HANDBOOK OF MECHANICS, MATERIALS, AND STRUCTURES

HANDBOOK OF MECHANICS, MATERIALS, AND STRUCTURES

Edited by
ALEXANDER BLAKE
Engineer at Large
Lawrence Livermore National Laboratory
University of California
Livermore, California

WILEY SERIES IN MECHANICAL ENGINEERING PRACTICE

CONSULTING EDITOR

Marvin D. Martin
President, Marvin D. Martin, Inc., Consulting Engineers, Tucson, Arizona

A Wiley-Interscience Publication
JOHN WILEY & SONS
New York • Chichester • Brisbane • Toronto • Singapore

Library of Congress Cataloging in Publication Data:

Main entry under title:

Handbook of mechanics, materials, and structures.
 (Wiley series in mechanical engineering practice,
ISSN 0749-0216)
 "A Wiley-Interscience publication."
 Includes index.
 1. Mechanics, Applied—Handbooks, manuals, etc.
2. Strength of materials—Handbooks, manuals, etc.
3. Structures, Theory of—Handbooks, manuals, etc.

I. Blake, Alexander. II. Martin, Marvin D. III. Series.
TA350.H23 1985   620.1   85-5373
ISBN 0-471-86239-8

Printed in the United States of America

10 9 8 7 6 5 4 3 2 1

CONTRIBUTORS

Raymond D. Ciatto, *Manager of Design, Teledyne Engineering Services, Waltham, Massachusetts*

Stephen C. Cowin, *Professor, Tulane University, New Orleans, Louisiana*

W. W. Feng, *Mechanical Engineer, Lawrence Livermore National Laboratory, Livermore, California*

A. Goldberg, *Senior Metallurgist, Lawrence Livermore National Laboratory, Livermore, California*

Okan Gurel, *IBM Cambridge Scientific Center, Cambridge, Massachusetts*

W. F. Kirkwood, *Consultant, Experimental Stress Analysis, Livermore, California*

Donald F. Landers, *Senior Vice President, Teledyne Engineering Services, Waltham, Massachusetts*

L. D. Mardis, *Senior Engineer, Pomona Division of General Dynamics, Pomona, California*

Joe W. McKinley, *President, Joe W. McKinley, Inc., Laguna Beach, California Former Professor, California State Polytechnic University, Pomona, California*

Frederick C. Nelson, *Dean of Engineering, Tufts University, Medford, Massachusetts*

R. G. Scott, *Supervisor, Lawrence Livermore National Laboratory, Livermore, California*

R. D. Streit, *Section Head, Lawrence Livermore National Laboratory, Livermore, California*

William C. Van Buskirk, *Professor, Tulane University, New Orleans, Louisiana*

SERIES PREFACE

The Wiley Series in Mechanical Engineering Practice is written for the practicing engineer. Students and academicians may find it useful, but its primary thrust is for the working engineer who needs a convenient and comprehensive reference on hand.

Two kinds of information are contained in the several volumes:

1. Numerical information such as strengths of materials, thermodynamic properties of fluids, standard pipe sizes, thread systems, and so on.
2. Descriptive and mathematical information typical of the state-of-the-art of the many facets and specialties encompassed by the broad term "mechanical engineering."

The profession has expanded to cover such a broad range of engineering activities that no one can be knowledgeable in more than a fraction of the whole field. Yet, in day-to-day work, practicing engineers frequently have to use, or at least interface with, specialty areas outside their normal sphere of competence. This book was written to provide readers with the state of the art information and standard practices in these other areas.

The task of covering such a vast amount of material has dictated the decision to split the series into five separate volumes:

Design and Manufacturing
Fluids and Fluid Machinery
Mechanics, Materials, and Structures
Power and Energy Systems
Instrumentation and Control

Each volume is designed to stand alone but the five complement each other in providing the broad coverage mentioned above. Within each volume chapter and section headings are designed to help the user find the material being sought.

A serious attempt was made to provide state-of-the-art material at the time of writing. Since many of the areas are in a state of rapid change, there will be some obsolescence by the time printing is complete. It is planned to revise and update at reasonable intervals so that users may purchase newer editions and keep their references up to date.

The many editors and contributors who have made this series possible join me in the hope that the several volumes will turn out to be really useful tools for the practicing engineer.

Marvin D. Martin

Tucson, Arizona
January 1985

PREFACE

This handbook is intended to be a design-oriented tool for the practicing engineer who often needs state-of-the-art information on diversified facets of mechanics, materials, and structures. These elements enter the decision-making process in hardware design, development, and research in industry subjected to a good deal of pressure from deadlines, economic constraints, and competitive forces. For this reason it is particularly fitting to provide a quick reference to the basic facts, simplified formulas, and essential materials data suitable for "ball park" estimates within reasonable bracketing assumptions of good engineering practice. Since the profession has expanded to cover such a wide spectrum of engineering topics, no one person or volume can relate to more than a fraction of the whole field. Hence this handbook emphasizes the most frequently needed portions of the body of knowledge of engineering design, including advanced areas of practice such as finite-element modeling, developments in fracture control, and basic regulatory aspects of pressure vessel design.

Although a number of topics selected for this volume must of necessity parallel the traditional lines of development, the handbook is not designed to be a reader's digest of abridged textbooks in the field but rather is an authoritative compendium of fundamentals accompanied by a minimum of derivation and descriptive material. In general, the essential topics of solid mechanics and design can be represented by a relatively small number of working formulas and models that straddle the disciplines of mechanical and civil engineering. Such rules are primarily concerned with the elastic response phenomena upon which nearly all the existing structures are based.

The transformation of engineering practice, particularly during the past 15 years, from the predominantly linear, static, and deterministic to the more dynamic, digital, and stochastic has been deeply felt by educational and industrial institutions. So far the effects of modern trends seem to be reflected in less emphasis on the conventional, design-oriented training and experience. The philosophy and manner of presentation of this handbook material have been aimed at promoting the necessary balance between highly theoretical and more practical problem solutions without losing sight of modern developments.

The division of this handbook material into discrete chapters may appear to be somewhat arbitrary since the various individual topics are crosslinked in many ways and venture into the neighboring areas of interest. Common foundations involved in the fields of mechanics, materials, and structures selected for this volume are bound to cross certain boundaries of each complicated topic. However, individual chapters presented here are expected to largely stand alone with emphasis on ready-to-use material as well as references to more advanced topics for the reader who is prepared to think beyond the elementary facts and formulas.

Essentially, Chapters 1 and 2 cover the more conventional subjects of engineering mathematics, statics, and dynamics in the form of a practical text that the working engineer should have on hand. The presentation of mathematics, for instance, emphasizes simple notation, graphical illustration, ready-to-use formulas, and numerical examples. However, standard data such as trigonometric tables, logarithms, powers, roots, and similar material are not included since this information is easily found on the conventional calculator. The chapter on statics and dynamics contains the basics of equilibrium, kinematics, frictional effects, principal laws of kinetics, and plane motion of rigid bodies. Chapter 3 provides a comprehensive treatment of practical strength of materials directly applicable to mechanical and structural design. Chapter 4 includes a brief review of conventional testing of materials together with the principles of metallurgy of metals and alloys, creep, fatigue, and the applications of fracture mechanics to characterization of metals. The topics of experimental stress analysis have often received considerable attention with no lack of good publications presenting this field in its entirety. For this reason Chapter 5 can only serve as an introduction. Similarly, Chapter 6 has been developed around the idea of providing the essential details of well-established theories in the areas of elasticity, elastic stability, and plasticity. Chapters 7, 8, and 9, however, contain a wealth of practical design information, covering straight members, curved members, and plates, written for the users rather than specialists in the field. This feature alone should place this handbook on the desk rather than on the bookshelf.

Because of the unique challenges and responsibilities associated with the design and quality control of pressurized components, Chapter 10 deals authoritatively with the technical background of the ASME Boiler and Pressure Vessel Code. The chapter is intended to guide the design engineer through the intricacies of the code and it provides an extensive list of key references essential for further study.

The material for this handbook has been prepared by a number of experts and practitioners in the field representing universities, private industry, and government research. An effort has been made to reduce unnecessary duplication of subject matter and to limit the extent of mathematical derivations. Therefore, the emphasis is placed on the ready-to-use formulas, design rules, and guides within the state-of-the-art boundaries in order to assist engineers in developing immediate answers to a variety of questions arising from their practical work.

The editor wishes to take this opportunity to express his gratitude to all major contributors listed in this volume, whose cooperation and understanding allowed completion of this design reference for engineers. This acknowledgment is extended to Walter F. Arnold, former Head of the Mechanical Engineering Department of the Lawrence Livermore National Laboratory, for his encouragement and professional interest in the project.

ALEXANDER BLAKE

Livermore, California
June 1985

CONTENTS

2 STATICS AND DYNAMICS   117

William C. Van Buskirk and Stephen C. Cowin

4 MECHANICAL PROPERTIES AND SCIENCE OF ENGINEERING MATERIALS 279

W. F. Kirkwood, W. W. Feng, R. G. Scott, R. D. Streit, and A. Goldberg

7 STRAIGHT MEMBERS   525

Raymond D. Ciatto

8 CURVED MEMBERS   575

Frederick C. Nelson

HANDBOOK OF MECHANICS, MATERIALS, AND STRUCTURES

CHAPTER 1
ENGINEERING MATHEMATICS

OKAN GUREL
IBM Cambridge Scientific Center
Cambridge, Massachusetts

1.1 BASIC ALGEBRAIC EQUATIONS

1.1-1 Algebraic Operations

If a set of elements $x, y, \ldots$, is given with *algebraic (binary) operations* such as addition and multiplication involving its elements in pairs, then the following laws are obeyed.

The Laws of Algebraic Operations

(a) Commutative law: $x + y = y + x$, $xy = yx$.
(b) Associative law: $x + (y + z) = (x + y) + z$, $x(yz) = (xy)z$.
(c) Distributive law: $z(x + y) = zx + zy$.

1.1-2 Progression

The sequence of numbers with a special relationship between the consecutive numbers forms a *progression*. Following are the most common progressions.

Arithmetic Progression

The *arithmetic progression* is a sequence of n numbers $x_1, \ldots, x_n$, such that the difference between two consecutive numbers is a constant, called a *common difference*, d. The nth number is given by

$$x_n = x_1 + (n - 1)d$$

The *sum* of n numbers, S_n, is given by

$$S_n = (n/2)(x_1 + x_n)$$

The *arithmetic mean* between x and y is $(x + y)/2$.

Example: $x_1 = 6$, $n = 4$, $d = 3$

$$x_1, x_2, x_3, x_4 \text{ become } 6, 9, 12, 15$$

$$S_4 = (4/2)(6 + 15) = 42$$

The arithmetic mean between 9 and 12 is $(9 + 12)/2 = 10.5$.

Geometric Progression

The *geometric progression* is a sequence of n numbers $x_1, \ldots, x_n$, such that the ratio between two consecutive numbers is a constant, called a *common ratio*, r. The nth number is given by

$$x_n = x_1 r^{n-1}$$

The *sum* of n numbers, S_n, is given by

$$S_n = x_1(1 - r^n)/(1 - r)$$

or

$$S_n = (x_1 - rx_n)/(1 - r)$$

The *geometric mean* between x and y is $\sqrt{xy}$.

Example: $x_1 = 6,\quad n = 4,\quad r = 3$

$$x_1, x_2, x_3, x_4 \text{ become } 6, 18, 54, 162$$

$$S_4 = 6(1 - 3^4)/(1 - 3) = 240$$

The geometric mean between 18 and 54 is $\sqrt{18 \cdot 54} = 31.2$.

Harmonic Progression

The *harmonic progression* is a sequence of n numbers $x_1, \ldots, x_n$, such that their *reciprocals* form an arithmetic progression. The nth number is given by

$$\frac{1}{x_n} = \frac{1}{x_1 + (n - 1)d}$$

The *harmonic mean* between x and y is $2xy/(x + y)$.

Example: $x_1 = 6,\quad n = 4,\quad d = 3$

$$x_1, x_2, x_3, x_4 \text{ become } 1/6, 1/9, 1/12, 1/15$$

The harmonic mean between $1/9$ and $1/12$ is $[2(1/9)(1/12)]/[(1/9) + (1/12)] = 1/10.5$.

1.1-3   Permutations and Combinations

An arrangement of n objects in groups of s $(< n)$ in all possible orders is called a *permutation*. For example, if there are four numbers 1, 2, 3, and 4, the following permutations of groups of three numbers are possible:

1 2 3	1 2 4	1 3 4	2 3 4
1 3 2	1 4 2	1 4 3	2 4 3
2 1 3	2 1 4	3 1 4	3 2 4
2 3 1	2 4 1	3 4 1	3 4 2
3 1 2	4 1 2	4 1 3	4 2 3
3 2 1	4 2 1	4 3 1	4 3 2

The general formula for the number of permutations of n objects in groups of s is

$$_nP_s = n(n - 1)(n - 2) \cdots (n - s + 1) = \frac{n!}{(n - s)!}$$

where $k! = 1 \cdot 2 \cdots k$ is called the *factorial*.

For $n = 4$ and $s = 3$ the above formula gives

$$_4P_3 = 4 \cdot 3 \cdot 2 = \frac{1 \cdot 2 \cdot 3 \cdot 4}{(4 - 3)!} = 24$$

The two special cases are

$$_nP_1 = n \quad \text{and} \quad _nP_n = n! \quad \text{(note that } 0! = 1\text{)}$$

Therefore the factorial $n!$ is also the number of permutations of all n objects.

A collection of n objects in groups of s $(< n)$ where the order of the objects is not relevant is called a *combination*. For example, combinations of four numbers, $1, 2, 3, 4$, in groups of 3 are

$$1\ 2\ 3 \quad 1\ 2\ 4 \quad 1\ 3\ 4 \quad 2\ 3\ 4$$

The general formula for the number of combinations possible is

$$_nC_s = \frac{_nP_s}{s!} = \frac{n!}{(n - s)!s!}$$

Thus for $n = 4$, $s = 3$, $_4C_3 = 4!/1!3! = 4$.

1.1-4 Exponentials and Logarithms

Definition

The *exponential*, x^n, is defined as the product of n x's, that is,

$$\underbrace{x \cdot x \cdot x \cdots x}_{n \text{ terms}}$$

where n is a positive integer and x is a real number. If $y = x^n$, y is called the nth *power* of x while x is the nth *root* of y.

Example: $x = 3$, $n = 4$,

$$x^n = 3^4 = 3 \cdot 3 \cdot 3 \cdot 3 = 81$$

The Laws of Exponents

1. $x^m \cdot x^n = x^{m+n}$;   for example, $x^3 \cdot x^4 = x^{3+4} = x^7$.
2. $x^m/x^n = x^{m-n}$;   for example, $x^4/x^1 = x^{4-1} = x^3$.
3. $(x^m)^n = x^{mn}$;   for example, $(x^2)^4 = x^{2 \cdot 4} = x^8$.

Also,

$$
\begin{array}{ll}
x^0 = 1 & \text{for example, for any } x \\
x^{m/n} = (x^m)^{1/n} & \text{for example, } x^{6/3} = (x^6)^{1/3} \\
x^{-m} = 1/x^m & \text{for example, } x^{-2} = 1/x^2
\end{array}
$$

Definition

The *logarithm to the base a*, $\log_a$, is the inverse function to the exponential function a^x, that is,

$$y = a^x \qquad x = \log_a y$$

Example: $100 = 10^2$, $y = 100$, $x = 2$, $a = 10$, thus $2 = \log_{10} 100$.

The Laws of Logarithms

1. $\log_a(xy) = \log_a x + \log_a y$; for example, $\log_{10}(3 \cdot 2) = \log_{10} 3 + \log_{10} 2$.
2. $\log_a(x/y) = \log_a x - \log_a y$; for example, $\log_{10}(6/2) = \log_{10} 6 - \log_{10} 2$.
3. $\log_a(x^n) = n \log_a x$; for example, $\log_{10}(3^4) = 4 \log_{10} 3$.

The common bases for logarithms are

$$10 \qquad \text{for common, or Briggsian logarithm}$$
$$e = 2.71828 \qquad \text{for natural, Naperian, hyperbolic logarithms}$$

The formulas for base changes of logarithms are:

$$\log_a x = \log_b x / \log_b a = (\log_b x) \cdot (\log_a b)$$

$$\log_{10} x = \log_e x / \log_e 10 = (\log_{10} e) \cdot (\log_e x)$$

$$= 0.43429\,44819 \log_e x$$

$$\log_e x = \log_{10} x / \log_{10} e = (\log_e 10) \cdot (\log_{10} x)$$

$$= 2.30258\,50930 \log_{10} x$$

$$\text{antilog } x = 1/\log x = \log^{-1} x$$

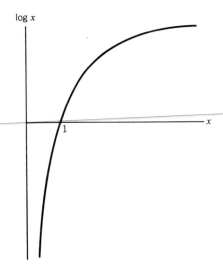

1.1-5 Polynomials

The left-hand-side of an *algebraic equation* in n unknowns, $x_1, \ldots, x_n$, expressed in the summation form of

$$\sum C_{k_1, \ldots, k_n} x_1^{k_1} \cdots x_n^{k_n} = 0$$

where $k_1, \ldots, k_n$ are *powers* of variables and $C_{k_1, \ldots, k_n}$ are *coefficients*, is called a *polynomial*. A single element is called a *term* and the sum (Σk_i) its *degree*. The highest degree is called the *degree of the polynomial*. A polynomial with elements of the same degree is called *homogeneous*. Polynomials with one term are *monomials* and two terms *binomials*. Two of the special homogeneous polynomials are *quadratic*, the second degree, and *cubic*, the third degree.

Examples:

$$6x_1^3x_2^2 \qquad \text{a } \textit{monomial } (\textit{degree} = 5)$$

$$5x_1^4 + 2x_2 \qquad \text{a } \textit{binomial } (\textit{degree} = 4)$$

$$5x_1^2 + 2x_1x_2 + 4x_2^2 \qquad \text{a } \textit{homogeneous polynomial } (\textit{degree} = 2)$$

$$x_1^2 + x_2 \qquad \text{a } \textit{quadratic } (\textit{degree} = 2)$$

$$2x_1^3 + x_2^2 + x_1x_2 \qquad \text{a } \textit{cubic } (\textit{degree} = 3)$$

Discriminant of a Polynomial

The *discriminant of a polynomial*,

$$C_0x^n + C_1x^{n-1} + \cdots + C_n = 0$$

with roots $x_1, \ldots, x_n$ is, by definition,

$$\prod_{i>k}(x_i - x_k)^2$$

Examples:

$$x^3 - 6x^2 + 11x - 6 = 0 \quad \textit{the polynomial}$$

$$(x - 1)(x - 2)(x - 3) = 0$$

$$x_1 = 1, \quad x_2 = 2, \quad x_3 = 3$$

$$\prod(x_3 - x_2)^2(x_3 - x_1)^2(x_2 - x_1)^2 \quad \textit{the discriminant}$$

$$\prod(3 - 2)^2(3 - 1)^2(2 - 1)^2 = 1 \cdot 4 \cdot 1 = 4$$

Infinite Series

A *series* is a sum of n terms, such as $1 + 5 + 9 + 13$ is a series of four terms. A series of n terms is

$$x_1 + x_2 + \cdots + x_n = \sum_{i=1}^{n} x_i$$

An *infinite series* is a series with infinitely many terms:

$$x_1 + x_2 + \cdots + x_n + \cdots = \sum_{i=1}^{\infty} x_i$$

A *partial sum* of series is given as

$$S_n = x_1 + x_2 + \cdots + x_n$$

If $\lim S_n = S$, as n goes to infinity, then the series is called *convergent* to the *sum S*. If the limit does not exist, it is said to *diverge*. Since

$$x_n = S_n - S_{n-1} \quad \text{as } n \to \infty$$

a series with the general term x_n approaching zero is a *convergent* series. Otherwise it is *divergent*. For example, the *geometric series*

$$1 + x + x^2 + \cdots + x^n + \cdots$$

is convergent if $|x| < 1$ and divergent if $|x| > 1$. If a series is convergent and also $\Sigma|x_i|$ is convergent, then Σx_i is said to be *absolutely convergent*.

The two tests for convergence of infinite series are Cauchy's test and D'Alembert's test.

Cauchy's Test. If $\lim |x_n|^{1/2} < 1$, as n goes to infinity, Σx_i is an *absolutely convergent* series.
D'Alembert's Test. If $|x_{n+1}/x_n| < \phi$, where ϕ is positive, less than 1, and independent of n, the series Σx_i is *absolutely convergent*.

Binomial Series

A polynomial expansion of the nth power of the sum of two quantities results in a *binomial series*. A theorem related to the computation of the coefficients of this series is known as the *binomial theorem*.

Binomial Theorem

The polynomial expansion of the nth (n is a positive integer) power of the sum of two quantities is

$$(x + y)^n = \sum_{s=0}^{n} \binom{n}{s} x^s y^{n-s}$$

The coefficients

$$_nC_s = \binom{n}{s} = \frac{n(n-1)\cdots(n-s+1)}{s!} = \frac{n!}{s(n-s)!}$$

are called *binomial coefficients*, and $s!$, *factorial s*, is the product $1 \cdot 2 \cdots s$. It should also be noted that the combination $_nC_s$ is equal to $\binom{n}{s}$. The binomial coefficients have the relationships

$$\binom{n}{0} = 1 \qquad \binom{n}{s} = \binom{n}{n-s} \qquad \binom{n}{s} + \binom{n}{n-s} = \binom{n+1}{s}$$

Based on the last relationship, coefficients for the $n + 1$ case can be computed from the preceding one. For example, for $n = 1$,

$$_1C_0 = 1, \qquad _1C_1 = \frac{1!}{1!0!} = 1$$

$$_1C_1 + _1C_0 = _2C_1$$

Similarly, $1 + 2 = 3$, $1 + 3 = 4$, $3 + 3 = 6$, and so on. Thus we have the following coefficients $\binom{n}{s}$ for corresponding values of n:

$$
\begin{array}{ll}
n = 1 & 1\ 1 \\
n = 2 & 1\ 2\ 1 \\
n = 3 & 1\ 3\ 3\ 1 \\
n = 4 & 1\ 4\ 6\ 4\ 1
\end{array}
$$

which is called the *Pascal triangle*. In fact, the Pascal triangle provides coefficients of the expansion of $(x + y)^n$. For example, $(x + y)^1 = x + y$, thus the coefficients are 1 and 1, $(x + y)^2 = x^2 + 2xy + y^2$, the coefficients are $1, 2, 1$, and so on. Some other properties of binomial coefficients are:

$$\sum_{s=0}^{n} \binom{n}{s} = 2^n$$

$$\sum_{s=0}^{n} (-1)^s \binom{n}{s} = 0$$

$$\sum_{s=0}^{n} \binom{n}{s}^2 = \binom{2n}{n}$$

Series Expansion of Common Functions

A series expansion of a function, f, of a variable, x, as $f(x)$ may be given as a polynomial. The series expansion of some of the common functions are discussed in Section 1.6, and a table of selected functions are tabulated in Section 1.11.

1.1-6   Equations and Their Roots

The general algebraic equation with special n (degree) values are:

Equation of degree 2, $n = 2$, quadratic equation.
Equation of degree 3, $n = 3$, cubic equation.
Equation of degree 4, $n = 4$, quartic or biquadratic equation.

Quadratic Equation

This may be reduced to the form

$$ax^2 + bx + c = 0$$

The two roots of a quadratic equation are

$$x = \frac{-b \pm (b^2 - 4ac)^{1/2}}{2a}$$

For real a, b, and c, $b^2 - 4ac > 0$ and the two roots are real and unequal; for $b^2 - 4ac = 0$, the two real roots are equal; for $b^2 - 4ac < 0$, both roots are imaginary.

Cubic Equation

By substituting $y = x - b/3a$, the equation $ay^3 + by^2 + cy + d = 0$ may be reduced to the form

$$x^3 + ex + f = 0$$

where the coefficients e and f are calculated as

$$e = \frac{c}{a} - \frac{b^2}{3a^2} \qquad f = \frac{b}{a^2}\left(\frac{2b^2}{27a} - \frac{c}{3}\right) + \frac{d}{a}$$

The three roots of a cubic equation are

$$x_1 = E + F$$

$$x_2 = -\frac{(E + F)}{2} + \frac{(E - F)(-3)^{1/2}}{2}$$

$$x_3 = -\frac{(E + F)}{2} - \frac{(E - F)(-3)^{1/2}}{2}$$

where

$$E = \left[-\frac{f}{2} + \left(\frac{f^2}{4} + \frac{e^3}{27}\right)^{1/2}\right]^{1/3}$$

$$F = -\left[-\frac{f}{2} + \left(\frac{f^2}{4} + \frac{e^3}{27}\right)^{1/2}\right]^{1/3}$$

If a, b, c, d are real, then

For $f^2/4 + e^3/27 > 0$ one real, two conjugate;
$= 0$ three real, at least two equal;
< 0 three real and unequal roots.

Quartic or Biquadratic Equation

An equation of the form

$$ay^4 + by^3 + cy^2 + dy + e = 0 \tag{1.1-1}$$

by substituting $y = x - b/4a$, may be transformed to

$$x^4 + fx^2 + gx + h = 0 \tag{1.1-2}$$

where the coefficients f, g, and h are calculated to be

$$f = \frac{c}{a} - \frac{3b^2}{8a^2} \qquad g = \frac{d}{a} - \frac{bc}{2a^2} + \frac{b^3}{8a^2}$$

$$h = -\frac{3b^4}{256a^4} + \frac{b^2c}{16a^3} - \frac{bd}{4a^2} + \frac{c}{a}$$

If the three roots of

$$z^3 + fz^2 + (f^2 - 4h)z - g^2 = 0 \tag{1.1-3}$$

are z_1, z_2, and z_3, then the four roots of the transformed equation are

$$x_1 = \tfrac{1}{2}\left[+ (z_1)^{1/2} + (z_2)^{1/2} + (z_3)^{1/2} \right]$$

$$x_2 = \tfrac{1}{2}\left[+ (z_1)^{1/2} - (z_2)^{1/2} - (z_3)^{1/2} \right]$$

$$x_3 = \tfrac{1}{2}\left[- (z_1)^{1/2} + (z_2)^{1/2} - (z_3)^{1/2} \right]$$

$$x_4 = \tfrac{1}{2}\left[- (z_1)^{1/2} - (z_2)^{1/2} + (z_3)^{1/2} \right]$$

The discriminants, Δ, of (1.1-1) and (1.1-3) are the same. If the equation is real, the following cases are possible:

For $\Delta < 0$ a pair of conjugate complex roots and a pair of real roots;
$\Delta > 0$ $(f^2 - 4h) > 0$ four real roots;
$\Delta > 0$ $(f^2 - 4h) < 0$ four complex roots;
$\Delta = 0$ four roots are not distinct.

Partial Fractions

If a function is expressed as

$$\phi(s) = \frac{\sum\limits_0^m a_r s^r}{\sum\limits_0^n b_r s^r} = \frac{a_0 + a_1 s + a_2 s^2 + \cdots + a_m s^m}{b_0 + b_1 s + b_2 s^2 + \cdots + b_n s^n} \qquad m < n$$

and the denominator is known to be factorized as

$$\sum_0^n b_r s^r = B(s - b_1)^{n_1}(s - b_2)^{n_2} \cdots$$

then $\phi(s)$ can be expressed as a sum in terms of the factorized elements as an expansion in the form of n partial fractions

$$\phi(s) = \sum_1^{n_1} \frac{B_{1,r}}{(s - b_1)^r} + \sum_1^{n_2} \frac{B_{2,r}}{(s - b_2)^r} + \cdots$$

Example:

$$\phi(s) = \frac{2s^2 + 1}{s^3 - s}$$

The denominator, $s^3 - s$, is factorized as $s(s + 1)(s - 1)$, then

$$\phi(s) = \frac{B_{1,1}}{s} + \frac{B_{2,1}}{s + 1} + \frac{B_{3,1}}{s - 1}$$

By the method of undetermined coefficients $B_{1,1} \ldots$ can be determined. The numerator of the left-hand side is equal to the numerator of the right-hand side:

$$2s^2 + 1 = B_{1,1}(s^2 - 1) + B_{2,1}(s^2 - s) + B_{3,1}(s^2 + s)$$

which results in

$$2s^2 + 1 = (B_{1,1} + B_{2,1} + B_{3,1})s^2 - (B_{2,1} - B_{3,1})s - B_{1,1}$$

Equating the coefficients of equal powers of s:

$$2 = B_{1,1} + B_{2,1} + B_{3,1}$$

$$0 = B_{2,1} - B_{3,1}$$

$$1 = -B_{1,1}$$

resulting in $B_{1,1} = -1$, $B_{2,1} = B_{3,1} = 3/2$.

$$\phi(s) = -\frac{1}{s} + \frac{3/2}{s + 1} + \frac{3/2}{s - 1}$$

The factorization of the denominator may consist of different forms: such as all factors may be linear and as in the above example, that is, $r = 1$. In some cases $r > 1$. Or the factors of the denominator may not be powers of real terms such as $s^2 + 1$. Then the numerator of each factor would be one less power than the power of the factor. For example, if

$$\phi(s) = \frac{2s^2 + 1}{(s^2 + 1)(s - 1)}$$

$$\phi(s) = \frac{B_{1,1}s + B_{2,1}}{s^2 + 1} + \frac{B_{3,1}}{s - 1}$$

$$2s^2 + 1 = B_{1,1}s^2 - B_{1,1}s + B_{2,1}s - B_{2,1} + B_{3,1}s^2 + B_{3,1}$$

$$= (B_{1,1} - B_{3,1})s^2 - (B_{1,1} - B_{2,1})s - (B_{2,1} - B_{3,1})$$

$$B_{1,1} + B_{3,1} = 2$$

$$B_{1,1} - B_{2,1} = 0$$

$$B_{2,1} - B_{3,1} = -1$$

$$B_{1,1} = B_{2,1} \qquad B_{1,1} = 1/2 \qquad B_{3,1} = 3/2$$

Applications of partial fractions are found in various examples. For example, if r is an integral an application is found in Laplace transforms (see Section 1.7). See also Section 1.6 for an application in such cases where the L'Hôpital rule fails.

Complex Numbers

A *complex number* consists of a *real number* and an *imaginary number*:

$$z = x + iy \qquad i = \sqrt{-1}$$

Here x is the *real* part of the complex number while y is the imaginary part. A complex number may be represented as a vector in the x-y plane (complex plane z) as shown in the figure.

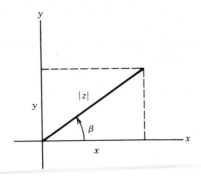

$$z = x + iy \qquad \bar{z} = x - iy$$

are *conjugate* complex numbers.

Addition of Complex Numbers

$$z_1 + z_2 = (x_1 + x_2) + i(y_1 + y_2)$$

We define $|z|$ as the *absolute value* (*modulus*) of the complex number and β as the *angle* of z. Using the trigonometric relations, see the figure above,

$$z = x + iy = |z|(\cos \beta + i \sin \beta) = |z|e^{i\beta}$$

$$\bar{z} = x - iy = |z|(\cos \beta - i \sin \beta) = |z|e^{-i\beta}$$

Multiplication of Complex Numbers

$$z_1 z_2 = (x_1 + iy_1)(x_2 + iy_2)$$

$$= x_1 x_2 - y_1 y_2 + i(x_1 y_2 + x_2 y_1)$$

Power of Complex Numbers

$$z^n = (x + iy)^n = [|z|(\cos \beta + i \sin \beta)]^n = |z|^n e^{in\beta}$$

$$z^n = (x - iy)^n = [|z|(\cos \beta - i \sin \beta)]^n = |z|^n e^{-in\beta}$$

$$\sqrt[n]{z} = \sqrt[n]{x + iy}$$

$$= \sqrt[n]{|z|}\left(\cos \frac{\beta + 2k}{n} + i \sin \frac{\beta + 2k}{n}\right)$$

$$= \sqrt[n]{|z|} e^{(\beta + 2k)/n}$$

1.2   PLANE AND SOLID GEOMETRY: MENSURATION FORMULAS

1.2-1   Plane Geometry

Triangles: Definitions

S = area
r (*or* R) = radius of the inscribed (or circumscribed) circle
A, B, C = angles and a, b, c the corresponding sides
t_a, t_b, t_c = lengths of the bisectors of angles A, B, C

m_a, m_b, m_c = lengths of the medians of sides a, b, c

h_a, h_b, h_c = lengths of the altitudes on sides a, b, c

$s = \frac{1}{2}(a + b + c)$

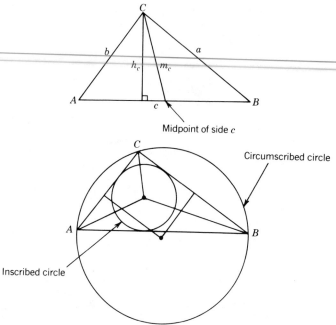

Midpoint of side c

Circumscribed circle

Inscribed circle

Triangles: Formulas

General Triangle

$$A + B + C = 180°$$

$$c^2 = a^2 + b^2 - 2ab \cos C \quad \text{(Law of cosine)}$$

$$S = \frac{1}{2}h_c c = \frac{1}{2}ab \sin C$$

$$= \frac{c^2 \sin A \sin B}{2 \sin C}$$

$$= rs = \frac{abc}{4R}$$

$$= \sqrt{s(s-a)(s-b)(s-c)} \quad \text{(Heron's formula)}$$

Equilateral Triangle

$$A = B = C = 60° \qquad a = b = c$$

$$S = \frac{1}{4}a^2\sqrt{3}$$

$$r = \frac{1}{6}a\sqrt{3} \qquad R = \frac{1}{3}a\sqrt{3}$$

$$h = \frac{1}{2}a\sqrt{3}$$

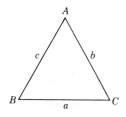

Right Triangle

$$A + B = C = 90°$$

$$c^2 = a^2 + b^2 \quad \text{(Pythagorean relation)}$$

$$a = \sqrt{(c + b)(c - b)}$$

$$S = \tfrac{1}{2}ab$$

$$r = \frac{ab}{a + b + c} \qquad R = \tfrac{1}{2}c$$

$$h = \frac{ab}{c} \qquad m = \frac{b^2}{c} \qquad n = \frac{a^2}{c}$$

$$m + n = c$$

m on b, n on a side of c

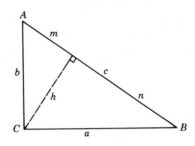

Quadrilaterals: Definitions

S = area
p, q = diagonals
A, B, C, D = angles
a, b, c, d = sides
$r(R)$ = radius of the inscribed (circumscribed) circles

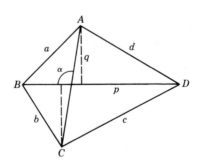

Quadrilaterals: Formulas

General Quadrangle or Quadrilateral

$$S = \tfrac{1}{2}pq \sin \alpha$$

Theorem. *If $a^2 + c^2 = d^2 + b^2$, then p and q are perpendicular to each other.*

Trapezoid. If any pair of opposite sides, for example, a and c, are parallel to each other, the quadrilateral is called a *trapezoid*.

$$m = \tfrac{1}{2}(a + c)$$

$$S = \tfrac{1}{2}(a + c)h = mh$$

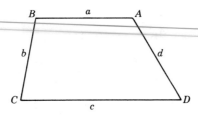

Parallelogram. If both pairs of opposite sides are parallel to each other, the quadrilateral is called a *parallelogram*.

$$A = C, \quad B = D, \quad A + B = 180°$$
$$a = c, \quad b = d$$
$$S = ah \quad h = \text{the distance between } a \text{ and } c$$
$$\quad = ab \sin A = ab \sin B$$
$$h = b \sin A = b \sin B$$
$$p^2 = a^2 + b^2 - 2ab \cos A \quad \text{if } A \text{ is an acute angle}$$
$$q^2 = a^2 + b^2 - 2ab \cos B = a^2 + b^2 + 2ab \cos A$$

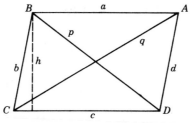

Rhombus. A parallelogram with $a = b = c = d$ is called a *rhombus*. By the theorem above p is perpendicular to q.

$$p^2 + q^2 = 4a^2$$

$$S = \tfrac{1}{2}pq$$

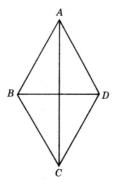

Rectangle. A parallelogram with $A = B = C = D = 90°$ is called a *rectangle*, thus $a = c$, $b = d$, and

$$S = ab$$

$$p = q = \sqrt{a^2 + b^2}$$

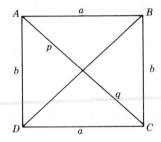

Polygons

A polygon with n equal sides of length s is called a *regular polygon*. The angles of a regular polygon with n sides is calculated as

$$\alpha = \left(\frac{n-2}{n}\right)180°, \quad \text{for example, } n = 6, \ \alpha = \frac{6-2}{6}180° = 120°$$

S = area
r (*or* R) = radius of the inscribed (or circumscribed) circle
p = perimeter

$$s = 2r\tan\frac{180°}{n} = 2R\sin\frac{180°}{n}$$

$$p = ns$$

$$S = \tfrac{1}{4}ns^2\cot\frac{180°}{n}$$

$$r = \tfrac{1}{2}s\cot\frac{180°}{n} \qquad R = \tfrac{1}{2}s\csc\frac{180°}{n}$$

Some regular polygons are:

$n = 5$ pentagon	$n = 6$ hexagon	$n = 7$ heptagon
$n = 8$ octagon	$n = 9$ nonagon	$n = 10$ decagon

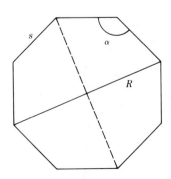

Circles

R = radius
D = diameter
C = circumference
S = area

$$C = 2\pi R = \pi D \qquad (\pi = 3.14159)$$

$$S = \pi R^2 = \tfrac{1}{4}\pi D^2$$

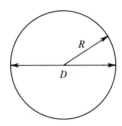

Sector of a Circle. Here s is the length of the arc, and $\alpha(< \pi)$ is the central angle in radians. We have

$$s = R\alpha$$

$$h = R - d$$

$$d = R\cos(\alpha/2)$$

$$\alpha = \frac{s}{R}$$

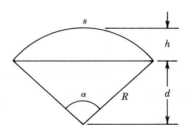

Conic Sections

Ellipse

a, b = the halves of the major axes
C = circumference = $2\pi\sqrt{(a^2 + b^2)/2}$ (approximate)
S = area = πab

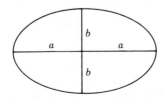

1.2-2 Solid Geometry

Definitions

S = area of lateral surface
T = total surface area
V = volume

Formulas

Cube

a = length of each edge
$T = 6a^2$
$V = a^3$
Diagonal of face = $a\sqrt{2}$
Diagonal of cube = $a\sqrt{3}$

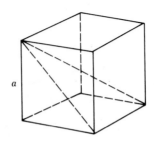

Rectangular Parallelepiped

a, b, c = the lengths of edges
$T = 2(ab + bc + ca)$
$V = abc$

Diagonal = $\sqrt{a^2 + b^2 + c^2}$

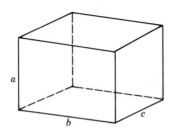

Prism

S = (perimeter of right section) × (lateral edge) = $(a + b + c + d)l$
V = (area of right section) × (lateral edge)
  = (area of base) × (altitude) = Al

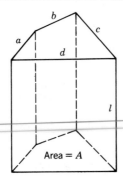

Pyramid

$S = \frac{1}{2}$(perimeter of base) × (slant height)
$V = \frac{1}{3}$(area of base) × (altitude)

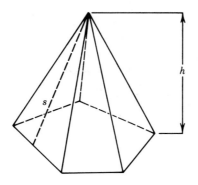

Frustum of Pyramid

B_1 = area of lower base
B_2 = area of upper base
h = altitude

$$V = \frac{1}{3}h\left(B_1 + B_2 + \sqrt{B_1 B_2} \right)$$

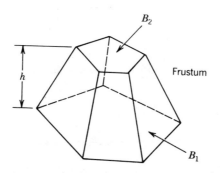

Regular Polyhedra

v = number of vertices

e = number of edges

f = number of faces

r (R) = radius of inscribed (circumscribed) sphere

$v - e + f = 2$ (Euler–Descartes formula for complex polyhedra)

If A is the area of each face, then

$$T = fA$$

$$V = \tfrac{1}{3}rfA = \tfrac{1}{3}rT$$

	v	e	f
Tetrahedon	4	6	4
Hexahedron	8	12	6
Octahedron	6	12	8
Dodecahedron	20	30	12
Icosahedron	12	30	20

Tetrahedron   Octahedron   Icosahedron

Cylinders and Cones

B_1 = area of lower base

B_2 = area of upper base

h = altitude

Cylinder

R = radius of base

$S = 2\pi R\ (h)$

$T = 2\pi R\ (R + h)$

$V = \pi R^2 h$

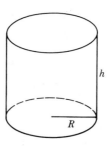

Cone

$$V = \tfrac{1}{3}B_1 h$$

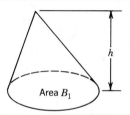

Area B_1

Right Circular Cone

s = slant length

$s = \sqrt{R^2 + h^2}$

$S = \pi R s = \pi\sqrt{R^2 + h^2}$

$T = \pi R(R + s) = \pi R\left(R + \sqrt{R^2 + h^2}\right)$

$V = \frac{1}{3}\pi r^2 h$

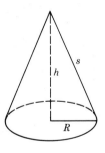

Frustrum of Cone

$$V = \frac{1}{3}h\left(B_1 + B_2\sqrt{B_1 B_2}\right)$$

If the cone is a right circular cone with radii R_1 and R_2 of the two bases, then

$$V = \frac{1}{3}\pi h\left(R_1^2 + R_2^2 + R_1 R_2\right)$$

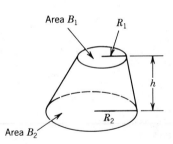

Area B_1 R_1

h

R_2

Area B_2

Sphere

$$D = 2R$$
$$S = 4\pi R^2 = \pi D^2$$
$$V = \tfrac{4}{3}\pi R^3 = \tfrac{1}{6}\pi D^2$$

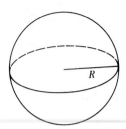

Circular Torus

r = radius of the circular cross-section
R = radius of rotating center of the cross-section circle

$$S = 4\pi^2 Rr$$
$$V = 2\pi^2 Rr^2.$$

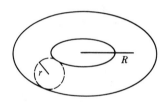

1.3   TRIGONOMETRY

1.3-1   Plane Trigonometry

Definitions

Angles

1 degree = 1/360 of one complete rotation
$180° = \pi$ radians
1 radian = the angle at the center of a circle corresponding to an arc of length equal to the radius of the circle.

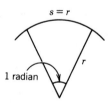

A *straight angle* is an angle of 180° (180 degrees).
A *right angle* is an angle of 90°.
An *acute angle* is an angle between 0 and 90°.
An *obtuse angle* is an angle between 90 and 180°.

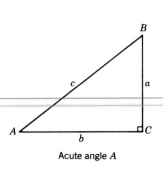

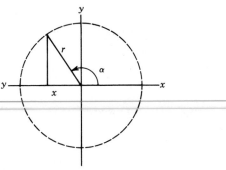

Acute angle A

Arbitrary angle α

Acute Angle, A

$$\sin A = a/c \qquad \cos A = b/c$$
$$\tan A = a/b \qquad \cot A = b/a$$
$$\sec A = c/b \qquad \csc A = c/a$$

Arbitrary Angle, α

$$\sin \alpha = y/r \qquad \cos \alpha = x/r$$
$$\tan \alpha = y/x \qquad \cot \alpha = x/y$$
$$\sec \alpha = r/x \qquad \csc \alpha = r/y$$

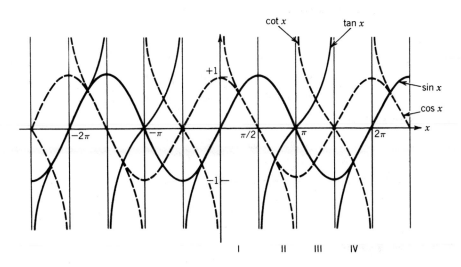

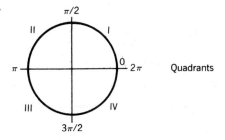

Quadrants

The values of trigonometric functions of some angles are as follows:

Function	0°	30°	45°	60°	90°	180°	270°	360°
$\sin \alpha$	0	$1/2$	$1/\sqrt{2}$	$\sqrt{3}/2$	1	0	-1	0
$\cos \alpha$	1	$\sqrt{3}/2$	$1/\sqrt{2}$	$1/2$	0	-1	0	1
$\tan \alpha$	0	$1/\sqrt{3}$	1	$\sqrt{3}$	∞	0	$-\infty$	0

Fundamental Identities

Relations

Product	Quotient	Reciprocal
$\sin \alpha = \tan \alpha \cos \alpha$	$\tan \alpha / \sec \alpha$	$1/\csc \alpha$
$\cos \alpha = \cot \alpha \sin \alpha$	$\cot \alpha / \csc \alpha$	$1/\sec \alpha$
$\tan \alpha = \sin \alpha \sec \alpha$	$\sin \alpha / \cos \alpha$	$1/\cot \alpha$
$\cot \alpha = \cos \alpha \csc \alpha$	$\cos \alpha / \sin \alpha$	$1/\tan \alpha$
$\sec \alpha = \csc \alpha \tan \alpha$	$\csc \alpha / \cot \alpha$	$1/\cos \alpha$
$\csc \alpha = \sec \alpha \cot \alpha$	$\sec \alpha / \tan \alpha$	$1/\sin \alpha$

Pythagorean Relations

$$\sin^2\alpha + \cos^2\alpha = 1$$

$$1 + \tan^2\alpha = \sec^2\alpha$$

$$1 + \cot^2\alpha = \csc^2\alpha$$

Angle Sum (and Difference) Relations

$$\sin(\alpha + \beta) = \sin \alpha \cos \beta + \cos \alpha \sin \beta$$

$$\sin(\alpha - \beta) = \sin \alpha \cos \beta - \cos \alpha \sin \beta$$

$$\cos(\alpha + \beta) = \cos \alpha \cos \beta - \sin \alpha \sin \beta$$

$$\cos(\alpha - \beta) = \cos \alpha \cos \beta + \sin \alpha \sin \beta$$

Double-Angle Relations

$$\sin 2\alpha = 2 \sin \alpha \cos \alpha = \frac{2 \tan \alpha}{1 + \tan^2\alpha}$$

$$\cos 2\alpha = \cos^2\alpha - \sin^2\alpha = \frac{1 - \tan \alpha}{1 + \tan^2\alpha}$$

$$\tan 2\alpha = \frac{2 \tan \alpha}{1 - \tan^2\alpha}$$

Multiple-Angle Relations

$$\sin 3\alpha = 3 \sin \alpha - 4 \sin^3 \alpha$$

$$\cos 3\alpha = 4 \cos^3\alpha - 3 \cos \alpha$$

$$\tan 3\alpha = \frac{3 \tan \alpha - \tan^3\alpha}{1 - 3 \tan^2\alpha}$$

$$\sin n\alpha = 2 \sin(n - 1)\alpha \cos \alpha - \sin(n - 2)\alpha$$

$$\cos n\alpha = 2 \cos(n - 1)\alpha \cos \alpha - \cos(n - 2)\alpha$$

$$\tan n\alpha = \frac{\tan(n - 1)\alpha + \tan \alpha}{1 - \tan(n - 1)\alpha \tan \alpha}$$

Function Product Relations

$$\sin \alpha \sin \beta = \tfrac{1}{2}\cos(\alpha - \beta) - \tfrac{1}{2}\cos(\alpha + \beta)$$

$$\cos \alpha \cos \beta = \tfrac{1}{2}\cos(\alpha - \beta) + \tfrac{1}{2}\cos(\alpha + \beta)$$

$$\sin \alpha \cos \beta = \tfrac{1}{2}\sin(\alpha + \beta) + \tfrac{1}{2}\sin(\alpha - \beta)$$

$$\cos \alpha \sin \beta = \tfrac{1}{2}\sin(\alpha + \beta) - \tfrac{1}{2}\sin(\alpha - \beta)$$

Function Sum and Difference Relations

$$\sin \alpha + \sin \beta = 2 \sin\tfrac{1}{2}(\alpha + \beta)\cos\tfrac{1}{2}(\alpha - \beta)$$

$$\sin \alpha - \sin \beta = 2 \cos\tfrac{1}{2}(\alpha + \beta)\sin\tfrac{1}{2}(\alpha - \beta)$$

$$\cos \alpha + \cos \beta = 2 \cos\tfrac{1}{2}(\alpha + \beta)\cos\tfrac{1}{2}(\alpha - \beta)$$

$$\cos \alpha - \cos \beta = -2 \sin\tfrac{1}{2}(\alpha + \beta)\sin\tfrac{1}{2}(\alpha - \beta)$$

$$\tan \alpha + \tan \beta = \frac{\sin(\alpha + \beta)}{\cos \alpha \cos \beta}$$

$$\tan \alpha - \tan \beta = \frac{\sin(\alpha - \beta)}{\cos \alpha \cos \beta}$$

$$\cot \alpha + \cot \beta = \frac{\sin(\alpha + \beta)}{\sin \alpha \sin \beta}$$

$$\cot \alpha - \cot \beta = \frac{\sin(\alpha - \beta)}{\sin \alpha \sin \beta}$$

$$\tan\tfrac{1}{2}(\alpha + \beta) = \frac{\sin \alpha + \sin \beta}{\cos \alpha + \cos \beta}$$

$$\tan\tfrac{1}{2}(\alpha - \beta) = \frac{\sin \alpha - \sin \beta}{\cos \alpha + \cos \beta}$$

Half-Angle Relations

$$\sin(\alpha/2) = -\sqrt{\frac{1 - \cos \alpha}{2}}$$

$$\cos(\alpha/2) = -\sqrt{(1 + \cos \alpha)/2}$$

$$\tan(\alpha/2) = -\sqrt{\frac{1 - \cos \alpha}{1 + \cos \alpha}}$$

$$= \frac{\sin \alpha}{1 + \cos \alpha} = \frac{1 - \cos \alpha}{\sin \alpha}$$

Power Relations

$$\sin^2\alpha = \tfrac{1}{2}(1 - \cos 2\alpha) \qquad \cos^2\alpha = \tfrac{1}{2}(1 + \cos 2\alpha)$$

$$\sin^3\alpha = \tfrac{1}{4}(3 \sin \alpha - \sin 3\alpha) \qquad \cos^3\alpha = \tfrac{1}{4}(3 \cos \alpha + \cos 3\alpha)$$

Exponential Relations: Euler Equation

$$e^{i\alpha} = \cos \alpha + i \sin \alpha \qquad i = \sqrt{-1}$$

$$\sin \alpha = \frac{e^{i\alpha} - e^{-i\alpha}}{2i}$$

$$\cos \alpha = \frac{e^{i\alpha} + e^{-i\alpha}}{2}$$

Relations Between Trigonometric Functions

	$\sin x = a$	$\cos x = b$	$\tan x = c$
$\sin x$	a	$\pm\sqrt{1 - b^2}$	$\pm c/\sqrt{1 + c^2}$
$\cos x$	$\pm\sqrt{1 - a^2}$	b	$\pm 1/\sqrt{1 + c^2}$
$\tan x$	$\pm a/\sqrt{1 - a^2}$	$\pm\sqrt{1 - a^2}/a$	c

Plane Triangle Formulas. For a given triangle, see Section 1.2, taking $s = (a + b + c)/2$, the following relations can be given:

Radius of inscribed circle: $r = \sqrt{(s - a)(s - b)(s - c)/s}$

Radius of circumscribed circle: $R = a/2 \sin A = b/2 \sin B = c/2 \sin C$

Law of sines: $a/\sin A = b/\sin B = c/\sin C$

Law of cosines: $a^2 = b^2 + c^2 - 2bc \cos A$ (similarly for b and c)

Law of tangents: $\dfrac{b - c}{b + c} = \dfrac{\tan(B - C)/2}{\tan(B + C)/2}$ (similarly for a, b and c, a)

1.3-2   Spherical Trigonometry

Oblique Spherical Angles

A, B, C = angles and a, b, c the sides measured as angles at the center of the sphere

$s = \frac{1}{2}(a + b + c)$

$S = \frac{1}{2}(A + B + C)$

Δ = area of triangle surface

E = spherical excess of triangle, see formulas below

R = radius of the sphere on which the triangle lies

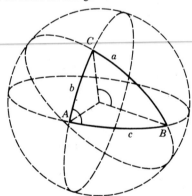

Formulas. Taking the sums of the angles defined above, either $a, b,$ and c or $A, B,$ and C varying as shown below, it can be seen that a point, a circle, and a plane triangle are all expressable in spherical coordinates.

$$0° \qquad < a + b + c < \qquad 360°$$
$$\uparrow \qquad\qquad\qquad\qquad\qquad \uparrow$$
$$\text{Point} \qquad\qquad\qquad\qquad \text{Circle}$$

$$180° \qquad < A + B + C < \qquad 540°$$
$$\uparrow \qquad\qquad\qquad\qquad\qquad \uparrow$$
$$\text{Plane triangle} \qquad\qquad\qquad \text{Circle}$$

$$E = A + B + C - 180°$$
$$\Delta = \pi R^2 E / 180°$$

Law of sines: $\sin a / \sin A = \sin b / \sin B = \sin c / \sin C$

Law of cosines for sides: $\cos a = \cos b \cos c + \sin b \sin c \cos A$ (similarly for b and c)

Law of cosines for angles: $\cos A = -\cos B \cos C + \sin B \sin C \cos a$ (similarly for B and C)

Law of tangents: $\dfrac{\tan\frac{1}{2}(B-C)}{\tan\frac{1}{2}(B+C)} = \dfrac{\tan\frac{1}{2}(b-c)}{\tan\frac{1}{2}(b+c)}$ (similarly for C, A and A, B pairs)

Half-angle formulas: $\tan\frac{1}{2}A = \dfrac{K}{\sin(s-a)}$ (similarly for B and C)

where $K^2 = \dfrac{\sin(s-a)\sin(s-b)\sin(s-c)}{\sin c}$

Half-side formulas: $\tan\frac{1}{2}a = K\cos(S-A)$ (similarly for b and c)

where $K^2 = \dfrac{-\cos S}{\cos(S-A)\cos(S-B)\cos(S-C)}$

Right Spherical Triangles

$C = 90°$, thus $\cos C = 0, \sin C = 1$.

$$\sin a = \tan b \cot B \qquad \sin a = \sin A \sin c$$
$$\sin b = \tan a \cot A \qquad \sin b = \sin B \sin c$$
$$\cos A = \tan b \cot c \qquad \cos A = \cos a \sin B$$
$$\cos B = \tan a \cot c \qquad \cos B = \cos b \sin A$$
$$\cos c = \cot A \cot B \qquad \cos c = \cos a \cos b$$

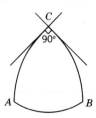

Hyperbolic Trigonometry

Definitions

$x = OM$

$y = MP$

$a = OA$

$v = $ shaded area$/a^2$

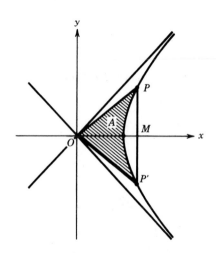

Hyperbolic sine: $\sinh v = y/a$

Hyperbolic cosine: $\cosh v = x/a$

Hyperbolic tangent: $\tanh v = \dfrac{\sinh v}{\cosh v}$

$$\operatorname{csch} v = \dfrac{1}{\sinh v}$$

$$\operatorname{sech} v = \dfrac{1}{\cosh v}$$

$$\coth v = \dfrac{1}{\tanh v}$$

Exponential Relations

$$e^{v} = \cosh v + \sinh v$$

$$e^{-v} = \cosh v - \sinh v$$

$$\sinh v = \tfrac{1}{2}\left(e^{v} - e^{-v}\right)$$

$$\cosh v = \tfrac{1}{2}\left(e^{v} + e^{-v}\right)$$

Identities

$$\sinh v = -\sinh(-v)$$

$$\cosh v = \cosh(-v)$$

$$\tanh v = -\tanh(-v)$$

$$\coth v = -\coth(-v)$$

$$\cosh^{2}v - \sinh^{2}v = 1$$

Sum and Difference Relations

$$\sinh(u + v) = \sinh u \cosh v + \cosh u \sinh v$$

$$\sinh(u - v) = \sinh u \cosh v - \cosh u \sinh v$$

$$\cosh(u + v) = \cosh u \cosh v + \sinh u \sinh v$$

$$\cosh(u - v) = \cosh u \cosh v - \sinh u \sinh v$$

Multiple-Argument Relations

$$\sinh 2v = 2 \sinh v \cosh v$$

$$\cosh 2v = \operatorname{conh}^{2}v + \sinh^{2}v$$

$$= 1 + 2 \sinh^{2}v$$

$$= -1 + 2 \cosh^{2}v$$

$$\sinh 3v = 3 \sinh v + 4 \sinh^{3}v$$

$$\cosh 3v = 3 \cosh^{3}v - 3 \cosh v$$

Inverse Hyperbolic Functions

$$\sinh^{-1}v = \log_e\left(v + \sqrt{v^2 + 1}\,\right)$$

$$\cosh^{-1}v = \log_e\left(v \pm \sqrt{v^2 + 1}\,\right) \qquad x \geq 1 \,(+\text{for principal value})$$

$$\tanh^{-1}v = \tfrac{1}{2}\log_e\left(\frac{1 + v}{1 - v}\right) \qquad v^2 < 1$$

Relations Between Trigonometric and Hyperbolic Functions

$$\sinh iv = i \sin v$$

$$\sinh v = -i \sin iv$$

$$\cosh iv = \cos v$$

$$\cosh v = \cos iv$$

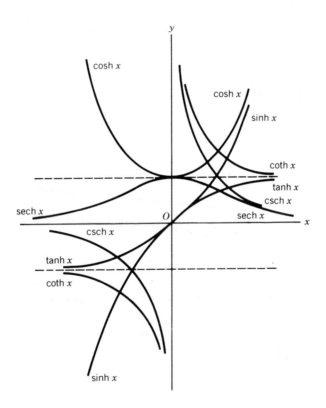

1.4   ANALYTIC GEOMETRY

Any point in a space is located by a set of quantities called its *coordinates* with respect to the *origin* whose coordinates are of the value zero. Depending on the *system of coordinates* the same point may be located by different sets of such quantities. Some of the types of coordinate system are introduced in this section.

1.4-1   Coordinate Systems on a Plane

Cartesian Coordinate System

On a plane any two straight lines OX and OY, called *axes*, intersecting each other at the *origin* form a *cartesian* coordinate system. If the point P lies on the same plane, the distance from P to the axes measured on the lines drawn parallel to the axes are called the *coordinates* of P in this cartesian coordinate system.

If the axes intersect each other at any angle, the coordinate system is called an *oblique cartesian coordinate system*. If the angle is 90°, the axes are said to be *perpendicular* (*orthogonal*) to each other, thus the special system is called a *rectangular coordinate system*.

Usually in a two-dimensional cartesian coordinate system the x axis is called the *abscissa* and the y axis is called the *ordinate*.

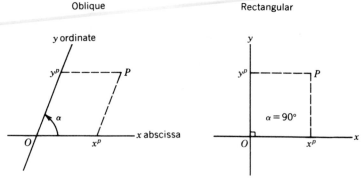

Oblique Coordinate System ($\alpha \neq 90°$)
Distance between any two points P_1, P_2:

$$\sqrt{(x_2 - x_1)^2 + (y_2 - y_1)^2 + 2(x_2 - x_1)(y_2 - y_1)\cos \alpha}$$

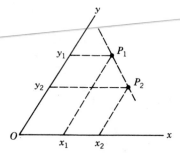

Three points, P_1, P_2, P_3, are *colinear* (lying on a line) if the determinant vanishes, that is,

$$\begin{vmatrix} x_1 & y_1 & 1 \\ x_2 & y_2 & 1 \\ x_3 & y_3 & 1 \end{vmatrix} = 0$$

Point P_3 dividing $P_1 P_2$ in the ratio of a/b has coordinates

$$\frac{ax_2 + bx_1}{a + b}, \qquad \frac{ay_2 + by_1}{a + b}$$

$$\overset{a \qquad b}{\underset{P_1 \quad P_3 \quad P_2}{\bullet\!-\!-\!\bullet\!-\!-\!\bullet}}$$

Line passing through points $P_1(c, 0)$, $P_2(0, d)$:

$$\frac{x}{c} + \frac{y}{d} = 1 \qquad \text{(intercept form)}$$

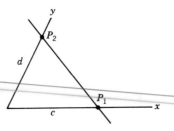

Line passing through any two points P_1, P_2:

$$\frac{y - y_1}{x - x_1} = \frac{y - y_2}{x - x_2} \quad \text{(two-point form)}$$

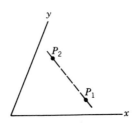

Line equation: $\qquad\qquad Cx + Dy + E = 0 \quad$ (general form)

Two lines: $\qquad$ parallel $\qquad\quad C_1 D_2 = C_2 D_1$

$\qquad\qquad\qquad$ perpendicular $\quad C_1 C_2 + D_1 D_2 = (C_1 D_2 + C_2 D_1)\cos \alpha$

Three lines concurrent: (Overlaping) if the determinant vanishes, that is,

$$\begin{vmatrix} C_1 & D_1 & E_1 \\ C_2 & D_2 & E_2 \\ C_3 & D_3 & E_3 \end{vmatrix} = 0$$

Rectangular Coordinate System ($\alpha = 90°$)

Distance between any two points: $\quad \sqrt{(x_2 - x_1)^2 + (y_2 - y_1)^2}$

Slope of $P_1 P_2$: $\qquad\qquad\qquad m = \tan \beta = \dfrac{y_2 - y_1}{x_2 - x_1}$

Line equation: $\qquad\qquad\qquad y = mx + b$

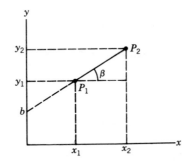

Angle between two lines with slopes m_1 and m_2:

$$\tan\theta = \frac{(m_2 - m_1)}{(1 + m_1 m_2)}$$

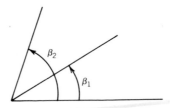

Parallel lines: $m_1 = m_2$

Perpendicular lines: $m_1 m_2 = -1$

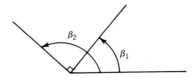

Plane Curves

Conics.   For a given fixed point F, called the *focus*, and a given fixed line L, called the *directrix*, the locus of points P at distance from F and L as f, l such that the ratio f/l is constant, the ratio f/l is called the *eccentricity*, e, v is the *vertex* line, and VF is the *axis*.

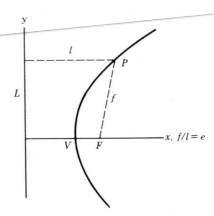

The general form of conics whose *axis* is oblique to the coordinate axes is

$$Ax^2 + Bxy + Cy^2 + Dx + Ey + F = 0$$

Ellipses: $e < 1$, $e = \sqrt{a^2 - b^2}/2$.   Two directrices, $2a$ = major axis, $2b$ = minor axis, center is at the origin, foci on x axis:

$$\frac{x^2}{a^2} + \frac{y^2}{b^2} = 1$$

In the general form, $B^2 - 4AC < 0$.

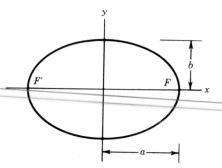

Circles: $a = b$, $e = 0$. Center is at the origin, radius $r = a$:

$$x^2 + y^2 = r^2$$

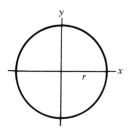

Parabolas: $e = 1$. One directrix, vertex at the origin, focus at $(p, 0)$:

$$y^2 = 4px$$

In the general form, $B^2 - 4AC = 0$.

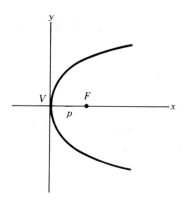

Hyperbolas: $e > 1$, $e = \sqrt{a^2 + b^2}\,/2$. Two directrices, $2a$ = transfer axis, $2b$ = conjugate axis, center is at origin, foci on x axis:

$$\frac{x^2}{a^2} - \frac{y^2}{b^2} = 1$$

Regular hyperbola, $a = b$, $e = \sqrt{2}$, and asymptotes are perpendicular.

In the general form, $B^2 - 4AC => 0$.

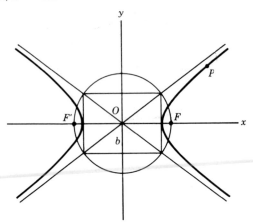

Polar Coordinate System

O = pole, origin
Ox = initial line
θ = vectorial angle
r = radius of vector
r, θ = polar coordinates of the point P

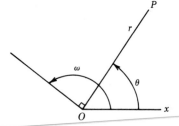

Distance between P_1 and P_2:

$$\sqrt{r_1^2 + r_2^2 - 2r_1 r_2 \cos(\theta_1 - \theta_2)}$$

Points P_1, P_2, P_3 are colinear if

$$r_2 r_3 \sin(\theta_3 - \theta_2) - r_3 r_1 \sin(\theta_3 - \theta_1) + r_1 r_2 \sin(\theta_2 - \theta_1) = 0$$

Straight line:

$$r\cos(\theta - \omega) = p \quad (normal\ form)$$

$$r(r_1 \sin(\theta - \theta_1) - r_2 \sin(\theta - \theta_2)) = \sin(\theta_2 - \theta_1) \quad (two\text{-}point\ form)$$

Relations between polar and rectangular coordinates:

$$x = r\cos\theta \qquad y = r\sin\theta$$

$$r = \sqrt{x^2 + y^2}$$

$$\theta = \arctan\frac{y}{x}$$

$$\sin\theta = \frac{y}{\sqrt{x^2 + y^2}} \qquad \cos\theta = \frac{x}{\sqrt{x^2 + y^2}}$$

Archimedean Spiral

$$r = a\theta$$

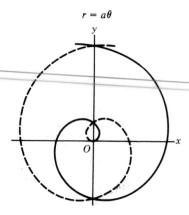

Astroid (Hypocycloid of Four Cusps)

$$x^{2/3} + y^{2/3} = a^{2/3}$$

$$x = a\cos^3\phi \qquad y = a\sin^3\phi$$

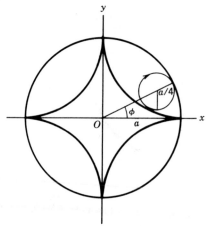

Deltoid (Hypocycloid of Three Cusps)

$$x = 2a\cos\phi + a\cos 2\phi \qquad y = 2a\sin\phi - a\sin 2\phi$$

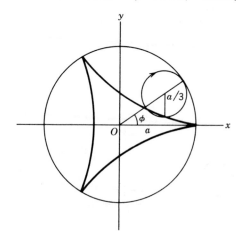

Catenary (Hyperbolic Cosine)

$$y = a \cosh(x/a)$$

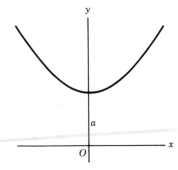

Cissoid of Diocles

$$y^2(a - x) = x^3$$

$$r = a \sin\theta \tan\theta$$

$$OB = AP$$

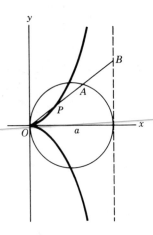

Cycloid with Vertex at the Origin
a = radius of the circle rolling on x axis

$$x = 2a \arcsin\sqrt{y/2a} + \sqrt{2ay - y^2}$$

$$x = a(\phi + \sin\phi)$$

$$y = a(1 - \cos\phi)$$

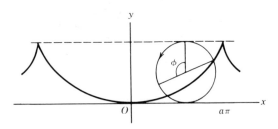

Cycloid, Curtate

$$x = a\phi - b \sin\phi$$

$$y = a - b \cos\phi \qquad a > b$$

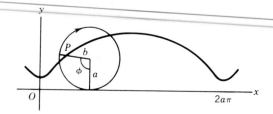

Cycloid, Prolate

$$x = a\phi - b \sin\phi$$

$$y = a - b \cos\phi \qquad a < b$$

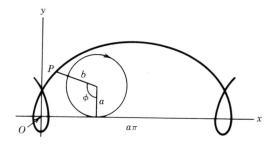

Epicycloid. Here b is the radius of the circle rolling on a circle of the radius a.

$$x = (a + b)\cos\phi - b \cos\left(\frac{a + b}{b}\phi\right)$$

$$y = (a + b)\sin\phi - b \sin\left(\frac{a + b}{b}\phi\right)$$

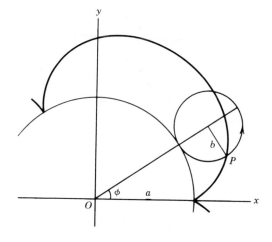

Folium of Descartes

$$x^3 + y^3 - 3axy = 0$$

$$x = \frac{3a\phi}{1 + \phi^3}$$

$$y = \frac{3a\phi^2}{1 + \phi^3}$$

$$r = \frac{3a \sin\theta \cos\theta}{\sin^3\theta + \cos^3\theta}$$

Asymptote:
$$x + y + a = 0$$

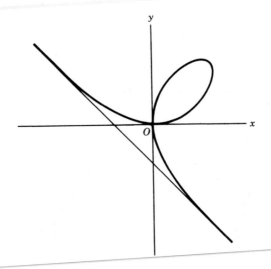

Lemniscate of Bernoulli, Two-Level Rose

$$\left(x^2 + y^2\right)^2 = a^2\left(x^2 - y^2\right)$$

$$r^2 = a^2\cos 2\theta$$

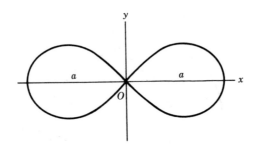

Ovals of Cassini

$$\left(x^2 + y^2 + b^2\right)^2 - 4b^2x^2 = k^4$$

$$F'P \cdot FP = k^2$$

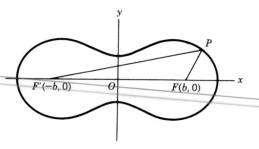

1.4-2 Coordinate Systems in a Space

Rectangular Coordinate System

The dimension of the space is $n = 3$. The rectangular coordinates of a point P are x, y, z, the distance of P from the yz, xz, and xy planes, respectively. If there are three points, P_1, P_2, P_3, the following definitions can be given:

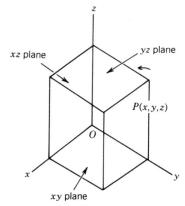

Distance between P_1 and P_2:

$$\sqrt{(x_2 - x_1)^2 + (y_2 - y_1)^2 + (z_2 - z_1)^2}$$

P_1, P_2, P_3 are *colinear* if and only if

$$x_2 - x_1 : y_2 - y_1 : z_2 - z_1 = x_3 - x_1 : y_3 - y_1 : z_3 - z_1$$

P_1, P_2, P_3, P_4 are *coplanar* if and only if the determinant vanishes, that is,

$$\begin{vmatrix} x_1 & y_1 & z_1 & 1 \\ x_2 & y_2 & z_2 & 1 \\ x_3 & y_3 & z_3 & 1 \\ x_4 & y_4 & z_4 & 1 \end{vmatrix} = 0$$

Point P_3 dividing P_1, P_2 in the ratio a/b has coordinates

$$\frac{ax_2 + bx_1}{a + b}, \quad \frac{ay_2 + by_1}{a + b}, \quad \frac{az_2 + bz_1}{a + b}$$

Direction Cosines

The angles between $P_1 P_2$ and the three axes x, y, z are called *direction angles*, α, β, γ.
If the distance between P_1 and P_2 is d, then the *direction cosines* of $P_1 P_2$ are

$$\cos \alpha = \frac{x_2 - x_1}{d} \qquad \cos \beta = \frac{y_2 - y_1}{d} \qquad \cos \gamma = \frac{z_2 - z_1}{d}$$

The sum of squares of direction cosines is equal to 1:

$$\cos^2\alpha + \cos^2\beta + \cos^2\gamma = 1$$

For parallel lines,

$$\alpha_1 = \alpha_2 \qquad \beta_1 = \beta_2 \qquad \gamma_1 = \gamma_2$$

For perpendicular lines,

$$\cos\alpha_1\cos\alpha_2 + \cos\beta_1\cos\beta_2 + \cos\gamma_1\cos\gamma_2 = 0$$

Line equations are as follows:

Point direction form:
$$\frac{x - x_1}{a} = \frac{y - y_1}{b} = \frac{z - z_1}{c}$$

Two point form:
$$\frac{x - x_1}{x_2 - x_1} = \frac{y - y_1}{y_2 - y_1} = \frac{z - z_1}{z_2 - z_1}$$

General form:
$$A_1 x + B_1 y + C_1 z + D_1 = 0$$
$$A_2 x + B_2 y + C_2 z + D_2 = 0$$

Parametric form:
$$x = x_1 + ta$$
$$y = y_1 + tb$$
$$z = z_1 + tc$$

Plane equation:
$$Ax + By + Cz + D = 0$$

Intercept form:
$$\frac{x}{a} + \frac{y}{b} + \frac{z}{c} = 1$$

Plane through point P_1 and perpendicular to direction (a, b, c):

$$a(x - x_1) + b(y - y_1) + c(z - z_1) = 0$$

Taking the general form $Ax + By + Cz + D = 0$, we have:

Parallel planes:
$$A_1 : B_1 : C_1 = A_2 : B_2 : C_2$$

Perpendicular planes:
$$A_1 A_2 + B_1 B_2 + C_1 C_2 = 0$$

Quadric Surfaces

Ellipsoid

$$\frac{x^2}{a^2} + \frac{y^2}{b^2} + \frac{z^2}{c^2} = 1$$

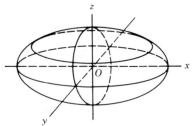

Sphere

$$x^2 + y^2 + z^2 = 1$$

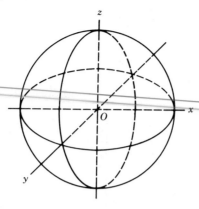

Elliptic Cone

$$\frac{x^2}{a^2} + \frac{y^2}{b^2} - \frac{z^2}{c^2} = 0$$

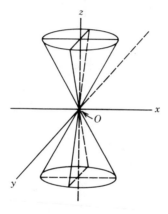

Elliptic Paraboloid

$$\frac{x^2}{a^2} + \frac{y^2}{b^2} = cz$$

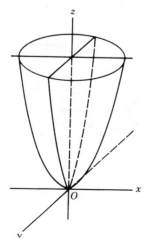

Elliptic Cylinder

$$\frac{x^2}{a^2} + \frac{y^2}{b^2} = 1$$

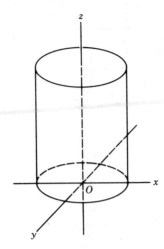

Hyperboloid of One Sheet

$$\frac{x^2}{a^2} + \frac{y^2}{b^2} - \frac{z^2}{c^2} = 1$$

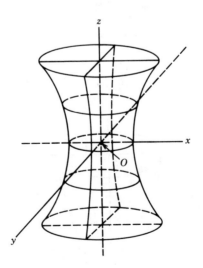

Hyperboloid of Two Sheets

$$\frac{x^2}{a^2} - \frac{y^2}{b^2} - \frac{z^2}{c^2} = 1$$

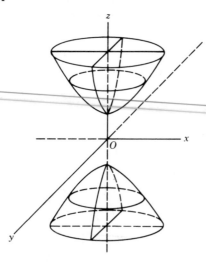

Hyperbolic Paraboloid

$$\frac{x^2}{a^2} - \frac{y^2}{b^2} = cz$$

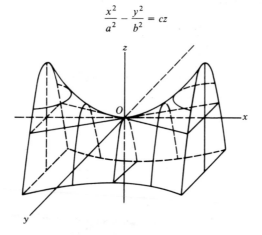

Transformation of Rectangular Coordinates

The rectangular coordinates of the old system (x, y, z) and of the new system (x', y', z'), of the origin of the new system, (h, k, l):

Translation

$$x = x' + h$$
$$y = y' + k$$
$$z = z' + l$$

Rotation About the Origin. (λ, μ, ν) are direction cosines of x', y', z' axes with respect to the old system.

$$x = \lambda_1 x' + \lambda_2 y' + \lambda_3 z'$$

$$y = \mu_1 x' + \mu_2 y' + \mu_3 z'$$

$$z = \nu_1 x' + \nu_2 y' + \nu_3 z'$$

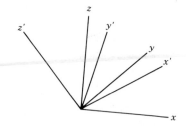

Cylindrical Coordinate System

For (x, y, z) cartesian (rectangular) coordinates,

$$x = r \cos \theta \qquad r = \sqrt{x^2 + y^2}$$

$$y = r \sin \theta \qquad r = \arctan(y/x)$$

$$z = z \qquad\qquad z = z$$

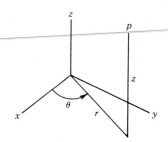

Spherical Coordinate System

For (x, y, z) cartesian (rectangular) coordinates,

$$x = \rho \cos \theta \sin \phi$$

$$y = \rho \sin \theta \sin \phi$$

$$z = \rho \cos \phi$$

$$\phi = \arccos \frac{z}{\sqrt{x^2 + y^2 + z^2}}$$

$$\theta = \arctan \frac{y}{z}$$

$$\rho^2 = x^2 + y^2 + z^2$$

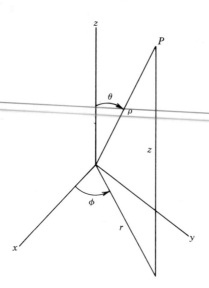

1.5 MATRIX ALGEBRA

1.5-1 Definitions

A *matrix* is an array of elements consisting of m rows and n columns:

$$\mathbf{A} = \begin{bmatrix} a_{11} & a_{12} & \cdots & a_{1n} \\ a_{21} & a_{22} & \cdots & a_{2n} \\ \vdots & \vdots & \cdots & \vdots \\ a_{m1} & a_{m2} & \cdots & a_{mn} \end{bmatrix}$$

Here a_{ij} indicates the element in the ith row and jth column.

If $m = n$, then the matrix is called a *square* matrix, otherwise it is a *rectangular* matrix. The diagonal spanning between the upper left corner and the lower right corner is called the *principal diagonal*.

The *transpose of a matrix* $\mathbf{A}$, denoted by $\mathbf{A}'$, or $\mathbf{A}^{\mathrm{T}}$, is the matrix with elements defined as

$$a_{ij} = a'_{ji}$$

Special forms of square matrix are:

The *symmetric* square matrix is the matrix with the property $\mathbf{A} = \mathbf{A}'$, that is, $a_{ij} = a_{ji}$.
The *skew-symmetric* (or *antisymmetric*) square matrix is the matrix with the property $\mathbf{A} = -\mathbf{A}'$, that is, $a_{ij} = -a_{ji}$.
A *lower triangular matrix* has all zeros as elements above the principal diagonal.
An *upper triangular matrix* has all zeros as elements below the principal diagonal.
A *diagonal matrix* has all its elements zero except those along the principal diagonal, and is denoted by $\mathbf{D}$.

Example:

$$\mathbf{D} = \begin{bmatrix} a_{11} & 0 & 0 \\ 0 & a_{22} & 0 \\ 0 & 0 & a_{33} \end{bmatrix}$$

Special forms of the rectangular matrix are:

A *column vector* is a matrix with m rows and 1 column.

A *row vector* is a matrix with n columns and 1 row.

A *scalar* is a matrix with 1 column and 1 row.

1.5-2 Operations

Addition of two matrices with the same number of columns and rows is

$$\mathbf{A} + \mathbf{B} = \mathbf{C}$$

where the elements are calculated as

$$a_{ij} + b_{ij} = c_{ij} \qquad i = 1, 2, \ldots, m \quad and \quad j = 1, 2, \ldots, n$$

Multiplication of a matrix by a scalar,

$$\mathbf{B} = \gamma\mathbf{A}$$

is calculated by the relation

$$b_{ij} = \gamma a_{ij}$$

Multiplication of two matrices,

$$\mathbf{AB} = \mathbf{C}$$

is possible only if the number of columns in **A** is equal to the number of rows of **B**. The elements are calculated as

$$c_{ij} = \sum_{k=1}^{n} a_{ik} b_{kj}$$

Example:

$$\mathbf{A} = \begin{bmatrix} 3 & 2 \\ 1 & 2 \\ 4 & 3 \end{bmatrix} \qquad \mathbf{B} = \begin{bmatrix} 2 & 3 & 3 & 1 \\ 5 & 4 & 2 & 3 \end{bmatrix}$$

A is a 3×2 matrix, **B** is a 2×4 matrix, thus **AB** is defined and equal to

$$\mathbf{C} = \mathbf{AB} = \begin{bmatrix} 3\cdot2+2\cdot5 & 3\cdot3+2\cdot4 & 3\cdot3+2\cdot2 & 3\cdot1+2\cdot3 \\ 1\cdot2+2\cdot5 & 1\cdot3+2\cdot3 & 1\cdot3+2\cdot2 & 1\cdot1+2\cdot3 \\ 4\cdot2+3\cdot5 & 4\cdot3+3\cdot4 & 4\cdot3+3\cdot2 & 4\cdot1+3\cdot3 \end{bmatrix}$$

$$\mathbf{C} = \begin{bmatrix} 16 & 17 & 13 & 9 \\ 12 & 9 & 7 & 7 \\ 23 & 24 & 18 & 13 \end{bmatrix}$$

The Laws of Matrix Operations

(a) Commutative law: $\mathbf{AB} \neq \mathbf{BA}$.

(b) Associative law: $\mathbf{A(BC)} = \mathbf{(AB)C}$.

(c) Distributive law: $\mathbf{(A + B)C} = \mathbf{AC} + \mathbf{BC}$.

Transpose of product of two matrices is equal to the product of the transpose of individual matrices, however in reverse order as follows:

$$\mathbf{(AB)'} = \mathbf{B'A'}$$

1.5-3 Special Forms

A *null matrix*, **0**, is a matrix with all its elements equal to zero.

The *identity matrix*, **I**, is a diagonal matrix with all its diagonal elements equal to one.

A *diagonal matrix*, $\gamma\mathbf{I}$, where γ is a scalar, is called a *scalar matrix*.

If a square matrix **Q** is given, and **x** is a column vector, and **x**′ its transpose, a row vector, and **y**, a column vector with equal number of elements, then:

x′**Qx** is called a *quadratic form*.
x′**Qy** is called a *bilinear form*.
$\mathbf{x}'\mathbf{x} = \Sigma x_i^2$ is the sum of *squares* of all elements.
$\mathbf{x}'\mathbf{y} = \Sigma x_i y_i$ is the sum of *products* of elements in **x** by those in **y**.

If the elements of a matrix are complex, that is,

$$a_{kj} = b_{kj} + ic_{kj} \quad \text{where } i = \sqrt{-1},$$

the matrix with elements conjugate to a_{kj}, that is,

$$a_{kj} = b_{kj} - ic_{kj}$$

is called the *conjugate matrix*, **A**.

$\mathbf{A}^H$ is called a *Hermitian conjugate matrix* which is obtained by transposing **A** and replacing its elements by conjugate complex. Since a symmetric (square) matrix is by definition $\mathbf{A} = \mathbf{A}'$, a square matrix is called Hermitian if

$$\mathbf{A} = \mathbf{A}^H$$

If $\mathbf{A}' = \mathbf{A}^{-1}$, then **A** is a *unitary* matrix. Here the inverse of a matrix $\mathbf{A}^{-1}$ is defined later in this section.

1.5-4 Determinants

The determinant of a square matrix, denoted as $|\mathbf{A}|$ or det(**A**), is defined as a sum of the products, and it is a scalar:

$$
\begin{vmatrix}
a_{11} & a_{12} & \cdots & a_{1n} \\
a_{21} & a_{22} & \cdots & a_{2n} \\
\cdot & \cdot & \cdots & \cdot \\
\cdot & \cdot & \cdots & \cdot \\
a_{n1} & a_{n2} & \cdots & a_{nn}
\end{vmatrix}
= \Sigma (-1)^{\delta} a_{1i_1} a_{2i_2} \cdots a_{ni_n}
$$

where $i_1, i_2, \ldots, i_n$ indicate n different columns and δ is the number of exchanges to put $i_1, i_2, \ldots, i_n$ in order or $1, 2, \ldots, n$. As a consequence, if any two rows (or columns) of a matrix are interchanged, the sign of the determinant changes.

For example, a 3×3 matrix has the determinant

$$
\begin{vmatrix}
a_{11} & a_{12} & a_{13} \\
a_{21} & a_{22} & a_{23} \\
a_{31} & a_{32} & a_{33}
\end{vmatrix}
= (-1)^0 a_{11} a_{22} a_{33} + (-1)^1 a_{11} a_{23} a_{32}
$$

$$+ (-1)^1 a_{12} a_{21} a_{33} + (-1)^4 a_{12} a_{23} a_{31}$$

$$+ (-1)^2 a_{13} a_{21} a_{32} + (-1)^3 a_{13} a_{22} a_{31}$$

$$= a_{11}(a_{22} a_{33} - a_{23} a_{32})$$

$$- a_{12}(a_{21} a_{33} - a_{23} a_{31})$$

$$+ a_{13}(a_{21} a_{32} - a_{22} a_{31})$$

A numerical example of determinant is calculated below for illustration:

$$
\begin{vmatrix}
3 & 3 & 2 \\
4 & 1 & 5 \\
3 & 4 & 1
\end{vmatrix}
= 3(1 \cdot 1 - 5 \cdot 4) - 3(4 \cdot 1 - 3 \cdot 5) + 2(4 \cdot 4 - 3 \cdot 1)
$$

$$= 3(-19) - 3(-11) + 2(13) = -57 + 33 + 26 = +2$$

The definition also implies that if a row or a column of a matrix is multiplied by a scalar α then the determinant is multiplied by the same scalar. If an (n, n) matrix is multiplied by a scalar, αA, then its determinant becomes $|\alpha A| = \alpha^n |A|$.

The product of the determinants of two matrices is

$$|A||B| = |AB|$$

The determinants of a matrix and of its transpose are equal, $|A| = |A'|$.

The matrix obtained by removing a row (rth) and a column (sth) is called the *minor* of the element a_{rs}, and denoted by M_{rs}. The *cofactor* of a_{rs} is the scalar found as $A_{rs} = (-1)^{r+s}|M_{rs}|$. The *cofactor matrix* is the matrix with A_{rs} as it elements. Based on the cofactors another definition of the determinant of a matrix is given as

$$|A| = \sum_{r=1}^{n} a_{rs}(-1)^{r+s}A_{rs}$$

An example for a 3×3 matrix is

$$\begin{vmatrix} a_{11} & a_{12} & a_{13} \\ a_{21} & a_{22} & a_{23} \\ a_{31} & a_{32} & a_{33} \end{vmatrix} = a_{11}\begin{vmatrix} a_{22} & a_{23} \\ a_{32} & a_{33} \end{vmatrix} - a_{12}\begin{vmatrix} a_{21} & a_{23} \\ a_{31} & a_{33} \end{vmatrix} + a_{13}\begin{vmatrix} a_{21} & a_{22} \\ a_{31} & a_{33} \end{vmatrix}$$

With the numerical values given, the determinant is evaluated as

$$\begin{vmatrix} 3 & 3 & 2 \\ 4 & 1 & 5 \\ 3 & 4 & 1 \end{vmatrix} = 3\begin{vmatrix} 1 & 5 \\ 4 & 1 \end{vmatrix} - 3\begin{vmatrix} 4 & 5 \\ 3 & 1 \end{vmatrix} + 2\begin{vmatrix} 4 & 1 \\ 3 & 4 \end{vmatrix}$$

The *transpose* of the matrix whose elements are A_{rs} is called the *adjoint matrix*, adj A.

Rank and Singularity

A matrix is called *singular* if it satisfies the relation, $Ax = 0$ or $A'x = 0$ where x is a nonzero column vector. Otherwise it is *nonsingular*.

The *rank* of a matrix A is the maximum number of rows or columns, r of the nonsingular square submatrix obtained from the matrix A with n rows or columns. The nonsingular submatrix of a singular matrix is called a *basis* of the matrix A.

If $r = n$ the matrix is nonsingular, and its determinant is nonzero. However, if $r < n$ then the matrix is singular and its determinant is zero.

Inverse of a Matrix

For a nonsingular matrix A, the matrix A^{-1} satisfying the relation, $AA^{-1} = A^{-1}A = I$ is called the *inverse* matrix. The inverse matrix is unique.

Example:

$$A = \begin{bmatrix} 1 & 2 & 1 \\ 3 & 1 & 4 \\ 2 & 1 & 3 \end{bmatrix}$$

$$A^{-1}A = I$$

$$\begin{bmatrix} a_{11} & a_{12} & a_{13} \\ a_{11} & a_{12} & a_{13} \\ a_{11} & a_{12} & a_{13} \end{bmatrix}\begin{bmatrix} 1 & 2 & 1 \\ 3 & 1 & 4 \\ 2 & 1 & 3 \end{bmatrix} = \begin{bmatrix} 1 & 0 & 0 \\ 0 & 1 & 0 \\ 0 & 0 & 1 \end{bmatrix}$$

By matrix multiplication,

$$a_{11} + 3a_{12} + 2a_{13} = 1$$

$$2a_{11} + a_{12} + a_{13} = 0$$

$$a_{11} + 4a_{12} + 3a_{13} = 0$$

Solving for a_{11}, a_{12}, a_{13}; $a_{11} = 1/2$, $a_{12} = 5/2$, $a_{13} = -7/2$:

$$a_{21} + 3a_{22} + 2a_{23} = 0$$

$$2a_{21} + a_{22} + a_{23} = 1$$

$$a_{21} + 4a_{22} + 3a_{23} = 0$$

Solving for a_{21}, a_{22}, a_{23}; $a_{21} = 1/2$, $a_{22} = -1/2$, $a_{23} = 1/2$:

$$a_{31} + 3a_{32} + 2a_{33} = 0$$

$$2a_{31} + a_{32} + a_{33} = 0$$

$$a_{31} + 4a_{32} + 3a_{33} = 1$$

Solving for a_{31}, a_{32}, a_{33}; $a_{31} = -1/2$, $a_{32} = -3/2$, $a_{33} = 5/2$. Thus the inverse matrix is

$$\mathbf{A}^{-1} = \begin{bmatrix} 1/2 & 5/2 & -7/2 \\ 1/2 & -1/2 & 1/2 \\ -1/2 & -3/2 & 5/2 \end{bmatrix}$$

If the inverse exists,

$$(\mathbf{A}^{-1})' = (\mathbf{A}')^{-1}$$

$$(\mathbf{ABC})^{-1} = \mathbf{C}^{-1}\mathbf{B}^{-1}\mathbf{C}^{-1}$$

$$|\mathbf{A}^{-1}| = \frac{1}{|\mathbf{A}|}$$

Some Special Determinants

A *Wronskian* is the determinant

$$|f_i^{(k)}| \quad \text{where} \quad f_i^{(k)} = \frac{d^k f_i}{dx^k}, \quad k = 0, 1, \ldots, n-1$$

Here $f_i^{(k)}$ is the kth derivative of f_i. The definition of derivative is given in Section 1.6. If the Wronskian vanishes the functions f_i are *linearly dependent*, or there is a relation such that

$$c_1 f_1 + \cdots + c_n f_n = 0$$

where c_i are nonzero constants.

Example: Let us assume that the three solutions of an equation are found as $f_1(x) = x^2$, $f_2(x) = x^3$ and $f_3(x) = x^2 + x^3$. The Wronskian is written as the determinant:

$$\begin{vmatrix} f_1 & f_2 & f_3 \\ f_1' & f_2' & f_3' \\ f_1'' & f_2'' & f_3'' \end{vmatrix}$$

Substituting the functions and derivatives in their positions in the Wronskian, one finds

$$\begin{vmatrix} x^2 & x^3 & x^2 & x^3 \\ 2x & 3x^2 & 2x & 3x^2 \\ 2 & 6x & 2 & 6x \end{vmatrix} = 0$$

Thus these three functions must be linearly dependent and the coefficients C_1, C_2, and C_3 below must be nonzero:

$$C_1 f_1 + C_2 f_2 + C_3 f_3 = 0$$

Substituting the functions in this equation,

$$C_1 x^2 + C_2 x^3 + C_3 (x^2 + x^3) = 0$$

Equating the coefficients of the equal powers of x

$$C_1 = -C_3 \qquad C_2 = -C_3$$

Thus $C_1 = C_2 = -C_3$, and there can be infinitely many sets of nonzero C values.

The *Jacobian* is the determinant

$$\left| \frac{\partial f_i}{\partial x_k} \right| = \left| \frac{\partial (f_1, \ldots, f_n)}{\partial (x_1, \ldots, x_n)} \right|$$

Here $\partial f_i / \partial x_k$ is a partial derivative of f_i. The definition of partial derivative is given in Section 1.6. If the Jacobian vanishes, there is a *functional dependence* between f_i, that is, there is a function as $F(f_1, \ldots, f_n) = 0$ for all values of the variables, x_i.

Example: The Jacobian determinant is used in solving implicit functions for determining transformations. For example, let us assume that w is a function of x and y as

$$w = x(x, y)$$

If we apply the transformations

$$f_1 = f_1(x, y)$$

$$f_2 = f_2(x, y)$$

then x and y may be obtained in terms of f_1 and f_2 as

$$x = x(f_1, f_2)$$

$$y = y(f_1, f_2)$$

Thus by substituting these in w,

$$w = \Omega(f_1, f_2)$$

However, this may not be easy. Instead the following technique is used. First the function w is differentiated with respect to x and y,

$$\frac{\partial w}{\partial x} = \frac{\partial w}{\partial f_1} \frac{\partial f_1}{\partial x} + \frac{\partial w}{\partial f_2} \frac{\partial f_2}{\partial x}$$

$$\frac{\partial w}{\partial y} = \frac{\partial w}{\partial f_1} \frac{\partial f_1}{\partial y} + \frac{\partial w}{\partial f_2} \frac{\partial f_2}{\partial y}$$

To find the derivatives of w with respect to f_1 and to f_2, that is, $\partial w / \partial f_1$ and $\partial w / \partial f_2$, the above set of linear equations is solved (see Simultaneous Linear Equations and particularly Cramer's rule in this section):

$$\frac{\partial w}{\partial f_1} = \frac{\begin{vmatrix} \partial w / \partial x & \partial f_2 / \partial x \\ \partial w / \partial y & \partial f_2 / \partial y \end{vmatrix}}{\begin{vmatrix} \partial f_1 / \partial x & \partial f_2 / \partial x \\ \partial f_1 / \partial y & \partial f_2 / \partial y \end{vmatrix}} \qquad \frac{\partial w}{\partial f_2} = \frac{\begin{vmatrix} \partial f_1 / \partial x & \partial w / \partial x \\ \partial f_1 / \partial y & \partial w / \partial y \end{vmatrix}}{\begin{vmatrix} \partial f_1 / \partial x & \partial f_2 / \partial x \\ \partial f_1 / \partial y & \partial f_2 / \partial y \end{vmatrix}}$$

The denominators are the Jacobian. If they vanish than there is no solution; otherwise the solutions exist.

1.5-5 Simultaneous Linear Equations

The simultaneous linear equations can be represented in a matrix form as

$$\mathbf{Ax} = \mathbf{b}$$

where $\mathbf{A}$ is a (n, n) matrix, and $\mathbf{x}$ and $\mathbf{b}$ are column vectors. The solution of this equation is unique and expressed in terms of the inverse of $\mathbf{A}$,

$$\mathbf{x} = \mathbf{A}^{-1}\mathbf{b}$$

Example: Let $\mathbf{A}$ be a 3×3 matrix,

$$\mathbf{A} = \begin{bmatrix} 1 & 2 & 1 \\ 3 & 1 & 4 \\ 2 & 1 & 3 \end{bmatrix}$$

$$\mathbf{x} = \begin{bmatrix} x_1 \\ x_2 \\ x_3 \end{bmatrix} \qquad \mathbf{b} = \begin{bmatrix} 3 \\ 5 \\ 4 \end{bmatrix}$$

For $\mathbf{Ax} = \mathbf{b}$, the solution is

$$\mathbf{x} = \mathbf{A}^{-1}\mathbf{b}$$

Therefore the inverse is searched. Since $\mathbf{A}$ is the same matrix as the one given in section on inverse,

$$\mathbf{A}^{-1} = \begin{bmatrix} 1/2 & 5/2 & -7/2 \\ 1/2 & -1/2 & 1/2 \\ -1/2 & -3/2 & 5/2 \end{bmatrix}$$

$$\mathbf{x} = \begin{bmatrix} 1/2 & 5/2 & -7/2 \\ 1/2 & -1/2 & 1/2 \\ -1/2 & -3/2 & 5/2 \end{bmatrix} \begin{bmatrix} 3 \\ 5 \\ 4 \end{bmatrix}$$

$$\mathbf{x} = \begin{bmatrix} 3/2 + 25/2 - 28/2 \\ 3/2 - 5/2 + 2 \\ -3/2 - 15/2 + 20/2 \end{bmatrix} = \begin{bmatrix} 0 \\ 1 \\ 1 \end{bmatrix}$$

Thus the solution is $x_1 = 0$, $x_2 = 1$, $x_3 = 1$.

Gauss–Jordan Elimination

If a set of simultaneous linear equations are given as

$$a_{11}x_1 + a_{12}x_2 + \cdots + a_{1n}x_n = b_1$$

$$a_{21}x_1 + a_{22}x_2 + \cdots + a_{2n}x_n = b_2$$

$$\vdots$$

$$a_{n1}x_1 + a_{n2}x_2 + \cdots + a_{nn}x_n = b_n$$

by dividing the first equation by the coefficient of its first term, a_{11}, and eliminating the x_1 terms from the remaining equations, and continuing this elimination process for the second, third, and the other equations, the system can be reduced to the *canonical form*:

$$\begin{aligned} x_1 \qquad\quad &= b_1^* \\ x_2 \quad\ &= b_2^* \\ \vdots \\ x_n &= b_n^* \end{aligned}$$

where the new values of the right-hand sides correspond directly to the solution of the system of equations.

Example:

$$\mathbf{A} = \begin{bmatrix} 1 & 2 & 1 \\ 3 & 1 & 4 \\ 2 & 1 & 3 \end{bmatrix}$$

$$\mathbf{b} = \begin{bmatrix} 3 \\ 5 \\ 4 \end{bmatrix}$$

Divide the first equation by the coefficients of x_1, and then eliminate x_1 terms from the other two:

$$x_1 + 2x_2 + x_3 = 3$$

$$-5x_2 + x_3 = -4$$

$$-3x_2 + x_3 = -2$$

Divide the new second equation by the coefficients of x_2, that is, -5, then eliminate the x_2 terms from the first and third equations:

$$x_1 \quad + \tfrac{7}{5}x_3 = \tfrac{7}{5}$$

$$x_2 - \tfrac{1}{5}x_3 = \tfrac{4}{5}$$

$$\tfrac{2}{5}x_3 = \tfrac{2}{5}$$

Divide the third equation by the coefficients of x_3, that is, $\tfrac{2}{5}$, and eliminate the x_3 terms from the first and second equations:

$$x_1 \qquad = 0$$

$$x_2 \quad = 1$$

$$x_3 = 1$$

Cramer's Rule

Both sides of the simultaneous linear equations, $\mathbf{Ax} = \mathbf{b}$ can be multiplied by the inverse of $\mathbf{A}$ from left to obtain

$$\mathbf{A}^{-1}\mathbf{Ax} = \mathbf{A}^{-1}\mathbf{b}$$

where $\mathbf{A}^{-1}\mathbf{A} = \mathbf{I}$, and $\mathbf{Ix} = \mathbf{x}$. Therefore,

$$\mathbf{x} = \mathbf{A}^{-1}\mathbf{b} = \frac{\text{adj } \mathbf{A}}{\det \mathbf{A}}\mathbf{b}$$

which is called *Cramer's Rule*.

Example:

$$\mathbf{x} = \mathbf{A}^{-1}\mathbf{b} = \frac{\text{adj } \mathbf{A}}{\det \mathbf{A}}\mathbf{b} \qquad \mathbf{b} = \begin{bmatrix} 3 \\ 5 \\ 4 \end{bmatrix}$$

$$\mathbf{A} = \begin{bmatrix} 1 & 2 & 1 \\ 3 & 1 & 4 \\ 2 & 1 & 3 \end{bmatrix}$$

$$\text{adj } \mathbf{A} = \begin{bmatrix} A_{11} & A_{21} & A_{31} \\ A_{12} & A_{22} & A_{23} \\ A_{31} & A_{32} & A_{33} \end{bmatrix}$$

Elements of the adj **A** are calculated as follows:

$$A_{11} = (-1)^2 \begin{vmatrix} 1 & 4 \\ 1 & 3 \end{vmatrix} = -1$$

$$A_{12} = (-1)^3 \begin{vmatrix} 3 & 4 \\ 2 & 3 \end{vmatrix} = -1$$

$$A_{13} = (-1)^4 \begin{vmatrix} 3 & 1 \\ 2 & 1 \end{vmatrix} = +1$$

$$A_{21} = (-1)^3 \begin{vmatrix} 2 & 1 \\ 1 & 3 \end{vmatrix} = -5$$

$$A_{22} = (-1)^4 \begin{vmatrix} 1 & 1 \\ 2 & 3 \end{vmatrix} = +1$$

$$A_{23} = (-1)^5 \begin{vmatrix} 1 & 2 \\ 2 & 1 \end{vmatrix} = +3$$

$$A_{31} = (-1)^4 \begin{vmatrix} 2 & 1 \\ 1 & 4 \end{vmatrix} = +7$$

$$A_{32} = (-1)^5 \begin{vmatrix} 1 & 1 \\ 3 & 4 \end{vmatrix} = -1$$

$$A_{33} = (-1)^6 \begin{vmatrix} 1 & 2 \\ 3 & 1 \end{vmatrix} = -5$$

$$\text{adj } \mathbf{A} = \begin{bmatrix} -1 & -5 & +7 \\ -1 & +1 & -1 \\ +1 & +3 & -5 \end{bmatrix}$$

$$\det \mathbf{A} = \begin{bmatrix} 1 & 2 & 1 \\ 3 & 1 & 4 \\ 2 & 1 & 3 \end{bmatrix}$$

$$= 1(3 - 4) - 2(9 - 8) + 1(3 - 2)$$

$$= -1 - 2 + 1 = -2$$

$$\frac{\text{adj } \mathbf{A}}{\det \mathbf{A}} \mathbf{b} = \frac{\begin{bmatrix} -1 & -5 & +7 \\ -1 & +1 & -1 \\ +1 & +3 & -3 \end{bmatrix} \begin{bmatrix} 3 \\ 5 \\ 4 \end{bmatrix}}{\begin{vmatrix} 1 & 2 & 1 \\ 3 & 1 & 4 \\ 2 & 1 & 3 \end{vmatrix}} = \frac{\begin{bmatrix} -3 - 25 + 28 \\ -3 + 5 - 4 \\ +3 + 15 - 20 \end{bmatrix}}{-2} = \begin{bmatrix} 0 \\ 1 \\ 1 \end{bmatrix}$$

Traces

If **A** is a square matrix, the *trace* of **A** is the sum of the diagonal elements, that is

$$\text{tr } \mathbf{A} = \sum_i a_{ii}$$

Example:

$$\mathbf{A} = \begin{bmatrix} 1 & 2 & 1 \\ 3 & 1 & 4 \\ 2 & 1 & 3 \end{bmatrix}$$

$$\text{tr } \mathbf{A} = 1 + 1 + 3 = 5$$

For the matrices **A**, **B**, and **C** with proper orders so that the matrix multiplications are defined,

$$\text{tr}(\mathbf{AB}) = \text{tr}(\mathbf{BA})$$

$$\text{tr}(\mathbf{ABC}) = \text{tr}(\mathbf{BCA}) = \text{tr}(\mathbf{CAB})$$

The k th order trace of a square matrix is the sum of determinants of all $\binom{n}{k}$ matrices of order $k \times k$,

$$\text{tr}_k \mathbf{A} = \sum \begin{vmatrix} a_{i_1 i_1} & a_{i_1 i_2} & \cdots & a_{i_1 i_k} \\ a_{i_2 i_1} & a_{i_2 i_2} & \cdots & a_{i_2 i_k} \\ & & \vdots & \\ a_{i_k i_1} & a_{i_k i_2} & \cdots & a_{i_k i_k} \end{vmatrix}$$

The sum is over all combinations of n elements taken k at a time in the order $i_1 < i_2 < \cdots < i_k$. Examples of k th order traces are:

$$k = 1; \quad \binom{n}{k} = \binom{3}{1} = \frac{3!}{1!(3-1)!} = \frac{1 \cdot 2 \cdot 3}{1 \cdot 2} = 3 \text{ matrices}$$

of order 1×1.

$$\text{tr}_1 \mathbf{A} = |a_{11}| + |a_{22}| + |a_{33}|$$

$$= |1| + |1| + |3|$$

$$= 1 + 1 + 3 = 5 = \text{tr} \mathbf{A}$$

$$k = 2; \quad \binom{n}{k} = \binom{3}{2} = \frac{3!}{2!(3-2)!} = \frac{1 \cdot 2 \cdot 3}{2 \cdot 1} = 3 \text{ matrices}$$

of order 2×2.

$$\text{tr}_2 \mathbf{A} = \begin{vmatrix} a_{11} & a_{12} \\ a_{21} & a_{22} \end{vmatrix} + \begin{vmatrix} a_{22} & a_{23} \\ a_{32} & a_{33} \end{vmatrix} + \begin{vmatrix} a_{33} & a_{31} \\ a_{13} & a_{11} \end{vmatrix}$$

$$= \begin{vmatrix} 1 & 2 \\ 3 & 1 \end{vmatrix} + \begin{vmatrix} 1 & 4 \\ 1 & 3 \end{vmatrix} + \begin{vmatrix} 3 & 2 \\ 1 & 1 \end{vmatrix}$$

$$= (1 - 6) + (3 - 4) + (3 - 2)$$

$$= -5 - 1 + 1 = -5$$

Characteristic Roots and Vectors

If $\mathbf{A}$ is a matrix of order n,

$$|\mathbf{A} - \lambda \mathbf{I}| = 0$$

is called the *characteristic equation* of the matrix $\mathbf{A}$, and it is a polynomial of degree n in λ:

$$\lambda^n - (\text{tr}_1 \mathbf{A})\lambda^{n-1} + (\text{tr}_2 \mathbf{A})\lambda^{n-2} - \cdots + (-1)^n |\mathbf{A}| = 0$$

The roots of this polynomial equation are called the *characteristic roots* of $\mathbf{A}$. Another name of these roots is the *eigenvalues* of $\mathbf{A}$.

The solution of $\mathbf{A} - \lambda \mathbf{x} = 0$ corresponding to λ values above are called *characteristic vectors*. Examples of the characteristic roots and characteristic vectors are natural frequencies of multidegree of freedom vibrating systems and the corresponding characteristic solutions.

Example:

$$\mathbf{A} = \begin{bmatrix} 1 & 2 & 1 \\ 3 & 1 & 4 \\ 2 & 1 & 3 \end{bmatrix}$$

$$|\mathbf{A} - \lambda \mathbf{I}| = \begin{vmatrix} \begin{bmatrix} 1 & 2 & 1 \\ 3 & 1 & 4 \\ 2 & 1 & 3 \end{bmatrix} - \lambda \begin{bmatrix} 1 & 0 & 0 \\ 0 & 1 & 0 \\ 0 & 0 & 1 \end{bmatrix} \end{vmatrix} = 0$$

$$= \begin{vmatrix} 1-\lambda & 2 & 1 \\ 3 & 1-\lambda & 4 \\ 2 & 1 & 3-\lambda \end{vmatrix} = 0$$

$$= (1-\lambda)\,[(1-\lambda)(3-\lambda) - 4]$$

$$-2\,[3(3-\lambda) - 2 \cdot 4]$$

$$+1\,[3 \cdot 1 - 2(1-\lambda)]$$

$$= -\lambda^3 + 5\lambda^2 + 5\lambda - 2 = 0$$

Changing the signs to make the first coefficient $+1$,

$$\lambda^3 - 5\lambda^2 - 5\lambda + 2 = 0$$

In the following formula

$$\lambda^3 - (\text{tr}_1\mathbf{A})\lambda^2 + (\text{tr}_2\mathbf{A})\lambda - (-1)^3|\mathbf{A}| = 0$$

$$\text{tr}_1\mathbf{A} = +5$$

$$\text{tr}_2\mathbf{A} = -5$$

$$|\mathbf{A}| = +2$$

which yields the same result for the characteristic equations obtained above.

Matrix Differentiation

If the elements of a matrix are functions of a scalar variable x, $\partial\mathbf{A}/\partial x$ is the matrix with elements $\partial a_{ij}/\partial x$. For definition of differentiation, see Section 1.6. If a scalar x is a function of a matrix, $\partial x/\partial\mathbf{A}$ is the matrix with elements $\partial x/\partial a_{ij}$.

The Eigenvalue Problem

The characteristic vector with unit length, $\mathbf{x}'\mathbf{x} = 1$, is the *eigenvector*. Finding characteristic values of a set of linear equations and their corresponding eigenvectors is called the eigenvalue problem.

Example: To illustrate the concept a simple example is given. The set of linear equations is

$$-x_1 + 2x_2 = 6$$

$$-3x_1 - 4x_2 = 8$$

with $x_1 = 4$ and $x_2 = 5$ as solutions. The *characteristic equation* is found by

$$\begin{vmatrix} -1-\lambda & 2 \\ -3 & 4-\lambda \end{vmatrix} = \lambda^2 - 3\lambda + 2 = 0$$

The *characteristic roots* are the solutions of this equation:

$$\lambda_1 = 1, \qquad \lambda = 2$$

The *characteristic (eigen) vectors* can be found by substituting the characteristic values into the equation:

$$(-1-\lambda)x_1 + 2x_2 = 0$$

$$-3x_1 + (4-\lambda)x_2 = 0$$

For $\lambda = 1$,

$$-2x_1 + 2x_2 = 0$$

$$-3x_1 + 3x_2 = 0$$

thus $x_1 = x_2$ satisfies these two equations. There are infinitely many such solutions. Since $x'x = 1$, the characteristic solution with unit length, the eigenvector is found by

$$[x_1 \quad x_1]\begin{bmatrix} x_1 \\ x_1 \end{bmatrix} = x_1^2 + x_1^2 = 1, \qquad 2x_1^2 = 1, \qquad x_1 = \sqrt{1/2} = \sqrt{2}/2$$

Therefore a set of eigenvectors is

$$(x_1, x_2) = (\sqrt{2}/2, \sqrt{2}/2)$$

Similarly for $\lambda = 2$,

$$-3x_1 + 2x_2 = 0$$

$$-3x_1 + 2x_2 = 0$$

thus $x_1 = \frac{2}{3}x_2$ satisfies these two equations. There are infinitely many such solutions. Since $x'x = 1$, the characteristic solution with unit length, the eigenvector is found by

$$[x_1 \quad \tfrac{3}{2}x_1]\begin{bmatrix} x_1 \\ \tfrac{3}{2}x_1 \end{bmatrix} = x_1^2 + \tfrac{9}{4}x_1^2 = 1, \qquad \tfrac{13}{4}x_1^2 = 1$$

This results in $x_1 = 2\sqrt{13}/13$, therefore $x_2 = 3\sqrt{13}/13$. Therefore a set of eigenvectors is

$$(x_1, x_2) = (2\sqrt{13}/13, 3\sqrt{13}/13)$$

In summary

Eigenvalues	Eigenvectors
$\lambda = 1$	$(\sqrt{2}/2, \sqrt{2}/2)$
$\lambda = 2$	$(2\sqrt{13}/13, 3\sqrt{13}/13)$

Application to Linear Differential Equations

For an application of matrices to the solution of linear differential equations (see Section 1.6 for definitions on differential equations), we consider a set of two linear differential equations of the first order given as

$$\frac{dx_1}{dt} = -x_1 + 2x_2$$

$$\frac{dx_2}{dt} = -3x_1 + 4x_2$$

At the origin where $x_1 = x_2 = 0$ the right-hand sides of these differential equations vanish, thus the derivatives of the variables are zero. This corresponds to a singular solution of the differential equation by definition. The coefficient matrix of the equations is the same as the example given above. The eigenvalues are those found above, namely, 1 and 2, two real positive eigenvalues. The question of stability of the singular solution, $x_1, x_2 = 0$, is answered by the sign of these eigenvalues. Since they are both greater than zero, that is, they are positive, this characteristic solution is an *unstable* solution.

1.6 DIFFERENTIAL AND INTEGRAL CALCULUS

1.6-1 Real Functions

Definitions

Given two sets, X and Y, of numbers x and y, respectively, a correspondence relating any x in X to one or more y in Y is called a *function*, f. The correspondence is shown as $y = f(x)$. Here X is called the *domain* of f. Also, x is then the *independent variable* and y the *dependent variable*. In the case of $y = f(x)$, the function is said to be *explicitly* given while, if $f(x, y) = 0$, it is *implicitly* given. The *inverse function* relates y to x: $x = f^{-1}(y)$.

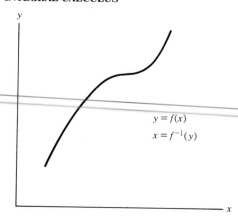

$$y = f(x)$$
$$x = f^{-1}(y)$$

1.6-2   Differentiation

If a function $f(x)$ is defined in the neighborhood of $x = x_0$, and the limit

$$\lim_{h \to 0} \frac{f(x_0 + h) - f(x_0)}{h}$$

exists, it is called the *derivative* or *differential coefficient* of $f(x)$ at x_0 and denoted by

$$f'(x_0) \quad \text{or} \quad (df/dx)_{x_0} \quad \text{or} \quad Df(x_0)$$

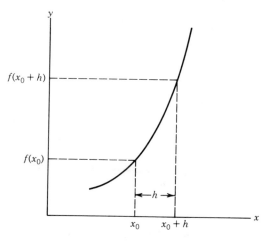

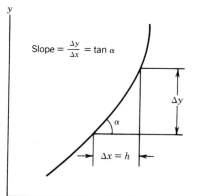

Therefore, while the mathematical meaning of derivative is *slope* of f, the physical meaning of derivative is the *rate of change* of function f as the variable x changes.

Similarly, the derivative of a derivative is called the *second derivative* and denoted by

$$f''(x_0) \quad \text{or} \quad (d^2f/dx^2)_{x_0} \quad \text{or} \quad D^2f(x_0).$$

On the other hand, the *operation* that derives $f'(x)$ from $f(x)$ is called the *differentiation* of $f(x)$.

The Differentiation Rules

The derivative of the sum of two functions equals the sum of derivatives:

$$(f + g)' = f' + g'$$

A constant multiplier, a is factored out of the derivative,

$$(af)' = a(f')$$

For given constants a and b and functions f and g:

Addition of functions: $(af + bg)' = (af' + bg')$

Product of functions: $(fg) = f'g + fg'$

Division of functions: $\left(\dfrac{f}{g}\right)' = \dfrac{f'g - fg'}{g^2}$

Function of a function:
(composite function) $\dfrac{d}{dx}f[g(x)] = \dfrac{df(g)}{dg}\dfrac{dg(x)}{dx}$

Leipnitz formula: $\dfrac{d^n}{dx^n}[fg] = \sum_{j=0}^{n} \binom{n}{j} f^{(j)} g^{(n-j)}$

$$\frac{d}{dx}[f(x)]^n_{x=x_0} = n[f(x_0)]^{n-1}f'(x_0)$$

Inverse function: If $y'(x)$, then $x'(y) = 1/y'(x)$.

Maxima and Minima

If the derivative of $f(x)$, df/dx, has the value 0 at some value of x, x_0,

$$|df/dx|_{x_0} = 0,$$

then f has a *stationary value* at x_0. There are three types of such points with stationary values of f.

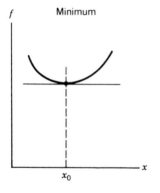
Minimum

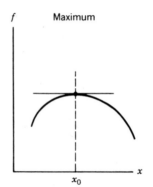
Maximum

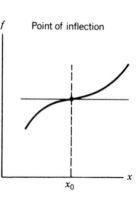
Point of inflection

If the f value everywhere in the neighborhood of x_0 is *greater* than the value at x_0, the function is said to have a *minimum* at x_0. If it is *lower* than the value at x_0, then function has a *maximum* at x_0. If neither is true, then x_0 is called a *point of inflection*.

Minimum occurs at $d^2f/dx^2 > 0$.
Maximum occurs at $d^2f/dx^2 < 0$.
Point of inflection is at $d^2f/dx^2 = 0$.

$d^3f/dx^3 > 0$

$d^3f/dx^3 < 0$

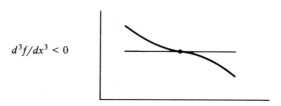

Example: A function is given as $f = x^3 - 3x$. The stationary points lie where

$$\frac{df}{dx} = 3x^2 - 3 = 0$$

This gives a solution $x^2 = 1$, thus $x_1 = +1$, $x_2 = -1$.

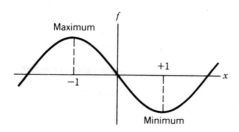

At $x_1 = +1$, $d^2f/dx^2 = 6x = 6 > 0$, thus f reaches its *minimum*.

At $x_1 = -1$, $d^2f/dx^2 = 6x = -6 < 0$, thus f reaches its *maximum*.

Example: If $f = x(10 - x)$, a rectangular area with width x and height $10 - x$, what should be the x value to have the maximum area?

$$\frac{df}{dx} = -2x + 10 \qquad x = 5$$

$$\frac{d^2f}{dx^2} = -2 < 0 \quad \text{thus } f \text{ is maximum at } x = 5$$

It is clear that this is a square. Thus, as expected, the square has the maximum area among all the rectangular areas with equal peripheries (edges). In this example the sum of edges is $2 \times 10 = 20$ units.

Multiple Variables

In the case of functions with multiple independent variables, $f(x_1, x_2, \ldots, x_n)$, the derivative is called the *partial derivative* and denoted by, for the case of two variables $x_1 = x$, $y_2 = y$,

$$\frac{\partial f(x, y)}{\partial x} \equiv \lim_{h \to 0} \frac{f(x + h, y) - f(x, y)}{h}$$

In general we have $\partial f(x_i)/\partial x_k$. There are as many *first* partial derivatives of f as the number of independent variables.

Higher derivatives would be

$$\frac{\partial^2 f}{\partial x_i^2} \quad \text{and} \quad \frac{\partial^2 f}{\partial x_i \, \partial x_k}$$

Similarly, for $\partial^n/\partial x_1 \cdots \partial x_j \cdots$, all the combinations of the independent variables must be considered.

The *total* derivative is defined as

$$df = \left(\frac{\partial f}{\partial x}\right) dx + \left(\frac{\partial f}{\partial y}\right) dy$$

In general

$$df = \left(\frac{\partial f}{\partial x_1}\right) dx_1 + \left(\frac{\partial f}{\partial x_2}\right) dx_2 + \cdots$$

From which we have

$$\frac{df}{dx_1} = \left(\frac{\partial f}{\partial x_1}\right) + \left(\frac{\partial f}{\partial x_2}\right)\left(\frac{dx_2}{dx_1}\right) + \cdots$$

where $x_2, \ldots$ can be considered as functions g_2 of x_1, and so on.

Maxima and Minima of Functions of Multiple Variables

Maxima and minima for the functions of multiple variables can similarly be defined. Here for a function of two variables, x, y,

$$\frac{\partial f}{\partial x} = 0 \qquad \frac{\partial f}{\partial y} = 0$$

corresponds to a stationary point. To determine the maximum or minimum, one refers to the Taylor expansion (see the section of Taylor series below). Two observations are that since the first derivatives vanish those terms with the first derivatives are dropped from the expansion. In addition, any term with derivatives higher than the second can be made small relative to this term, thus negligible, by taking small enough intervals. Intervals are taken as h_1 along the x axis, h_2 along the y axis. Therefore,

$$f(x, y) = f(x_0, y_0) + \frac{1}{2!}\left[h_1^2 \frac{\partial^2 f}{\partial x^2} + 2 h_1 h_2 \frac{\partial^2 f}{\partial x \, \partial y} + h_2^2 \frac{\partial^2 f}{\partial y_2}\right]$$

In a simpler form

$$f = f_0 + (1/2!) h_1^2 A + 2 h_1 h_2 B + h_2^2 C$$

where A, B, C are the corresponding derivatives evaluated at the stationary point. Rewriting the above equation as

$$f - f_0 = A\left(h_1 + \frac{Bh_2}{A}\right)^2 + \left(C - \frac{B^2}{A}\right) h_2^2$$

we have the following:

Minimum: $f - f_0 > 0$ for all h_1, h_2: $A > 0$, $C - (B^2/A) > 0$.
Maximum: $f - f_0 < 0$ for all h_1, h_2: $A < 0$, $C - (B^2/A) < 0$.

Otherwise it is a *saddle point*.

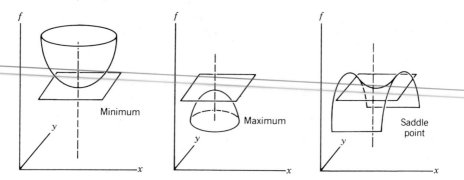

Minimum   Maximum   Saddle point

Therefore

$$\frac{\partial^2 f}{\partial x^2} > 0 \qquad\qquad \text{for a minimum}$$

$$\frac{\partial^2 f}{\partial x^2} < 0 \qquad\qquad \text{for a maximum}$$

$$\frac{\partial^2 f}{\partial x^2}\frac{\partial^2 f}{\partial y^2} > \left(\frac{\partial^2 f}{\partial x \partial y}\right)^2 \qquad \text{in either case}$$

$$\frac{\partial^2 f}{\partial x^2}\frac{\partial^2 f}{dy^2} < \left(\frac{\partial^2 f}{\partial x \, \partial y}\right)^2 \qquad \text{for a saddle point}$$

This can be generalized such that for the case of multiple variables,

$$\frac{\partial f}{\partial x_i} = 0 \qquad i = 1, 2, \ldots, n$$

$$h_i h_j \left[\frac{\partial^2 f}{\partial x_i \partial x_j}\right] > 0 \quad \text{for minimum}$$

$$< 0 \quad \text{for maximum}$$

Series Expansion of Functions

Using either the Taylor or Maclaurin theorem, discussed below, the series expansion of functions can be obtained. Illustrative examples of series expansions of some common functions, such as exponential functions, trigonometric functions, and so on are given here:

$$e^x = 1 + \frac{x}{1!} + \frac{x^2}{2!} + \cdots + \frac{x^n}{n!} + \cdots \qquad \text{for all } x$$

$$\lim(1 + x) = x - \frac{x^2}{2} + \frac{x^3}{3} - \cdots + (-1)^{n-1}\frac{x^n}{n} + \cdots \qquad |x| < 1$$

$$\sin x = x - \frac{x^3}{3!} + \frac{x^5}{5!} - \frac{x^7}{7!} + \cdots$$

$$\cos x = 1 - \frac{x^2}{2!} + \frac{x^4}{4!} - \frac{x^6}{6!} + \cdots$$

$$\sinh x = x + \frac{x^3}{3!} + \frac{x^5}{5!} + \frac{x^7}{7!} + \cdots$$

$$\cosh x = 1 + \frac{x^2}{2!} + \frac{x^4}{4!} + \frac{x^6}{6!} + \cdots$$

Indeterminate Forms

Series expansion of functions may be used in evaluating these functions. For example, if two functions are $f(x)$ and $g(x)$, defining another function as their ratio, $f(x)/g(x)$, and evaluating this ratio at some x values may be *indeterminate*. An example of such a case is $F(x) = 1 - e^x$, and $g(x) = \lim(1 + x)$; thus

$$\left. \frac{1 - e^x}{\lim(1 + x)} \right|_{x=0} = \frac{0}{0} \quad \text{which is indeterminate}$$

In such cases the *L'Hôpital rule* is applied which states that the actual value is

$$\left. \frac{f'(x)}{g'(g)} \right|_{x=0} = -1$$

This can easily be seen from the above series expansion

$$f(x) = 1 - e^x = 1 - 1 - \frac{x}{1!} - \frac{x^2}{2!} - \cdots = -\frac{x}{1!} - \frac{x^2}{2!} - \cdots$$

$$g(x) = \lim(1 + x) = x - \frac{x^2}{2} + \frac{x^3}{3} - + \cdots$$

$$f'(x) = -1 - x - \cdots$$

$$g'(x) = +1 - x + \cdots$$

L'Hôpital's rule may be applied successively such that derivatives of the functions forming the quotient may be taken successively until it is determinate.

Taylor's Theorem

In expanding functions in power series, Taylor's expansion is used. Series expansions are useful in mathematical analysis of various problems. A function, $f(x)$, n times differentiable in the interval $(x_0, x_0 + h)$, can be expanded about the value x_0 as

$$f(x_0 + h) = f(x_0) + \frac{h}{1!}f'(x_0) + \frac{h^2}{2!}f''(x_0) + \cdots + \frac{h^{n-1}}{(n-1)!}f^{n-1}(x_0) + R_n$$

where the remainder R_n is

$$R_n = \frac{h^n(1 - q)^{n-m}}{(n-1)!m!}f^n(x_0 + qh)$$

Lagrange remainder $(m = n)$: $R_n = \frac{h^n}{n!}f^{(n)}(x_0 + q_1 h)$

Cauchy remainder $(m = 1)$: $R_n = \frac{h^n(1 - q_2)^{n-1}}{(n-1)!}f^{(n)}(x_0 + q_2 h)$

where q_1 and q_2 are numbers with values between 0 and 1.

Example:

$$f(x) = e^x$$

$$f'(x) = e^x$$

$$f''(x) = e^x$$

$$\vdots$$

$$e(x_0 + h) = e^{x_0} + \frac{h}{1!}e^{x_0} + \frac{h^2}{2!}e^{x_0} + \cdots$$

$$= e^{x_0}\left(1 + \frac{h}{1!} + \frac{h^2}{2!} + \cdots + \frac{h^{n-1}}{(n-1)!} + R_n\right)$$

$$e^{x_0}e^h = e^{x_0}\left(1 + \frac{h}{1!} + \frac{h^2}{2!} + \cdots\right)$$

Therefore the Taylor expansion of e^h is

$$e^h = 1 + \frac{h}{1!} + \frac{h^2}{2!} + \cdots$$

$$e = 1 + \frac{1}{1!} + \frac{1}{2!} + \cdots$$

would be a method of calculating the value of e.

Maclaurin's Theorem

The Taylor expansion for a special case, $x_0 = 0$ and $h = x_1$, gives the series expansion obtained by Maclaurin's theorem:

$$f(x) = f(0) + \frac{x}{1!}f'(0) + \frac{x^2}{2!}f''(0) + \cdots + \frac{x^{n-1}}{(n-1)!}f^{n-1}(0) + R_n$$

where the remainder R_n is

$$R_n = x^n \frac{f^n(\phi x)}{n!} \qquad \phi \text{ between 0 and 1}$$

For example, for $F(x) = e^x$,

$$e^x = e^0 + \frac{x}{1!}e^0 + \frac{x^2}{2!}e^0 + \cdots \qquad e^0 = 1$$

$$e^x = 1 + \frac{x}{1!} + \frac{x^2}{2!} + \cdots \qquad \text{the Maclaurin expansion of } e^x$$

1.6-3 Integration

Integration is an operation inverse to differentiation. For a differentiation given as

$$f(x) = \frac{d}{dx}F(x)$$

its *inverse*, $F(x)$, is an *indefinite integral* of $f(x)$,

$$F(x) = \int f(x)\,dx$$

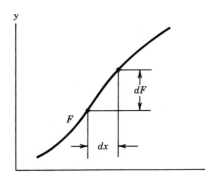

In general this function is not unique, and it is expressed as $F(x) + C$, where C is a constant.

Similar to the operation of differentiation the *operation* which results in $\int(\cdots)\,dx$ is called *integration*.

General Rules for Indefinite Integrals

$$\int cf(x)\,dx = c\int f(x)\,dx \quad \text{where } c \text{ is a constant}$$

$$\int[f_1(x) + f_2(x)]\,dx = \int f_1(x)\,dx + \int f_2(x)\,dx$$

Integration by Parts

$$\int f'(x)g(x)\,dx = f(x)g(x) - \int f(x)g'(x)\,dx$$

$$\int f(x)\,dx = \int f(u)\phi'(u)\,du \qquad x = \phi(u)$$

$$\int f'(x)/f(x)\,dx = \log f(x)$$

The *definite integral* is

$$\int_a^b f(x)\,dx = F(b) - F(a)$$

which defines the area between the curve $y = f(x)$ and the x axis, between the ordinates $x = a$ and $x = b$. This is known as the *Newton–Leipnitz formula*.

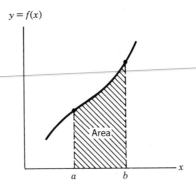

The *infinite integral*, if it exists, is defined by

$$\lim_{X \to \infty} \int_a^X f(x)\,dx = \int_a^\infty f(x)\,dx$$

which is called a Cauchy–Riemann integral.

Multiple integral is defined as limits of multiple sums as an extension of the integral defined above. For example, the *double integral* is

$$\iint_S f(x, y)\,dx\,dy$$

which can be shown to be the volume between the surface $z = f(x, y)$ and the plane $z = 0$, and lines parallel to z axis through the boundary of the S region. The evaluation of multiple integral may be

$$\int dy\left[\int f(x, y)\,dx\right] \qquad \int dx\left[\int f(x, y)\,dy\right]$$

Special Integrals

Some of the special integrals have been solved yielding closed solutions. These functions are useful in that they are analytical and available in tabulated form, thus, for engineering applications there are sufficiently comprehensive tables. Some of these functions are discussed briefly in this section.

Gamma Function

$$\Gamma(z) = \int_0^\infty e^{-t} t^{z-1}\, dt \qquad R(z) > 0$$

where z is the complex variable, $z = \text{Re}(z) + i\,\text{Im}(z)$ and $i = \sqrt{-1}$.

$$\Gamma(z) = z^{-1} e^{-\gamma z} \prod \left[(1 + z/n)^{-1} e^{z/n} \right] \qquad \text{all } z$$

$$\Gamma(1/2) = (\pi)^{1/2}$$

$$\Gamma(n + 1) = n! \qquad n = 1, 2, 3, \ldots$$

$$\Gamma(z + 1) = z\Gamma(z)$$

$$\Gamma(1 - z) = \pi \csc \pi z$$

Stirling's formula. Asymptotic behavior of the gamma function leads to Stirling's formula:

$$\ln \Gamma(z) \sim \left(z - \tfrac{1}{2} \right) \ln z - z + \tfrac{1}{2} \ln 2\pi$$

$$n! \sim \left(\frac{n}{e} \right)^n \sqrt{2\pi n}$$

Beta Function

The Beta function is related to the Gamma function:

$$B(z, \omega) = \int_0^1 t^{z-1} (1 - t)^{\omega - 1}\, dt$$

$$= \frac{\Gamma(z)\Gamma(\omega)}{\Gamma(z + \omega)} \qquad R(z) > 0,\ R(\omega) > 0$$

Elliptic Integrals

The general form of an elliptic integral is

$$\int R(x, \sqrt{X})\, ds$$

Here R denotes a rational function,

$$X = a + bx + cx^2 + dx^3 + ex^4$$

an algebraic function of third or fourth degree. The special forms are listed below, where k is called the *modulus* and usually lies between 0 and 1.

Incomplete Elliptic Integrals

1. *Elliptic integral of the first kind.*

$$F(k, x) = \int_0^t \frac{dx}{\left[(1 - x^2)(1 - k^2 x^2) \right]^{1/2}} \tag{1.6-1}$$

2. *Elliptic integral of the second kind.*

$$E(k, x) = \int_0^t \left(\frac{1 - k^2 x^2}{(1 - x^2)} \right)^{1/2} \tag{1.6-2}$$

3. *Elliptic integral of the third kind.*

$$\pi(k, \alpha, x) = \int_0^t \frac{dx}{(1 + \alpha^2 x^2)[(1 - x^2)(1 - k^2 x^2)]^{1/2}} \tag{1.6-3}$$

where $-\infty < \alpha^2 < \infty$, and integer $k^2 < 1$.

These integrals can be transformed by putting $x = \theta$, the replacing the upper limit of the integral, t, by $\phi = \sin^{-1} t$, thus resulting in functions

$$F(k, \phi) = \int_0^\phi \frac{d\phi}{(1 - k^2 \sin^2 \phi)^{1/2}} \tag{1.6-1'}$$

$$E(k, \phi) = \int_0^\phi (1 - k^2 \sin^2 \phi)^{1/2} \, d\phi \tag{1.6-2'}$$

$$\pi(k, \alpha^2, \phi) = \int_0^\phi \frac{d\phi}{(1 + \alpha^2 \sin^2 \phi)(1 - k^2 \sin^2 \phi)^{1/2}} \tag{1.6-3'}$$

Complete Elliptic Integrals. Putting $\phi = \pi/2$, these equations become complete elliptic integrals. Thus

$$F(k) \equiv F(k, \pi/2) = \pi/2 \left[1 + (1/2)^2 k^2 + (3/2 \cdot 4)^2 k^4 + \cdots \right] \tag{1.6-1''}$$

$$E(k) \equiv E(k, \pi/2) = \pi/2 \left[1 - (1/2)^2 k^2 - (3/2 \cdot 4)^2 k^4 - \cdots \right] \tag{1.6-2''}$$

$$\pi(k, \alpha^2) \equiv \pi(k, \alpha^2, \pi/2) \tag{1.6-3''}$$

1.6-4 Differential Equations

Ordinary Differential Equations

Definitions. The differential equations are equations containing derivatives of the *dependent variable* y with respect to the *independent variable* x. If D is the *differential operator* acting on a function y,

$$Dy = \frac{dy}{dx} \qquad D^2 y = \frac{d^2 y}{dx^2} \qquad D^n y = \frac{d^n}{dx^n}$$

An equation such as

$$P(D)y = f(x) \tag{1.6-4}$$

is called an *ordinary differential equation*. Here if P denotes a polynomial in D (see Section 1.1).

Example: $3 \left(\dfrac{d^2 y}{dx^2} \right) + \dfrac{dy}{dx} = x^3.$

The highest derivative in a differential equation indicates the *order* of the differential equation. If, for example, D^n is the highest derivative, it is the nth *order differential equation*. The *degree* of a differential equation is the power of the highest derivative when the differential equation is rational and integral in terms of the derivatives involved.

If the power of the function y and of its derivatives is 1, the equation is called *linear*. If any one of these is of more power than 1, the equation is called *nonlinear*. Equation (1.6-4) is a *linear ordinary differential equation with constant coefficients*. The equation where $f(x) = 0$ is called the *homogeneous* equation.

$$P(D)y = 0 \tag{1.6-5}$$

Or, we can explicitly write

$$\left[a_1 D^n + a_2 D^{n-1} + \cdots + a_n D + a_{n+1} \right] y = 0$$

$$a_1 \frac{d^n y}{dx^n} + \cdots + a_n \frac{dy}{dx} + a_{n+1} = 0$$

Solution of Differential Equations. The general solution of (1.6-5) is called the *complementary solution*, y_c. If y_p is a *particular solution* of (1.6-4), then the *general solution* of (1.6-4) is

$$y = y_c + y_p$$

If P is a polynomial in D of degree n, it can be *factorized* (see Section 1.11) to obtain

$$\prod_{i=1}^{n} (D - a_i) = 0 = (D - a_1)(D - a_2) \cdots (D - a_n) \tag{1.6-5'}$$

Based on the a_i values, the following forms of the exponential functions correspond to the complementary solutions of (1.6-4).

For nonrepeated real linear factors, that is, $(D - a_1), (D - a_2), \ldots,$

$$c_1 e^{a_1 x} + \cdots + c_n e^{a_n x} \tag{1.6-6}$$

where the c_i's are arbitrary constants. For repeated real linear factors, $(D - a)^k$, the k terms of the form

$$\left(c_1 + c_2 x + \cdots + c_k x^{k-1} \right) e^{ax}$$

are taken.

For a conjugate complex pair of factors, $a + ib, a - ib,$

$$c_1 e^{ax} \cos bx + c_2 e^{ax} \sin bx$$

And for the repeated ones,

$$\left[(c_1 + c_3 x + \cdots) \cos bx + (c_2 + c_4 x + \cdots) \sin bx \right] e^{ax}$$

The sum of all these terms form the complementary function.

The *particular solution* can be found by the *method of variation of parameters* which is explained below. The complementary solution is taken and the coefficients $c_1, c_2, \ldots$ are replaced by $c_1(x), c_2(x), \ldots$ functions of x. This becomes the particular solution with unknown coefficients:

$$y_p = c_1(x) e^{a_1 x} + c_2(x) e^{a_2 x} + \cdots + c_n(x) e^{a_n x} \tag{1.6-7}$$

Then y_p is differentiated to obtain y_p',

$$dy_p/dx = c_1'(x) e^{a_1 x} + c_2'(x) e^{a_2 x} + \cdots$$

$$+ c_1(x) a_1 e^{a_1 x} + c_2(x) a_2 e^{a_2 x} + \cdots$$

Here the term with the first derivatives, $c_1'(x), c_2'(x), \ldots,$ are set to zero,

$$c_1'(x) e^{a_1 x} + c_2'(x) e^{a_2 x} + \cdots + c_n'(x) e^{a_n x} = 0 \tag{1.6-8}$$

In subsequent derivatives the same condition is imposed until the nth derivative, $d^n y_p/dx^n$, where the terms with $c_1'(x), c_2'(x), \ldots$ are maintained.

Substituting the y_p and its derivatives in (1.6-4) and considering all the conditions of the type (1.6-8) imposed as equations, for n coefficients $c_i'(x)$, n equations are obtained from which $c_i'(x)$ can be solved. Integrating $c_i'(x)$,

$$c_i(x) = \int c_i'(x) \, dx$$

coefficients $c_i(x)$ are obtained.

The complete solution is thus determined as the sum of (1.6-6) and (1.6-7) as

$$y = y_c + y_p$$

$$= c_1 e^{a_1 x} + c_2 e^{a_2 x} + \cdots + c_n e^{a_n x}$$

$$+ c_1(x) e^{a_1 x} + c_2(x) e^{a_2 x} + \cdots + c_n(x) e^{a_n x}$$

Example:

$$\frac{d^2y}{dx^2} - 5\frac{dy}{dx} + 6y = x^2$$

is a differential equation of the form (1.6-4). The corresponding homogeneous equation can be written in terms of the nonrepeated linear factors as

$$(D - 3)(D - 2)y = 0$$

where $a_1 = 3$, $a_2 = 2$. Thus the *complementary solution* is

$$y_c = c_1 e^{3x} + c_2 e^{2x}$$

To find the *particular solution* we set the coefficients as functions of the independent variable x (variations of parameters) as

$$y_p = c_1(x)e^{3x} + c_2(x)e^{2x} \qquad (y_p)$$

After differentiating

$$y_p' = c_1'e^{3x} + c_2'e^{2x} + c_1 3e^{3x} + c_2 2e^{2x} + \cdots$$

We impose the condition

$$c_1'e^{3x} + c_2'e^{2x} = 0 \qquad (1.6\text{-}9)$$

Thus

$$y_p' = c_1 3e^{3x} + c_2 2e^{2x} \qquad (y_p')$$

Differentiating once again,

$$y_p'' = c_1' 3e^{3x} + c_2' 2e^{2x} + c_1 9e^{3x} + c_2 4e^{2x} \qquad (y_p'')$$

Substituting y_p, y_p', and y_p'' in the original differential equation gives

$$3c_1'e^{3x} + 2c_2'e^{2x} + 9c_1 e^{3x} + 4c_2 e^{2x} - 5(3c_1 e^{3x} + 2c_2 e^{2x}) + 6(c_1 e^{3x} + c_2 e^{2x}) = x^2$$

Eliminating the cancelling terms this relation reduces to

$$3c_1'e^{3x} + 2c_2'e^{2x} = x^2 \qquad (1.6\text{-}10)$$

The two equations (1.6-9) and (1.6-10) in c_1', c_2' are

$$c_1'e^{3x} + c_2'e^{2x} = 0$$

$$3c_1'e^{3x} + 2c_2'e^{2x} = x^2$$

From these equations c_1' and c_1' can be obtained as

$$c_1'e^{3x} = x^2$$

$$-c_2'e^{2x} = x^2$$

By integrating these equations,

$$dc_1 = x^2 e^{-3x}\, dx$$

$$c_1 = \int x^2 e^{-3x}\, dx$$

$$c_1 = -e^{-3x}\left[x^2/3 + 2x/9 + 2/27\right]$$

$$dc_2 = -x^2 e^{-2x}\,dx$$

$$c_2 = -\int x^2 e^{-2x}\,dx$$

$$c_2 = +e^{-2x}\left[x^2/2 + x/2 + 1/4\right]$$

Therefore the complete solution is

$$y = y_c + y_p$$

$$y = c_1 e^{3x} + c_2 e^{2x}$$

$$-\left[x^2/3 + 2x/9 + 2/27\right]$$

$$+\left[x^2/2 + x/2 + 1/4\right]$$

$$y = c_1 e^{3x} + c_2 e^{2x} + \left[x^2/6 + 5x/18 + 19/108\right]$$

Solution in Series by the Method of Frobenius. A homogeneous linear equation can be transformed into

$$x^n y^{(n)} + x^{n-1} y^{(n-1)} p_1(x) + \cdots + p_n(x)y = 0$$

The solution is assumed to be of the form

$$y = x^s \sum_{r=0}^{\infty} a_r x^r$$

By substituting it in the differential equations the numbers $s, a_0, a_1, \ldots$ are determined by equating coefficients of each power of x to zero. The solution is valid in the range of convergence of the obtained series. The same method can be extended to the solution about a given point x_0 (not origin) by replacing x by $x - x_0$ in the above series and the equation.

Example: $y'' - 5y' + 6y = 0$, multiplying by x^2,

$$x^2 y'' - 5x^2 y' + 6x^2 = 0$$

$$y = x^s \sum_{r=0}^{\infty} a_r x^r$$

$$y' = \sum_{r=0}^{\infty} a_r (r + s) x^{s+r-1}$$

$$y'' = \sum_{r=0}^{\infty} a_r (r + s)(r + s - 1) x^{s+r-2}$$

Substituting in the equation

$$\sum a_r (r + s)(r + s - 1) x^{s+r-2} - 5\sum a_r (r + s) x^{s+r-1} + 6\sum a_r x^r = 0$$

In a more explicit form,

$$a_0 s(s - 1) x^{s-2} - 5a_0 s x^{s-1} + 6a_0 x^s + a_1(s + 1)s x^{s-1} - 5a_1(s + 1) x^s + 6a_1 x^{s+1}$$

$$+ a_2(s + 2)(s + 1) x^s - \cdots = 0$$

The coefficients of like powers of x must be equal to zero, that is,

$$a_0 s(s-1) = 0$$

$$-5a_0 s + a_1(s+1)s = 0$$

$$6a_0 - 5a_1(s+1) + a_2(s+2)(s+1) = 0$$

Solving the first equality, $s = 0$, $s = 1$, and $a_0 = 0$ are found. However, for $s = 1$, from the second equality,

$$a_1 = 5a_0/2$$

from the third equality,

$$a_2 = 19a_0/2$$

can be found. Therefore the solution of the differential equation becomes

$$y = a_0\left[x + \tfrac{5}{2}x^2 + \tfrac{19}{6}x^3 + \cdots \right]$$

Series Solution in Descending Powers of x is valid for large values of the independent variable,

$$y = x^{-s}\sum_{r=0}^{\infty} a_r x^{-r}$$

Solution by Laplace Transformations is obtained by applying Laplace transforms, see Section 1.7, to linear differential equations. From the resulting Laplace transform of the variable y the function y can be obtained by inverse transforms.

Example: For an illustrative example see Section 1.7 on Laplace transformations.

As an example of *Perturbation Theory for Nonlinear Differential Equations* we use van der Pol's equation:

$$y'' + \varepsilon f(y, y') + y = F(t)$$

where $y' = dy/dt$ and ε is small.

$$y(t) = y_0(t) + \varepsilon y_1(t) + \varepsilon^2 y_2(t) + \cdots$$

is used as the perturbed solution about y_0. The coefficients of the powers of ε are equated to zero and the solution is obtained.

Partial Differential Equations

Definition.   In the above section there was only one dependent variable, y, and one independent variable, x. However, some differential equations might involve multiple *dependent variables*, $u_1, u_2, \ldots, u_m$, as well as multiple *independent variables*, $x_1, x_2, \ldots, x_r$. The derivatives of dependent variables with respect to one or more independent variables are called *partial derivatives*. For example, if $u_1(x_1, x_2)$ is given as a function of two independent variables, partial derivatives of u_1 with respect to x_1 alone are, $\partial u_1/\partial x_1, \partial^2 u_1/\partial x_1^2, \ldots$.

Any differential equation containing partial derivatives is called a *partial differential equation*. The *order* is determined by the highest partial derivative appearing in the equation. For example, if the highest partial derivative is $\partial^2 u/\partial x_1^2$, then the partial differential equation is said to be of the second order. Some well known examples are given below for illustration.

For a *Linear System of Equations*, in most general form,

$$\sum_{j=1}^{m} P_{ij}(D_1, \ldots, D_n) u_j = f_i(x_j) \qquad i = 1, 2, \ldots, r$$

where $D_i = \partial/\partial x_i$ and P_{ik} are polynomials in D_i with coefficients depending on x_j. Some well known examples are discussed briefly below.

Example: $r = 2$, $m = 3$.

$$P_{11}(D_1, D_2, D_3)u_1 + P_{12}(D_1, D_2, D_3)u_2 = f_1(x_1, x_2, x_3)$$

$$P_{21}(D_1, D_2, D_3)u_1 + P_{22}(D_1, D_2, D_3)u_2 = f_2(x_1, x_2, x_3)$$

given as

$$\left(\frac{\partial}{\partial x_1} - \frac{\partial}{\partial x_2} + \frac{\partial}{\partial x_3}\right)u_1 + \left(\frac{\partial}{\partial x_1} + \frac{\partial^2}{\partial x_1^2}\right)u_2 = x_1^2 - x_3$$

$$\frac{\partial}{\partial x_3}u_1 + \left[\frac{\partial}{\partial x_2} + \left(\frac{\partial}{\partial x_3}\right)^3\right]u_2 = x_2^3$$

However, in the case of *a linear system* of equations the dependent variables and all their derivatives are not higher than the first order. An example is

$$\left(\frac{\partial}{\partial x_1} - \frac{\partial}{\partial x_2} + \frac{\partial}{\partial x_3}\right)u_1 + \left(\frac{\partial}{\partial x_1} + \frac{\partial^2}{\partial x_1^2}\right)u_2 = x_1^2 - x_3$$

$$\frac{\partial}{\partial x_3}u_1 + \left(\frac{\partial}{\partial x_2} + \frac{\partial}{\partial x_3}\right)u_2 = x_2^3$$

Second-Order Equations.

$$A\frac{\partial^2 u}{dx^2} + 2B\frac{\partial^2 u}{dx\,dy} + C\frac{\partial^2 u}{dy^2} = f(x, y, \partial u/\partial x, \partial u/\partial y)$$

Boundary conditions:

Dirichlet: u is given at the boundary, for example, $u = 1$.

Neumann: $\partial u/\partial n$ is given at boundary, for example $\partial u/\partial n = 0$. Normal component of the gradient of u at each point of boundary.

Cauchy: u and $\partial u/\partial n$ at each point; $u(x_0, y_0) = 1$, $\partial u(x_0, y_0)/\partial n = 0$.

Types of equations:

Hyperbolic	$B^2 > AC$
Parabolic	$B^2 = AC$
Elliptic	$B^2 < AC$

Some examples of these three types of equation are given below.

Elliptic Equations

The Laplace Equation. In n dimensional space the Laplace equation (potential equation) is given as

$$\nabla^2 u = \sum_{i=1}^{n} u_{x_i x_i} = 0$$

Here u_y is the derivative of u with respect to y, and u_{yy} is the second derivative with respect to y. The notation ∇^2, called the *Laplacian*, represents the summation of second derivatives of u with respect to the variables, $x_1, x_2, \ldots, x_n$.

For example, for $n = 3$, the Laplace equation, $\nabla^2 u = 0$, becomes

$$\frac{d^2 u}{dx_1^2} + \frac{d^2 u}{dx_2^2} + \frac{d^2 u}{dx_3^2} = 0$$

The Laplace equation is obtained in formulating various physical problems such as heat flow problems, some problems in electrostatics, and so on. The solutions of this equation are called *harmonic* or *potential* functions. The *harmonic functions* are real functions $u(x_1, \ldots)$ with continuous second derivative which satisfy the Laplace equation in a given region. For example, trigonometric functions are such functions for the one dimensional Laplace equation as illustrated in examples below.

The two special problems are the *Dirichlet problem* and the *Neumann problem*. The difference between them is the boundary conditions: for the first one, u is given at the boundary (i.e., temperature defined for the heat flow problem), and for the second one, u_x is given at the boundary (i.e., flow of heat at the boundary for the heat flow problem).

Poisson's Equation. While the Laplace equation is homogeneous in that $\nabla^2 u = 0$, the Poisson equation has a nonvanishing right-hand side as follows:

$$\nabla^2 u = f(x_1, \ldots, x_n)$$

which is satisfied by

$$u = \int vf \, dx_1 \cdots dx_n$$

where v is the *fundamental solution*.

Wave Equation.

$$\nabla^2 u + \lambda u = 0$$

The *one dimensional wave equation* is

$$\frac{\partial^2 u}{\partial x^2} = \frac{1}{c^2} \frac{\partial^2 u}{\partial t^2}$$

where c is the speed of the wave. An example is the vibrating string. For solution of the wave equation see the section on the Methods of Separation of Variables below.

Example:

$$\frac{\partial^2 u}{\partial t^2} = c^2 \frac{\partial^2 u}{\partial x^2}$$

A solution is $u = \cos x \cos ct$.

Parabolic Equation of Second Order

The Heat Equation.

$$u_t = \nabla^2 u = u_{xx} + u_{yy} + u_{zz}$$

A simple example of the heat equation for a one dimensional problem, where the only independent variable is x, is discussed in the next section on the diffusion equation.

The Diffusion Equation.

$$u_t = k u_{xx} \quad \text{(one dimensional)}$$

Example: The temperature $u(x, t)$ of an isolated bar at x at time t satisfies the diffusion equation:

$$u_t = a^2 u_{xx}$$

The boundary conditions are

$$u_x(0, t) = u_x(l, t) = 0 \quad \text{at the insulated ends}$$

$$u(x, 0) = f(x) \quad \text{initial condition}$$

By separation of variables (see the Method of Separation of Variables),

$$u = X(x)T(t), \qquad T'/T = -p^2, \qquad a^2 X''/X = -p^2$$

$$T = e^{-p^2 t}$$

$$X = \cos(p/a)x \qquad X = \sin(p/a)x$$

However, with the given boundary conditions, $u(0, t) = 0$, $\cos(p/a)x$ is the only one satisfying the diffusion equation, and the boundary condition $u_x(l, t) = 0$ results in $X'(l) = 0$, which yields

$$p = n\pi a/l \quad \text{where } n \text{ an integer}$$

Therefore the solution is

$$T(t)X(x) = e^{-(n\pi a/l)^2 t}\cos(n\pi x/l)$$

Hyperbolic Equation in Two Independent Variables

$$Lu = au_{xx} + 2bu_{xy} + cu_{yy} + 2du_x + 2eu_y + fu = g$$

The coefficients $a, \dots, g$ are known functions of x, y. The hyperbolicity condition is satisfied as $ac - b^2 < 0$.

Higher-Order Equations

Plate Equation.

$$\frac{\partial^4 u}{\partial x^4} + 2\frac{\partial^4 u}{\partial x^2 \partial y^2} + \frac{\partial^4 u}{\partial y^4} = 0$$

Solution of Partial Differential Equations

Equations with Constant Coefficients. If the order of each term is the same, the specific solutions can be determined by setting

$$u(x, y) = f(p)$$

$$p = ax + by$$

and substituting in the equations. An example is the *wave equation*.

$$\frac{\partial^2 \psi}{\partial x^2} - \frac{1}{c^2}\frac{\partial^2 \psi}{\partial t^2} = \sin(x + t)$$

Let $p = ax + bt$, $\psi(x, t) = f(p)$

$$\frac{\partial \psi}{\partial x} = \frac{\partial f}{\partial p}a; \qquad \frac{\partial^2 \psi}{\partial x^2} = \frac{\partial^2 f}{\partial p^2}a^2$$

$$\frac{\partial \psi}{\partial t} = \frac{\partial f}{\partial p}b; \qquad \frac{\partial^2 \psi}{\partial t^2} = \frac{\partial^2 f}{\partial p^2}b^2$$

$$\frac{\partial^2 f}{\partial p^2}a^2 - \frac{b^2}{c^2}\frac{\partial^2 f}{\partial p^2} = \sin p$$

$$\frac{\partial^2 f}{\partial p^2}\left(a^2 - \frac{b^2}{c^2}\right) = \sin p$$

$$f = -\sin p\frac{c^2}{a^2 c^2 - b^2}$$

$$\psi(x, y) = -\frac{c^2}{a^2 c^2 - b^2}\sin(ax + bt)$$

Since $a = b = 1$,

$$\psi(x, y) = -\frac{c^2}{c^2 - 1} \sin(x + t)$$

However, in the case of the diffusion equation, the order of various terms is different, thus the above method is not applicable. Laplace transformation can be used.

Method of Separation of Variables

A solution is sought in terms of

$$u(x_1, x_2, \ldots, x_n) = X_1(x_1) X_2(x_2) \cdots X_n(x_n)$$

If a solution is of this form, it is said to be *separable* in $x_1, \ldots, x_n$. By substituting in the partial differential equation, for example, wave equation,

$$\frac{X_1''}{X_1} = -\mu_1^2, \ldots, \frac{X_n''}{X_n} = -\mu_n^2$$

where the μ_i's are *separation constants*. The solution is obtained as $X_1(x_1) + \exp(i\mu_1 x_1), \ldots, X_n(x_n) = \exp(i\mu_n x_n)$ and

$$u(x_1, \ldots, x_n) = \exp[i(\mu_1 x_1 + \cdots + \mu_n x_n)]$$

Example:

$$\frac{\partial^2 u}{\partial x^2} - \frac{1}{c^2} \frac{\partial^2 u}{\partial t^2} = 0$$

Let

$$u(x, t) = X(x) T(t)$$

and substitute in the equation to obtain

$$TX'' - \frac{1}{c^2} XT'' = 0$$

Denoting

$$\frac{X''}{X} = -\mu_1^2 \qquad \frac{T''}{T} = -\mu_2^2$$

$$X'' + \mu_1^2 X = 0 \qquad T'' + \mu_2^2 T = 0$$

Solving these equations for X and T,

$$X = e^{i\mu_1 x} \qquad T = e^{i\mu_2 t}$$

By the Euler equations, given in Section 1.3, e^{ix} are expressed in terms of trigonometric functions to obtain

$$X = \cos \mu_1 x + \sin \mu_1 x$$

$$T = \cos \mu_2 t + \sin \mu_2 t$$

Thus, substituting in the above relation,

$$u(x, t) = (\cos \mu_1 x + \sin \mu_1 x)(\cos \mu_2 t + \sin \mu_2 t)$$

Here $\mu_2^2 = c\mu_1^2$ can be obtained by substituting the above solution in the original equation, thus expressing μ_2 in terms of μ_1.

Numerical Solutions of Partial Differential Equations

See Finite Difference Methods in Section 1.11.

1.6-5   Integral Equations

Integral Equations of the Second Kind: Fredholm Equation

$$\phi(x) - \lambda \int_a^b K(x,t)\phi(t)\, dt = f(x) \tag{1.6-11}$$

where $K(x,t)$ and $f(x)$ are given functions, is an *integral equation of the second kind* for the unknown function $\phi(x)$. Here $K(x,t)$ is called the *kernel*, λ is the parameter.

Corresponding to this equation are two additional equations: the *homogeneous equation*, where

$$f(x) = 0 \tag{1.6-12}$$

and the *transposed integral equation* where $K(x,t)$ is replaced by $K(t,x)$,

$$\psi(x) - \lambda \int_a^b K(t,x)\psi(t)\, dt = f(x) \tag{1.6-13}$$

The number of linearly independent solutions of (1.6-12), $\phi^{(i)}(x)$, $i = 1, 2, \ldots, n$, is the same as for (1.6-13). The general solution of the homogeneous equation is

$$\sum_{i=1}^n c_i \phi^{(i)}(x) \qquad c_i = \text{constant}$$

If $n = 0$ and $f(x)$ is continuous, (1.6-11) and (1.6-13) have one solution:

$$\phi(x) = f(x) + \int_a^b L(x,t)f(t)\, dt$$

$$\psi(x) = f(x) + \int_a^b L(t,x)f(t)\, dt$$

where $L(x,t)$ is called the *resolvent kernel*.

If $n > 0$, and $\sum_a^b \psi^{(i)}(t)f(t)\, dt = 0$, $i = 1, 2, \ldots, n$, then (1.6-11) has a unique solution. The general solution of (1.6-11) is the sum of the particular solution of (1.6-11) and the general solution of (1.6-12).

There are various methods of solution of linear integral equations of the second kind. These methods result in different forms of solutions.

1.   The *Method of Successive Substitutions* due to Neumann, Liouville, and Volterra gives $\phi(x)$ as an integral series in λ, where the coefficients of various powers of λ are functions of x. In this case ϕ is replaced in the integrand and the process is continued successively.

2.   The *Method of Successive Approximations* takes $\phi_0(x)$ as a first approximation and substitutes it in the equation obtaining successive approximations as $\phi_1, \phi_2, \ldots$.

3.   The *Method of Frobenious* gives $\phi(x)$ as the ratio of two integral series in λ. The numerator only has coefficients of the various powers of λ as functions of x.

4.   The *Method of Hilbert and Schmidt* gives $\phi(x)$ in terms of a set of *fundamental* functions that are the solutions of the corresponding homogeneous equation for the characteristic values of λ. This is applicable to symmetric kernels only (see below for the definition of symmetric kernels).

Symmetric Kernel

If $K(x,t) = K(t,x)$, that is, *symmetric*, the integral equation has solutions for only those (real) values which are called *eigenvalues*. The solutions found are *eigenfunctions* belonging to $K(x,t)$.

The method of successive approximations can be used for solving this equation. The sequence of functions

$$q_1(x) = f(x)$$

$$q_n(x) = f(x) + \lambda \int_a^b K(x,t)q^{n-1}(t)\, dt \qquad n = 2, 3, \ldots$$

converges to the solution $\phi(x)$, if

$$\lambda^2 < \left[\iint K^2(s,t)\,ds\,dt\right]^{-1}$$

Degenerate Kernel

If

$$K(x,t) = \sum_{i=1}^{n} \phi_i(x)\psi_i(t)$$

then the kernel is called *degenerate*. The integral equation can be solved by solving a system of algebraic equations.

Example:

$$\phi(x) - \lambda \int_a^b xt\phi(t)\,dt = x$$

Integral Equation of the First Kind

In Equation (1.6-11) we replace the unknown function $\phi(x)$ outside the integral sign by zero, and the integral equation obtained,

$$f(x) = \int_a^b K(x,t)d(t)\,dt$$

is called the equation of the first kind. An example is in the theory of Laplace transforms, where the kernel is $K(x,t) = e^{-xt}$. For Fourier transforms the kernel becomes, $K(x,t) = e^{-ixt}$ (see Section 1.7).

Volterra Equation

$$\phi(x) - \lambda \int_0^x K(x,\xi)\phi(\xi)\,d\xi = f(x)$$

is equivalent to ordinary differential equations, and solution can be obtained by Laplace transforms.

Example: The differential equation is given as

$$d\phi/dx = \phi(x) + x \quad \text{with } \phi(0) = \phi_0$$

By integration,

$$\phi(x) = \int_0^x (\phi + t)\,dt + \phi_0$$

is a linear Volterra equation of the second kind.

Nonlinear Integral Equation

$$\phi(x) = f(x) + \int_0^x K[x,\xi;\phi(\xi)]\,d\xi$$

which can be solved by successive approximations.

Example: The differential equation is given as

$$d\phi/dx = \phi^2(x) + x \quad \text{with } \phi(0) = \phi_0$$

By integration,

$$\phi(x) = \int_0^x (\phi^2(t) + t)\,dt + \phi_0$$

is a nonlinear integral equation.

1.7   TRANSFORMS AND OPERATIONAL MATHEMATICS

For a given function $f(x)$ one can associate an integral transform $\phi(p)$ given by

$$\phi(p) = \int \Omega(x, p) f(x)\, dx$$

where the function $\Omega(x, p)$ is the operator. Depending on the operator and the limits of the integration there may be different types of transforms that are commonly used in applications. Some special transforms are listed below. Tables of transforms of some functions are given in Section 1.10.

1.7-1   Fourier Transform

If $f(x)$ is defined for $x > 0$, and piecewise continuous over any finite interval, and

$$\int_0^\infty f(x)\, dx \quad \text{is absolutely convergent,}$$

for the *Fourier cosine transform* of $f(x)$, $\Omega(x, p) = \sqrt{2/\pi}\, \cos px$:

$$\phi_c(p) = \sqrt{2/\pi} \int_0^\infty f(x) \cos(px)\, dx$$

for the *Fourier sine transform* of $f(x)$, $\Omega(x, p) = \sqrt{2/\pi}\, \sin px$:

$$\phi_s(p) = \sqrt{2/\pi} \int_0^\infty f(x) \sin(px)\, dx$$

See Section 1.10 for tables of Fourier cosine and Fourier sine transforms.

1.7-2   Laplace Transform

For *Laplace transforms*, $\Omega(x, s) = e^{-sx}$:

$$Lf(x) = \phi(s) = \int_0^\infty e^{-sx} f(x)\, dx$$

For the *inverse Laplace transform*,

$$f(x) = \frac{1}{2\pi i} \int_{a-i\infty}^{a+i\infty} e^{sx} \phi(s)\, ds$$

or

$$L^{-1} \frac{1}{(s-\beta)^r} = \frac{x^{r-1}}{(r-1)!} e^{\beta x}$$

which is the sum of n such functions of x (see partial fractions in Section 1.1). Various functions and their transforms are given in a table form in Section 1.10.

Example: Let us take the given function as $f(x) = 1$.

$$\phi(s) = \Omega[f(x)] = \int_0^\infty e^{-sx} 1\, dx$$

$$= \frac{1}{-s} e^{-sx} \Big|_0^\infty = -\frac{1}{s} \left[\frac{1}{e^\infty} - \frac{1}{e^0} \right] = -\frac{1}{s}[0 - 1]$$

$$= \frac{1}{s}$$

See the tables in Section 1.11 for other Laplace transforms.

The inverse Laplace transform of $1/s$ can be found by using the formula above as illustrated below:

$$L^{-1}\frac{1}{(s-\beta)^r} = \frac{x^{r-1}}{(x-1)!}e^{\beta x} \quad \text{where } b = 0, r = 1$$

$$L^{-1}\frac{1}{s} = \frac{x^0}{(r-1)!}1 = \frac{1}{1}\cdot 1 = 1$$

1.7-3   Solution of Linear Differential Equations

Consider the differential equation given as

$$y'' + y = f(t) \quad \text{with initial conditions } y(0) = y'(0) = 0$$

This is the vibration equation with an external forcing function, $f(t)$, given as

$$f(t) = 0 \quad \text{for } t < 0$$

$$f(t) = 1 \quad \text{for } t > 0 \text{ and } = 0$$

Laplace transform of the elements of this equation are

$$L(y'') = s^2L(y)$$

$$L(1) = 1/s$$

thus,

$$s^2L(y) + L(y) = 1/s$$

$$L(y)[s^2 + 1] = 1/s$$

$$L(y) = \frac{1}{s(s^2 + 1)}$$

By partial fractions,

$$L(y) = \frac{1}{s} - \frac{s}{s^2 + 1} \qquad \text{,}$$

By the inverse transforms (see the tables in Section 1.10),

$$y = x - \cos x \quad \text{for } x > 0$$

$$y = 0 \quad \text{for } x < 0$$

1.7-4   Solution of Integral Equations

The equation of the curve defined by a particle starting from rest and sliding down, frictionless under gravity, leads to *tautoschrone* which is a cycloid (see Section 1.4). The problem is formulated as

$$\int_0^y f(z)(y - z)^{-1/2} \, dx = C_0$$

where C_0 is a constant. Taking the Laplace transform

$$L(f)(-1/2)!s^{-1/2} = C_0 s^{-1}$$

results in

$$L(f) = C_1 s^{-1/2}$$

Here C_1 is also a constant. Thus

$$f(y) = Cy^{-1/2}$$

Denoting the arc along the curve by u, $du^2 = dx^2 + dy^2$, the corresponding differential equation becomes

$$f(y) = \frac{du}{dy} = \left[1 + \left(\frac{dx}{dy} \right)^2 \right]^{1/2} = cy^{-1/2}$$

Setting $y = c^2 \sin^2(\phi/2)$, the parametric equations of a cycloid are obtained as follows:

$$x = \tfrac{1}{2}c^2 (\phi + \sin\phi)$$

$$y = \tfrac{1}{2}c^2 (1 - \cos\phi)$$

1.8   VECTOR ANALYSIS

1.8-1   Definitions

A *vector* is a quantity determined by its magnitude and direction, thus represented by a directed line segment. In application, quantities such as velocity and force are vectors:

$v = |v|$ the magnitude of a vector v
v/v unit vector
$-v$ negative vector, in opposite direction to v

Any two intersecting vectors V_1 and V_2 define a *plane*. Any other vector on the same plane can be expressed as a linear combination of these two *coplanar* vectors:

$$V = aV_1 + bV_2$$

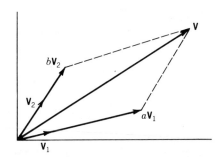

Here a and b are *scalar*. If A defines point A and B point B, any point P lying on the line segment AB is determined by its *position vector*,

$$V = sA + (1 - s)B$$

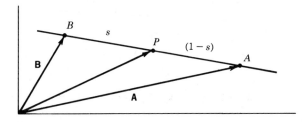

where the scalar s varies between 0, for $V = B$ (Point B), and 1, for $V = A$ (Point A).

Linear Dependence. n vectors are linearly dependent if scalars $s_1, \ldots, s_n$ exist such that

$$s_1V_1 + s_2V_2 + \cdots + s_nV_n = 0$$

Any vector **V** linearly dependent on $V_1 \cdots V_n$ can be expressed as

$$\mathbf{V} = s_1\mathbf{V}_1 + s_2\mathbf{V}_2 + \cdots + s_n\mathbf{V}_n$$

If the $v_1, \ldots, v_n$ are unit vectors along the coordinate axes $x_1, \ldots, x_n$, any vector **A** in this n-dimensional space can be represented as

$$\mathbf{A} = a_1\mathbf{v}_1 + \cdots + a_n\mathbf{v}_n$$

where $a_1, \ldots, a_n$ are respective magnitudes of the projections of **v** on the coordinate axes. The magnitude of **V** is v:

$$v = \sqrt{a_1^2 + \cdots + a_n^2}$$

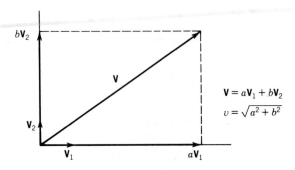

$$\mathbf{V} = a\mathbf{V}_1 + b\mathbf{V}_2$$
$$v = \sqrt{a^2 + b^2}$$

The *direction cosines* of **v** are

$$\cos a_1 = a_1/v, \cdots, \cos a_n = a_n/v$$

1.8-2   Vector Operations

The two vectors **A** and **B** and the angle ψ from **A** to **B** are given. The following relations are relevant to the vector operations.

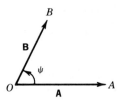

Addition of Vectors

$$\mathbf{A} + \mathbf{B} = (a_1 + b_1)\mathbf{v}_1 + \cdots + (a_n + b_n)\mathbf{v}_n$$

Commutative: $$\mathbf{A} + \mathbf{B} = \mathbf{B} + \mathbf{A}$$

Distributive: $$(s_1 + s_2)\mathbf{A} = s_1\mathbf{A} + s_2\mathbf{A}$$

$$s_1(\mathbf{A} + \mathbf{B}) = s_1\mathbf{A} + s_1\mathbf{A}$$

Associative: $$\mathbf{A} + (\mathbf{B} + \mathbf{C}) = (\mathbf{A} + \mathbf{B}) + \mathbf{C} = \mathbf{A} + \mathbf{B} + \mathbf{C}$$

Product of Vectors

Scalar (Dot or Inner) Product.

$$\mathbf{A} \cdot \mathbf{B} = ab \cos \psi$$

$$= a_1b_1 + a_2b_2 + \cdots + a_nb_n = \text{a scalar}$$

Commutative:
$$\mathbf{A} \cdot \mathbf{B} = \mathbf{B} \cdot \mathbf{A}$$

$$(\mathbf{A} + \mathbf{B}) \cdot \mathbf{C} = \mathbf{A} \cdot \mathbf{C} + \mathbf{B} \cdot \mathbf{C}$$

$$\mathbf{A} \cdot (\mathbf{B} + \mathbf{C}) = \mathbf{A} \cdot \mathbf{B} + \mathbf{A} \cdot \mathbf{C}$$

If **A** and **B** are perpendicular to each other, $\mathbf{A} \cdot \mathbf{B} = 0$. If **A** and **B** are parallel to each other, $\mathbf{A} \cdot \mathbf{B} = ab$

For unit vectors
$$\mathbf{v}_i \cdot \mathbf{v}_j = 0 \qquad i, j = 1, 2, \dots, n$$

$$\mathbf{v}_i \cdot \mathbf{v}_i = 1 \qquad i = 1, 2, \dots, n$$

Vector (Cross or Skew) Product.

$$\mathbf{A} \times \mathbf{B} = ab(\sin \psi)\mathbf{c}$$

where **c** is the unit vector perpendicular to the plane of **A** and **B** (the direction of **c** as shown).

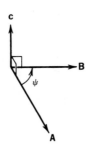

$$\begin{vmatrix} \mathbf{v}_1 & \mathbf{v}_2 & \mathbf{v}_3 \\ a_1 & a_2 & a_3 \\ b_1 & b_2 & b_3 \end{vmatrix} = (a_2 b_3 - a_3 b_2)\mathbf{v}_1 + (a_3 b_1 - a_1 b_3)\mathbf{v}_2 + (a_1 b_2 - a_2 b_1)\mathbf{v}_3$$

$$\tan \psi = \frac{|\mathbf{A} \times \mathbf{B}|}{\mathbf{A} \times \mathbf{B}}$$

Noncommutative:
$$\mathbf{A} \times \mathbf{B} = -\mathbf{B} \times \mathbf{A}$$

Distributive:
$$\mathbf{A} \times (\mathbf{B} + \mathbf{C}) = \mathbf{A} \times \mathbf{B} + \mathbf{A} \times \mathbf{C}$$

$$(\mathbf{A} + \mathbf{B}) \times \mathbf{C} = \mathbf{A} \times \mathbf{C} + \mathbf{B} \times \mathbf{C}$$

For $\mathbf{v}_1, \mathbf{v}_2, \mathbf{v}_3$ orthogonal unit vectors,

$$\mathbf{v}_1 \times \mathbf{v}_2 = \mathbf{v}_3 \qquad \mathbf{v}_2 \times \mathbf{v}_3 = \mathbf{v}_1 \qquad \mathbf{v}_3 \times \mathbf{v}_1 = \mathbf{v}_2$$

$$\mathbf{v}_1 \times \mathbf{v}_1 = 0 \qquad \mathbf{v}_2 \times \mathbf{v}_2 = 0 \qquad \mathbf{v}_3 \times \mathbf{v}_3 = 0$$

Scalar Triple Product.

$$\mathbf{A} \cdot (\mathbf{B} \times \mathbf{C}) = (\mathbf{A} \times \mathbf{B}) \cdot \mathbf{C} = |ABC|$$

This corresponds to the volume of the parallelepiped with edges A, B, C.

Vector Triple Product.

$$\mathbf{A} \times (\mathbf{B} \times \mathbf{C}) = (\mathbf{A} \cdot \mathbf{B})\mathbf{B} - (\mathbf{A} \cdot \mathbf{B})\mathbf{C}$$

This results in a vector perpendicular to **A**, lying in the plane of **B** and **C**.

Differentiation of Vectors

$$\mathbf{A} = a_1\mathbf{v}_1 + a_2\mathbf{v}_2 + a_3\mathbf{v}_3$$

$$\mathbf{B} = b_1\mathbf{v}_1 + b_2\mathbf{v}_2 + b_3\mathbf{v}_3$$

and **A** and **B** are functions of a scalar, t,

$$\frac{d\mathbf{A}}{dt} = \frac{da_1}{dt}\mathbf{v}_1 + \frac{da_2}{dt}\mathbf{v}_2 + \frac{da_3}{dt}\mathbf{v}_3$$

$$\frac{d}{dt}(\mathbf{A} + \mathbf{B}) = \frac{d\mathbf{A}}{dt} + \frac{d\mathbf{B}}{dt}$$

$$\frac{d}{dt}(\mathbf{A} \cdot \mathbf{B}) = \frac{d\mathbf{A}}{dt}\mathbf{B} + \mathbf{A}\frac{d\mathbf{B}}{dt}$$

$$\frac{d}{dt}(\mathbf{A} \times \mathbf{B}) = \frac{d\mathbf{A}}{dt} \times \mathbf{B} + \mathbf{A} \times \frac{d\mathbf{B}}{dt}$$

$$\mathbf{A} \cdot \frac{d\mathbf{A}}{dt} = a\frac{da}{dt}$$

Triple Product.

$$\frac{d}{dt}\mathbf{ABC} = \frac{d\mathbf{A}}{dt}\mathbf{BC} + \mathbf{A}\frac{d\mathbf{B}}{dt}\mathbf{C} + \mathbf{AB}\frac{d\mathbf{C}}{dt}$$

$$\frac{d}{dt}\mathbf{A} \times (\mathbf{B} \times \mathbf{C}) = \frac{d\mathbf{A}}{dt} \times (\mathbf{B} \times \mathbf{C}) + \mathbf{A} \times \left(\frac{d\mathbf{B}}{dt} \times \mathbf{C}\right) + \mathbf{A} \times \left(\mathbf{B} \times \frac{d\mathbf{C}}{dt}\right)$$

Differential Operators

If F is a scalar function of x, y, z, then

$$dF = \frac{\partial F}{\partial x}\,dx + \frac{\partial F}{\partial y}\,dy + \frac{\partial F}{\partial z}\,dz$$

$$\nabla \equiv \text{del} \equiv \mathbf{v}_1\frac{\partial}{\partial x} + \mathbf{v}_2\frac{\partial}{\partial y} + \mathbf{v}_3\frac{\partial}{\partial z}$$

$$\nabla^2 \equiv \text{Laplacian} \equiv \frac{\partial^2}{\partial x^2} + \frac{\partial^2}{\partial y^2} + \frac{\partial^2}{\partial z^2}$$

The *gradient* is the direction and magnitude of the maximum rate of increase of F at any point:

$$\nabla F \equiv \text{grad } F \equiv \frac{\partial F}{\partial x}\mathbf{v}_1 + \frac{\partial F}{\partial y}\mathbf{v}_2 + \frac{\partial F}{\partial z}\mathbf{v}_3$$

The Distributive law: $\qquad \nabla(F + G) = \nabla F + \nabla G$

The Associative law: $\qquad \nabla(FG) = F\nabla G + G\nabla F$

A is a vector function with A_1, A_2, A_3 as magnitudes of the components in coordinate axes.

The *divergence* of **A** is given by

$$\nabla \cdot \mathbf{A} \equiv \text{div } \mathbf{A} \equiv \frac{\partial A_1}{\partial x} + \frac{\partial A_2}{\partial y} + \frac{\partial A_3}{\partial z}$$

The Distributive law: $\quad \nabla \cdot (\mathbf{A} + \mathbf{B}) = \nabla \cdot \mathbf{A} + \nabla \cdot \mathbf{B}$

$$\nabla \cdot (F\mathbf{A}) = (\nabla F) \cdot \mathbf{A} + F(\nabla \cdot \mathbf{A})$$

$$\nabla \cdot (\mathbf{A} \times \mathbf{B}) = \mathbf{B} \cdot (\nabla \times \mathbf{A}) - \mathbf{A}(\nabla \times \mathbf{B})$$

The *Curl of* **A** is given by

$$\nabla \times \mathbf{A} \equiv \text{curl } \mathbf{A} \equiv \left(\frac{\partial A_3}{\partial y} - \frac{\partial A_2}{\partial z} \right) \mathbf{v}_1 + \left(\frac{\partial A_1}{\partial z} - \frac{\partial A_3}{\partial x} \right) \mathbf{v}_2 + \left(\frac{\partial A_2}{\partial y} - \frac{\partial A_3}{\partial x} \right) \mathbf{v}_3$$

$$= \begin{vmatrix} \mathbf{v}_1 & \mathbf{v}_2 & \mathbf{v}_3 \\ \dfrac{\partial}{\partial x} & \dfrac{\partial}{\partial y} & \dfrac{\partial}{\partial z} \\ A_1 & A_2 & A_3 \end{vmatrix}$$

For **A** given as

$$\mathbf{A} = A_1 \mathbf{v}_1 + A_2 \mathbf{v}_2 + A_3 \mathbf{v}_3$$

$$\nabla \cdot \mathbf{A} = \nabla A_1 \cdot \mathbf{v}_1 + \nabla A_2 \cdot \mathbf{v}_2 + \nabla A_3 \cdot \mathbf{v}_3$$

$$\nabla \times \mathbf{A} = \nabla A_1 \times \mathbf{v}_1 + \nabla A_2 \times \mathbf{v}_2 + \nabla A_3 \times \mathbf{v}_3$$

$$\text{div grad } F = \nabla \cdot (\nabla F) \equiv \text{Laplacian } F \equiv \nabla^2 F$$

$$\text{curl grad } F \equiv 0$$

$$\text{div curl } \mathbf{A} \equiv 0$$

Integration of Vectors and Integrals

F = a vector function
V = volume
S = a surface bounding a closed region with volume *V*
C = a curve, closed and bounds *S*
n = the unit vector normal to *S* outward at point *A*
r = the unit vector tangent to *C* at point *A*

Green's (Gauss') Theorem.

$$\iiint_{(v)} (\nabla \cdot \mathbf{F}) \, dV = \iint_{(s)} (\mathbf{F} \cdot \mathbf{n}) \, dS$$

Stokes' Theorem.

$$\iint_{(s)} \mathbf{n} \cdot (\nabla \times \mathbf{F}) \, dS = \int_{(c)} \mathbf{F} \cdot d\mathbf{r}$$

1.9 STATISTICS AND PROBABILITY

1.9-1 Statistics

Definitions

Statistical data consist of *N* values of variables x_i, called *class marks*. One of these classes, x_0 is taken to be the *origin*. The *intervals*, $x_{i+1} - x_i = c$, are called *class intervals*. The *frequency distribution* of the data is a tabulation by classes where the *frequencies* or *weights*, f_i, of a class x_i are given.

Since the total number of observations, *N*, must be equal to the sum of frequencies of various classes, $N = \Sigma f_i$.

Example: Ten students ($N = 10$) receive five different grades, $n = 5$, as follows:

3	grades are	50
1	grade is	70
3	grades are	80
2	grades are	90
1	grade is	100

Classes: $x_1 = 50,\ x_2 = 70,\ x_3 = 80,\ x_4 = 90,\ x_5 = 100$

Frequencies: $f_1 = 3,\ f_2 = 1,\ f_3 = 3,\ f_4 = 2,\ f_2 = 1$

Mean (Arithmetic Mean).

$$\bar{x} = \frac{1}{n} \sum_{i=1}^{n} x_i = \frac{x_1 + \cdots + x_n}{n}$$

For the above example, the mean becomes

$$\bar{x} = \frac{50 + 70 + 80 + 90 + 100}{5} = 78$$

Weighted Mean. If each value x_i has an associated weight, $\omega_i > 0$, then $\sum_{i=1}^{n} \omega_i$ is the total weight and

$$\bar{x} = \frac{\displaystyle\sum_{i=1}^{n} \omega_i x_i}{\displaystyle\sum_{i=1}^{n} \omega_i} = \frac{\omega_1 x_1 + \cdots + \omega_n x_n}{\omega_1 + \cdots + \omega_n}$$

For the above example, $\omega_i = f_i$, thus $\Sigma \omega_i = 3 + 1 + 3 + 2 + 1 = 10$ and the weighted mean becomes

$$\bar{x} = \frac{50.3 + 70.1 + 80.3 + 90.2 + 100.1}{10} = 64$$

Geometric Mean.

$$\bar{x}_G = \sqrt[n]{x_1 \cdots x_n}$$

Geometric Mean in Logarithmic Form.

$$\log \bar{x}_G = \frac{1}{n} \sum \log x_i = \frac{\log x_i + \cdots + \log x_n}{n}$$

Harmonic Mean.

$$\bar{x}_H = \frac{n}{\displaystyle\sum_{i=1}^{n} (1/x_i)} = \frac{n}{(1/x_1) + \cdots + (1/x_n)}$$

Here $\bar{x}_H < \bar{x}_G < \bar{x}$. If sample values are identical $\bar{x}_H = \bar{x}_G = \bar{x}$.

Median. If the sample is arranged in an ascending order of values, then the median, M_d, of n such values is the value at the $[(n + 1)/2]$th position of the sequence:

$$M_d = \frac{n + 1}{2}$$

that is, when n is odd, the middle value (if even)—the mean of the two middle values of the above ordered set of values—is the median.

For the above example the classes are 50, 70, 80, 90, 100. Therefore the median is 80, that is, two grades 50 and 70 are less than the median and two grades 90 and 100 are more than the median, thus equal number of grades are below and above the median value, 80.

Quartiles. For an ascending order, $i(n + 1)/4$, where $i = 1, 2, 3$ corresponds to the first, second, and third quartiles, respectively.

Deciles. They are every $[(n + 1)/10]$th value.

Percentile. They are every $[(n + 1)/100]$th value.

The sample example is shown below with various notions identified for illustration.

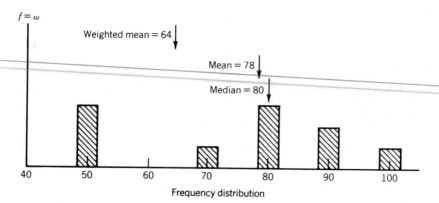

Frequency distribution

The example given above is a *discrete* distribution. There are various types of distributions used in statistical analysis. One of the important frequency distributions is the *normal distribution* which is a *continuous* one in that class values continuously vary as shown in the figure below. The normal distribution is symmetric and bell shaped.

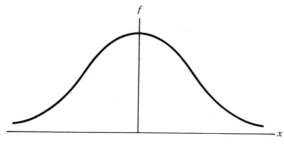

Normal distribution

Mean Deviation

$$\frac{1}{n} \sum_{i=1}^{n} |x_i - \bar{x}|$$

Standard Deviation

$$s = \sqrt{\frac{\sum_{i=1}^{n} (x_i - \bar{x})^2}{n - 1}}$$

Variance

$$V = s^2$$

Standardized variable: $z = \dfrac{x_i - \bar{x}}{s}$

Moments

The rth moment about the *origin* (see above definitions) is given by

$$m_r' = \frac{1}{n} \sum_{i=1}^{n} x_i^r$$

The rth moment about the *mean* $\bar{x}$ (see above definitions) is given by

$$m_r = \frac{1}{n} \sum_{i=1}^{n} (x_i - \bar{x}) r$$

Curve Fitting

If (x_i, y_i) form a set of ordered pairs where x_i are fixed variables and y_i are dependent variables, a curve can be fitted to find $y = f(x)$ in terms of a *polynomial function*, $y = b_0 + b_1 x + \cdots + b_n x_n$. Here, by the method of least squares, the coefficients $b_0, \ldots, b_n$ can be calculated.

For the special case of $n = 1$ the polynomial becomes a line equation. Thus the curve fitted to the data is a *straight line*.

Similarly, an *exponential curve*, $y = ab^x$, can also be fitted to given data. Here $\log y = \log a + x \log b$ becomes the straight line and $(x_i, \log y_i)$ are fitted by the method of mean squares.

A *power function*, $y = ax^b$, can also be fitted to data by taking logarithms, $\log y = \log a + b \log x$. Thus the ordered pair is $(\log x_i, \log y_i)$, and the fit is a straight line fit.

Regression

In a *regression* problem *one particular variable* is studied in terms of the *remaining variables*. For example, if the two variables are x and y, y can be chosen as the particular variable. For the *linear regression* of y on x, we have

$$E(y/x) = b_0 + b_1 x$$

where $E(y/x)$ is the mean of the distribution of y for a given x.

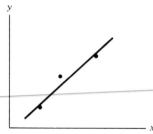

Linear relationship of y to x

After fitting a regression line to a given set of points, the accuracy of predicting y values can be determined. For this the standard error estimate is defined.

The *standard error estimate* is given by

$$s_e = \sqrt{\frac{\sum [y_i - (b_0 + b_1 x_i)]^2}{n - 2}}$$

where b_0 and b_1 are calculated by straight line fitting using the least squares method.

In the case of multiple x's the techniques for predicting the y variable are similar to the above.

Correlation

In a *correlation* problem, however, *several variables* are studied *simultaneously* to show their interrelation. For example, for the case of two variables, x and y, an estimate of the population *correlation coefficient* is given by and defined the desired *measure of relationship*.

$$r = \sqrt{\frac{\sum (x_i - \bar{x})(y_i - \bar{y})}{\left[\sum (x_i - x)^2\right]\left[\sum (y_i - y)^2\right]}}$$

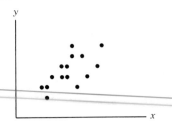

Scatter diagram of
two variables used in correlation

1.9-2   Probability

Definitions

In an experiment, any outcome corresponds to one *element* (an *event*) E, while the set of all such events forms the *sample space*, S. Therefore, the *probability* of an event is

$$P(E) = m/n$$

where n is the number of mutually exclusive and likely *outcomes* of an experiment and m is the number of those corresponding to the event E.

Example: Let us experiment by throwing a dice with six sides, numbered $1, 2, 3, 4, 5, 6$. The probability of any side being an *outcome* is equal to $1/6$. Here $m = 1$ and $n = 6$. If the even E represents any one of the six sides being an outcome,

$$P(E) = m/n = 1/6$$

A *set* is defined as a collection of *elements*. The elements are objects with particular characteristics, for example, a set of integers, a set of square areas, and so on. The following notations are frequently used in operations with sets.

$S = \{s_1, s_2, \ldots, s_n\}$ is a set.

$s_1 \in S$ defines s_1 as an element of S.

$S_1 \subset S_2$ indicates that all the elements of S_1 are also elements of S_2; thus S_1 is a *subset* of S_2.

If $S_1 \subset S_2$ and $S_2 \subset S_1$, then $S_1 = S_2$.

If $S_2 \subset S_3$ but at least one element in S_3 is not in S_2, then S_2 is a *proper subset* of S_3.

$S_1 \cap S_2$ is the *intersection* of two sets, S_1 and S_2 such that the elements in this intersection are elements of both S_1 and S_2.

$S_1 \cup S_2$ is the union of two sets, S_1 and S_2 such that the elements in this union are elements of either S_1 or S_2.

$E \cup F$ $E \cap F$

Theorems

1. If $\varnothing$ is the *null set*, $P(\varnothing) = 0$.
2. If S is the *sample set*, $P(S) = 1$.
3. If E and F are two events,

$$P(E \cup F) = P(E) + P(F) - P(E \cap F)$$

4. If E and F are *mutually exclusive* events,

$$P(E \cup F) = P(E) + P(F)$$

5. If E and E' are *complementary* events,

$$P(E) = 1 - P(E')$$

6. The *conditional* probability of an event E given an event F, where conditional means that the event E occurs only if the event F occurs, is defined as

$$P(E/F) = \frac{P(E \cap F)}{P(F)}$$

where $P(F) \neq 0$.

7. Two events E and F are *independent* if and only if

$$P(E \cap F) = P(E)P(F)$$

E is *statistically independent* of F if

$$P(E/F) = P(E) \quad \text{and} \quad P(F/E) = P(F)$$

Bayes' Theorem. If $E_1, \ldots, E_n$ are mutually exclusive events and the sample space is the union of these events, for $P(E) \neq 0$,

$$P(E_j/E) = \frac{P(E_j)P(E/E_j)}{\sum_{i=1}^{n} [P(E_i)P(E/E_i)]}$$

Random Variables

A function between the sample space, S (domain), and the set of real numbers, x (range), is called a *random variable*, X. A random variable can be *discrete* if only a finite number of values on the real axis is assumed. It is *continuous* if a continuum of values on the real axis is represented.

Discrete Case: Probability Function. The function $f = P[X = x]$ with properties

$$f(x_i) \geq 0, \qquad \sum_i f(x_i) = 1 \qquad i = 1, 2, \ldots$$

and for every event E

$$P(E) = P[X \text{ is in } E] = \sum_E f(x)$$

where the summation implies for the values x that are in E, is called the *probability function* of the discrete random variable X.

Discrete Case: Cumulative Distribution Function. The probability of the value of a random variable X less than or equal to some real number x, is defined as the *cumulative distribution function*:

$$F(x) = P(X \leq x) = \sum f(x_i)$$

for $-\infty < x < \infty$ and $x_i \leq x$.

Continuous Case: Probability Density. The function $f(x)$ with properties $f(x) \geq 0$, and

$$\int_{-\infty}^{\infty} f(x)\, dx = 1 \quad \text{for all } x, \ -\infty < x < \infty$$

also for any event E

$$P(E) = P(X \text{ is in } E) = \int_E f(x)\, dx$$

is called the *probability density* of the continuous random variable X.

Continuous Case: Cumulative Distribution Function. The probability of the value of a random variable X less than or equal to some real number x is defined as the *cumulative distribution function*:

$$F(x) = P(X \leq x) = \int_{-\infty}^{x} f(x)\, dx \qquad \text{for } -\infty < x < \infty$$

From this the *density* can be found as

$$f(x) = \frac{dF(x)}{dx}$$

and

$$P(a \leq X \leq b) = P(X \leq b) - P(X \leq a)$$
$$= F(b) - F(a)$$

Expected Value

$$E(X) = \sum_{x} xf(x) \quad \text{(discrete case)}$$

$$= \int_{-\infty}^{\infty} xf(x)\, dx \quad \text{(continuous case)}$$

Theorems:

1. $E[aX + bY] = aE(X) + bE(Y)$.
2. $E[XY] = E(X)E(Y)$, if X and Y are statistically independent.

Moments

1. **The rth moment about the origin.**

$$\mu_r = E(X^r) = \sum_{x} x^r f(x) \quad \text{(discrete case)}$$

$$= \int_{-\infty}^{\infty} x^r f(x)\, dx \quad \text{(continuous case)}$$

2. **The first moment (mean).**

$$\mu = \sum_{x} x^r f(x) \quad \text{(discrete case)}$$

$$= \int_{-\infty}^{\infty} xf(x)\, dx \quad \text{(continuous case)}$$

3. **The rth moment about the mean.**

$$\mu_r = E\big[(X - \mu)^r\big] = \sum_{x} (x - \mu)^r f(x) \quad \text{(discrete case)}$$

$$= \int_{-\infty}^{\infty} (x - \mu)^r f(x)\, dx \quad \text{(continuous case)}$$

4. **The second moment about the mean.**

$$\mu_2 = E\big[(x - \mu)^2\big] = \mu_2' - \mu^2$$

This is called the *variance* and denoted by σ^2. Here σ is the *standard deviation*.

Generating Function

1. **Moment generating function.**

$$m_x(t) = E(e^{iX}) = \sum_x e^{ix}f(x) \quad \text{(discrete)}$$

$$= \int_{-\infty}^{\infty} e^{ix}f(x)\,dx \quad \text{(continuous)}$$

2. **Fractional moment generating function.**

$$E(t^X) = \sum_x t^x f(x) \quad \text{(discrete)}$$

$$= \int_{-\infty}^{\infty} t^x f(x)\,dx \quad \text{(continuous)}$$

Characteristic Function

$$d(t) = E(e^{itX}) = \sum_x e^{itx}f(x) \quad \text{(discrete)}$$

$$= \int_{-\infty}^{\infty} e^{itx}f(x)\,dx \quad \text{(continuous)}$$

Probability Distributions: Discrete Case

The probability distribution possessed by the discrete random variable X depends on the probability function. These functions and corresponding distributions are listed below.

1. **Discrete uniform distribution of X.**

$$P(X = x) = f(x) = 1/n \qquad x = x_1, x_2, \ldots, x_n$$

2. **Binomial distribution of X** (see Section 1.1).

$$P(X = x) = f(x) = \binom{n}{x}\theta^x(1 - \theta)^{n-x} \qquad x = 0, 1, 2, \ldots, n$$

where $f(x)$ is the general term of the expansion of $[\theta + (1 - \theta)]^n$.

3. **Geometric distribution of X.**

$$P(X = x) = f(x) = \theta^x(1 - \theta)^{x-1} \qquad x = 0, 1, 2, \ldots, n$$

4. **Multinomial distribution of X.**

$$P(X_1 = x_1, \ldots, X_n = x_n) = f(x_1, \ldots, x_n) = \frac{N!}{\prod_{i=1}^{n} x_i!} \prod_{i=1}^{n} \theta_i x_i$$

5. **Poisson distribution of X.**

$$P(X = x) = f(x) = \frac{e^{-\lambda}\lambda^x}{x!} \qquad \lambda > 0, \quad x = 0, 1, \ldots$$

6. **Hypergeometric distribution of X.**

$$P(X = x) = f(x) = \frac{\binom{k}{x}\binom{N-k}{n-x}}{\binom{N}{n}} \qquad x = 0, 1, \ldots, [n, k] \text{ (smaller of } n \text{ or } k)$$

Probability Distributions: Continuous Case

The probability distribution possessed by the continuous random variable X depends on the density function. These functions and corresponding distributions are listed below.

1. Uniform distribution of X.

$$f(x) = \frac{1}{\beta - \alpha} \qquad \lambda < x < \beta$$

where α and β are parameters.

2. Normal distribution of X.

$$f(x) = \left(1/\sigma\sqrt{2}\,\pi\right) e^{-(x-\mu)^2/2\sigma^2} \qquad -\infty < x < \infty$$

where μ, the mean, and σ, the standard deviation, are parameters.

3. Gamma distribution of X.

$$f(x) = \frac{1}{\Gamma(\alpha + 1)\beta^{\alpha+1}} x^{\alpha} e^{-x/\beta}$$

where α and β are parameters, $\alpha > -1$, $\beta > 0$.

4. Exponential distribution of X.

$$f(x) = (1/\theta) e^{-x/\theta}, \qquad x > 0, \quad \theta > 0$$

5. Beta distribution of X.

$$f(x) = \frac{\Gamma(\alpha + \beta + 2)}{\Gamma(\alpha + 1)\Gamma(\beta + 1)} x^{\alpha}(1 - x)^{\beta}$$

1.10 MATHEMATICAL TABLES

1.10-1 Series Expansion of Common Functions

See Section 1.1.

Binomial Functions

$$(x + y)^n = x^n + nx^{n-1}y + \frac{n(n-1)}{2!}x^{n-2}y^2 + \frac{n(n-1)(n-2)}{3!}x^{n-3}y^3 + \cdots \qquad (y^2 < x^2)$$

$$(1 \pm x)^n = 1 \pm nx + \frac{n(n-1)x^2}{2!} \pm \frac{n(n-1)(n-2)x^3}{3!} + \cdots \qquad (x^2 < 1)$$

$$(1 \pm x)^{-n} = 1 \mp nx + \frac{n(n+1)x^2}{2!} \mp \frac{n(n+1)(n+2)x^3}{3!} + \cdots \qquad (x^2 < 1)$$

$$(1 \pm x)^{-1} = 1 \mp x + x^2 \mp x^3 + x^4 \mp x^5 + \cdots \qquad (x^2 < 1)$$

$$(1 \pm x)^{-2} = 1 \mp 2x + 3x^2 \mp 4x^3 + 5x^4 \mp 6x^5 + \cdots \qquad (x^2 < 1)$$

Exponential Functions

$$e = 1 + \frac{1}{1!} + \frac{1}{2!} + \frac{1}{3!} + \frac{1}{4!} + \cdots$$

$$e^x = 1 + x + \frac{x^2}{2!} + \frac{x^3}{3!} + \frac{x^4}{4!} + \cdots \qquad \text{(all real values of } x \text{)}$$

$$a^x = 1 + x \log_e a + \frac{\left(x \log_e a \right)^2}{2!} + \frac{\left(x \log_e a \right)^3}{3!} + \cdots$$

$$e^x = e^a \left[1 + (x - a) + \frac{(x - a)^2}{2!} + \frac{(x - a)^3}{3!} + \cdots \right]$$

Logarithmic Functions

$$\log_e x = \frac{x - 1}{x} + \frac{1}{2}\left(\frac{x - 1}{x} \right)^2 + \frac{1}{3}\left(\frac{x - 1}{x} \right)^3 + \cdots \qquad \left(x > \tfrac{1}{2} \right)$$

$$\log_e x = (x - 1) - \tfrac{1}{2}(x - 1)^2 + \tfrac{1}{3}(x - 1)^3 - \cdots \qquad (2 \geq x > 0)$$

$$\log_e x = 2\left[\frac{x - 1}{x + 1} + \frac{1}{3}\left(\frac{x - 1}{x + 1} \right)^3 + \frac{1}{5}\left(\frac{x - 1}{x + 1} \right)^5 + \cdots \right] \qquad (x > 0)$$

$$\log_e (1 + x) = x - \tfrac{1}{2}x^2 + \tfrac{1}{3}x^3 - \tfrac{1}{4}x^4 + \cdots \qquad (-1 < x < 1)$$

$$\log_e (n + 1) - \log_e (n - 1) = 2\left[\frac{1}{n} + \frac{1}{3n^3} + \frac{1}{5n^5} + \cdots \right]$$

$$\log_e (a + x) = \log_e a + 2\left[\frac{x}{2a + x} + \frac{1}{3}\left(\frac{x}{2a + x} \right)^3 + \frac{1}{5}\left(\frac{x}{2a + x} \right)^5 + \cdots \right]$$

$$(a > 0, \, -a < x < +\infty)$$

$$\log_e \frac{1 + x}{1 - x} = 2\left[x + \frac{x^3}{3} + \frac{x^5}{5} + \cdots + \frac{x^{2n-1}}{2n - 1} + \cdots \right] \qquad -1 < x < 1$$

$$\log_e x = \log_e a + \frac{(x - a)}{a} - \frac{(x - a)^2}{2a^2} + \frac{(x - a)^3}{3a^3} - + \cdots \qquad 0 < x \leq 2a$$

Trigonometric Functions

$$\sin x = x - \frac{x^3}{3!} + \frac{x^5}{5!} - \frac{x^7}{7!} + \cdots \qquad \text{(all real values of } x \text{)}$$

$$\cos x = 1 - \frac{x^2}{2!} + \frac{x^4}{4!} - \frac{x^6}{6!} + \cdots \qquad \text{(all real values of } x \text{)}$$

$$\tan x = x + \frac{x^3}{3} + \frac{2x^5}{15} + \frac{17x^7}{315} + \frac{62x^9}{2835} + \cdots + \frac{2^{2n}(2^{2n} - 1) B_n}{(2n)!} x^{2n-1} + \cdots$$

$$\left[x^2 < \pi^2/4, \text{ and } B_n \text{ represents the } n \text{th Bernoulli number} \right]$$

$$\cot x = \frac{1}{x} - \frac{x}{3} - \frac{x^2}{45} - \frac{2x^5}{945} - \frac{x^7}{4725} - \cdots - \frac{2^{2n} B_n}{(2n)!} x^{2n-1} - \cdots$$

$$\left[x^2 < \pi^2, \text{ and } B_n \text{ represents the } n \text{th Bernoulli number} \right]$$

$$\sin^{-1} x = x + \frac{x^3}{2 \cdot 3} + \frac{1 \cdot 3}{2 \cdot 4 \cdot 5}x^5 + \frac{1 \cdot 3 \cdot 5}{2 \cdot 4 \cdot 6 \cdot 7}x^7 + \cdots \qquad \left(x^2 < 1, \, -\frac{\pi}{2} < \sin^{-1} x < \frac{\pi}{2} \right)$$

$$\cos^{-1}x = \frac{\pi}{2} - \left(x + \frac{x^3}{2 \cdot 3} + \frac{1 \cdot 3}{2 \cdot 4 \cdot 5}x^5 + \frac{1 \cdot 3 \cdot 5x^7}{2 \cdot 4 \cdot 6 \cdot 7} + \cdots \right) \qquad (x^2 < 1, 0 < \cos^{-1}x < \pi)$$

$$\tan^{-1}x = x - \frac{x^3}{3} + \frac{x^5}{5} - \frac{x^7}{7} + \cdots \qquad (x^2 < 1)$$

$$\tan^{-1}x = \frac{\pi}{2} - \frac{1}{x} + \frac{1}{3x^2} - \frac{1}{5x^5} + \frac{1}{7x^7} - \cdots \qquad (x > 1)$$

$$\tan^{-1}x = -\frac{\pi}{2} - \frac{1}{x} + \frac{1}{3x^2} - \frac{1}{5x^5} + \frac{1}{7x^7} - \cdots \qquad (x < -1)$$

$$\cot^{-1}x = \frac{\pi}{2} - x + \frac{x^3}{3} - \frac{x^5}{5} + \frac{x^7}{7} - \cdots \qquad (x^2 < 1)$$

$$\log_e \sin x = \log_e x - \frac{x^2}{6} - \frac{x^4}{180} - \frac{x^6}{2835} - \cdots \qquad (x^2 < \pi^2)$$

$$\log_e \cos x = -\frac{x^2}{2} - \frac{x^4}{12} - \frac{x^6}{45} - \frac{17x^8}{2520} - \cdots \qquad \left(x^2 < \frac{\pi^2}{4} \right)$$

$$\log_e \tan x = \log_e x + \frac{x^2}{3} + \frac{7x^4}{90} + \frac{62x^6}{2835} + \cdots \qquad \left(x^2 < \frac{\pi^2}{4} \right)$$

$$e^{\sin x} = 1 + x + \frac{x^2}{2!} - \frac{3x^4}{4!} - \frac{8x^5}{5!} - \frac{3x^6}{6!} + \frac{56x^7}{7!} + \cdots$$

$$e^{\cos x} = e\left(1 - \frac{x^2}{2!} + \frac{4x^4}{4!} - \frac{31x^6}{6!} + \cdots \right)$$

$$e^{\tan x} = 1 + x + \frac{x^2}{2!} + \frac{3x^3}{3!} + \frac{9x^4}{4!} + \frac{37x^5}{5!} + \cdots \qquad \left(x^2 < \frac{\pi^2}{4} \right)$$

$$\sin x = \sin a + (x - a)\cos a - \frac{(x - a)^2}{2!}\sin a - \frac{(x - a)^3}{3!}\cos a + \frac{(x - a)^4}{4!}\sin a + \cdots$$

Hyperbolic and Inverse Functions

$$\sinh x = x + \frac{x^3}{3!} + \frac{x^5}{5!} + \frac{x^7}{7!} + \cdots + \frac{x^{2n+1}}{(2n+1)!} + \cdots \qquad |x| < \infty$$

$$\cosh x = 1 + \frac{x^2}{2!} + \frac{x^4}{4!} + \frac{x^6}{6!} + \cdots + \frac{x^{2n}}{(2n)!} + \cdots \qquad |x| < \infty$$

$$\tanh x = x - \frac{1}{3}x^3 + \frac{2}{15}x^5 - \frac{17}{315}x^7 - \frac{62}{2835}x^9 - \cdots$$

$$+ \frac{(-1)^{n+1}2^{2n}(2^{2n} - 1)}{(2n)!}B_n x^{2n-1} \pm \cdots \qquad |x| < \frac{\pi}{2}$$

$$\coth x = \frac{1}{x} + \frac{x}{3} - \frac{x^3}{45} + \frac{2x^5}{945} - \frac{x^7}{4725} + \cdots + \frac{(-1)^{n+1}2^{2n}}{(2n)!}B_n x^{2n-1} \pm \cdots \qquad 0 < |x| < \pi$$

$$\operatorname{sech} x = 1 - \frac{1}{2!}x^2 + \frac{5}{4!}x^4 - \frac{61}{6!}x^6 + \frac{1385}{8!}x^8 - \cdots + \frac{(-1)^n}{(2n)!}E_n x^{2n} \pm \cdots \qquad |x| < \frac{\pi}{2}$$

$$\operatorname{cosech} x = \frac{1}{x} - \frac{x}{6} + \frac{7x^3}{360} - \frac{31x^5}{15{,}120} + \cdots + \frac{2(-1)^n(2^{2n-1} - 1)}{(2n)!}B_n x^{2n-1} + \cdots \qquad 0 < |x| < \pi$$

Additional Relations for Fourier Series

$$1 = \frac{4}{\pi}\left[\sin\frac{\pi x}{k} + \frac{1}{3}\sin\frac{3\pi x}{k} + \frac{1}{5}\sin\frac{5\pi x}{k} + \cdots\right] \qquad [0 < x < k]$$

$$x = \frac{2k}{\pi}\left[\sin\frac{\pi x}{k} - \frac{1}{2}\sin\frac{2\pi x}{k} + \frac{1}{3}\sin\frac{3\pi x}{k} - \cdots\right] \qquad [-k < x < k]$$

$$x = \frac{k}{2} - \frac{4k}{\pi^2}\left[\cos\frac{\pi x}{k} + \frac{1}{3^2}\cos\frac{3\pi x}{k} + \frac{1}{5^2}\cos\frac{5\pi x}{k} + \cdots\right] \qquad [0 < x < k]$$

$$x^2 = \frac{2k^2}{\pi^3}\left[\left(\frac{\pi^2}{1} - \frac{4}{1}\right)\sin\frac{\pi x}{k} - \frac{\pi^2}{2}\sin\frac{2\pi x}{k} + \left(\frac{\pi^2}{3} - \frac{4}{3^3}\right)\sin\frac{3\pi x}{k} - \frac{\pi^2}{4}\sin\frac{4\pi x}{k}\right.$$

$$\left. + \left(\frac{\pi^2}{5} - \frac{4}{5^3}\right)\sin\frac{5\pi x}{k} + \cdots\right] \qquad [0 < x < k]$$

$$x^2 = \frac{k^2}{3} - \frac{4k^2}{\pi^2}\left[\cos\frac{\pi x}{k} - \frac{1}{2^2}\cos\frac{2\pi x}{k} + \frac{1}{3^2}\cos\frac{3\pi x}{k} - \frac{1}{4^2}\cos\frac{4\pi x}{k} + \cdots\right] \qquad [-k < x < k]$$

$$1 - \frac{1}{3} + \frac{1}{5} - \frac{1}{7} + \cdots = \frac{\pi}{4}$$

$$1 + \frac{1}{2^2} + \frac{1}{3^2} + \frac{1}{4^2} + \cdots = \frac{\pi^2}{6}$$

$$1 - \frac{1}{2^2} + \frac{1}{3^2} - \frac{1}{4^2} + \cdots = \frac{\pi^2}{12}$$

$$1 + \frac{1}{3^2} + \frac{1}{5^2} + \frac{1}{7^2} + \cdots = \frac{\pi^2}{8}$$

$$\frac{1}{2^2} + \frac{1}{4^2} + \frac{1}{6^2} + \frac{1}{8^2} + \cdots = \frac{\pi^2}{24}$$

Some Special Limits

$$\lim_{x \to 0}\frac{\sin x}{x} = 1, \quad \lim_{x \to 0}(1 + x)^{1/x} = e = 2.71828\cdots = 1 + 1 + \frac{1}{2!} + \frac{1}{3!} + \cdots$$

1.10-2   Trigonometric Relationships

See Section 1.3.

$$\sin x = \frac{1}{2i}(e^{ix} - e^{-ix}) \qquad\qquad\qquad \sinh x = \frac{1}{2}(e^x - e^{-x})$$

$$= -i\sinh ix \qquad\qquad\qquad\qquad = -i\sin(ix)$$

$$\cos x = \frac{1}{2}(e^{ix} + e^{-ix}) \qquad\qquad\qquad \cosh x = \frac{1}{2}(e^x + e^{-x})$$

$$= \cosh ix \qquad\qquad\qquad\qquad = \cos ix$$

$$\tan x = \frac{\sin x}{\cos x} = \frac{1}{i}\tanh ix \qquad\qquad \tanh x = \frac{\sinh x}{\cosh x} = \frac{1}{i}\tan ix$$

$$\cot x = \frac{\cos x}{\sin x} = \frac{1}{\tan x} = i\coth ix \qquad \coth x = \frac{\cosh x}{\sinh x} = \frac{1}{\tanh x} = i\cot i$$

$$\cos^2 x + \sin^2 x = 1 \qquad\qquad\qquad \cosh^2 x - \sinh^2 x = 1$$

$$\sin(x \pm y) = \sin x \cos y \pm \sin y \cos x$$

$$\sinh(x \pm y) = \sinh x \cosh y \pm \sinh y \cosh x$$

$$\sin(x \pm iy) = \sin x \cosh y \pm i \sinh y \cos x$$

$$\sinh(x \pm iy) = \sinh x \cos y \pm i \sin y \cosh x$$

$$\cos(x \pm y) = \cos x \cos y \mp \sin x \sin y$$

$$\cosh(x \pm y) = \cosh x \cosh y \pm \sinh x \sinh y$$

$$\cos(x \pm iy) = \cos x \cosh y \mp i \sin x \sinh y$$

$$\cosh(x \pm iy) = \cosh x \cos y \pm i \sinh x \sin y$$

$$\tan(x \pm y) = \frac{\tan x \pm \tan y}{1 \mp \tan x \tan y}$$

$$\tanh(x \pm y) = \frac{\tanh x \pm \tanh y}{1 \pm \tanh x \tanh y}$$

$$\tan(x \pm iy) = \frac{\tan x + i \tanh y}{1 \mp i \tan x \tanh y}$$

$$\tanh(x \pm iy) = \frac{\tanh x \pm i \tan y}{1 \pm i \tanh x \tan y}$$

$$\sin x \pm \sin y = 2 \sin\tfrac{1}{2}(x \pm y)\cos\tfrac{1}{2}(x \mp y)$$

$$\sinh x \pm \sinh y = 2 \sinh\tfrac{1}{2}(x \pm y)\cosh\tfrac{1}{2}(x \mp y)$$

$$\cos x + \cos y = 2 \cos\tfrac{1}{2}(x + y)\cos\tfrac{1}{2}(x - y)$$

$$\cosh x + \cosh y = 2 \cosh\tfrac{1}{2}(x + y)\cosh\tfrac{1}{2}(x - y)$$

$$\cos x - \cos y = 2 \sin\tfrac{1}{2}(x + y)\sin\tfrac{1}{2}(y - x)$$

$$\cosh x - \cosh y = 2 \sinh\tfrac{1}{2}(x + y)\sinh\tfrac{1}{2}(x - y)$$

$$\tan x \pm \tan y = \frac{\sin(x \pm y)}{\cos x \cos y}$$

$$\tanh x \pm \tanh y = \frac{\sinh(x \pm y)}{\cosh x \cosh y}$$

$$\sin^2 x - \sin^2 y = \sin(x + y)\sin(x - y) = \cos^2 y - \cos^2 x$$

$$\sinh^2 x - \sinh^2 y = \sinh(x + y)\sinh(x - y) = \cosh^2 x - \cosh^2 y$$

$$\cos^2 x - \sin^2 y = \cos(x + y)\cos(x - y) = \cos^2 y - \sin^2 x$$

$$\sinh^2 x + \cosh^2 y = \cosh(x + y)\cosh(x - y) = \cosh^2 x + \sinh^2 y$$

$$\sin^2 x = \tfrac{1}{2}(-\cos 2x + 1)$$

$$\sin^3 x = \tfrac{1}{4}(-\sin 3x + 3 \sin x)$$

$$\sin^4 x = \tfrac{1}{8}(\cos 4x - 4\cos 2x + 3)$$

$$\sin^5 x = \tfrac{1}{16}(\sin 5x - 5\sin 3x + 10\sin x)$$

$$\sin^6 x = \tfrac{1}{32}(-\cos 6x + 6\cos 4x - 15\cos 2x + 10)$$

$$\sin^7 x = \tfrac{1}{64}(-\sin 7x + 7\sin 5x - 21\sin 3x + 35\sin x)$$

$$\sinh^2 x = \tfrac{1}{2}(\cosh 2x - 1)$$

$$\sinh^3 x = \tfrac{1}{4}(\sinh 3x - 3\sinh x)$$

$$\sinh^4 x = \tfrac{1}{8}(\cosh 4x - 4\cosh 2x + 3)$$

$$\sinh^5 x = \tfrac{1}{16}(\sinh 5x - 5\sinh 3x + 10\sinh x)$$

$$\sinh^6 x = \tfrac{1}{32}(\cosh 6x - 6\cosh 4x + 15\cosh 2x - 10)$$

$$\sinh^7 x = \tfrac{1}{64}(\sinh 7x - 7\sinh 5x + 21\sinh 3x - 35\sinh x)$$

$$\cos^2 x = \tfrac{1}{2}(\cos 2x + 1)$$

$$\cos^3 x = \tfrac{1}{4}(\cos 3x + 3\cos x)$$

$$\cos^4 x = \tfrac{1}{8}(\cos 4x + 4\cos 2x + 3)$$

$$\cos^5 x = \tfrac{1}{16}(\cos 5x + 5\cos 3x + 10\cos x)$$

$$\cos^6 x = \tfrac{1}{32}(\cos 6x + 6\cos 4x + 15\cos 2x + 10)$$

$$\cos^7 x = \tfrac{1}{64}(\cos 7x + 7\cos 5x + 21\cos 3x + 35\cos x)$$

$$\cosh^2 x = \tfrac{1}{2}(\cosh 2x + 1)$$

$$\cosh^3 x = \tfrac{1}{4}(\cosh 3x + 3\cosh x)$$

$$\cosh^4 x = \tfrac{1}{8}(\cosh 4x + 4\cosh 2x + 3)$$

$$\cosh^5 x = \tfrac{1}{16}(\cosh 5x + 5\cosh 3x + 10\cosh x)$$

$$\cosh^6 x = \tfrac{1}{32}(\cosh 6x + 6\cosh 4x + 15\cosh 2x + 10)$$

$$\cosh^7 x = \tfrac{1}{64}(\cosh 7x + 7\cosh 5x + 21\cosh 3x + 35\cosh x)$$

$$(\cos x + i\sin x)^n = \cos nx + i\sin nx \qquad\qquad (\cosh x + \sinh x)^n = \sinh nx = \cosh ny \qquad (n \text{ is an integer})$$

$$\sin\frac{x}{2} = \pm\sqrt{\tfrac{1}{2}(1 - \cos x)} \qquad\qquad \sinh\frac{x}{2} = \pm\sqrt{\tfrac{1}{2}(\cosh x - 1)}$$

$$\cos\frac{x}{2} = \pm\sqrt{\tfrac{1}{2}(1 + \cos x)} \qquad\qquad \cosh\frac{x}{2} = \sqrt{\tfrac{1}{2}(\cosh x + 1)}$$

$$\tan\frac{x}{2} = \frac{1 - \cos x}{\sin x} = \frac{\sin x}{1 + \cos x} \qquad\qquad \tanh\frac{x}{2} = \frac{\cosh x - 1}{\sinh x} = \frac{\sinh x}{\cosh x + 1}$$

1.10-3   Derivatives

See Section 1.6.

$$\frac{dc}{dx} = 0, \qquad dc = 0 \qquad \frac{d(x)}{dx} = 1, \qquad d(x) = dx$$

$$\frac{d}{dx}(u + v - w) = \frac{du}{dx} + \frac{dv}{dx} - \frac{dw}{dx}, \qquad d(u + v - w) = du + dv - dw$$

$$\frac{d}{dx}(cv) = c\frac{dv}{dx}, \qquad d(cv) = c\,dv$$

$$\frac{d}{dx}(uv) = u\frac{dv}{dx} + v\frac{du}{dx}, \qquad d(uv) = u\,dv + v\,du$$

$$\frac{d}{dx}(v^n) = nv^{n-1}\frac{dv}{dx}, \quad \frac{d}{dx}(x^n) = nx^{n-1}, \quad d(v^n) = nv^{n-1}\,dv$$

$$\frac{d}{dx}\left(\frac{u}{v}\right) = \frac{v\dfrac{du}{dx} - u\dfrac{dv}{dx}}{v^2}, \qquad d\left(\frac{u}{v}\right) = \frac{v\,du - u\,dv}{v^2}$$

$$\frac{d}{dx}\left(\frac{c}{v}\right) = -\frac{c\dfrac{dv}{dx}}{v^2}, \qquad d\left(\frac{c}{v}\right) = -\frac{c\,dv}{v^2}$$

$$\frac{d(v_1 v_2 \cdots v_n)}{v_1 v_2 \cdots v_n} = \frac{dv_1}{v_1} + \frac{dv_2}{v_2} + \cdots + \frac{dv_n}{v_n} = d\log(v_1 v_2 \cdots v_n)$$

$$\frac{dy}{dx} = \frac{dy}{dv}\frac{dv}{dx} \qquad \frac{dy}{dx} = \frac{1}{\dfrac{dx}{dy}} \qquad \frac{dy}{dx} = \frac{\dfrac{dy}{dt}}{\dfrac{dx}{dt}}$$

$$\frac{d^n}{dx^n}(uv) = \frac{d^n u}{dx^n}v + n\frac{d^{n-1}u}{dx^{n-1}}\frac{dv}{dx} + \frac{n(n-1)}{2!}\frac{d^{n-2}u}{dx^{n-2}}\frac{d^2 v}{dx^2}$$

$$+ \frac{n(n-1)(n-2)}{3!}\frac{d^{n-3}u}{dx^{n-3}}\frac{d^3 v}{dx^3} + \cdots + u\frac{d^n v}{dx^n}$$

$$\frac{d}{dx}(\sin u) = \cos u\frac{du}{dx} \qquad\qquad \frac{d}{dx}(\cos u) = -\sin u\frac{du}{dx}$$

$$\frac{d}{dx}(\tan u) = \sec^2 u\frac{du}{dx} \qquad\qquad \frac{d}{dx}(\cot u) = -\csc^2 u\frac{du}{dx}$$

$$\frac{d}{dx}(\sec u) = \sec u \tan u\frac{du}{dx} \qquad\qquad \frac{d}{dx}(\csc u) = -\csc u \cot u\frac{du}{dx}$$

$$\frac{d}{dx}\left(\arcsin\frac{u}{a}\right) = \frac{1}{\sqrt{a^2 - u^2}}\frac{du}{dx} \qquad \frac{d}{dx}\left(\arccos\frac{u}{a}\right) = -\frac{1}{\sqrt{a^2 - u^2}}\frac{du}{dx}$$

$$\frac{d}{dx}\left(\arctan\frac{u}{a}\right) = \frac{a}{a^2 + u^2}\frac{du}{dx} \qquad \frac{d}{dx}\left(\operatorname{arccot}\frac{u}{a}\right) = -\frac{a}{a^2 + u^2}\frac{du}{dx}$$

$$\frac{d}{dx}(\text{arc sec} \frac{u}{a}) = \frac{a}{u\sqrt{u^2 - a^2}} \frac{du}{dx} \qquad \frac{d}{dx}(\text{arc csc} \frac{u}{a}) = -\frac{a}{u\sqrt{u^2 - a^2}} \frac{du}{dx}$$

$$\frac{d}{dx}(\text{arc vers} \frac{u}{a}) = \frac{1}{\sqrt{2au - u^2}} \frac{du}{dx}$$

$$\frac{d}{dx} \log_a u = \frac{1}{\log a} \frac{1}{u} \frac{du}{dx} \qquad\qquad \frac{d}{dx} \log u = \frac{1}{u} \frac{du}{dx}$$

$$\frac{d}{dx}(a^u) = (\log a) a^u \frac{du}{dx} \qquad\qquad \frac{d}{dx}(e^u) = e^u \frac{du}{dx}$$

$$\frac{d}{dx}(u^v) = u^v \log u \frac{dv}{dx} + v u^{v-1} \frac{du}{dx}$$

1.10-4   Integrals

See Section 1.6.

Elementary Integral Forms

$$\int a \, dx = ax$$

$$\int a f(x) \, dx = a \int f(x) \, dx$$

$$\int \phi(y) \, dy = \int \frac{\phi(y)}{y'} \, dy \qquad \text{where } y' = \frac{dy}{dx}$$

$$\int (u + v) \, dv = \int u \, dx + \int v \, dx \qquad \text{where } u \text{ and } v \text{ are any functions of } x$$

$$\int u \, dv = u \int dv - \int v \, du = uv - \int v \, du$$

$$\int u \frac{dv}{dx} \, dx = uv - \int v \frac{du}{dx} \, dx$$

$$\int x^n \, dx = \frac{x^{n+1}}{n + 1} \qquad \text{except } n = -1$$

$$\int \frac{f'(x) \, dx}{f(x)} = \log f(x) \qquad (df(x) = f'(x) \, dx)$$

$$\int \frac{dx}{x} = \log x$$

$$\int \frac{f'(x) \, dx}{2\sqrt{f(x)}} = \sqrt{f(x)} \qquad (df(x) = f'(x) \, dx)$$

$$\int e^x \, dx = e^x$$

$$\int e^{ax} \, dx = e^{ax}/a$$

$$\int b^{ax} \, dx = \frac{b^{ax}}{a \log b} \qquad (b > 0)$$

$$\int \log x \, dx = x \log x - x$$

$$\int a^x \log a \, dx = a^x \qquad (a > 0)$$

$$\int \frac{dx}{a^2 + x^2} = \frac{1}{a} \tan^{-1} \frac{x}{a}$$

$$\int \frac{dx}{a^2 - x^2} = \begin{cases} \dfrac{1}{a} \tanh^{-1} \dfrac{x}{a} \\ \dfrac{1}{2a} \log \dfrac{a + x}{a - x} \qquad (a^2 > x^2) \end{cases}$$

$$\int \frac{dx}{x^2 - a^2} = \begin{cases} -\dfrac{1}{a} \coth^{-1} \dfrac{x}{a} \\ \quad\text{or} \\ \dfrac{1}{2a} \log \dfrac{x - a}{x + a} \qquad (x^2 > a^2) \end{cases}$$

$$\int \frac{dx}{\sqrt{a^2 - x^2}} = \begin{cases} \sin^{-1} \dfrac{x}{|a|} \\ \quad\text{or} \\ -\cos^{-1} \dfrac{x}{|a|} \qquad (a^2 > x^2) \end{cases}$$

$$\int \frac{dx}{\sqrt{x^2 \pm a^2}} = \log\left(x + \sqrt{x^2 \pm a^2}\right)$$

$$\int \frac{dx}{x\sqrt{x^2 - a^2}} = \frac{1}{|a|} \sec^{-1} \frac{x}{a}$$

$$\int \frac{dx}{x\sqrt{a^2 \pm x^2}} = -\frac{1}{a} \log\left(\frac{a + \sqrt{a^2 \pm x^2}}{x}\right)$$

Integrals Containing Hyperbolic Functions

$$\int (\sinh x) \, dx = \cosh x$$

$$\int (\cosh x) \, dx = \sinh x$$

$$\int (\tanh x) \, dx = \log \cosh x$$

$$\int (\coth x) \, dx = \log \sinh x$$

$$\int (\operatorname{sech} x) \, dx = \tan^{-1}(\sinh x)$$

$$\int \operatorname{csch} x \, dx = \log \tanh\left(\frac{x}{2}\right)$$

$$\int x (\sinh x) \, dx = x \cosh x - \sinh x$$

$$\int x^n (\sinh x) \, dx = x^n \cosh x - n \int x^{n-1} (\cosh x) \, dx$$

$$\int x (\cosh x) \, dx = x \sinh x - \cosh x$$

$$\int x^n (\cosh x) \, dx = x^n \sinh x - n \int x^{n-1} (\sinh x) \, dx$$

Integrals Containing Trigonometric Functions

$$\int \sin x \, dx = -\cos x \qquad \int \cos x \, dx = \sin x$$

$$\int \sin^2 x \, dx = \tfrac{1}{2}x - \tfrac{1}{2}\sin x \cos x = \tfrac{1}{2}x - \tfrac{1}{4}\sin 2x$$

$$\int \cos^2 x \, dx = \tfrac{1}{2}x + \tfrac{1}{2}\sin x \cos x = \tfrac{1}{2}x + \tfrac{1}{4}\sin 2x$$

$$\int \sin^3 x \, dx = -\tfrac{1}{3}(\sin^2 x + 2)\cos x$$

$$\int \cos^3 x \, dx = \tfrac{1}{3}(\cos^2 x + 2)\sin x$$

$$\int \sin^n x \, dx = -\frac{\sin^{n-1}x \cos x}{n} + \frac{n-1}{n}\int \sin^{n-2}x \, dx$$

$$\int \cos^n x \, dx = \frac{\cos^{n-1}x \sin x}{n} + \frac{n-1}{n}\int \cos^{n-2}x \, dx$$

$$\int \sin^m x \cos x \, dx = \frac{\sin^{m+1}x}{m+1}$$

$$\int \sin x \cos^m x \, dx = -\frac{\cos^{m+1}x}{m+1}$$

$$\int \sin^2 x \cos^2 x \, dx = -\tfrac{1}{8}\left(\tfrac{1}{4}\sin 4x - x\right)$$

$$\int \cos^m x \sin^n x \, dx = \frac{\cos^{m-1}x \sin^{n+1}x}{m+n} + \frac{m-1}{m+n}\int \cos^{m-2}x \sin^n x \, dx$$

$$\int \cos^m x \sin^n x \, dx = -\frac{\sin^{n-1}x \cos^{m+1}x}{m+n} + \frac{n-1}{m+n}\int \cos^m x \sin^{n-2}x \, dx$$

$$\int \frac{dx}{\sin^m x} = -\frac{1}{m-1}\frac{\cos x}{\sin^{m-1}x} + \frac{m-2}{m-1}\int \frac{dx}{\sin^{m-2}x}$$

$$\int \frac{dx}{\cos^m x} = \frac{1}{m-1}\frac{\sin x}{\cos^{m-1}x} + \frac{m-2}{m-1}\int \frac{dx}{\cos^{m-2}x}$$

$$\int \frac{dx}{\sin x \cos x} = \log \tan x$$

$$\int \frac{dx}{\sin^m x \cos^n x} = \frac{1}{n-1}\frac{1}{\sin^{m-1}x \cos^{n-1}x} + \frac{m+n-2}{n-1}\int \frac{dx}{\sin^m x \cos^{n-2}x}$$

$$\int \frac{dx}{\sin^m x \cos^n x} = -\frac{1}{m-1}\frac{1}{\sin^{m-1}x \cos^{n-1}x} + \frac{m+n-2}{m-1}\int \frac{dx}{\sin^{m-2}x \cos^n x}$$

$$\int \frac{\cos^m x \, dx}{\sin^n x} = -\frac{\cos^{m+1}x}{(n-1)\sin^{n-1}x} - \frac{m-n+2}{n-1}\int \frac{\cos^m x \, dx}{\sin^{n-2}x}$$

$$\int \frac{\cos^m x \, dx}{\sin^n x} = \frac{\cos^{m-1}x}{(m-n)\sin^{n-1}x} + \frac{m-1}{m-n}\int \frac{\cos^{m-2}x \, dx}{\sin^n x}$$

$$\int \frac{\sin^n x \, dx}{\cos^m x} = -\int \frac{\cos^n\left(\tfrac{1}{2}\pi - x\right)d\left(\tfrac{1}{2}\pi - x\right)}{\sin^m\left(\tfrac{1}{2}\pi - x\right)}$$

$$\int \tan x \, dx = -\log \cos x \qquad \int \cot x \, dx = \log \sin x$$

$$\int \tan^2 x \, dx = \tan x - x \qquad \int \cot^2 x \, dx = -\cot x - x$$

$$\int \tan^n x \, dx = \frac{\tan^{n-1} x}{n-1} - \int \tan^{n-2} x \, dx$$

$$\int \cot^n x \, dx = -\frac{\cot^{n-1} x}{n-1} - \int \cot^{n-2} x \, dx$$

$$\int \sec x \, dx = \log(\sec x + \tan x) \quad \text{or} \quad \log \tan\left(\tfrac{1}{4}\pi + \tfrac{1}{2}x\right)$$

$$\int \csc x \, dx = \log(\csc x - \cot x) \quad \text{or} \quad \log \tan\tfrac{1}{2}x$$

$$\int \sec^2 x \, dx = \tan x \qquad \int \csc^2 x \, dx = -\cot x$$

$$\int \sec^n x \, dx = \int \frac{dx}{\cos^n x} \qquad \int \csc^n x \, dx = \int \frac{dx}{\sin^n x}$$

$$\int \frac{dx}{a^2 \cos^2 x + b^2 \sin^2 x} = \frac{1}{ab} \arctan \frac{b \tan x}{a}$$

$$\int x \sin x \, dx = \sin x - x \cos x$$

$$\int x \cos x \, dx = \cos x + x \sin x$$

$$\int x^2 \sin x \, dx = 2x \sin x - (x^2 - 2)\cos x$$

$$\int x^2 \cos x \, dx = 2x \cos x + (x^2 - 2)\sin x$$

$$\int x^m \sin x \, dx = -x^m \cos x + m \int x^{m-1} \cos x \, dx$$

$$\int x^m \cos x \, dx = x^m \sin x - m \int x^{m-1} \sin x \, dx$$

$$\int \frac{\sin x \, dx}{x} = x - \frac{x^3}{3 \cdot 3!} + \frac{x^5}{5 \cdot 5!} - \frac{x^7}{7 \cdot 7!} + \cdots$$

$$\int \frac{\cos x \, dx}{x} = \log x - \frac{x^2}{2 \cdot 2!} + \frac{x^4}{4 \cdot 4!} - \frac{x^6}{6 \cdot 6!} + \cdots$$

Integrals Containing Exponential Functions

$$\int e^{ax} \, dx = \frac{e^{ax}}{a} \qquad \int a^x \, dx = \frac{a^x}{\log a}.$$

$$\int x e^{ax} \, dx = \frac{e^{ax}}{a^2}(ax - 1)$$

$$\int x^m e^{ax} \, dx = \frac{x^m e^{ax}}{a} - \frac{m}{a} \int x^{m-1} e^{ax} \, dx$$

$$\int \frac{e^{ax} \, dx}{x^m} = -\frac{1}{m-1} \frac{e^{ax}}{x^{m-1}} + \frac{a}{m-1} \int \frac{e^{ax} \, dx}{x^{m-1}}$$

$$\int e^x \sin x \, dx = \tfrac{1}{2} e^x (\sin x - \cos x)$$

$$\int e^x \cos x \, dx = \tfrac{1}{2} e^x (\sin x + \cos x)$$

$$\int e^{ax} \sin bx \, dx = \frac{e^{ax}(a \sin bx - b \cos bx)}{a^2 + b^2}$$

$$\int e^{ax} \cos bx \, dx = \frac{e^{ax}(b \sin bx + a \cos bx)}{a^2 + b^2}$$

$$\int e^{ax} \cos^n x \, dx = \frac{e^{ax} \cos^{n-1} x (a \cos x + n \sin x)}{a^2 + n^2} + \frac{n(n-1)}{a^2 + n^2} \int e^{ax} \cos^{n-2} x \, dx$$

$$\int e^{ax} \sin^n x \, dx = \frac{e^{ax} \sin^{n-1} x (a \sin x - n \cos x)}{a^2 + n^2} + \frac{n(n-1)}{a^2 + n^2} \int e^{ax} \sin^{n-2} x \, dx$$

$$\int \frac{e^x}{x} \, dx = \log x + x + \frac{x^2}{2 \cdot 2!} + \frac{x^3}{3 \cdot 3!} + \frac{x^4}{4 \cdot 4!} + \cdots$$

Integrals Containing Logarithmic Functions

$$\int \log x \, dx = x \log x - x$$

$$\int x^n \log x \, dx = x^{n+1} \left[\frac{\log x}{n+1} - \frac{1}{(n+1)^2} \right]$$

$$\int x^n (\log x)^m \, dx = \frac{x^{n+1}}{n+1} (\log x)^m - \frac{m}{n+1} \int x^n (\log x)^{m-1} \, dx$$

$$\int \frac{(\log x)^n}{x} \, dx = \frac{1}{n+1} (\log x)^{n+1}$$

$$\int \frac{dx}{x \log x} = \log(\log x)$$

$$\int \frac{dx}{x (\log x)^n} = - \frac{1}{(n-1)(\log x)^{n-1}}$$

$$\int \frac{x^n \, dx}{(\log x)^m} = - \frac{x^{n+1}}{(m-1)(\log x)^{m-1}} + \frac{n+1}{m-1} \int \frac{x^n \, dx}{(\log x)^{m-1}}$$

$$\int e^{ax} \log x \, dx = \frac{e^{ax} \log x}{a} - \frac{1}{a} \int \frac{e^{ax}}{x} \, dx$$

Integrals Containing (a + bx)

$$\int \frac{dx}{a + bx} = \frac{1}{b} \log(a + bx)$$

$$\int \frac{dx}{(a + bx)^n} = \frac{1}{b(1-n)(a+bx)^{n-1}} \quad \text{if } n \neq 1$$

$$\int (a + bx)^n \, dx = \frac{(a+bx)^{n+1}}{b(n+1)} \quad \text{if } n \neq -1$$

$$\int \frac{x \, dx}{a + bx} = \frac{1}{b^2} [a + bx - a \log(a + bx)]$$

$$\int \frac{x^2\,dx}{a + bx} = \frac{1}{b^3}\left[\tfrac{1}{2}(a + bx)^2 - 2a(a + bx) + a^2\log(a + bx)\right]$$

$$\int \frac{x\,dx}{(a + bx)^2} = \frac{1}{b^2}\left[\log(a + bx) + \frac{a}{a + bx}\right]$$

$$\int \frac{x^2\,dx}{(a + bx)^2} = \frac{1}{b^3}\left[a + bx - 2a\log(a + bx) - \frac{a^2}{a + bx}\right]$$

$$\int \frac{x\,dx}{(a + bx)^3} = \frac{1}{b^2}\left[\frac{a}{2(a + bx)^2} - \frac{1}{a + bx}\right]$$

$$\int \frac{dx}{x(a + bx)} = -\frac{1}{a}\log\frac{a + bx}{x}$$

$$\int \frac{dx}{x(a + bx)^2} = \frac{1}{a(a + bx)} - \frac{1}{a^2}\log\frac{a + bx}{x}$$

$$\int \frac{dx}{x^2(a + bx)} = -\frac{1}{ax} + \frac{b}{a^2}\log\frac{a + bx}{x}$$

$$\int \frac{dx}{x^2(a + bx)^2} = \frac{-1}{ax(a + bx)} - \frac{2b}{a}\int \frac{dx}{x(a + bx)^2}$$

Integrals Containing $(x^2 - a^2)^{1/2}$

$$\int (x^2 - a^2)^{1/2}\,dx = \tfrac{1}{2}x(x^2 - a^2)^{1/2} - \tfrac{1}{2}a^2\log\left[x + (x^2 - a^2)^{1/2}\right]$$

$$\int x(x^2 - a^2)^{1/2}\,dx = \tfrac{1}{3}(x^2 - a^2)^{3/2}$$

$$\int x^2(x^2 - a^2)^{1/2}\,dx = \frac{x}{8}(2x^2 - a^2)(x^2 - a^2)^{1/2} - \frac{a^4}{8}\log\left[x + (x^2 - a^2)^{1/2}\right]$$

$$\int x^3(x^2 - a^2)^{1/2}\,dx = \tfrac{1}{5}(x^2 - a^2)^{5/2} + \frac{a^2}{3}(x^2 - a^2)^{3/2}$$

$$\int (x^2 - a^2)^{3/2}\,dx = \frac{x}{8}(2x^2 - 5a^2)(x^2 - a^2)^{1/2} + \frac{3a^4}{8}\log\left[x + (x^2 - a^2)^{1/2}\right]$$

$$\int x(x^2 - a^2)^{3/2}\,dx = \tfrac{1}{5}(x^2 - a^2)^{5/2}$$

Integrals Containing $(a + bx)^{1/2}$

$$\int (a + bx)^{n/2}\,dx = \frac{2(a + bx)^{(n+2)/2}}{b(n + 2)}$$

$$\int x(a + bx)^{n/2}\,dx = \frac{2}{b^2}\left[\frac{(a + bx)^{(n+4)/2}}{n + 4} - \frac{a(a + bx)^{(n+2)/2}}{n + 2}\right]$$

$$\int x^{-1}(a + bx)^{n/2}\,dx = b\int (a + bx)^{(n-2)/2}\,dx + a\int x^{-1}(a + bx)^{(n-2)/2}\,dx$$

$$\int x^{-1}(a + bx)^{-n/2}\,dx = \frac{1}{a}\int x^{-1}(a + bx)^{-(n-2)/2}\,dx - \frac{b}{a}\int (a + bx)^{-n/2}\,dx$$

$$\int \frac{x^m \, dx}{(a+bx)^{1/2}} = \frac{2x^m(a+bx)^{1/2}}{(2m+1)b} - \frac{2ma}{(2m+1)b} \int \frac{x^{m-1} \, dx}{(a+bx)^{1/2}} \, dx$$

$$\int \frac{dx}{x^n(a+bx)^{1/2}} = -\frac{(a+bx)^{1/2}}{(n-1)ax^{n-1}} - \frac{(2n-3)b}{(2n-2)a} \int \frac{dx}{x^{n-1}(a+bx)^{1/2}}$$

Integrals Containing $(a^2 + x^2)^{1/2}$

$$\int (a^2+x^2)^{1/2} \, dx = \tfrac{1}{2}x(a^2+x^2)^{1/2} + \tfrac{1}{2}a^2 \log\left[x + (a^2+x^2)^{1/2} \right]$$

$$\int x(a^2+x^2)^{1/2} \, dx = \tfrac{1}{3}(a^2+x^2)^{3/2}$$

$$\int x^2(a^2+x^2)^{1/2} \, dx = \frac{x}{8}(2x^2+a^2)(a^2+x^2)^{1/2} - \frac{a^4}{8}\log\left[x + (a^2+x^2)^{1/2} \right]$$

$$\int x^3(a^2+x^2)^{1/2} \, dx = \tfrac{1}{5}(a^2+x^2)^{5/2} - \frac{a^2}{3}(a^2+x^2)^{3/2}$$

$$\int (a^2+x^2)^{3/2} \, dx = \frac{x}{8}(2x^2+5a^2)(a^2+x^2)^{1/2} + \frac{3a^4}{8}\log\left[x + (a^2+x^2)^{1/2} \right]$$

$$\int x(a^2+x^2)^{3/2} \, dx = \tfrac{1}{5}(a^2+x^2)^{5/2}$$

Integrals Containing $(a^2 - x^2)^{1/2}$

$$\int (a^2-x^2)^{1/2} \, dx = \tfrac{1}{2}x(a^2-x^2)^{1/2} + \tfrac{1}{2}a^2 \arcsin\frac{x}{a}$$

$$\int x(a^2-x^2)^{1/2} \, dx = -\tfrac{1}{3}(a^2-x^2)^{3/2}$$

$$\int x^2(a^2-x^2)^{1/2} \, dx = \frac{x}{8}(2x^2-a^2)(a^2-x^2)^{1/2} + \frac{a^4}{8}\arcsin\frac{x}{a}$$

$$\int x^3(a^2-x^2)^{1/2} \, dx = \tfrac{1}{5}(a^2-x^2)^{5/2} - \frac{a^2}{3}(a^2-x^2)^{3/2}$$

$$\int (a^2-x^2)^{3/2} \, dx = \frac{x}{8}(5a^2-2x^2)(a^2-x^2)^{1/2} + \frac{3a^4}{8}\arcsin\frac{x}{a}$$

$$\int x(a^2-x^2)^{3/2} \, dx = -\tfrac{1}{5}(a^2-x^2)^{5/2}$$

Integrals Containing $(a + bx^n)$

$$\int \frac{dx}{a^2+x^2} = \frac{1}{a}\arctan\frac{x}{a} = \frac{1}{a}\arcsin\frac{x}{(a^2+x^2)^{1/2}}$$

$$\int \frac{dx}{x^2-a^2} = \frac{1}{2a}\log\frac{x-a}{x+a} \quad \text{if } x^2 > a^2$$

$$= \frac{1}{2a}\log\frac{a-x}{a+x} \quad \text{if } x^2 < a^2$$

$$\int \frac{dx}{a+bx^2} = \frac{1}{\sqrt{ab}}\arctan\left(x\sqrt{\frac{b}{a}} \right) \quad \text{if } a > 0, \, b > 0$$

$$= \frac{1}{2}\frac{1}{\sqrt{-ab}}\log\frac{\sqrt{a}+x\sqrt{-b}}{\sqrt{a}-x\sqrt{-b}} \quad \text{if } a > 0,\ b < 0$$

$$\int\frac{dx}{a^2 - b^2 x^2} = \frac{1}{2ab}\log\frac{a+bx}{a-bx}$$

$$\int\frac{dx}{(a+bx^2)^{m+1}} = \frac{x}{2ma(a+bx^2)^m} + \frac{2m-1}{2ma}\int\frac{dx}{(a+bx^2)^m}$$

$$\int\frac{x\,dx}{(a+bx^2)^{m+1}} = \frac{1}{2}\int\frac{dz}{(a+bz)^{m+1}} \quad \text{if } z = x^2$$

$$\int\frac{x\,dx}{a+bx^2} = \frac{1}{2b}\log\left(x^2 + \frac{a}{b}\right)$$

$$\int\frac{x^2\,dx}{a+bx^2} = \frac{x}{b} - \frac{a}{b}\int\frac{dx}{a+bx^2}$$

$$\int\frac{dx}{x(a+bx^2)} = \frac{1}{2a}\log\frac{x^2}{a+bx^2}$$

$$\int\frac{dx}{x^2(a+bx^2)} = -\frac{1}{ax} - \frac{b}{a}\int\frac{dx}{a+bx^2}$$

$$\int\frac{dx}{(a+bx^2)^2} = \frac{x}{2a(a+bx^2)} + \frac{1}{2a}\int\frac{dx}{a+bx^2}$$

$$\int\frac{x^2\,dx}{(a+bx^2)^m} = \frac{1}{b}\int\frac{dx}{(a+bx^2)^{m-1}} - \frac{a}{b}\int\frac{dx}{(a+bx^2)^m}$$

$$\int\frac{dx}{x^2(a+bx^2)^m} = \frac{1}{a}\int\frac{dx}{x^2(a+bx^2)^{m-1}} - \frac{b}{a}\int\frac{dx}{(a+bx^2)^m}$$

Definite Integrals

$$\int_0^\infty\frac{dx}{a^2+x^2} = \frac{\pi}{a}$$

$$\int_0^\infty x^{n-1}e^{-x}\,dx = \Gamma(n)$$

where

$$\Gamma(n+1) = n\Gamma(n) \quad \text{if } n > 0$$

$$\Gamma(n+1) = n! \quad \text{if } n \text{ is a positive integer}$$

$$\Gamma(2) = \Gamma(1) = 1$$

$$\Gamma\left(\tfrac{1}{2}\right) = \pi^{1/2}$$

$$\int_0^1\left(\log\frac{1}{x}\right)^n dx = \Gamma(n+1)$$

$$\int_0^\infty e^{-zx}z^n x^{n-1}\,dx = \Gamma(n)$$

$$\int_0^1 x^{m-1}(1-x)^{n-1}\,dx = \int_0^\infty\frac{x^{m-1}\,dx}{(1+x)^{m+n}} = \frac{\Gamma(m)\Gamma(n)}{\Gamma(m+n)}$$

$$\int_0^\infty \frac{x^{n-1}}{1+x}\, dx = \frac{\pi}{\sin n\pi}$$

$$\int_0^\infty e^{-a^2 x^2}\, dx = \frac{\pi^{1/2}}{2a}$$

$$\int_0^\infty \frac{\sin^2 x\, dx}{x^2} = \frac{\pi}{2}$$

$$\int_0^\infty \cos(x^2)\, dx = \int_0^\infty \sin(x^2)\, dx = \tfrac{1}{2}\left(\tfrac{1}{2}\pi\right)^{1/2}$$

$$\int_0^\infty \frac{\sin mx\, dx}{x} = \frac{\pi}{2} \quad \text{if } m > 0.$$

$$\int_0^\infty e^{-ax}\cos bx\, dx = \frac{a}{a^2 + b^2}$$

$$\int_0^\infty e^{-ax}\sin bx\, dx = \frac{b}{a^2 + b^2}$$

$$\int_0^\infty e^{-a^2 x^2}\cos bx\, dx = \frac{\pi^{1/2}}{2a} e^{-b^2/4a^2}$$

$$\int_0^1 \frac{x^b - x^a}{\log x}\, dx = \log\frac{1+b}{1+a}$$

$$\int_0^1 \frac{\log x}{1-x}\, dx = -\frac{\pi^2}{6}$$

$$\int_0^1 \frac{\log x}{1+x}\, dx = -\frac{\pi^2}{12}$$

$$\int_0^1 \frac{\log x}{1-x^2}\, dx = -\frac{\pi^2}{8}$$

$$\int_0^1 \log\left(\frac{1+x}{1-x}\right)\frac{dx}{x} = \frac{\pi^2}{4}$$

$$\int_0^\infty \log\left(\frac{e^x + 1}{e^x - 1}\right) dx = \frac{\pi^2}{4}$$

1.10-5   Laplace Transforms

See Section 1.7.

$\phi(s)$	$F(t)$
$\dfrac{1}{s}$	1
$\dfrac{1}{s^2}$	t
$\dfrac{1}{s^n} \quad (n = 1, 2, \ldots)$	$\dfrac{t^{n-1}}{(n-1)!}$
$\dfrac{1}{\sqrt{s}}$	$\dfrac{1}{\sqrt{\pi T}}$
$s^{-3/2}$	$2\sqrt{\dfrac{t}{\pi}}$

(*Continued*)

$\phi(s)$	$F(t)$
$s^{-[n+(1/2)]}$ $(n = 1, 2, \ldots)$	$\dfrac{2^n t^{n-(1/2)}}{1 \cdot 3 \cdot 5 \cdots (2n-1)\sqrt{\pi}}$
$\dfrac{\Gamma(k)}{s^k}$ $(k \geq 0)$	t^{k-1}
$\dfrac{1}{s-a}$	e^{at}
$\dfrac{1}{(s-a)^2}$	te^{at}
$\dfrac{1}{(s-a)^n}$ $(n = 1, 2, \ldots)$	$\dfrac{1}{(n-1)!} t^{n-1} e^{at}$
$\dfrac{\Gamma(k)}{(s-a)^k}$ $(k \geq 0)$	$t^{k-1} e^{at}$
$\dfrac{1}{(s-a)(s-b)}$	$\dfrac{1}{a-b}(e^{at} - e^{bt})$
$\dfrac{s}{(s-a)(s-b)}$	$\dfrac{1}{a-b}(ae^{at} - be^{bt})$
$\dfrac{1}{(s-a)(s-b)(s-c)}$	$-\dfrac{(b-c)e^{at} + (c-a)e^{bt} + (a-b)e^{ct}}{(a-b)(b-c)(c-a)}$
$\dfrac{1}{s^2 + a^2}$	$\dfrac{1}{a}\sin at$
$\dfrac{s}{s^2 + a^2}$	$\cos at$
$\dfrac{1}{s^2 - a^2}$	$\dfrac{1}{a}\sinh at$
$\dfrac{s}{s^2 - a^2}$	$\cosh at$
$\dfrac{1}{s(s^2 + a^2)}$	$\dfrac{1}{a^2}(1 - \cos at)$
$\dfrac{1}{s^2(s^2 + a^2)}$	$\dfrac{1}{a^3}(at - \sin at)$
$\dfrac{1}{(s^2 + a^2)^2}$	$\dfrac{1}{2a^3}(\sin at - at \cos at)$
$\dfrac{s}{(s^2 + a^2)^2}$	$\dfrac{t}{2a}\sin at$
$\dfrac{s^2}{(s^2 + a^2)^2}$	$\dfrac{1}{2a}(\sin at + at \cos at)$
$\dfrac{s^2 - a^2}{(s^2 + a^2)^2}$	$t \cos at$
$\dfrac{s}{(s^2 + a^2)(s^2 + b^2)}$ $(a^2 \neq b^2)$	$\dfrac{\cos at - \cos bt}{b^2 - a^2}$
$\dfrac{1}{(s-a)^2 + b^2}$	$\dfrac{1}{b}e^{at}\sin bt$
$\dfrac{s-a}{(s-a)^2 + b^2}$	$e^{at}\cos bt$

(Continued)

$\phi(s)$	$F(t)$
$\dfrac{4a^3}{s^4 + 4a^4}$	$\sin at \cosh at - \cos at \sinh at$
$\dfrac{s}{s^4 + 4a^4}$	$\dfrac{1}{2a^2}\sin at \sinh at$
$\dfrac{1}{s^4 - a^4}$	$\dfrac{1}{2a^3}(\sinh at - \sin at)$
$\dfrac{s}{s^4 - a^4}$	$\dfrac{1}{2a^2}(\cosh at - \cos at)$
$\dfrac{8a^3 s^2}{(s^2 + a^2)^3}$	$(1 + a^2 t^2)\sin at - \cos at$
$\dfrac{1}{s}\left(\dfrac{s-1}{s}\right)^n$	$L_n(t) = \dfrac{e^t}{n!}\dfrac{d^n}{dt^n}(t^n e^{-t})$
$\dfrac{s}{(s-a)^{3/2}}$	$\dfrac{1}{\sqrt{\pi t}}e^{at}(1 + 2at)$
$\sqrt{s-a} - \sqrt{s-b}$	$\dfrac{1}{2\sqrt{\pi t^3}}(e^{bt} - e^{at})$
$\dfrac{1}{\sqrt{s}}e^{-k/s}$	$\dfrac{1}{\sqrt{\pi t}}\cos 2\sqrt{kt}$
$\dfrac{1}{\sqrt{s}}e^{k/s}$	$\dfrac{1}{\sqrt{\pi t}}\cosh 2\sqrt{kt}$
$\dfrac{1}{s^{3/2}}e^{-k/s}$	$\dfrac{1}{\sqrt{\pi k}}\sin 2\sqrt{kt}$
$\dfrac{1}{s^{3/2}}e^{k/s}$	$\dfrac{1}{\sqrt{\pi k}}\sinh 2\sqrt{kt}$

1.10-6 Fourier Transforms

See Section 1.7.

Fourier Transforms

$f(x)$	$\phi(\alpha)$
$\dfrac{\sin ax}{x}$	$\begin{cases}\sqrt{\dfrac{\pi}{2}} & \|\alpha\| < a \\ 0 & \|\alpha\| > a\end{cases}$
$\begin{cases}e^{iwx} & (p < x < q) \\ 0 & (x < p,\ x > q)\end{cases}$	$\dfrac{i}{\sqrt{2\pi}}\dfrac{e^{ip(w+\alpha)} - e^{iq(w+\alpha)}}{(w+\alpha)}$
$\begin{cases}e^{-cx+iwx} & (x > 0) \\ 0 & (x < 0)\end{cases}(c > 0)$	$\dfrac{i}{\sqrt{2\pi}\,(w + \alpha + ic)}$
$e^{-px^2}R(p) > 0$	$\dfrac{1}{\sqrt{2p}}e^{-\alpha^2/4p}$
$\cos px^2$	$\dfrac{1}{\sqrt{2p}}\cos\left[\dfrac{\alpha^2}{4p} - \dfrac{\pi}{4}\right]$
$\sin px^2$	$\dfrac{1}{\sqrt{2p}}\cos\left[\dfrac{\alpha^2}{4p} + \dfrac{\pi}{4}\right]$

(*Continued*)

$f(x)$	$\phi(\alpha)$
$\|x\|^{-p} \quad (0 < p < 1)$	$\sqrt{\dfrac{2}{\pi}} \dfrac{\Gamma(1-p)\sin\dfrac{p\pi}{2}}{\|\alpha\|^{(1-p)}}$
$\dfrac{e^{-\alpha\|x\|}}{\sqrt{\|x\|}}$	$\dfrac{\sqrt{\sqrt{(a^2+\alpha^2)}+a}}{\sqrt{a^2+\alpha^2}}$
$\dfrac{\cosh ax}{\cosh \pi x} \quad (-\pi < a < \pi)$	$\sqrt{\dfrac{2}{\pi}} \dfrac{\cos\dfrac{a}{2}\cosh\dfrac{\alpha}{2}}{\cosh\alpha + \cos a}$
$\dfrac{\sinh ax}{\sinh \pi x} \quad (-\pi < a < \pi)$	$\dfrac{1}{\sqrt{2\pi}} \dfrac{\sin a}{\cosh\alpha + \cos a}$

Finite Fourier Sine Transforms $\phi_s(n) = \displaystyle\int_0^\pi f(x)\sin nx\,dx\ (n = 1, 2, \dots)$

$\phi_s(n)$		$F(x)$
$\dfrac{1}{n}$	$\dots$	$\dfrac{\pi - x}{\pi}$
$\dfrac{(-1)^{n+1}}{n}$		$\dfrac{x}{\pi}$
$\dfrac{1-(-1)^n}{n}$		1
$\dfrac{2}{n^2}\sin\dfrac{n\pi}{2}$		$\begin{cases} x & \text{when } 0 < x < \pi/2 \\ \pi - x & \text{when } \pi/2 < x < \pi \end{cases}$
$\dfrac{(-1)^{n+1}}{n^3}$		$\dfrac{x(\pi^2 - x^2)}{6\pi}$
$\dfrac{1-(-1)^n}{n^3}$		$\dfrac{x(\pi - x)}{2}$
$\dfrac{\pi^2(-1)^{n-1}}{n} - \dfrac{2[1-(-1)^n]}{n^3}$		x^2
$\pi(-1)^n\left(\dfrac{6}{n^3} - \dfrac{\pi^2}{n}\right)$		x^3
$\dfrac{n}{n^2 + c^2}[1 - (-1)^n e^{c\pi}]$		e^{cx}
$\dfrac{n}{n^2 + c^2}$		$\dfrac{\sinh c(\pi - x)}{\sinh c\pi}$
$\dfrac{n}{n^2 - k^2} \quad (k \neq 0,1,2,\dots)$		$\dfrac{\sin k(\pi - x)}{\sin k\pi}$
$\begin{cases} \dfrac{\pi}{2} & \text{when } n = m \\ 0 & \text{when } n \neq m \end{cases} \quad (m = 1,2,\dots)$		$\sin mx$
$\dfrac{n}{n^2 - k^2}[1 - (-1)^n\cos k\pi] \quad (k \neq 1,2,\dots)$		$\cos kx$
$\begin{cases} \dfrac{n}{n^2 - m^2}[1 - (-1)^{n+m}] \\ \quad \text{when } n \neq m = 1,2,\dots \\ 0 \quad \text{when } n = m \end{cases}$		$\cos mx$

Finite Fourier Cosine Transforms $\phi_c(n) = \int_0^\pi f(x)\cos nx\,dx \ (n = 0, 1\,2, \ldots)$

$\phi_c(n)$	$f(x)$
$(-1)^n\phi_c(n)$	$f(\pi - x)$
$\left.\begin{array}{l}0 \quad\text{when } n = 1, 2, \ldots \\ \phi_c(0) = \pi\end{array}\right\}$	1
$\left.\begin{array}{l}-\dfrac{1-(-1)^n}{n^2} \\ \phi_c(0) = \pi^2/2\end{array}\right\}$	x
$\left.\begin{array}{l}\dfrac{(-1)^n}{n^2} \\ \phi_c(0) = \pi^2/6\end{array}\right\}$	$\dfrac{x^2}{2\pi}$
$\left.\begin{array}{l}\dfrac{1}{n^2}; \\ \phi_c(0) = 0\end{array}\right\}$	$\dfrac{(\pi - x)^2}{2\pi} - \dfrac{\pi}{6}$
$\left.\begin{array}{l}3\pi^2\dfrac{(-1)^n}{n^2} - 6\dfrac{1-(-1)^n}{n^4} \\ \phi_c(0) = \pi^4/4\end{array}\right\}$	x^3
$\dfrac{(-1)^n e^c\pi - 1}{n^2 + c^2}$	$\dfrac{1}{c}e^{cx}$
$\dfrac{1}{n^2 + c^2}$	$\dfrac{\cosh c(\pi - x)}{c\sinh c\pi}$
$\begin{array}{l}\dfrac{k}{n^2 - k^2}[(-1)^n\cos \pi k - 1] \\ (k \neq 0, 1, 2, \ldots)\end{array}$	$\sin kx$
$\begin{array}{l}\dfrac{(-1)^{n+m} - 1}{n^2 - m^2} \\ \phi_c(m) = 0 \quad (m = 1, 2, \ldots)\end{array}$	$\dfrac{1}{m}\sin mx$
$\dfrac{1}{n^2 - k^2} \quad (k \neq 0, 1, 2, \ldots)$	$-\dfrac{\cos k(\pi - x)}{k\sin k\pi}$
$\left.\begin{array}{l}0 \quad (n = 1, 2, \ldots) \\ \phi_c(m) = \pi/2 \quad (m = 1, 2, \ldots)\end{array}\right\}$	$\cos mx$

1.10-7 Conversion Factors
Metric to English

To Obtain	Multiply	By
Inches	Centimeters	0.3937007874
Feet	Meters	3.280839895
Yards	Meters	1.093613298
Miles	Kilometers	0.6213711922
Ounces	Grams	$3.527396195 \times 10^{-2}$
Pounds	Kilograms	2.204622622
Gallons (U.S. Liquid)	Liters	0.2641720524
Fluid ounces	Milliliters (cc)	$3.381402270 \times 10^{-2}$
Square inches	Square centimeters	0.1550003100
Square feet	Square meters	10.76391042
Square yards	Square meters	1.195990046
Cubic inches	Milliliters (cc)	$6.102374409 \times 10^{-2}$
Cubic feet	Cubic meters	35.31466672
Cubic yards	Cubic meters	1.307950619

English to Metric

To Obtain	Multiply	By
Microns	Mils	25.4
Centimeters	Inches	2.54
Meters	Feet	0.3048
Meters	Yards	0.9144
Kilometers	Miles	1.609344
Grams	Ounces	28.34952313
Kilograms	Pounds	0.45359237
Liters	Gallons (U.S. Liquid)	3.785411784
Milliliters (cc)	Fluid ounces	29.57352956
Square centimeters	Square inches	6.4516
Square meters	Square feet	0.09290304
Square meters	Square yards	0.83612736
Milliliters (cc)	Cubic inches	16.387064
Cubic meters	Cubic feet	$2.831684659 \times 10^{-2}$
Cubic meters	Cubic yards	0.764554858

1.11 NUMERICAL METHODS

Solutions of equations are obtained by applying *analytical methods*. However, if there is no analytical method to solve an equation, rather than an *analytical* a *numerical* solution is sought by using a *numerical method*. Such a method approximates the true solution by a numerical value. Various methods have been developed to solve specific types of equations. Since a numerical solution is an *approximation* at best, by using a numerical method based on these approximations of solutions, an *error* is always introduced as the difference between the *true* value of a solution and the approximate solution obtained which deviates from the solution. Also, to evaluate the error introduced by such a numerical approximation, methods have been devised. Therefore, together with numerical analysis the concept of *error analysis* has developed. An error analysis of a numerical method results in the limits of errors (*bounds of error*) introduced, rather than the actual error for each calculated solution. For example, an error introduced would be between 0.01 and 0.005% less than or more than the approximate value obtained by a numerical method.

Another important issue in numerical analysis is the *stability* of the numerical method. If the approximate solution is obtained by an *iterative* numerical method, for *unstable* methods the rounding error exponentially increases and each time it runs further away from the actual solution. In the case of *stable* methods the solution settles in an approximate value which is within the error bounds mentioned in the paragraph above.

Some of the numerical methods for finding approximate solutions are illustrated for classes of problems. The most common problem requiring a numerical solution is obtaining the roots of an algebraic equation which is discussed here first. Another important class of problems where numerical methods are widely used is the class of equations involving derivatives, namely, differential equations (see Section 1.6). Also, the inverse of differentiation—integration—can also be solved by numerical methods. Such problems and some of their solution methods are also discussed in this section.

Some of these numerical methods can easily be implemented by performing the necessary calculations by hand. However, with the advent of computers most of the methods are designed to be implemented on such *digital computers*. The use of computers as a tool in assisting numerical analysis of analytical problems brought about the two important improvements:

1. The *speed of computation* is *increased*. Because of the high speed with which the computer can perform algebraic operations, such as addition and multiplication, the solution to a numerical problem requiring a large number of operations can be accomplished in a minute fraction of a second, compared to an excessive amount of time if done without a computer.

2. The *error* introduced is *decreased* by implementing solution methods which are more refined than those simpler methods that give approximate solutions with larger errors.

To implement the numerical methods on a computer an additional step is required—writing a *program* that the computer can understand. This may be done by *programming languages* developed exactly for this purpose. Since these programs express the *discrete* formulas representing the numerical approximation of the analytical expressions to the computer, the computer languages have been accordingly designed. Among the common languages BASIC, FORTRAN, PL/I, and APL may be named. Each language has its own *syntax*, thus a program written in one computer programming language differs from the others. Use of some programming languages is simpler than others. Some are more powerful in expressing even the complicated mathematical expressions. FORTRAN (FORmula TRANslation) has been designed for translating formulas to computer language and is still widely used among scientists and engineers.

For example, the FORTRAN program corresponding to the expression

$$y = x^2 - (1/2x^3)$$

is simply

$$Y = X**2 - (1/(2**3)).$$

Similar to the mathematical concepts, X and Y are *variables*, * indicates the *multiplication* operation, while ** represents the *exponential*, and / represents the *division*. Although when executed by the computer there is a hierarchy among the operations such as exponential, multiplication, division, and then addition, the parentheses used above indicate how the operations will be done, thus eliminating any ambiguity. A simple example of this question of expressing the operations correctly is the following:

The FORTRAN expression is given as $10 - 2 \times 3$. Since the multiplication comes before subtraction, the computer interprets this as $10 - (2 \times 3)$ and calculates it as $10 - 6 = 4$. However, if it is meant to be $(10 - 2)$ and then multiplied by 3, $(10 - 2) \times 3 = 8 \times 3 = 24$ would be the answer which is not equal to 4 obtained above.

This simple illustration shows that each programming language should be properly applied in solving the mathematical expressions by numerical methods. Therefore each language must be learned so that the computer representation of the expressions are correct.

1.11-1   Roots of Algebraic Equations

An algebraic equation in one unknown x, given as

$$c_0 x^n + c_1 x^{n-1} + \cdots + c^n = 0$$

can be *factorized* such that the polynomial becomes

$$c_0 (x - x_1)(x - x_2) \cdots (x - x_n) = 0$$

where $x_1, \ldots, x_n$ are the n roots of the algebraic equations. If there are *multiple roots*, then such a root appears in more than one place in the factorized form above.

Multiple Roots. A *root of multiplicity k* of a function $f(x)$ is a *root of multiplicy $n - 1$* of its derivative $f'(x)$.

Negative Real Roots. These (or *negative real part* of complex roots) are important for answering the stability question of singular solutions of the differential equations (see Section 1.5).

Hurwitz and Routh Criterion. A necessary and sufficient condition for the real part of all the roots to be negative is that all the determinants Δ_i are positive:

$$\Delta_0 = c_0 \qquad \Delta_1 = c_1$$

$$\Delta_2 = \begin{vmatrix} c_1 & c_0 \\ c_3 & c_2 \end{vmatrix} \qquad \Delta_3 = \begin{vmatrix} c_1 & c_0 & 0 \\ c_3 & c_2 & c_1 \\ c_5 & c_4 & c_3 \end{vmatrix} \quad \text{and so on}$$

Methods of Obtaining Roots of an Algebraic Equation

Rearrangement of the equation. If the algebraic equation, $g(x) = 0$ is rearranged as

$$x = f(x)$$

starting with an initial value for x, x_1, by a recursive relation

$$x_n = f(x_{n-1})$$

$x_1, x_2, \ldots$ can be calculated until the difference between x_n and x_{n-1} is within the acceptable tolerance.

Linear Interpolation. The two values are chosen as $x_{10} < x_{20}$ such that $f(x_{10})$ and $f(x_{20})$ are of opposite sign. The line segment between $[x_{10}, f(x_{10})]$ and $[x_{20}, f(x_{20})]$ intersects the x axis at

$$x_1 = \frac{x_{10}f(x_{20}) - x_{20}f(x_{10})}{f(x_{20}) - f(x_{10})}$$

Next x_1 is taken as x_{10} and the same iteration is followed for x_1 and x_{20} to find x_2, then $x_3, \ldots$, until x_n is obtained close to the x_{n-1} value within the acceptable tolerance.

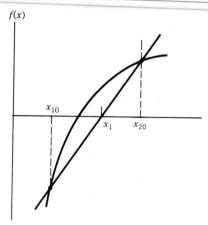

Binary Chopping. Same as linear interpolation, $x_{10} < x_{20}$ are chosen such that $f(x_{10})$ and $f(x_{20})$ are of opposite sign. $x_1 = \frac{1}{2}(x_{10} + x_{20})$ is found. $f(x_1)$ is calculated. Here x_1 replaces x_{10} (or x_{20}) if $f(x_{10})$ (or $f(x_{20})$) is of the same sign as $f(x_1)$. Similarly, the iteration is carried out.

Newton's Method. If x_0 is close to a root, x_r, then

$$x_{n+1} = x_n - \frac{f(x_n)}{f'(x_n)} \qquad n = 0, 1, 2, \ldots$$

and $x_n \to x_r$.

1.11-2 Simultaneous Linear Equations

Simultaneous linear equations may be solved numerically by such methods as Gauss–Jordan Elimination or Cramer's rule. See examples given in Section 1.5 on Matrix Algebra.

1.11-3 Least Square Method

If a set of n ordered pairs (x_i, y_i), $i = 1, 2, \ldots, n$, is given where x's are the fixed variables and y's are dependent variables, a curve can be fitted by *the method of least squares*.

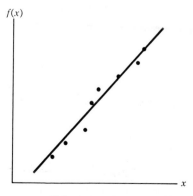

Fitting by a straight line

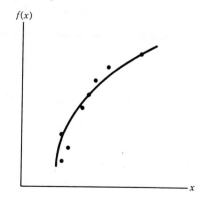

Fitting by a polynomial

For fitting the given set of ordered pairs by a *straight line*

$$y = b_0 + b_1 x$$

the difference, $\delta_i = y_i - b_0 - b_1 x_i$, can be minimized. For this the sum of the squares of the differences must be minimized:

$$E = \sum \delta_i^2$$

This means (see maxima and minima in Section 1.6) that

$$\partial E / \partial b_0 = 0$$

$$\partial E / \partial b_1 = 0$$

should be satisfied resulting in the following two equations in two unknowns.

$$\sum y_i = n b_0 + b_1 \sum x_i$$

$$\sum x_i y_i = b_0 \sum x_i + b_1 \sum x_i^2$$

The above two equations can be solved for b_0, b_1.
    For fitting the given set of ordered pairs by a *polynomial*,

$$y = b_0 + b_1 x + b_2 x^2 + \cdots + b_n x^n$$

the similar equations are obtained as

$$\sum y_i = n b_0 + b_1 \sum x_i + b_2 \sum x_i^2 + \cdots + b_n \sum x_i^n$$

$$\sum x_i y_i = b_0 \sum x_i + b_1 \sum x_i^2 + \cdots + b_n \sum x_i^{n+1}$$

$$\vdots$$

$$\sum x_i^n y_i = b_0 \sum x_i^n + b_1 \sum x_i^{n+1} + \cdots + b_n \sum x_i^{2n}$$

These $n + 1$ equations can be solved for $b_0, \ldots, b_n$.

1.11-4 Numerical Differentiation

The first derivative of f

$$f_i^{(1)} = (df/dx)|_{x_i}$$

can be calculated by referring to the Taylor expansion.

Central Difference.   By subtracting the two Taylor expansions about x_{i-1} and x_{i+1},

$$= \frac{f_{i+1} - f_{i-1}}{2h} - \frac{h^3}{3!} \frac{d^3 f}{dx^3} - \cdots$$

Forward Difference.   By expanding about x_i and x_{i+1},

$$= \frac{f_{i+1} - f_i}{h} - \frac{h^2}{2!} \frac{d^2 f}{dx^2} - \cdots$$

Backward Difference.   By expanding about x_i and x_{i-1},

$$= \frac{f_i - f_{i-1}}{h} - \frac{h^2}{2!} \frac{d^2 f}{dx^2} - \cdots$$

The error is clearly greater in the last two, being of the order h^2, while smaller for the first where the order is h^3.

1.11-5 Finite Difference Methods

Using the numerical differentiation schemes above, differential equations can be solved numerically. In the case of ordinary differential equations, the difference is taken along one independent variable only. For the partial differential equations, finite difference is taken along each variable. In general, a mesh is formed at each intersection of coordinate mesh. The differential equations are then expressed as *difference equations*, corresponding to various mesh points, based on the *finite difference method* used. As seen above, in numerical differentiation, various schemes are proposed to obtain higher accuracy by minimizing the error introduced by the approximation.

The values of a given function f, at equally spaced intervals of length h is denoted by $f_n = f(a + nh)$. Therefore, the *first, second,..., forward differences* of f_n are found as

$$\Delta f_n = f_{n+1} - f_n$$

$$\Delta^2 f_n = \Delta \Delta f_n = (f_{n+2} - f_{n+1}) - (f_{n+1} - f_n)$$

$$= f_{n+2} - 2f_{n+1} + f_n$$

Similarly we can define

$$\delta f_{n+1}, \delta^2 f_{n+1}, \ldots \quad \text{as central differences}$$

$$\Delta f_{n+1}, \Delta^2 f_{n+1}, \ldots \quad \text{as backward differences}$$

1.11-6 Numerical Solution of Differential Equations

An application of finite differences is used in the solution of differential equations. The purpose of these methods is, starting from an initial value, to calculate values of the dependent variable by the relationship given as the differential equation.

First-Order Methods

Point-Slope Formula:

$$y_{n+1} = y_n + hy_n' + O(h^2)$$

$$y_{n+1} = y_{n-1} + 2hy_n' + O(h^2)$$

Example: The differential equation is given as

$$dy/dx = x^2$$

The point slope formula is applied to have

$$\Delta y_n = y_{n+1} - y_n$$

$$\Delta x = h$$

Thus the above equation is written to calculate y_{n+1} from the previous y_n value as

$$y_{n+1} = y_n + x_n^2 h$$

In FORTRAN language,

```
    I = 1
    DO 20 I = 1, N
    Y(I + 1) = Y(I) + (X(I) * * 2) * H
20  CONTINUE
```

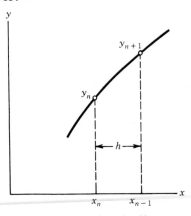

Point-slope formula with
two points only

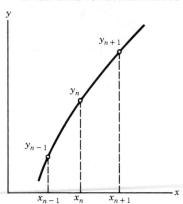

Point-slope formula with
three points

Trapezoidal Formula.   Another first-order method is the trapezoidal formula given as

$$y_{n+1} = y_n + (h/2)(y'_{n+1} + y'_n) + O(h^3)$$

Second-Order Methods

Runge–Kutta Methods:

$$y_{n+1} = y_n + (1/2)(k_1 + k_2) + O(h^3)$$

where $k_1 = hf(x_n, y_n)$ and $k_2 = hf(x_n + h, y_n + k_1)$.

Higher-Order Methods

There are other methods in which the order of the error term $O(h^n)$ is improved.

Predictor–Corrector Methods

Milne's Method: These formulas are used first to *predict* a value by a *predictor* formula and then to *correct* by the *corrector* formula.

System of Differential Equations

The above methods have also been extended to system of differential equations in which there are multiple equations corresponding to multiple dependent variables.

Partial Differential Equations

In the case of partial differential equations, since there are multiple independent variables the geometry of the space of independent variables is partitioned as a mesh to apply finite difference methods.

1.11-7   Integration

The integration

$$I = \int_a^b f(x)\, dx$$

can be expressed in a discrete form in various ways. Since it is basically the area under the $f(x)$ several methods of approximating this area have been suggested.

Trapezium Rule.   By taking the *linear* approximation for the function $f(x)$ between x_i and x_{i+1}, the area of the trapezoid between these two points is

$$A_i = (h/2)(f_i + f_{i+1})$$

thus if $a = x_0$, $b = x_n$

$$I = \sum_{i=0}^{n-1} A_i = (h/2)(f_0 + 2f_1 + 2f_2 + \cdots + 2f_{n-1} + f_n)$$

Example: The shaded area which is the integral of the shown curve is calculated by the following Trapezium rule.

$$A = (h/2)(f_{n-1} + 2f_n + f_{n+1})$$

A FORTRAN program would be

```
I = 1
DO 30 I = 1, N
A(I) = (H/2)*(F(I − 1) + F(I))
A(I + 1) = A(I) + (H/2)*(F(I) + F(I + 1))
30   CONTINUE
```

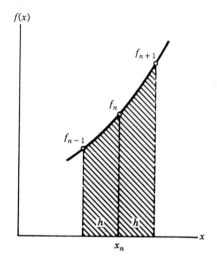

Simpson's Rule.   This is a *parabolic* approximation such that in each interval $(0, 2h)$, $(2h, 4h), \ldots$, the function $f(x)$ is approximated by a parabola:

$$y = ax^2 + bx + c$$

Therefore the numerical approximation formulas become

$$f_{i+1} = f(x_i + h) = f_i + bh + ah^2$$

$$f_{i-1} = f(x_i - h) = f_i - bh + ah^2$$

$$A_i = (h/3)(4f_i + f_{i+1} + f_{i-1})$$

$$I = (h/3)\left(f_0 + f_n + 4 \sum_{k \text{ odd}} f_k + 2 \sum_{k \text{ even}} f_k \right)$$

BIBLIOGRAPHY

Abramowitz, M., and I. A. Stegun (Eds.), *Handbook of Mathematical Functions*, Dover Publications, New York, 1965.

Beyer, W. H. (Ed.), *CRC Standard Mathematical Tables*, 26th ed., CRC Press, Boca Raton, Fla., 1981.

Beyer, W. H. (Ed.), *CRC Handbook of Mathematical Sciences*, 5th ed., CRC Press, West Palm Beach, Fla., 1978.

Beyer, W. H. (Ed.), *CRC Handbook of Tables for Probability and Statistics*, 4th ed., CRC Press, Boca Raton, Fla., 1979.

Byrd, P. F., and M. D. Friedman, *Handbook of Elliptic Integrals for Engineers and Physicists*, Springer-Verlag, Berlin, 1954.

Carmichael, R. D., and E. R. Smith (Compilers), *Mathematical Tables and Formulas*, Dover Publications, New York, 1962.

Condon, E. U., and H. Odishaw (Eds.), *Handbook of Physics*, McGraw-Hill, New York, 1958.

Eshbach, O. W., and M. Souders (Eds.), *Handbook of Engineering Fundamentals*, 3rd ed., Wiley, New York, 1975.

Flugge, S. (Ed.), *Handbuch der Physik, Mathematische Methoden I und II*. Springer-Verlag, Berlin, 1955.

Gradshteyn, I. S., and I. M. Ryzhik, *Table of Integrals, Series, and Products*, Academic Press, New York, 1980.

Klerer, M., and G. A. Korn, *Digital Computer User's Handbook*, McGraw-Hill, New York, 1967.

Korn, G. A., and T. M. Korn, *Mathematical Handbook for Scientist and Engineers*, 2nd ed., McGraw-Hill, New York, 1968.

Ryshik, I. M., and I. S. Gradstein, *Tables of Series, Products and Integrals*, VEB Deutscher Verlag der Wissenschaften, Berlin, 1963.

Weast, R. C., and S. M. Selby (Eds.), *Handbook of Tables for Mathematics*, 5th ed., The Chemical Rubber Co., Cleveland, Ohio, 1978.

CHAPTER 2

STATICS AND DYNAMICS

WILLIAM C. VAN BUSKIRK
STEPHEN C. COWIN

Tulane University
New Orleans, Louisiana

2.1 FUNDAMENTALS

2.1-1 Mechanics

Mechanics is the physical science that deals with the effects of forces on objects. Traditionally mechanics has been divided into four subdisciplines:

1. Particle mechanics.
2. Rigid-body mechanics.
3. Deformable solid-body mechanics.
4. Fluid mechanics.

In this chapter we consider the statics and dynamics of particles and rigid bodies. A particle is an idealized object in which all the mass is concentrated in a single point that can move and interact with other objects idealized as particles. The particle model works well even for very large objects if one is only interested in the translational movement of the object in space and not the spinning of the object. A rigid body is one in which no dimension changes as a consequence of the applied load. Using rigid-body or particle statics one can determine the reaction forces for a body constrained from moving. Using rigid-body or particle dynamics one can determine the motion of the body from a knowledge of the applied forces.

When forces are applied to a body there are both external and internal effects. The external effects are the reaction forces or acceleration of the body. The internal effects of a solid body are the strain and deformation in the body. The rigid-body assumption precludes the possibility of internal deformation. Thus, there is no internal effect in a rigid body. Two sets of forces that have the same external effect on a body are said to be rigid-body equivalent.

2.1-2 Newton's Three Laws

The following three laws of motion given by Sir Isaac Newton form the basis for classical mechanics:

1. A body remains at rest or moves at a constant speed in a straight line unless acted on by a force. This is a statement of the principle of inertia.
2. The total force acting on a body is equal to the mass of the body times its acceleration; that is, $F = ma$, where F and a are in the same direction.
3. If a body exerts a force on a second body, the second body exerts a force on the first body that is equal in magnitude and opposite in direction to the first force. This is the law of action and reaction.

2.1-3 Dimensions and Units

Dimensions and units are not the same thing. The physical dimensions of importance in mechanics are mass [M], length [L], time [T], and force [F]. These dimensions can be measured using any convenient

**TABLE 2.1-1   PREFIXES
FOR SI UNITS**

$$10^{-9} = \text{nano (n)}$$
$$10^{-6} = \text{micro } (\mu)$$
$$10^{-3} = \text{milli (m)}$$
$$10^{3} = \text{kilo (k)}$$
$$10^{6} = \text{mega (M)}$$
$$10^{9} = \text{giga (G)}$$

unit system. Every equation must be dimensionally homogeneous; that is, the dimensions in every term of the equation must be the same. Thus, Newton's second law implies the following relationship for the four dimensions:

$$[F] = \left[\frac{ML}{T^2} \right] \qquad (2.1\text{-}1)$$

The units used most by engineers working in the United States today are U.S. customary units and the International System of Units (abbreviated as SI). Since engineers in current practice are likely to encounter both sets of units, both sets are used in this chapter. In U.S. customary units the base units for force, length, and time are the pound (lb), the foot (ft), and the second (s). The unit for mass is the slug. The slug is related to the base units using Eq. (2.1-1). That is, 1 slug = 1 lb·s^2/ft. Other U.S. customary units for length are the inch (in.) and the mile. Force is sometimes expressed in kilopounds (kip). One kilopound is equal to 1000 lb.

In SI the base units for mass, length, and time are the kilogram (kg), the meter (m), and the second (s). The unit for force is the newton (N). The newton is related to the base units through Eq. (2.1-1). That is, 1 N = 1 kg·m/s^2. Additional units of time used both in SI and U.S. customary units are the minute and the hour.

New SI units are created using certain prefixes. The most important prefixes are shown in Table 2.1-1. Numerical values in SI are written whenever possible to lie between the values 0.1 and 1000. When that is not possible, spaces rather than commas are used to separate digits in groups of three. Thus, one might write the number 13 800, but 13 800 N would be written as 13.8 kN. Spaces are also used to the right of the decimal; for example, 0.000 323. Units are combined using dots or slashes; that is, a newton-meter is written as N·m and the unit for acceleration is m/s. Prefixes are never used in the denominator of a combined unit. Thus, a possible unit for stress or pressure is MN/m^2 but never N/mm^2.

2.1-4   Weight, Mass, and the Acceleration of Gravity

Weight and mass are different physical entities and should never be confused with one another. Weight is the force of gravitational attraction of a body by the earth. It is properly given in pounds or newtons and not in kilograms. The mass of a body is a measure of its resistance to acceleration. It is properly given in kilograms or slugs and not in pounds. The equation that relates these two quantities at the earth's surface is $W = mg$ where g is the acceleration of gravity at the surface of the earth. In U.S. customary units, $g = 32.2$ ft/s^2, and in SI, $g = 9.81$ m/s^2.

2.1-5   Conversion from One Unit System to Another

To convert from U.S. customary units to SI or vice versa, one uses conversion ratios. A conversion ratio is a ratio of units equal to one. Conversion ratios between U.S. customary units and SI are based on the following equalities:

$$1 \text{ ft} = 0.3048 \text{ m}$$
$$1 \text{ lb} = 4.4482 \text{ N}$$
$$1 \text{ slug} = 14.59 \text{ kg}$$

Example 2.1-1.   Express 25 N·m in U.S. customary units.

$$(25 \text{ N} \cdot \text{m}) \left(\frac{1 \text{ lb}}{4.4482 \text{ N}} \right) \left(\frac{1 \text{ ft}}{0.3048 \text{ m}} \right) = 18.44 \text{ ft} \cdot \text{lb}$$

Example 2.1-2.   Express 50 psi (pounds per square inch) in SI units.

$$\left(50 \text{ lb/in.}^2 \right) \left(\frac{12 \text{ in.}}{0.3048 \text{ m}} \right)^2 \left(\frac{4.448 \text{ N}}{1 \text{ lb}} \right) = 345\,000 \text{ N/m}^2$$

or 345 kN/m². In the SI system 1 newton per square meter is defined as 1 pascal (Pa). Hence the above result may be stated as 345 kilopascals.

2.2 VECTORS

2.2-1 Forces and Vectors

The primary physical entity in mechanics is force. A force is represented mathematically by a vector; that is, a quantity that possesses both magnitude and direction. A quantity that possesses magnitude only is a scalar. Forces and other vectors will be represented in this chapter by a boldface letter, for example, **F**. For handwritten work we suggest that the vector be indicated by a wavy underline. A vector is shown in a sketch by a directed line segment. The magnitude of the vector **F** is a scalar and is indicated by the length of the line and is denoted simply as F. The direction of the vector is indicated by the orientation of the directed line segment.

2.2-2 Addition of Vectors

The sum or resultant of two vectors **A** and **B** acting at a point P is denoted by **A** + **B** and may be found by forming a parallelogram as shown in Fig. 2.2-1. The magnitude and direction of the resultant vector **C** are given by the magnitude and direction of the diagonal of the parallelogram that passes through the point P. This procedure for finding the resultant is known as the parallelogram law. The triangle rule is a procedure that gives an equivalent result. Using this rule the tail of one of the two vectors is joined with the tip of the other as shown in Fig. 2.2-2. The resultant vector is then found by drawing a line from the tail of the first vector to the tip of the second.

2.2-3 Law of Sines and Law of Cosines

In performing a vector addition, the parallelogram law or triangle rule can be applied rigorously with good graphical tools or can be applied loosely to obtain an approximate sketch of the result. In the

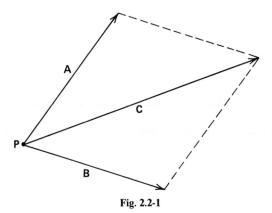

Fig. 2.2-1

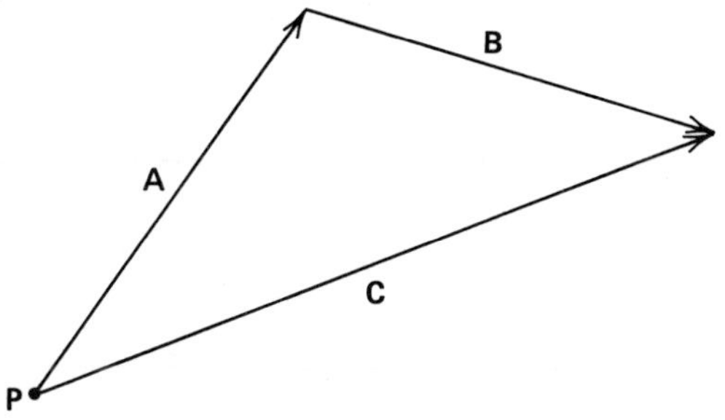

Fig. 2.2-2

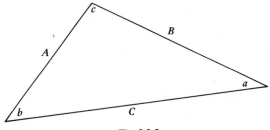

Fig. 2.2-3

latter case the magnitude and orientation of the resultant is usually found using trigonometry. Consider the triangle shown in Fig. 2.2-3. For that triangle the law of cosines states

$$C^2 = A^2 + B^2 - 2AB \cos c$$
$$A^2 = B^2 + C^2 - 2BC \cos a$$
$$B^2 = A^2 + C^2 - 2AC \cos b$$

The law of sines for this triangle is

$$\frac{\sin a}{A} = \frac{\sin b}{B} = \frac{\sin c}{C}$$

These laws relate the magnitudes of **A**, **B**, and **C**, and the included angles between them: a, b, and c.

2.2-4   Resolution of a Vector into Components

The inverse operation of vector addition is the resolution of a vector into components. The vector **C** in Fig. 2.2-4 can be resolved into unique components in any set of two directions; for example, the directions indicated by the lines aa and bb. These components when added together give **C**. Let the components in this example be denoted by **A** and **B**. The magnitudes of **A** and **B** are found by forming the parallelogram shown in Fig. 2.2-5. Note that **C** is along the diagonal of the parallelogram and **A** and **B** are directed along the lines of action of aa and bb, respectively. The magnitude of the two components may be found using the law of sines.

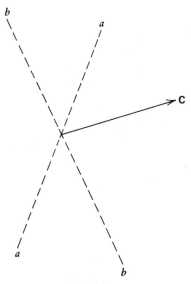

Fig. 2.2-4

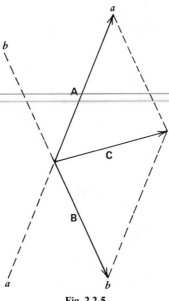

Fig. 2.2-5

2.2-5 Unit Vectors

A vector of magnitude or length one is a unit vector. Any vector may be written as the product of its magnitude and a unit vector in the direction of the vector; for example, if $\mathbf{n}$ is a unit vector in the direction of $\mathbf{A}$, then $\mathbf{A} = A\mathbf{n}$. The unit vector in the direction of a particular vector can be found by dividing the expression for the vector by its magnitude. Three unit vectors that are frequently used in mechanics are the unit vectors in the directions of the x, y, and z coordinate axes. These are denoted by $\mathbf{i}$, $\mathbf{j}$, and $\mathbf{k}$, respectively.

2.2-6 Rectangular Components

Vectors are often resolved into components that are perpendicular to one another. These are called rectangular components. Consider the vector $\mathbf{F}$ shown in Fig. 2.2-6. It is resolved into rectangular

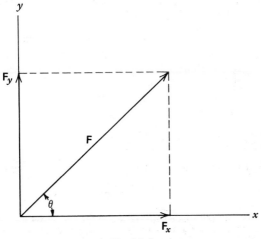

Fig. 2.2-6

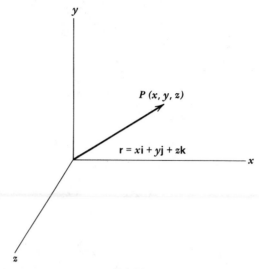

Fig. 2.2-7

components $\mathbf{F}_x$ and $\mathbf{F}_y$. These are the vector components of $\mathbf{F}$. Note that $\mathbf{F}_x + \mathbf{F}_y = \mathbf{F}$. Using unit vectors along the x and y axes we can write $\mathbf{F} = F_x \mathbf{i} + F_y \mathbf{j}$. F_x and F_y are the scalar components of $\mathbf{F}$. If there is no ambiguity, they are simply called the components of $\mathbf{F}$. Note that $F_x = F\cos\theta$ and that $F_y = F\sin\theta$. Also note that $F^2 = F_x^2 + F_y^2$.

In three-dimensional space a vector can be resolved into three rectangular components; that is, the vector $\mathbf{F}$ can be written as $\mathbf{F} = F_x \mathbf{i} + F_y \mathbf{j} + F_z \mathbf{k}$. We know from the Pythagorean theorem that $F^2 = F_x^2 + F_y^2 + F_z^2$.

2.2-7   Addition of Vectors by Adding Components

If two vectors are given in the form $\mathbf{S} = S_x \mathbf{i} + S_y \mathbf{j} + S_z \mathbf{k}$ and $\mathbf{T} = T_x \mathbf{i} + T_y \mathbf{j} + T_z \mathbf{k}$, then the resultant of the two vectors is given by $\mathbf{R} = R_x \mathbf{i} + R_y \mathbf{j} + R_z \mathbf{k}$ where $R_x = S_x + T_x$, $R_y = S_y + T_y$, and $R_z = S_z + T_z$.

2.2-8   Position Vectors

The position vector of a point P with coordinates (x, y, z) with respect to the origin of the reference frame is the vector from the origin of the reference frame to the point P and is given by $\mathbf{r} = x\mathbf{i} + y\mathbf{j} + z\mathbf{k}$ (see Fig. 2.2-7). The position vector of P (x_2, y_2, z_2) with respect to Q (x_1, y_1, z_1) is denoted as $\mathbf{r}_{P/Q}$ and is given by

$$\mathbf{r}_{P/Q} = (x_2 - x_1)\mathbf{i} + (y_2 - y_1)\mathbf{j} + (z_2 - z_1)\mathbf{k}$$

(see Fig. 2.2-8). The unit vector along the line from Q to P is found by dividing the vector $\mathbf{r}_{P/Q}$ by its magnitude, which is found from

$$|\mathbf{r}_{P/Q}| = \sqrt{(x_2 - x_1)^2 + (y_2 - y_1)^2 + (z_2 - z_1)^2}$$

2.2-9   Cross Product

The cross product of two vectors $\mathbf{S}$ and $\mathbf{T}$ is denoted by $\mathbf{S} \times \mathbf{T}$ and is defined as

$$\mathbf{S} \times \mathbf{T} = ST\sin\theta\,\mathbf{n}$$

where θ is the angle less than 180° between $\mathbf{S}$ and $\mathbf{T}$ and the unit vector $\mathbf{n}$ is perpendicular to both $\mathbf{S}$ and $\mathbf{T}$ in the direction given by the right-hand rule. The right-hand rule may be stated as follows: The fingers of the right hand are extended in the direction of $\mathbf{S}$. The hand is then rotated through the angle

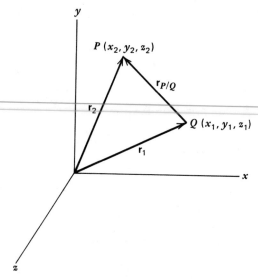

Fig. 2.2-8

θ to point in the direction of **T**. The direction of **n** is indicated by the thumb of the right hand if the thumb is held perpendicular to the fingers during the rotation of the hand.

The cross product of two vectors given in terms of their components may be expressed as a determinant:

$$\mathbf{S} \times \mathbf{T} = \begin{vmatrix} \mathbf{i} & \mathbf{j} & \mathbf{k} \\ S_x & S_y & S_z \\ T_x & T_y & T_z \end{vmatrix}$$

A straightforward way to expand a determinant is to write the first two columns again as shown below. The value of the determinant is found by multiplying along the diagonals as shown by the arrows and then adding the products. The negative signs mean that the result of that multiplication should be multiplied by -1.

$$\mathbf{S} \times \mathbf{T} = \begin{array}{ccccc} \mathbf{i} & \mathbf{j} & \mathbf{k} & \mathbf{i} & \mathbf{j} \\ S_x & S_y & S_z & S_x & S_y \\ T_x & T_y & T_z & T_x & T_y \end{array}$$

$$\quad - \quad - \quad - \quad + \quad + \quad +$$

The result of the expansion of this determinant is

$$\mathbf{S} \times \mathbf{T} = \left(S_y T_z - S_z T_y \right)\mathbf{i} + \left(S_z T_x - S_x T_z \right)\mathbf{j} + \left(S_x T_y - S_y T_x \right)\mathbf{k}$$

2.2-10 Moment of a Force

A force **F** acts on a body as shown in Fig. 2.2-9. The moment of F about O is a measure of the turning effect of the force about the point O. The moment of a force possesses both magnitude and direction and is a vector. The moment is defined as

$$\mathbf{M}_O = \mathbf{r} \times \mathbf{F}$$

where **r** is a position vector from O to any point on the line of action of **F**. The magnitude of a moment can often be more easily found using the scalar formula, $M_O = Fd$, where d is the perpendicular distance from O to the line of action of F. The direction of the turning effect is usually obvious, but can always be determined by the right-hand rule (Section 2.2-9).

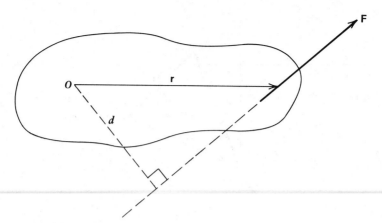

Fig. 2.2-9

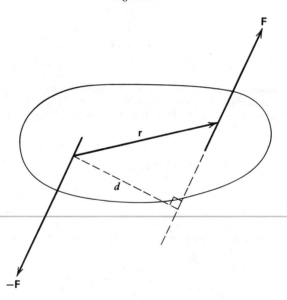

Fig. 2.2-10

2.2-11 Moment of a Couple

Consider two forces equal in magnitude, opposite in direction, and a perpendicular distance d apart as shown in Fig. 2.2-10. Two such forces constitute a couple. One of the forces is denoted as $\mathbf{F}$ and the other as $-\mathbf{F}$. The moment of a couple about any point is given by $\mathbf{M} = \mathbf{r} \times \mathbf{F}$, where $\mathbf{r}$ is a position vector from any point on the line of action of $-\mathbf{F}$ to any point on the line of action of $\mathbf{F}$. Since the moment of a couple is the same about every point in the body, a couple provides a pure torque—nothing but a turning effect. Sometimes the moment of a couple is itself called a couple.

The moment of a couple can often be more easily calculated using the scalar formula, $M = Fd$, where d is the perpendicular distance between the two forces which constitute the couple.

2.2-12 Dot Product

The dot product between two vectors $\mathbf{S}$ and $\mathbf{T}$ is denoted by $\mathbf{S} \cdot \mathbf{T}$ and is defined as $ST\cos\theta$ where θ is the angle between the two vectors. Note that the result of the dot product is a scalar. For that reason the dot product is sometimes called the scalar product. Suppose the vectors $\mathbf{S}$ and $\mathbf{T}$ are given in terms

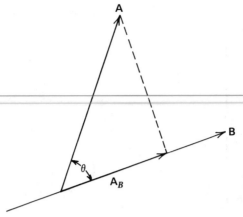

Fig. 2.2-11

of their components: $\mathbf{S} = S_x\mathbf{i} + S_y\mathbf{j} + S_z\mathbf{k}$ and $\mathbf{T} = T_x\mathbf{i} + T_y\mathbf{j} + T_z\mathbf{k}$. In that case the dot product is given by

$$\mathbf{S} \cdot \mathbf{T} = S_xT_x + S_yT_y + S_zT_z$$

One of the uses of the dot product is to find the angle between two vectors. From the definition of the dot product we see that

$$\theta = \cos^{-1}\frac{\mathbf{S} \cdot \mathbf{T}}{ST}$$

Another use is to find the projection of one vector on a second vector. The projection of the vector $\mathbf{A}$ on the vector $\mathbf{B}$ is defined as $A_B = \mathbf{A} \cdot \mathbf{B}/B$ (see Fig. 2.2-11). The projection of a vector is the rectangular component of the vector in the direction of the second vector. Thus $\mathbf{A} \cdot \mathbf{i} = A_x$, $\mathbf{A} \cdot \mathbf{j} = A_y$, and $\mathbf{A} \cdot \mathbf{k} = A_z$.

2.3 EQUIVALENT FORCE SYSTEMS

2.3-1 Resultant Force and Resultant Couple

A system of forces and couples is rigid-body equivalent to a second system if the two systems have the same external effect on the body. The external effect of a system of forces consists of the reaction forces if the body is in static equilibrium or the motion of the body if it is free to move. Any system of forces and couples acting on a body can be replaced by a rigid-body equivalent system consisting of a single resultant force applied at a point and a resultant couple. The resultant force is simply the sum of the forces:

$$\mathbf{R} = \sum_{i=1}^{n} \mathbf{F}_i$$

where $\mathbf{R}$ is the resultant force and $\mathbf{F}_i$ represents the ith applied force in a set of n applied forces. The resultant couple is the moment of the forces and couples about the point of application of the resultant force:

$$\mathbf{M} = \sum_{i=1}^{n} \mathbf{r}_i \times \mathbf{F}_i + \sum_{j=1}^{m} \mathbf{C}_j$$

where $\mathbf{r}_i$ is a position vector from the point of application of the resultant force to the line of action of $\mathbf{F}_i$ and $\mathbf{C}_j$ is the jth applied couple in a set of m applied couples.

Example 2.3-1. The plate illustrated in Fig. 2.3-1 is subjected to the forces and couples shown. Replace this system of forces by a rigid-body equivalent system consisting of a force at O and a couple.

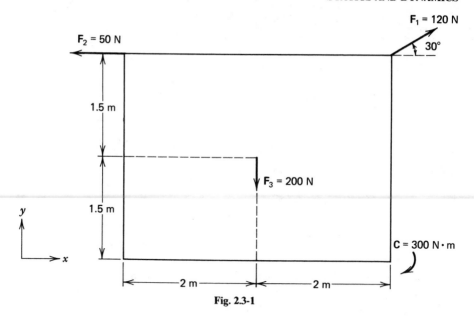

Fig. 2.3-1

The resultant force is given by

$$\mathbf{R} = \sum_{i=1}^{3} \mathbf{F}_i = (103.9\mathbf{i} + 60\mathbf{j} - 50\mathbf{i} - 200\mathbf{j}) \text{ N}$$

or

$$\mathbf{R} = (53.9\mathbf{i} - 140\mathbf{j}) \text{ N}$$

The resultant couple is found by

$$\mathbf{M} = \sum_{i=1}^{3} \mathbf{r}_i \times \mathbf{F}_i + \mathbf{C} = (4\mathbf{i} + 3\mathbf{j}) \times (103.9\mathbf{i} + 60\mathbf{j}) + (3)(50)\mathbf{k} - (2)(200)\mathbf{k} - 300\mathbf{k}$$

or

$$\mathbf{M} = -622\mathbf{k} \text{ N} \cdot \text{m}$$

Note that two systems of forces that have the same resultant force and couple are rigid-body equivalent. We also observe that two forces with the same magnitude, direction, and line of action are rigid-body equivalent. This concept is known as the principle of transmissibility.

2.3-2   Reduction of a Force and a Coplanar Couple to a Single Equivalent Resultant Force

Consider a force acting in the xy plane and a couple acting in that same plane (see Fig. 2.3-2). These may be replaced by a single resultant force which is rigid-body equivalent. This resultant force has the same magnitude and direction as the original force, but it has a new line of action. The new line of action is a perpendicular distance $d = M/F$ away from the original line in a direction so as to produce the same turning effect as the original couple. This process is perhaps best illustrated by an example.

Example 2.3-2. The resultant force and couple found in Example 2.3-1 are shown in Fig. 2.3-3. Replace these with a single equivalent resultant force.

The magnitude of **R** is

$$R = \sqrt{(53.9)^2 + (140)^2} = 150 \text{ N}$$

The single equivalent force acts along the dashed line that is parallel to the original force, R. The

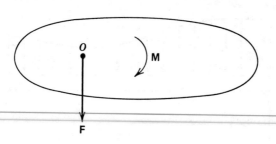

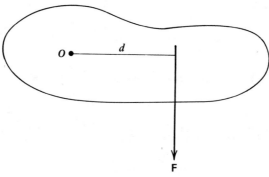

Fig. 2.3-2

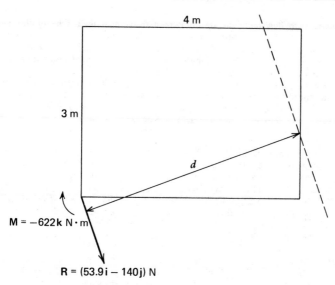

Fig. 2.3-3

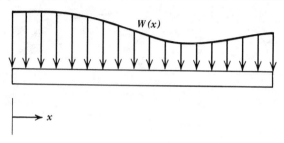

Fig. 2.3-4

distance d is

$$d = \frac{M}{F} = \frac{622}{150} = 4.15 \text{ m}$$

2.3-3 Reduction of a Parallel Force System to a Single Equivalent Resultant Force

A system of forces consisting only of parallel forces can be reduced to a single resultant force that is rigid-body equivalent. The direction of the resultant force is of course the same as the direction of the parallel forces. The magnitude is simply the sum of the magnitudes of the parallel forces. The determination of the line of action is illustrated here through an example.

Example 2.3-3. Consider a parallel system of forces. Without loss of generality, we let the y axis lie in the direction of the forces. We shall replace this system with a single resultant force. The resultant force is found from $\mathbf{R} = R\mathbf{j} = \Sigma \mathbf{F}$. To find the line of action of the resultant force, first find the resultant couple at the origin. In terms of components, the resultant couple is $\mathbf{M} = M_x \mathbf{i} + M_z \mathbf{k}$. The line of action of $\mathbf{R}$ is parallel to the y axis. The coordinates of the point in the xz plane through which it passes are denoted by (x', z'). These coordinates are given by $x' = M_z/R$ and $z' = -M_x/R$. Note that a force of magnitude R acting in the y direction passing through the point (x', z') has the same resultant couple about the origin as does the original system of forces.

2.3-4 Resultant Force of a Distributed Load

It is often necessary to determine the resultant of a distributed load. One of the most common distributed loads is that associated with the weight of a body. In that case the resultant force is the total weight of the body acting at the center of mass (sometimes called the center of gravity) of the body. For a uniform body the center of mass is the same as the centroid of volume. Determination of the center of mass of a body is discussed in Section 2.10-1.

In two-dimensional problems one often encounters the type of loading illustrated in Fig. 2.3-4. For this case the resultant force acts in the direction of the distributed load and is equal to the area under the load curve. The line of action of the load passes through the centroid of the area under the load curve. The centroids of a number of common areas are shown in Table 2.3-1.

Example 2.3-4. The beam shown in Fig. 2.3-5 is subjected to a distributed load. We wish to replace this load with a single equivalent resultant force. To find the resultant force divide this load into two parts: a uniform load over the left half and a triangular load over the right half. The area under the uniform load curve is 600 N and the area under the triangular curve is 300 N. These forces act through the centroids of their respective areas. Thus a rigid-body equivalent load is that shown in Fig. 2.3-6.

This system may now be replaced by a resultant force at A and a couple. The resultant force is $\mathbf{R} = -900\mathbf{j}$ N and the resultant couple is $\mathbf{M} = -2100\mathbf{k}$ N · m. This system can in turn be replaced by a single resultant force $\mathbf{R} = -900\mathbf{j}$ N acting at a point $x = M/F = 2.33$ m from the left end.

It is important to note that when one seeks the stress or strain in a deformable body, rigid-body equivalent loadings are no longer applicable because the body is no longer rigid. For a deformable body the exact nature of the distribution of loading is significant whereas, for a rigid body, it is not.

TABLE 2.3-1   PROPERTIES OF PLANE ARCS AND AREAS

SHAPE	AREA	AREA MOMENT OF INERTIA FOR CENTROIDAL AXES $(J_{\hat{z}} = I_{\hat{x}} + I_{\hat{y}})$
Rectangle	$\mathscr{A} = ab$	$I_{\hat{x}} = \frac{1}{12}ab^3$
		$I_{\hat{y}} = \frac{1}{12}a^3b$

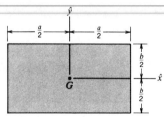

Right triangle	$\mathscr{A} = \frac{1}{2}ab$	$I_{\hat{x}} = \frac{1}{36}ab^3$
		$I_{\hat{y}} = \frac{1}{36}a^3b$
		$I_{\hat{x}\hat{y}} = -\frac{1}{72}a^2b^2$

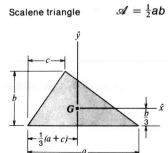

Scalene triangle	$\mathscr{A} = \frac{1}{2}ab$	$I_{\hat{x}} = \frac{1}{36}ab^3$
		$I_{\hat{y}} = \frac{1}{36}ab(a^2 + c^2 - ac)$
		$I_{\hat{x}\hat{y}} = \frac{1}{72}ab^2(2c - a)$

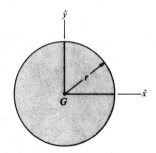

Circle	$\mathscr{A} = \pi r^2$	$I_{\hat{x}} = I_{\hat{y}} = \frac{1}{4}\pi r^4$

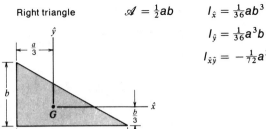

TABLE 2.3-1 (*Continued*)

SHAPE	AREA	AREA MOMENT OF INERTIA FOR CENTROIDAL AXES $(J_{\hat{z}} = I_{\hat{x}} + I_{\hat{y}})$

Quarter circle $\quad\quad \mathscr{A} = \tfrac{1}{4}\pi r^2 \quad\quad I_{\hat{x}} = I_{\hat{y}} = \left(\dfrac{9\pi^2 - 64}{144\pi}\right)r^4$

$$I_{\hat{x}\hat{y}} = \left(\dfrac{9\pi - 32}{72\pi}\right)r^4$$

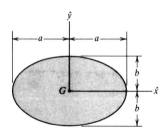

Circular sector $\quad\quad \mathscr{A} = \tfrac{1}{2}\alpha r^2 \quad\quad I_{\hat{x}} = \tfrac{1}{8}(\alpha - \sin\alpha)r^4$

$$I_{\hat{y}} = \left[\dfrac{\alpha + \sin\alpha}{8} - \dfrac{4}{9\alpha}(1 - \cos\alpha)\right]r^4$$

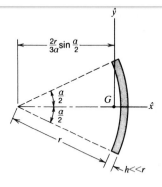

Circular arc $\quad\quad \mathscr{A} = \alpha r h \quad\quad I_{\hat{x}} = \tfrac{1}{2}(\alpha - \sin\alpha)r^3 h$

$$I_{\hat{y}} = \left[\dfrac{\alpha + \sin\alpha}{2} - \dfrac{2}{\alpha}(1 - \cos\alpha)\right]r^3 h$$

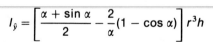

Ellipse $\quad\quad \mathscr{A} = \pi ab \quad\quad I_{\hat{x}} = \dfrac{\pi}{4}ab^3$

$$I_{\hat{y}} = \dfrac{\pi}{4}a^3 b$$

TABLE 2.3-1   (*Continued*)

SHAPE	AREA	AREA MOMENT OF INERTIA FOR CENTROIDAL AXES ($J_{\hat{z}} = I_{\hat{x}} + I_{\hat{y}}$)
Quarter ellipse	$\mathscr{A} = \frac{1}{4}\pi ab$	$I_{\hat{x}} = \left(\dfrac{9\pi^2 - 64}{144\pi}\right)ab^3$
		$I_{\hat{y}} = \left(\dfrac{9\pi^2 - 64}{144\pi}\right)a^3 b$
		$I_{\hat{x}\hat{y}} = \left(\dfrac{9\pi - 32}{72\pi}\right)a^2 b^2$

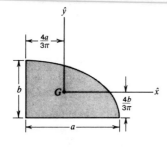

Parabolic section	$\mathscr{A} = \frac{2}{3}ab$	$I_{\hat{x}} = \frac{8}{175}ab^3$
		$I_{\hat{y}} = \frac{19}{480}a^3 b$
		$I_{\hat{x}\hat{y}} = -\frac{1}{60}a^2 b^2$

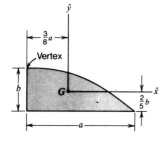

Parabolic spandrel	$\mathscr{A} = \frac{1}{3}ab$	$I_{\hat{x}} = \frac{19}{1050}ab^3$
		$I_{\hat{y}} = \frac{1}{80}a^3 b$
		$I_{\hat{x}\hat{y}} = -\frac{1}{120}a^2 b^2$

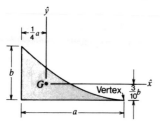

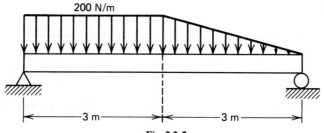

Fig. 2.3-5

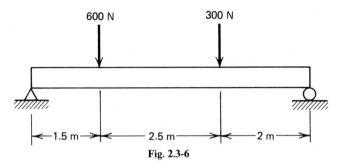

Fig. 2.3-6

2.4 EQUILIBRIUM

2.4-1 Equations of Equilibrium

A rigid body is said to be in equilibrium if the resultant force and resultant couple acting at any point O are each equal to zero:

$$R = \sum_{i=1}^{n} F_i = 0$$

$$M_O = \sum_{i=1}^{n} r_i \times F_i + \sum_{j=1}^{m} C_j = 0$$

where r_i is a position vector from O to any point on the line of action of F_i. The six scalar equations of equilibrium are

$$\sum F_x = 0 \qquad \sum M_x = 0$$

$$\sum F_y = 0 \qquad \sum M_y = 0$$

$$\sum F_z = 0 \qquad \sum M_z = 0$$

For a coplanar system of forces; for example, forces in the xy plane only, there are three equations of equilibrium:

$$\sum F_x = 0$$

$$\sum F_y = 0$$

$$\sum M_O = 0$$

where $\sum M_O$ is the sum of the moments about an axis parallel to the z axis through any point O.

2.4-2 Free-Body Diagrams

One of the keys to the solution of a static equilibrium problem is a proper free-body diagram. A free-body diagram is a sketch of the body or part of the body isolated from its surroundings. Interactions with other bodies are represented by appropriate forces. There are two kinds of interactions: surface contact interactions and interactions at a distance. Surface contact interactions are associated with bodies that touch or are contiguous. There is only one interaction at a distance that is of great interest in mechanics, and that is the gravitational force which we call the weight of an object.

If parts of the body have been isolated from other parts, the force exerted by a part taken away must be represented on the imaginary internal plane that has been exposed. The weight of the body or part of body being represented in the sketch must also be included.

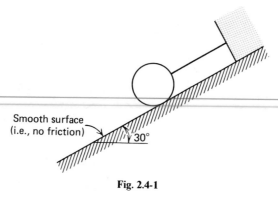

Smooth surface
(i.e., no friction)

30°

Fig. 2.4-1

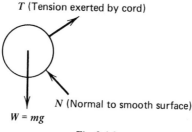

T (Tension exerted by cord)

N (Normal to smooth surface)

$W = mg$

Fig. 2.4-2

Example 2.4-1. In Fig. 2.4-1 an object is supported by a cord and a smooth (frictionless) surface. The free-body diagram of this object is shown in Fig. 2.4-2. Note that two of the interactions are represented by contact forces and the interaction at a distance is represented by the weight of the object.

2.4-3 Example of Two-Dimensional Equilibrium

The truss illustrated in Fig. 2.4-3 is subjected to the forces shown. We wish to determine the reaction forces at A and B. The free-body diagram for the truss is shown in Fig. 2.4-4. The weight of the truss has been neglected. The triangle at A represents a smooth pin, which prevents motion in the x and y directions but provides no resistance to rotation. The circle shown at B represents a smooth roller, which prevents motion normal to the surface upon which it acts. In this case it prevents motion in the x direction.

After sketching the free-body diagram, we apply the equations of equilibrium. If we let a counterclockwise moment be positive, we have

$$\sum M_A = 0: \quad 8B + (4)(2000) - (8)(1500) - (16)(1000) = 0$$

Therefore

$$B = 2500 \text{ lb}$$

$$\sum F_x = 0: \quad A_x + B + 2000 = 0; \quad A_x = -4500 \text{ lb}$$

$$\sum F_y = 0: \quad A_y - 1500 - 1000 = 0; \quad A_y = 2500 \text{ lb}$$

2.4-4 Two- and Three-Force Bodies

A two-force body is a body in equilibrium subjected to two forces only. Since it is in equilibrium, the resultant force is zero. Thus, the two forces must be equal in magnitude, opposite in direction, and collinear.

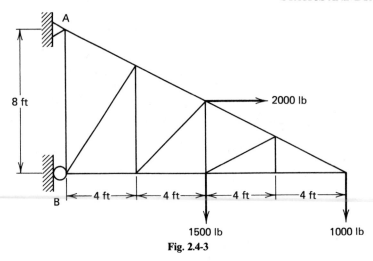

Fig. 2.4-3

Fig. 2.4-4

A three-force body is a body in equilibrium subjected to three forces. Consider the body shown in Fig. 2.4-5. Note that F_1 and F_2 can be added together at the point of intersection to form a resultant vector R. The body is now a two-force body with forces F_3 and R. These forces must be collinear; that is, F_3 passes through the intersection of F_1 and F_2. Therefore, the three forces of a three-force body must all pass through the same point. An exception to this rule occurs if the three forces are parallel.

Example 2.4-2. The 20-ft long ladder shown in Fig. 2.4-6 rests on a rough surface at B and on a smooth surface at A. A 200-lb person stands at the midpoint of the ladder. Determine the reaction forces at A and B.

A free-body diagram for the situation is shown in Fig. 2.4-7. Note that B_x and B_y are the components of a single force, and thus we have a three-force body. The three forces must intersect at the common point shown in Fig. 2.4-8. The resultant of the three forces acting on the ladder is zero. Thus, one can form the force triangle shown in Fig. 2.4-9. The magnitudes of A and B are then found using trigonometry: $A = 200 \tan 15° = 53.6$ lb and $B = 200/\cos 15° = 207$ lb.

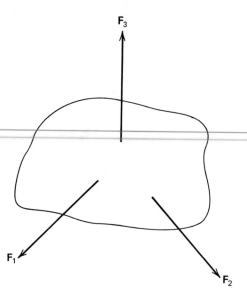

Fig. 2.4-5

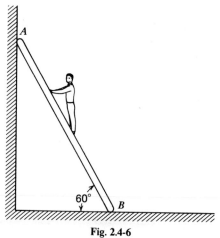

Fig. 2.4-6

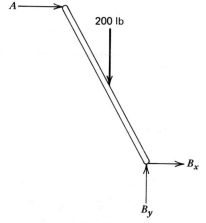

Fig. 2.4-7

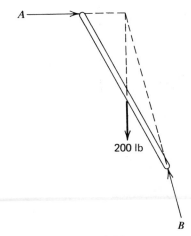

Fig. 2.4-8

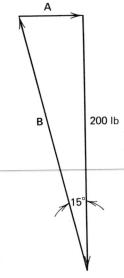

Fig. 2.4-9

2.4-5 Three-Dimensional Equilibrium

A three-dimensional equilibrium problem is solved in the same way as a two-dimensional problem. The steps are:

1. Draw an appropriate free-body diagram.
2. Apply the vector or scalar equations of equilibrium.
3. Solve the resulting set of algebraic equations.

As an example consider the 300-lb uniform bar 10 ft in length supported as shown in Fig. 2.4-10. We wish to determine the force carried in each of the cables.

As a consequence of symmetry, we know that the force carried by each cable is the same. We denote that force by T. A free-body diagram of the bar is shown in Fig. 2.4-11. The two unit vectors

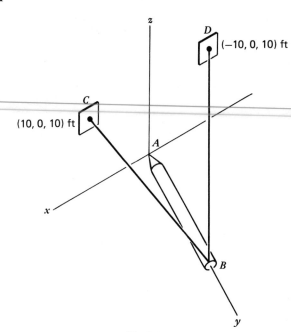

Fig. 2.4-10

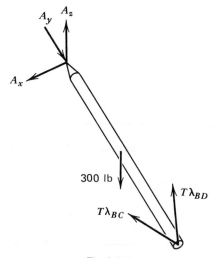

Fig. 2.4-11

are found by

$$\lambda_{BC} = \frac{\mathbf{BC}}{BC} = \frac{10\mathbf{i} - 10\mathbf{j} + 10\mathbf{k}}{17.32}$$

$$\lambda_{BD} = \frac{\mathbf{BD}}{BD} = \frac{-10\mathbf{i} - 10\mathbf{j} + 10\mathbf{k}}{17.32}$$

The equations of equilibrium are

$$\sum M_A = 0: \quad 10\mathbf{j} \times T\frac{(10\mathbf{i} - 10\mathbf{j} + 10\mathbf{k})}{17.32} + 10\mathbf{j} \times T\frac{(-10\mathbf{i} - 10\mathbf{j} + 10\mathbf{k})}{17.32} + 5\mathbf{j} \times (-300\mathbf{k}) = 0$$

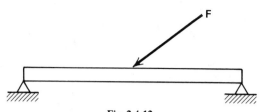

Fig. 2.4-12

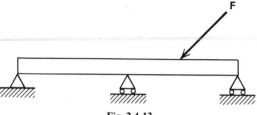

Fig. 2.4-13

or

$$\frac{200}{17.32} T\mathbf{i} - 1500\mathbf{i} = 0$$

therefore

$$T = 129.9 \text{ lb}$$

$$\sum F_x = 0: \qquad A_x + \frac{10}{17.32} T - \frac{10}{17.32} T = 0: \quad A_x = 0$$

$$\sum F_y = 0: \qquad A_y - \frac{10}{17.32} T - \frac{10}{17.32} T = 0: \quad A_y = 150 \text{ lb}$$

$$\sum F_z = 0: \quad A_z - 300 + \frac{10}{17.32} T + \frac{10}{17.32} T = 0: \quad A_z = 150 \text{ lb}$$

In this case the algebra was straightforward. More complicated problems can result in as many as five or six equations with five or six unknowns.

2.4-6   Static Indeterminacy

Some static equilibrium problems cannot be solved using rigid-body mechanics because the number of equations obtained from equilibrium considerations is less than the number of unknowns. An algebraic rule ensures a unique solution to a system of linear algebraic equations only if the number of equations and number of unknowns is equal. If the number of unknowns exceeds the number of equations, the system of equations is said to be indeterminate.

Consider the simple examples shown in Figs. 2.4-12 and 2.4-13. In each of these cases there are four unknown force components but only three equations of equilibrium. These types of problems are said to be statically indeterminate. One must use the principles of mechanics of materials in addition to statics to solve for the reaction forces.

2.5   STRUCTURES AND MACHINES

2.5-1   Trusses

One of the most important engineering structures is the truss. It has many applications, but it is found, in particular, in bridges and buildings. The elementary analysis of a truss is based on the following

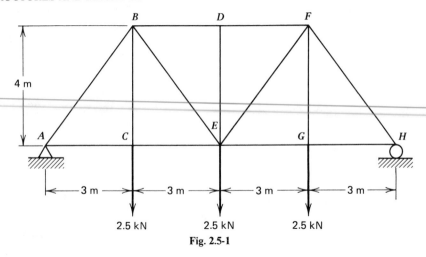

2.5 kN   2.5 kN   2.5 kN

Fig. 2.5-1

three idealizations:

1. Forces are applied only at the joints.
2. Members are joined together by smooth pins.
3. The weight of a member is either negligible or can be divided in two and applied at the two ends.

As a consequence of these idealizations, each member of a truss can be treated as a two-force body. Since the member is in equilibrium, the two forces acting on the body must be equal in magnitude, opposite in direction, and collinear; that is, they must lie along the axis of the bar and oppose one another. If the forces tend to elongate the bar, the bar is said to be in tension. If they tend to shorten the bar, the bar is said to be in compression. There are two principal ways of solving for the forces carried by the members of a truss: the method of joints and the method of sections. These two methods are described in the next two sections.

2.5-2 Method of Joints

The method of joints will be illustrated by way of an example. Consider the truss shown in Fig. 2.5-1. We wish to determine the force carried by each member of the truss.

The usual first step is to determine the reaction force at each of the external supports by a consideration of a free-body diagram of the entire truss such as the one shown in Fig. 2.5-2. By applying the equations of equilibrium we find $A = H = 3.75$ kN.

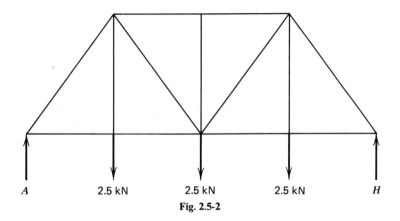

A 2.5 kN   2.5 kN   2.5 kN H

Fig. 2.5-2

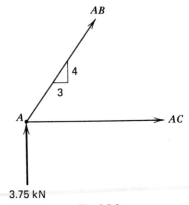

Fig. 2.5-3

The next step is to isolate and draw the free-body diagram of a joint with only two unknown member forces; for example, joint A (see Fig. 2.5-3). In this example the internal force in each bar has been labeled with the end points of the bar. The bars are assumed to be in tension. A bar in tension "pulls" on the joint. If in the application of the equations of equilibrium we obtain a negative number for the force in a member, we then know that the member is, in fact, in compression. Applying the equations of equilibrium to joint A we obtain

$$\sum F_y = 0: \quad 3.75 + 0.8AB = 0; \quad AB = -4.69 \text{ kN}$$

$$\sum F_x = 0: \quad AC + 0.6AB = 0; \quad AC = 2.81 \text{ kN}$$

Thus we find that AB carries a compressive load of 4.69 kN and that AC carries a tensile load of 2.81 kN.

The next step is to isolate an adjacent joint with only two unknown member forces. We cannot move to B because there are three unknowns. We therefore next consider joint C. A free-body diagram of C is shown in Fig. 2.5-4. By inspection (i.e., a quick mental application of the laws of equilibrium) we see that $CE = 2.81$ kN and $BC = 2.5$ kN.

We next consider joint B (see Fig. 2.5-5) for a free-body diagram). An application of the equations of equilibrium yields

$$\sum F_y = 0: \quad 0.8(4.69) - 2.5 - 0.8BE = 0; \quad BE = 1.565 \text{ kN}$$

$$\sum F_x = 0: \quad 0.6(4.69) + 0.6BE + BD = 0; \quad BD = -3.75 \text{ kN}$$

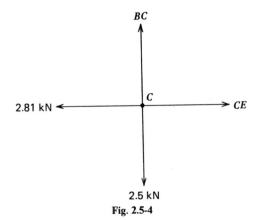

Fig. 2.5-4

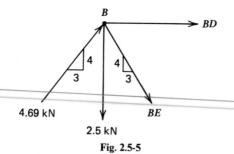

Fig. 2.5-5

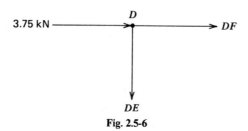

Fig. 2.5-6

We now isolate joint D. A free-body diagram is sketched in Fig. 2.5-6. The equations of equilibrium show that $DF = -3.75$ kN and $DE = 0$. DE is called a zero-force member. Zero-force members are often included in a truss to prevent buckling of long compression members (in this case, BF).

Since this truss is symmetric in both geometry and loading (i.e., there is a mirror image about the center line), the force carried by each of the other members can be determined by symmetry:

$$EG = CE = 2.81 \text{ kN}$$

$$EF = BE = 1.565 \text{ kN}$$

$$GH = AC = 2.81 \text{ kN}$$

$$FH = AB = -4.69 \text{ kN}$$

$$FG = BC = 2.5 \text{ kN}$$

2.5-3 Method of Sections

The method of joints is a laborious procedure employed when one wishes to know the force carried by every member of a truss. If the problem is to determine the force carried by only a few members of the truss, a more straightforward method is the method of sections. We demonstrate the method of sections by way of an example.

Consider again the truss shown in Fig. 2.5-1. Suppose we wish to know the force carried by BD only. After determining the reaction forces by applying the equations of equilibrium to a free-body diagram of the entire truss, we divide the truss into two sections by making an imaginary cut through the truss. The imaginary cut must, of course, pass through BD. A free-body diagram of either section is then drawn (see Fig. 2.5-7). Each of the three unknowns on this free-body diagram could be found using the three equations of two-dimensional equilibrium. In this case since we are interested in the force carried by BD only, we need only one equation of equilibrium:

$$\sum M_E = 0: \quad 4BD + (6)(3.75) - (3)(2.5) = 0; \quad BD = -3.75 \text{ kN}$$

2.5-4 Frames

A frame is a structure designed to carry loads. It differs from a truss in that it has multiforce members. The forces acting on the members of a frame are determined by dividing the frame into component

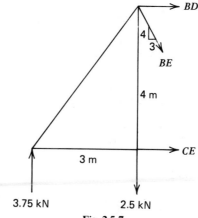

Fig. 2.5-7

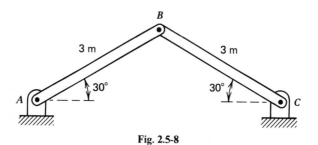

Fig. 2.5-8

parts and analyzing those parts using free-body diagrams and the equations of equilibrium. The process is illustrated here by an example.

Consider the simple frame shown in Fig. 2.5-8. Bar AB is uniform and has a mass of 20 kg, and bar BC is uniform and has a mass of 30 kg. We wish to determine the forces acting on each of the bars.

In drawing the free-body diagrams (see Fig. 2.5-9), it is important to remember the law of action and reaction. Note that the forces at B on BC are drawn as equal in magnitude but opposite in direction to the forces at B on bar AB. Note that for each bar there are four unknowns but only three equations of equilibrium to solve for them. Between the two, however, there are six unknowns and six

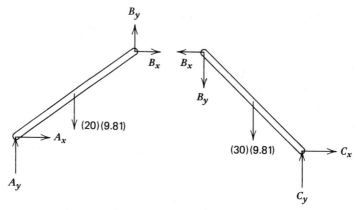

Fig. 2.5-9

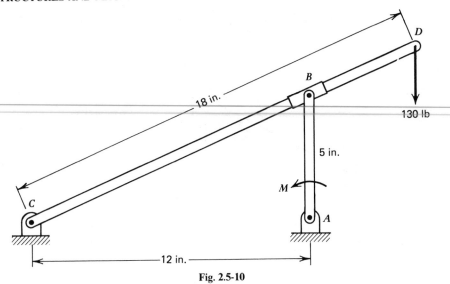

Fig. 2.5-10

equations. For AB we have

$$\sum F_x = 0: \quad A_x + B_x = 0$$

$$\sum F_y = 0: \quad A_y + B_y - (20)(9.81) = 0$$

$$\sum M_A = 0: \quad (20)(9.81)(1.5 \cos 30°) + B_x (3 \sin 30°) - B_y (3 \cos 30°) = 0$$

For the bar BC

$$\sum F_x = 0: \quad -B_x + C_x = 0$$

$$\sum F_y = 0: \quad C_y - B_y - (30)(9.81) = 0$$

$$\sum M_C = 0: \quad (30)(9.81)(1.5 \cos 30°) + B_y (3 \cos 30°) + B_x (3 \sin 30°) = 0$$

The simultaneous solution of these six equations gives:

$$A_x = 212 \text{ N}, \ A_y = 221 \text{ N},$$

$$B_x = -212 \text{ N}, \ B_y = -24.5 \text{ N}, \ C_x = -212 \text{ N}, \quad \text{and} \quad C_y = 270 \text{ N}.$$

2.5-5 Machines

A machine transmits and alters forces. The forces acting on the components of a machine are determined by isolating those components and analyzing them using free-body diagrams and equations of equilibrium.

Consider the quick return mechanism shown in Fig. 2.5-10. We wish to determine the applied couple M. We assume there is no friction between the slider B and the rod CD. A free-body diagram is drawn for each component (see Fig. 2.5-11). Considering the bar CD we can write

$$\sum M_C = 0: \quad 13B - (18)(12/13)(130) = 0; \quad B = 166.15 \text{ lb}$$

For bar AB

$$\sum M_A = 0: \quad M - (5/13)(5)(166.15) = 0; \quad M = 319.5 \text{ in.} \cdot \text{lb}$$

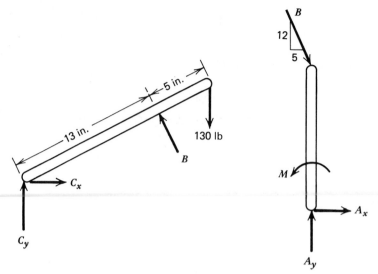

Fig. 2.5-11

2.6 FRICTION

2.6-1 Dry Friction

When a force is applied to an object in contact with a surface, a friction force that resists the motion is produced between the object and the surface. For a dry surface the maximum force that can be developed by friction is

$$f = \mu_s N$$

where μ_s is the coefficient of static friction and N is the normal force. The normal force is the component of the force of contact between the object and surface that is perpendicular or normal to the surface. When motion does not impend, $f < \mu_s N$, and the friction force must be determined using the equations of equilibrium.

For an object sliding on a surface the friction force is

$$f = \mu_k N$$

where μ_k is the coefficient of kinetic friction. In general, $\mu_k < \mu_s$. Some approximate dry friction coefficients are shown in Table 2.6-1.

2.6-2 Angle of Friction

The friction force and normal force may be added together to give the total resultant force between the object and surface. The angle between the resultant force and normal force is known as the angle

TABLE 2.6-1 APPROXIMATE DRY FRICTION COEFFICIENTS

Materials	μ_s	μ_k
Steel on steel	0.5	0.4
Steel on wood	0.5	0.4
Steel on Teflon	0.04	0.04
Wood on wood	0.4	0.3
Wood on leather	0.4	0.3
Auto tire on pavement	0.9	0.8

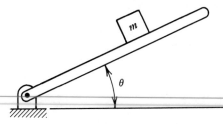

Fig. 2.6-1

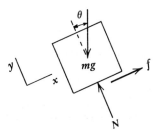

Fig. 2.6-2

of friction. If the angle of friction is denoted by ϕ, we have the relationship

$$\tan\phi = \frac{f}{N}$$

If the object is sliding on the surface or at impending motion, $f = \mu N$, and

$$\phi = \tan^{-1}\mu$$

2.6-3 Example

A block of mass m rests on an inclined plane as illustrated in Fig. 2.6-1. We wish to find the angle θ at which the motion of the block down the plane will impend. From the free-body diagram in Fig. 2.6-2, the equations of equilibrium are expressible as

$$\sum F_x = 0: \quad f = mg\sin\theta$$

$$\sum F_y = 0: \quad N = mg\cos\theta$$

The ratio of f to N is then $\tan\theta$; that is, $f/N = \tan\theta$. It follows that motion impends when $\mu_s = \tan\theta$ or, equivalently, $\theta = \phi$, the angle of friction. For $\theta \leq \phi$ the block does not move. For $\theta > \phi$ it slides down the plane resisted by a friction force, $f = \mu_k mg\cos\theta$.

2.7 KINEMATICS OF PARTICLES

2.7-1 Position, Velocity, and Acceleration

Kinematics is concerned with describing the motion of a body without reference to the forces which produce that motion. The important parameters in the kinematics of particles are the position, velocity, and acceleration of the particle as a function of time. If the position of the particle is known as a function of time, the velocity and acceleration can be determined as shown below.

2.7-2 Rectilinear Motion

Rectilinear motion is motion along a straight line. We denote the axis along which the particle moves as the x axis. Let x be positive to the right. The position of the particle is given as $x(t)$. The magnitude of the velocity of the particle, v, is given by

$$v = \frac{dx}{dt} = \dot{x}$$

The dot notation indicates differentiation with respect to time. A positive value of v means that the particle is moving to the right. The magnitude of the acceleration of the particle, a, is given by

$$a = \frac{dv}{dt} = \dot{v} = \ddot{x}$$

Both v and a may be functions of time. The term *deceleration* is often used to denote the slowing down of a particle. Deceleration does not imply negative acceleration. If the particle were moving to the left, a negative value for acceleration would indicate that the particle was speeding up.

2.7-3 Constant Acceleration

Consider the rectilinear motion of a particle subjected to constant acceleration. A typical situation is that of a particle moving under the influence of gravity with the acceleration of gravity, $a = g = 9.81$ m/s². For constant acceleration the location of the particle as a function of time is given by

$$x = x_0 + v_0 t + \tfrac{1}{2} a t^2$$

where x_0 and v_0 are the initial position and velocity of the particle and t is the elapsed time. The velocity as a function of time is

$$v = v_0 + at$$

The velocity as a function of position can be found from

$$v^2 = v_0^2 + 2a(x - x_0)$$

2.7-4 Curvilinear Motion

A particle traveling along a curved path or trajectory is said to be in curvilinear motion. The position of the particle as a function of time is given by the position vector $\mathbf{r}(t)$, as shown in Fig. 2.7-1. Another way of specifying the position of the particle is in terms of the arc length $s(t)$ along the trajectory from an arbitrary point designated as $s = 0$. The velocity of a particle moving along a curved

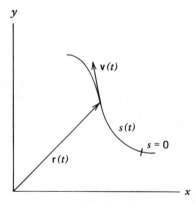

Fig. 2.7-1

trajectory is always tangential to the trajectory and is given by

$$\mathbf{v} = \frac{d\mathbf{r}}{dt} = \dot{\mathbf{r}}$$

The magnitude of the velocity vector is known as the speed. If $s(t)$ is known, the speed can be found from

$$v = \frac{ds}{dt} = \dot{s}$$

The acceleration of a particle in curvilinear motion is given by

$$\mathbf{a} = \frac{d\mathbf{v}}{dt} = \frac{d^2\mathbf{r}}{dt^2} = \ddot{\mathbf{r}}$$

For three-dimensional motion the position vector of a particle can be expressed in terms of its rectangular components as a function of time: $\mathbf{r}(t) = x(t)\mathbf{i} + y(t)\mathbf{j} + z(t)\mathbf{k}$. The velocity and acceleration in terms of rectangular components are given by

$$\mathbf{v}(t) = \dot{\mathbf{r}}(t) = \dot{x}(t)\mathbf{i} + \dot{y}(t)\mathbf{j} + \dot{z}(t)\mathbf{k}$$

$$\mathbf{a}(t) = \ddot{\mathbf{r}}(t) = \ddot{x}(t)\mathbf{i} + \ddot{y}(t)\mathbf{j} + \ddot{z}(t)\mathbf{k}$$

2.7-5 Motion of a Projectile

Often the acceleration of a particle is known in one or more directions, and from the acceleration we wish to determine the motion of the body, that is, we wish to know $\mathbf{r}(t)$. A typical example is the motion of a projectile. Consider the motion of a particle with an initial velocity v_0 launched from the point $x = 0$ and $y = 0$ at an angle θ from the x axis as illustrated in Fig. 2.7-2. If we neglect air resistance, the acceleration in the x direction is the $\ddot{x} = 0$. Integrating this expression once, we find the velocity in the x direction is a constant, $\dot{x} = v_0 \cos \theta$. Integrating one more time we find x as a function of time

$$x = v_0 t \cos \theta \qquad (2.7\text{-}1)$$

where we have required that $x = 0$ when $t = 0$.

The acceleration of the particle in the y direction is the acceleration of gravity, $\ddot{y} = -g$. The velocity in the y direction is $\dot{y} = -gt + v_0 \sin \theta$. The y location as a function of time is

$$y = -\tfrac{1}{2}gt^2 + v_0 t \sin \theta \qquad (2.7\text{-}2)$$

where we have set $y = 0$ when $t = 0$.

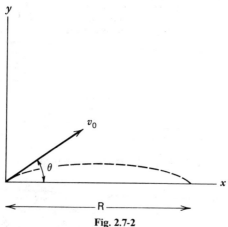

Fig. 2.7-2

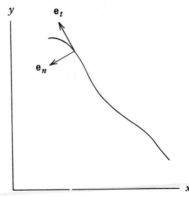

Fig. 2.7-3

Knowing $x(t)$ and $y(t)$, we can determine anything we wish to know about the motion of the particle. For example, by eliminating t between Equations (2.7-1) and (2.7-2) we can find y as a function of x,

$$y = \frac{-gx^2}{2v_0^2\cos^2\theta} + x\tan\theta$$

We see from this equation that the trajectory is parabolic. If the projectile is moving across a flat surface, the time of flight, t_f, can be found by setting $y = 0$ in Equation (2.7-2). The result is

$$t_f = \frac{2v_0\sin\theta}{g}$$

The range of the projectile R can be found by introducing the time of flight into Equation (2.7-1). We find

$$R = \frac{2v_0^2\sin\theta\cos\theta}{g} = \frac{v_0^2}{g}\sin 2\theta$$

2.7-6 Normal and Tangential Components

Consider the motion of a particle along a path as shown in Fig. 2.7-3. Sometimes it is useful to express the velocity and acceleration of the particle in terms of components that are normal and tangential to the path. We define two perpendicular unit vectors which move with the particle: The unit tangent vector $(\mathbf{e}_t)$, which is tangent to the path and is in the direction of motion of the particle, and the unit normal vector $(\mathbf{e}_n)$, which is normal to the path and points toward the instantaneous center of curvature of the path. In terms of these two unit vectors the velocity and acceleration of the particle may be written as

$$\mathbf{v} = v\mathbf{e}_t$$

$$\mathbf{a} = \dot{v}\mathbf{e}_t + \frac{v^2}{\rho}\mathbf{e}_n$$

where ρ is the radius of curvature of the path.

2.7-7 Polar Coordinates and Radial and Transverse Components

The location of a particle in two dimensions can be expressed not only in terms of its x and y coordinates but also in terms of its polar coordinates r and θ where r is the magnitude of the position vector from O to the particle and θ is the counterclockwise angle between the x axis and the position

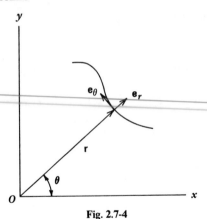

Fig. 2.7-4

vector (see Fig. 2.7-4). When polar coordinates are being used to express the location of the particle, it is usually most convenient to express the velocity and acceleration of the particle in terms of radial and transverse components. Two unit vectors are defined that move with the particle: A unit radial vector, e_r, which is in the same direction as the position vector r, and the unit transverse vector, e_θ, which is perpendicular to e_r and which points in the direction the particle would move if θ were increased. The velocity and acceleration of the particle in terms of radial and transverse components is given by

$$v = \dot{r} e_r + r\dot{\theta} e_\theta$$

$$a = (\ddot{r} - r\dot{\theta}^2) e_r + (r\ddot{\theta} + 2\dot{r}\dot{\theta}) e_\theta$$

2.7-8 Cylindrical Coordinates

The location of a particle in three-dimensional space can be given in terms of its polar coordinates and z coordinate. These three coordinates (r, θ, z) are known as the cylindrical coordinates of the particle. The velocity and acceleration of a particle in terms of cylindrical components are

$$v = \dot{r} e_r + r\dot{\theta} e_\theta + \dot{z} k$$

$$a = (\ddot{r} - r\dot{\theta}^2) e_r + (r\ddot{\theta} + 2\dot{r}\dot{\theta}) e_\theta + \ddot{z} k$$

2.7-9 Circular Motion

Consider the motion of a particle on a circular path. The tangential acceleration of the particle is

$$a_t = \alpha r$$

where α is the angular acceleration of the line between the particle and the center of the circle and r is the radius of the circle. The normal acceleration is directed toward the center of the circle and is known as the centripetal (center seeking) acceleration. The centripetal acceleration may be written in either of the following forms:

$$a_n = \omega^2 r = \frac{v^2}{r}$$

where v is the velocity along the circular path and ω is the angular velocity of the line between the particle and the center of the circle. The velocity of the particle is given by

$$v = \omega r$$

2.8 KINETICS OF PARTICLES

2.8-1 Newton's Second Law

The basic equation for the analysis of the motion of particles is Newton's second law, which may be expressed in the form

$$\mathbf{F} = m\mathbf{a}$$

where m is the mass of the object. This vectoral equation can be written in a scalar form using any convenient coordinate system. For rectangular coordinates

$$F_x = ma_x$$

$$F_y = ma_y$$

$$F_z = ma_z$$

where the subscripts denote direction. In normal and tangential components, Newton's Second Law may be written as

$$F_n = ma_n = m\frac{v^2}{\rho}$$

$$F_t = ma_t$$

where v is the velocity of the particle and ρ is the radius of curvature. In radial and transverse components

$$F_r = ma_r = m(\ddot{r} - r\dot{\theta}^2)$$

$$F_\theta = ma_\theta = m(r\ddot{\theta} + 2\dot{r}\dot{\theta})$$

where r and θ are the polar coordinates.

Example 2.8-1. Consider the system shown in Fig. 2.8-1. We wish to determine the tension in the cord as well as the acceleration of the two blocks. The coefficient of friction between the block and the inclined plane is 0.3. The mass and friction in the pulley is negligible.

The approach to dynamics problems is very much the same as the approach to statics problems. The first step is to draw a free-body diagram. We then apply the appropriate equations of motion. The last step is to solve the resulting set of algebraic equations. Kinematics is employed either in the application of the equations of motion or in a separate step to provide the additional equations needed to solve for the unknowns in the problem.

To solve this problem we first draw two free-body diagrams as shown in Fig. 2.8-2. The acceleration a of each of the particles is shown in the diagrams. Applying the equations of motion to

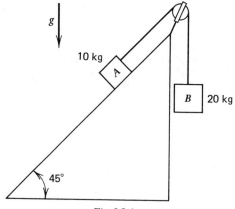

Fig. 2.8-1

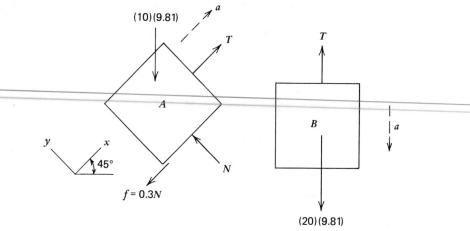

Fig. 2.8-2

block A we find

$$\sum F_y = 0: \quad N = (10)(9.81)\cos 45°$$

$$\sum F_x = ma: \quad T - (10)(9.81)\sin 45° - 0.3N = 10a$$

Applying Newton's second law to block B leads to

$$F = ma: \quad (20)(9.81) - T = 20a$$

Note that in applying the equations of motion, forces in the direction of the acceleration are positive. Solving the above equations simultaneously we obtain

$$T = 125.5 \text{ N}$$

$$a = 3.54 \text{ m/s}$$

Example 2.8-2. A 5-kg mass moves at the end of a 2-m cord with a speed of 3 m/s (see Fig. 2.8-3). Determine the tension in the cord.

A free-body diagram of the mass is shown in Fig. 2.8-4. The mass is traveling in a circular path. Applying Newton's second law normal to the path we find

$$F_n = ma_n = m\frac{v^2}{\rho}: \quad T - (5)(9.81) = 5(3)^2/2$$

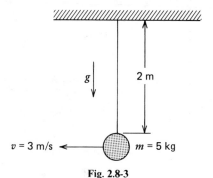

Fig. 2.8-3

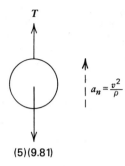

(5)(9.81)

Fig. 2.8-4

therefore

$$T = 71.55 \text{ N}$$

2.8-2   Kinetic Energy of a Particle

The kinetic energy of a particle is denoted by T and is given by

$$T = \tfrac{1}{2} m v^2$$

The units for kinetic energy are foot-pound in U.S. customary units and newton-meter in SI. In discussions of energy, a newton-meter is defined as a joule for which the symbol is J. Although the unit in SI for moment or torque is also a newton-meter one never uses the joule as a unit for torque.

2.8-3   Definition of Work

Consider the motion of a particle along a trajectory as shown in Fig. 2.8-5. The particle is being acted on by a force **F**. The differential work done by the force **F** as the particle moves through a differential distance $d\mathbf{r}$ is defined as $dU = \mathbf{F} \cdot d\mathbf{r}$. If the angle between **F** and $d\mathbf{r}$ is denoted by α then $dU = F \, ds \cos \alpha$. Note that $F \cos \alpha = F_t$, where F_t is the component of force F tangent to the path. Using this term, the differential work may be written as $dU = F_t \, ds$, where ds is a differential length along the trajectory (i.e., the magnitude of $d\mathbf{r}$). Another expression for the differential work which follows from the definition is

$$dU = F_x \, dx + F_y \, dy + F_z \, dz$$

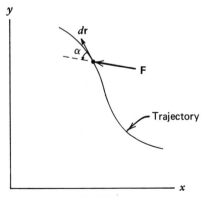

Fig. 2.8-5

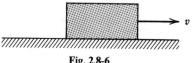

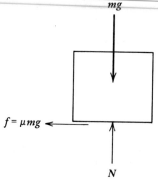

Fig. 2.8-6

Fig. 2.8-7

The total work done on a particle as it moves form position 1 to position 2 on a trajectory is found by integrating the expression for differential work:

$$U_{1 \to 2} = \int_1^2 \mathbf{F} \cdot d\mathbf{r}$$

The units used for work are foot-pound in U.S. customary units and joules (J) in SI.

Example 2.8-3 Work Due to a Frictional Force. A frictional force generally opposes the motion of the body and thus does negative work. Consider the body shown in Fig. 2.8-6. Suppose that it moves with a speed v. A free-body diagram of the body is shown in Fig. 2.8-7. The work done by the frictional force μN as it moves through the distance x is $U = -\mu N x$, where $N = mg$.

Example 2.8-4 Work Due to the Force of Gravity. Consider the work due to a force $F_y = -mg$. The differential work associated with this gravitational force is $dU = -mg\,dy$. The work done as the body is moved from a position y_1 to a position y_2 is given by $U = -mg(y_2 - y_1)$.

Example 2.8-5 Work Due to the Force of a Spring. Consider a spring with a spring constant k. When deformed, the spring exerts a restoring force which is given by $F = -kx$, where x is the amount of deflection from the unstretched position. The differential work associated with the spring force is $dU = -kx\,dx$. If the stretch or compression of a spring is changed from x_1 to x_2, the work done by the spring force is $U = -\frac{1}{2}k(x_2^2 - x_1^2)$.

2.8-4 Principle of Work and Energy

The principle of work and energy is derived from Newton's second law. It states that the work done on a body is equal to the change in the kinetic energy or

$$U_{1 \to 2} = T_2 - T_1$$

A more useful form for problem solving may be obtained through rearrangement:

$$T_1 + U_{1 \to 2} = T_2$$

The kinetic energy at a second position is equal to the initial kinetic energy plus the work done on the particle as it moves from position 1 to position 2.

The principle of work and energy is most useful in solving problems in which one wishes to calculate the change in velocity of a particle associated with a change in position. Often problems of this sort are even more readily solved using conservation of mechanical energy when it is appropriate. They can also be solved using Newton's second law since that law is the basis for both the principle of work and energy and conservation of mechanical energy.

Example 2.8-6. A 2500-lb automobile is traveling at 60 mph (88 ft/s). The driver applies the brakes. If the coefficient of friction between the tires and the road is 0.8, determine the stopping distance.
    The initial kinetic energy is

$$T_1 = \tfrac{1}{2}mv^2 = \tfrac{1}{2}\left(\frac{2500}{32.2}\right)(88)^2 = 300{,}600 \text{ ft} \cdot \text{lb}$$

The final kinetic energy is zero. The work due to a frictional force was found in Example 2.8-3 to be $-\mu N x$. Therefore

$$300{,}600 - (0.8)(2500)x = 0$$

or $x = 150.3$ ft.

2.8-5   Potential Energy

The work done by gravity in moving a body from a position y_1 to y_2 was shown in Example 2.8-4 to be $U = -mg(y_2 - y_1)$. The work done depends only on the initial and final locations of the particle. It is entirely independent of the path taken. We define the quantity mgy as the potential energy of the body with respect to gravity. We denote the potential energy by V. Note that a change in the potential energy is equal to the work done on the body.

$$U_{1\rightarrow 2} = V_1 - V_2$$

A decrease in potential energy indicates that positive work has been done. Indeed the phrase "potential energy with respect to gravity" means that gravity has the potential of doing work on the body. As that work is done, the potential energy decreases. The term y in the definition of potential energy with respect to gravity is the height above an entirely arbitrary datum. Potential energy may be positive or negative depending on whether the body is above or below the datum selected.
    In Example 2.8-5 we demonstrated that the work done by a spring undergoing a change in deflection from x_1 to x_2 is given by $U = -\tfrac{1}{2}k(x_2^2 - x_1^2)$. The work done by the spring is dependent only on the end points and is independent of the path taken (i.e., x can vary in any conceivable way between x_1 and x_2 without affecting the work done). If we define the potential energy with respect to the spring as $V = \tfrac{1}{2}kx^2$, we see that the work done is equal once again to the change in the potential energy

$$U_{1\rightarrow 2} = V_1 - V_2$$

In the case of a spring the datum is not arbitrary. The distance x must be the amount of stretch or compression of the spring from its unstretched length. Note that the potential energy with respect to a spring can never be negative. Whether a spring is stretched or compressed it has the potential for doing work.

2.8-6   Conservation of Mechanical Energy

Consider the work done on a body as it moves from position 1 to position 2 if the only forces acting on the body are those associated with a spring or with gravity.

$$U_{1\rightarrow 2} = V_1 - V_2 = T_2 - T_1$$

where V is the sum of the potential energy with respect to gravity and with respect to the spring. This equation can be rearranged so that it states

$$T_1 + V_1 = T_2 + V_2$$

The mechanical energy of a body is the sum of its kinetic and potential energy. The above equation shows that if the only forces acting are those associated with a spring or gravity then mechanical energy is conserved. Forces for which mechanical energy is conserved are said to be conservative forces. A potential energy can be defined for any conservative force. A typical nonconservative force is that associated with friction. A frictional force usually takes mechanical energy out of the system. If only conservative forces are acting, mechanical energy is conserved. The above equation is a statement of the principle of conservation of mechanical energy. For problems in which it is applicable it is often easier to use this principle than the principle of work and energy from which it is derived.

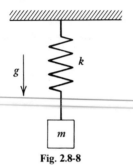

Fig. 2.8-8

Example 2.8-7.   A mass of 5 kg attached to a spring with a spring constant of 3000 N/m is supported such that the spring is not extended (see Fig. 2.8-8). Determine the maximum extension of the spring if the support is suddenly withdrawn.

The kinetic energy of the mass in the initial position and at the point of maximum extension of the spring is zero. If we let the gravitational datum be at the point of maximum extension of the spring and if we denote that maximum extension by x, conservation of mechanical energy yields

$$mgx = \tfrac{1}{2}kx^2$$

$$(5)(9.81)x = \tfrac{1}{2}(3000)x^2$$

therefore

$$x = 32.7 \text{ mm}$$

2.8-7   Linear Momentum of a Particle

The linear momentum of a particle is denoted by $\mathbf{L}$ and is defined as

$$\mathbf{L} = m\mathbf{v}$$

where m is the mass of the particle and $\mathbf{v}$ is its velocity. With this definition of linear momentum, Newton's second law may be written as

$$\mathbf{F} = \frac{d\mathbf{L}}{dt}$$

where m is constant.

2.8-8   Principle of Impulse and Momentum

The linear impulse acting on a particle is defined as

$$\mathbf{Imp}_{1 \to 2} = \int_{t_1}^{t_2} \mathbf{F}\, dt$$

By integrating Newton's second law with respect to time we obtain the principle of impulse and momentum

$$\mathbf{Imp}_{1 \to 2} = m\mathbf{v}_2 - m\mathbf{v}_1$$

That is, the impulse acting on a particle between times 1 and 2 is equal to the change in the linear momentum of the particle. A form of the equation that is more convenient for problem solving is

$$m\mathbf{v}_1 + \mathbf{Imp}_{1 \to 2} = m\mathbf{v}_2$$

Note that this is a vectoral expression. This equation can also be written in terms of its three

Fig. 2.8-9

components:

$$mv_{x_1} + \int_{t_1}^{t_2} F_x \, dt = mv_{x_2}$$

$$mv_{y_1} + \int_{t_1}^{t_2} F_y \, dt = mv_{y_2}$$

$$mv_{z_1} + \int_{t_1}^{t_2} F_z \, dt = mv_{z_2}$$

2.8-9   Impact

The impact between two objects is governed by the principle of conservation of linear momentum; that is, the total linear momentum of the two objects is not changed by the impact. Consider the two particles shown in Fig. 2.8-9 with masses m_1 and m_2 and speeds v_1 and v_2 in the x direction. Note that $v_1 > v_2$ and that an impact is about to occur. An impact such as this is known as a direct central impact. A central impact is one in which the objects impact at a point along the line of impact, which is defined as the line joining their centers of mass. In a direct central impact the motion of the objects is along the line of impact. The principle of conservation of linear momentum states that

$$m_1 v_1 + m_2 v_2 = m_1 v_1' + m_2 v_2' \tag{2.8-1}$$

where the primes denote the speeds after impact. It should be noted that linear momentum is a vectoral quantity and the expression for the conservation of linear momentum is a vectoral equation. In this case since all of the vectors are in the x direction, we have written a scalar expression.

The usual problem is to determine the speeds after the impact if we know the speeds prior to impact. In that event there are two unknowns in Equation (2.8-1) and we need another equation. This second equation is provided by the concept of the coefficient of restitution. The coefficient of restitution, denoted by e, is the ratio of the separation speed of the two objects after impact to the approach speed of the two objects prior to impact. For the case we are discussing,

$$e = \frac{v_2' - v_1'}{v_1 - v_2} \tag{2.8-2}$$

The value of e is always between 0 and 1 and is dependent on the nature of the two impacting bodies. When $e = 1$, energy as well as linear momentum is conserved, and the impact is said to be perfectly elastic. When $e = 0$, the objects adhere to one another, and the impact is said to be perfectly plastic.

In an oblique central impact such as the one illustrated in Fig. 2.8-10, the velocity of the two bodies is not along the line of impact. The velocities of the two bodies after an oblique central impact are found using the following assumptions:

1. Equations (2.8-1) and (2.8-2) hold for those velocity components along the line of impact.
2. Velocity components perpendicular to the line of impact are unchanged by the impact.

Example 2.8-8.   Consider the impact of two spheres as shown in Fig. 2.8-11. The coefficient of restitution for the two spheres is 0.8. We wish to determine the velocity of each sphere after impact.
In vectoral form the velocity of each of the spheres is

$$\mathbf{v}_1 = 10\mathbf{j} \text{ m/s}$$

$$\mathbf{v}_2 = \frac{8}{\sqrt{2}} (-\mathbf{i} - \mathbf{j}) \text{ m/s}$$

The line of impact is along the y axis. Conservation of linear momentum in the y direction results in

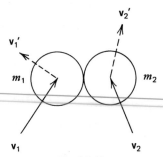

Fig. 2.8-10

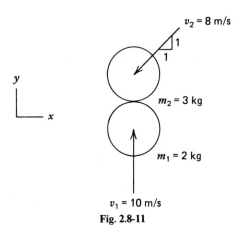

Fig. 2.8-11

the equation

$$(2)(10) + (3)\left(-\frac{8}{\sqrt{2}}\right) = 2v'_{1y} + 3v'_{2y}$$

Using the coefficient of restitution we obtain

$$0.8 = \frac{v'_{2y} - v'_{1y}}{10 - \left(-8/\sqrt{2}\right)}$$

We solve these two equations simultaneously to find

$$v'_{2y} = 5.62 \text{ m/s}$$

$$v'_{1y} = -6.91 \text{ m/s}$$

Velocity components in the x direction are unchanged; that is, $v'_{1x} = 0$ and $v'_{2x} = -8/\sqrt{2}$. Thus, we find

$$\mathbf{v}'_1 = -6.91\mathbf{j} \text{ m/s}$$

$$\mathbf{v}'_2 = (-5.66\mathbf{i} + 5.62\mathbf{j}) \text{ m/s}$$

2.9 RIGID-BODY KINEMATICS

2.9-1 Basic Equations

There are two basic equations that describe the general motion of a rigid body. The first of these relates the velocity of any point A on the body to the velocity of a second point B:

$$\mathbf{v}_A = \mathbf{v}_B + \boldsymbol{\omega} \times \mathbf{r}_{A/B}$$

where $\boldsymbol{\omega}$ is the angular velocity of the body and $\mathbf{r}_{A/B}$ is the position vector of A with respect to B. The second equation relates the acceleration of any point A to the acceleration of a second point B:

$$\mathbf{a}_A = \mathbf{a}_B + \boldsymbol{\alpha} \times \mathbf{r}_{A/B} + \boldsymbol{\omega} \times (\boldsymbol{\omega} \times \mathbf{r}_{A/B})$$

where $\boldsymbol{\alpha}$ is the angular acceleration of the body. The angular acceleration is equal to

$$\boldsymbol{\alpha} = \frac{d\boldsymbol{\omega}}{dt}$$

Example 2.9-1. Consider the bar shown in Fig. 2.9-1. A and B are constrained to move along the surfaces with which they are in contact. The velocity and acceleration of B are $\mathbf{v}_B = -3\mathbf{j}$ m/s and $\mathbf{a}_B = 0.5\mathbf{j}$ m/s^2. We wish to determine $\mathbf{v}_A$ and $\mathbf{a}_A$ and the acceleration of the center of the rod G.
   We find the velocity of A from the equation

$$\mathbf{v}_A = \mathbf{v}_B + \boldsymbol{\omega} \times \mathbf{r}_{A/B}$$

which leads to

$$v_A \mathbf{i} = -3\mathbf{j} + \omega \mathbf{k} \times (2\cos 60°\mathbf{i} - 2\sin 60°\mathbf{j})$$

or

$$v_A \mathbf{i} = -3\mathbf{j} + 1.732\,\omega\mathbf{i} + \omega\mathbf{j}$$

Equating $\mathbf{j}$ components we obtain

$$0 = -3 + \omega$$

That is, $\omega = 3$ rad/s. Equating $\mathbf{i}$ components we find

$$v_A = 1.732\,\omega = 5.20 \text{ m/s}$$

The acceleration of A is given by

$$\mathbf{a}_A = \mathbf{a}_B + \boldsymbol{\alpha} \times \mathbf{r}_{A/B} + \boldsymbol{\omega} \times (\boldsymbol{\omega} \times \mathbf{r}_{A/B})$$

Since $\boldsymbol{\omega}$ and $\mathbf{r}_{A/B}$ are perpendicular to one another (as always occurs in a plane-motion problem), the

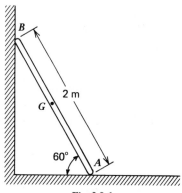

Fig. 2.9-1

last term can be written as $-\omega^2 r_{A/B}$. Thus,

$$a_A i = 0.5j + \alpha k \times (2 \cos 60°i - 2 \sin 60°j)$$

$$- (3)^2 (2 \cos 60°i - 2 \sin 60°j)$$

or

$$a_A i = 0.5j + 1.732 \alpha i + \alpha j - 9i + 15.59j$$

Equating **j** components we find

$$0 = 0.5 + \alpha + 15.59$$

or $\alpha = -16.09$ rad/s. Equating **i** components we obtain

$$a_A = 1.732\alpha - 9 = -36.9 \text{ m/s}^2$$

The acceleration of G is found from

$$\mathbf{a}_G = \mathbf{a}_B + \boldsymbol{\alpha} \times \mathbf{r}_{G/B} - \omega^2 \mathbf{r}_{G/B}$$

which leads to

$$\mathbf{a}_G = 0.5j - 16.09k \times (\cos 60°i - \sin 60°j)$$

$$- (3)^2 (\cos 60°i - \sin 60°j)$$

or

$$\mathbf{a}_G = 0.5j - 13.93i - 8.05j - 4.5i + 7.79j$$

Therefore

$$\mathbf{a}_G = (-18.43i + 0.24j) \text{ m/s}^2$$

2.9-2 Rotation About a Fixed Axis

When a body rotates about a fixed axis, every point in the body is moving in a circular path about that axis. Thus, we can use the kinematics of circular motion to describe the motion of any point in the body. The velocity of any point is given by

$$v = \omega r$$

where ω is the angular velocity of the body and r is the distance from the fixed axis to the point. The tangential acceleration of the point is

$$a_t = \alpha r$$

where α is the angular acceleration of the body. The centripetal acceleration or acceleration of the point toward the axis is given by

$$a_n = \omega^2 r$$

If the body is rotating at a constant angular acceleration α, the following relationships hold. The angular position as a function of time is

$$\theta = \theta_0 + \omega_0 t + \tfrac{1}{2}\alpha t^2$$

where θ_0 is the initial angular position and ω_0 is the initial angular velocity. The angular velocity as a function of time is

$$\omega = \omega_0 + \alpha t$$

Another useful relationship is

$$\omega^2 = \omega_0^2 + 2\alpha(\theta - \theta_0)$$

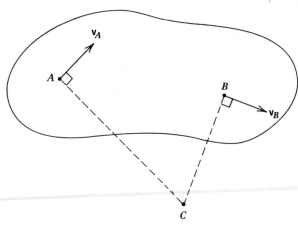

Fig. 2.9-2

2.9-3   Instantaneous Center of Rotation

Consider a body moving in such a way that at a particular instant you know the velocity of at least two points in the body. Lines are drawn at the two points perpendicular to the velocity vectors (see Fig. 2.9-2). The point of intersection of those two lines is the instantaneous center, denoted by C in the figure. At the instant being considered, the body is rotating about the instantaneous center as if it were a fixed axis. The instantaneous center can, of course, move from instant to instant. The velocity of any point in the body at the particular instant being considered is in a direction perpendicular to a line joining the point and the instantaneous center, and is given by

$$v = \omega r$$

where r is the distance from the instantaneous center to the point in question and ω is the angular velocity of the body.

Example 2.9-2.   Consider again the bar shown in Example 2.9-1. The velocity of B is $v_B = -3\mathbf{j}$ m/s, and A and B are constrained to move along the surfaces with which they are in contact. We wish to find the instantaneous center and to use the concept of the instantaneous center to find the velocity of A and the angular velocity of the bar.

A sketch of the kinematics of the bar is shown in Fig. 2.9-3. Lines have been drawn perpendicular to the velocity vectors at A and B. The intersection of the lines at C is the instantaneous center. We

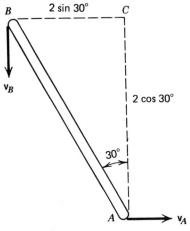

Fig. 2.9-3

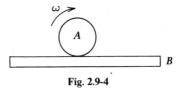

Fig. 2.9-4

apply $\omega = v/r$ to the point B and find

$$\omega = \frac{v_B}{r_B} = \frac{3}{2 \sin 30°} = 3 \text{ rad/s}$$

We then use $v = \omega r$ to find the velocity of A:

$$v_A = \omega r_A = (3)(2 \cos 30°) = 5.20 \text{ m/s}$$

2.9-4   Rolling Without Slip

Consider an object rolling without slip on a second object as shown in Fig. 2.9-4. In such a situation, the velocity of the point of contact on A is the same as the velocity of the point of contact on B. If the object shown as B is a nonmoving surface, the point of contact is the instantaneous center for A.

2.10   PLANE MOTION OF SYMMETRIC RIGID BODIES

2.10-1   Center of Mass

The location of the center of mass of a body is given by the volume integral over the body

$$\bar{r} = \frac{1}{m} \int_B r \, dm$$

where r is the position vector to a generic point of mass dm. If the density of the body is uniform, then the center of mass is at the centroid of volume. The centroid of volume for a number of common geometric shapes is shown in Table 2.10-1.

For a body that is a composite of a number of bodies, the center of mass is found from

$$\bar{r} = \frac{1}{m} \sum \bar{r}_i m_i$$

where m is the total mass, $\bar{r}_i$ is the position vector to the center of mass of the ith part, and m_i is the mass of the ith component part. The coordinates of the center of mass of the composite are given by

$$\bar{x} = \frac{1}{m} \sum \bar{x}_i m_i$$

$$\bar{y} = \frac{1}{m} \sum \bar{y}_i m_i$$

$$\bar{z} = \frac{1}{m} \sum \bar{z}_i m_i$$

where $(\bar{x}_i, \bar{y}_i, \bar{z}_i)$ are the coordinates of the center of mass of the ith particle.

Example 2.10-1.   A body is composed of a rod of length l and mass m_1 and a sphere of radius r and mass m_2 as shown in Fig. 2.10-1. The center of mass of the composite body taken from the end of the rod is given by

$$\bar{x} = \frac{1}{m_1 + m_2} \left[\frac{l}{2} m_1 + (r + l) m_2 \right]$$

TABLE 2.10-1   MOMENTS OF INERTIA AND CENTER OF MASS FOR COMMON GEOMETRIC SHAPES OF UNIFORM DENSITY

SHAPE	VOLUME	MOMENTS OF INERTIA
Uniform slender bar	$\mathscr{V} = L\mathscr{A}$	$I_{xx} = I_{zz} = \frac{1}{12}mL^2$
		$I_{yy} = 0$

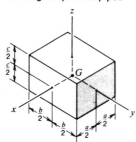

Cross sectional area $\mathscr{A}$

Rectangular parallelipiped	$\mathscr{V} = abc$	$I_{xx} = \frac{1}{12}m(b^2 + c^2)$
		$I_{yy} = \frac{1}{12}m(a^2 + c^2)$
		$I_{zz} = \frac{1}{12}m(a^2 + b^2)$

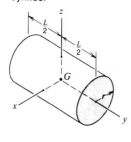

Cylinder	$\mathscr{V} = \pi r^2 L$	$I_{xx} = I_{zz} = \frac{1}{12}m(3r^2 + L^2)$
		$I_{yy} = \frac{1}{2}mr^2$

Semicylinder	$\mathscr{V} = \frac{1}{2}\pi r^2 L$	$I_{xx} = \left(\frac{9\pi^2 - 64}{36\pi^2}\right)mr^2 + \frac{1}{12}mL^2$
		$I_{yy} = \left(\frac{9\pi^2 - 32}{18\pi^2}\right)mr^2$
		$I_{zz} = \frac{1}{12}m(3r^2 + L^2)$

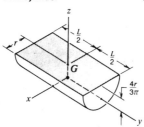

TABLE 2.10-1 (*Continued*)

SHAPE	VOLUME	MOMENTS OF INERTIA
Semicylindrical Shell 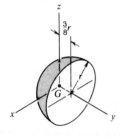	$\mathcal{V} = \pi r L h$	$I_{xx} = \left(\dfrac{1}{2} - \dfrac{4}{\pi^2}\right)mr^2 + \dfrac{1}{12}mL^2$ $I_{yy} = \left(1 - \dfrac{4}{\pi^2}\right)mr^2$ $I_{zz} = \dfrac{1}{2}mr^2 + \dfrac{1}{12}mL^2$
Sphere 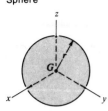	$\mathcal{V} = \dfrac{4}{3}\pi r^3$	$I_{xx} = I_{yy} = I_{zz}$ $= \dfrac{2}{5}mr^2$
Hemisphere 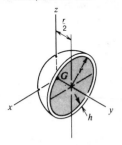	$\mathcal{V} = \dfrac{2}{3}\pi r^3$	$I_{xx} = I_{zz} = \dfrac{83}{320}mr^2$ $I_{yy} = \dfrac{2}{5}mr^2$
Hemispherical shell	$\mathcal{V} = 2\pi r^2 h$	$I_{xx} = I_{zz} = \dfrac{5}{12}mr^2$ $I_{yy} = \dfrac{2}{3}mr^2$
Semiellipsoid 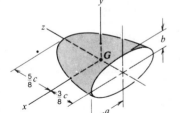	$\mathcal{V} = \dfrac{2}{3}\pi abc$	$I_{xx} = \dfrac{m}{320}(64b^2 + 19c^2)$ $I_{yy} = \dfrac{m}{320}(64a^2 + 19c^2)$ $I_{zz} = \dfrac{1}{5}m(a^2 + b^2)$

TABLE 2.10-1 *(Continued)*

SHAPE	VOLUME	MOMENTS OF INERTIA
Cone	$\mathcal{V} = \frac{1}{3}\pi r^2 L$	$I_{xx} = I_{zz} = \frac{3}{80}m(4r^2 + L^2)$ $I_{yy} = \frac{3}{10}mr^2$

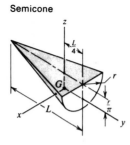

Semicone $\mathcal{V} = \frac{1}{6}\pi r^2 L$

$$I_{xx} = \left(\frac{3}{20} - \frac{1}{\pi^2}\right)mr^2 + \frac{3}{80}mL^2$$

$$I_{yy} = \left(\frac{3}{10} - \frac{1}{\pi^2}\right)mr^2$$

$$I_{zz} = \frac{3}{80}m(4r^2 + L^2)$$

$$I_{yz} = -\frac{1}{20\pi}\,mrL$$

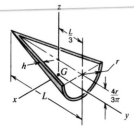

Semiconical shell $\mathcal{V} = \frac{\pi}{2}hr(r^2 + L^2)^{1/2}$

$$I_{xx} = \left(\frac{1}{4} - \frac{4}{9\pi^2}\right)mr^2 + \frac{1}{18}mL^2$$

$$I_{yy} = \left(\frac{1}{2} - \frac{4}{9\pi^2}\right)mr^2$$

$$I_{zz} = \frac{1}{36}m(9r^2 + 2L^2)$$

$$I_{yz} = -\frac{1}{9\pi}mrL$$

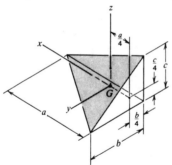

Orthogonal tetrahedron $\mathcal{V} = \frac{1}{6}abc$

$$I_{xx} = \frac{3}{80}m(b^2 + c^2)$$

$$I_{yy} = \frac{3}{80}m(a^2 + c^2)$$

$$I_{zz} = \frac{3}{80}m(a^2 + b^2)$$

$$I_{xy} = -\frac{1}{80}mab$$

$$I_{xz} = -\frac{1}{80}mac$$

$$I_{yz} = -\frac{1}{80}mbc$$

TABLE 2.10-1 (*Continued*)

SHAPE	VOLUME	MOMENTS OF INERTIA

Right triangular prism $\quad \mathscr{V} = \frac{1}{2}abc$

$I_{xx} = \frac{1}{36}m(3b^2 + 2a^2)$

$I_{yy} = \frac{1}{18}m(b^2 + c^2)$

$I_{zz} = \frac{1}{36}m(2a^2 + 3b^2)$

$I_{xz} = -\frac{1}{36}mac$

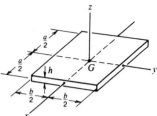

Rectangular plate $\quad \mathscr{V} = abh$

$I_{xx} = \frac{1}{12}mb^2$

$I_{yy} = \frac{1}{12}ma^2$

$I_{zz} = \frac{1}{12}m(a^2 + b^2)$

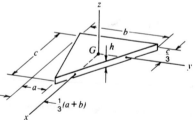

Triangular plate $\quad \mathscr{V} = \frac{1}{2}bch$

$I_{xx} = \frac{1}{18}m(a^2 + b^2 - ab)$

$I_{yy} = \frac{1}{18}mc^2$

$I_{zz} = \frac{1}{18}m(a^2 + b^2 + c^2 - ab)$

$I_{xy} = \frac{1}{36}mc(2a - b)$

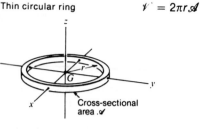

Thin circular ring $\quad \mathscr{V} = 2\pi r \mathscr{A}$

$I_{xx} = I_{yy} = \frac{1}{2}mr^2$

$I_{zz} = mr^2$

Cross-sectional area $\mathscr{A}$

Circular plate $\quad \mathscr{V} = \pi r^2 h$

$I_{xx} = I_{yy} = \frac{1}{4}mr^2$

$I_{zz} = \frac{1}{2}mr^2$

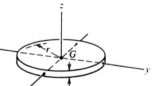

TABLE 2.10-1 *(Continued)*

SHAPE	VOLUME	MOMENTS OF INERTIA
Quarter circular plate 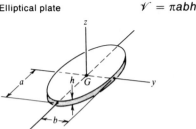	$\mathscr{V} = \dfrac{\pi}{4} r^2 h$	$I_{xx} = I_{yy} = \left(\dfrac{9\pi^2 - 64}{36\pi^2}\right) mr^2$ $I_{zz} = \left(\dfrac{9\pi^2 - 64}{18\pi^2}\right) mr^2$ $I_{xy} = \left(\dfrac{9\pi - 32}{18\pi}\right) mr^2$
Elliptical plate	$\mathscr{V} = \pi abh$	$I_{xx} = \tfrac{1}{4} mb^2$ $I_{yy} = \tfrac{1}{4} ma^2$ $I_{zz} = \tfrac{1}{4} m(a^2 + b^2)$

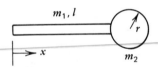

Fig. 2.10-1

2.10-2 Mass Moment of Inertia

The mass moment of inertia of a body about an axis through that body is given by the volume integral over the body:

$$I = \int_B r^2 \, dm$$

where r is the distance from the axis to a generic point of mass dm. The mass moment of inertia about axes through the center of mass for a number of common geometric shapes is given in Table 2.10-1.

Sometimes the inertial properties of a body are given in terms of the radius of gyration, k. If so, the mass moment of inertia is given by $I = mk^2$.

2.10-3 Parallel-Axis Theorem

The mass moment of inertia about any axis through a body can be found using the parallel-axis theorem which states

$$I = \bar{I} + md^2$$

where $\bar{I}$ is the mass moment of inertia about an axis through the center of mass, I is the mass moment of inertia about any axis through the body parallel to the first axis, m is the mass of the body, and d is the distance between the two axes.

Example 2.10-2. The mass moment of inertia about an axis perpendicular to and through the center of mass of a rod of mass m and length l is found in Table 2.10-1 to be $ml^2/12$. The mass moment of inertia about a parallel axis through the end of the rod is

$$I = \tfrac{1}{12}ml^2 + m\left(\tfrac{1}{2}l\right)^2 = \tfrac{1}{3}ml^2$$

The parallel-axis theorem can also be used to find the moment of inertia of a composite body as shown in Example 2.10-3.

Example 2.10-3. Consider the body shown in Fig. 2.10-1 with the center of mass at $\bar{x}$ as found in Example 2.10-1. The mass moment of inertia of the composite body about an axis through its center of mass and perpendicular to the rod is found in this way. Using the parallel-axis theorem the mass moment of inertia of the rod alone about the center of mass of the composite body is

$$I_r = \tfrac{1}{12}m_1 l^2 + m_1\left(\bar{x} - \tfrac{1}{2}l\right)^2$$

The mass moment of inertia of the sphere alone about the center of mass of the composite body is

$$I_s = \tfrac{2}{5}m_2 r^2 + m_2\left(r + l - \bar{x}\right)^2$$

The total mass moment of inertia for the composite body is the sum of these two moments of inertia.

2.10-4 Equations of Motion

The equations of motion for a symmetric rigid body in plane motion are

$$\mathbf{F} = m\bar{\mathbf{a}}$$

$$M = I\alpha$$

where $\mathbf{F}$ is the resultant force acting on the body, m is its mass, and $\bar{\mathbf{a}}$ is the acceleration of the center of mass. In the second equation, M is the resultant moment of the applied force system about the center of mass, I is the mass moment of inertia about an axis through the center of mass perpendicular to the plane of motion, and α is the angular acceleration of the body. For the second expression to be valid, the body must be symmetric about the plane of motion. In the case of fixed-axis rotation, $M = I\alpha$ can be used with M as the resultant moment of the force system about the fixed axis and I as the mass moment of inertia about that axis.

The solution of rigid-body kinetics problems involving plane motion of a symmetric body is very similar to the solution of a two-dimensional equilibrium problem. The first step is to draw a free-body diagram. The equations of motion are then applied. Kinematics is employed to relate $\bar{\mathbf{a}}$ and α. Finally, the resulting algebraic equations are solved.

Example 2.10-4. Consider the rod of mass m and length l as shown in Fig. 2.10-2. The cord at B is cut and the rod is free to swing down about an axis through A. We wish to determine the reaction force at A immediately after the string at B is cut.

A free-body diagram of the rod is shown in Fig. 2.10-3. The center of mass is shown as G. This is a fixed-axis rotation problem, so we can use $M = I\alpha$, with M and I about point A. Application of this equation yields

$$M = I\alpha: \quad mg\frac{l}{2} = \left(\tfrac{1}{3}ml^2\right)\alpha$$

Fig. 2.10-2

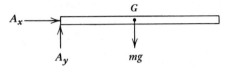

Fig. 2.10-3

Thus,

$$\alpha = \frac{3}{2}\frac{g}{l}$$

Denoting the x and y components of $\bar{\mathbf{a}}$ by $\bar{a}_x$ and $\bar{a}_y$, respectively, and applying Newton's second law in the x and y directions, we obtain

$$\sum F_x = m\bar{a}_x: \quad A_x = m\bar{a}_x = m\frac{v^2}{\rho} = 0$$

$$\sum F_y = m\bar{a}_y: \quad mg - A_y = m\bar{a}_y$$

in which y has been taken as positive downward. To find the relationship between a and α we use the kinematics of circular motion (see Section 2.7-9). In this situation the radius is $l/2$ and $a_t = \bar{a}_y$. Thus,

$$\bar{a}_y = \tfrac{1}{2}l\alpha$$

Solving the above equations simultaneously we find

$$A_y = \tfrac{1}{4}mg$$

Example 2.10-5. The 6-m uniform rod of mass 30 kg shown in Fig. 2.10-4 is released from rest. We wish to determine the reaction forces at A and B given that the surfaces at those two points are smooth.

A free-body diagram of the rod is given in Fig. 2.10-5. Application of the moment equation gives us

$$M_G = \bar{I}\alpha: \quad (3\cos 60°)B - (3\sin 60°)A = \tfrac{1}{12}(30)(6)^2\alpha$$

Since we do not have a fixed-axis rotation in this problem, the only axis about which we can apply this equation is through the center of mass. Applying Newton's second law in the x and y directions we

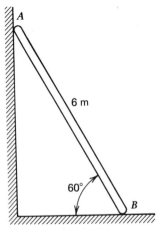

Fig. 2.10-4

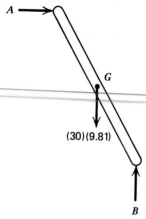

Fig. 2.10-5

find

$$\sum F_x = m\bar{a}_x: \quad A = 30\bar{a}_x$$

$$\sum F_y = m\bar{a}_y: \quad B - (30)(9.81) = 30\bar{a}_y$$

At this point in our solution we have three equations and five unknowns: A, B, α, $\bar{a}_x$, and $\bar{a}_y$. We need two additional equations. We obtain these using kinematics (see Section 2.9-1):

$$\mathbf{a}_A = \mathbf{a}_B + \boldsymbol{\alpha} \times \mathbf{r}_{A/B} + \boldsymbol{\omega} \times (\boldsymbol{\omega} \times \mathbf{r}_{A/B})$$

At the instant being considered $\omega = 0$. Therefore,

$$a_A \mathbf{j} = a_B \mathbf{i} + \alpha \mathbf{k} \times (-6\cos 60°\mathbf{i} + 6\sin 60°\mathbf{j})$$

or

$$a_A \mathbf{j} = a_B \mathbf{i} - 5.196\alpha \mathbf{i} - 3\alpha \mathbf{j}$$

Equating $\mathbf{j}$ components we find

$$a_A = -3\alpha$$

To find the relationship between the acceleration of the center of mass and the angular acceleration of the rod, we again employ kinematics:

$$\mathbf{a}_G = \mathbf{a}_A + \boldsymbol{\alpha} \times \mathbf{r}_{G/A}$$

or

$$\bar{a}_x \mathbf{i} + \bar{a}_y \mathbf{j} = -3\alpha \mathbf{j} + \alpha \mathbf{k} \times (3\cos 60°\mathbf{i} - 3\sin 60°\mathbf{j})$$

That is,

$$\bar{a}_x \mathbf{i} + \bar{a}_y \mathbf{j} = -3\alpha \mathbf{j} + 2.598\alpha \mathbf{i} + 1.5\alpha \mathbf{j}$$

Equating like components we have

$$\bar{a}_x = 2.598\alpha$$

$$\bar{a}_y = -1.5\alpha$$

These two equations combined with the three equations of equilibrium give us five equations and five unknowns. Solving simultaneously we obtain $A = 95.6$ N and $B = 239$ N.

2.10-5   Kinetic Energy of a Symmetric Body in Plane Motion

The kinetic energy of a symmetric body in plane motion is

$$T = \tfrac{1}{2}m\bar{v}^2 + \tfrac{1}{2}\bar{I}\omega^2$$

where m is the mass of the body, $\bar{v}$ is the velocity of the center of mass, $\bar{I}$ is the mass moment of inertia about an axis through the center of mass, and ω is the angular velocity of the body.
   For rotation about a fixed axis

$$T = \tfrac{1}{2}I_0\omega^2$$

where I_0 is the mass moment of inertia about the fixed axis.

2.10-6   Principle of Work and Energy

To the work done by forces as described in Section 2.8-3, we must add the work done by a moment M as a body rotates. The work done by M, the resultant moment about the center of mass, is

$$U_{1\rightarrow 2} = \int_1^2 M\,d\theta$$

where $d\theta$ is the differential angle through which the body rotates. If M is not a function of θ, then

$$U_{1\rightarrow 2} = M(\theta_2 - \theta_1)$$

   Using the definition of kinetic energy given in Section 2.10-5 and the definition of work given in this section and in Section 2.8-3, we have the principle of work and energy which holds that

$$T_1 + U_{1\rightarrow 2} = T_2$$

   If there are only conservative forces acting on the body (e.g., springs and gravity), then the principle of conservation of mechanical energy holds:

$$T_1 + V_1 = T_2 + V_2$$

The potential energy of a spring is $kx^2/2$ where x is the extension or compression of the spring from its undeflected position. The potential energy due to gravity is mgh where h is the height of the center of mass of the body above an arbitrary datum.

Example 2.10-6.   A 6-m uniform rod of mass 30 kg is released from rest in the position shown in Fig. 2.10-6. We wish to determine the angular velocity of the bar as it swings through the vertical position.
   Since the only force that does work in this problem is gravity, we can use the principle of conservation of mechanical energy. In this case the initial kinetic energy, T_1, is zero. If we let the initial position of the center of mass be the gravitational datum, then the initial potential energy, V_1, is

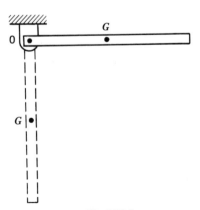

Fig. 2.10-6

also zero. The final kinetic energy is

$$T_2 = \tfrac{1}{2}I_0\omega^2 = \tfrac{1}{2}\left[\tfrac{1}{3}(30)(6)^2\right]\omega^2$$

and the final potential energy is

$$V_2 = mgh = (30)(9.81)(-3)$$

Therefore,

$$0 = 180\omega^2 - 882.9$$

or $\omega = 2.21$ rad/s.

CHAPTER 3

STRENGTH OF MATERIALS

JOE W. McKINLEY

Consultant
Laguna Beach, California

L. D. MARDIS

Pomona Division of General Dynamics
Pomona, California

3.1 INTRODUCTION TO STRESS AND STRAIN

3.1-1 Normal and Shear Stress

A structural member that is loaded by surface forces or body forces (inertial, gravitational, or electromagnetic) develops an internal force distribution in opposition to the load, according to Newton's third law. *Stress* is the intensity of the internal forces, that is, force per unit area; it varies with location in the member and is therefore a point function.

To completely define the general (three-dimensional) state of stress at a point, it is necessary to specify components on three orthogonal planes passing through the point. Positive stresses are most easily represented as shown acting on the infinitesimal cube in Fig. 3.1-1, with three orthogonal components shown on each of the three positive faces. A cube face is positive if it faces a positive coordinate system direction. A *normal* stress component (σ) carries a single subscript that identifies the coordinate axis to which it is parallel and the cube face to which it is perpendicular. A normal stress is positive (tensile) if directed outward from the cube and is negative (compressive) if directed inward. A *shear* stress component (τ) carries two subscripts that indicate, first, the face to which it acts tangential and, second, the coordinate axis toward which it is directed. A shear stress is positive if it acts on a positive face in the direction of a positive coordinate direction or if it acts on a negative face in the axis of a negative coordinate direction.

For static loading, equilibrium requires that each normal stress be equally opposed of a cube face, it must be accompanied by three others, which act in the same plane. These shear stresses balance statically by being directed head-to-head and tail-to-tail as shown in Fig. 3.1-1. The magnitudes are equal unless surface couples or body couples are present.[1,2]

Stress is a tensor of second order, represented by the array:

$$T = \begin{bmatrix} \sigma_x & \tau_{yx} & \tau_{zx} \\ \tau_{xy} & \sigma_y & \tau_{zy} \\ \tau_{xz} & \tau_{yz} & \sigma_z \end{bmatrix}$$

In most cases surface and body couples are not significant and may be ignored. Equilibrium then reduces from nine to six the number of independent stress components necessary to define the general state of stress, and the tensor array is symmetric:

$$T = \begin{bmatrix} \sigma_x & \tau_{xy} & \tau_{xz} \\ \tau_{xy} & \sigma_y & \tau_{yz} \\ \tau_{xz} & \tau_{yz} & \sigma_z \end{bmatrix} \tag{3.1-1}$$

since $\tau_{xy} = \tau_{yx}$, $\tau_{yz} = \tau_{zy}$, and $\tau_{xz} = \tau_{zx}$.

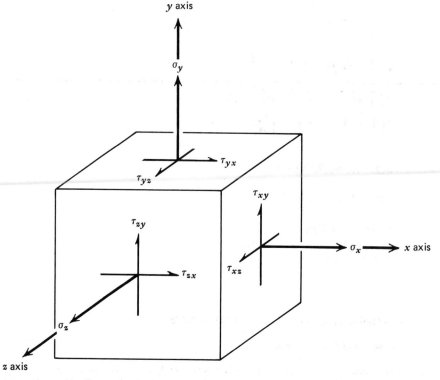

Fig. 3.1-1 Stresses on infinitesimal cubic element.

The coordinate system and its orientation are selected for convenience, based on the geometry of the structural member and loads. The state of stress may be quantified using any coordinate system or set of dimensional units. The components of one system will differ from another system. However, the state of stress is not dependent on the system used to quantify it; it is *invariant*. Sections 3.1-2 and 3.1-5 show how reorientation of the cartesian coordinate system modifies the component magnitudes. Loading conditions that are simpler than the general (three-dimensional) case may reduce the stress to a biaxial[3] (see Section 3.1-5) or uniaxial (see Section 3.1-2) state. There is a corresponding reduction of nonzero components in the stress tensor array.

3.1-2 Uniaxial Stress and Strain

In Fig. 3.1-2 a rod is shown, subject to a static externally applied tensile load (P). The load is applied along the longitudinal axis of the rod, acting through the centroid of each lateral cross section (centric loading). If the material is homogeneous and isotropic, no bending will occur, and both the loading

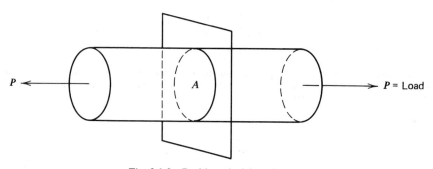

Fig. 3.1-2 Rod in uniaxial tension.

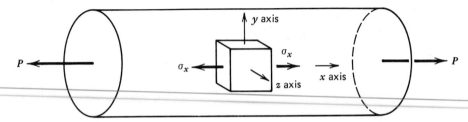

Fig. 3.1-3   Imaginary cubic element of rod in tension (x, y, z are cube axes).

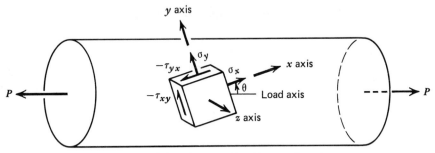

Fig. 3.1-4   Cubic element with arbitrary orientation.

and stress states are uniaxial. For these conditions the normal (tensile) stress across a plane perpendicular to the load is correctly assumed to be uniform across the cross section. The stress is not uniform at the rod ends where point loads are applied, or near any abrupt change in the cross section, where *stress concentrations* will result (see Section 3.9). A long, slender rod may buckle (see Section 3.6) when loaded in compression, and the presence of bending stress causes the stress distribution to be nonuniform.

An infinitesimal cubic element of the rod subject to the above restrictions is shown in Fig. 3.1-3, oriented with the cube x axis aligned with the rod longitudinal axis. Only normal (tensile) stress appears on the two cube faces that are normal to the longitudinal axis of the rod. No shear stress acts on any face of the cube. Select a second cube that is oriented an angle θ about a lateral z axis at the same location (Fig. 3.1-4). Equilibrium shows that shear components appear on the x and y faces of the second cube. A normal stress component exists on the y face, and the magnitude of normal stress on the x face is less than for the original cube. Magnitudes of the stress components may be found from the equations:

$$\sigma_x = \frac{dP(\cos^2\theta)}{dA} = \frac{P(\cos^2\theta)}{A}$$

$$\tau_{xy} = \frac{dP(\cos\theta\sin\theta)}{dA} = \frac{P(\cos\theta\sin\theta)}{A}$$

The differential is deleted since there is a uniform stress distribution across plane A, and therefore the local stress is equal to the average stress across this plane. These functions are normalized to the ratio P/A and are graphed with θ as the abscissa in Fig. 3.1-5. The figure confirms that shear stress has a maximum ($\tau_{max} = P/2A$) on planes 45° from the load axis, and normal stress has a maximum ($\sigma_{max} = P/A$) on the plane perpendicular to the load axis. The significance of these relationships becomes apparent during failure analysis (see Section 3.8). Mohr's circle for uniaxial stress is an alternative method of showing how shear and normal stress components vary on an arbitrary plane (Fig. 3.1-6).

The stress magnitudes described above are average values based on the assumption that the material is a continuum, and not influenced by the orientations of atomic lattices. Lattice stress magnitudes differ from average values. For metals the structure is usually polycrystalline, of random size and crystallographic orientation, and the continuum assumption is permissible for structural members of greater than microscopic size. Structures the size of electronic "chip" circuits may require special treatment.

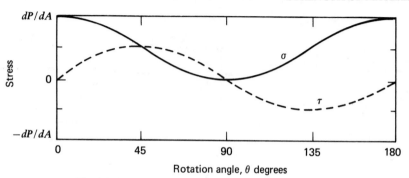

Fig. 3.1-5 Stress components versus θ for uniaxial load.

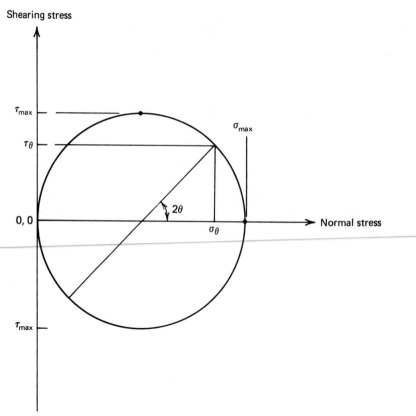

Fig. 3.1-6 Mohr's circle for uniaxial stress.

Eccentric axial loading applied to a structural member results in a stress distribution that is not uniform over the cross section (Fig. 3.1-7). The eccentricity (e) is the distance between the line of action of the load to the centroid of the cross section of the member. Bending stress (see Section 3.3-3) and uniaxial stress components are additive according to superposition (see Section 3.1-5) for linear elastic materials. When the eccentricity is large, the bending stress will exceed the uniaxial component over a portion of the cross section, leading to a sign reversal for stress over that portion. Part of the remaining area will have much larger than average stresses.

Materials that are weak in tension are threatened when loaded eccentrically in compression. Similarly, when loaded eccentrically in tension, thin shell or frame structures may buckle on one side due to local compression.

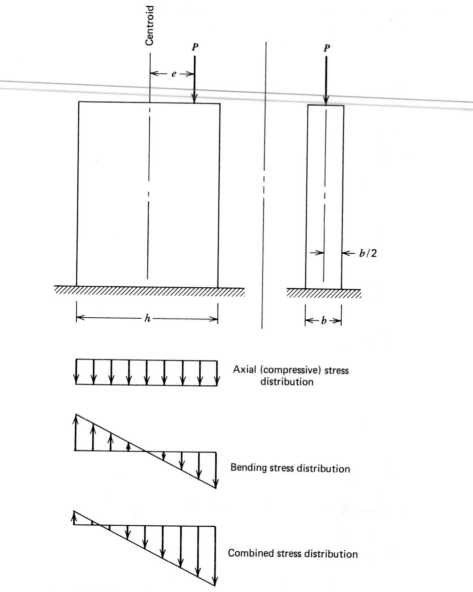

Fig. 3.1-7 Eccentric loading on short compression block.

To prevent stress reversal of this kind, the line of action of the force must be restricted to an area close to the centroid of the cross section, called the *core* or *kern*. The *kern limit* is the perimeter of the kern. Table 3.1-1 shows kern limits for several cross sections.[4]

The rod in Fig. 3.1-2 will elongate due to the external tensile load until the internal force distribution balances the external load or until the strength of the constituent material is exceeded, causing failure of the rod (see Sections 3.8 and 6.2-1). *Hooke's law* implies that the elongation will be linearly proportional to the load, and for most structural materials this is a close approximation for both tension and compression. The *proportional limit* is a material stress (or strength) limit, above which the linear relationship no longer applies. A material may be *elastic*, without being linear, if it returns to its original dimensions once the load is released. For some materials this property is time or temperature dependent. Despite these qualifications, materials that deform linearly are given the more general title "elastic," and an "elastic constant" is used to approximate the relationship between load and displacement (see Section 3.4).

TABLE 3.1-1   KERN LIMITS FOR COMPRESSION MEMBERS[a]

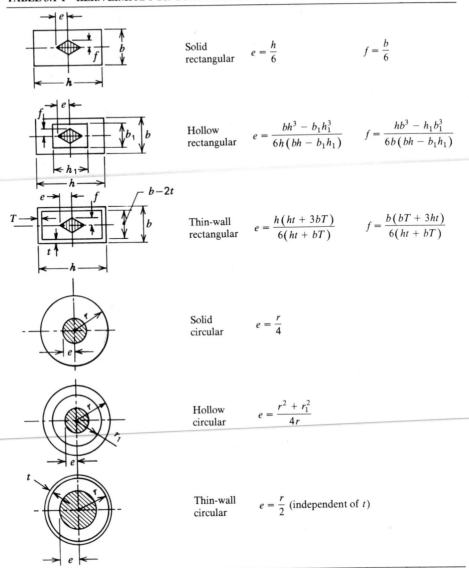

Solid rectangular	$e = \dfrac{h}{6}$	$f = \dfrac{b}{6}$
Hollow rectangular	$e = \dfrac{bh^3 - b_1 h_1^3}{6h(bh - b_1 h_1)}$	$f = \dfrac{hb^3 - h_1 b_1^3}{6b(bh - b_1 h_1)}$
Thin-wall rectangular	$e = \dfrac{h(ht + 3bT)}{6(ht + bT)}$	$f = \dfrac{b(bT + 3ht)}{6(ht + bT)}$
Solid circular	$e = \dfrac{r}{4}$	
Hollow circular	$e = \dfrac{r^2 + r_1^2}{4r}$	
Thin-wall circular	$e = \dfrac{r}{2}$ (independent of t)	

[a] Kern limit areas are defined by shaded portions of sections. (From Reference 4.)

Axial strain (ε), also called normal strain, is the longitudinal deformation of the rod divided by the original length of the rod. The units for axial strain are length/length.

The *average* axial strain is defined $\varepsilon_{ave} = \delta/L$, where δ is the change in length and L is the initial length. The local strain in the longitudinal axis is $\varepsilon_{local} = d\delta/dl$, and it may differ from the average strain along rods of nonuniform cross section.

Through the definitions of stress and strain, the elastic behavior of a material is expressed such that it is independent of the geometry of the rod: Stress and strain are essentially the force and displacement normalized to the initial cross-sectional area and length of the rod. The elastic behavior is then $\sigma = E\varepsilon$, where E is the *modulus of elasticity* of the material, also called *Young's modulus* (dimensions are force/area). Values of Young's modulus are tabulated for several materials in Section 3.1-4. Increased temperature generally reduces Young's modulus. The linear axial deflection of a straight, isotropic, centrically loaded rod is, integrating over the length,

$$\delta = \int_0^L \frac{P(l)}{A(l)E(l)}\, dl$$

where load, modulus of elasticity, or cross-sectional area may be functions of position along the rod (but must be constant over each cross section). For constant stiffness and concentrated end loads, the deflection is simply $\delta = PL/AE$.

The force–displacement relationship for a uniaxial rod element of constant stiffness in matrix form is

$$\{P\} = [K]\{\delta\}$$

or

$$\begin{Bmatrix} P_1 \\ P_2 \end{Bmatrix} = \frac{AE}{L} \begin{bmatrix} 1 & -1 \\ -1 & 1 \end{bmatrix} \begin{Bmatrix} \delta_1 \\ \delta_2 \end{Bmatrix}$$

where subscripts 1 and 2 relate to the rod ends.

In a uniaxial member axial strain is accompanied by *lateral strain*. Elongation of the rod leads to a reduction of the cross-sectional area for most materials.

Lateral strain is the lateral deflection divided by the initial lateral dimension. *Poisson's ratio* (ν or μ) is defined as $\nu = |\varepsilon_{lateral}/\varepsilon_{axial}|$. For isotropic materials Poisson's ratio is independent of orientation. A perfectly plastic material will have a constant volume under stress and attains the theoretical limit of 0.5. A perfectly inelastic (brittle) material has no lateral deformation under stress, and the volume change is due only to deformation in the loading axis. The value of ν is then zero.

Material volume change is usually calculated for hydrostatic (triaxial) stress, and is expressed by the *bulk modulus*, also called the modulus of volume expansion (K), which is the ratio of applied normal stress to unit volume change.

3.1-3 Direct Shear Stress and Strain

Transverse loading tends to cause material to deform much like a stack of paper that is pushed over. *Shear stress* (τ) and *strain* (γ) are shown in Fig. 3.1-8. Shear strain is angular in form. The geometry of loading has led to the nomenclature described in Fig. 3.1-9. In the first two cases a circular rod is loaded (P) perpendicular to its longitudinal axis. The *average shear stress* (τ) acts over the cross-sectional area (A) of the rod so that $\tau = P/A$. In the last two cases the flat plate and bolt head are under shear stress at the circumference of the shaft and hole. The area upon which the stress acts is $A = \pi t d$, and the average shear stress is $\tau = P/\pi t d$. Although the actual shear stress distribution is not uniform, the average shear stress is commonly used when bending stress is not significant. If close tolerances are not kept, and gaps appear between these members, *bending stresses* may be introduced. A combined stress condition then exists which lowers the load-carrying capability of the structures (see Section 3.1-5).

An example is shown in Fig. 3.1-10 in which the loading *eccentricity e* causes the member under shear to behave as a very short beam (cf. Section 3.7-3).

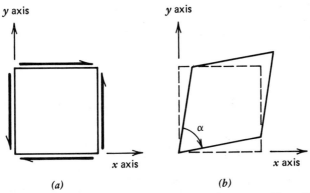

(a)   *(b)*

Fig. 3.1-8 Shear stress and strain sign conventions. (*a*) Positive shear stress; (*b*) positive shear strain $\gamma = \pi/2 - \alpha$ (radians).

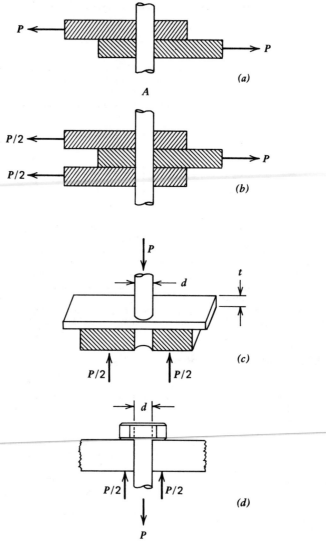

Fig. 3.1-9 Shear loading nomenclature. (a) Single direct shear; (b) double direct shear; (c) punching shear; (d) punching shear.

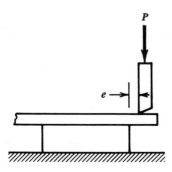

Fig. 3.1-10 Shear with accidental bending component.

TABLE 3.1-2 ELASTIC CONSTANTS[a]

Material	Modulus of Elasticity in Tension $\times 10^6$ (lb/in.^{2})	Modulus of Rigidity $\times 10^6$ (lb/in.^{2})	Poisson's Ratio
Beryllium	42.5	20	0.1
Chromium	42		
Gold	12		0.42
Platinum	25		0.39
Silver	11		0.37
Tungsten	59		0.28
Uranium (depleted)	20–30		0.21
Beryllium copper	18.5	7.3	0.27
Brasses	15	5.6	
Bronzes (including phosphor B)	14–16	6.5	
Cobalt and alloys	30–33.6		
Coppers and alloys	17–18.7	7	0.34
Invar	21	8.1	
Iron alloys	26.5–30	10–11	0.30–0.32
Lead and alloys	2		0.43
Molybdenum and alloys	46–53		.32
Nickel and alloys	23–35		
Tantalum and alloys	26–27		
Titanium and alloys	11–18.5	5.3–6.7	0.31–0.34
Carbon and graphite, fibrous reinforced	0.03–1.8		
Beryllium carbide	30–46		
Boron carbide	42–65		
Columbium carbide	49		
Silicon carbide	13–95		
Tantalum carbide	55		
Titanium carbide	36–65		
Tungsten carbide	65–102		

Basic Material	Percent Glass Fiber Content	Modulus of Elasticity in Tension $\times 10^6$ (lb/in.^{2})
Acetal, copolymer	25	1.25
Epoxy, GL laminates		3.3–5
Nylons	30	1–1.4
Phenylene oxide-based resins	20–30	0.93–1.3
Polycarbonate	40	0.86–1.7
Polyester, thermoplastic	30, 45	1.3–2.1
Polyetherimide	10, 30	0.65, 1.3
Polyimides		4.5
Polyphenylene sulfide	40	1.12
Polystyrene	30	1.2
Polysulfone	30–40	1.1–1.5
Styrene acylonitrile	30	1.8

[a] References 4–7 and 11–13.

The *modulus of elasticity in shear* (G; also called modulus of rigidity) relates shear stress and strain: $\tau = G\gamma$. The shear modulus of elasticity is listed for several materials in Section 3.1-4 and has dimensions force/length^2.

Torsional loading is not considered a form of direct shear and is discussed in Section 3.4-1.

3.1-4 Elastic Constants

Linear stress–strain relationships were presented for conventional structural materials in Sections 3.1-2 and 3.1-3. The four elastic constants are related by the equations

$$E = 2G(1 + \nu) \qquad \text{Modulus of elasticity}$$
$$G = E/2(1 + \nu) \qquad \text{Modulus of rigidity}$$
$$K = E/3(1 - 2\nu) \qquad \text{Bulk modulus}$$
$$\nu = (E/2G) - 1 \qquad \text{Poisson's ratio}$$

Elastic constants for selected materials are provided in Table 3.1-2. For a given material substantial variations in elastic behavior exist, and typical bounds are shown for some of the materials listed

TABLE 3.1-3 TYPICAL PROPERTIES OF STRUCTURAL MATERIALS AT 70° F

Material	Form and Condition	Yield Strength 0.2% Offset (1000 psi)	Tensile Strength (1000 psi)	Elongation in 2 in.(%)	Density (lb/in.^{3})	Coefficient of Expansion (in./in. °F $\times 10^{-6}$)	Modulus of Elasticity (psi $\times 10^{-6}$)
Carbon Steels							
SAE 1020	Annealed	40	60	35	0.284	6.7	30
	W-200° F^a	80	104	6	0.284	6.7	30
	W-1000° F	62	90	22	0.284	6.7	30
SAE 1045	W-600° F	114	150	8	0.284	6.7	30
	W-1000° F	89	120	19	0.284	6.7	30
SAE 4340	0-400° F^b	225	290	10	0.284	6.7	30
	0-1000° F	160	185	15	0.284	6.7	30
Stainless Steel							
Type 304	Annealed	30	85	50	0.286	9.6	29
Type 410	Heat-treated	115	150	15	0.277	6.1	28
17-7 PH	TH-1050	182	193	10	0.281	6.0	29
Nonferrous Metals							
Aluminum 1100-0	Annealed	5	13	45	0.098	12.9	10
Aluminum 2017-T4	Heat-treated	40	62	22	0.101	12.9	10
Beryllium copper	Heat-treated	130	175	5	0.297	9.2	18
Naval brass	Annealed	22	56	40	0.304	11.2	15
Admirality brass	Annealed	20	55	65	0.308	10.2	15
Nickel 200	Annealed	25	65	45	0.321	7.4	30
Magnesium AZ80A-T5	Heat-treated	40	55	7	0.064	14.0	6
Titanium 8-1-1	Heat-treated	150	160	15	0.170	4.7	17
Wood (W-Grain, 12% Moisture)							
Birch		8.3^c	2.0^d	—	0.026	1.1	2.1
White Oak		7.0^c	1.9^d	—	0.028	2.7	1.6

^{a} Water quench, tempered at 200° F.
^{b} Oil quench, tempered at 400° F.
^{c} Compressive strength.
^{d} Shear strength.

TABLE 3.1-4 COMPARATIVE PROPERTIES AT 70° F

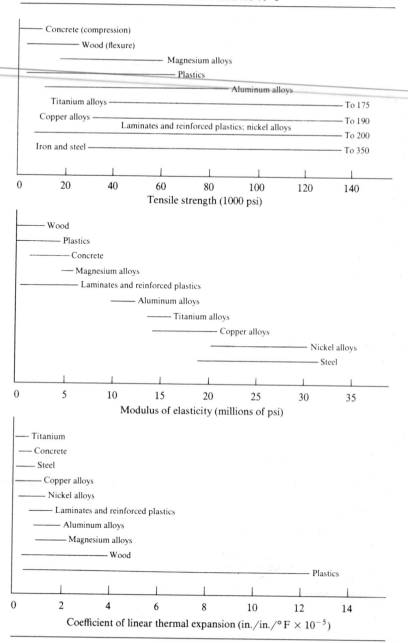

Concrete (compression)

Wood (flexure)

Magnesium alloys

Plastics

Aluminum alloys

Titanium alloys ————————————————————————— To 175

Copper alloys ———— ———————————————————— To 190

Laminates and reinforced plastics; nickel alloys

To 200

Iron and steel ————————————————————————— To 350

| 0 | 20 | 40 | 60 | 80 | 100 | 120 | 140 |

Tensile strength (1000 psi)

Wood

Plastics

Concrete

Magnesium alloys

Laminates and reinforced plastics

Aluminum alloys

Titanium alloys

Copper alloys

Nickel alloys

Steel

| 0 | 5 | 10 | 15 | 20 | 25 | 30 | 35 |

Modulus of elasticity (millions of psi)

Titanium

Concrete

Steel

Copper alloys

Nickel alloys

Laminates and reinforced plastics

Aluminum alloys

Magnesium alloys

Wood

Plastics

| 0 | 2 | 4 | 6 | 8 | 10 | 12 | 14 |

Coefficient of linear thermal expansion (in./in./°F $\times$ 10^{-5})

(Tables 3.1-3–3.1-6). Fabrication methods influence elastic behavior, and testing is recommended for critical applications, when samples of the specific material are available. The data in Tables 3.1-2–3.1-6 are generally restricted to static loading of homogeneous, isotropic materials below the proportional limit (see Section 6.2-1). The properties are for room temperatures only.

For materials not within these limitations, Hooke's law is approximated by a more general form whereby stress and strain are related by as many as 36–81 independent coefficients, depending on the

TABLE 3.1-5   ELASTIC CONSTANTS OF METALS[a]

Metal	Modulus of Elasticity (Young's Modulus) (1,000,000 psi)	Modulus of Rigidity (Shearing Modulus) (1,000,000 psi)	Bulk Modulus (1,000,000 psi)	Poisson's Ratio
Cast steel	28.5	11.3	20.2	0.265
Cold-rolled steel	29.5	11.5	23.1	0.287
Stainless steel 18-8	27.6	10.6	23.6	0.305
All other steels, including high-carbon, heat-treated	28.6–30.0	11.0–11.9	22.6–24.0	0.283–0.292
Cast iron	13.5–21.0	5.2–8.2	8.4–15.5	0.211–0.299
Malleable iron	23.6	9.3	17.2	0.271
Copper	15.6	5.8	17.9	0.355
Brass, 70-30	15.9	6.0	15.7	0.331
Cast brass	14.5	5.3	16.8	0.357
Tobin bronze	13.8	5.1	16.3	0.359
Phosphor bronze	15.9	5.9	17.8	0.350
Aluminum alloys, various	9.9–10.3	3.7–3.9	9.9–10.2	0.330–0.334
Monel metal	25.0	9.5	22.5	0.315
Inconel	31	11		
Z-nickel	30	11		
Elektron (magnesium alloy)	6.3	2.5	4.8	0.281
Titanium (99.0 Ti), annealed bar	15–16			
Zirconium, crystal bar	11–14			

[a] Mostly from tests of R. W. Vose.

degree of symmetry of the material.[1] Typical examples of materials that do not behave the same in all directions are rolled metals and fiber-reinforced composites. The introduction of fibers causes the elastic modulus in tension to be much higher than in compression.

3.1-5   COMBINED STRESSES

The maximum load a structure can bear may be estimated if the material strength and the state of stress of the structure are known (see Section 3.8). Material properties are readily available.[6,7] Stress components are easily identified if the type of loading is only normal, or only shear (see Sections 3.1-2–3.4-1). But for the general state of stress, Fig. 3.1-1, the most important stresses are not necessarily aligned with the applied load, and so are more difficult to resolve.

Of primary interest are the maximum normal (σ_1) and shear (τ_{max}) stresses because materials tend to fail through tensile or shear mechanisms. The uniaxial case (see Section 3.1-2) revealed that maximum shear and normal stresses at a point act on planes oriented 45° apart. This is also applicable to the general case.

For any combination of loads, there exist at any point three orthogonal planes upon which only normal stresses act. These are the *principal* (normal) *stresses*, designated in declining order of magnitude: σ_1, σ_2, and σ_3. The maximum shear stresses act on planes that are inclined 45° from the σ_1 and σ_3 planes. Orthogonal to the maximum shear planes are the minimum (τ_{min}) and intermediate shear planes, which are of no consequence. Normal stresses may act on these shear planes, but shear stresses do not act on principal normal planes. The principal stresses are found by diagonalizing the general stress tensor, Equation (3.1-1), so that the shear components are zero:

$$T = \begin{bmatrix} \sigma_1 & 0 & 0 \\ 0 & \sigma_2 & 0 \\ 0 & 0 & \sigma_3 \end{bmatrix} \qquad (3.1\text{-}2)$$

Several hand-held computers have solution routines for cubic roots, but a common procedure is based

TABLE 3.1-6 TYPICAL PHYSICAL PROPERTIES OF AND ALLOWABLE STRESSES FOR SOME COMMON STRUCTURAL MATERIALS[a]

Material	Unit Weight (lb/in.³)	Ultimate Strength (ksi)			Yield Strength[g] (ksi)		Allow. Stresses[i], psi		Elastic Moduli (×10⁶ psi)		Coef. of Thermal Expans. (×10⁻⁶ per °F)
		Tens.	Comp.[c]	Shear	Tens.[h]	Shear	Tens. or Comp.	Shear	Tens. or Comp.	Shear	
Aluminum alloy (extruded) 2024-T4	0.100	60	—	32	44	25			10.6	4.00	12.9
6061-T6		38	—	24	35	20			10.0	3.75	13.0
Cast iron Gray	0.276	30	120	—[e]	—	—			13	6	5.8
Malleable		54	—	48	36	24			25	12	6.7
Concrete[b] 8 gal/sack	0.087	—	3	—[e]	—	—	−1,350[j]	66	3	—	6.0
6 gal/sack		—	5	—	—	—	−2,250[j]	86	5	—	
Magnesium alloy, AM100A	0.065	40	—	21	22	—			6.5	2.4	14.0
Steel 0.2% Carbon (hot-rolled)	0.283	65	—	48	36	24	±24,000	14,500	30[k]	12	6.5
0.6% Carbon (hot-rolled)		100	—	80	60	36					
0.6% Carbon (quenched)		120	—	100	75	45					
3½% Ni, 0.4% C		200	—	150	150	90					
Wood Douglas fir (coast)	0.018	—	7.4[d]	1.1[f]	—	—	±1,900[j]	120[j]	1.76	—	—
Southern pine (longleaf)	0.021	—	8.4[d]	1.5[f]	—	—	±2,250[j]	135[j]	1.76	—	—

[a] Mechanical properties of metals depend not only on composition but also on heat treatment, previous cold working, etc. Data for wood are for clear 2-in.-by-2-in. specimens at 12% moisture content. True values vary.

[b] 8 gal/sack means 8 gallons of water per 94-lb sack of Portland cement. Values for 28-day-old concrete.

[c] For short blocks only. For ductile materials the ultimate strength in compression is indefinite; may be assumed to be the same as that in tension.

[d] Compression parallel to grain on short blocks. Compression perpendicular to grain at proportional limit 950 psi, 1,190 psi, respectively. Values from Wood Handbook, U.S. Dept. of Agriculture.

[e] Fails in diagonal tension.

[f] Parallel to grain.

[g] For most materials at 0.2% set.

[h] For ductile materials compressive yield strength may be assumed the same.

[i] For ductile materials much lower stresses required in machine design because of fatigue properties and dynamic loadings.

[j] In bending only. No tensile stress is allowed in concrete. Timber stresses are for select or dense grade.

[k] AISC recommends the value of 29 × 10⁶ psi.

on the solution of the determinant:

$$\begin{bmatrix} (\sigma_x - \sigma) & \tau_{xy} & \tau_{xz} \\ \tau_{xy} & (\sigma_y - \sigma) & \tau_{yz} \\ \tau_{xz} & \tau_{yz} & (\sigma_z - \sigma) \end{bmatrix} = 0 \qquad (3.1\text{-}3)$$

which expands to

$$\sigma^3 - B\sigma^2 + C\sigma - D = 0 \qquad (3.1\text{-}4)$$

where

$$B = \sigma_x + \sigma_y + \sigma_z$$

$$C = F + H + J - M - N - P$$

$$D = (F - N)\sigma_z - N\sigma_x - P\sigma_y + 2Q$$

and

$$F = \sigma_x\sigma_y \qquad M = \tau_{xy}^2$$

$$H = \sigma_y\sigma_z \qquad N = \tau_{yz}^2$$

$$J = \sigma_x\sigma_z \qquad P = \tau_{xz}^2$$

$$Q = \tau_{xy}\tau_{yz}\tau_{xz}$$

Again, the roots of Equation (3.1-4) are the *principal normal stresses* (σ_1, σ_2, and σ_3). Texts are available that have programmed solutions to cubic equations, such as the Newton–Raphson, successive substitution, and interval-halving methods.[8,9]

The *maximum shear stress* (τ_{max}) will be the greatest calculated from the principal shears: $\tau_1 = (\sigma_1 - \sigma_2)/2$, $\tau_2 = (\sigma_2 - \sigma_3)/2$, and $\tau_3 = (\sigma_3 - \sigma_1)/2$. If τ_1 does not have the greatest magnitude, the shear subscripts are reassigned so that $\tau_1 > \tau_2 > \tau_3$. The sign convention is defined in Section 3.1-1. For example, a compressive normal stress is negative and should carry its sign into the above equations. The *direction cosines* for the principal planes may be found by substituting σ_1, σ_2, and σ_3 for σ of Equation (3.1-4) and solving for l, m, and n, with the condition that $l^2 + m^2 + n^2 = 1$.[10]

$$l(\sigma_x - \sigma) + m\tau_{xy} + n\tau_{xz} = 0$$

$$l\tau_{xy} + m(\sigma_y - \sigma) + n\tau_{yz} = 0 \qquad (3.1\text{-}5)$$

$$l\tau_{xz} + m\tau_{yz} + n(\sigma_z - \sigma) = 0$$

Although the above analytical method is preferred to graphical techniques, *Mohr's circles* provide a simple aid to visualize stress relations, once principal stresses are known. The method of construction for the three-dimensional case is shown in Fig. 3.1-11. Observe that negative values are less than zero or positive values, regardless of the absolute values of magnitudes. The radius of the largest circle is τ_{max} (also called $\tau_{max\text{-}max}$ because lesser τ_{max} values exist for the minimum shear and intermediate shear planes). Table 3.1-7 shows selected Mohr's circles for common loading conditions. The state of stress at a point may only exist within the region inside the largest circle, and outside the smaller circles. When no shear stress acts on a plane, the circle representing shear stress on that plane is reduced to a point, located on the abscissa and having magnitude equal to the principal stress. When one of the principal stresses has zero magnitude, the condition is then two-dimensional, or *plane stress*, and at least one circle intersects the origin. The principal normal stresses and maximum shear stress for the plane stress condition are found from components within the *xy* plane:

$$\sigma_1 = \frac{\sigma_x + \sigma_y}{2} + \left[\left(\frac{\sigma_x - \sigma_y}{2} \right)^2 + \tau_{xy}^2 \right]^{1/2}$$

$$\sigma_2 = \frac{\sigma_x + \sigma_y}{2} - \left[\left(\frac{\sigma_x - \sigma_y}{2} \right)^2 + \tau_{xy}^2 \right]^{1/2}$$

$$\sigma_3 = 0$$

$$\tau_{max} = \left[\left(\frac{\sigma_x - \sigma_y}{2} \right)^2 + \tau_{xy}^2 \right]^{1/2} = \frac{\sigma_1 - \sigma_2}{2}$$

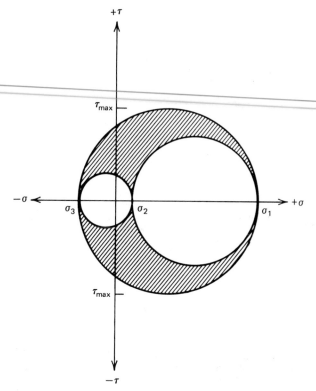

Fig. 3.1-11 Construction of three-dimensional Mohr's circle for stress.

The planes on which the maximum (in-plane) shear acts are 45° from the principal planes. The inclination of the principal normal stresses is found from

$$\theta = \tfrac{1}{2} \tan^{-1} \left(\frac{2\tau_{xy}}{\sigma_x - \sigma_y} \right)$$

The Mohr's circle is constructed as shown in Fig. 3.1-12, where the radius is τ_{max} and the center of the circle is located at $\sigma = (\sigma_1 + \sigma_2)/2$. Out-of-plane stresses are not shown. (The biaxial equation for τ_{max} and the two-dimensional Mohr's circle do not account for $\tau_{max\text{-}max}$, which does not act on the xy plane. It must be calculated from the triaxial equation for τ_1.)

An example of combined torsion and bending stresses is provided in Section 3.4-5.

A summary of stress–strain relations is shown in Table 3.1-8, for isotropic materials.

3.1-6 Thermal Effects

A temperature change (ΔT) causes the rod of Fig. 3.1-13 to expand or contract (ΔL) an amount which is often assumed to be linearly proportional to ΔT:

$$\Delta L = \alpha L \Delta T = \delta$$

where L is the initial dimension, such as length, and α is the *coefficient of thermal expansion*. This assumption is accurate for common structural materials, near room temperature, and for small temperature changes so that the material does not undergo a phase change, as is common for certain metals. Nitinol (a nickel and titanium alloy) is a noteworthy alloy which displays a phenomenal plastic memory during phase change.

The dimensional change will not occur if the rod ends are restrained, as in Fig. 3.1-14. Thermal stresses then develop and are for a uniform rod

$$\sigma = E \alpha \Delta T$$

Values of α are listed for selected materials in Table 3.1-9.[6,7]

TABLE 3.1-7 MOHR'S CIRCLE FOR SELECTED LOADS

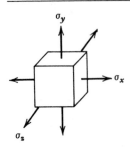

(A) Equal Triaxial Tension
$(\sigma_x = \sigma_y = \sigma_z = \sigma_1 = \sigma_2 = \sigma_3)$

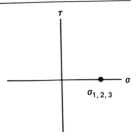

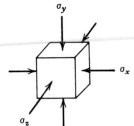

(B) Equal Triaxial Compression
$(\sigma_x = \sigma_y = \sigma_z = \sigma_1 = \sigma_2 = \sigma_3)$

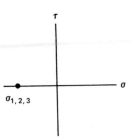

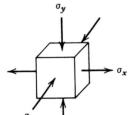

(C) Equal Compression (2 planes)
with Lateral Tension
$(\sigma_x = -\sigma_y = -\sigma_z = \sigma_1 = -\sigma_2 = -\sigma_3)$

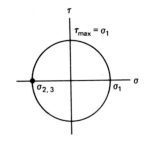

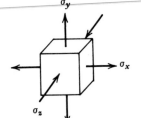

(D) Equal Tension (2 planes)
with Lateral Compression
$(\sigma_x = \sigma_y = -\sigma_z = \sigma_1 = -\sigma_2 = -\sigma_3)$

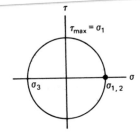

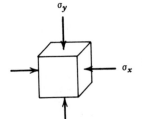

(E) Equal Biaxial Compression
$(\sigma_x = \sigma_y = \sigma_1 = \sigma_2)$

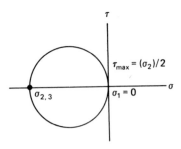

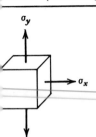

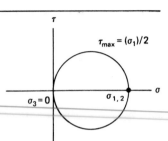

(F) Equal Biaxial Tension
$(\sigma_x = \sigma_y = \sigma_1 = \sigma_2)$

$\tau_{max} = (\sigma_1)/2$

$\sigma_3 = 0$ $\sigma_{1,2}$

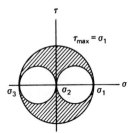

(G) Pure Shear
$(\tau_{xy} = \tau_{yx})$; $(\sigma_2 = 0)$

$\tau_{max} = \sigma_1$

σ_3 σ_2 σ_1

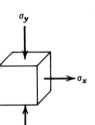

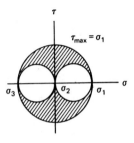

(H) Pure Shear $(\sigma_x = -\sigma_y)$
$(\sigma_x = -\sigma_y = \sigma_1 = -\sigma_2)$; $(\sigma_2 = 0)$

$\tau_{max} = \sigma_1$

σ_3 σ_2 σ_1

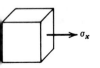

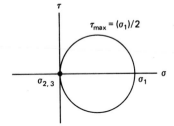

(I) Uniaxial Tension
$(\sigma_x = \sigma_1)$; $(\sigma_2 = \sigma_3 = 0)$

$\tau_{max} = (\sigma_1)/2$

$\sigma_{2,3}$ σ_1

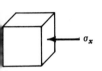

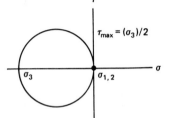

(J) Uniaxial Compression
$(\sigma_x = \sigma_3)$; $(\sigma_1 = \sigma_2 = 0)$

$\tau_{max} = (\sigma_3)/2$

σ_3 $\sigma_{1,2}$

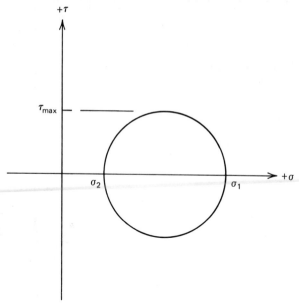

Fig. 3.1-12 Construction of two-dimensional Mohr's circle for stress. (Note: $\sigma_3 = 0$ and $\tau_{\text{max}-\text{max}}$ is not shown.)

TABLE 3.1-8 GENERALIZED STRESS–STRAIN RELATIONS

Triaxial Stress

$$\varepsilon_x = \frac{\sigma_x - \nu(\sigma_y + \sigma_z)}{E}$$

$$\varepsilon_y = \frac{\sigma_y - \nu(\sigma_x + \sigma_z)}{E}$$

$$\varepsilon_z = \frac{\sigma_z - \nu(\sigma_x + \sigma_y)}{E}$$

$$\sigma_x = \frac{E\left[(1 - \nu)\varepsilon_x + \nu(\varepsilon_y + \varepsilon_z)\right]}{(1 + \nu)(1 - 2\nu)}$$

$$\sigma_y = \frac{E\left[(1 - \nu)\varepsilon_y + \nu(\varepsilon_x + \varepsilon_z)\right]}{(1 + \nu)(1 - 2\nu)}$$

$$\sigma_z = \frac{E\left[(1 - \nu)\varepsilon_z + \nu(\varepsilon_x + \varepsilon_y)\right]}{(1 + \nu)(1 - 2\nu)}$$

$$\tau_{xy} = G\gamma_{xy} = \frac{E\gamma_{xy}}{2(1 + \nu)}$$

$$\tau_{yz} = G\gamma_{yz} = \frac{E\gamma_{yz}}{2(1 + \nu)}$$

$$\tau_{xz} = G\gamma_{xz} = \frac{E\gamma_{xz}}{2(1 + \nu)}$$

TABLE 3.1-8   (*Continued*)

Plane Stress ($\sigma_z = \tau_{xz} = \tau_{yz} = 0$)

$$\varepsilon_x = \frac{\sigma_x - \nu\sigma_y}{E}$$

$$\varepsilon_y = \frac{\sigma_y - \nu\sigma_x}{E}$$

$$\varepsilon_z = \frac{-\nu(\sigma_x + \sigma_y)}{E}$$

$$\sigma_x = \frac{E(\varepsilon_x + \nu\varepsilon_y)}{1 - \nu^2}$$

$$\sigma_y = \frac{E(\varepsilon_y + \nu\varepsilon_x)}{1 - \nu^2}$$

$$\tau_{xy} = \tau_{yx} = G\gamma_{xy} = G\gamma_{yx}$$

Plane Strain ($\varepsilon_z = \gamma_{xz} = \gamma_{yz} = 0$)

$$\varepsilon_x = \frac{\left[\sigma_x(1 - \nu) - \nu\sigma_y\right](1 + \nu)}{E}$$

$$\varepsilon_y = \frac{\left[\sigma_y(1 - \nu) - \nu\sigma_x\right](1 + \nu)}{E}$$

$$\sigma_x = \frac{E\left[(1 - \nu)\varepsilon_x + \nu\varepsilon_y\right]}{(1 + \nu)(1 - 2\nu)}$$

$$\sigma_y = \frac{E\left[(1 - \nu)\varepsilon_y + \nu\varepsilon_x\right]}{(1 + \nu)(1 - 2\nu)}$$

$$\sigma_z = \nu(\sigma_x + \sigma_y)$$

$$\tau_{xy} = \tau_{yx} = G\gamma_{xy} = G\gamma_{yx}$$

Uniaxial Stress ($\sigma_y = \sigma_z = 0$)

$$\varepsilon_x = \frac{\sigma_x}{E}$$

$$\varepsilon_y = \frac{-\nu\sigma_x}{E}$$

$$\varepsilon_z = \frac{-\nu\sigma_x}{E}$$

$$\sigma_x = E\varepsilon_x = \frac{-E\varepsilon_y}{\nu} = \frac{-E\varepsilon_z}{\nu}$$

Direct (Pure) Shear

$$\gamma = \frac{\tau}{G}$$

$$\tau = \gamma G$$

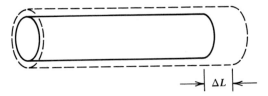

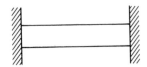

Fig. 3.1-13   Thermal strain.

Fig. 3.1-14   Thermal stress for fixed boundaries.

TABLE 3.1-9   COEFFICIENTS OF THERMAL EXPANSION[a]

Material	$\alpha\ (10^{-6}\ \text{in./in.})$
Beryllium (68° F)	6.4–9.
Chromium (68° F)	3.4
Gold (68° F)	7.9
Platinum (68° F)	4.9
Silver (32–212° F)	10.9
Tungsten (70° F)	2.5
Uranium (depleted) (70° F)	7.7
Aluminum and alloys (68–212° F)	10.8–13.7
Brasses, wrought (68–572° F)	9. –12.
Bronzes, wrought (68–572° F)	10. –11.8
Cobalt and alloys (70–1800° F)	6.8– 9.9
Copper and alloys (68–572° F)	9. –12.
Iron and alloys (70–212° F)	6. –10.8
Lead and alloys (68–212° F)	16. –16.3
Magnesium and alloys	13. –14.
Titanium and alloys (68–1000° F)	4.5– 6.
Beryllium carbide (77–1472° F)	5.8
Boron carbide (0–2250° F)	1.7
Tantalum carbide (77–1472° F)	4.6
Tungsten carbide	2.5–4.1
Alumina (77–1830° F)	0.4–4.5
Beryllia (68–2550° F)	5.8
Silica (68–2280° F)	0.28
Carbon, graphite	1. –9.4

[a] From References 6 and 7.

3.1-7   Symbols Used

Symbol	Meaning
A	Cross-sectional area of structural member
E	Modulus of elasticity, Young's modulus
e	Eccentricity of load to centroid of structure
G	Modulus of rigidity, shear modulus
K	Bulk modulus
$[K]$	Stiffness matrix
L	Initial length (of structural member)
dl	Differential length

(*Continued*)

Symbol	Meaning
l, m, n	Direction cosines
P	Externally applied force, load
T	Stress tensor
ΔT	Change in temperature
x, y, z	Coordinate axes
α	Coefficient of thermal expansion, or measurement angle (alpha)
γ	Shear strain (gamma)
δ	Normal deflection, change in length (delta)
ε	Normal strain (epsilon)
θ	Angle of rotation (theta)
ν, μ	Poisson's ratio (nu, mu)
σ	Normal stress (sigma)
τ	Shear stress (tau)

REFERENCES

3.1-1 A. P. Boresi et al., *Advanced Mechanics of Materials*, 3rd ed. Wiley, New York, 1978.

3.1-2 A. P. Boresi and P. P. Lynn, *Elasticity in Engineering Mechanics*, 2nd ed. Prentice-Hall, Englewood Cliffs, NJ, 1974.

3.1-3 J. W. McKinley, *Fundamentals of Stress Analysis*. Matrix Publishers, Portland, OR, 1979.

3.1-4 A. Blake, *Practical Stress Analysis in Engineering Design*. Marcel Dekker, New York, 1982.

3.1-5 A. Higdon et al., *Mechanics of Materials*. Wiley, New York, 1976.

3.1-6 J. C. Bittence (Ed.), "1983 Materials Selector," *Materials Engineering*, Vol. 96, No. 6, December 1982.

3.1-7 *Metallic Materials and Elements for Aerospace Vehicle Structures*. Department of Defense, Washington, D.C., 1978.

3.1-8 B. Carnahan et al., *Applied Numerical Methods*. Wiley, New York, 1969.

3.1-9 A. Ralston and H. S. Wilf, *Mathematical Methods for Digital Computers*, Vol. 2, Wiley, New York, 1967.

3.1-10 R. G. Budynas, *Advanced Strength and Applied Stress Analysis*, McGraw-Hill, New York, 1977.

3.1-11 J. H. Faupel and F. E. Fisher, *Engineering Design*, 2nd ed. Wiley, New York, 1981.

3.1-12 T. Baumeister and L. S. Marks, *Mechanical Engineer's Handbook*. McGraw-Hill, New York, 1958.

3.1-13 E. P. Popov, S. Nagarajan, and Z. A. Lu, *Mechanics of Materials*, 2nd ed. Prentice-Hall, New Jersey, 1976.

3.2 CENTROID, MOMENT OF INERTIA, AND PRODUCT OF INERTIA

3.2-1 Introduction

The area properties discussed in this section apply to area centroid, moments, and product of inertia and torsion factors. Figure 3.2-1 shows two areas and a set of reference axes for each area. The area of Fig. 3.2-1 can be described as a "continuous" shape. The composite shape is made up of two or more standard shapes. Standard shapes include rectangles, triangles, circles, or other shapes for which area properties are commonly tabulated.

For the area of Fig. 3.2-1*a* the centroid is defined as

$$\bar{x} = \frac{\int_A x \, dA}{\int_A dA}$$

$$\bar{y} = \frac{\int_A y \, dA}{\int_A dA}$$

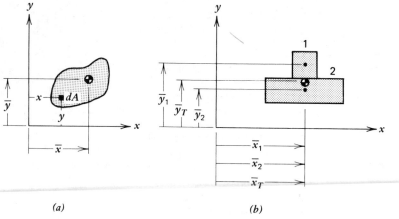

(a)   (b)

Fig. 3.2-1   General and composite areas and coordinates.

These definitions are commonly used to find the centroids of the standard shapes. The *moment of inertia* of this area is defined to be

$$I_x = \int_A y^2 \, dA$$

about the x axis and

$$I_y = \int_A x^2 \, dA$$

about the y axis.

The *product of inertia is defined* as

$$I_{xy} = \int_A xy \, dA$$

For the areas that are composites of standard shapes, such as the one shown in Fig. 3.2-1b, we assume that the centroid moment and product of inertia of the standard shapes can be looked up in tables. Then the total properties are found in terms of component properties.

For the centroid we have

$$\bar{x}_T = \frac{A_1 \bar{x}_1 + A_2 \bar{x}_2}{A_1 + A_2}$$

$$\bar{y}_T = \frac{A_1 \bar{y}_1 + A_2 \bar{y}_2}{A_1 + A_2}$$

The above equations apply to Fig. 3.2-1b, and other figures with two component areas. In general, these two equations become

$$\bar{x}_T = \frac{\sum\limits_{i=1}^{n} A_i \bar{x}_i}{\sum\limits_{i=1}^{n} A_i} \qquad \bar{y}_T = \frac{\sum\limits_{i=1}^{n} A_i \bar{y}_i}{\sum\limits_{i=1}^{n} A_i}$$

To calculate the moments and products of inertia of composite shapes, it is usually expedient to use the parallel axis, or transfer of axis formula. This formula relates the inertia properties of the areas about their *own centroidal axis* to some other axis that is parallel to the centroidal axis but is at some distance away. Refer to Fig. 3.2-2.

The formulas are

$$I_x = I_{\bar{x}-\bar{x}} + A d_y^2$$

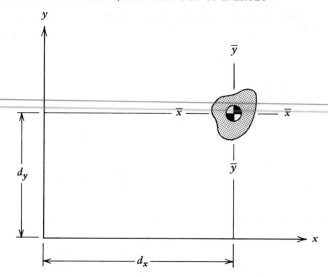

Fig. 3.2-2 Geometry for transformation of axes.

(inertia any || axis = inertia own centroid + area × transfer distance squared)

$$I_y = I_{\bar{y}-\bar{y}} + Ad_x^2$$

and

$$I_{xy} = I_{\overline{xy}} + Ad_x d_y$$

Note that the signs of the terms d_x and d_y are significant; they are plus as shown in Fig. 3.2-2. The statement applies to the formula for I_{xy}.

Another inertia relationship is the effect of a rotation of axes on the moments and product of inertia. In this case we are considering the variation of inertias for a given body with given inertia properties and centroid; for a given body with given inertia properties and centroid the *only* variable is the orientation of the axes with respect to an angle θ as shown in Fig. 3.2-3.

It is emphasized that the angle θ is the only variable. This is the angle of rotation of the centroidal reference axes and expresses the change of reference axes from the xy system to the $x'y'$ system.

As the angle θ varies, we can make several observations:

1. There will be a value of θ for which $I_{x'y'}$ is zero. At this value of θ the *principal axes* of inertia are defined. At the principal axis orientation, the values of $I_{x'}$ and $I_{y'}$ will be maximum and minimum.

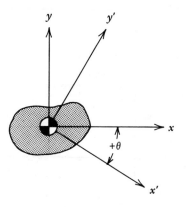

Fig. 3.2-3 Geometry for rotation of axes.

2. The value of θ at which the principal axes occur is given in terms of the known initial xy axis properties by

$$\tan 2\theta = \frac{2I_{xy}}{I_x - I_y}$$

3. The principal moments of inertia are given by

$$I_{x'} = I_{max} = \frac{I_x + I_y}{2} + \sqrt{\left(\frac{I_x - I_y}{2}\right)^2 + I_{xy}^2}$$

$$I_{y'} = I_{min} = \frac{I_x + I_y}{2} - \sqrt{\left(\frac{I_x - I_y}{2}\right)^2 + I_{xy}^2}$$

when the principal angle θ is known, it can be shown that

$$I_{x'} = \frac{I_x + I_y}{2} + \left(\frac{I_x - I_y}{2}\right)\cos 2\theta + I_{xy}\sin 2\theta$$

$$I_{y'} = \frac{I_x + I_y}{2} - \left(\frac{I_x - I_y}{2}\right)\cos 2\theta - I_{xy}\sin 2\theta$$

The following observations are frequently helpful:

1. An axis for which the product of inertia is zero is a principal axis.
2. If an area has an axis of symmetry, the symmetry axis will be a principal axis and an axis of zero product of inertia.

Example 3.2-1. Find the centroid of the following cross section of a beam.

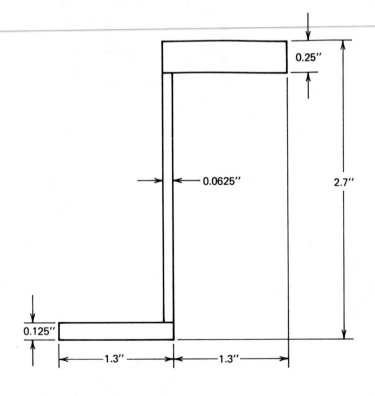

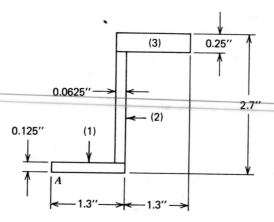

Solution. Start at *A* for the origin. Use the right-hand coordinate system axis.

Member	A (in.^{2})	$\bar{X}$ (in.)	$\bar{Y}$ (in.)	$A \cdot \bar{X}$	$A\bar{Y}$
1	0.1625	0.65	0.0625	0.106	0.0102
2	0.145	1.269	1.288	0.184	0.187
3	0.341	1.919	2.575	0.654	0.877
Total	0.648			0.944	1.074

$$\bar{X} = \frac{\sum Ax}{\sum A} = \frac{0.913}{0.6325} = \bar{X} = 1.457 \text{ in.}$$

$$\bar{Y} = \frac{\sum Ay}{\sum A} = \frac{1.032}{0.6325} = \bar{Y} = 1.657 \text{ in.}$$

Example 3.2-2. Compute I_{yy} and I_{xx}, the angle of inclination of principal axis, I_{max}, and I_{min}.

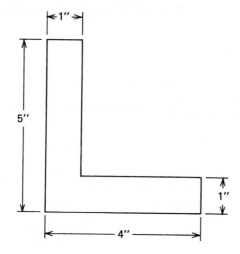

Solution.

1. The centroid was computed and is shown in the diagram.

2. Using a right-hand coordinate system axis at the centroid, one computes the following: For Part A,

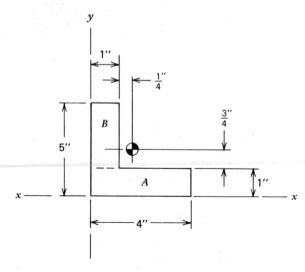

$$I_{xy_A} = \bar{I}_{xy} + dy\,dx\,A = 0 + \left(-\tfrac{5}{4}\right)\left(\tfrac{3}{4}\right)(4)$$

$$= -3.75 \text{ in.}^4$$

For Part B,

$$I_{xy_B} = \bar{I}_{xy} + dy\,dx\,A = 0 + \left(\tfrac{5}{4}\right)\left(-\tfrac{3}{4}\right)(4)$$

$$= -3.75 \text{ in.}^4$$

$$I_{xy_{total}} = I_{xy_A} + I_{xy_B} = -3.75 - 3.75$$

$$= -7.50 \text{ in.}^4$$

3. $I = \bar{I} + Ad^2$. For Part A,

$$I_x = \left(\tfrac{1}{12}\right)(4)(1^3) + \left(\tfrac{5}{4}\right)^2 4 = 6.583 \text{ in.}^4$$

$$I_y = \left(\tfrac{1}{12}\right)(1)(4^3) + \left(\tfrac{3}{4}\right)^2 4 = 7.583 \text{ in.}^4$$

For Part B,

$$I_x = 11.583 \text{ in.}^4$$

$$I_y = 2.583 \text{ in.}^4$$

For the entire structure,

$$I_{xx} = 6.583 + 11.583 = 18.17 \text{ in.}^4$$

$$I_{yy} = 7.583 + 2.583 = 10.167 \text{ in.}^4$$

4. This is with respect to the centroidal axis. Angle of inclination is given by

$$\tan 2\theta = \frac{2I_{xy}}{I_x - I_y} = \frac{-2(7.50)}{10.167 - 18.167} = 1.875$$

$$2\theta = 61°56'$$

$$\theta = 30°58'$$

TABLE 3.2-1 AREA PROPERTIES

Section	Moment of inertia	Section modulus	Radius of gyration
$d_m = \frac{1}{2}(D + d)$ $s = \frac{1}{2}(D - d)$	$I = \frac{\pi}{64}(D^4 - d^4)$ $= \frac{\pi}{4}(R^4 - r^4)$ $= \frac{1}{4}A(R^2 + r^2)$ $= 0.05(D^4 - d^4)$ (approx)	$\frac{I}{c} = \frac{\pi}{32}\frac{D^4 - d^4}{D}$ $= \frac{\pi}{4}\frac{R^4 - r^4}{R}$ $= 0.8d_m^2 s$ (approx) when $\frac{s}{d_m}$ is very small	$\frac{\sqrt{R^2 + r^2}}{2} = \frac{\sqrt{D^2 + d^2}}{4}$
	$I = r^4\left(\frac{\pi}{8} - \frac{8}{9\pi}\right)$ $= 0.1098r^4$	$\frac{I}{c_2} = 0.1908r^3$ $\frac{I}{c_1} = 0.2587r^3$ $c_1 = 0.4244r$	$\frac{\sqrt{9\pi^2 - 64}}{6\pi}r = 0.264r$
	$I = 0.1098(R^4 - r^4)$ $- \frac{0.283R^2r^2(R - r)}{R + r}$ $= 0.3tr_1^3$ (approx) when $\frac{t}{r_1}$ is very small	$c_1 = \frac{4}{3\pi}\frac{R^2 + Rr + r^2}{R + r}$ $c_2 = R - c_1$	$\sqrt{\frac{2I}{\pi(R^2 - r^2)}}$ $= 0.31r_1$ (approx)
	$I = \frac{\pi a^3 b}{4} = 0.7854a^3b$	$\frac{I}{c} = \frac{\pi a^2 b}{4} = 0.7854a^2b$	$\frac{a}{2}$
	$I = \frac{\pi}{4}(a^3b - a_1^3b_1)$ $= \frac{\pi}{4}a^2(a + 3b)t$ (approx)	$\frac{I}{c} = \frac{\pi}{4}a(a + 3b)t$ (approx)	$\sqrt{\frac{I}{(\pi ab - a_1b_1)}} =$ $\frac{a}{2}\sqrt{\frac{a + 3b}{a + b}}$ (approx)
	$I = \frac{1}{12}\left[\frac{3\pi}{16}d^4 + b(h^3 - d^3) + b^3(h - d)\right]$ $\frac{I}{c} = \frac{1}{6h}\left[\frac{3\pi}{16}d^4 + b(h^3 + d^3) + b^3(h - d)\right]$		$\sqrt{\frac{I}{\pi\frac{d^2}{4} + 2b(h - d)}}$ (approx)
	$I = \frac{t}{4}\left(\frac{\pi B^3}{16} + B^2h + \frac{\pi Bh^2}{2} + \frac{2}{3}h^3\right)$ $h = H - \frac{1}{2}B$ $\frac{I}{c} = \frac{2I}{H + t}$		$\sqrt{\frac{I}{2\left(\frac{\pi B}{4} + h\right)t}}$

TABLE 3.2-1   (*Continued*)

$(I$ = moment of inertia; I/c = section modulus; $r = \sqrt{I/A}$ = radius of gyration$)$

Section	Moment of inertia	Section modulus	Radius of gyration
 $I = \dfrac{bh^3}{12}$ $\dfrac{I}{c} = \dfrac{bh^2}{6}$ $r = \dfrac{h}{\sqrt{12}} = 0.289h$	 $\dfrac{bh^3}{3}$ $\dfrac{bh^2}{3}$ $\dfrac{h}{\sqrt{3}} = 0.577h$	 $\dfrac{b^3h^3}{6(b^2+h^2)}$ $\dfrac{b^2h^2}{6\sqrt{b^2+h^2}}$ $\dfrac{bh}{\sqrt{6(b^2+h^2)}}$	 $\dfrac{bh}{12}\,(h^2\cos^2 a + b^2\sin^2 a)$ $\dfrac{bh}{6}\left(\dfrac{h^2\cos^2 a + b^2\sin^2 a}{h\cos a + b\sin a}\right)$ $\sqrt{\dfrac{h^2\cos^2 a + b^2\sin^2 a}{12}}$
 $I = \dfrac{b}{12}\,(H^3 - h^3)$ $\dfrac{I}{c} = \dfrac{b}{6}\,\dfrac{H^3 - h^3}{H}$ $r = \sqrt{\dfrac{H^3 - h^3}{12(H - h)}}$	 $\dfrac{H^4 - h^4}{12}$ $\dfrac{1}{6}\,\dfrac{H^4 - h^4}{H}$ $\sqrt{\dfrac{H^2 + h^2}{12}}$	 $\dfrac{H^4 - h^4}{12}$ $\dfrac{\sqrt{2}}{12}\,\dfrac{H^4 - h^4}{H}$ $\sqrt{\dfrac{H^2 + h^2}{12}}$	 $\dfrac{bh^3}{36};\ c = \dfrac{2}{3}h$ $\dfrac{bh^2}{24}$ $\dfrac{h}{\sqrt{18}}$
 $I = \dfrac{bh^3}{12}$ $\dfrac{I}{c} = \dfrac{bh^2}{12}$ $r = \dfrac{h}{\sqrt{6}}$	 $\dfrac{5\sqrt{3}}{16}\,R^4$ $\tfrac{5}{8}R^3$ $\sqrt{\dfrac{5}{24}}\,R$	 $\dfrac{5\sqrt{3}}{16}\,R^4$ $\dfrac{5\sqrt{3}}{16}\,R^3$ $\sqrt{\dfrac{5}{24}}\,R$	 $\dfrac{1 + 2\sqrt{2}}{6}\,R^4$ $0.6906R^3$ $0.475R$

Square, axis same as first rectangle, side = h; $I = h^4/12$; $I/c = h^3/6$; $r = 0.289h$.
Square, diagonal taken as axis: $I = h^4/12$; $I/c = 0.1179h^3$; $r = 0.289h$.

TABLE 3.2-1 (*Continued*)

Section	Moment of inertia	Section modulus	Radius of gyration
Equilateral Polygon A = area, (see p. 1–39) R = rad circumscribed circle r = rad inscribed circle n = no. sides a = length of side Axis as in preceding section of octagon	$I = \dfrac{A}{24}(6R^2 - a^2)$ $= \dfrac{A}{48}(12r^2 + a^2)$ $= \dfrac{AR^2}{4}$ (approx)	$\dfrac{I}{c} = \dfrac{I}{r}$ $= \dfrac{I}{R\cos\dfrac{180°}{n}}$ $= \dfrac{AR}{4}$ (approx)	$\sqrt{\dfrac{6R^2 - a^2}{24}} \approx \dfrac{R}{2}$ $\sqrt{\dfrac{12r^2 + a^2}{48}}$

| | $I = \dfrac{6b^2 + 6bb_1 + b_1^2}{36(2b + b_1)}h^3$ $c = \dfrac{1}{3}\dfrac{3b + 2b_1}{2b + b_1}h$ | $\dfrac{I}{c} = \dfrac{6b^2 + 6bb_1 + b_1^2}{12(3b + 2b_1)}h^2$ | $\dfrac{h\sqrt{12b^2 + 12bb_1 + 2b_1^2}}{6(2b + b_1)}$ |

| | $I = \dfrac{BH^3 + bh^3}{12}$ $\dfrac{I}{c} = \dfrac{BH^3 + bh^3}{6H}$ | | $\sqrt{\dfrac{BH^3 + bh^3}{12(BH + bh)}}$ |

| | $I = \dfrac{BH^3 - bh^3}{12}$ $\dfrac{I}{c} = \dfrac{BH^3 - bh^3}{6H}$ | | $\sqrt{\dfrac{BH^3 - bh^3}{12(BH - bh)}}$ |

| | $I = \tfrac{1}{3}(Bc_1^3 - B_1h^3 + bc_2^3 - b_1h_1^3)$ $c_1 = \dfrac{1}{2}\dfrac{aH^2 + B_1d^2 + b_1d_1(2H - d_1)}{aH + B_1d + b_1d_1}$ | | $\sqrt{\dfrac{I}{(Bd + bd_1) + a(h + h_1)}}$ |

| | $I = \tfrac{1}{3}(Bc_1^3 - bh^3 + ac_2^3)$ $c_1 = \dfrac{1}{2}\dfrac{aH^2 + bd^2}{aH + bd}$ $c_2 = H - c_1$ $r = \sqrt{\dfrac{I}{[Bd + a(H - d)]}}$ | | |

| | $I = \dfrac{\pi d^4}{64} = \dfrac{\pi r^4}{4} = \dfrac{A}{4}r^2$ $= 0.05d^4$ (approx) | $\dfrac{I}{c} = \dfrac{\pi d^3}{32} = \dfrac{\pi r^3}{4} = \dfrac{A}{4}r$ $= 0.1d^3$ (approx) | $\dfrac{r}{2} = \dfrac{d}{4}$ |

TABLE 3.2-1 (*Continued*)

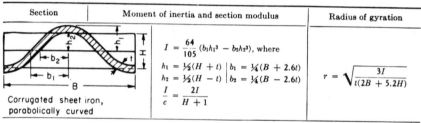

Section	Moment of inertia and section modulus	Radius of gyration
Corrugated sheet iron, parabolically curved	$I = \dfrac{64}{105}(b_1 h_1^3 - b_2 h_2^3)$, where $h_1 = \frac{1}{2}(H + t)$ $b_1 = \frac{1}{4}(B + 2.6t)$ $h_2 = \frac{1}{2}(H - t)$ $b_2 = \frac{1}{4}(B - 2.6t)$ $\dfrac{I}{c} = \dfrac{2I}{H + 1}$	$r = \sqrt{\dfrac{3I}{t(2B + 5.2H)}}$

Approximate values of *least* radius of gyration r

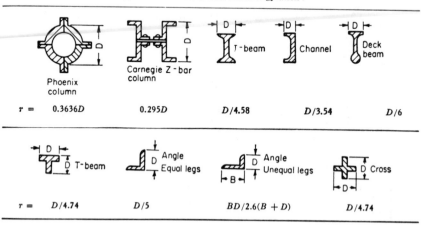

	Phoenix column	Carnegie Z-bar column	T-beam	Channel	Deck beam
$r =$	$0.3636D$	$0.295D$	$D/4.58$	$D/3.54$	$D/6$

	T-beam	Angle Equal legs	Angle Unequal legs	Cross
$r =$	$D/4.74$	$D/5$	$BD/2.6(B + D)$	$D/4.74$

This formula is for the **strength of beams**. For rectangular beams, $M = 1/6 Sbh^2$, where $b =$ breadth, and $h =$ depth; *i.e.*, the elastic **strength of beam sections** varies as follows: (1) for equal width, as the square of the depth; (2) for equal depth, directly as the width; (3) for equal depth and width, directly as the strength of the material; (4) if span varies, then for equal depth, width and material, inversely as the span.

If a beam is cut in halves horizontally, the two halves laid side by side will carry only one-half as much as the original beam.

The term **section modulus** is given to the value of I/c, where c is the distance to the fiber carrying greatest stress. Moment of inertia of cross section $= I$.

Oblique Loading. It should be noted that Table 3.2-1 includes certain cases for which the horizontal axis is not a neutral axis, assuming the common case of vertical loading. The rectangular section with the diagonal as a horizontal axis is such a case. These cases must be handled by the principles of oblique loading.

Every section of a beam has two principal axes passing through the center of gravity, and these two axes are always at right angles to each other. The principal axes are axes with respect to which the moment of inertia is, respectively, a maximum and a minimum, and for which the product of inertia is zero. For symmetrical sections, axes of symmetry are always principal axes. For unsymmetrical sections, like a rolled angle section, the inclination of the principal axis with the X-axis may be found from the formula $\tan 2\theta = 2I_{xy}/(I_y - I_x)$, in which $\theta =$ angle of inclination of the principal axis to the X-axis, $I_{xy} =$ the product of inertia of the section with respect to the X- and Y-axes, $I_y =$ moment of inertia of the section with respect to the Y-axis, $I_x =$ moment of inertia of the section with respect to the X-axis. When this principal axis has been found, the other principal axis is at right angles to it.

TABLE 3.2-1 (*Continued*)

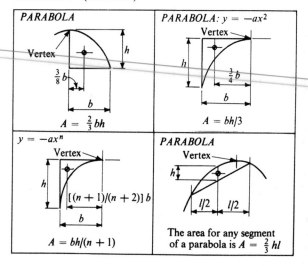

5. I_{max}, I_{min}

$$I_{x'} = \frac{I_x + I_y}{2} + \left(\frac{I_x - I_y}{2}\right)(\cos 2\theta) + I_{xy}\sin 2\theta$$

$$I_{y'} = \frac{I_x + I_y}{2} - \left(\frac{I_x - I_y}{2}\right)(\cos 2\theta) - I_{xy}\sin 2\theta$$

$$I_{x'} = \frac{18.167 + 10.167}{2} + \frac{18.167 - 10.167}{2}(0.4705) + (7.5)(0.8824)$$

$$= 22.67 \text{ in.}^4$$

$$I_{y'} = \frac{18.167 + 10.167}{2} - \frac{18.167 - 10.167}{2}(0.4705) - (7.5)(0.8824)$$

$$= 5.67 \text{ in.}^4$$

Because of the definition of the principal axis, one concludes that

$$I_{x'} = I_{max} = 22.67 \text{ in.}^4$$

$$I_{y'} = I_{min} = 5.67 \text{ in.}^4$$

One could also use the equivalent formulas to obtain

$$I_{x'} = I_x\cos^2\theta + I_y\sin^2\theta - 2I_{xy}\sin\theta\cos\theta$$

$$I_{y'} = I_x\sin^2\theta + I_y\cos^2\theta + 2I_{xy}\sin\theta\cos\theta$$

3.2-2 Tabulation of Area Properties

Area properties are widely tabulated. See References 1–3. Area properties tabulated here include area, centroid, moments of inertia, and torsion factors. The application of the torsion factors is discussed in

TABLE 3.2-2 AREA PROPERTIES FOR TORSIONAL STRESS AND DEFLECTION

General formulas: $\theta = \dfrac{TL}{KG}$, $s = \dfrac{T}{Q}$, where θ = angle of twist (rad); T = twisting moment (in.-lb); L = length (in.); s = unit shear stress (lb. per sq. in.); G = modulus of rigidity (lb. per sq. in.); K (in.^{4}) and Q (in.^{3}) are functions of the cross section.

Form and dimensions of cross sections, other quantities involved, and case number	Formula for K in $\theta = \dfrac{TL}{KG}$	Formula for shear stress
1. Solid circular section	$K = \tfrac{1}{2}\pi r^4$	$\text{Max } s = \dfrac{2T}{\pi r^3}$ at boundary
2. Solid elliptical section	$K = \dfrac{\pi a^3 b^3}{a^2 + b^2}$	$\text{Max } s = \dfrac{2T}{\pi a b^2}$ at ends of minor axis
3. Solid square section	$K = 0.1406 a^4$	$\text{Max } s = \dfrac{T}{0.208 a^3}$ at mid-point of each side
4. Solid rectangular section	$K = a b^3\left[\dfrac{16}{3} - 3.36\dfrac{b}{a}\left(1 - \dfrac{b^4}{12a^4}\right)\right]$	$\text{Max } s = \dfrac{T(3a + 1.8b)}{8a^2 b^2}$ at mid-point of each longer side

5. Solid triangular section (equilateral)

$K = \frac{a^4\sqrt{3}}{80}$

Max $s = \frac{20T}{a^3}$ at mid-point of each side

---

6. Hollow concentric circular section

$K = \frac{1}{2}\pi(r_1^4 - r_0^4)$

Max $s = \frac{2Tr_1}{\pi(r_1^4 - r_0^4)}$ at outer boundary

---

7. Hollow elliptical section, outer and inner boundaries similar ellipses

$q = \frac{a_0}{a} = \frac{b_0}{b}$

$K = \frac{\pi a^3 b^3}{a^2 + b^2}(1 - q^4)$

Max $s = \frac{2T}{\pi ab^2(1 - q^4)}$ at ends of minor axis on outer surface

---

8. Hollow, thin-walled elliptical section of uniform thickness. U = length of median boundary, shown dotted

$U = \pi(a + b - t)\left[1 + 0.27\frac{(a - b)^2}{(a + b)^2}\right]$ approx.

$K = \frac{4\pi^2 t[(a - \frac{1}{2}t)^2(b - \frac{1}{2}t)^2]}{U}$

Average $s = \frac{T}{2\pi t(a - \frac{1}{2}t)(b - \frac{1}{2}t)}$ (stress nearly uniform if t is small)

---

9. Any thin tube of uniform thickness. U = length of median boundary, A = mean of areas enclosed by outer and inner boundaries, or (approx.) area within median boundary

$K = \frac{4A^2t}{U}$

Average $s = \frac{T}{2tA}$ (stress nearly uniform if t is small)

TABLE 3.2-2 (Continued)

Form and dimensions of cross sections, other quantities involved, and case number	Formula for K in $\theta = \dfrac{TL}{KG}$	Formula for shear stress
10. Any thin tube. U and A as for Case 9; t = thickness at any point	$K = \dfrac{4A^2}{\int \dfrac{dU}{t}}$	Average s on any thickness $AB = \dfrac{T}{2tA}$ (Max s where t is a minimum)
11. Hollow rectangle	$K = \dfrac{2t t_1 (a-t)^2 (b-t_1)^2}{at + bt_1 - t^2 - t_1^2}$	Average $s = \dfrac{T}{2t(a-t)(b-t_1)}$ near mid-length of short sides $\quad$ Average $s = \dfrac{T}{2t_1(a-t)(b-t_1)}$ near mid-length of long sides $\quad$ (There will be higher stresses at inner corners unless fillets of fairly large radius are provided)
12. Thin circular open tube of uniform thickness. r = mean radius	$K = \tfrac{2}{3}\pi r t^3$	Max $s = \dfrac{T(6\pi r + 1.8t)}{4\pi^2 r^2 t^2}$, along both edges remote from ends (this assumes t small compared with mean radius; otherwise use formulas given for Cases 14 to 20)
13. Any thin open tube of uniform thickness. U = length of median line, shown dotted	$K = \tfrac{1}{3} U t^3$	Max $s = \dfrac{T(3U + 1.8t)}{U^2 t^2}$, along both edges remote from ends (this assumes t small compared with least radius of curvature of median line; otherwise use formulas given for Cases 14 to 20)

For all solid sections of irregular form (Cases 14 to 20, inclusive) the max shear stress occurs at or very near one of the points where the largest inscribed circle touches the boundary,* and of these, at the one where the curvature of the boundary is algebraically least. (Convexity represents positive, concavity negative, curvature of the boundary.) At a point where the curvature is positive (boundary of section straight or convex) this max stress is given approximately by: $s = G\frac{\theta}{L}c$ or $s = \frac{T}{K}c$ where

$$c = \frac{D}{1 + \frac{\pi^2 D^4}{16 A^2}}\left[1 + 0.15\left(\frac{\pi^2 D^4}{16 A^2} - \frac{D}{2r}\right)\right], \text{ where}$$

D = diameter of largest inscribed circle
r = radius of curvature of boundary at the point (positive for this case)
A = area of the section
At a point where the curvature is negative (boundary or section concave, or reentrant) this max stress is given approximately by $s = G\frac{\theta}{L}c$ or $s = \frac{T}{K}c$

$$\text{where } c = \frac{D}{1 + \frac{\pi^2 D^4}{16 A^2}}\left[1 + \left\{0.118 \log_e\left(1 - \frac{D}{2r}\right) - 0.238\frac{D}{2r}\right\} \tanh \frac{2\phi}{\pi}\right]$$

where D, A, and r have same meaning as before and ϕ = angle through which a tangent to the boundary rotates in turning or traveling around the reentrant portion, measured in radians. (Here r is negative.) The above formulas should also be used for Cases 12 and 13, when t is relatively large compared with radius of median line

* Unless at some other point on boundary there is a sharp reentrant angle, causing high local stress.

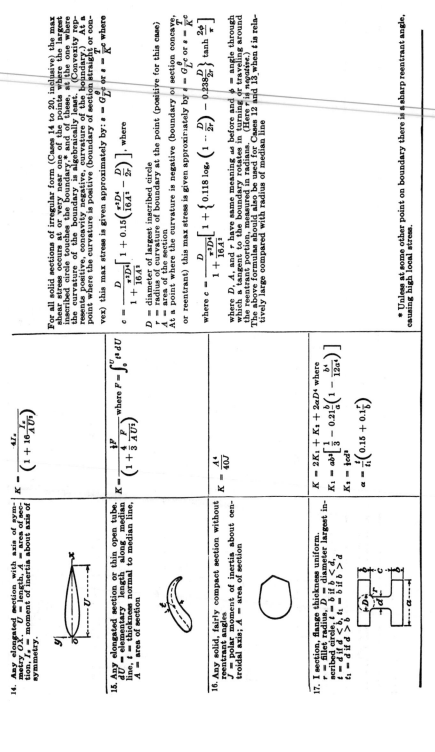

14. Any elongated section with axis of symmetry OX. U = length, A = area of section, I_x = moment of inertia about axis of symmetry.

$$K = \frac{4I_x}{\left(1 + 16\frac{I_x}{AU^2}\right)}$$

15. Any elongated section or thin open tube. dU = elementary length along median line, t = thickness normal to median line, A = area of section

$$K = \frac{\frac{1}{3}F}{\left(1 + \frac{4}{3}\frac{F}{AU^2}\right)} \text{ where } F = \int_0^U t^3\, dU$$

16. Any solid, fairly compact section without reentrant angles.
J = polar moment of inertia about centroidal axis; A = area of section

$$K = \frac{A^4}{40J}$$

17. I section, flange thickness uniform. r = fillet radius, D = diameter largest inscribed circle, t = b if $b < d$, $t = d$ if $d < b$, $t_1 = b$ if $b > d$, $t_1 = d$ if $d > b$

$$K = 2K_1 + K_2 + 2\alpha D^4 \text{ where}$$
$$K_1 = ab^3\left[\frac{1}{3} - 0.21\frac{b}{a}\left(1 - \frac{b^4}{12a^4}\right)\right]$$
$$K_2 = \frac{1}{3}cd^3$$
$$\alpha = \frac{t}{t_1}\left(0.15 + 0.1\frac{r}{b}\right)$$

TABLE 3.2-2 (*Continued*)

Form and dimensions of cross sections, other quantities involved, and case number	Formula for K in $\theta = \dfrac{TL}{KG}$	Formula for shear stress
18. T section, flange thickness uniform; r, D, t and t_1 as for Case 17	$K = K_1 + K_2 + \alpha D^4$ where $K_1 = ab^3\left[\dfrac{1}{3} - 0.21\dfrac{b}{a}\left(1 - \dfrac{b^4}{12a^4}\right)\right]$ $K_2 = cd^3\left[\dfrac{1}{3} - 0.105\dfrac{d}{c}\left(1 - \dfrac{d^4}{192c^4}\right)\right]$ $\alpha = \dfrac{t}{t_1}\left(0.15 + 0.10\dfrac{r}{b}\right)$	
19. L section, r and D as for Cases 17 and 18, $b \geq d$	$K = K_1 + K_2 + \alpha D^4$ where $K_1 = ab^3\left[\dfrac{1}{3} - 0.21\dfrac{b}{a}\left(1 - \dfrac{b^4}{12a^4}\right)\right]$ $K_2 = cd^3\left[\dfrac{1}{3} - 0.105\dfrac{d}{c}\left(1 - \dfrac{d^4}{192c^4}\right)\right]$ $\alpha = \dfrac{d}{b}\left(0.07 + 0.076\dfrac{r}{b}\right)$	
20. U section or Z section	$K = $ sum of K's of constituent L sections, computed as for Case 19	
21. Eccentric hollow circular section	$K = \pi(D^4 - d^4)/32Q$ where $Q = 1 + \left[\dfrac{16n^2}{(1 - n^2)(1 - n^4)}\right]\lambda^2$ $+ \left[\dfrac{384n^4}{(1 - n^2)^2(1 - n^4)^4}\right]\lambda^4$	$\text{Max } S = 16TDF/\pi(D^4 - d^4)$ where $F = 1 + \left[\dfrac{4n^2}{1 - n^2}\right]\lambda + \left[\dfrac{32n^2}{(1 - n^2)(1 - n^4)}\right]\lambda^2$ $+ \left[\dfrac{48n^4(1 + 2n^2 + 3n^4 + 2n^6)}{(1 - n^2)(1 - n^6)}\right]\lambda^3$ $+ \left[\dfrac{64n^2(2 + 12n^2 + 19n^4 + 28n^4 + 18n^6 + 14n^{10} + 3n^{12})}{(1 - n^2)(1 - n^6)(1 - n^8)}\right]\lambda^4$

23. Circular sector

$K = Cr^4$
where C depends on α and has values as follows:

α	45°	60°	90°	120°
C	0.0181	0.0349	0.0825	0.148
α	180°	270°	300°	360°
C	0.296	0.528	0.686	0.878

Max $s = \dfrac{T}{Q}$ on radial boundary. Here $Q = Cr^3$, where C has values as follows:

α	60°	120°	180°
C	0.0712	0.227	0.35

24. Isosceles triangle

$K = \dfrac{a^3b^3}{15a^2 + 20b^2}$
For $\alpha = 90°$, $K = 0.0261c^4$
For $\alpha = 60°$, $K = 0.0216c^4$

$Q = 0.0666ab^2$
For $\alpha = 90°$, $Q = 0.0554c^3$
For $\alpha = 60°$, $Q = 0.050c^3$
Max S_s at center longest side

25. Trapezoid

$K = \tfrac{1}{12}b(m + n)(m^2 + n^2) - V_L m^4 - V_s n^4$
where
$V_L = 0.10504 - 0.10s + 0.0848s^2 - 0.06746s^3 + 0.0515s^4$
$V_s = 0.10504 + 0.10s + 0.0848s^2 + 0.06746s^3 + 0.0515s^4$
and $s = \dfrac{m - n}{b}$

Max S as for Cases 14 to 20

26. Circular shaft with opposite sides flattened

$K = cr^4$
where c depends on ratio w/r and has values as follows:

$\dfrac{w}{r}$	$\dfrac{7}{8}$	$\dfrac{3}{4}$	$\dfrac{5}{8}$	$\dfrac{1}{2}$
c	1.357	1.076	0.733	0.438

$Q = cr^3$
where c depends on ratio w/r and has values as follows:

$\dfrac{w}{r}$	$\dfrac{7}{8}$	$\dfrac{3}{4}$	$\dfrac{5}{8}$	$\dfrac{1}{2}$
c	1.155	0.912	0.638	0.471

Section 3.4; the application of moment of inertia is discussed in Section 3.3. See Tables 3.2-1 and 3.2-2.

REFERENCES

3.2-1 Theodore Baumeister, *Marks' Mechanical Engineers' Handbook*, 6th ed. McGraw-Hill, New York, 1958.
3.2-2 Raymond J. Roark, *Formulas for Stress and Strain*, 4th ed. McGraw-Hill, New York, 1965.
3.2-3 E. P. Popov, *Mechanics of Materials*, 2nd ed. Prentice-Hall, Englewood Cliffs, NJ, 1976.

3.3   BEAM ELEMENT IN BENDING AND SHEAR: STRESS AND DEFLECTION

3.3-1   Introduction

A beam is a relatively long, slender member that carries loads that are perpendicular to its longitudinal axis and that develops internal stresses due to its internal bending moment and shear force. These internal moments and shear forces result from externally applied loads as will be shown shortly. A typical beam is depicted in Fig. 3.3-1. The study of beams and their behavior has a long and fascinating history. Beams were first studied by Galileo.[1]

Further developments in beam theory include the efforts of Jacob Bernoulli, Daniel Bernoulli, and L. Euler.[2] The beam is a very important structural element. The first-approximation model of many structural elements is that they behave as a beam in bending. Some examples are steel beams in a building, wooden floor joists in a home, freeways, freeway bridges and overpasses, railroad rails, power transmission shafting, airplane wings, and fuselage and tail sections. An understanding of beam theory and behavior is essential to the understanding of many other kinds of structures.

The shear and moment diagrams are developed as a result of considering internal equilibrium of the beam as shown in Fig. 3.3-2. The usual sign conventions for positive bending and shear are shown in Fig. 3.3-3.

The principle of static equivalent load can be used on the beam of Fig. 3.3-4*b* as follows.

1. The net load is equal to the area under the loading curve.
2. The line of action of the net load is at the centroid of the loading curve area.

This replacement is shown in Fig. 3.3-4*c*. Note that this step works *only* for *external* equilibrium. The beams of Figs. 3.3-4*a* and 3.3-4*c* are statically equivalent externally; that is, they have identical reactions. Their internal forces are different. We shall discuss how to treat the internal forces in a moment.

Once the reactions have been calculated by the use of the statically equivalent external load, we can then proceed to calculate the internal shear and moment by taking a slice of the beam and considering the real (distributed) load that acts on this portion. The distributed load *and* the previously calculated reactions will be in equilibrium with the internal shear and moment. It is convenient to assume that the shear and moment are positive. It is also convenient, when working with continuous functions, to work out the results for a general coordinate *x*. This is shown in Fig. 3.3-5. Now we can replace the distributed loading acting on the appropriate *piece* of the beam with its static equivalent as shown.

3.3-2   Basic Relationship

There is a convenient differential and integral calculus relationship that relates load, shear, and bending moment. It is especially convenient to use these relationships for continuous loading

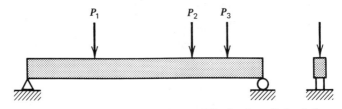

Fig. 3.3-1   A typical beam: long and slender; laterally loaded.

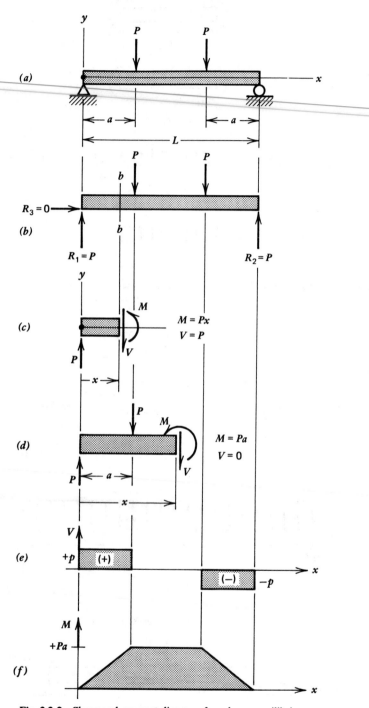

Fig. 3.3-2 Shear and moment diagrams from beam equilibrium.

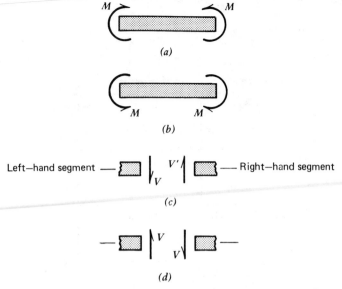

Fig. 3.3-3 Sign convention for positive shear and moment. (*a*) Positive bending moment; (*b*) negative bending moment; (*c*) positive internal shear; (*d*) negative internal shear.

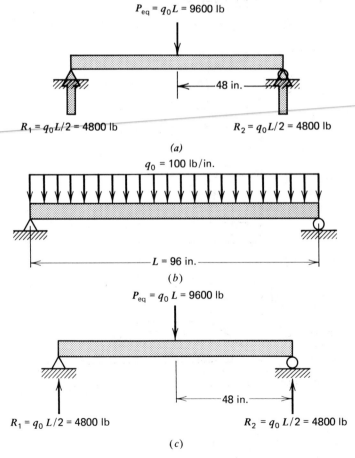

Fig. 3.3-4 Illustration of static equivalent load.

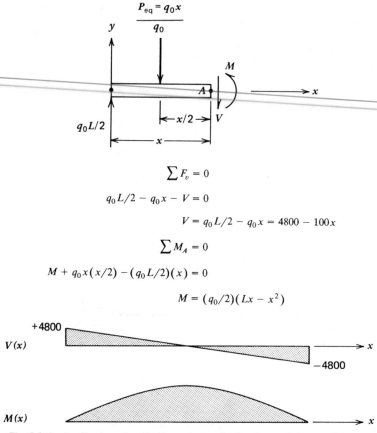

$$\sum F_v = 0$$

$$q_0 L/2 - q_0 x - V = 0$$

$$V = q_0 L/2 - q_0 x = 4800 - 100x$$

$$\sum M_A = 0$$

$$M + q_0 x (x/2) - (q_0 L/2)(x) = 0$$

$$M = (q_0/2)(Lx - x^2)$$

Fig. 3.3-5   Application of statically equivalent load to internal shear and moment.

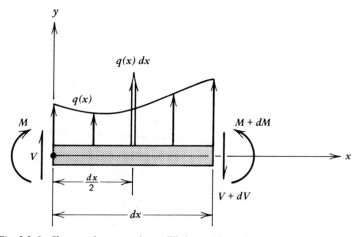

Fig. 3.3-6   Shear and moment in equilibrium with continuous loading function.

functions. Consider Fig. 3.3-6 where the loading function is shown, along with the shear and moment relations. All quantities are shown acting in their *positive* sense.

The load is shown as a varying function, but as $dx \to 0$ in the limit, the static equivalent force will be $q(x)\, dx$. This force acts at $dx/2$ as shown.

The equilibrium of forces in the vertical direction will give

$$\sum F_v = 0$$

$$V + q(x)\, dx - (V + dV) = 0$$

$$q(x)\, dx - dV = 0$$

$$q(x) = \frac{dV(x)}{dx}$$

Beam element

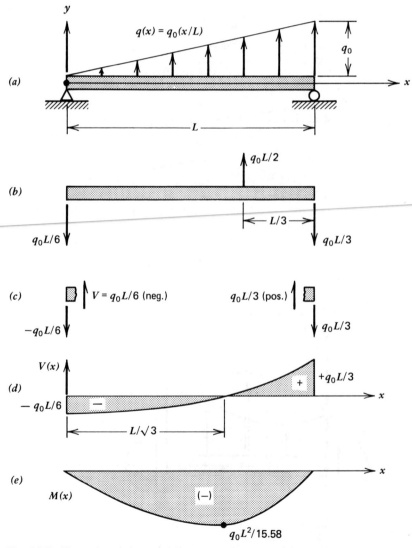

Fig. 3.3-7   Illustration of the use of differential relations among $q(x)$, $v(x)$, and $m(x)$.

The summation of moments around the left end of the segment gives the following:

$$\sum M = 0$$

$$- M + q(x)\, dx \left(\frac{dx}{2} \right) + (M + dM) - (V + dV)\, dx = 0$$

If we ignore higher-order terms

$$dM - V\, dx = 0$$

$$V(x) = \frac{dM(x)}{dx}$$

An application of these two equations is shown in Fig. 3.3-7.

3.3-3 Beam in Symmetric Bending

At any point along the span of the beam the bending moment gives rise to internal stresses. Let us consider a cross section of the beam shown in Fig. 3.3-8a.

If we examine the end view of the cross section, we will view the moment vector and the beam cross section as shown in Fig. 3.3-8b. The fibers are shown in end view in Fig. 3.3-8b. Figure 3.3-8c shows an isometric view. (Only a few of the "fibers" are shown.)

It can be seen that the bending moment causes the upper fibers to be compressed and the bottom fibers to be extended. Thus the bending moment causes beam tensile and compressive stresses, which are uniaxial stresses along the longitudinal axis of the beam. These are not necessarily the only stresses that take place, but these are the ones we will examine in detail now.

We have considered the cross section to be a rectangle for the sake of simplicity. The remarks of this section apply equally well to sections meeting the requirements for *symmetric bending*. There are two requirements for symmetric bending:

1. The cross section must have an axis of symmetry.
2. The moment vector, as depicted in Fig. 3.3-8b, must be perpendicular to the axis of symmetry.

Some other examples of cross sections that meet the two requirements for symmetric bending are shown in Fig. 3.3-9.

It is especially important to ensure that the formulas developed in this cross section for symmetric bending are only used when appropriate. The methods for treating unsymmetric bending will be discussed in the next section.

We now turn our attention to the study of the *geometry* of bending. Let us consider the action of bending alone. For example, we can consider the midportion of the beam of Fig. 3.3-2. In this region of the span the bending moment is constant. The first observation that we make is that bending of this portion of the beam will cause the top to go in compression and the bottom to go in tension. Furthermore, the curvature of the beam will be circular. This is shown in Fig. 3.3-10a. (In this figure, the curvature is exaggerated for purpose of illustration.) The radius of curvature, R, is measured to the line which, in end view, passes through the centroid of the cross section. Line aa is called the neutral axis. All fibers above line aa are in compression; all fibers below it are in tension. Line aa and line bb (which is coincident with moment vector $\mathbf{M}$) form a plane called the neutral plane.

A second important observation is that lines such as cc, dd, ee, and ff, which were initially straight and vertical, *remain straight during bending*. The preceding statement has some important meaning to the deformation of the entire cross section. This is depicted in Fig. 3.3-10b. The plane of the cross section rotates about line bb. The top of the plane moves inward and the bottom moves outward. All the particles on a line such as gg move the same distance to form line $g'g'$. This is also true of any set of particles on a line parallel to bb. All of the particles on such a line will move the same distance (but not the same as the particles on line gg). This geometric behavior is customarily described by the statement that during bending *plane sections remain plane*.

Let us now return to Fig. 3.3-10a and look at what happens with respect to the y axis (depth) as far as strain and deformation are concerned. Consider a portion of length dx as shown. At some distance y up from the neutral axis, the material is undergoing compression δ as shown. (We have previously defined positive moment as one that causes compression in the top of the beam.) Before bending, all of the longitudinal fibers in the shaded trapezoid were of length dx. Therefore, the axial strain of any one of these fibers, may be written as a function of y, which is $\varepsilon(y) = -\delta(y)/dx$. (Positive y causes compressive strain that is negative.)

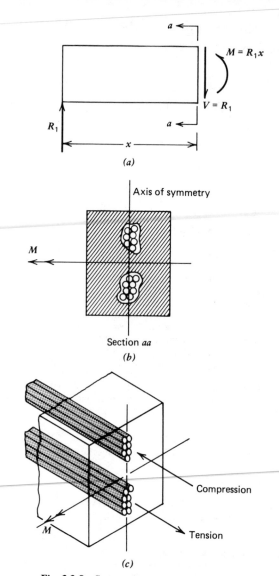

Fig. 3.3-8   Stresses in a beam cross section.

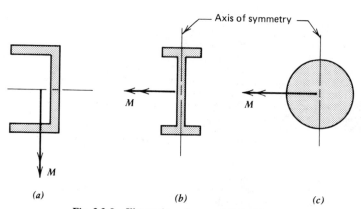

Fig. 3.3-9   Illustration of symmetric bending.

216

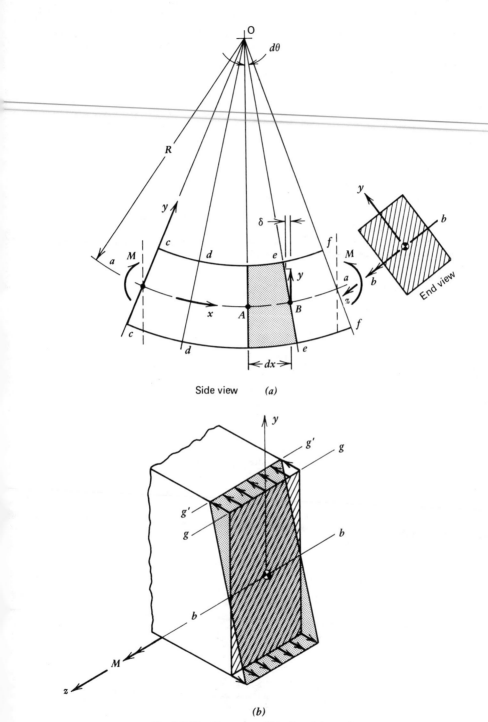

Side view   (a)

End view

(b)

Fig. 3.3-10 Geometry of bending deformation.

We have already discussed the fact that $\delta(y)$ is a linear function. To get $\delta(y)$, note that for small deflections triangle OAB is similar to the small triangle defined by y and δ, so

$$\frac{\delta(y)}{y} = \frac{dx}{R}$$

or

$$\delta(y) = \frac{y}{R}\, dx$$

Substituting gives

$$\varepsilon(y) = -\frac{y}{R}$$

For an elastic material the stress–strain relation is $\sigma(y) = E\varepsilon(y)$. For equilibrium we must consider two things. Refer now to Fig. 3.3-11.

The first is that the sum of all the forces in the x direction must be zero. These are the forces down the span of the beam, which must be zero at every cross section, since no such forces are present. Thus

$$\int_A \sigma(y)\, dA = 0$$

Since

$$\int_A \frac{-Ey}{R}\, dA = -\frac{E}{R}\int_A y\, dA = 0$$

we obtain

$$\int y\, dA = 0$$

The $\int_A y\, dA$ is called the first moment of area.

The second equilibrium condition is that the internal moments on any cross section must all sum up to the internal moment, M, which acts at that cross section.

$$-\int_A \sigma(y)\, dA\, y = M$$

or

$$-\int_A E\varepsilon(y)\, dA\, y = M$$

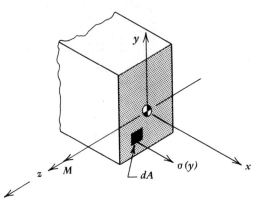

Fig. 3.3-11 Force on element da is $\sigma\, da$.

or

$$\int_A \frac{E}{R} y^2 \, dA = M$$

or

$$\frac{E}{R} \int_A y^2 \, dA = M$$

In this equation the term $\int_A y^2 \, dA$ is the *centroidal moment of inertia = I*. Thus,

$$\frac{EI}{R} = M$$

or

$$\frac{\sigma(y)I}{\varepsilon(y)R} = M$$

or

$$\sigma(y) = \frac{-My}{I}$$

3.3-4 Unsymmetric Bending

We now consider the methods for treating unsymmetric bending. The question of symmetry is, as previously stated, really concerned with the relationship of the moment vector to any axis of symmetry that the cross section in bending may have. Let us distinguish two categories of unsymmetric bending:

1. The cross section has symmetry, but the orientation of the moment vector is incorrect for symmetric bending (Fig. 3.3-12a).
2. The cross section has no axis of symmetry (Fig. 3.3-12b).

The sign conventions are:

1. Positive axes are shown in Fig. 3.3-12a.
2. Moment components are positive when their direction is coincident with the positive x or y coordinate (Fig. 3.3-12a).
3. Tensile stresses are positive; compressive stresses are negative.

Then we can superimpose the stresses due to M_x and M_y by adding the two symmetric bending cases. (Note: In Fig. 3.3-12a we have a cross section with double symmetry. We only need one axis of

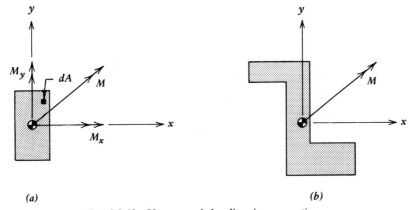

(a)   (b)

Fig. 3.3-12 Unsymmetric bending sign conventions.

symmetry to use this method, however, because one axis of symmetry will be enough to give a zero product of inertia. This will be explained more thoroughly later in this section.) The formula is obtained by using the equation for symmetric bending twice and watching the signs.

$$\sigma = \frac{M_x y}{I_x} - \frac{M_y x}{I_y} \qquad (3.3\text{-}1)$$

The small element dA in Fig. 3.3-12 experiences positive values of M_x, M_y, x, and y. M_x will cause tensile (plus) stress and M_y will cause compressive (negative) stress.

An important consideration in this kind of bending is the location of the neutral axis in the face of the cross section. This can be better understood by contrasting the symmetric and unsymmetric cases of Fig. 3.3-13a and b. In Fig. 3.3-13a line bb forms the neutral plane and is coincident with the moment vector. This was mentioned previously in our discussion of symmetric bending. The deflection of the cross section will be perpendicular to both the moment vector and line bb as shown in Fig. 3.3-13. In the case of unsymmetric bending, Fig. 3.3-13b, we can observe experimentally and mathematically that line bb, which forms the neutral plane, is no longer coincident with the moment vector $\mathbf{M}$. Let us denote the orientation of line bb with respect to the xy coordinate system by the angle β, as shown in the figure. In general, the angle β will be different from the angle α. Angle α is the orientation of the moment vector in the xy coordinate system.

The orientation of line bb can be determined mathematically by realizing that the neutral plane is the focus of all points in the cross section that have zero stress. So we can set Equation (3.3-1) equal to zero:

$$\frac{M_x y}{I_x} - \frac{M_y x}{I_y} = 0$$

In this coordinate system $\tan \beta$ is defined to be y/x, so

$$\tan \beta = \frac{y}{x} = \frac{M_y I_x}{I_y M_x}$$

This equation will permit us to solve for angle β. The direction of deflection of the beam will be as shown in Fig. 3.3-13b. This direction is perpendicular to axis bb. It is interesting to note that the orientation of the neutral axis is now a function of the magnitude of loading and the x- and y-inertia properties of the area.

The preceding method may be named the *principal axis method*, and this method is one of the ways to solve the problem of Fig. 3.3-12b. The problem is simply resolved into principal coordinates as shown in Fig. 3.3-14.

When the principal axes x' and y' are found, the moments $M_{x'}$ and $M_{y'}$ are resolved in this system. For a point such as A, Fig. 3.3-14, the coordinates x'_A and y'_A are measured in the $x'y'$ system.

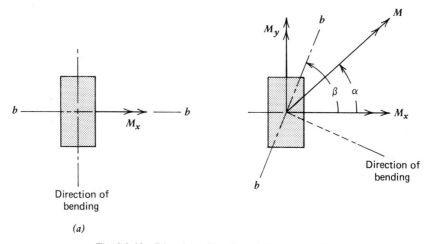

Fig. 3.3-13   Direction of bending of the cross sections.

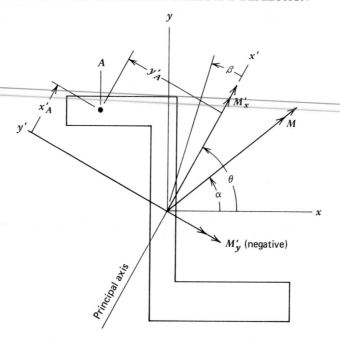

Fig. 3.3-14 Coordinates for unsymmetric bending.

Equation (3.3-1) is then written as

$$\sigma = \frac{M_{x'} y'}{I_{x'}} - \frac{M_{y'} x'}{I_{y'}}$$

Also

$$\tan \beta = \frac{M_{y'} I_{x'}}{I_{y'} M_{x'}}$$

Angle β is now referenced to the $x'y'$ system as shown in Fig. 3.3-14. Note that $I_{x'}$ and $I_{y'}$ are principal moments of inertia and that $I_{x'y'} = 0$ in the $x'y'$ system.

The *general method* for unsymmetric bending eliminates the requirement for transferring to the principal axes. This is best understood by deriving the equation in xy coordinates. In Fig. 3.3-15 assume that the unsymmetic cross section is bent in positive total curvature R, as shown.

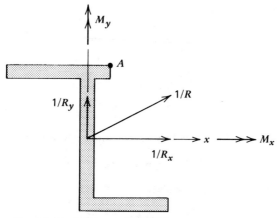

Fig. 3.3-15 Curvatures of unsymmetric cross section.

The strain at a typical point such as A is given as a superposition of the linear strains by

$$\varepsilon = \left(\frac{1}{R_x}\right) y - \left(\frac{1}{R_y}\right) x$$

The result is shown to be

$$\sigma(x, y) = \left(\frac{M_x I_y + M_y I_{xy}}{I_x I_y - I_{xy^2}}\right) y - \left(\frac{M_x I_{xy} + M_y I_x}{I_x I_y - I_{xy^2}}\right) x$$

This means that, unlike symmetric bending, the application of M_x alone on an unsymmetric cross section will induce a twisting about the y axis. This comes as a result of I_{xy} being nonzero for the unsymmetric section.

Example 3.3-1   Unsymmetric Bending.   For the beam cross section as shown and loaded below, calculate the bending stress at point A, and calculate the orientation of the neutral axis:

$$I_x = 693.3 \text{ in.}^4$$

$$I_y = 173.3 \text{ in.}^4$$

$$I_{xy} = -240 \text{ in.}^4$$

Solution.   The stress at any point on the beam is given by the unsymmetrical bending formula

$$\sigma(x, y) = \left(\frac{M_x I_y + M_y I_{xy}}{I_x I_y - I_{xy^2}}\right) y - \left(\frac{M_x I_{xy} + M_y I_x}{I_x I_y - I_{xy^2}}\right) x$$

Substituting in the proper values we have

$$\sigma = \frac{-(100,000)(-240) - (10,000)(693.3)}{(693.3)(173.3) - (-240)^2} x + \frac{(10,000)(-240) + (100,000)(173.3)}{(693.3)(173.3) - (-240)^2} y$$

which yields the following equation for bending stress at any particular point on the cross section.

$$\sigma = 272.9x + 238.7y \qquad\qquad (3.3\text{-}2)$$

Point A is located by coordinates

$$x_A = 5 \text{ in.} \qquad y_A = -4 \text{ in.}$$

Hence,

$$\sigma_A = 272.9(5) + 238.7(-4)$$

$$= 409.7 \text{ psi tension}$$

The orientation of the neutral axis is obtained from Equation (3.3-2) by substituting $\sigma = 0$. Namely,

$$0 = 272.9x + 238.7y$$

or

$$\frac{y}{x} = \tan \beta = -\frac{272.9}{238.7}$$

which leads to

$$\beta = -48.8° \quad \text{(see diagram below)}$$

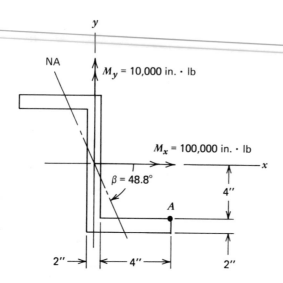

3.3-5 Design of Beams of Constant Strength

In general, a beam of uniform strength[2] will be dependent on bending moment along the span and general cross-section shape. It should be apparent from the preceding discussion that the selection of a prismatic beam is based only on the stresses at the critical sections. At all other sections through the beam the stresses will be below the allowable level. Therefore, the potential capacity of a given material is not fully utilized. This situation may be improved by designing a beam of variable cross section, that is, by making the beam nonprismatic. Since flexural stresses control the design of most beams, as has been shown, the cross sections may everywhere be made just strong enough to resist the corresponding moment. Such beams are called beams of constant strength. Shear governs the design at sections through these beams where the bending moment is small.

Example 3.3-2. Design a cantilever of constant strength for resisting a concentrated force applied at the end. Neglect the beam's own weight.

Solution (see Fig. a). A cantilever with a concentrated force applied at the end is shown in Fig. *a*; the corresponding moment diagram is plotted in Fig. *b*. Basing the design on the bending moment, the required section modulus at an arbitrary section is given by

$$Z = \frac{M}{\sigma_{allow}} = \frac{Px}{\sigma_{allow}}$$

A great many cross-sectional areas satisfy this requirement; so first, it will be assumed that the beam will be of rectangular cross section and of constant height h. The section modulus for this beam is given by $bh^2/6 = S$, hence

$$b = \left[\frac{6P}{h^2\sigma_{allow}}\right]x = \frac{b_0}{L}x$$

where the bracketed expression is a constant and is set equal to b_0/L, so that when $x = L$ the width is b_0. A beam of constant strength with a constant depth in a plane view looks like the wedge* shown in

---

*Since this beam is not of constant cross-sectional area, the use of the elementary flexure formula is not entirely correct. When the angle included by the sides of the wedge is small, little error is involved. As this angle becomes large, the error may be considerable. An exact solution shows that when the total included angle is 40° the solution is in error by nearly 10%.

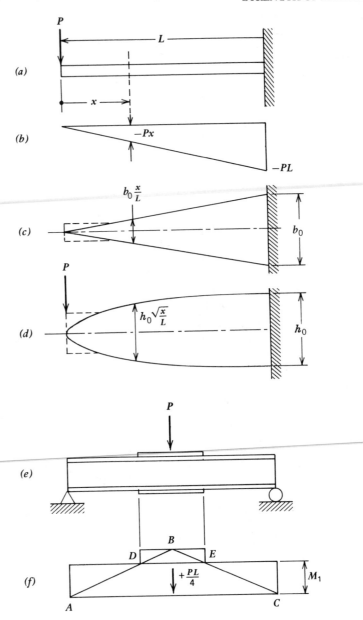

Fig. *c*. Near the free end this wedge must be modified to be of adequate strength to resist the shearing force $V = P$.

If the width or breadth b of the beam is constant,

$$h = h_0 \sqrt{\frac{x}{L}}$$

this expression indicates that a cantilever of constant width loaded at the end is also of constant strength if its height varies parabolically from the free end (Fig. *d*).

Beams of constant strength are used in leaf springs and in many machine parts that are cast or forged. In structural work an approximation of a beam of constant strength is frequently made. For example, the moment diagram for the beam loaded as shown in Fig. *e* is given by the lines AB and BC in Fig. *f*. By selecting a beam of flexural capacity equal only to M_1, the middle portion of the beam is overstressed. However, cover plates can be provided near the middle of the beam to boost the flexural capacity of the composite beam to the required value of the maximum moment. For the case shown the cover plates must extend at least over the length *DE* of the beam and in practice they are made somewhat longer.

3.3-6   Shear Stresses in Beams

We know that at any point along the span of the beam where the moment is changing that a shear force exists since

$$V = \frac{dM}{dx}$$

The manner in which this shear force is carried over the cross section of the beam is the point of study of this section. The shear stress is not distributed across the cross section uniformly, so that $\tau \neq P/A$, as in bolt or rivet shear. Consider the segment of the beam shown in Fig. 3.3-16. We will only consider the shear stresses in symmetric bending.

The change in moment along the portion of span dx will cause an unbalanced force along the span. The shaded area on the front face of the cross section will see the force $P + dP$ while the force on the corresponding area of the back face is P. This requires a shear force V down the span as shown.

Application of equilibrium leads to the result

$$\tau = \frac{VQ}{Ib}$$

A significant property of the beam cross section is the *shear center*. A beam that is loaded vertically through the centroid of the cross section as shown in Fig. 3.3-17 will twist. This will happen if the load is applied at the centroid of the cross section or at any other point except the shear center.

Each cross section will have a shear center. This can be calculated by balancing out the shear stresses that "flow" in the cross section. This is shown in Fig. 3.3-18 for the channel section. Shear centers are tabulated in Table 3.3-1.[3]

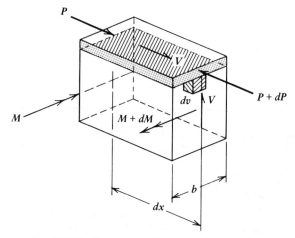

Fig. 3.3-16   Shear stresses in symmetric bending.

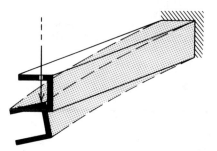

Fig. 3.3-17   Illustration of shear center.

TABLE 3.3-1   SOME SHEAR CENTERS OF TYPICAL CROSS SECTIONS[a]

—Position of Flexural Center Q for Different Sections

Form of section	Position of Q
1. Any narrow section symmetrical about the x axis. Centroid at $x = 0, y = 0$	$e = \dfrac{1 + 3\nu}{1 + \nu} \dfrac{\int \chi t^3\, dx}{\int t^3\, dx}$ For narrow triangle (with $\nu = 0.25$), $e = 0.187a$ For any equilateral triangle, $e = 0$
2. Sector of thin circular tube	$e = \dfrac{2R}{(\pi - \theta) + \sin\theta\cos\theta}[(\pi - \theta)\cos\theta + \sin\theta]$ For complete tube split along element ($\theta = 0$), $e = 2R$
3. Semicircular area	$e = \left(\dfrac{8}{15\pi}\dfrac{3 + 4\nu}{1 + \nu}\right) R$ (Q is to right of centroid)
4. Angle	Leg 1 = rectangle $w_1 h_1$; leg 2 = rectangle $w_2 h_2$ I_1 = moment of inertia of leg 1 about Y_1 (central axis) I_2 = moment of inertia of leg 2 about Y_2 (central axis) $e_x = \tfrac{1}{2}h_2\left(\dfrac{I_2}{I_1 + I_2}\right)$ $e_y = \tfrac{1}{2}h_1\left(\dfrac{I_1}{I_1 + I_2}\right)$ If w_1 and w_2 are small, $e_x = e_y = 0$ (practically) and Q is at 0
5. Channel	$e = h\left(\dfrac{H_{xy}}{I_x}\right)$ where H_{xy} = product of inertia of the half section (above X) with respect to axes X and Y, and I_x = moment of inertia of whole section with respect to axis X If t is uniform, $e = \dfrac{b^2 h^2 t}{4 I_x}$
6. Tee	$e = \tfrac{1}{2}(t_1 + t_2)\left[\dfrac{1}{1 + \dfrac{d_1^3 t_1}{d_2^3 t_2}}\right]$ For a T-beam of ordinary proportions, Q may be assumed to be at 0
7. I with unequal flanges and thin web	$e = b\left(\dfrac{I_2}{I_1 + I_2}\right)$ where I_1 and I_2, respectively, denote moments of inertia about X-axis of flange 1 and flange 2
8. Beam composed of n elements, of any form, connected or separate, with common neutral axis (*e.g.*, multiple-spar airplane wing)	$e = \dfrac{E_2 I_2 x_2 + E_3 I_3 x_3 \ldots + E_n I_n x_n}{E_1 I_1 + E_2 I_2 + E_3 I_3 \ldots + E_n I_n}$ where I_1, I_2, etc., are moments of inertia of the several elements about the X-axis (*i.e.*, Q is at the centroid of the products EI for the several elements)

226

TABLE 3.3-1 (Continued)

—POSITION OF FLEXURAL CENTER Q.—

Form of section	Position of Q

9. Lipped channel (t small) — Values of e/h

c/h \ b/h	1.0	0.8	0.6	0.4	0.2
0	0.430	0.330	0.236	0.141	0.055
0.1	0.477	0.380	0.280	0.183	0.087
0.2	0.530	0.425	0.325	0.222	0.115
0.3	0.575	0.470	0.365	0.258	0.138
0.4	0.610	0.503	0.394	0.280	0.155
0.5	0.621	0.517	0.405	0.290	0.161

10. Hat section (t small) — Values of e/h

c/h \ b/h	1.0	0.8	0.6	0.4	0.2
0	0.430	0.330	0.236	0.141	0.055
0.1	0.464	0.367	0.270	0.173	0.080
0.2	0.474	0.377	0.280	0.182	0.090
0.3	0.453	0.358	0.265	0.172	0.085
0.4	0.410	0.320	0.235	0.150	0.072
0.5	0.355	0.275	0.196	0.123	0.056
0.6	0.300	0.225	0.155	0.095	0.040

11. D-section (A = enclosed area) — Values of $e(h/A)$

t_1/t_s \ S/h	1	1.5	2	3	4	5	6	7
0.5	.....	.....	1.0	0.800	0.665	0.570	0.500	0.445
0.6	.....	.....	0.910	0.712	0.588	0.498	0.434	0.386
0.7	.....	0.980	0.831	0.641	0.525	0.443	0.384	0.338
0.8	.....	0.910	0.770	0.590	0.475	0.400	0.345	0.305
0.9	.....	0.850	0.710	0.540	0.430	0.360	0.310	0.275
1.0	1.0	0.800	0.662	0.500	0.400	0.330	0.285	0.250
1.2	0.905	0.715	0.525	0.380	0.304	0.285	0.244	0.215
1.6	0.765	0.588	0.475	0.345	0.270	0.221	0.190	0.165
2.0	0.660	0.497	0.400	0.285	0.220	0.181	0.155	0.135
3.0	0.500	0.364	0.285	0.200	0.155	0.125	0.106	0.091

[a] From Reference 3.

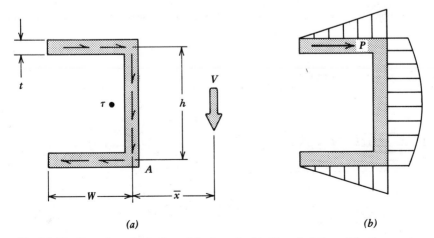

(a) (b)

Fig. 3.3-18 Shear flows in the face of the beam lead to the calculation of the shear center.

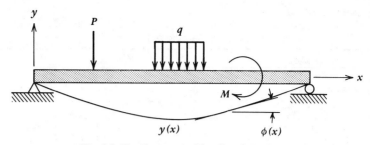

Fig. 3.3-19 Geometry of bending deflection.

3.3-7 Bending Deflection of Beams

The loading of a beam transverse to its longitudinal axis will produce a lateral deflection as shown in Fig. 3.3-19.

The slope of the *elastic curve* $y(x)$ is $dy/dx = \phi(x)$. The relationship between moment and curvature is given by

$$\frac{1}{R} = \frac{M(x)}{EI}$$

Recall from calculus the definition of curvature

$$\frac{1}{R} = \frac{(d^2y/dx^2)}{\left[1 + (dy/dx)^2\right]^{3/2}}$$

which, for small dy/dx becomes

$$\frac{1}{R} = \frac{d^2y}{dx^2}$$

The beam deflections are found by integrating and applying the boundary conditions.

It can be noted that for small deflections and a linear elastic material, the method of superposition can be used to superimpose the elastic curves $y(x)$ or the maximum deflections, and so forth.

The deflections for beams are frequently tabulated in tables such as Table 3.3-2. Extensive sources are available; see, for example, Reference 3. See also Section 3.7.

See Section 3.7 for the graphical method for deflection due to shear and for tables of deflections.

TABLE 3.3-2   BEAM DEFLECTION

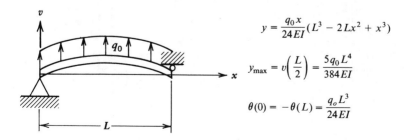

$$y = \frac{q_0 x}{24EI}(L^3 - 2Lx^2 + x^3)$$

$$y_{max} = v\left(\frac{L}{2}\right) = \frac{5q_0 L^4}{384EI}$$

$$\theta(0) = -\theta(L) = \frac{q_0 L^3}{24EI}$$

When $0 \le x \le a$, then

$$y = \frac{Pb}{6EIL}[(L^2 - b^2)x - x^3]$$

When $a = b = L/2$, then

$$y = \frac{Px}{48EI}(3L^2 - 4x^2) \qquad (0 \le x \le L/2)$$

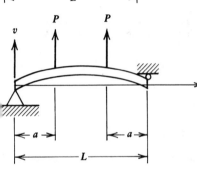

$$y_{max} = v\left(\frac{L}{2}\right) = \frac{PL^3}{48EI}$$

$$\theta(0) = -\theta(L) = \frac{PL^2}{16EI}$$

$$y = -\frac{M_0 x}{6EIL}(L^2 - x^2)$$

$$y_{max} = v\left(\frac{L}{\sqrt{3}}\right) = -\frac{M_0 L^2}{9\sqrt{3}\,EI}$$

$$\theta(0) = -\frac{\theta(L)}{2} = -\frac{M_0 L}{6EI}$$

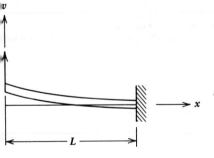

$$y_a = v(a) = \frac{Pa^2}{6EI}(3L - 4a)$$

$$y_{max} = v(L/2) = \frac{Pa}{24EI}(3L^2 - 4a^2) \qquad \theta(0) = \frac{Pa}{2EI}(L - a)$$

$$y = \frac{P}{6EI}(2L^3 - 3L^2x - x^3)$$

$$v_{max} = v(0) = \frac{PL^3}{3EI} \qquad \theta(0) = -\frac{PL^3}{2EI}$$

$$y = \frac{q_0}{24EI}(x^4 - 4L^3x - 3L^4)$$

$$v_{max} = v(0) = \frac{q_0 L_4}{8EI} \qquad \theta(0) = -\frac{q_0 L^3}{6EI}$$

REFERENCES

3.1-1 Joe W. McKinley, *Fundamentals of Stress Analysis*. Matrix Publishers, Portland, OR, 1979.
3.3-2 E. P. Popov, *Mechanics of Materials*, 2nd ed. Prentice-Hall, Englewood Cliffs, NJ, 1976.
3.3-3 Raymond J. Roark, *Formulas for Stress and Strain*, 4th ed. McGraw-Hill, New York, 1965.

3.4 TORSION AND THE TORSION ELEMENT

3.4-1 Introduction

A structural element that is being twisted with respect to its longitudinal axis is in torsion. This is illustrated in Fig. 3.4-1, where the double arrowheads indicate twisting in accordance with the right-hand rule as shown.

Elements may be in torsion due to the direct application of torque; due to their presence in a structure where they are twisted due to the deformation of surrounding structure; or due to the fact that they are being used to transmit horsepower.

A shaft that is being used to transmit horsepower has internal torque given by the following formula, which is a handy one to have memorized:

$$T = \frac{63,000 \text{ hp}}{N}$$

where N = rpm and T = torque.

In analyzing element in torsion, it is quite useful to be able to classify them into one of three categories and then proceed in a definite method of analysis that depends on which category the *cross section* falls in.

The three categories are:

1. Circular (either solid or hollow).
2. Hollow, closed, thin wall.
3. Other.

The classification into one of these three categories is not difficult. However, some comments are in order so that doubtful areas may be overcome.

The choice of a circular cross section is obvious. The circular cross section may be solid or hollow as indicated in Fig. 3.4-2. If the cross section is hollow, the circles must be concentric.

The thin-wall hollow, closed cross section may be of any shape. The wall thickness does not have to be constant. The question of what a *thin* wall is frequently arises. As a guide, the ratio of thickness t to some significant size parameter D (see Fig. 3.4-2) should meet the requirement

$$t/D < 0.1$$

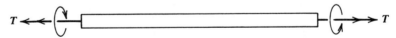

Fig. 3.4-1 A structural element in torsion.

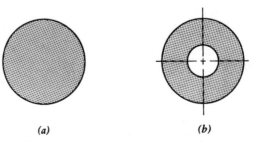

(a)   (b)

Fig. 3.4-2 Circular cross sections may be solid or hollow.

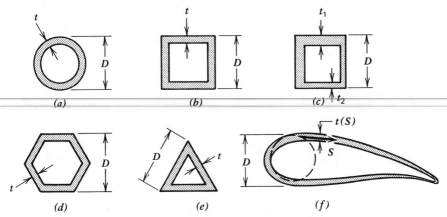

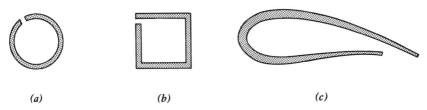

Fig. 3.4-3 Examples of thin-wall, hollow cross sections.

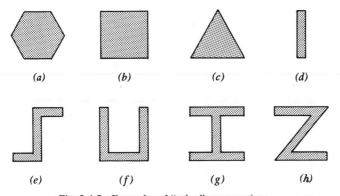

Fig. 3.4-4 Sections that appear to be thin-wall, hollow but are not.

Fig. 3.4-5 Examples of "other" cross sections.

The hollow, thin-wall tube must be *completely* closed (Fig. 3.4-3). If there is a break in the cross section, caused by a longitudinal cut, the section does *not* qualify under category 2. Some examples are shown in Fig. 3.4-4. They are category 3, "other."

Any section not classified as category 1 or 2, will be (automatically) category 3. Some examples of category 3, in addition to those shown in Fig. 3.4-4, are shown in Fig. 3.4-5.

3.4-2 Torsional Stress and Deflection in the Circular Shaft

Consider a shaft in torsion as shown in Fig. 3.4-1. If we analyze a small square, such as the shaded one, we can postulate that the change from square to diamond shape is caused by shear stress, as shown by Fig. 3.4-6a–c.

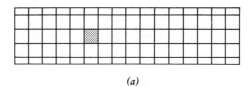

(a)

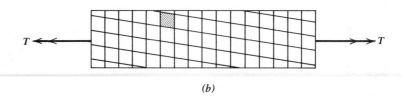

(b)

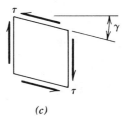

(c)

Fig. 3.4-6 Behavior of a circular bar in torsion.

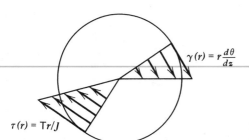

$$\gamma(r) = r\frac{d\theta}{dz}$$

$$\tau(r) = Tr/J$$

Fig. 3.4-7 Stresses and strains in the circular cross section.

The results for angular deflection and torsional stress are

$$\frac{d\theta}{dz} = \frac{T}{GJ} \quad \text{and} \quad \tau = \frac{Tr}{J}$$

The shear stress as a function of radius is given in terms of applied torque and polar moment of inertia.[1]

The variation of stress and the angle of twist with respect to radius is shown in Fig. 3.4-7. This same distribution applies all the way around the circumferential coordinate of the circle.

Example 3.4-1   Circular Shaft in Torsion.   A 4-ft length of 7075-T6 aluminum bar has a diameter of 1.25 in. and carries a torque of 1000 in.· lb. What is the angle of twist? What is the maximum shear stress?

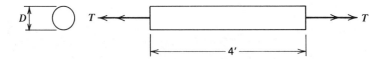

Solution. The angle of twist per unit length is

$$\frac{d\theta}{dz} = \frac{T}{JG}$$

So

$$d\theta = \frac{T\,dz}{JG}$$

$$T = 1000 \text{ in.} \cdot \text{lb}$$

$$G \text{ (Table 3.1-2)} = 3.9 \times 10^6 \text{ psi}$$

$$J \text{ for a circle} = \frac{\pi D^4}{32}$$

$$= \frac{\pi(1.25)^4}{32} = 0.2397 \text{ in.}^4$$

$$\theta = \int_0^L \frac{T}{JG}\,dz = \frac{TL}{JG} = \frac{1000(4 \times 12)}{0.2397 \times 3.9 \times 10^6} = 0.0513 \text{ rad}$$

$$= 2.94°$$

Note that

$$\theta = \frac{TL}{JG}$$

is the solution for the twist of a uniform shaft of length L. The shear stress is given by

$$\tau_{max} = \frac{T r_0}{J}$$

$$T = 1000 \text{ in.} \cdot \text{lb}$$

$$r_0 = \text{outside radius} = 0.625 \text{ in.}$$

$$J = 0.2397 \text{ in.}^4$$

$$\tau_{max} = \frac{1000(0.625)}{0.2397} = 2607 \text{ lb/in.}^2$$

Example 3.4-2   Hollow Circular Shaft.   If a 0.5-in. diameter hole is drilled in the shaft of Example 3.4-1 calculate:

1. The percent reduction of volume (and thus weight).
2. The percent increase in angle of twist.
3. The percent increase in stress.

Solution.   The percent reduction in area is

$$\frac{A_0 - A_f}{A_0} \times 100$$

where A_0 is original area and A_f is final area, and

$$A_0 = \frac{\pi D_0^2}{4} = \frac{\pi(1.25)^2}{4} = 1.227 \text{ in.}^2 \quad \text{(total outside area)}$$

$$A_i = \frac{\pi D_i^2}{4} = \frac{\pi(0.5)^2}{4} = 0.196 \text{ in.}^2 \quad \text{(area of hole)}$$

$$A_f = 1.227 - 0.196 - 1.031 \text{ in.}^2$$

And

$$\text{percent reduction in area} = \frac{1.227 - 1.031}{1.227} \times 100 = 15.9\% \text{ reduction}$$

The increase in angle of twist and shear stress is governed by the new polar moment of inertia

$$J = \frac{1}{32}\pi\left(D_0^4 - D_i^4\right)$$

$$= \frac{1}{32}\pi(1.25^4 - 0.5^4) = 0.2335 \text{ in.}^4$$

The new angle of twist is

$$\theta = \frac{TL}{JG} = \frac{1000(48)}{0.2335 \times 3.9 \times 10^6} = 0.0527 \text{ rad}$$

The percent increase in angle of twist is

$$100 \times \frac{\theta_{new} - \theta_{old}}{\theta_{new}} = \frac{0.0527 - 0.0513}{0.0527} \times 100$$

$$= 2.66\%$$

The new stress is given by

$$\tau_{max} = \frac{Tr_0}{J}$$

(The stress distribution for a hollow shaft is shown in the sketch.)

$$\tau_{max} = \frac{1000(0.625)}{0.2335} = 2676 \text{ psi}$$

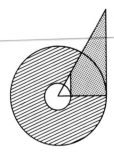

The percent increase in stress is

$$\frac{\tau_{new} - \tau_{old} \times 100}{\tau_{new}} = \frac{2676 - 2607}{2676} \times 100 = 2.58\%$$

The above numbers show why shafts in actual machinery are often hollow. A significant weight saving can be obtained with a small deflection and stress penalty.

3.4-3 Torsion of Thin-Wall, Closed Hollow Tubes

Earlier, the guideline $t/D < 0.1$ was given for a thin-wall tube. A more accurate description of a thin-wall tube is that the shear stress variation across the wall is negligible. Consider Fig. 3.4-8 which shows a hollow circular tube with a thick wall compared to a thin wall. In the thick-wall tube the stress variation from inside to outside is significant. In the thin wall the stress variation is small.

The decision as to whether or not something is thin wall may require some judgment, but basically, the wall is thin enough with respect to the rest of the structure so that the shear stress variation across the wall thickness is essentially constant as shown in Fig. 3.4-8.

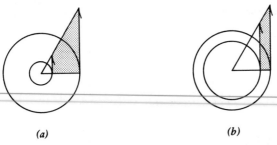

(a)   (b)

Fig. 3.4-8   Stress distribution in the hollow circular cross section.

For thin-wall structures it is customary to work with the shear flow (in lb/in.) defined by

$$q = \tau t \qquad lb/in.$$

The shear flow is assumed to act at the midline of the wall. The coordinate S runs around the midline of the wall as shown in Fig. 3.4-9.
   Imposition of *equilibrium* of the *in-plane moments* gives

$$\underbrace{q\,ds}_{\text{(force)}} \quad \times \quad \underset{\text{(arm)}}{r} \quad = \quad \underset{\text{= (applied torque)}}{T}$$

Now note that

$$r\,ds = 2\,dA$$

where dA is the shaded area in Fig. 3.4-9. Letting ds cover section gives

$$q = \frac{T}{2A}$$

where A is the *entire* area enclosed by the centerline of the tube (*not* the area of actual material). Note that even though the thickness t (and thus the shear stress) may vary around the circumferential coordinate of the tube, the shear flow q *remains constant.*

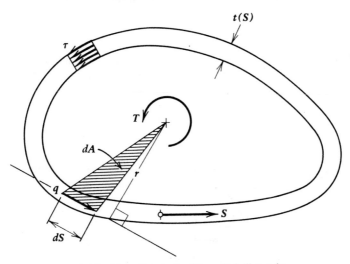

Fig. 3.4-9   Shear flow in the thin-wall, hollow tube.

Example 3.4-3. See figure below. For the tube shown calculate the shear flow. Calculate the shear stress in each leg. $T = 1000$ in. · lb.

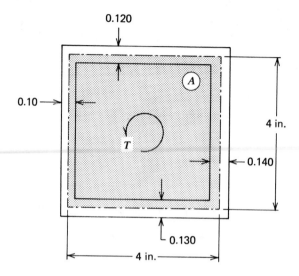

Solution. The area A is total area enclosed by the centerline of the tube. This is shown as shaded in the figure.

$$A = 4 \times 4 = 16 \text{ in.}^2$$

Then

$$q = \frac{T}{2A} = \frac{1000}{(2)16} = 31.25 \text{ lb/in.}$$

In the 0.10 wall: $\tau = \dfrac{q}{t} = \dfrac{31.25}{0.10} = 312.5$ psi

In the 0.120 wall: $\tau = \dfrac{q}{t} = \dfrac{31.25}{0.120} = 260.4$ psi

In the 0.140 wall: $\tau = \dfrac{q}{t} = \dfrac{31.25}{0.140} = 223.2$ psi

In the 0.130 wall: $\tau = \dfrac{q}{t} = \dfrac{31.25}{0.130} = 240.4$ psi

When we consider the angular deflection of a hollow, closed, thin-wall structure, it is convenient to use Castigliano's theorem. The strain energy of a structure in shear, per unit volume, is shown in Fig. 3.4-10.

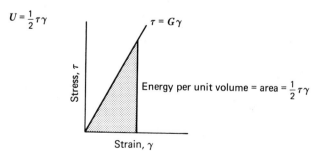

Fig. 3.4-10 Strain energy in torsion.

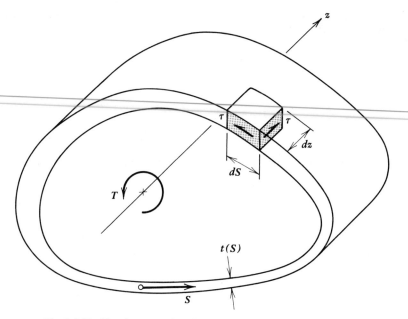

Fig. 3.4-11 Development of angle of twist for thin-wall hollow tube.

Since we will restrict ourselves, as usual, to the elastic range, we have $\tau = G\gamma$ and

$$U = \frac{\tau^2}{2G}$$

Consider now the element of tube along the Z axis (longitudinal axis) as shown in Fig. 3.4-11. The shear strain energy on the elemental volume is

$$dU = \frac{\tau^2 t}{2G} \, ds \, dz$$

If we *assume* that we have a uniform tube and that the angle of twist per unit length is a constant, then we can write the total energy as an integration over the length and around the circumference.

$$U = \int_0^L \oint_s \frac{\tau^2 t}{2G} \, ds \, dz$$

Then

$$U = \int_0^L \oint_s \frac{T^2}{8A^2 t(s)G} \, ds \, dz$$

and

$$\theta = \frac{TL}{4A^2 G} \oint_s \frac{ds}{t(s)}$$

If we compare this with

$$\theta = \frac{TL}{JG}$$

then we see that the area property for this tube is given by

$$J_H = \frac{4A^2}{\displaystyle\oint_s \frac{ds}{t(s)}}$$

Example 3.4-4. For the tube of the previous example, find the angle of twist if the length is 6 ft and the tube is made of AISI 1025 steel.

Solution.

$$\theta = \frac{TL}{GJ_H}$$

$$T = 1000 \text{ in.} \cdot \text{lb}$$

$$L = 6 \times 12 = 72 \text{ in.}$$

$$G = 11 \times 10^6 \text{ psi (Table 3.1-2)}$$

To calculate J_H, we can replace the integration by a summation along each leg of the tube, since t is constant in each leg.

$$J_H = \frac{4A^2}{\displaystyle\sum_{i=1}^{4} \frac{s_i}{t_i}}$$

$$A^2 = 16^2 = 256 \text{ in.}^4$$

$$\frac{s_1}{t_1} = \frac{4}{0.10} = 40 \qquad \frac{s_3}{t_3} = \frac{4}{0.140} = 28.6$$

$$\frac{s_2}{t_2} = \frac{4}{0.120} = 33.3 \qquad \frac{s_4}{t_4} = \frac{4}{0.130} = 30.8$$

$$\sum \frac{s_i}{t_i} = 132.7$$

$$J_H = \frac{4A^2}{\displaystyle\sum \frac{s_i}{t_i}} = \frac{4 \times 256}{132.7} = 7.71 \text{ in.}^4$$

$$\theta = \frac{1000 \times 72}{11 \times 10^6 \times 7.71} = 0.000848 \text{ rad}$$

$$= 0.04864°$$

3.4-4 Torsion of the "Other" Cross Sections

For practical purposes the method for stress analysis of cross sections classified as "other" is simply stated: Find the proper formula in a handbook or other reference. Reference 2 is an excellent source.

In this section we will list a few of the more common other shapes that are used in practice. One common shape is the rectangle, shown in Fig. 3.4-12.

The ratio of b/t is unity for a square cross section and approaches infinity for a long, thin rectangular cross section. The appropriate formulas, presented without derivation, are

$$\tau = \frac{T}{\alpha b t^2}$$

$$\theta = \frac{TL}{\beta b t^3 G}$$

where α and β are given in Table 3.4-1

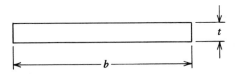

Fig. 3.4-12 Rectangular cross section in torsion.

TABLE 3.4-1 VALUES OF α AND β FOR b/t FOR RECTANGULAR CROSS SECTIONS[a]

b/t[b]	1.00	1.50	1.75	2.00	2.50	3.00	4	6	8	10	∞
α	0.208	0.231	0.239	0.246	0.258	0.267	0.282	0.299	0.307	0.313	0.333
β	0.141	0.196	0.214	0.229	0.249	0.263	0.281	0.299	0.307	0.313	0.333

[a]From Reference 3.
[b]Rectangular shape, b = longer side; t = shorter side.

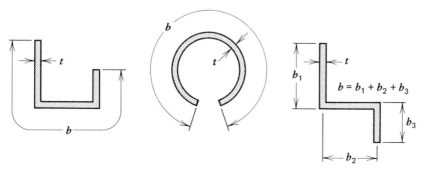

Fig. 3.4-13 J equivalent for "rectangular" shapes.

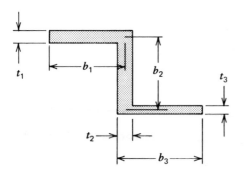

Fig. 3.4-14   "Rectangles" of unequal thickness.

The equations for τ and θ can be used for rectangular shapes that are not flat, as shown in Fig. 3.4-13. The dimension b is now taken to be the running length of the centerline.

For shapes made up of thin rectangles with legs of different thickness, such as shown in Fig. 3.4-14, the formula

$$\theta = \frac{TL}{KG}$$

is used, with

$$K = \beta_1 b_1 t_1^3 + \beta_2 b_2 t_2^3 + \beta_3 b_3 t_3^3$$

Sections of other shapes are infinite in variety. Table 3.4-2 provides a sample. For more complex shapes additional tables may be found in various handbooks and publications. References 1 and 3–5 contain useful information.

Shape factors and solutions for bars of "other" cross sections are given in Section 3.2, which is adapted from Reference 2.

TABLE 3.4-2 SOME EQUATIONS FOR THE TORSION OF IRREGULAR CROSS SECTIONS[a]

Any elongated section with axis of symmetry $0X$. U = length, A = area of section, I_x = moment of inertia about axis of symmetry.

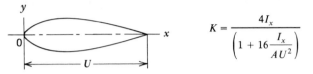

$$K = \frac{4I_x}{\left(1 + 16\dfrac{I_x}{AU^2}\right)}$$

Any elongated section or thin open tube. dU = elementary length along median line, t = thickness normal to median line, A = area of section.

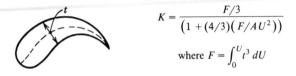

$$K = \frac{F/3}{\left(1 + (4/3)\left(F/AU^2\right)\right)}$$

$$\text{where } F = \int_0^U t^3\, dU$$

Any solid, fairly compact section without reentrant angles. J = polar moment of inertia about centroidal axis; A = area of section.

$$K = \frac{A^4}{40J}$$

I section, flange thickness uniform. r = fillet, radius, D = diameter largest inscribed circle, $t = b$ if $b < d$, $t = d$ if $d < b$, $t_1 = b$ if $b > d$, $t_1 = d$ if $d > b$.

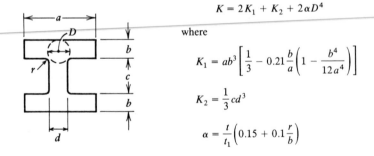

$$K = 2K_1 + K_2 + 2\alpha D^4$$

where

$$K_1 = ab^3\left[\frac{1}{3} - 0.21\frac{b}{a}\left(1 - \frac{b^4}{12a^4}\right)\right]$$

$$K_2 = \frac{1}{3}cd^3$$

$$\alpha = \frac{t}{t_1}\left(0.15 + 0.1\frac{r}{b}\right)$$

For all solid section of irregular form the maximum shear stress occurs at or very near one of the points where the largest inscribed circle touches the boundary,[b] and of these, at the one where the curvature of the boundary is algebraically least. (Convexity represents positive, concavity negative, curvature of the boundary.) At a point where the curvature is positive (boundary of section straight or convex) this maximum stress is given approximately by $s = G(\theta/L)c$ or $s = (T/K)c$ where

$$c = \frac{D}{1 + \dfrac{\pi^2 D^4}{16A^2}}\left[1 + 0.15\left(\frac{\pi^2 D^4}{16A^2} - \frac{D}{2r}\right)\right]$$

where D = diameter of largest inscribed circle
   r = radius of curvature of boundary at the point (positive for this case)
   A = area of the section

At a point where the curvature is negative (boundary of section concave, or reentrant) this maximum

TABLE 3.4-2 (*Continued*)

stress is given approximately by $s = G(\theta/L)c$ or $s = (T/K)c$ where

$$c = \frac{D}{1 + \dfrac{\pi^2 D^4}{16A^2}}\left[1 + \left\{0.118\log_e\left(1 - \frac{D}{2r}\right) - 0.238\frac{D}{2r}\right\}\tanh\frac{2\phi}{\pi}\right]$$

where D, A, and r have the same meaning as before and ϕ = angle through which a tangent to the boundary rotates in turning or traveling around the reentrant portion, measured in radians. (Here r is negative).

Circular sector
$K = Cr^4$ where C depends on α and has values as follows:

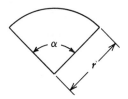

α	45°	60°	90°	120°
C	0.0181	0.0349	0.0825	0.148
α	180°	270°	300°	360°
C	0.296	0.528	0.686	0.878

Max $s = T/Q$ on radial boundary. Here $Q = Cr^3$, where C has values as follows:

α	60°	120°	180°
C	0.0712	0.227	0.35

Isosceles triangle

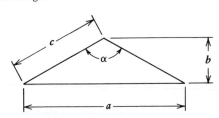

$$K = \frac{a^3 b^3}{15a^2 + 20b^2}$$

For $\alpha = 90°$, $K = 0.0261c^4$
For $\alpha = 60°$, $K = 0.0216c^4$

$Q = 0.0666\,ab^2$
For $\alpha = 90°$, $Q = 0.0554c^3$
For $\alpha = 60°$, $Q = 0.050c^3$
Max S_s at center longest side

Trapezoid

$$K = \tfrac{1}{12}b(m + n)(m^2 + n^2) - V_L m^4 - V_s n^4$$

where

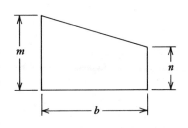

$$V_L = 0.10504 - 0.10s + 0.0848s^2 - 0.06746s^3 + 0.0515s^4$$

$$V_s = 0.10504 + 0.10s + 0.0848s^2 + 0.0674s^3 + 0.0515s^4$$

and

$$s = \frac{m - n}{b}$$

Max S as for elongated section cases.

TABLE 3.4-2   *(Continued)*

Circular shaft with opposite sides flattened
$K = cr^4$ where c depends on ratio w/r and has values as follows:

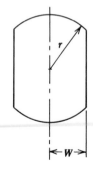

$\dfrac{w}{r}$	$\dfrac{7}{8}$	$\dfrac{3}{4}$	$\dfrac{5}{8}$	$\dfrac{1}{2}$
c	1.357	1.076	0.733	0.438

$Q = cr^3$ where c depends on ratio w/r and has values as follows:

$\dfrac{w}{r}$	$\dfrac{7}{8}$	$\dfrac{3}{4}$	$\dfrac{5}{8}$	$\dfrac{1}{2}$
c	1.155	0.912	0.638	0.471

[a] From Reference 2.
[b] Unless at some other point on boundary there is a sharp reentrant angle, causing high local stress.
Note: $\theta = TL/KG$.

REFERENCES

3.4-1 F. R. Shanley, *Mechanics of Materials*, McGraw-Hill, New York, 1967.
3.4-2 Raymond J. Roark, *Formulas for Stress and Strain*, 4th ed. McGraw-Hill, New York, 1965.
3.4-3 E. F. Bruhn, *Analysis and Design of Flight Vehicle Structures*. Tri-State Offset Co., 1973.
3.4-4 David J. Peery, *Aircraft Structures*, McGraw-Hill, New York, 1950.
3.4-5 E. P. Popov, *Mechanics of Materials*, 2nd ed. Prentice-Hall, Englewood Cliffs, NJ, 1976.

3.5 METHOD OF VIRTUAL WORK AND STATICALLY INDETERMINATE STRUCTURES

3.5-1 Virtual Work and the Deflection of Trusses

Virtual work is one of a number of energy methods that forms an intriguing approach to static, dynamic, and stability analysis of structures. References 1 and 2 provide an interesting summary of this topic.

To do this, consider the truss of Fig. 3.5-1. A truss, 1-2-3, with members 1-3 and 2-3, is loaded by load P at node 3. What is the *structural* deflection at node 3? How does the structure, the assembly of elements, displace?

It is important to distinguish between structural displacements and element deformations. The *structural displacement* in this case is the movement of 3 to 3′ during the loading. The *element deformations* are the respective lengthening or shortening of the two elements due to their own internal loads which result from the application of load P.

It is also very important to distinguish between *external*, or structural, *loads* and *internal member forces*. In Fig. 3.5-1, P is an external load. The two members develop internal forces due to P. The two internal member forces may be found by considering the static equilibrium of node 3. For members 1 and 2, we will denote the internal forces, due to P, as f_1 and f_2.

Look now at Fig. 3.5-2. The element force–deformation relationship is shown graphically in Fig. 3.5-2. The element stress–strain relationship is also shown. The load–deflection (P–Δ) curve for the structure will also be linear.

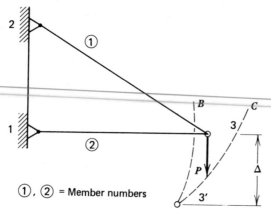

Fig. 3.5-1 Structural deflections of a truss.

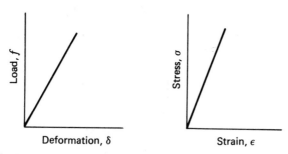

Fig. 3.5-2 Load–deflection curve for structure and stress–strain curve for element.

The external work done by a slowly applied load is

$$We = \frac{P\Delta}{2}$$

The internal work done (strain energy stored in members)

$$u_i = \sum_{i=1}^{2} \frac{f_i \delta_i}{2}$$

The structural deflection Δ is calculated as follows:

$$\Delta = \sum_{i=1}^{2} \frac{u_i f_i L_i}{A_i E_i}$$

It is essential to understand the meaning of each term in the equation.

Symbol	Meaning
Δ	Structural deflection at the desired node and in the desired direction. In this case the vertical deflection of node 3.
u_i	Force in the ith member resulting from the application of a unit load at the desired node and in the direction of the desired deflection.
f_i	Force in the ith member due to the applied external load. In this case P.
L_i, A_i, E_i	The ith member properties.

Generally, then, the equation for calculating the deflection of any determinate truss is

$$\Delta = \sum_{i=1}^{n} \frac{u_i f_i L_i}{A_i E_i}$$

Example 3.5-1. Calculate the displacement of point C in a direction $45°$ with the horizontal as shown below.

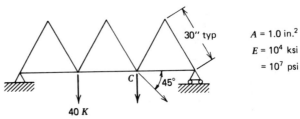

$A = 1.0$ in.^{2}
$E = 10^4$ ksi
$= 10^7$ psi

40 K

Solution. The member forces S resulting from the applied loads are calculated from equilibrium at each joint, values shown are in kips (1000 lb). The forces, u, (in the members) resulting from a unit virtual load at point C in the direction of the desired deflection are given below.

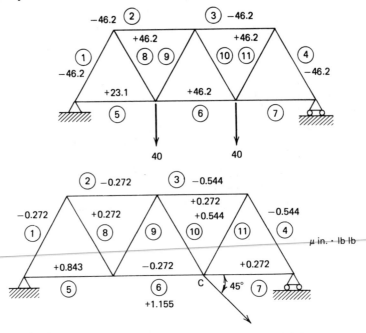

These results are summarized and final calculations are made in the table below.

Member No.	f (kip)	$\dfrac{L}{AE}$ (in./kip)	$\dfrac{fL}{AE}$ (in.)	u (lb/lb)	$\dfrac{ufL}{AE}$ (in.)
1	−46.2	0.003	−0.1386	−0.272	+0.0377
2	−46.2		−0.1386	−0.272	+0.0377
3	−46.2		−0.1386	−0.544	+0.0754
4	−46.2		−0.1386	−0.544	+0.0754
5	+23.1		+0.0693	+0.843	+0.0584
6	+46.2		+0.1386	+ .155	+0.1600
7	+23.1		+0.0693	+0.272	+0.0188
8	+46.2		+0.1386	+0.272	+0.0377
9	0		0	−0.272	0
10	0		0	+0.272	0
11	+46.2		+0.1386	+0.544	+0.0754

$$\Delta = \sum \frac{ufL}{AE} = +0.5765$$

3.5-2 Beam Deflections by the Method of Virtual Work

Before going further and treating beam, frame, and torsional deflections by the method of virtual work, it may be helpful to look at a simple illustration of the relation between virtual work and Castigliano's theorem.

Recall that the calculation for the structural deflections of a truss (which is composed of axial elements) involves two separate calculations, one for the real external loads and one for the appropriate unit load. Let us show the link between this operation and the use of Castigliano's theorem.

Castigliano's theorem says that the deflection of a structure at, and in the sense of, a load P is equal to the first partial derivative of the strain energy with respect to the load:

$$\Delta_p = \frac{\partial U}{\partial P}$$

A simple example of this is the deflection of an axial force element as shown in Fig. 3.5-3. For this type of structural element, the strain energy is

$$U = \frac{P^2 L}{2AE}$$

and

$$\Delta = \frac{\partial U}{\partial P} = \frac{PL}{AE}$$

It is instructive to look at the first partial of the above equation in the following form

$$\frac{\partial U}{\partial P} = \frac{2P(\partial P/\partial P)L}{2AE}$$

The term $\partial P/\partial P = 1$ is the response of this simple structure to a unit load at the proper node, and in the sense of the desired deflection. Thus

$$\Delta = \frac{PL}{AE}$$

In general, when a structure such as a truss is composed of many elements, the strain energy is the sum over all the elements:

$$U = \sum_{i=1}^{n} \frac{f_i^2 L_i}{2A_i E_i}$$

and Castigliano's theory says

$$\Delta = \frac{\partial U}{\partial P} = \sum_{i=1}^{n} \frac{f_i(\partial f_i/\partial P) L_i}{A_i E_i} = \sum_{i=1}^{n} \frac{f_i u_i L_i}{A_i E_i}$$

The term $\partial f_i/\partial P$ represents the u_i for each member. Thus we see the relation between Castigliano's theorem and virtual work.

Strain Energy of a Beam in Bending

A beam in bending stores strain energy. The energy relationship can be derived with the aid of Fig. 3.5-4.

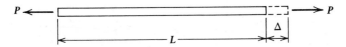

Fig. 3.5-3 Element load deformation.

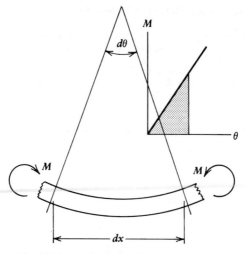

Fig. 3.5-4 $M-\theta$ relationship for pure bending.

We know that

$$y'' = \frac{d\theta}{dx} = \frac{M}{EI}$$

$$d\theta = \frac{M}{EI}\,dx$$

The moment–slope relationship for a beam in pure bending is

$$\theta = \frac{Mx}{EI}$$

If we associate the term $\partial M/\partial P$ with the response of a structure due to a unit load, we can write

$$\Delta = \int_0^L \frac{Mm\,dx}{EI}$$

To generalize this expression and relate it to the method of virtual work, consider a beam loaded by an external load P as shown in Fig. 3.5-5. The external work is $P\Delta_p/2$. Applying the unit load method gives

$$\Delta = \int_0^L \frac{Mm\,dx}{EI}$$

One term is the strain energy due to the external load and one is due to the product of the constant internal moment, m, due to the internal load, riding through the deflections due to P.

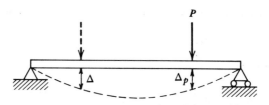

Fig. 3.5-5   Work done during a unit load application.

It is important to know precisely the meaning of each term of the equation. It is also instructive to see the relation between each of the terms:

$$\Delta = \int_0^L \frac{Mm\,dx}{EI}$$

Symbol	Meaning
Δ	Deflection of a beam or frame.
M	Bending moment in beam due to applied external loads.
m	Bending moment in beam due to unit load at, and in the sense of, the desired deflection.
EI	Section properties of beam.

Note that the terms in the equation for beam deflection have the same meaning, in the most general sense, as the terms for truss deflection. Bending moment M represents the response of the structure to external loading, and corresponds to the f_i of the truss. The response of the structure to the unit load is given by m, which corresponds to the u_i of the truss. A pattern is emerging whereby we begin to see that the response of the structure to the "real" external loading and also to the appropriate unit load are both important.

Example 3.5-2. For the beam shown, find the horizontal displacement of point B.

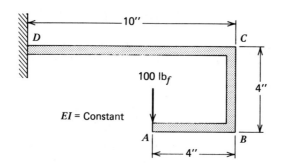

Solution. The moment distribution on the frame due to the 100-lbf load at point A is given below:

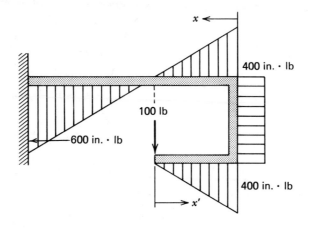

or

$$M = 100x' \qquad 400 \qquad 400-100x$$

$$A \text{ to } B \qquad B \text{ to } C \qquad C \text{ to } D$$

The moment distribution due to the horizontal virtual load f_e is given by:

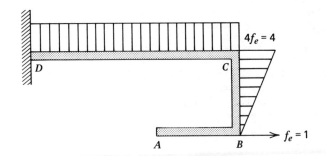

or

$$m = ly \qquad m = 4 \qquad m = 0$$

$$B \text{ to } C \qquad C \text{ to } D \qquad A \text{ to } B$$

Now

$$\delta = \int_A^B \frac{M_A - {}_BM_A - {}_Bds}{EI} + \int_B^C \frac{M_B - {}_CM_B - {}_Cds}{EI} + \int_C^D \frac{M_C - {}_DM_C - {}_Dds}{EI}$$

$$\delta = \int_0^4 \frac{(100x')}{EI}(0)\,dx' + \int_0^4 \frac{400(ly)}{EI}\,dy + \int_0^{10} \frac{(400 - 100x)}{EI}\,4\,dx$$

$$\delta = \frac{400}{EI}\frac{y^2}{2}\Big|_0^4 + \frac{1600x}{EI}\Big|_0^{10} - \frac{400}{EI}\frac{x^2}{2}\Big|_0^{10}$$

$$\delta = -\frac{800}{EI} \text{ in.}$$

Sign indicates deflection in opposite direction of applied unit load.

Example 3.5-3. A circular "arch" is loaded horizontally as shown below. Find the horizontal deflection, Δ_H, of the support.

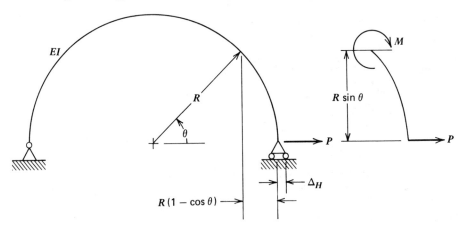

Solution. The appropriate equation for this deflection is

$$\Delta_H = \int_L \frac{Mm\,ds}{EI}$$

Bending moment M is the internal moment due to P. The free-body diagram shown in the small figure above shows

$$M = PR\sin\theta \quad (\text{compression outside} = +)$$

Bending moment m is the moment due to a unit load at, and in the sense of Δ_H. In this problem this is the same as setting $P = 1$. Then

$$m = R \sin \theta$$

In polar coordinates

$$ds = R\, d\theta$$

So that

$$\Delta_H = \int_0^\pi \frac{PR \sin \theta R \sin \theta R\, d\theta}{EI}$$

$$= \frac{PR^3}{EI} \int_0^\pi \sin^2 \theta\, d\theta$$

$$= \frac{PR^3}{EI} \left(\frac{\theta}{2} - \frac{\sin 2\theta}{4} \Big|_0^\pi \right)$$

$$\Delta_H = \frac{\pi PR^3}{2EI}$$

3.5-3 Virtual Work Method for Torsional Members

The derivation for the deflection of a member in torsion by the method of virtual work proceeds in a manner that is analogous to that for beam bending. The derivation will not be given here. The result has the form

$$\delta = \frac{TM_t}{GJ}\, ds$$

where $\delta =$ desired structural deflection
$T =$ internal torque distribution due to external loading
$M_t =$ internal torque distribution due to a unit load applied at, and in the sense of, the desired deflection
$GJ =$ member properties (need not be constant)

The similarity between these terms and those for truss and beam deflections should be noted. In particular note the following:

1. There is always a term that represents the internal response of the structure to external loading. For a truss, S_i; for a beam, M; for torsion, T.
2. There is always a term that represents the internal response of the structure to the appropriate unit load. For a truss, u_i; for a beam, m; for torsion, M_t.
3. A unit load is always put on the structure at, and in the sense of, the desired deflection. Remember that "load" is used in the general sense and may mean a force, moment, or torque.
4. Terms that give the member properties are always included. For a truss, AE; for a beam, EI; for torsion, GJ.

Example 3.5-4. Find the deflection of point A in the direction of the x axis by the virtual work method. Set up integral only.
Solution. The bending moment, torsional, unit bending, and unit torsion diagrams are as shown below for x-axis deflections of point A.

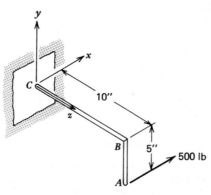

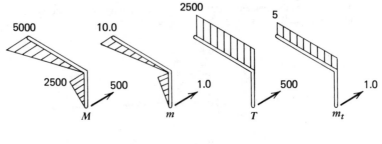

$$\delta_{Ax} = \int \frac{Mm}{EI}\,ds + \int \frac{TM_t}{JG}\,ds$$

$$= \int_0^5 \frac{(2500/5)\,y\,(y)\,dy}{EI} + \int_0^{10} \frac{(5000/10)\,z\,(z)\,dz}{EI} + \int_0^{10} \frac{2500(5)\,dz}{JG}$$

Example 3.5-5. Find the deflection of point A in the direction of the y axis by the method of virtual work. Set up the integrals and then integrate using the section properties given below.

Section Properties:

$E = 29 \times 10^6$ psi

$G = 11 \times 10^6$ psi

$I = 1$ in.^{4}

$J = 2$ in.^{4}

Solution.

$$\delta_{y_{total}} = \int \frac{Mm\,dx}{EI} + \int \frac{Mm\,dz}{EI} + \int \frac{Tt\,dz}{GJ}$$

$$= \int_0^5 \frac{1000x^2\,dx}{EI} + \int_0^{10} \frac{1000z^2\,dz}{EI} + \int_0^{10} \frac{5000(5)\,dz}{GJ}$$

$$= \left.\frac{1000x^3}{3EI}\right|_0^5 + \left.\frac{1000z^3}{3EI}\right|_0^{10} + \left.\frac{25000z}{GJ}\right|_0^{10}$$

Since $E = 29 \times 10^6$ psi, $I = 1$ in.^{4}, $G = 11 \times 10^6$ psi, and $J = 2$ in.^{4} we have

$$\delta_{y_{total}} = \frac{1000(5)^3}{3(29)10^6}(1) + \frac{1000(10)^3}{3(29)10^6}(1) + \frac{25000(10)}{11(10^6)(2)}$$

$$= 0.001437 + 0.011494 + 0.011364$$

$$= 0.024295 \text{ in.}$$

Therefore

$$\delta_{y_{total}} \approx 0.0243 \text{ in.} \quad \text{at point } A$$

3.5-4 Statically Indeterminate Structures

In the previous section we saw how the method of virtual work is used to calculate the deflections of various types of structures. We shall now extend this skill to a systematic way of calculating the forces in statically indeterminate structures. Most "real" structures are statically indeterminate.

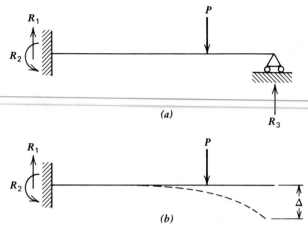

Fig. 3.5-6 Unit load method for a redundant beam.

A structure is statically indeterminate if the number of unknown forces (either internal member forces or reactions) exceeds the equations of static equilibrium available to calculate them. Thus the beam of Fig. 3.5-6 is indeterminate because it has three redundants, R_1, R_2, and R_3 and only two static equations of equilibrium. (Sum of forces in the horizontal direction does not help in this case.) Therefore, we have one excess force, or one redundant. The first step in understanding and solving statically indeterminate structures is to be able to determine the number of redundants. The number of redundants is the number of excess forces, internal or external, that the structures possess over those that can be solved for by using static equilibrium.

The basic plan for solving statically indeterminate structures can be illustrated with the beam of Fig. 3.5-6. The procedure goes as follows:

1. Remove the redundant forces by physically freeing the structure, as shown in Fig. 3.5-6b. In this figure R_3 has been chosen as the redundant force, and R_3 is removed from the structure. The structure will now be *statically determinate*. This statically determinate structure is called the *primary structure*.

2. Calculate the amount that the primary structure displaces in the sense of the (removed) redundant. This is illustrated as displacement Δ in Fig. 3.5-6b.

3. Calculate the amount of redundant force required to restore the structure to its original configuration. In the example of Fig. 3.5-6 this is the magnitude of R_3 that is required to restore Δ to zero. Note that this calculation is also for the primary (determinate) structure. This step enforces structural compatibility.

General Comments

The above example has been used to illustrate a procedure that will be used for all statically indeterminate structures problems. First, the redundants are removed, thus forming a primary structure. The deflections of the primary structure due to the external loading are calculated. Then the magnitude of redundant forces required to restore those displacements back to the original structure configuration are calculated. We shall see that the method of virtual work will be useful for these calculations.

If we have a structure that has n redundants (is n degrees statically indeterminate), then we will end up with a system of n linear algebraic equations in n unknowns, with the n redundant forces as the unknowns. Because it is the redundant forces that are the unknowns in this method of analyzing indeterminate structures, the method is called the force method. The force method is an ancestor of the matrix force method, and the matrix force method was one of the first computer methods for stress analysis.

We can now generalize the notation for deflection.

Recall the steps used to obtain reaction X_a:

1. Cut the structure back to a determinant structure (Fig. 3.5-7a).

2. Calculate deflection, at point of redundant, in direction of redundant, for cut-back structure.

$$\delta_{a0} = \frac{PL_1^3}{3EI} + \frac{PL_1^2 L_2}{2EI} \quad \text{(Fig. 3.5-7b and c)}$$

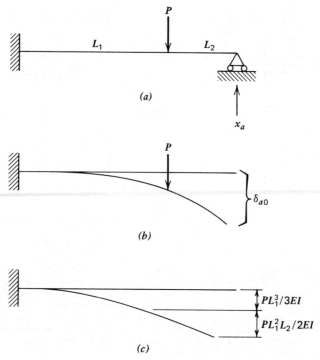

(a)

(b)

(c)

Fig. 3.5-7 Meaning of δ_{a0}.

3. Apply unit load on cut-back structure, in direction of redundant, at redundant.

$$\delta_{aa} = \frac{1(L_1 + L_2)^3}{3EI} \quad \text{(Fig. 3.5-8)}$$

4. X_a must be just enough to push δ_{a0} back to zero.

$$\delta_{a0} + X_a\delta_{aa} = 0$$

$$X_a = \frac{-\delta_{a0}}{\delta_{aa}}$$

Now in general, we denote

δ_{a0}

    — Due to external loads

      — Displacement at the redundant X_a and in the direction of X_a

$\delta_{ab} \ (\delta_{aa}, \delta_{ba}, \ldots, \delta_{mn})$

    — Deflection due to a unit load at redundant X_b and in the sense of X_b

    — Deflection is at the redundant X_a and in the direction of X_a

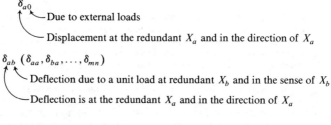

Fig. 3.5-8 Calculation of δ_{aa} and the meaning of the primary structure.

One-Degree Redundant Trusses

A truss with one internal redundant is shown in Fig. 3.5-9. We wish to find the distribution of forces (the internal member forces) for this truss. It is important to realize that—in contrast to a determinate truss—the distribution of internal forces in the truss members is a function of the size and material properties of each truss member. This is, for a determinate truss, the distribution of internal member forces is not affected by member size, but only by static equilibrium. For an indeterminate truss, the distribution of internal forces is affected by member size (really, by member stiffness). To emphasize this, the truss members have been given different properties, as shown in Fig. 3.5-9.

The primary structure is shown in Fig. 3.5-10. Member number 5 has been selected as the redundant and is shown as being cut, but still being present. We solve for the reactions and internal forces and apply the method of virtual work to calculating the "gap" developed in the primary structure; this is δ_{a0}, the deflection of the primary structure due to the external loading. The results are shown in Table 3.5-1.

The results of Table 3.5-1 show the primary structure deflection δ_{a0} to be -77.46×10^{-4} in. This is the amount of gap, or structural incompatibility, that would develop if the determinant primary structure were loaded by the real external structural loads. Note that in using the method of virtual work, we have calculated the u_i as the set of internal forces in the primary structure resulting from the application of a unit, virtual load which is applied at, and in the sense of, the desired deflection. This amounts to the prescription of a unit force in member 5. This is shown in Fig. 3.5-10.

It is important to note that the appropriate unit load in this case is a value of unity in the truss member, and not a 1-lb force put on the nodes of the truss as external loads, as shown in Fig. 3.5-11.

Let us now return to our original objective, which was to calculate the value of the redundant force in member 5. Let us call this value of force X_a. The question is: How much is X_a? To answer this

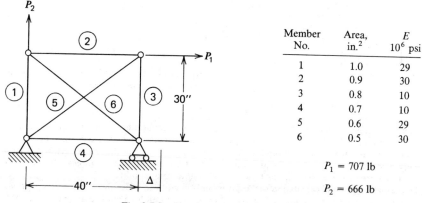

Member No.	Area, in.^{2}	E 10^6 psi
1	1.0	29
2	0.9	30
3	0.8	10
4	0.7	10
5	0.6	29
6	0.5	30

$P_1 = 707$ lb

$P_2 = 666$ lb

Fig. 3.5-9   One degree redundant truss.

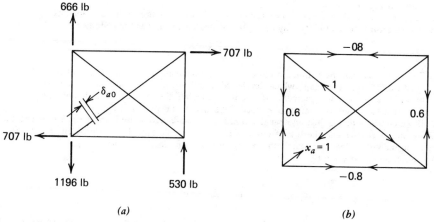

(a)   (b)

Fig. 3.5-10   Responses of the primary structure. (a) To real loads; (b) to unit x_a.

TABLE 3.5-1   SOLUTION FOR INDETERMINATE TRUSS[a]

Member No.	f_i (lb)	u_i (lb)	$\dfrac{L_i}{A_i E_i}$ $(10^{-6}$ in./lb)	$\dfrac{f_i u_i L_i}{A_i E_i}$ $(10^{-4}$ in.)
1	+1196	−0.6	1.03	−7.39
2	+707	−0.8	1.48	−8.37
3	0	−0.6	3.75	0
4	+707	−0.8	5.71	−32.30
5	0	+1.0	2.87	0
6	−883	+1.0	3.33	−29.40

$$\delta_{a0} = \sum \frac{f_i u_i L_i}{A_i E_i} = -77.46 \times 10^{-4} \text{ in.}$$

[a] Tension = (+), Compression = (−)

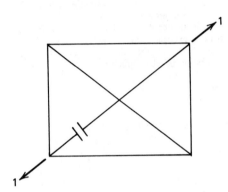

Fig. 3.5-11 Unit load must be introduced directly into the cut member.

question, we must reflect on the meaning of structural compatibility. The *primary* structure has member 5 cut, and develops displacement δ_{a0}. The *real* structure does not have member 5 cut and develops internal force X_a in member 5 instead. The amount of X_a in the *real* structure is therefore, logically, just the amount of force required to pull δ_{a0} back to zero in the *primary* structure. (X_a is the force required to close the gap, or to restore structural compatibility.)

To perform the actual calculation for X_a, it is convenient to consider the displacement response of the primary structure due to a *unit* value of X_a. That is, to a unit load applied at, and in the sense of, the redundant force. It is interesting, and it saves a lot of work, to note that this is the same load shown in Fig. 3.5-10, and already used to calculate the u_i of Table 3.5-1. Note the structure is not loaded as in Fig. 3.5-11.

If we denote the displacement response of the *primary* structure to this unit load as δ_{aa}, then the logic of structural compatibility may be expressed mathematically as the condition

$$X_a \delta_{aa} + \delta_{a0} = 0$$

That is to say that the redundant force X_a, times the displacement due to a unit redundant force δ_{aa}, just cancels δ_{a0}, the "gap" which developed in the primary structure due to the external loads.

Now, let us think for a moment. We want to calculate the response of a determinate truss (the primary structure) due to a unit-applied load. This, again, is depicted in Fig. 3.5-10.

If we consider this as just another determinate truss calculation, and forget for a moment about indeterminate trusses and the rest of our worries, we will note that this is a problem where the real load and the virtual load are exactly the same, namely, a unit load as shown in Fig. 3.5-10. Therefore, the deflection δ_{aa} is given by

$$\delta_{aa} = \sum_{i=1}^{6} \frac{u_i^2 L_i}{A_i E_i}$$

where the u_i have already been calculated in Table 3.5-1. If we perform the sum indicated in the

TABLE 3.5-2   FINAL FORCES FOR INDETERMINATE TRUSS

Member No.	Final Force, f_i (lb)
1	824.84
2	212.12
3	−371.16
4	212.12
5	618.6
6	−264.4

above equation we get

$$\delta_{aa} = 1.252 \times 10^{-5} \text{ in./lb}$$

where δ_{aa} is deflection per unit load and the units are in./lb. Now, using the equation for X_a we get

$$X_a = \frac{-\delta_{a0}}{\delta_{aa}} = 618.6 \text{ lb}$$

Note that we have had to be careful about signs throughout the calculations, with tension being plus and compression being minus.

Once we know redundant force X_a, we can calculate the rest of the member forces from statics. But it is much easier to note that the force in each member is just $X_a u_i + f_i$ since the u_i are the response of the structure to a unit value of redundant. The final forces are shown in Table 3.5-2.

This concludes the calculation for the member forces in the truss. All trusses of one-degree redundancy may be treated with the same general approach. Let us now turn our attention to a different, but related, problem: that of calculating structural deflection, Δ, of the real structure. An example is shown in Fig. 3.5-12, where Δ is the horizontal structural displacement of the right support due to the external loads on the real structure. We note that δ_{a0} and δ_{aa} were not real structure displacements of the primary structure which were calculated for the purpose of determining X_a.

Considering now the structure of Fig. 3.5-12 and the calculation of Δ, we may treat this as another problem of truss deflection by the method of virtual work with

$$\Delta = \sum_{i=1}^{n} \frac{f_i u_i L_i}{A_i E_i}$$

where f_i are the internal member forces in the *real* structure due to the external load. (Just the solution to our original statically determinant problem.) Results are in Table 3.5-2.

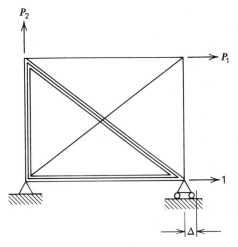

Fig. 3.5-12 Stable substructure.

u_i are the internal member forces in the *real* structure due to a unit load applied at the point of desired displacement and in the direction of desired displacement. L_i, A_i, and E_i are member properties.

Looking at Fig. 3.5-12, the situation at first looks grim. The unit load is shown in the figure, and it appears that we will have to solve another statically indeterminate problem since we must have the u_i for the real structure. However, one feature of the method of virtual work will save us much effort. It is presented here without proof.

For the u_i (the internal response to the external unit load) *any* set of internal forces that is geometrically compatible with the real structure and that is in equilibrium with the external unit load may be used.

This is extremely important because it permits us to select a geometrically compatible subtruss which is statically determinate by selecting particular members from our *real* truss. An example is shown in Fig. 3.5-12, where the subtruss is indicated by heavy lines.

The u_i for this subtruss is simply calculated as $u_4 = 1$. All others equal 0. So the displacement Δ is given by

$$\Delta = \frac{f_4 u_4 L_4}{A_4 E_4} = 1.22 \times 10^{-3} \text{ in.}$$

It is necessary to elaborate slightly on what is meant by a geometrically compatible subtruss. This means that a stable, determinant subtruss must be formed from actual members of the real truss. We cannot introduce new members or create a set of internal equilibrating forces that is not consistent with the real truss.

Example 3.5-6. Truss with Single Redundancy.

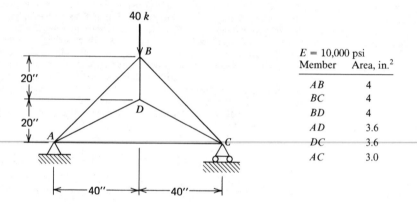

Member	Area, in.^{2}
AB	4
BC	4
BD	4
AD	3.6
DC	3.6
AC	3.0

$E = 10,000$ psi

Select as a redundant the member AC. Cut this member.

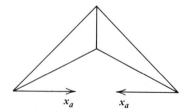

Real loads on cut-back structure may be found from statics. Deflection δ_{a0} may be found from virtual work:

$$\delta_{a0} = \sum \frac{SuL}{AE}$$

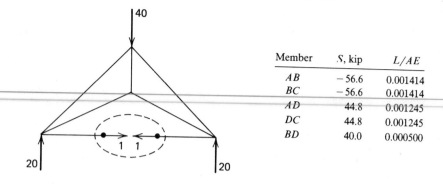

Member	S, kip	L/AE
AB	−56.6	0.001414
BC	−56.6	0.001414
AD	44.8	0.001245
DC	44.8	0.001245
BD	40.0	0.000500

Now the deflection δ_{aa} may be calculated; the appropriate unit force is the one already shown. But the deflection of member AC must be included

$$\delta_{aa} = \frac{u \cdot uL}{AE} = \frac{u^2 L}{AE}$$

Member	u	$\dfrac{SuL}{AE}$	$\dfrac{u^2 L}{AE}$
AB	1.414	−0.1130	0.00283
BC	1.414	−0.1130	0.00283
AD	−2.233	−0.1245	0.00621
DC	−2.233	−0.1245	0.00621
BD	−2.0	−0.0400	0.00200
		−0.5150	0.00267 AC!!!!!
			0.02275

Now

$$X_a = -\frac{\delta_{a0}}{\delta_{aa}} = \frac{0.5150}{0.02275} = 22.6 \text{ kips}$$

The total members forces are

$$S_T = S_0 + X_a u$$

REFERENCES

3.5-1 David J. Perry, *Aircraft Structures*. Chapter 16, McGraw-Hill, New York, 1950.
3.5-2 H. L. Langhaar, *Energy Methods in Applied Mechanics*. Wiley, New York, 1962.

3.6 METHODS OF DEFLECTION ANALYSIS, AND BEAM FORMULAS

3.6-1 Introduction

Section 3.3 discusses the analysis of straight and curved beams and develops the shear and moment relationship and the equation for the elastic axis. Section 3.5 develops the application of virtual work and energy methods to straight and curved beams. In the present are tabulated results for shear and moment diagrams as well as deflection curves for straight and curved beams.

3.6-2 Summary of Deflection Analysis Methods

The deflection analysis methods are:

1. Integration of the equation $d^2y/dx^2 = M/EI$ (see Section 3.3).
2. Integration of the equation $d^4y/dx^4 = q/EI$ (see Section 3.3).

3. The method of virtual work; See Section 3.5.
4. The method of Castigliano; See Section 3.5.
5. The method of graphical integration; See Section 3.6-4.
6. The area–moment method; see Reference 1.
7. The finite element method; see Section 6.5.

If tabulated results cannot be found for a particular problem, one of the above methods should be effective.

3.6-3 Table for Beam Deflection

Table 3.6-1 shows deflection formulas for beams.

TABLE 3.6-1   DEFLECTION FORMULAS FOR BEAMS[a]

1. Cantilever with end load.

$$y = \frac{Fx^2}{6EI}(x - 3l) \qquad y_{max} = -\frac{Fl^3}{3EI}$$

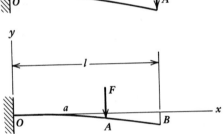

2. Cantilever with intermediate load.

$$y_{OA} = \frac{Fx^2}{6EI}(x - 3a) \qquad y_{AB} = \frac{Fa^2}{6EI}(a - 3x)$$

$$y_{max} = \frac{Fa^2}{6EI}(a - 3l)$$

3. Cantilever with uniform load.

$$y = \frac{wx^2}{24EI}(4lx - x^2 - 6l^2) \qquad y_{max} = -\frac{wl^4}{8EI}$$

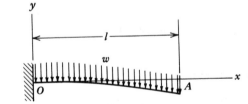

4. Cantilever with moment load.

$$y = \frac{M_A x^2}{2EI} \qquad y_{max} = \frac{M_A l^2}{2EI}$$

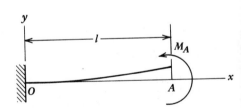

5. Simply supported with center load.

$$y_{OA} = \frac{Fx}{48EI}(4x^2 - 3l^2) \qquad y_{max} = -\frac{Fl^3}{48EI}$$

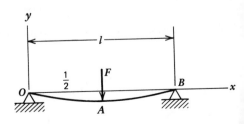

TABLE 3.6-1 (*Continued*)

6. Simply supported with intermediate load.

$$y_{OA} = \frac{Fbx}{6EIl}(x^2 + b^2 - l^2)$$

$$y_{AB} = \frac{Fa(l-x)}{6EIl}(x^2 + a^2 - 2lx)$$

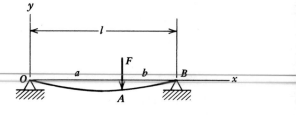

7. Simply supported with uniform load.

$$y = \frac{wx}{24EI}(2lx^2 - x^3 - l^3)$$

$$y_{max} = -\frac{5wl^4}{384EI}$$

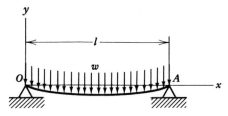

8. Simply supported with moment load.

$$y_{OA} = \frac{M_A x}{6EIl}(x^2 + 3a^2 - 6al + 2l^2)$$

$$y_{AB} = \frac{M_A}{6EIl}\left[x^3 - 3lx^2 + x(2l^2 + 3a^2) - 3a^2l\right]$$

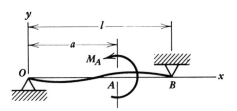

9. Simply supported with twin loads.

$$y_{OA} = \frac{F_x}{6EI}(x^2 + 3a^2 - 3la)$$

$$y_{AB} = \frac{Fa}{6EI}(3x^2 + a^2 - 3lx)$$

$$y_{max} = \frac{Fa}{24EI}(4a^2 - 3l^2)$$

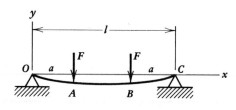

10. Simply supported with overhanging load.

$$y_{OA} = \frac{Fax}{6EIl}(l^2 - x^2)$$

$$y_{AB} = \frac{F(x-l)}{6EI}\left[(x-l)^2 - a(3x-l)\right]$$

$$y_B = -\frac{Fa^2}{3EI}(l+a)$$

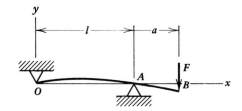

11. One fixed and one simple support with center load.

$$y_{OA} = \frac{Fx^2}{96EI}(11x - 9l)$$

$$y_{AB} = \frac{F(l-x)}{96EI}(5x^2 + 2l^2 - 10lx)$$

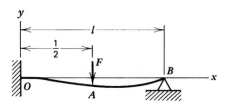

259

TABLE 3.6-1 (*Continued*)

12. One fixed and one simple support with intermediate load.

$$y_{OA} = \frac{Fbx^2}{12EIl^3}[3l(b^2 - l^2) + x(3l^2 - b^2)]$$

$$y_{AB} = y_{OA} - \frac{F(x-a)^3}{6EI}$$

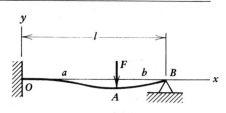

13. One fixed and one simple support with uniform load.

$$y = \frac{wx^2}{48EI}(l-x)(2x-3l)$$

$$y_{max} = -\frac{wl^4}{185EI}$$

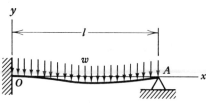

14. Both ends fixed with center load.

$$y_{OA} = \frac{Fx^2}{48EI}(4x - 3l) \qquad y_{max} = -\frac{Fl^3}{192EI}$$

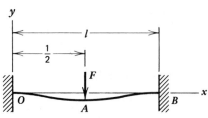

15. Both ends fixed with intermediate load.

$$y_{OA} = \frac{Fb^2x^2}{6EIl^3}[x(3a+b) - 3al]$$

$$y_{AB} = \frac{Fa^2(l-x)^2}{6EIl^3}[(l-x)(3b+a) - 3bl]$$

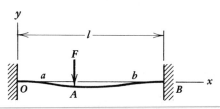

16. Both ends fixed with uniform load.

$$y = -\frac{wx^2}{24EI}(l-x)^2 \qquad y_{max} = -\frac{wl^4}{384EI}$$

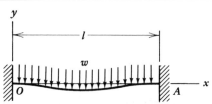

17.

$$M_A = \frac{2EI}{L}\theta_B \qquad R_A = \frac{6EI}{L^2}\theta_B$$

$$M_B = \frac{4EI}{L}\theta_B \qquad R_B = -\frac{6EI}{L^2}\theta_B$$

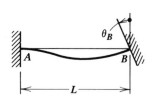

18.

$$M_A = M_0\left(-1 + 4\frac{a}{L} - \frac{3a^2}{L^2}\right) \qquad R_A = \frac{6M_0a}{L^2}\left(1 - \frac{a}{L}\right)$$

$$M_B = \frac{M_0a}{L}\left(2 - 3\frac{a}{L}\right) \qquad R_B = -\frac{6M_0a}{L^2}\left[1 - \left(\frac{a}{L}\right)\right]$$

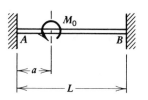

---

[a] From References 1 and 2.

3.6-4 Deflections by Graphical Integration

Beginning with the bending moment diagram, two integrations will yield the slope and deflection diagrams.[2] When this method is used, we usually begin with the M/EI diagram rather than the M diagram. One advantage of the method of graphical integration is that the beam need not have the same moment of inertia throughout its length. Because the slope at the ends of simply supported beams is unknown, we shall illustrate the method with the following example.

Example 3.6-1. The problem selected for demonstration of the method is illustrated in Fig. 3.6-1*a*. This is a shouldered steel shaft, commonly found in rotating machinery. The shoulders are used to

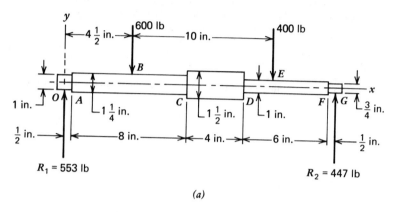

(a)

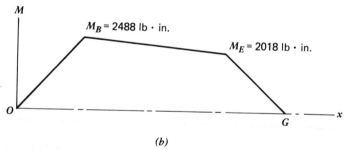

(b)

Shaft Segment	Moment, lb · in.	Moment of inertia, in.^{4}	M/EI, rad/in.
OA	$M_0 = 0$	0.049	0
	$M_A = 276$		$188(10)^{-6}$
AB	$M_A = 276$	0.120	$76.7(10)^{-6}$
	$M_B = 2488$		$691(10)^{-6}$
BC	$M_B = 2488$	0.120	$691(10)^{-6}$
	$M_C = 2300$		$639(10)^{-6}$
CD	$M_C = 2300$	0.248	$309(10)^{-6}$
	$M_D = 2112$		$284(10)^{-6}$
DE	$M_D = 2112$	0.049	$1437(10)^{-6}$
	$M_E = 2018$		$1373(10)^{-6}$
EF	$M_E = 2018$	0.049	$1373(10)^{-6}$
	$M_F = 224$		$152(10)^{-6}$
FG	$M_F = 224$	0.0156	$479(10)^{-6}$
	$M_G = 0$		0

(c)

Fig. 3.6-1 Shaft for graphical integration example.

locate gears, bearings, pulleys, and the like in an axial direction and to resist thrust loads when they are present. In this case the effect of the bearings and gears, say, have been replaced by the forces that they exert on the shaft. This means, as shown, that the beam is assumed to be simply supported and that the distributed forces exerted by the parts on the shaft have been replaced by their resultants. The bending moment diagram is shown in Fig. 3.6-1*b*. Find the deflection diagram, its scale, and the maximum deflection with its location.

   Solution. The first step is to compute the M/EI values corresponding to every point of discontinuity on the diagram that results. The computation is most convenient if made in tabular form. This has been done, and the results are displayed in Fig. 3.6-1*c*. The moment of inertia is computed from the equation $I = \pi d^4/64$. Then, M/EI is computed using $E = 30(10)^6$ psi.

   In Fig. 3.6-2, the shaft drawing is reproduced for reference purposes using the scale $S_x = 4$ in./in. Note that these are two kinds of inches, and so they cannot be canceled. Thus, 4 in./in. means every 4 in. of shaft is represented by a 1-in. space on the drawing. The M/EI diagram is now plotted in line with the shaft drawing using any convenient scale, but, of course, using the same scale for x. In this case, a scale $S_{M/EI} = 800(10)^{-6}$ rad/in.· in. means that an $M/EI = 1437(10)^{-6}$ rad/in. at D occupies about 1.8 in. of drawing, which gives a moderate-sized diagram.

   Before integrating, we divide the diagram into vertical strips, using more strips for the steep portions in order to get a smoother integral. Now choose the pole distance H_2 and place with the

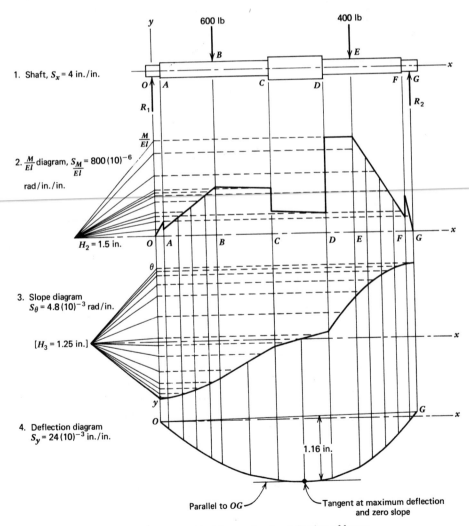

Fig. 3.6-2 Graphical integration for deflection of beam.

graphical integration, as described in Section 3.6-5. The result is the slope diagram. The scale computed is

$$S_\theta = H_2 S_{M/EI} S_\theta = (1.5)(800)(10)^{-6}(4) = 4800(10)^{-6} \text{ rad/in.}$$

or $4.8(10)^{-3}$ rad/in.

Before integrating the slope diagram, choose a place to draw the x axis. Usually, this will be chosen so as to cross the slope diagram near the middle of the beam. This does not mean that the slope is zero at this crossing at all, but only that we think the slope is close to zero at the slope–curve crossing. In other words we do not know where this crossing should be, so we guess at it.

Next, choose a pole distance H_3 and integrate again. This yields the deflection diagram as shown. Note that more vertical strips were used to obtain a smoother integral. The closing chord, OG in this case, will seldom be a horizontal line. The location of the point of maximum deflection, and hence of zero slope, is found by drawing a tangent to the deflection curve parallel to the closing chord. In this case the scale of the deflection diagram is

$$S_y = H_3 S_0 S_x = (1.25)(4.8)(10)^{-3}(4) = 24(10)^{-3} \text{ in./in.}$$

To get the maximum deflection, measure the vertical distance between the tangent and the closing chord. This is 1.16 in. of drawing. Therefore, the maximum deflection is

$$y_{max} = 1.16 S_y = 1.16(24)(10)^{-3} = 0.0278 \text{ in.}$$

This is located, as shown, at $x = 10.5$ in.

3.6-5 Graphical Integration

By using instruments and a reasonably large drawing, the various beam diagrams can be obtained with good accuracy by the use of graphical integration.[1] And these same techniques also constitute a useful means of estimating the shapes of the diagrams by free-hand sketching.

Graphical integration can be explained by reference to Fig. 3.6-3 where function $y = f(x)$ is plotted in part a and the integral in part b. As in graphical differentiation, three scales are necessary —one for the dependent variable, one for the independent variable, and one for the integral. These scales are related in the same manner as for differentiation. The scale formula is

$$S_{yi} = H S_y S_x$$

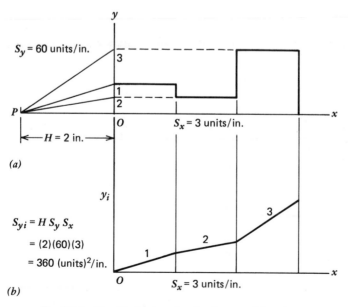

Fig. 3.6-3 Graphical integration for the general function.

where H = pole distance, in.
    S_x = scale of x in units of x per inch
    S_y = scale of y in units of y per inch
    S_{yi} = scale of integral in units of x times units of y per inch

The integral of a function $y = f(x)$ between $x = a$ and $x = b$ is simply the area below the curve between ordinates erected at $x = a$ and $x = b$. To integrate the function of Fig. 3.6-3 choose the line OP, called the pole distance, at any convenient length, say 2 in. Project the tops of the various rectangles horizontally to the y axis and draw the ray $P1$, $P2$, and $P3$ to each of these intersections. Each ray has a slope proportional to the height of the corresponding rectangle. The middle rectangle, for example, is short, and the ray $P2$ has only a small slope; but the last rectangle is tall, and so the ray $P3$ is steep.

The integral in Fig. 3.6-3 is drawn by drawing lines parallel to $P1$, $P2$, and $P3$. Thus, line 1 is parallel to the ray $P1$, line 2 is parallel to $P2$, and line 3 is parallel to $P3$. The scale of the integral is shown in the figure.

When the function to be integrated is a curve, it is divided into a number of vertical strips or segments. Each of these segments is then treated as if it were a rectangle. The integration can then be performed as in Fig. 3.6-3. Of course, the result is not exact, but the approximation is often quite satisfactory for many practical purposes.

3.6-6   Deflection Due to Shear

The shear force in a beam will contribute significantly to its deflection if the beam is relatively short.[1] For long, slender beams the deflections are governed by bending. Consider the following example.

Example 3.6-2.   Find the maximum deflection due to a force P applied at the end of a cantilever having a rectangular cross section, Fig. 3.6-4. Consider the effect of the flexural and shearing deformations.

    Solution.   If the force P is gradually applied to the beam, the external work $W_e = \frac{1}{2} P\Delta$, where Δ is the total deflection of the end of the beam. The internal strain energy consists of two parts. One part is due to the bending stresses; the other is caused by the shearing stresses. These strain energies may be directly superposed.

The strain energy in pure bending is obtained from $U = M^2 \, dx / (2EI)$, by noting that $M = -Px$. The strain energy in shear is found from $dU_{shear} = [\tau^2/(2G)] \, dV$. In this particular case the shear at every section is equal to the applied force P, while the shearing stress τ, is distributed parabolically, as

$$\tau = \left[\frac{P}{2I} \right] \left[\left(\frac{h}{2} \right)^2 - y^2 \right]$$

At any one level y, this shearing stress does not vary across the breadth b and the length L of the beam. Therefore the infinitesimal volume dV in the shear energy expression is taken as $Lb \, dy$. By equating the sum of these two internal strain energies to the external work, the total deflection is obtained:

$$U_{bending} = \int_0^L \frac{M^2 \, dx}{2EI} = \int_0^L \frac{(-Px)^2 \, dx}{2EI} = \frac{P^2 L^3}{6EI}$$

$$U_{shear} = \int_{vol} \frac{\tau^2}{2G} \, dV = \frac{1}{2G} \int_{-h/2}^{+h/2} \left\{ \frac{P}{2I} \left[\left(\frac{h}{2} \right)^2 - y^2 \right] \right\}^2 Lb \, dy$$

$$= \frac{P^2 Lb}{8GI^2} \frac{h^5}{30} = \frac{P^2 Lbh^5}{240G} \left(\frac{12}{bh^3} \right)^2 = \frac{3P^2 L}{5AG}$$

where $A = bh$ is the cross section of the beam. Then

$$W_e = U = U_{bending} + U_{shear}$$

$$\frac{P\Delta}{2} = \frac{P^2 L^3}{6EI} + \frac{3P^2 L}{5AG} \quad \text{or} \quad \Delta = \frac{PL^3}{3EI} + \frac{6PL}{5AG}$$

The first term in this answer, $PL^3/(3EI)$, is the deflection of the beam due to the flexure. The second term is the deflection due to shear. The factor, such as $6/5$ in this term, varies for different shapes of the cross section, since it depends on the nature of the shearing-stress distribution.

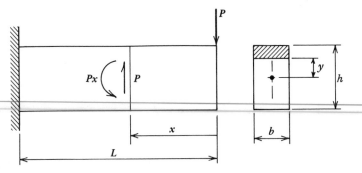

Fig. 3.6-4   Cantilever for shear deflections.

It is instructive to recast the expression for the total deflection Δ as

$$\Delta = \frac{PL^3}{3EI}\left(1 + \frac{3E}{10G}\frac{h^2}{L^2}\right)$$

where as before, the last term gives the deflection due to shear.

To gain further insight into this problem, replace in the last expression the ratio E/G by 2.5, a typical value for steels. Then

$$\Delta = \left(\frac{1 + 0.75h^2}{L^2}\right)\Delta_{\text{bending}}$$

From this equation it can be seen that for a short beam (e.g., one with $L = h$) the total deflection is 1.75 times that due to bending. Hence shear deflection is very important in comparable cases. On the other hand, if $L = 10h$, the deflection due to shear is less than 1%. Small deflections due to shear are typical for ordinary, slender beams. This fact can be noted further from the original equation for Δ. There, the deflection due to bending increases as the cube of the span length, whereas the deflection due to shear increases directly. Hence, as beam length increases, the bending deflection quickly becomes dominant. For this reason it is usually possible to neglect the deflection due to shear.

REFERENCES

3.6-1   E. P. Popov, *Mechanics of Materials*, 2nd ed. Prentice-Hall, Englewood Cliffs, NJ, 1976, 1952.
3.6-2   Joseph Edward Shigley, *Applied Mechanics of Materials*. McGraw-Hill, New York, 1976.

3.7   THEORIES OF STRENGTH AND FAILURE

3.7-1   Introduction

What stresses will cause a part to fail? This is a vital question. For the present, let us define failure as *yielding* or *fracture*. In the case of uniaxial tension (see Section 3.1) the failure criteria are simple. The stress is given by $\sigma = P/A$. The part will *yield* if the uniaxial stress equals the yield strength of the material. The part will *fracture* if the uniaxial stress equals the ultimate tensile strength.

When the stresses are biaxial or triaxial, the situation is more complicated. Various theories of failure have been adopted because they agree with experimental observation. See References 1–3.

3.7-2   Maximum Normal Stress Theory

When various stress components act at a point in a body, it is always possible to use Mohr's circle methods to obtain the principal stresses.[1] In studying failure theories, we shall assume that this has been done. The resulting stress state at a point can then be diagrammed as in Fig. 3.7-1. The three principal stresses σ_1, σ_2, and σ_3 are all shown, with the third, σ_3, shown to emphasize that it must not be forgotten.

No matter what values the principal stresses have, the subscripts can always be interchanged. Thus, we shall assume that $\sigma_1 > \sigma_2 > \sigma_3$ in all cases, with σ_1 being the greatest positive stress and σ_3 being the least (or most negative) stress.

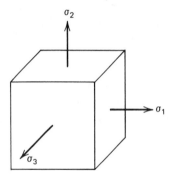

Fig. 3.7-1 Triaxial stress element.

The maximum normal stress theory states that failure occurs when the largest principal normal stress equals the strength. All other principal stresses are ignored in using the theory. And either the yield strength or the ultimate strength can be used. Thus, for tension, failure occurs whenever

$$\sigma_1 = F_{ty} \quad \text{or} \quad \sigma_1 = F_{tu}$$

And a pure compressive failure would occur when

$$\sigma_3 = -F_{cy} \quad \text{or} \quad \sigma_3 = -F_{cu}$$

provided, of course, that σ_1 is a tensile stress and σ_3 a compressive stress.

The maximum normal stress theory is an easy one to use, but it may give results that are on the unsafe side. In fact, it predicts that the yield strength in shear is the same as the yield strength in tension, which is contrary to experimental evidence.

3.7-3 Maximum Shear Stress Theory

This is an easy theory to use; it is always on the safe side of test results, and it is in use in many construction and safety codes. Since it is used only to predict when yielding occurs, it applies only to ductile materials, that is, materials that have a yield strength.

Figure 3.7-2 shows Mohr's circle diagram for the simple tension test. For this test, σ_2 and σ_3 are both zero and yielding occurs when $\sigma_1 = F_{yt}$, as shown. For this circle the maximum shear stress theory states that failure occurs whenever the maximum shear stress at a point equals the maximum shear stress in the tension-test specimen when that specimen begins to yield. Thus, for a general stress state, failure occurs whenever

$$\tfrac{1}{2}(\sigma_1 - \sigma_3) = \tfrac{1}{2}F_{tu}$$

because, with $\sigma_1 > \sigma_2 > \sigma_3$, the maximum shear stress is $(\sigma_1 - \sigma_3)/2$.

According to this theory, the yield strength in shear is

$$F_{ys} = \tfrac{1}{2}F_{tu}$$

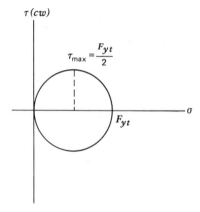

Fig. 3.7-2 Mohr's circle for simple tension.

All three equations in this section can also be written in terms of the yield strength in compression, F_{yc}.

3.7-4 Distortion-Energy Theory

This theory is an important one because it comes closest of all to verifying experimental results. The derivation is based on the assumption that Hooke's law (stress is proportional to strain) applies, and so this theory is valid only for ductile materials, that is, materials that may yield.

In the literature the distortion-energy theory is also called the shear-energy theory, as well as the von Mises–Hencky theory.

The distortion-energy theory originated because of the observation that ductile materials stressed hydrostatically (equal tension or equal compression along three coordinate axes) had yield strengths greatly in excess of the values given by the simple tension test. It was postulated that yielding was not a simple tensile or compressive phenomenon at all, but rather, that it was related somehow to the angular distortion of the stressed element. One of the earlier theories of failure, now abandoned, predicted that yielding would begin whenever the total energy stored in the stressed element became equal to the energy stored in an element of the tension-test specimen at the yield point. This theory was called the maximum strain energy theory; it was a forerunner of the distortion-energy theory. The distortion-energy theory is based on subtracting from the total energy that which produces only a simple volume change. Then, the energy left produces only angular distortion. We compare this energy with the distortion energy in an element of the tension-test specimen to predict yielding.

Let us assume that an element is acted upon by stresses arranged so that $\sigma_1 > \sigma_2 > \sigma_3$.

According to the distortion-energy theory, yielding will occur when

$$\sigma_1^2 - \sigma_1\sigma_3 + \sigma_3^2 = F_y^2$$

For analytical purposes it is convenient to define a von Mises stress as

$$\sigma' = \sqrt{\sigma_1^2 - \sigma_1\sigma_3 + \sigma_3^2}$$

Then, failure is predicted when

$$\sigma_1 = F_y$$

For pure torsion, $\sigma_1 = \tau$ and $\sigma_3 = -\sigma_1$. Failure will occur in torsion when

$$\tau = \frac{S_y}{\sqrt{3}}$$

Since $1/\sqrt{3} = 0.577$, this means that the yield strength in shear, as predicted by the distortion-energy theory, is

$$F_{sy} = 0.577 F_y$$

Example 3.7-1. A ductile steel has a yield strength of 40 kpsi. Find factors of safety, n, corresponding to failure by the maximum normal stress theory, the maximum shear stress theory, and the distortion-energy theory, for the following stress state

$$\text{(a)} \qquad \sigma_x = 10 \text{ kpsi}, \qquad \sigma_y = -4 \text{ kpsi}$$

Solution. (a) A Mohr's circle diagram will reveal that $\sigma_1 = 10$ kpsi, $\sigma_3 = -4$ kpsi, and $\tau_{max} = 7$ kpsi. For the maximum normal stress theory,

$$n = \frac{F_{ty}}{\sigma_1} = \frac{40}{10} = 4$$

For the maximum shear stress theory, we have

$$n = \frac{F_{ty}/2}{\tau_{max}} = \frac{40/2}{7} = 2.86$$

To use the distortion-energy theory, first compute the von Mises stress:

$$\sigma' = \sqrt{\sigma_1^2 - \sigma_1\sigma_3 + \sigma_3^2} = \sqrt{(10)^2 - (10)(-4) + (-4)^2} = 12.5 \text{ kpsi}$$

$$n = \frac{F_{ty}}{\sigma'} = \frac{40}{12.5} = 3.20$$

3.7-5   Failure Theories for Ductile Materials

A ductile material is usually classified as a material that has a yield strength and that exhibits more than 5% elongation in the standard tension test. Such materials have about the same yield strength in compression as in tension.

Let us now select a general plane stress situation composed of the stresses σ_1 and σ_2, which may have any magnitude and be either positive or negative. Thus, $\sigma_3 = 0$ because it is a plane stress state. In order to compare the failure theories with each other, we create a rectangular reference system (Fig. 3.7-3) in which σ_1 is plotted on the horizontal axis and σ_2 on the vertical axis. The positive (tensile) directions are taken upward on the σ_2 axis, and to the right on the σ_1 axis.

The tension test gives the tensile yield strength F_{ty} and the compression test provides F_{cy}. As shown, these points can be located on each axis. Now, by specifying that $\sigma_2 = a\sigma_1$ and letting a vary from 0 to 1, we can calculate sets of values for each failure theory and plot the results in the first and third quadrants. A similar procedure using $\sigma_2 = -b\sigma_1$, with b varying from 0 to 1, will provide the necessary results for completing the paths in the second and fourth quadrants. Figure 3.7-4 shows some of the results from this procedure for the three theories of failure and experimental points.

To use such a diagram, consider the two stress components σ_x and σ_y, which have been plotted to obtain point A on the diagram. Point A falls inside the limits defined by the maximum-normal-stress theory and the distortion-energy theory. So, by these two theories, the stress state σ_x, σ_y is considered safe. But, A falls outside the limit defined by the maximum-shear-stress theory, and so this theory predicts failure. By plotting the principal stresses, the example on this or a similar diagram, it is easy to see the relations between the various factors of safety that were obtained.

Figure 3.7-4 is a chart where only the first and fourth quadrants have been used. This is permissible because σ_1 and σ_2 can always be interchanged. A number of actual test points are shown on the chart with the purpose of discovering which theory is the best one to use.

An examination of the first quadrant, where σ_1 and σ_2 have the same sign, shows all points to be outside the failure lines for the maximum-normal-stress theory and the maximum-shear-stress theory. Thus, both of these theories predict smaller values for the strength, and so they are on the conservative side and are safe to use. Of course, both of these theories give the same results in this quadrant. The first quadrant also shows that the distortion-energy theory comes closest to predicting actual failure.

Examination of the data in the fourth quadrant, where σ_1 and σ_2 have opposite signs, again shows that the distortion-energy theory comes closest to predicting actual failure. Note, too, that all data points fall inside the maximum-normal-stress trajectory. This means that the maximum-normal-stress

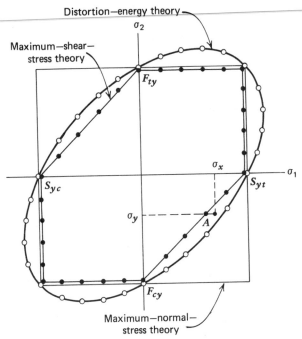

Fig. 3.7-3   Comparison of theories of failure.

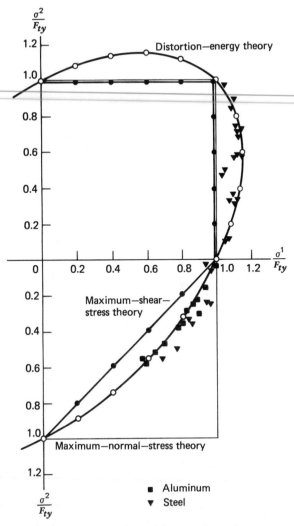

Fig. 3.7-4 Comparison of theory and experiment.

theory is a dangerous one to use when the principal stresses have opposite signs, because it will predict safety when, in fact, none exists. The fourth quadrant plot also shows that the maximum shear stress theory is always on the safe side. Of course, in evaluating any set of experimental evidence, the statistical nature of the data should be considered because there will usually be dome data points that vary considerably from the normal.

3.7-6 Failure Theories for Brittle Materials

In selecting a failure theory for use with brittle materials we first observe that most brittle materials have the following characteristics:

1. The stress–strain diagram is a smooth continuous line to the failure point; failure occurs by fracture, and these materials do not have a yield point.
2. The compressive strength is usually many times greater than the tensile strength.
3. The modulus of rupture F_{su}, sometimes called the ultimate shear strength, is approximately the same as the tensile strength.

The Coulomb–Mohr theory, also called the internal-friction theory, is based upon the results of two tests, the standard tension and compression tests. These experiments give the tensile and

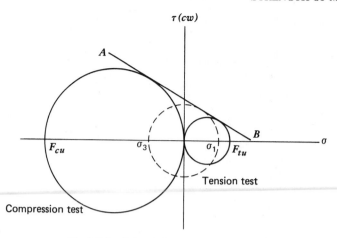

Fig. 3.7-5 Culomb–Mohr theory of failure.

compressive strengths F_{ut} and F_{uc}. The theory is based upon plotting two Mohr's circles on the same diagram, one for each test. These are shown in Fig. 3.7-5. Now draw line AB tangent to both circles. The Coulomb–Mohr theory states that failure occurs for any stress state that produces a circle tangent to the envelope formed by the two test circles and the tangent line. As an example, suppose a particular stress state in which the principal stresses are arranged such that $\sigma_1 > \sigma_2 > \sigma_3$. Then, the largest circle will be formed by σ_1 and σ_3, since σ_2 would lie between σ_1 and σ_3. If this circle is

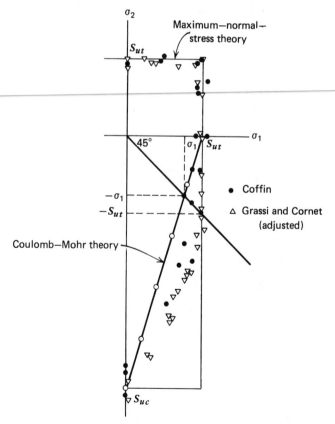

Fig. 3.7-6 Culomb–Mohr theory versus test data.

tangent to the envelope, as shown by the dashed lines in Fig. 3.7-5, then failure will occur. The stresses σ_1 and σ_3 and the two strengths can be related by the equation

$$\frac{\sigma_1}{F_{tu}} + \frac{\sigma_3}{F_{cu}} = 1$$

where both σ_3 and F_{cu} are negative quantities.

Figure 3.7-6 is a plot of the maximum-normal-stress theory and the Coulomb–Mohr theory (with test results for comparison purposes). Note that σ_3 is assumed to be zero, that σ_1 is always tension, and that σ_2 may be either a tensile or a compressive stress. It is seen that the maximum-normal-stress theory is satisfactory when both stresses are tension. Then, with one compressive stress, the Coulomb–Mohr theory is seen to be on the safe side of test results.

For pure torsion $\sigma_2 = -\sigma_1$. This is shown on Fig. 3.7-6 as a 45° line, and results in two predictions for the modulus of rupture F_{su}. Based on the maximum-normal-stress theory, we see that $F_{su} = F_{ut}$. And based on the Coulomb–Mohr theory, the above equation can be solved to give the modulus of rupture as

$$F_{su} = \frac{F_{tu}F_{cu}}{F_{tu} + F_{cu}}$$

3.7-7 Brittle Failures in Ductile Materials

An extremely insidious form of failure is the brittle fracture (failure) in a material that was presumed to be ductile. Ductile materials are more "forgiving" than brittle ones, and they usually give more warning before they fracture. Examples of the occurrence of this in engineering practice are: (i) material becoming embrittled by use at a temperature below its transition *temperature*; (ii) impact loading of notched materials; and (iii) metal fatigue and stress concentration.

Metal fatigue and stress concentration are discussed in Chapter 4. The transition temperature is observed to have an embrittling effect on materials. It should be noted that low temperatures, high rates of strain, and/or notch effects may result in the brittle failure of ductile materials (Reference 2, p. 34). Some transition temperatures for steels are illustrated in Fig. 3.7-7 (Reference 3, p. 335). Note that the transition temperature is related to energy.

The results of notching members subjected to impact loading are similar to those encountered upon static loading (Reference 2, p. 71). Notching results in a concentration of stress at the root of the notch and produces a complex stress pattern which serves to restrain flow. Thus, notching tends to make otherwise ductile materials behave in a brittle fashion. The tendency toward embrittlement is known as notch sensitivity. Notch sensitivity varies considerably among metals and is dependent on heat treatment.

The size and shape of the notch exert a pronounced effect on impact toughness. In general, the deeper the notch, the less the sensitivity of the test to differences in behavior of ductile and brittle metals. The effect of notch sharpness on impact toughness at different temperatures is shown in Fig. 3.7-8. This compares the effect of a standard V notch with a V notch having a crack at its base. The sharper the notch, the smaller the area of ductile fracture under the notch, that is, the sharper the notch, the greater the restraint.

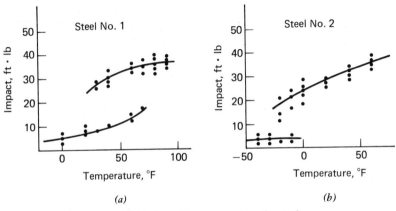

(a)   (b)

Fig. 3.7-7 Transition temperatures for steels.

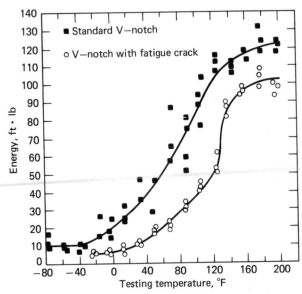

Fig. 3.7-8 Effects of notch sharpness on impact toughness.

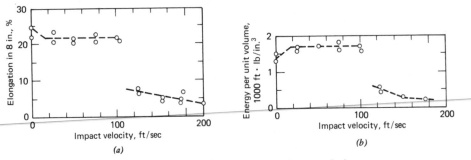

Fig. 3.7-9 Elongation and energy versus impact velocity.

Velocity of Loading

In connection with impact tests velocity has little if any effect on behavior when the velocities vary between 10 and 20 ft/sec. A comparison of standard unnotched tension impact test results with static tests indicates that low and intermediate velocities (up to 200 ft/sec) will usually (but not always) produce somewhat higher values for strength, toughness, and reduction of area, than static loading. However, above some critical velocity of loading the value of toughness decreases rapidly with increasing speed. An example of this is shown in Fig. 3.7-9, which shows energy absorption for a steel that happens to have a critical velocity less than 200 ft/sec.

REFERENCES

3.7-1 Joseph Edward Shigley, *Applied Mechanics of Materials*. McGraw-Hill, New York, 1976.
3.7-2 Carl A. Keyser, *Materials of Engineering*. Prentice-Hall, Englewood Cliffs, NJ, 1956.
3.7-3 Lawrence H. Van Vlack, *Elements of Materials Science*. Addison-Wesley Reading, MA, 1959.

3.8 STRESS CONCENTRATIONS

3.8-1 Introduction

Stress concentrations are the local peak stresses caused by any geometric discontinuity or irregularity in the structure. Stress concentrations are virtually inevitable in "real" structure due to grooves, fillets, oil holes, set screw holes, machining, and just about anything else. The worst stress concentrations are

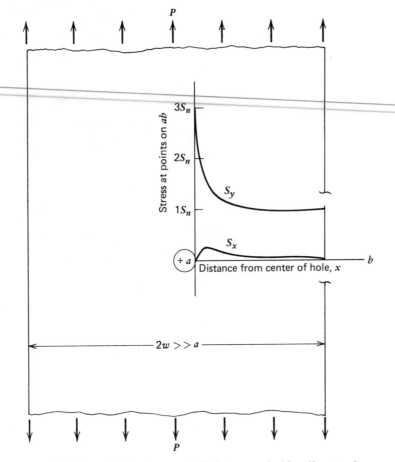

Fig. 3.8-1 Stress distribution around a hole in a panel with uniform tension.

frequently those due to unforseen situations such as machining errors, gravel nicks in propellers, microvoids in microstructures, nonmetallic inclusions, poor workmanship, inadequate detection of damage, and so forth.

Figure 3.8-1 shows the distribution of stress around a circular hole in a flat panel. Let us consider carefully what is happening to the stresses in the vicinity of the hole, for the situation is *worse* than it seems at first.

First, consider a flat panel with no hole at all. The stress in the panel is

$$\sigma = \frac{P}{A} = \frac{P}{tw}$$

When a hole is drilled in the panel, it is natural to assume that the panel has a higher stress due to the fact that there is now less cross-sectional area to carry the load. Thus, one would think that

$$\sigma_{\text{hole}} = \frac{P}{A_{\text{net}}}$$

where A_{net} is the net cross-sectional area left when hole is drilled out.

This equation is NOT correct. This is because the hole causes *severe* local disturbances in the "flow" of stress through the panel. In the region near the hole the peak stress is quite high, as is shown in Fig. 3.8-1.

The stress at the maximum point is given by

$$\sigma_p = \sigma_{\text{peak}} = K_t \frac{P}{A_{\text{net}}}$$

where K_t is the theoretical stress concentration factor.

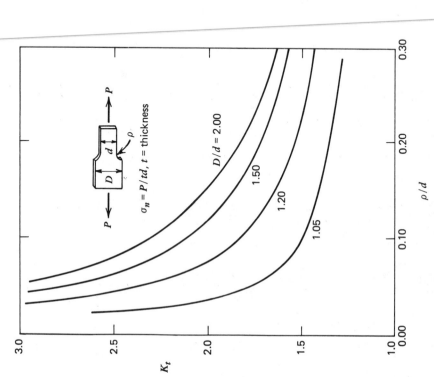

Fig. 3.8-3 Stress concentration factors for a round bar with fillet under tension.

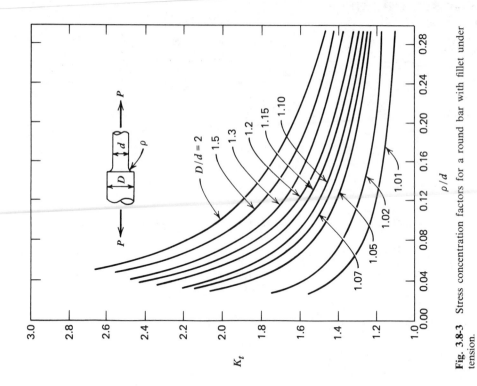

Fig. 3.8-2 Stress concentration factors for flat bar with fillet under tension (based on photoelastic data).

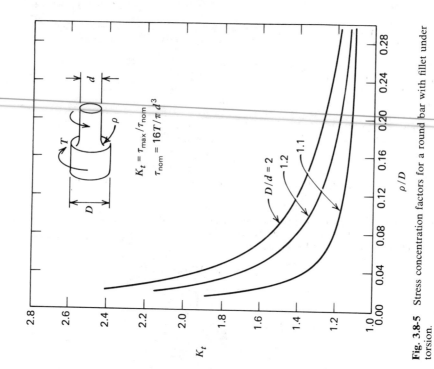

Fig. 3.8-5 Stress concentration factors for a round bar with fillet under torsion.

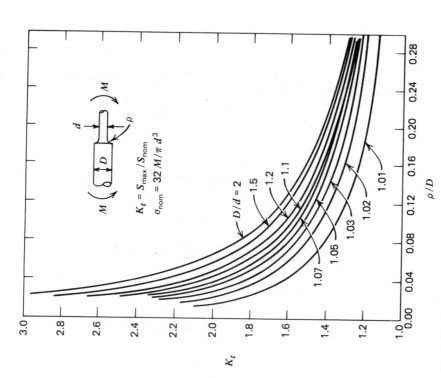

Fig. 3.8-4 Stress concentration factors for a round bar with fillet under bending.

275

K_t is a function of geometry (shape, size) *only*, so long as the material is elastic. Thus, K_t is not a function of load or materials; it is the same for all magnitudes of the same loading state and for all different (elastic) materials.

The usual way of presenting stress concentrations is shown in the figures of this section. Many strength of materials books, civil and mechanical handbooks, company stress manuals, and so on present such charts in great volume. Reference 1 is an entire book devoted to the subject of stress concentration. References 2 and 3 are also recommended. Important stress concentration charts are given in Figs. 3.8-2 through 3.8-5.

The stress concentration factor K_t is usually found by electrical strain gauges or photoelasticity. When K_t is known, the peak stress in a given situation can be found from

$$\sigma_{peak} = K_t \sigma_{nom}$$

where σ_{nom} is the nominal stress calculated from the usual theory ($\sigma = P/A$, etc.) for the case in question.

Example 3.8-1.   The flat coupon shown is loaded in tension. What is the peak stress at the root of the fillet? $D = \frac{3}{4}$ in.; $d = \frac{5}{8}$ in.; $\rho = \frac{1}{8}$ in.; $t = 0.10$.
   Solution.   Refer to Fig. 3.8-2.

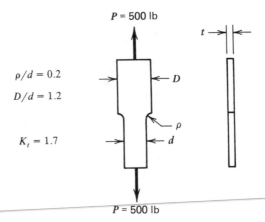

$P = 500$ lb

$t \rightarrow ||\leftarrow$

$\rho/d = 0.2$

$D/d = 1.2$

From the figure

$K_t = 1.7$

D

ρ

d

$P = 500$ lb

3.8-2   Plastic Reduction of the Elastic Stress Concentration Factor

If a part is made of a ductile material, a relatively small load or stress may result in peak stresses that exceed the yield stress. If the plastic zone is small, then the material (ductile) may have no problem tolerating it. The actual magnitude of the peak stress

$$\sigma_p = K_t \sigma_{nom}$$

will lead to a ficticiously high value if it predicts stresses above the elastic limit since the local yielding will result in a lowering of stress magnitude and a redistribution of stress. For these reasons, the stress concentration factor, is often ignored for ductile materials and static loads.

If the material is brittle, the stress concentration factor should be considered. Recall that *ductile materials may fail in brittle fashion due to cyclic loading, shock loading, low temperatures, hydrogen embrittlement, or welding.*

There are instances where plastic behavior may be intentionally permitted. Refer to Fig. 3.8-6.[4] If the load is increased until the peak stress equals F_{ty}, the resulting stress distribution is as shown in Fig. 3.8-6a. Let P_e represent the tensile load on the plate when this occurs. The load can be increased beyond P_e; however, the stress will increase only slightly or not at all, depending on the shape of the plastic range of the stress–strain curve. For an elastic–plastic material with a stress–strain curve depicted in Fig. 3.8-7, the stress will be limited to F_{ty} and the plastic zone will extend as shown in Figs. 3.8-6b and 3.8-6c. *During this time the plastic strain will increase* and fracture will occur at strain value ε_f. Thus, care must be taken in plastic design. If the stress concentration factor for the hole is 2.5, then potentially

$$P_p = 2.5 P_e$$

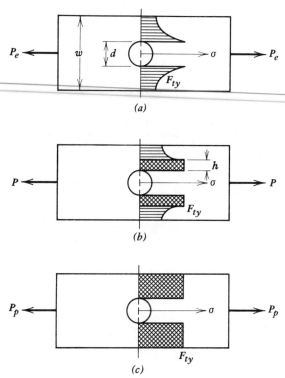

Fig. 3.8-6 Stress concentration beyond elastic limit. (*a*) Elastic limit; (*b*) $P > P_e h$ = plastic zone; (*c*) fully plastic.

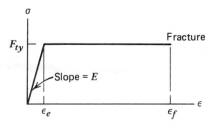

Elastic–perfectly plastic (*E–PP*)

Fig. 3.8-7 Stress–strain curve for rigid-plastic material.

or the plate can tolerate 2.5 times the load that first initiates plastic behavior before the condition of Fig. 3.8-6*c* occurs.

The theoretical stress concentration factor, K_t, depends only on the geometry of the part. Considerations as to its use in full or as to using some reduced values depends on the type of material and the loading conditions in each specific case.[5-7]

The stress concentration in the plastic range for flat plates containing notches and fillets has been given as

$$K_p = 1 + [(K_e - 1) E_S] E_\infty \tag{3.8-1}$$

where E_S is the secant modulus (local at the notch) and E_∞ the secant modulus at a location far removed from the notch. (See Reference 6, which contains an extensive discussion and bibliography.) Reference 7 suggests extending Eq. (3.8-1) to any type of stress concentration factor by

$$K_p = 1 + (K_t - 1) \frac{E_S}{E} \tag{3.8-2}$$

where E is Young's modulus. In addition, it gives the formula for a circular hole in a very wide plate as

$$K_p = 1 + 2\frac{E_S}{E}$$

(3.8-3)

REFERENCES

3.8-1 R. E. Peterson, *Stress Concentration Design Factors*. Wiley, New York, 1953.
3.8-2 Raymond J. Roark, *Formulas for Stress and Strain*, 4th ed. McGraw-Hill, New York, 1965.
3.8-3 Joseph Edward Shigley, *Applied Mechanics of Materials*. McGraw-Hill, New York, 1976.
3.8-4 Richard G. Budynas, *Applied Strength and Stress Analysis*. McGraw-Hill, New York, 1977.
3.8-5 Joseph E. Shigley, *Mechanical Engineering Design*. McGraw-Hill, New York, 1963.
3.8-6 Joseph H. Faupel and Franklin E. Fisher, *Engineering Design*, 2nd ed. Wiley, New York, 1981.
3.8-7 J. A. Collins, *Failure of Materials in Mechanical Design: Analysis, Prediction, Prevention*. Wiley, New York, 1981.

CHAPTER **4**

MECHANICAL PROPERTIES AND SCIENCE OF ENGINEERING MATERIALS

W. F. KIRKWOOD

Consultant, Experimental Stress Analysis
Livermore, California

W. W. FENG
R. G. SCOTT
R. D. STREIT
A. GOLDBERG

Lawrence Livermore National Laboratory
Livermore, California

4.1   BASIC PROPERTIES AND TESTING

W. F. Kirkwood

In Chapter 3 the reader was introduced to the principles of stress and strain and the fundamental methods of stress analysis. By careful scrutiny of the particular stress analysis problems, it may be noted that in some cases it was possible to perform analysis without reference to the elastic constants. When an experimental stress analysis employing elastic strain measurements is performed, the elastic constants must be known. If the analysis is to cover the plastic range, involving cold working, metal forming, or creep of the material, mechanical properties beyond the plastic range will be required.

4.1-1   Stress, Strain, and Displacement

Stress is the internal force per unit area by which a body resists a change in shape. The accepted unit of stress per the American National Standards Institute is the pascal, and these standards will also apply to all variables and constants relative to characterizing engineering materials. The types of stress that are of primary interest are the normal and shear stresses that have been previously discussed in Chapter 3.

Strain is the unit change in length of a body. This change may be produced by external forces, body forces, or a variation in the temperature of the body. As discussed in Chapter 3, strains may be of the normal or shear type. The normal strain in any direction is the displacement in that direction divided by a fixed length over which that displacement was measured as shown in Fig. 4.1-1*a*. In Fig. 4.1-1*b* the shearing strain is the tangent of the angle γ. As can be seen from the strain equations, strain is a dimensionless quantity. In plotting stress–strain curves, the numerical values of strain may be expressed in percent or strain times 10^{-6}. In experimental stress analysis the favored term is microstrain.

Displacement is the linear dimensional change in a body due to external forces. In this chapter displacement will conform to the ANSI standards for length.

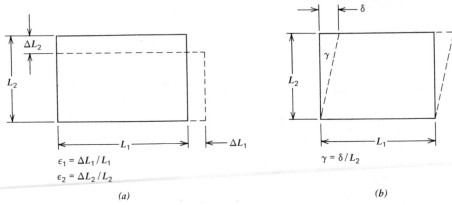

Fig. 4.1-1 Strain and displacement relations. (a) Normal strains; (b) shear strain.

4.1-2 Stress–Strain Diagrams

Uniaxial tension and compression tests on materials are very useful for determining mechanical properties. The response of the instrumented test specimens to axial forces are recorded as force and strain from which a stress versus strain diagram may be plotted. The modern commercially available mechanical properties characterization systems have software written to completely characterize the material.

Stress–strain diagrams may be presented in many forms, but in this section the discussion will cover the ordinary stress–strain diagram and the true stress–strain diagram.

The *ordinary stress–strain diagram* is plotted from a series of simultaneous load and strain readings as they are monitored during the test by a suitable load cell and extensometer, respectively. The stress solution is simply the axial load P divided by the original cross-sectional area A_0 of the specimen. Strain is usually read directly as microstrain from a digital voltmeter or may be recorded simultaneously with the load in digital or analog form as a function of time. From this data the ordinary stress–strain curve may be hand plotted or programmed to draw a graph as shown in Fig. 4.1-2. This diagram provides the information for the basic mechanical properties of the material.

The proportional limit and the elastic limit of the material are often used synonymously and should not be confused. The *proportional limit* is the maximum stress the material will maintain a constant stress–strain relationship. In Fig. 4.1-2b this corresponds to the stress at point A.

The *elastic limit* is the maximum stress applied to the test specimen that will not deform the material upon releasing the load. Determining the elastic limit is time consuming, and the proportional limit is usually accepted as equivalent to the elastic limit.

The *Johnson's apparent elastic limit*, although not generally used anymore, was devised for materials for which the point of departure from the straight line is poorly defined. It is defined as the

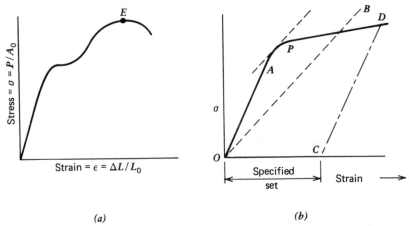

Fig. 4.1-2 Ordinary stress–strain curves derived from uniaxial test specimens.

stress where the stress–strain curve slope is 50% less than the slope at O in Fig. 4.1-2b. This is determined by drawing the slope OB 50% less than the linear portion of OA and then translating itself parallel through point P. This value will then be the Johnson's apparent elastic limit.

The *yield point* as defined by ASTM E6 is that stress where a material exhibits a limiting permanent set. This arbitrary limit is based on an allowable strain devised to fit stress versus strain curves that exhibit a gradual departure from the elastic to the plastic state of the material such as low-carbon steel. This specified set will range between 0.10–0.20% of the original gauge length. Figure 4.1-2b shows the graphical construction for determining the yield point. Draw CD parallel to OA and displace the distance of the specified set $\overline{OC}$. Position D on the stress–strain curve determines the yield stress.

The *tensile strength* is defined by the location on the stress–strain curve where the load on the specimen is a maximum as indicated at E in Fig. 4.1-2a.

Elongation of the material using modern electrical extensometers may be measured until the material fails, provided failure occurs within the gauge length of the extensometer. ASTM A370-77 recommends that specimens be gauge marked with a center punch, scribe multiple divided, or drawn with ink. Elongation is determined after uniaxial testing by fitting the ends of the fractured specimen together and measuring the distance between the original gauge marks. This elongation is the increase of length of the gauge length expressed as a percentage of the original gauge length.

Reduction of area is determined after fracture and is defined as the ratio of the change in the original cross-section area at the smallest cross section divided by the original area of cross section. This ratio is then multiplied by 100 in line with the procedure outlined in ASTM A370.

4.1-3 The True Stress–Strain Diagram

The true stress–strain diagram is a curve of the true stress versus the measured strain measured on a uniaxial tensile specimen. This requires the experimenter to continually monitor axial load and specimen cross-sectional area during the testing history. The true stress is the load divided by the cross-sectional area A at that instant. This stress is then plotted versus the true strain, defined as follows:

$$\varepsilon_{\text{natural}} = \log_e\left(\frac{L}{L_0}\right) = \log_e\left(\frac{A_0}{A}\right) \tag{4.1-1}$$

where $A_0 =$ original cross-sectional area of the specimen
$A =$ instantaneous cross-sectional area of the specimen
$L =$ length of a small element at a given load
$L_0 =$ the original element length

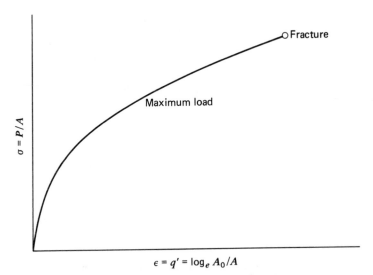

$$\epsilon = q' = \log_e A_0/A$$

Fig. 4.1-3   The true stress–strain diagram.

For round test specimens the true axial strain ε_{nat} is equal to the true change in area:

$$q' = \frac{A_0 - A}{A_0} = 2 \log_e\left(\frac{d_0}{d}\right) \tag{4.1-2}$$

Here d_0 and d refer to the original and the instantaneous diameters, respectively.

Figure 4.1-3 shows a true stress–strain diagram based on these measurements and equations. Although it is assumed that the true stress is based on the known axial load and actual cross-sectional area at any instant in time, this stress in the specimen may be affected by volumetric changes along the necked-down region as failure is imminent. Therefore, at the instant of failure, the stress across the fracture may not be uniform and should be considered the average stress.

It has been shown through experiments on many materials subjected to various environments that the true stress versus strain diagram behaves linearly from the vicinity of the maximum load value to fracture. These tests have also shown that the P/A true stress is really an average stress as noted above but varies very little from the stress that would be present if the specimen had not necked.

4.1-4   Stiffness, Elastic Modulus, and Poisson's Ratio

Whenever a uniaxial specimen is loaded in tension or compression, axial deformation occurs as a function of load. The ratio of stress to strain is a measure of the material stiffness and under tensile stress is expressed as Young's modulus designated by the letter E. If the specimen is tested in torsion, the ratio of the shear stress to shear strain is called the shearing modulus of elasticity or the modulus of rigidity and is expressed by the letter G. The unit for modulus is pascals or pounds per square inch (psi) in the English system of units.

Figure 4.1-4 shows two ordinary stress–strain diagrams from which the modulus of elasticity may be determined. In Fig. 4.1-4a the elastic portion of the curve is straight, producing a constant value of E along $\overline{OA}$. Figure 4.1-4b exhibits a material where the stress–strain relationship is constantly changing. The slope of the line $\overline{OA}$ is called the original tangent modulus at point O. If B is the point at which the modulus is desired, then the slope of the tangent at B is also the tangent modulus at B. If point C is selected and a line $\overline{OC}$ drawn through C, then this slope is the secant modulus of elasticity at C. All these values hold true provided the material still exhibits elastic behavior.

Poisson's Ratio

When a tension or compression specimen is uniaxially loaded, the principal strain e_1 in the axial direction is accompanied by the second principal strain e_2 90° away in the transverse direction. Simultaneous measurements of these two strains will yield a ratio of $e_2/e_1 = \mu = $ Poisson's ratio, a dimensionless quantity.

These simultaneous measurements may be made for e_1 by extensometer or bonded strain gauge and by micrometer or bonded strain gauge for e_2.

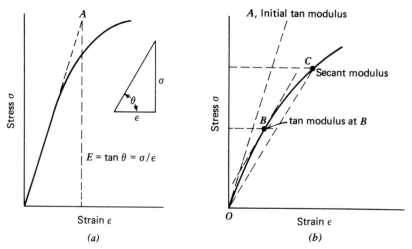

Fig. 4.1-4   Moduli of elasticity. (a) Stress–strain diagram with a linear region for determining E; (b) Stress–strain diagram continually curved.

4.2   TESTING MACHINES

W. F. Kirkwood

Mechanical testing of materials is conducted in commercial, academic, and research laboratories using state-of-the-art equipment designed for specific mechanical properties evaluations. This equipment or testing machines can be classified into two types.

1.   Machines for characterizing mechanical properties of materials.
2.   Machines or integrated structural test components for testing assemblies or structures.

The discussion in this section will be devoted to machines for mechanical properties investigations. All of these mechanical testing machines require a means of loading the specimen and recording the response for evaluation of the results. The principal methods of accomplishing this are discussed as each machine is introduced. Table 4.2-1 classifies several material test machines as to load rate, type of stress, and temperature environment.

4.2-1   Loading Applications

The various configurations of test machines employed for determining mechanical properties of materials may employ one or a combination of the following loading methods.

Weights

Weights may be applied directly to load a specimen in tension, compression, or bending. Some early test machines employed this method of loading. The great disadvantage with this method is that for most specimens, cross-sectional area would be too small to obtain representative material properties.

Weights and Levers

Weights and lever systems similar to commercial scales was another means of applying static loads to specimens for materials testing. But even with this approach, the system does not meet the requirements for efficient accurate testing.

Mechanical Systems

Mechanical gear systems have been employed for loading uniaxial specimens. The loading mechanism employs a motor-driven horizontal shaft with a screw and gear mechanism to transfer rotary motion to a translatory motion of the test machine head.

Hydraulic Pistons

Hydraulic systems employing a pump, pressure regulator, valves, controls, and hydraulic ram for loading the specimen have become very popular in static and dynamic testing for materials and structures.

TABLE 4.2-1   MACHINES FOR MECHANICAL PROPERTIES TESTING

Load Rate	Temperature	Stress Category
Static	Cold, normal, high	*Simple*
		1. Tension
		2. Compression
		3. Bending
		4. Torsion
		5. Shear
		Combined
		1. Biaxial
		a. Tension–tension
		b. Tension–compression
		c. Compression–compression

4.2-2 Static Testing Machines

Simple static testing machines are used for the characterization of mechanical properties of materials. The simple modes of loading are uniaxial tension and compression, bending, shear, and torsion. These machines are of two types:

1. Universal test machines test specifically designed specimens for tension, compression, transverse shear, and bending properties. Some of the later designed universal test systems are capable of torsion tests and multiaxial tests.
2. Special test frames: Some of these machines are loading frames and force systems devoted to specific torsion, compression, or flexure tests.

4.2-3 Universal Test Machines

The modern design test machine has two basic functions: load application system and load measuring system.

In addition, the test laboratory will have various accessories for gripping the specimen, maintaining alignment of the load through the specimen, devices for specimen strain and displacement measurements, and a power supply.

The early commercial, universal test machines employed a single lever for the loading and measuring functions. This design provided no means of compensating for specimen deformation. With the introduction of screw and hydraulic driven machines, the load measuring system became independent, and these are the types that are discussed in this section.

4.2-4 Screw-Gear Machines

Screw-gear driven universal test machines apply the load mechanically by a screw and gear mechanism. The test machine is usually a rectangular frame consisting of three horizontal members;

1. Fixed crosshead at the top.
2. Movable crosshead.
3. Platen or bedplate.

The vertical members are:

1. Two or four columns for connecting the fixed crosshead and platen.
2. Two or more screws for driving the movable crosshead.

Fig. 4.2-1 120,000 lbf screw-driven universal test machine.

Tension loading of the specimen is performed with the specimen mounted between the top crosshead and the screw-driven movable crosshead. Compression testing is accomplished with the specimen mounted between the movable crosshead and the platen or bedplate.

A screw-gear machine is shown in Fig. 4.2-1 with the movable and fixed members clearly evident.

4.2-5 Hydraulic Machines

The introduction of hydraulic rams into the universal machine test frame has broadened the capabilities of mechanical properties testing. Although the term *static testing* is generally used in materials characterization with these machines, hydraulic power provides means of higher strain rates and many waveshape functions.

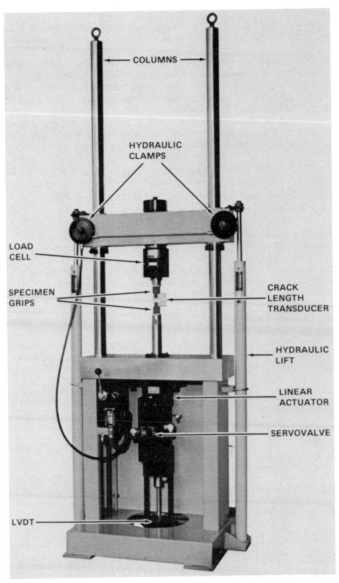

Fig. 4.2-2 Hydraulic universal testing machine with closed-loop control. (Courtesy Koehring, Pegasus Division.)

The test frame for the hydraulic loading universal test machine has a movable crosshead that is positioned vertically and fixed into position prior to testing by mechanical or hydraulic clamps.

Figure 4.2-2 shows a modern hydraulic loading universal test machine. The system shown here consists of a basic load frame with hydraulic lifts for moving the crosshead, a linear actuator (hydraulic ram) with a LVDT (linear variable displacement transducer) attached to it, a load cell, and suitable grips.

The hydraulic lifts raise and lower the crosshead to accommodate various size test specimens and are secured by hydraulic locks. The load cell attached to the crosshead measures the applied force during the test. At the bottom of the test frame the LVDT is attached to the lower platen or bedplate and measures the actuator displacement.

4.2-6 Basic Control Systems

Conventional universal test machines operate on the basis of open-loop control for testing specimens. The control modes available with modern equipment are crosshead or ram displacement, load control using a load cell, or strain control with a transducer on the specimen.

Figure 4.2-3 shows a hydraulic open-loop system using displacement control. If the valve is left uncontrolled, the position is unregulated. This requires a human operator to observe and control the position.

Figure 4.2-4 shows a typical closed-loop control system. Lacking a human operator, there is no need for a visual indicator of piston position. Instead a transducer is substituted, providing a signal proportional to piston position. Signals from this transducer are transmitted to a controller that compares it with the signal from the manually adjusted command control. This signal difference drives the control valve regulating the fluid flow in a direction to balance the signals.

For an explanation of closed-loop testing, refer to Fig. 4.2-5. Here the control mode of the test is load, and the linear programmer has been set for a function of tensile load versus time. The three-bridge load cell sends out three individual signals:

1. Through the load control mode switch to the servovalve which regulates the oil to the upper end of the double-acting ram. This will maintain the programmed load rate.
2. To the dial indicator where the load is visually monitored.
3. To the *XY* recorder.

The strain transducer mounted on the specimen is also sending a signal to the *XY* recorder. This provides the second signal, enabling the operator to simultaneously record load and strain and to make a load versus strain plot during the test.

The machines that have been discussed may be employed for testing in tension, compression, and bending. Structural testing of machine components and assemblies may be accomplished with fixturing devised to secure and load the particular subject.

4.2-7 Torsion Testing Machines

Testing a specimen to determine the shear strength is the primary function of torsion testing machines. These machines are produced as screw powered or hydraulic powered. In the past it was not practical to adapt the universal test machines to torsion testing. However, with closed-loop control systems

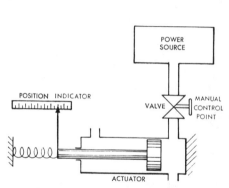

Fig. 4.2-3 Open-loop control system.

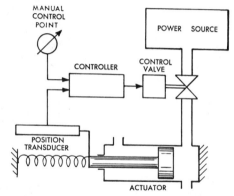

Fig. 4.2-4 Closed-loop control system.

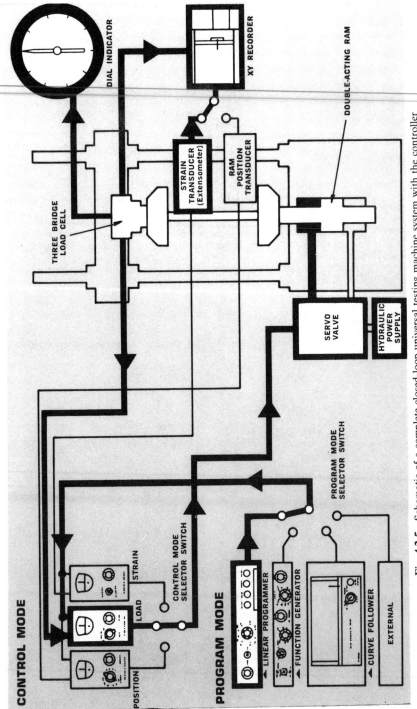

Fig. 4.2-5 Schematic of a complete closed-loop universal testing machine system with the controller set in the load control mode. (Courtesy Timius Olsen Testing Machine Co.)

DIAL INDICATOR

XY RECORDER

DOUBLE-ACTING RAM

STRAIN TRANSDUCER (Extensometer)

RAM POSITION TRANSDUCER

THREE BRIDGE LOAD CELL

SERVO VALVE

HYDRAULIC POWER SUPPLY

CONTROL MODE

STRAIN

LOAD

POSITION

CONTROL MODE SELECTOR SWITCH

PROGRAM MODE

LINEAR PROGRAMMER

FUNCTION GENERATOR

CURVE FOLLOWER

EXTERNAL

PROGRAM MODE SELECTOR SWITCH

some of the manufacturers have produced multiaxial test machines that can simultaneously load and control test specimens in three loading modes:

1. The conventional uniaxial tension or compression test.
2. Torsion—independent actuator.
3. Pressurize the specimen—independent system.

A properly designed torsion specimen may then be mounted in this machine. The control modes selected would be torque load (newton-meters versus time) or angular displacement versus time. A second parameter would be load control from the universal test machine hydraulic actuator. During the conduct of the test and as the torque function is applied, the axial forces have been programmed for zero load.

4.2-8 Multiaxial Testing

When machines and structures are in service, they are usually subject to forces that impose combined stresses on the components. Some designers and stress analysts require the mechanical properties of materials subject to two principal stresses in order to certify a design. Some special combined stress testing apparatus have been built.

The early experiments on combined stresses were conducted on solid right circular cylindrical specimens. These specimens were subject to combined stresses of torsion with axial loading or torsion combined with bending. This develops a nonuniform stress distribution, and in order to eliminate this difficulty thin-wall tubular specimens were designed. The thin-wall tube provides three test options:

1. Torsion and axial tension.
2. Internal pressure and axial tension.
3. Internal pressure and axial compression.

The principal stresses in a thin-wall tube subjected to combined torsion and axial tension loads are as follows and they have opposite signs:

$$\left.\begin{matrix}\sigma_1 \\ \sigma_2\end{matrix}\right\} = \frac{F}{2A} \pm \sqrt{\frac{F^2}{4A^2} + \frac{M_T^2 r^2}{I_p^2}} \tag{4.2-1}$$

Fig. 4.2-6 Closed-loop universal test machine providing axial, torsion, and internal pressure capability for biaxial testing.

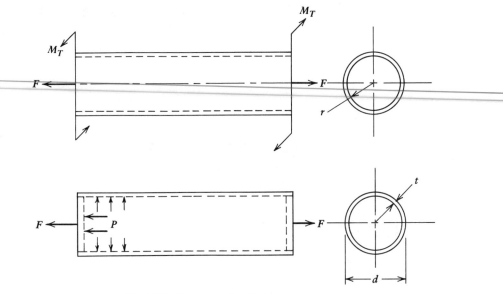

Fig. 4.2-7 Specimens for biaxial stress tests.

where A = cross-sectional area of the tube
F = axial load
r = outer tube radius
M = twisting moment
I_p = polar moment of inertia

The principal stresses for the hollow tube subject to axial tensile load and internal pressure are both positive in sign:

$$\sigma_1 = \frac{F}{A} + \frac{Pd}{4t} \tag{4.2-2}$$

$$\sigma_2 = \frac{Pd}{2t} \tag{4.2-3}$$

where F = axial load
P = internal pressure
t = wall thickness
d = internal tube diameter

In combined stress tests the load rates should be programmed to maintain a constant ratio between the principal stresses. This keeps the stress–strain plots relatively smooth and the biaxial properties such as yield, ultimate strength, and ductility may be obtained.

Figure 4.2-6 shows a modern closed-loop universal test machine capable of performing the three multiaxial modes of testing already discussed. Axial and torsion loading is simultaneously accomplished by initiating the test programmed to a common time base. The torsion loading is in a secondary closed control loop.

A combined pressure and axial loading test is also accomplished on a common time base with independent closed-loop control. Specimens for biaxial stress tests are illustrated in Fig. 4.2-7.

4.3 CHARACTERIZING MECHANICAL PROPERTIES

4.3-1 Tension Tests—Metals

W. F. Kirkwood

The tension test is the most common method of investigating the mechanical properties of metals. These tests are usually conducted in a universal test machine with specially designed specimens. The specimen may be flat or cylindrical, and special grips hold them aligned as accurately as possible to ensure central application of the load. The standards for the mechanical testing of metals in the

TABLE 4.3-1 TYPICAL PROPERTIES OF STRUCTURAL MATERIALS

Metals

Material	δ, Weight, lb/in.^{3}	γ, Coefficient of Thermal Exp., °F $\times 10^5$	E, Million lb/in.^{2}	G, Million lb/in.^{2}	Poisson's Ratio	Strength Properties, thousand lb. per sq in.								
						Tensile Properties			Compressive Properties		Shear Properties		Modulus of Rupture in Cross-bending	Endurance Strength (Rot. beam, 10^7 cycles)
						Ultimate Strength (σ_u)	Elastic Limit	Yield Pt. or Yield Strength	Ultimate Strength	Yield Strength	Ultimate Strength	Yield Strength		
Aluminum, cast, pure	0.0976	1.30	9	3.7	0.36	11	5	—	—	—	—	—	—	11
Aluminum, cast, 220-T4	0.093	1.36	9.5	3.55	0.33	42	—	22	—	23	30	—	—	—
Aluminum, wrought, 2014-T6	0.101	1.28	10.6	4	0.33	68	—	60	—	62	39	35	—	20
Aluminum, wrought, 6061-T6	0.098	1.30	10	3.75	0.3	38	—	35	—	35	24	20	—	17
Beryllium copper	0.297	0.93	19	7	—	100–200	110–150	140	—	—	100–130	70–100	—	40
Brass, naval	0.304	1.18	15	5.5	—	57–75	—	25–50	—	—	40–45	—	—	0.35σ_u
Bronze, phosphor, ASTM B159	0.320	0.99	15	6.5	—	100–150	60–110	—	70–110	50–85	70–110	50–85	—	0.32σ_u
Cast iron, gray, no. 20	0.251	0.60	14	—	0.25	20	—	—	90	—	32	—	46	10
Cast iron, gray, no. 30	0.260	0.60	15.2	—	0.25	30	—	—	115	—	44	—	57	14.5
Cast iron, gray. no. 40	0.260	0.60	18.3	—	0.25	40	—	—	130	—	51	—	66	19
Cast iron, gray. no. 60	0.270	0.60	19	—	0.25	60	—	—	180	—	72	—	100	24
Cast iron, malleable	0.266	0.75	26	8.8	0.25	50–65	—	32–45	200	—	49	—	62	32
Cast iron, nodular	0.257	0.66	23.5	—	0.25	60–100	—	45–65	200	—	—	—	—	—
Magnesium, AZ80A-T5	0.065	1.60	6.5	2.4	0.34	55	—	38	—	17	24	—	—	16
Titanium, pure	0.163	0.53	15.5	5.8	0.34	65–80	—	55–70	—	—	—	—	—	0.6σ_u
Titanium, alloy, 5 Al, 2.5 Sn	0.161	0.57	17	6.2	0.33	115	—	110	—	110	100	—	—	0.6σ_u

Steel for bridges and buildings, ASTM A7-61T														
All shapes	0.283	0.65	29	11.5	0.27	60–75	—	33	—	33	—	17	—	$0.5\sigma_u$
Plates $t < 1.5$	0.283	0.65	29	11.5	0.27	60–72	—	33	—	33	—	17	—	$0.5\sigma_u$
Plates $t > 1.5$	0.283	0.65	29	11.5	0.27	60–75	—	33	—	33	—	17	—	$0.5\sigma_u$
High-strength low-alloy structural steel, ASTM A242-63T														
Most shapes	0.283	0.65	29	11.5	0.27	70	—	50	—	50	—	25	—	$0.5\sigma_u$
Plates $t < 0.75$	0.283	0.65	29	11.5	0.27	70	—	50	—	50	—	25	—	$0.5\sigma_u$
Plates $0.75 < t < 1.5$	0.283	0.65	29	11.5	0.27	67	—	46	—	46	—	23	—	$0.5\sigma_u$
Plates $1.5 < t < 4$	0.283	0.65	29	11.5	0.27	63	—	42	—	42	—	21	—	$0.5\sigma_u$
High-strength steel castings for structural purposes, ASTM A148-60 (7 grades)	0.283	0.83	29	11.5	0.27	80–175	—	40–145	—	—	—	—	—	$0.4\sigma_u$
Steel, spring, carbon, SAE 1095	0.28	—	30	—	—	170–220	125–170	—	—	—	—	—	—	$0.36\sigma_u$
Steel, spring, alloy, SAE 4068	0.28	—	30	—	—	200–270	175–240	—	—	—	—	—	—	
Steel, ball bearings, SAE 52100	0.28	—	30	—	—	326	—	—	—	—	—	—	—	
Steel, stainless (0.08–0.2 C, 17 Cr, 7 Ni) $\frac{1}{4}$ hard	0.28	0.96	28	12.5	—	125	—	78	—	67	—	—	—	$0.35\sigma_u$
Same, full hard	0.28	0.96	26.6	12.0	—	185	—	150	—	99	—	—	—	$0.35\sigma_u$

TABLE 4.3-1 (*Continued*)

Wood Products[a]

| Species | Weight lb/ft³ | γ, Coefficient Thermal Exp., °F | | E,[b] lb/in.² | Compressive Strength | | | Shear Strength With Grain | Bending Strength | |
| | | With Grain | Across Grain | | With Grain | | Across Grain Elastic Limit | | Modulus of Rupture (rect. section) | Fiber Stress at Elastic Limit |
					Ultimate	Elastic limit				
Ash (white)	41	0.0000053	—	1,680,000	7280	5580	1510	1920	14,600	8,900
Birch (sweet, yellow)	44	0.0000011	0.000016	2,070,000	8310	6200	1250	2020	16,700	10,100
Elm (American)	35	—	—	1,340,000	5520	4030	850	1510	11,800	7,600
Hickory (true)	51	—	—	2,180,000	8970		2310	2140	19,700	10,900
Maple (sugar)	44	0.0000012	—	1,830,000	7830	5390	1810	2430	15,800	9,500
Oak (red)	44	0.0000019	0.00002	1,810,000	6920	4610	1260	1830	14,400	8,400
Oak (white)	48	0.0000027	0.00003	1,620,000	7040	4350	1410	1890	13,900	7,900
Fir (Douglas)	36	—	—	1,920,000	7420	6450	910	1140	11,700	8,100
Hemlock (Eastern)	30	—	—	1,200,000	5410	4020	800	1060	8,900	6,100
Spruce (Sitka)	26	—	—	1,570,000	5610	4780	710	1150	10,200	6,700
Cypress (southern)	32	—	—	1,440,000	6360	4740	900	1000	10,600	7,200
Pine (southern long-leaf)	40	0.000003	0.000019	1,990,000	8440	6150	1190	1500	14,700	9,300

Brick and Masonry

Material	Weight, lb/ft³	γ, Coefficient, Thermal Exp., °F.	E, lb/in.²	ν	Ultimate Strength Values			
					Compression	Tension	Shear	Modulus of Rupture in Bending
Concrete	150	0.0000060	—	—	σ_c	$\frac{1}{10}\sigma_c$	$\frac{1}{2}\sigma_c$	$200 + 0.09\,\sigma_c$
1:1½:3, w/c = 6.5	—	—	3,500,000	0.15	3,500			
1:2½:3½, w/c = 7.5	—	—	3,000,000	0.13	2,500			
1:3:5, w/c = 9.0	—	—	2,500,000	0.10	1,500			
(w/c = water/cement ratio, gal per sack)			(at 30 days)		(at 30 days)			
Brick	—	0.000003		—	σ_c			
Soft	120		1,500,000	—	2,000	—	—	400
Medium	—		—	—	3,500	—	—	600
Hard	144		3,500,000	—	5,000	—	—	900
Vitrified	120		—	—	10,000	—	—	1,800
Brick masonry								
1:3 Portland cement mortar	—		—	—	$0.30\,\sigma_c$	—	—	
1:3 lime mortar	—		—	—	$0.15\,\sigma_c$	—	—	
(wall or pier height = 15 × thickness)								
Granite	168	0.0000036	7,000,000	0.28	25,000	—	—	2,500
Limestone	166	0.0000028	6,000,000	0.21	8,000–16,000	—	—	700–1,500
Marble	175	0.0000038	8,000,000	0.26	12,000	—	—	1,200
Sandstone	156	0.0000052	2,500,000	0.28	6,000	—	—	600

[a] Values are taken from *Wood Handbook* and are based on tests of small specimens of select, clear, seasoned wood at 12% moisture content. The endurance limit in reversed bending is approximately 28% of the modulus of rupture.

[b] G is approximately one-sixteenth E.

United States are defined by ASTM E8. These standards establish the methods that will be most likely to assure quality results in materials testing. When these tests are performed with skilled investigators and qualified technicians using the proper equipment and test machines, the results will present a reliable measure of material performance. Table 4.3-1 is a compilation of data from U.S. government and industrial research laboratory tests on various structural materials using several ASTM standards.

4.3-2 Standard Tensile Specimens

W. F. Kirkwood

The geometry of test specimens is based on the material available for fabrication, usually the cross section will be round or rectangular. Cylindrical metal specimens are preferred over rectangular ones because by design the gripping ends may be easily aligned with the longitudinal center line. When the specimen is made from sheet or plate stock, a flat specimen is more conveniently fabricated.

Figure 4.3-1 shows a typical tensile specimen design where the ends may be modified to match the gripping device. It is desired that the specimen be designed to produce failure within the gauge length. This requires a generous fillet in either round or bar stock specimens to provide a smooth stress transition into the grip area (minimum stress concentration).

Table 4.3-2 shows several ASTM specimen designs with some basic dimensions. The model 1 specimen with a 12.5 mm diameter is generally used for testing metals. If the material diameter is less than that or lacks the required thickness, smaller-diameter models may be scaled down from this drawing. Figure 4.3-2 shows two standard ASTM E8 specimens of the 12.5-mm design. The top specimen in this figure has been mounted in the threaded grips ready for installation in a universal test machine.

4.3-3 Specimen Grips

W. F. Kirkwood

The ASTM E8 publication has illustrations of grip designs universally accepted by engineers and material scientists for mounting tensile specimens in the standard universal test machines. Figure 4.3-3 shows several of these configurations used for testing cylindrical and flat specimens.

The wedge grip illustrated in Fig. 4.3-3a is satisfactory for most commercial tests of ductile materials provided the specimen is long enough to minimize bending. As the wedge grips have no alignment adjustments, they are not satisfactory for testing brittle materials. The faces of the grips are serrated to reduce slippage during loading.

In the ideal uniaxial test the load should pass through the longitudinal centerline of the specimen to obtain uniform stress distribution across the minimum section. Bending of the specimen is minimized in cylindrical specimens by employing spherical seats as shown in Fig. 4.3-3b. Some machines also employ a device called a flexure, which performs like a universal joint. The distance between the spherical seats depends on the combined length of the test specimen and the load linkage, with the longest possible assembly producing the least bending.

4.3-4 Tension Test Procedures

W. F. Kirkwood

Tests on uniaxial tension specimens should be run in a universal test machine capable of operating in one of three control modes prescribed by the customer or principal investigator. Most commercial tests on metals are run in the displacement control mode until fracture, and the data obtained will

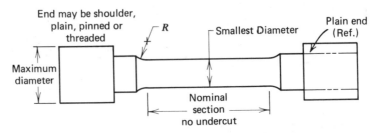

Fig. 4.3-1 Typical cylindrical tensile specimen.

Model No.	Specimen Sketch	Remarks
1	Threaded 12.7 140	For metals and alloys except brittle material Ref. ASTM E8
2	432 max. 60% Nom. dia. Nom. dia. 50.8 max.	For plastic rod. Use test machine JAWS 89 mm Ref. ASTM D638-82a
3	19 m 29 246 See D 638	For sheet, plate and molded plastic Ref. ASTM D638-82a
4	12.7 50.8 203.2	Recommended for metal plate with maximum thickness of 16 mm Ref. ASTM E8

^{a}Dimensions are for the largest models of each specimen.

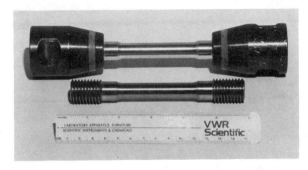

Fig. 4.3-2   Standard cylindrical 0.505-in. specimen with threaded ends. Upper specimen has grips attached.

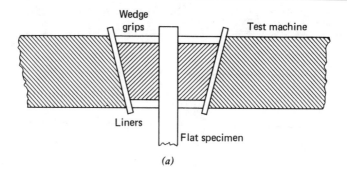

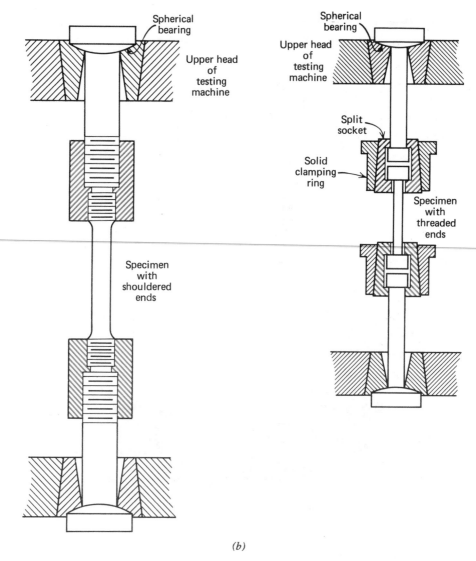

Fig. 4.3-3 Grips for tensile tests. (*a*) Flat specimen; (*b*) cylindrical specimens.

TABLE 4.3-3 DESIGNATED TEST MACHINE SPEEDS FOR SOME TENSILE SPECIMENS PER ASTM

Material	ASTM Ref.	Maximum Crosshead Speed (mm/min)		Stress Rate (mPa/min)
		To Yield	To Ultimate	
Metals	E8		12/24 mm of gauge length	
Steel	A370	1.57/25.4 mm of gauge length		690 to yield
Gray cast iron	A48		3.175 above 103.4 mPa	

		Specified Grip Speed (mm/min)	Nominal Strain Rate at Start of Test (mm/mm/min)
Plastics			
Type 1, 2, and 3	D638	5 ± 25%	0.1
Rods and tubes		50 ± 10%	1.0
Rigid and semirigid		500 ± 10%	10.0
Wood	D143		
Parallel to grain		1.27 to Ultimate	
Perpendicular to grain		2.54 to Ultimate	

provide the strength and elastic modulus properties. Table 4.3-3 provides a general guide for tensile testing requirements on crosshead speeds.

There are two other modes for testing specimens. Strain control enables the investigator to control the strain rate with an electrical extensometer attached to the specimen. Load control may be employed by the investigator if the test requirements are based on rate of stress. These modes of control are discussed in the section on test machines.

Uniaxial tension or compression tests will usually observe displacement and load (Table 4.3-4). If it is possible to record these data simultaneously as a function of time, it will facilitate drawing a stress–strain diagram. If an electrical extensometer is attached to the specimen, it is possible to plot a load versus strain diagram on a suitable recorder. Modern systems will completely program the test machine to control the input, record all the pertinent data, and plot the results.

TABLE 4.3-4 DESIGNATED TEST MACHINE SPEEDS FOR SOME COMPRESSION SPECIMENS PER ASTM

Material	ASTM Ref.	Recommended Crosshead Speed or Stress Rate	
Metals	E9	0.005/min to 0.003/min[a]	
Concrete	C39	Screw Machine 1.3 mm/min idling speed	Hydraulic Machine *Load Control* 0.14 to 0.34 mPa/s may use faster rate during first half of anticipated maximum load
Rigid plastics	D695	1.3 ± 0.3 mm/min until yield Increase to 5–6 mm/min until specimen breaks	
Wood	D143		
Parallel to grain		0.60 mm/min	
Perpendicular to grain		0.30 mm/min	

[a]Strain rate limits if material is rate sensitive.

4.3-5   Mechanical Characterization of Rigid Plastics

W. F. Kirkwood

Plastic materials tend to be temperature and strain rate sensitive, and material properties may show considerable variation. The tensile properties may vary with specimen preparation although if the material is reasonably homogeneous, isotropic, and uniform in size, the results between specimens will usually be comparable.

Tensile and compression tests on plastic materials are conducted to determine the fundamental mechanical properties. These properties are determined from the recorded data of crosshead displacement and load in order to develop the stress–strain curve.

The standard test method for tensile properties of plastics is covered in ASTM D638 and specimens are shown in Table 4.3-2. These methods may provide useful results for plastics engineering design purposes, but the determination of a true elastic limit in plastics is debatable. At low stresses and at room temperature (21°C) or colder, many plastics will exhibit a linear response for determining the slope for the elastic modulus. Figure 4.3-4 shows the response of an epoxy specimen from which the elastic modulus was readily determined. This specimen was run in strain control at 21°C with an averaging electrical extensometer attached to it.

The calculations for mechanical properties of plastics depends on the stress–strain curve to determine the values of strength, elongation, and initial modulus. These are patterned similar to the ordinary stress–strain diagrams (Fig. 4.1-2) and the appendix of ASTM D638.

Figure 4.3-5 shows a plastic tensile specimen installed in a universal test machine that includes an environmental chamber for temperature control. This specimen is a standard dog-bone with plastic caps telescoped over and epoxied to the specimen. The end caps have female threaded inserts that pick up the test machine components. Axial loading is maintained by the upper universal joint and the lower ball joint.

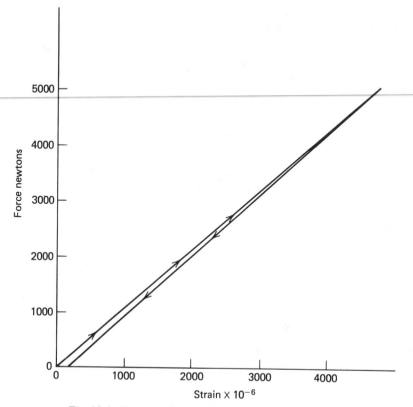

Fig. 4.3-4   Force–strain curve for an anhydride-cured epoxy.

Fig. 4.3-5 Rigid plastic tensile specimen in a universal test machine equipped with a thermal chamber for controlled temperature tests. The load train to the specimen employs an upper universal joint and lower ball joint for load alignment.

Another tensile test setup for plastics is shown in Fig. 4.3-6. This schematic shows a 12.7 mm diameter plastic specimen. The split cone seats center the specimen in the special designed fixtures at the specimen ends. Alignment is maintained by the lower universal joint and the commercial flexure at the top. Neither Fig. 4.3-6 nor 4.3-7 conforms to the exact procedures of ASTM D638 on specimen design, but they have proved very satisfactory for characterizing high-explosive plastic materials.

Compressions tests of rigid plastics are conducted in universal test machines, and standards are prescribed by ASTM D695. The preferred specimens are right-circular cylinders 12.7 mm diameter by 25.4 mm long. When determining modulus and offset yield stress data, it is suggested that the right-circular cylinder will have the dimensions 12.5 mm diameter by 50.8 mm long. Also cylinders of 20.26 mm (0.798 in.) diameter by 81 mm (3.2 in.) length are convenient for converting axial load to stress when it is acceptable to express stress in pounds per square inch.

4.3-6 Testing Viscoelastic Materials

W. F. Kirkwood

In order to cover the general behavior of viscoelastic materials, this section will consider the test philosophy applicable to uniaxial testing in both the tensile and compressive modes.

The mechanical behavior of these materials are functions of time, temperature, and load rate. Although it is not completely true, the only practical way to characterize these materials over the

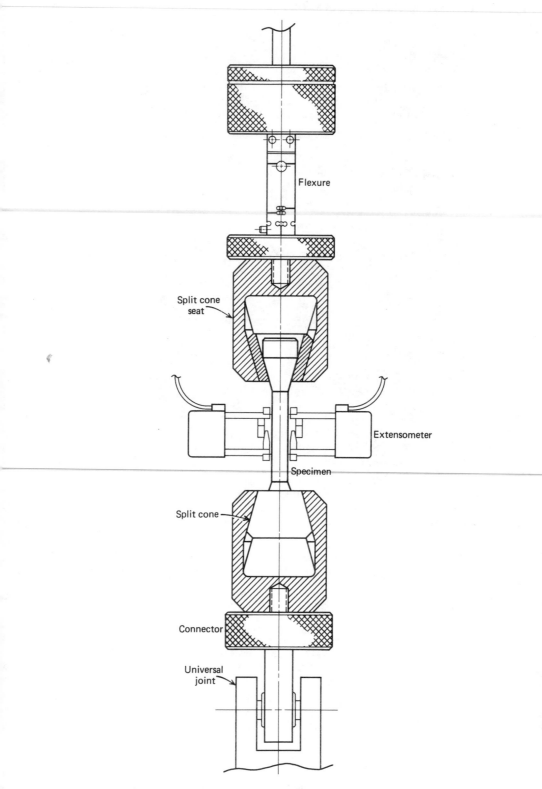

Fig. 4.3-6 Schematic of plastic tensile specimen in a universal test machine.

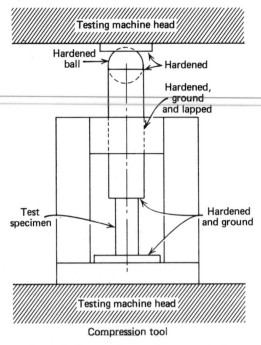

Fig. 4.3-7 Compression test fixture for plastics.

temperature range of -54 to $+74°C$ is to make the following assumptions:

The material is homogeneous.

The material is isotropic.

Linear elastic or linear viscoelastic theories apply to characterization.

Poisson's ratio is time and temperature independent.

The material is thermorheologically simple (the time–temperature postulate holds).

These assumptions have been explored and found to be reasonable for characterizing viscoelastic materials at the Lawrence Livermore National Laboratory.

The initial step for determining the appropriate mechanical properties data is to evaluate the time frame in which the structural components will be performing. A plot of stress magnitude versus time given in Fig. 4.3-8 is useful for indicating the type of data needed to characterize the material. If the component is to perform in a shock or impact environment, then elastic theory applies. All other regions slower than shock may combine elastic and viscoelastic behavior with all time scales above 1 sec requiring viscoelastic solutions.

Figure 4.3-9 shows the difference in response between elastic and viscoelastic materials when under constant stress. The elastic material when stressed below the yield point with a constant load will exhibit a constant strain response, while the viscoelastic material will have both an initial strain response and a time-dependent strain response called creep.

Consider another type of uniaxial test where the specimen is subjected to a programmed saw-tooth controlled strain input. Figure 4.3-10a shows a fast and slow strain rate input. Figures 4.3-10b and c show the response of elastic and viscoelastic materials, respectively, to these inputs. Notice that the elastic response is independent of strain rate and time, but the viscoelastic material response is rate dependent. Also, note the stress reversal occurring during the descending portion of the stress–time curve for the viscoelastic specimen.

The following section on viscoelastic modeling provides the basis for applying mechanical properties to engineering analysis.

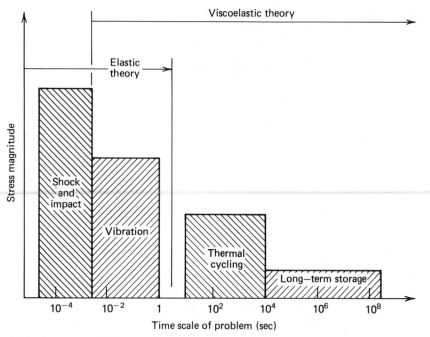

Fig. 4.3-8 Stress magnitude versus time for mechanical environments that influence the appropriate theory for stress analysis of viscoelastic materials.

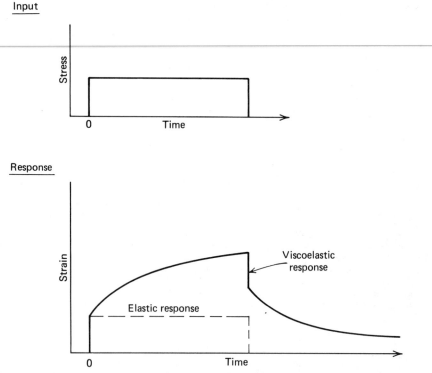

Fig. 4.3-9 Elastic and viscoelastic material responses to constant stress and unloading.

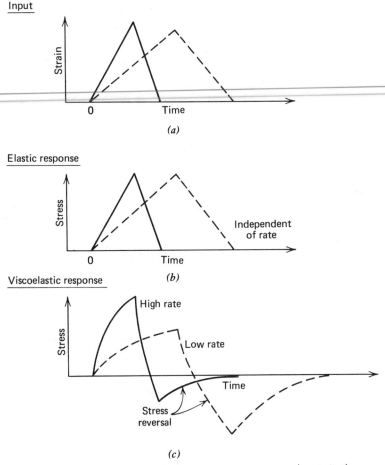

Fig. 4.3-10 Elastic and viscoelastic response to constant strain rate testing.

4.3-7 Viscoelastic Models

W. W. Feng

Introduction

The theory of elasticity accounts for materials that store energy while loading and release the stored energy while unloading. No energy is dissipated during the loading and unloading cycles. On the other hand the theory of Newtonian viscous fluids accounts for materials that do not store energy; all external work done on Newtonian viscous fluids is dissipated. The elastic solids and the Newtonian viscous fluids are two extreme cases. In reality, all materials possess the capacity to store some part of energy and dissipate the remaining part. These materials are categorized as viscoelastic or elastic-plastic materials. The deformation of elastic-plastic materials is independent of the time scale involved in loading and unloading, while the deformation of viscoelastic materials has a specific time or rate dependence. In this section the viscoelastic materials are considered.

Linear Viscoelastic Models

The models for obtaining one-dimensional stress–strain relationships can be obtained by various arrangements of springs and dashpots. The spring represents the part of the material that stores energy and the dashpot represents the part that dissipates energy. The most commonly known models are Maxwell fluid, Kelvin solid, three-parameter solid, and three-parameter fluid models. The governing differential equations that relate stress and strain, the creep compliance, and the relaxation modulus

TABLE 4.3-5 LINEAR VISCOELASTIC MODELS

Model	Name	Differential Equation / Inequalities	Creep Compliance $J(t)$	Relaxation Modulus $G(t)$
	Maxwell fluid	$\sigma + p_1\dot{\sigma} = q_1\dot{\varepsilon}$	$(p_1 + t)/q_1$	$\dfrac{q_1}{p_1} e^{-t/p_1}$
	Kelvin solid	$\sigma = q_0\varepsilon + q_1\dot{\varepsilon}$	$\dfrac{1}{q_0}(1 - e^{-\lambda t}), \quad \lambda = \dfrac{q_0}{q_1}$	$q_0 + q_1\delta(t)$
	Three-parameter solid	$\sigma + p_1\dot{\sigma} = q_0\varepsilon + q_1\dot{\varepsilon}$ $q_1 > p_1 q_0$	$\dfrac{p_1}{q_1}e^{-\lambda t} + \dfrac{1}{q_0}(1 - e^{-\lambda t}),$ $\lambda = q_0/q_1$	$\dfrac{q_1}{p_1}e^{-t/p_1} + q_0(1 - e^{-t/p_1})$
	Three-parameter fluid	$\sigma + p_1\dot{\sigma} = q_1\dot{\varepsilon} + q_2\ddot{\varepsilon}$ $p_1 q_1 > q_2$	$\dfrac{t}{q_1} + \dfrac{p_1 q_1 - q_2}{q_1^2}(1 - e^{-\lambda t})$ $\lambda = q_1/q_2$	$\dfrac{q_2}{p_1}\delta(t) + \dfrac{1}{p_1}\left(q_1 - \dfrac{q_2}{p_1}\right)e^{-t/p_1}$

for these models are shown in Table 4.3-5. The creep compliance is defined as the response of strain due to a unit step input of stress. The relaxation modulus is defined as the response of the stress due to a unit step input of strain. In Table 4.3-5 p_i and q_i $(i = 1, \ldots,)$ are material constants of the springs and the viscosity coefficients of the dashpots. The symbol $\delta(t)$ is the Kronecker delta. It is to be noted that with various combinations of arrangements of springs and dashpots, other models can be achieved.

Convolution Integrals

As the relaxation modulus is known, the stress response can be determined under any strain loading condition through a convolution integral. These equations are

$$\sigma(t) = \varepsilon(0)G(t) + \int_0^t G(t - \tau)\frac{d\varepsilon(\tau)}{d\tau}\,d\tau \tag{4.3-1}$$

or

$$\sigma(t) = \varepsilon(t)G(0) + \int_0^t \varepsilon(\tau)\frac{dG(t - \tau)}{d\tau}\,d\tau \tag{4.3-2}$$

with a similar equation for the strain response due to arbitrary stress input:

$$\varepsilon(t) = \sigma(0)J(t) + \int_0^t J(t - \tau)\frac{d\sigma(\tau)}{d\tau}\,d\tau \tag{4.3-3}$$

or

$$\varepsilon(t) = \sigma(t)J(0) + \int_0^t \sigma(\tau)\frac{dJ(t - \tau)}{d\tau}\,d\tau \tag{4.3-4}$$

Actually, one can bypass the spring dashpot models for determining the creep compliance and the relaxation modulus, and determine them directly from the integral Eqs. (4.3-1) and (4.3-3).

The viscoelastic material properties are determined when either the creep compliance or the relaxation modulus is given. In fact, they are related by the following integral equation:

$$\int_0^t G(t - \tau)\frac{dJ(\tau)}{d\tau}\,d\tau = \mu(t) \tag{4.3-5}$$

where $\mu(t)$ is the unit step function. Therefore, only one test, either creep or relaxation test, is required for determining the viscoelastic material properties.

Generalized Constitutive Equations

The one-dimensional constitutive equations can be extended to general three-dimensional constitutive equations. There are many forms of constitutive equations. A general and most common one are presented here:

$$\sigma_{ij}(t) = G_{ijkl}(t)\varepsilon_{kl}(0) + \int_0^t G_{ijkl}(t - \tau)\frac{d\varepsilon_{kl}(\tau)}{d\tau}\,d\tau \tag{4.3-6}$$

$$\sigma_{ij}(t) = G_{ijkl}(0)\varepsilon_{kl}(t) + \int_0^t \varepsilon_{kl}(t - \tau)\frac{dG_{ijkl}(t - \tau)}{d\tau}\,d\tau \tag{4.3-7}$$

In the above equations the indicial notations are used. The free indices take the value from 1 to 3. The repeated indices indicate summation from 1 to 3. The stress σ_{ij} and the strain ε_{ij} are second-order tensor quantities, and G_{ijkl} are the fourth-order tensor relaxation modulus. A set of similar constitutive equations for strains, when the stress history is known, can be obtained.

A Nonlinear Constitutive Equation

Viscoelasticity was mainly developed due to the large-scale development and utilization of polymeric materials. Utilization of the polymeric material often undergoes finite deformation; therefore, it is important to develop a nonlinear viscoelastic constitutive equation. The constitutive equations developed by Christensen for nonlinear viscoelastic incompressible materials and for finite deformations are presented here:

$$\sigma_{ij}(t) = -p\delta_{ij} + x_{i,K}(t)x_{j,L}(t)\left[g_0\delta_{KL} + \int_0^t g_1(t-\tau)\frac{\partial E_{KL}(t)}{\partial \tau}d\tau\right] \quad (4.3\text{-}8)$$

where again cartesian tensor notation is employed with x_K denoting the initial, underformed configuration, while $x_i(t)$ relates to the deformed configuration. Symbol p is the hydrostatic pressure, g_0 is an elastic constant, while $g_1(t)$ is the viscoelastic relaxation function with

$$\lim_{t\to\infty} g_1(t) = 0 \quad (4.3\text{-}9)$$

The strain E_{KL} is given by

$$E_{KL} = \tfrac{1}{2}(x_{i,K}x_{i,L} - \delta_{KL}) \quad (4.3\text{-}10)$$

For sufficiently slow processes, the integral term in (4.3-8) becomes negligibly small and the remaining terms constitute the theory of rubber elasticity.

The strain tensor (4.3-10) for an incompressible material involving simple extension is given by

$$(E_{KL}) = \frac{1}{2}\begin{bmatrix} \lambda^2 - 1 & 0 & 0 \\ 0 & \dfrac{1}{\lambda} - 1 & 0 \\ 0 & 0 & \dfrac{1}{\lambda} - 1 \end{bmatrix} \quad (4.3\text{-}11)$$

where λ is the stretch ratio and is defined by

$$\lambda = \frac{L}{L_0} \quad (4.3\text{-}12)$$

In the above equation L and L_0 are the deformed and undeformed length of the simple extension specimen, respectively. Using the condition $\sigma_{22} = \sigma_{33} = 0$ to evaluate p, it is found that the constitutive equation reduces to

$$\sigma_{11}(t) = g_0\left(\lambda^2 - \frac{1}{\lambda}\right) - \frac{1}{2\lambda}\int_0^t g_1(t-\tau)\frac{d}{d\tau}\frac{1}{\lambda(\tau)}d\tau + \frac{\lambda^2}{2}\int_0^t g_1(t-\tau)\frac{d\lambda^2(\tau)}{d\tau}d\tau \quad (4.3\text{-}13)$$

Equation (4.3-13) is a constitutive equation of an incompressible rubberlike material subjected to simple extension finite deformations.

BIBLIOGRAPHY

Flügge, W. *Viscoelasticity*. Blaisdell, Berkeley, CA, 1967.

Christensen, R. M. *An Introduction Theory of Viscoelasticity*, 2nd ed. Academic, New York, 1982.

4.3-8   Testing Wood Specimens

W. F. Kirkwood

As wood is a product of multitudinous purposes, it is necessary to discuss some of the basic procedures that are recommended for mechanical properties characterization. These recommendations are the results of extensive testing and evaluation by the U.S. Forest Service, the Forest Products Laboratories of Canada, and other similar organizations. The methods proposed here are detailed in ASTM D143 and are under the jurisdiction of ASTM Committee D7 on wood.

The principal mechanical tests are conducted to obtain mechanical properties both parallel and perpendicular to the grain. These are static bending, compression parallel to the grain of the wood, impact bending, toughness, compression perpendicular to the grain, hardness, shear parallel to the grain, cleavage, and tension parallel to the grain. Figure 4.3-11 shows a specimen designed for parallel to the grain tension testing. For tensile testing perpendicular to the grain, the specimen is designed as shown in Fig. 4.3-12. In each case special grips to load the shoulder or the circular cut out is necessary. These are from ASTM D143, which also has examples of load displacement curves of tests run on several wood varieties. Table 4.3-1 shows representative wood properties of tests run on clear straight-grained specimens with a 12% moisture content.

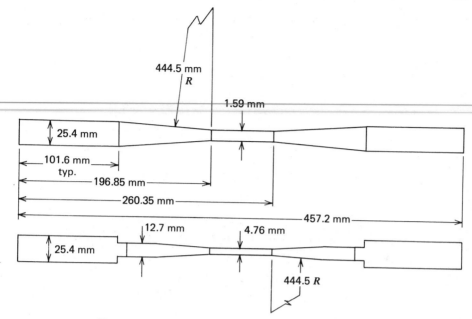

Fig. 4.3-11 Tension specimen—wood parallel to the grain.

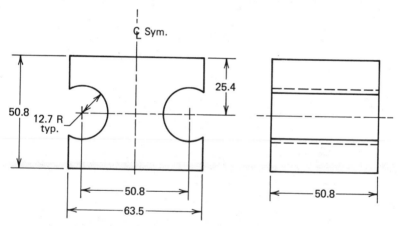

Fig. 4.3-12 Tensile-perpendicular to the grain wood test specimen.

Compression specimens are usually square (25.4 × 25.4 mm) cross-section blocks. Parallel to the grain blocks are 203 mm long and perpendicular to the grain blocks are 152 mm long.

4.3-9 Concrete Tests—Tensile Splitting of Cylinders

W. F. Kirkwood

The tensile testing of brittle materials is not easily accomplished because of the localized stresses generated in the grip areas. In the case of concrete this material may also have variations in strength due to two fundamental sources. These are variations in strength-producing properties of the materials mixture and apparent differences in strength due to discrepancies in test techniques.

The ASTM Standard Test Method for testing cylindrical concrete cores is designated C496-71. The specimens are standard 152.4 mm (6 in.) length by 304.8 mm (12 in.) diameter cylinders. Figure 4.3-13 shows the basic loading configuration and method for loading the specimen in a standard universal

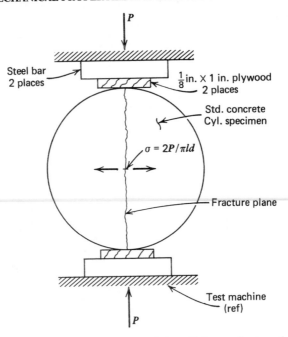

Fig. 4.3-13   Schematic for concrete cylinder splitting tensile strength test.

test machine. This test rig applies the load through the vertical diameter of the specimen. The splitting tensile strength at the center of the cylinder may be computed from

$$\sigma = \frac{2P}{\pi l d} \tag{4.3-14}$$

where σ = splitting tensile strength, kPa or psi
   P = maximum applied load indicated by the testing machine, N or lbf
   l = cylinder length, in. or m
   d = cylinder diameter, in. or m

4.3-10   General Compression Tests

W. F. Kirkwood

This mode of testing is principally employed for testing brittle materials such as stone, concrete, cast iron, ceramics, and some powdered metallurgy formulations. The general concepts pertaining to tension tests will apply to compression tests. However, in testing compression specimens between plane surfaces of the test machine, the following limitations may be experienced:

1. True uniaxial loading may not be achieved.
2. End effects due to surface friction between the specimen and bearing plates may influence the strain readings.
3. Specimens may require larger cross-sectional areas for load stability. This may require the use of a larger test machine or if the compromise is toward a smaller specimen, the instrumentation may be limited.
4. Compressive loading is relatively unstable and there is always the possibility for developing unwanted bending stresses.

   Most of these problems are overcome by proper specimen design, the use of spherical seats and bearing blocks, and finally, proper testing techniques.

4.3-11   Compression Specimen Requirements

W. F. Kirkwood

The desirable compression specimen is the right-circular cylinder with square or rectangular shapes employed in cases where it is only feasible to use the material in the manufactured state. This is also true for wood specimens where compression tests are run parallel and perpendicular to the grain.

The right-circular cylinder compression specimen performs best with a ratio of length to diameter of 2 or more, with 10 being the upper limit for practical experiments. These cylindrical specimens require the ends to be flat and perpendicular to the axis, and the gauge length should be one diameter or less than the specimen length.

4.3-12   Compression Tests of Concrete

W. F. Kirkwood

Concrete tests are conducted in laboratories and in the field to ensure uniform production of concrete that meets the desired strength and quality to satisfy structural integrity of the assigned project. The American Concrete Institute has established the recommended procedure for evaluating the compression test results of field concrete in AC1214-65.

The strength value of concrete is subject to the uniformity of the aggregates, cement, and admixtures of the region in which it is produced. Therefore, ASTM test methods and properly calibrated test machines are necessary for an accurate characterization of the product.

The standard test method for compressive strength of cylindrical concrete specimens is defined by ASTM C39-81. The scope of this document covers the procedures for the testing and structural evaluation of concrete under compressive stresses. The schematic for the compression test setup is depicted in Fig. 4.3-14.

Calculations on these specimens are usually related to determining the maximum compressive strength, which is the maximum load carried by the specimen divided by the average cross-sectional area. The average diameter will be used in calculating the cross-sectional area.

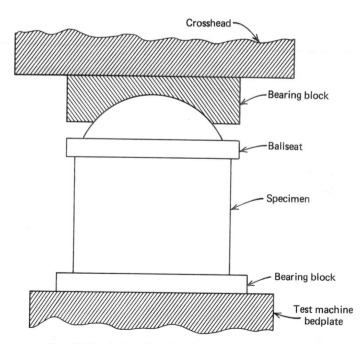

Fig. 4.3-14   Spherical bearing blocks for compression tests.

4.3-13   Special Features of Compression Testing

W. F. Kirkwood

The commercial testing laboratories usually conduct compression tests on a large number of a customer's specimens to determine the ultimate strength of the material. This ultimate strength will usually be expressed in terms of the stress at time of failure.

Some procedures that aid in producing reliable test results are as follows:

1. The specimens' dimensions must be recorded.
2. Cylindrical specimen ends should be parallel and perpendicular to the axis of the specimen (limits established by specimen design).
3. Specimen and bearing blocks should be centered and aligned.

Test machine crosshead speed is usually the method of loading for commercial tests. Table 4.3-3 specifies the recommended speeds to be used with standard compression specimens.

As many compression tests are conducted on brittle materials, they will commonly rupture along a diagonal plane or cone, with splitting according to the specimen type.

Figure 4.3-15 illustrates typical failure profiles of several brittle materials using square and cylindrical specimens.

For uniaxial compression specimens whose resistance to failure is due to a combination of internal friction and cohesion, the angle of rupture will not coincide with the maximum shear plane but is a function of the angle of internal friction ϕ. This is illustrated in Fig. 4.3-16 where the uniaxial specimen has the two principal stresses $\sigma_1 = \sigma_{max}$ and $\sigma_2 = 0$. From Mohr's circle, the angle of internal friction ϕ and the limiting shear stress may be determined.

Nonhomogeneous materials such as cast iron or concrete may not conform to the Mohr theory of rupture. Also, end effects due to bearing plate friction on the specimen ends may cause the angle of rupture to deviate from the theoretical values.

In the case of very short specimens a normal failure plane may not develop and other modes of failure such as crushing may occur. Some short brittle specimens may separate longitudinally into columnar fragments known as columnar fracture. Good quantitative results are difficult to achieve in specimens outside the prescribed ASTM standards.

4.3-14   Torsion Tests

W. F. Kirkwood

Torsional loading of properly designed test specimens is the most practical approach for easily characterizing the shear properties of engineering materials. The texts on strength of engineering

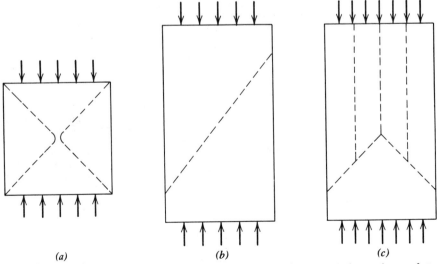

(a)   *(b)*   *(c)*

Fig. 4.3-15   Failure types of brittle compression specimens. (*a*) Shear cone (cube specimens of stone or mortar); (*b*) diagonal shear (cast iron or concrete); (*c*) shear cone with columnar splitting (concrete).

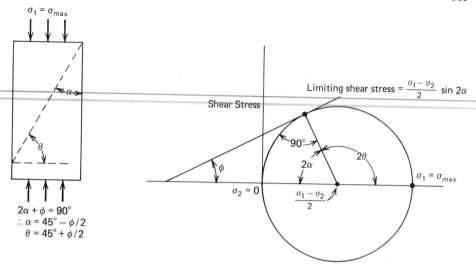

Fig. 4.3-16 Mohr's circle showing relationship of angle of rupture α and angle of internal friction ϕ in a uniaxially loaded compression specimen.

materials provide the theoretical equation for a solid cylindrical bar subjected to pure torsional load with a maximum shear stress at the outer fibers of

$$\tau = \frac{TR}{J} = \frac{2T}{\pi R^3} \tag{4.3-15}$$

If the specimen is tubular, the maximum shear stress at the outer fibers is written

$$\tau = \frac{2TR_0}{\pi\left(R_0^4 - R_i^4\right)} \tag{4.3-16}$$

where T = torsional moment
   J = polar moment of inertia
   R_0 = outer tube radius
   R_i = inner tube radius

Figure 4.3-17 shows the stress–strain relationships for a right-circular cylinder test specimen in pure torsion. Prior to testing, the distance L is carefully marked on the surface. After testing to failure the distance L' is determined by knowing the shear strain $\gamma = R\theta$, where θ is equal to the angle of twist per unit length of the cylinder. The elongation or ductility of the test specimen may then be expressed as a percentage of elongation of the outer fiber which is equal to

$$\left(\frac{L' - L}{L}\right) \times 100 = \text{percentage of elongation}$$

The results of torsion tests on cylinders and round tubes are usually plotted as torsion stress–strain diagrams. The modulus of elasticity in shear is the ratio of shear stress to shear strain at low loads, which is the initial slope of this diagram. The modulus of elasticity in shear is also called the modulus of rigidity, designated by the letter G. The relationship of this property in terms of torque and angle of twist is

$$G = \frac{TL}{J\theta} \tag{4.3-17}$$

However, G is usually expressed in terms of the modulus of elasticity in tension and Poisson's ratio by the following equation:

$$G = \frac{E}{2(1 + \mu)} \tag{4.3-18}$$

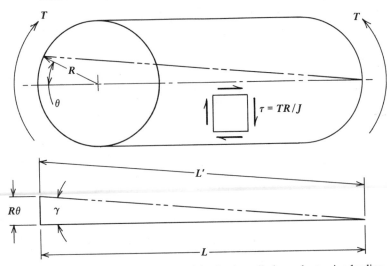

Fig. 4.3-17   Stress–strain relation for a right-circular cylinder under torsion loading.

G is a basic mechanical property derived from torsion tests. Other properties such as the proportional limit and ultimate shearing stress cannot be treated as a basic property due to "form factor" effects.

4.4   HARDNESS SCALES FOR STRUCTURAL MATERIALS

R. G. Scott

The hardness of materials is not a measure of any single fundamental property. It is largely a surface phenomenon influenced by yield strength, true tensile strength, work hardening, modulus of elasticity, and others. Tests for hardness are comparative tests, and evolved mainly from the need for some convenient measure of a materials resistance to scratching, wear or abrasion, and indentation. A hardness test of a material can be a guide to its overall strength or its resistance to deformation provided a sufficiently large amount of material is deformed during the test and if there is enough baseline information from previous experience. The hardness test can serve to grade similar materials and to check and control the quality level of materials and products. Many comparative tests could be and have been devised to give the necessary information. Of these, four tests have endured and are favored in industrial use. These are the Brinell, Vickers, Rockwell, and Shore scleroscope tests. The first three are all indentation tests, differing mainly in the shape of the indentor. In the Brinell and the Vickers tests the hardness number is defined as the mean pressure on the actual surface area of the indentation. The Rockwell test differs in that the hardness number is relative to the depth of penetration. The Shore scleroscope is completely different from the others in that the hardness of a surface is defined by the height of rebound of a small weight dropped onto it. The three hardness tests that rely on indentors are usually considered nondestructive because of the small size of the indentations made by the tests. The Shore scleroscope hardness indications depend on the resilience of a small pointed hammer and the resilience of the material being tested; however, the permanent deformation of the material is also an important factor. When the hammer falls onto a soft surface, it penetrates that surface to some extent before rebounding and produces a minute indentation. This absorbs part of the energy of the fall, and the energy of the rebound is comparatively small. If the surface is hard, the size of the indentation is much smaller, less energy is used in the deformation process, and the rebound is higher. In any case the indentation is slight and nearly indiscernible. Thus, the Shore scleroscope test is generally considered nondestructive.

4.4-1   Brinell Hardness Test

In the Brinell hardness test a known load or force is applied through a hardened metallic ball of a known diameter to the surface of the material being tested. The load or force is in kilograms, and the diameter of the hardened metallic ball is in millimeters. The resulting permanent indentation into the surface of the test piece is assumed to be in the shape of a spherical section surface, even though it is realized that the elastic behavior of both the ball and the material being tested prevent an exact duplication of the ball's unloaded surface. The diameter of the indentation where it intersects the test piece's surface is carefully measured in millimeters. From this dimension the surface area of the indentation can be calculated. The Brinell hardness number is the numerical value of the load or force supplied divided by this area. If P is the applied load or force (in kilograms), D is the diameter of the hardened metallic ball (in millimeters) and d is the diameter of the indentation (in millimeters).

$$\text{Brinell hardness number} = \text{HB} = \frac{\text{load or force on the ball}}{\text{area of the indentation}}$$

$$\text{HB} = \frac{2P}{\pi D \left(D - \sqrt{D^2 - d^2} \right)} \tag{4.4-1}$$

The standard test method for determining the Brinell hardness of metals is described in ASTM E10-78. This standard provides the methods for two general classes of standard tests, which are verification, laboratory or referee tests, where a high degree of accuracy is required, and routine tests, where an adequate but lower degree of accuracy is acceptable.

The method specifies for the standard test the following requirements. The hardened ball indentor is to have a diameter of 10.000 mm with a deviation from this value of not more than 0.005 mm in any diameter. A steel ball with a Vickers hardness (HV) of at least 850 using a 10-kgf load may be used on material having a HB not over 450. Above this hardness and to but not over 630, a carbide ball is used. It is recommended that the Brinell test not be used for material having a HB over 630. The load in the standard test is to be 3000, 1500, or 500 kgf.

The Brinell hardness machine for laboratory or referee tests is acceptable for this purpose over a loading range within which the machine error does not exceed $\pm 1\%$. For routine Brinell hardness testing, a machine error within $\pm 2\%$ is acceptable.

The choice of the load to be used is based on the desirability that the test load be of such a magnitude that the diameter of the impression be in the range of from 2.5 to 6.0 mm or expressed in percent of indentor diameter, 25–60%. This range of impression diameters is not mandatory, but it should be realized that different Brinell hardness numbers may be obtained for a given material by using different loads on a 10-mm ball. The time interval over which the load is applied can influence the indicated hardness. Except for certain soft metals, the full test load is applied for 10–15 s in the standard test. For soft metals the test load is applied for 30 s.

The Brinell hardness number followed by the symbol HB without any suffix numbers defines the following test conditions:

Ball diameter	10 mm
Load	300 kgf
Duration of load	10 to 15 s

Any other test conditions must denote the ball diameter, load, and duration of loading in that order. For example, 61 HB 10/500/30 indicates a Brinell hardness of 61 measured with a ball 10 mm in diameter and a load of 500 kgf applied for 30 s. The variation of test conditions is necessitated by the limitations of a single set of test conditions over the range of material dimensions and hardness. Hardness tests on thin specimens must not indicate that the deformation process was continuous through the thickness of the specimen when the reverse side is examined. A smaller ball and a lower applied load is sometimes necessary to develop the desired 25–60% relationship between the impression diameter and the indentor ball diameter. Such tests, which are not to be regarded as standard, will approximate the standard test more closely if the relation between the applied load, P, measured in kilograms force and the ball diameter, D, measured in millimeters is the same as in the standard test where

$P/D^2 = 30$ for 3000-kgf load and 10-mm ball
$P/D^2 = 15$ for 1500-kgf load and 10-mm ball
$P/D^2 = 5$ for 500-kgf load and 10-mm ball

For soft metals special tests that do not simulate the standard tests are made with P/D^2 ratios of 2.5, 1.25, and 1.0.

On tests where the ball size is less than 10 mm in diameter, the test load, ball size, and duration of the full-load application are reported. In any case balls used for hardness tests should conform to the requirements for the material and permissible variations in diameter specified for the standard ball.

Indentation diameters are usually measured with a low-power (approximately 20X) portable microscope capable of resolving the direct measurement to 0.1 mm and the estimation of the diameter to 0.02 mm. These specifications apply to any other device used for the measurement. In actual measurements of the indentation, the diameter is usually read to 0.05 mm. In the case of referee or standardization tests, a more precise instrument such as a micrometer microscope is used.

There are instances where the indentation might not be circular because the compressive properties of a flat specimen are not uniform, owing perhaps to the rolling direction or cooling stresses. In such cases the average Brinell hardness may be obtained if the diameter is taken as the average of four directions, approximately 45° apart. Brinell hardness determinations on material having a radius of curvature, even with uniform properties, will not exhibit an impression with a circular boundary unless the test material's surface is spherical. If the radius of curvature is not less than 25 mm, the diameter of the impression may be taken as the average of the maximum and minimum of diameters.

The appealing advantage of the Brinell hardness test is the fundamental simplicity of the equipment needed to perform the test. Several types of machines are available. The differences between them may be in the method of loading, method of measuring the load, and the size of the machine. Loads can be applied by hydraulic pressure, gear-driven screw, or by weights with levers. The machines can be hand operated or motor driven. Load measuring methods include a piston with weights, a bourdon gauge, dynamometer, or weights with lever. There are large size machines for laboratory use and small portable sizes for field work. For tests of thin sheet-metal products, a small hand-held plier device using a 1.2 mm ball and a 10-kg spring pressure has been employed. The Brinell test can be accomplished in a universal test machine with suitable load ranges if a fixture to hold the ball properly is used.

Machines used for Brinell hardness testing are verified over the range of loads to be used (ASTM methods E10-78). Besides the standard load of 3000, 1500 and 500 kgf, any other load to be used should be verified and checked periodically with a proving ring, deadweights and proving levers, or by an elastic calibration device or springs. ASTM methods E8 Verification of Test Machines provides for the guidance of the procedure.

If the machine is used only for routine hardness testing, standardized hardness blocks may be used. The machine is considered standardized if the mean diameter of any hardness impression made differs by no more than 3% of the mean diameter to the hardness value of the standard test block.

Measuring microscopes or other devices used for determining the diameter of the impressions made by the test should be verified at five intervals over the working range by the use of an accurate scale such as a stage micrometer. The measuring microscope should be adjusted so that, throughout

TABLE 4.4-1   APPROXIMATE BRINELL–ROCKWELL B HARDNESS CONVERSIONS FOR AUSTENITIC STAINLESS STEEL PLATE IN ANNEALED CONDITION

Rockwell Hardness Number B Scale (100-kgf load, $\frac{1}{16}$-in. (1.588-mm) ball)	Brinell Hardness Number (3000-kgf load, 10-mm ball)	Rockwell Hardness Number B Scale (100-kgf load, $\frac{1}{16}$-in. (1.588-mm) ball)	Brinell Hardness Number (3000-kgf load, 10-mm ball)
100	256	79	150
99	248	78	147
98	240	77	144
97	233	76	142
96	226	75	139
95	219	74	137
94	213	73	135
93	207	72	132
92	202	71	130
91	197	70	128
90	192	69	126
89	187	68	124
88	183	67	122
87	178	66	120
86	174	65	118
85	170	64	116
84	167	63	114
83	163	62	113
82	160	61	111
81	156	60	110
80	153		

the range covered, the difference between the scale divisions of the microscope and that of the calibrating scale does not exceed 0.01 mm.

There is no general method to convert accurately Brinell hardness number to other hardness scales or tensile values. ASTM E10-78 recommends that such comparisons, which are at best only approximations, be avoided. Exceptions to this are special cases where a reliable basis for the approximate conversion has been made by comparison tests.

Table 4.4-1 refers to approximate Brinell–Rockwell B hardness conversions for steel.

4.4-2 Vickers Hardness Test

The Vickers hardness test is included here for two reasons. First, the Brinell steel ball indentor is specified to have a Vickers hardness (HV) of at least 850 using a 10-kgf load and, second, the tests are somewhat similar in that an indentation is made and the hardness number is determined from the ration P/A of the load P in kilograms to the surface area A of the indentation in square millimeters. The indentor is a square-based diamond pyramid with an included angle of 136° (ASTM E92). The load is varied according to the behavior and/or the thickness of the material being tested over the range of 1–120 kg.

The Vickers hardness number (HV) is computed from

$$HV = \frac{2P \sin(\alpha/2)}{d^2} = \frac{1.844P}{d^2} \tag{4.4-2}$$

where P = load, kgf
$\quad\quad\quad d$ = mean diagonal of the impression, mm
$\quad\quad\quad \alpha$ = face angle of the diamond = 136°

The Vickers hardness testing machine consists of a support for the specimen and provides for the indentor and the specimen to be brought into contact gradually and smoothly, under a predetermined load. This load is applied for a fixed period of time. The machine is designed so that no rocking from side to side is allowed while the load is being applied or removed. A measuring microscope is usually mounted on the machine in a manner that allows it to readily locate the impression in its optical field.

The highly polished pointed indentors have face angles of $136° \pm \frac{1}{2}°$. Obviously, the indentor should be examined periodically for cracks, chips, and looseness in its mounting.

The advantage of the Vickers hardness test is in the measurement of the diagonals of the indentation. A much more accurate measurement can be obtained of the diagonal of a square than that of the diameter of a circle where the measurement must be made between two tangents to the circle. The Vickers hardness test can be accurately used for hardnesses up to 1300 (approximately 850 HB). The test can be conducted on material as thin as 0.00024 mm. It is often used to indicate the friability of nitrided-steel cases. The hardness so determined appears to be a good measure of the wearing qualities of nitrided steel.

4.4-3 Rockwell Hardness Test

The Rockwell test differs from the Brinell and Vickers tests in that the hardness number relates inversely to the depth of the impression made by a penetrator into the specimen under certain fixed conditions of the test. Hardness numbers are read from a dial on the test machine. The penetrator may be a steel ball or a diamond sphere-conical penetrator. Because the indentor is smaller and the applied load smaller than that in the Brinell test, the size of the indentation is smaller and thus less marring.

The Rockwell hardness test machine is designed so that the specimen rests on a smooth hardened steel anvil and can be brought into contact with the penetrator and brought gradually to a minor preload of 10 kgf with a hand-manipulated screw. The preloading process causes the pointer on the dial to make a certain number of complete revolutions coming to rest within ±5 minor scale divisions of the "set" position. So that the proper number of complete revolutions is made, the machine will have either a reference mark on the stem of the dial or an auxiliary hand on the dial. After the minor load has been applied, the dial pointer is set to the zero on the black scale. The major load is applied by tripping the operating lever and is removed within 2 s after the motion of the operating lever has stopped. In the case of material exhibiting little or no plastic flow, the pointer will come to rest before the operating lever motion stops. The major load is removed immediately in this case. The Rockwell hardness test has 15 scales which can cover a complete spectrum of hardness. Each scale has a specific penetrator and major load combination. Penetrators are of two types: steel balls of various specified sizes and a sphero-conical diamond point. The diamond penetrator has a 120° included angle, the apex of which is rounded to a spherical surface with a 0.200 mm radius. The usual penetrators are a $\frac{1}{16}$ steel ball and the sphero-conical diamond point. The 10-kgf minor load is used in all Rockwell tests. The usual major loads are customarily 60 or 100 kgf for the $\frac{1}{16}$-inch diameter steel ball and 150 kgf for the diamond. Table 4.4-2 gives the various Rockwell hardness scales and their typical applications.

TABLE 4.4-2 ROCKWELL HARDNESS SCALES

Scale Symbol	Penetrator	Major Load (kgf)	Dial Figures	Typical Applications of Scales
B	$\frac{1}{16}$-in. (1.588-mm) ball	100	Red	Copper alloys, soft steels, aluminum alloys, malleable iron, etc.
C	Diamond	150	Black	Steel, hard cast irons, pearlitic malleable iron, titanium, deep case-hardened steel, and other materials harder than B100.
A	Diamond	60	Black	Cemented carbides, thin steel, and shallow case-hardened steel.
D	Diamond	100	Black	Thin steel and medium case-hardened steel, and pearlitic malleable iron.
E	$\frac{1}{8}$-in. (3.175-mm) ball	100	Red	Cast iron, aluminum and magnesium alloys, bearing metals.
F	$\frac{1}{16}$-in. (1.588-mm) ball	60	Red	Annealed copper alloys, thin soft sheet metals.
G	$\frac{1}{16}$-in. (1.588-mm) ball	150	Red	Malleable irons, copper–nickel–zinc and cupro–nickel alloys. Upper limit G 92 to avoid possible flattening of ball.
H	$\frac{1}{8}$-in (3.175-mm) ball	60	Red	Aluminum, zinc, lead.
K	$\frac{1}{8}$-in. (3.175-mm) ball	150	Red	
L	$\frac{1}{4}$-in. (6.350-mm) ball	60	Red	
M	$\frac{1}{4}$-in. (6.350-mm) ball	100	Red	Bearing metals and other very soft or thin materials. Use smallest ball and heaviest load that does not give anvil effect.
P	$\frac{1}{4}$-in. (6.350-mm) ball	150	Red	
R	$\frac{1}{2}$-in. (12.70-mm) ball	60	Red	
S	$\frac{1}{2}$-in. (12.70-mm) ball	100	Red	
V	$\frac{1}{2}$-in. (12.70-mm) ball	150	Red	

4.4-4 Rockwell Superficial Hardness Test

The Rockwell superficial test is used where the marring of the specimen is desired to be minimized such as on hardened and polished finished-steel products. The tester for the Rockwell superficial test is a specialized form of the regular tester. It measures hardness by the same principles as the regular test but employs a smaller minor load, smaller major loads, and a more sensitive depth-measuring system. (In the regular Rockwell test one Rockwell number represents 0.002 mm while in the superficial test, one Rockwell number represents 0.001 mm movement of the penetrator.) The major loads (total loads) are 15, 30, or 45 kgf. The standard scales for the Rockwell superficial test are given in Table 4.4-3.

The N scales are used for materials similar to those tested on the Rockwell C, H, and D scales, but of thinner gauges or case depth, or where minute indentation is required. The W, X, and Y scales are used for very soft materials.

Besides causing less marring than the Brinell test, the advantage of the Rockwell test is that the hardness read from the dial is rapid and more positive. Hardness numbers are reported to nearest

TABLE 4.4-3 ROCKWELL SUPERFICIAL HARDNESS SCALES

	Scale Symbols				
Major Load, kgf (N)	N Scale Diamond Penetrator	T Scale, $\frac{1}{16}$-in. (1.588-mm) Ball	W Scale, $\frac{1}{8}$-in. (3.175-mm) Ball	X Scale, $\frac{1}{4}$-in. (6.350-mm) Ball	Y Scale, $\frac{1}{2}$-in. (12.70-mm) Ball
15(147)	15N	15T	15W	15X	15Y
30(294)	30N	30T	30W	30X	30Y
45(441)	45N	45T	45W	45X	45Y

whole number unless otherwise specified. The Rockwell hardness numbers when reported must indicate the scale used. There is no Rockwell hardness number designated by a figure alone.

Rockwell hardness testers, both the regular and the superficial, should be periodically checked for accuracy. Two methods are generally used. Separate verification of load application penetrator and the depth-measuring device is followed by a performance test involving a series of impressions on standardized hardness test blocks. This procedure is used on new or rebuilt machines. The standard test block method used in the above test as a performance test is used by itself to verify Rockwell hardness testers used in referee, laboratory, or routine tests. The extent of the standardized hardness block tests is detailed in ASTM E18.

4.5   FATIGUE OF METALS

R. D. Streit

4.5-1   Nature of Fatigue

Fatigue is the process by which the strength of a structural member is degraded due to the cyclic application of load or strain. The fatigue load that a structure can withstand is often significantly less than the load which it would be capable of if the load were applied only once. In traditional fatigue analysis no distinction is made between fatigue crack initiation and crack growth to failure. The fatigue life (i.e., the number of cycles to failure, N) is shown to be a function of the cyclic stress, S, or strain, ε, amplitude and is plotted on an $S-N$ or $\varepsilon-N$ curve for a given part. However, such $S-N$ type curves are not unique to the material or part, but rather, are influenced by a number of factors such as surface roughness, material heat treatment, mean stress or strain, environment, and details of the geometry. A number of empirical "modifying factors," which are based on extensive test data, are employed to shift the $S-N$ curve to account for the actual condition. The classical or traditional approach to fatigue testing and analysis is presented in Section 4.5-2.

A second approach to fatigue analysis is to consider crack growth from some preexisting or assumed initial crack size. Using a fracture mechanics type stress analysis (see Section 4.6) to calculate the cyclic stress intensity at the tip of the crack one can obtain the number of cycles to propagate the crack a specified amount. Generally, failure is predicted when the crack reaches a length such that the remaining uncracked ligament can no longer support the maximum load. Thus, while more accurate and quantitative in the prediction of the lifetime of a flawed component, this method ignores the initiation phase of crack growth, which in many cases is a large percentage of the life of the given component. This approach is discussed in Section 4.5-3.

Mechanisms of Fatigue Crack Initiation and Growth

The fatigue process can be divided into three phases: crack initiation, growth, and final failure. While there is no clear distinction between the various phases, some general guidelines delineate their behaviors. Initiation can be considered to consist of the creation of a macrocrack from a crack-free material. For a fatigue evaluation, a macrocrack can be considered to be of a size which is large compared to the microstructural size characteristics of the material, for example, many grain diameters. Based on the observed behavior of common structural materials, the crack initiation phase can consume from 10 to 95% of the life of a component depending on the material, loading, geometry, and environment.[1] As a general rule of thumb, for components which have a long fatigue life ($> 10^4$ cycles to failure) the crack initiation phase consumes $> 75\%$ of the life of the component. On the other hand, for fatigue life $< 10^3$ cycles, the crack initiation occurs in the first 50% of the fatigue life. If a part has an initial defect due to manufacturing (e.g., weld cracking, processing, etc.) or previous service, its fatigue life can be significantly shorter than a similar component that was initially free of flaws.

A number of theories exist as to the actual mechanisms of fatigue crack initiation: however, there is no one mechanism that is universally accepted. In the absence of a corrosive environment, all proposed mechanisms include some localized plastic deformation. Fatigue cracks are usually initiated at points of high stress concentration such as holes or fillets and are formed at or near the surface of the material. On a microscale, the local plastic deformation at structural discontinuities such as inclusions or grain boundaries are of a highly localized nature, and are believed to contribute to the initiation process. Typical mechanisms include the cyclic shearing of material at the surface of the component, or, at structural discontinuities. In either case, the deformation is a result of sequential dislocation motion, and as such, it is the maximum shear stress law that governs the initiation of fatigue cracks.[2] Further, since the surface is the prime location for fatigue crack initiation, the fatigue strength of the part can be increased by strengthening the surface of the part (e.g., by strain aging) or removing the damaged surface layer. Conversely, if the surface of the part is degraded due to corrosion pitting or brittle plating, for example, the fatigue life can be significantly reduced in cases where initiation is a large percentage of fatigue life.

Once the crack initiates by the process of linking many microcracks and is of a size which is large relative to the microstructure of the material, the process changes from shear deformation to a process controlled by the maximum tensile stress. Crack propagation becomes more readily visible and can be studied by either following the advance of the surface trace during stress cycling or by examination of the striation markings on the fracture surface using scanning electron microscopy (SEM). The direction of crack propagation is perpendicular to the direction of the maximum tensile stress.

Final failure occurs when the loading is such that the component fails catastrophically due to the application of a single cycle of load. In this case the remaining uncracked area, applied loading, geometry, and the mechanical properties of the material govern failure. The mechanism or failure model may be stress controlled (i.e., by exceeding the yield or ultimate tensile strength of the material) or due to a fracture-controlled model (i.e., plane strain crack propagation or elastic–plastic tearing).

Fatigue Surface Appearance

Fatigue surfaces are characteristically flat and are oriented perpendicular to the direction of applied tensile loading. On closer examination of the surface, a number of features can be discerned and related to the fatigue process. A high-magnification examination of the fatigue surface often reveals the presence of a fine array of parallel lines referred to as striations. Each striation is formed with each successive application of applied load and as such represents the crack front at the time of the load application. The orientation of the fatigue striations are therefore perpendicular to the direction of crack advance and can be used to determine the origin of crack initiation, see Fig. 4.5-1.[3] Because of the very fine nature of striations, they can be easily obliterated due to the rubbing of the fracture surfaces during the fatigue process, final failure and post fatigue material handling, or due to environmental attack of the surface.

Fatigue striations are not developed during all phases of fatigue. During the crack initiation phase or where crack growth is proceeding under very low loads such that the growth per cycle is on the

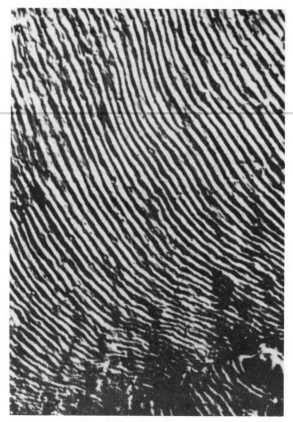

Fig. 4.5-1 Fractograph at 3500X, of a fracture surface showing stage II striations, typical of fatigure failure. (From *Metals Handbook*, Vol. 10, 8th ed., American Society for Metals, 1974, p. 17.)

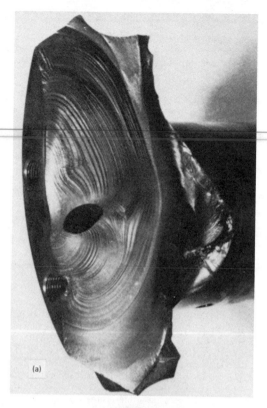

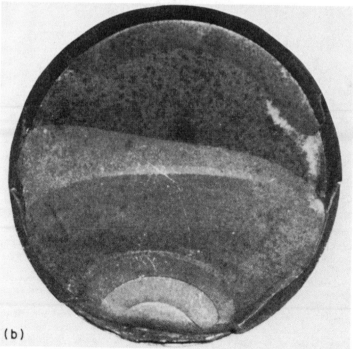

Fig. 4.5-2 A classic fatigue–fracture surface, showing progression marks (beach marks) that indicate successive positions of the advancing crack front. (From *Metals Handbook*, Vol. 10, 8th ed., American Society for Metals, 1974, p. 21.)

order of the atomic spacing, classical striations are not formed. Similarly, during the final stages of fatigue growth, where the number of cycles to failure is relatively low, striations are not formed. Under this condition the high crack tip cyclic stress intensity results in crack growth by microvoid coalescence.[4]

In addition to the fine-scale striations, a pattern of readily visible concentric "beach marks" develops as a result of changing loading or environmental conditions, Fig. 4.5-2.[3] These beach marks correspond to various positions of the crack front when the magnitude of the cyclic stress was changed, and in turn, the rate of propagation changed. Under constant stress cycling and when the operating environment is not changed, these markings are not observed. At high shear stress levels, many shiny, smeared areas may also be present in the fracture surface. The polishing action is due to the rubbing of the two fracture surfaces during the compressing half of the loading cycle prior to final fracture.

4.5-2   Traditional Approach to Fatigue Testing and Design

In traditional fatigue analysis and testing, no distinction is made between the crack initiation, crack growth, and the final fracture phases of fatigue failure. Within traditional fatigue testing, two regimes of material behavior are generally considered: high-cycle fatigue in which failure occurs in excess of 10^3 cycles, and low-cycle fatigue in which failure occurs in relatively few cycles, $< 10^3$. In high-cycle fatigue the deformations are small and can be characterized by a material's elastic behavior. In low-cycle fatigue crack growth is often accompanied by large-scale plastic deformation.

In high-cycle fatigue analysis one assumes that the component is initially crack-free and estimates the total fatigue life by means of an $S-N$ diagram (i.e., applied stress versus cycles to failure). While it would be best to obtain the $S-N$ diagram for a given component on material samples of that component, in general smooth tensile or rotating beam bend specimens are employed in the experimental evaluation. Results from the experimental $S-N$ curves are then modified for a particular application to account for part geometry, surface finish, environment, and so on.[5]

In low-cycle fatigue, materials testing is used to develop empirical models for the cyclic stress–strain behavior and fatigue life relationship for a given material. These models can then be used in the design analysis to estimate the system response to applied loads and the fatigue life of the component. As with high-cycle fatigue, to obtain the most accurate results for a given design application, the test conditions (e.g., temperature, environment) should approximate the actual design conditions as much as possible.

High-Cycle Fatigue Testing

The best results for predicting the life of a component are obtained by testing the actual component under conditions which approximate the actual service loading and environmental conditions. While this procedure can be very expensive and time consuming it is often warranted in critical applications because of the increased reliability of the life prediction. These types of tests are done routinely in the automobile and aircraft industry. The more usual approach is to test the material itself and then modify the results to allow for the shape of the component, and conditions of the part and loading. This method relies on empirical relations that have been formulated by doing extensive testing while isolating each fatigue life modifying effect.

For evaluating material properties, the usual starting point is a standardized specimen that has been machined and polished to eliminate all surface defects oriented perpendicular to the applied loading direction. Two types of fatigue test are readily used: constant-amplitude axial load fatigue tests and the rotating beam bend test. Standard practices for conducting constant-amplitude axial fatigue tests of metallic material have been developed by the American Society for Testing and Materials.[6] Per this ASTM Standard Practice E466-82 for high-cycle fatigue, the significance of the axial load test is to determine the effect of variations in material, geometry, surface condition, stress, and so forth, on the fatigue resistance of metallic materials. It has been standardized such that fatigue data can be compared, reproduced, and correlated among different laboratories. It is not intended for the specialized fatigue loading of components or parts. Results of these tests are useful in design analysis based on empirical relationships between test variables and actual design conditions.

The high-cycle test practice allows for a certain amount of flexibility in the specimen design. Specimens of either round or rectangular cross section may be employed, and, in addition, the specimens may be either notched or unnotched depending on the goals of the test program. For unnotched specimens a number of general guidelines are provided[6] to ensure that the sample fails in the test section and that the stress in the test section is not unduly influenced by the specimen geometry. Some typical unnotched specimen geometries are shown in Figs. 4.5-3 and 4.5-4. Since notched geometries are generally used for a specialized test program, no restrictions are placed on the design other than it be consistent with the objectives of the test program. In high-cycle fatigue testing the primary data recorded is the number of cycles to failure at a specified maximum and minimum

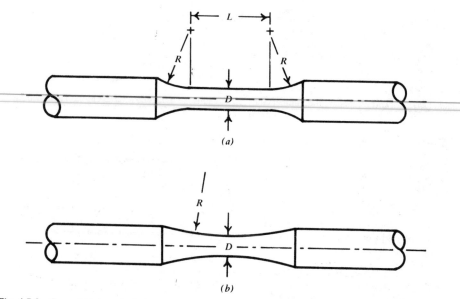

(a)

(b)

Fig. 4.5-3 Round high-cycle fatigue specimens per ASTM Standard Practice E466-82.[6] The radius R should be greater than eight times the specimen diameter D. In (a) the test section L should be greater than three times the specimen diameter D. (a) Specimens with tangentially blending fillets between the test section and the ends; (b) specimens with a continuous radius between ends. (Reprinted with permission from the *Annual Book of ASTM Standards.* Copyright 1983, ASTM, 1916 Race St., Philadelphia, PA 19103.)

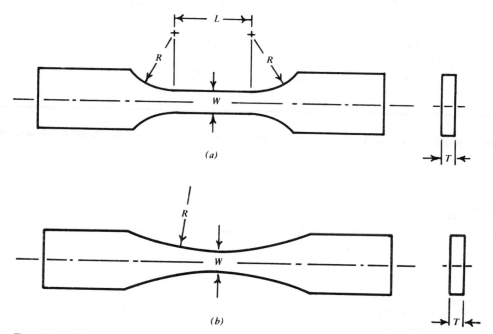

(a)

(b)

Fig. 4.5-4 Flat high-cycle fatigue specimens per ASTM specimen width W Standard Practice E466-82.[6] The radius R should be greater than eight times the specimen. In (a) the test section should be greater than three times W. (a) Specimens with tangentially blending fillets between the uniform test section and the ends; (b) specimens with continuous radius between ends. (Reprinted with permission from the *Annual Book of ASTM Standards.* Copyright 1983, ASTM, 1916 Race St., Philadelphia, PA 19103.)

stress level. These results are often used to determine a characteristic $S-N$ or $\varepsilon-N$ curve for the material as described by ASTM Standard E468.

In addition to the constant-amplitude, axial load tests standardized by the ASTM, rotating beam bending tests are also commonly used in fatigue testing. For these tests a polished, 0.3 in. (7.62 mm) diameter, smooth round specimen is subjected to a bending moment while rotating. This gives rise to a stress on the outer surface that varies between equal tension and compression given by Mr/I, where M is the bending moment, r is the radius of the specimen, and I is the moment of inertia given by $(\pi r^4/4)$. A counter registers the number of cycles to failure at a particular stress level. Then, similar to the axial fatigue test data, the data is plotted as stress versus log N.

Comments on High-Cycle Fatigue Testing

Two points relative to fatigue testing and data presentation should be noted. First, the two methods of fatigue testing described (i.e., axial fatigue and rotating beam bending) result in data that are slightly different. Some of this change may be due to the difficulty of eliminating all bending in the axial loading tests. But even the most rigid and carefully aligned testing machines seem to yield fatigue strength approximately 10% below that of a rotating beam type test. Differences between the two types of loading, uniform stress field (axial) versus nonuniform (bending) may also contribute to lower fatigue strength observed in the axial fatigue tests. These factors should be considered when using general empirical relations to relate fatigue tests to design analysis.

Secondly, fatigue tests generally exhibit appreciable scatter—even when evaluating a given material. This scatter may be the result of a number of factors such as specimen preparation (surface finish, tolerances, heat treating, etc.), specimen alignment, instrumentation, and material variability. Because of this scatter, a number of approaches to apply statistical methods to fatigue data have been attempted. While no "handbook" methods exist that cover all cases of statistical analysis, the reader is cautioned that such methods exist and may be necessary depending on the data and application.

High-Cycle Fatigue Analysis

In high-cycle fatigue analysis where the loading is alternate tension–compression (zero mean stress) as obtained in the rotating beam bending tests, for example, the $S-N$ diagram can be divided into two regimes, Fig. 4.5-5. For stresses less than the endurance limit of the material, S_e', the fatigue life becomes infinite. For applied stresses between the regime of low-cycle fatigue (approximately 10^3 cycles) and the endurance limit, the $S-N$ behavior can be estimated as a linear fit of stress versus log (N). If two points on this line are known, the entire $S-N$ behavior in that range can be estimated either graphically or based on an equation of that line. While structural materials such as steel and titanium exhibit a true endurance limit, S_e', at approximately 10^6 cycles, many other materials do not show a true endurance limit. These materials are usually rated, S_n', at a certain fatigue life, generally in

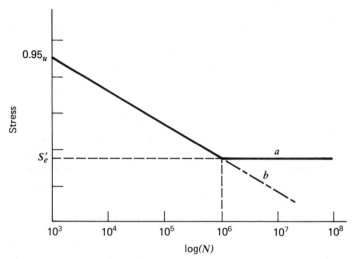

Fig. 4.5-5 Schematic representation of $S-N$ diagram. The solid line a models a material exhibiting a true endurance limit, S_e', while curve b does not show endurance limit behavior. Such materials b are usually rated at a specified stress level for a given number of cycles.

the range of 10^8 cycles. The endurance limit (or rated life) for a number of commonly used materials can generally be approximated by

Material	S'_e or S'_n $(\times S_u)^*$	Rated Life Endurance Limit (cycles)	Max (S'_n) (ksi)
Wrought steel	0.50	10^6	100
Titanium	0.45–0.65	10^6	—
Wrought aluminum	0.40	5×10^8	19
Magnesium	0.35	10^8	—
Copper	0.25–0.50	10^6	—
Nickel	0.25–0.50	10^8	—

$*S'_e$ and S'_n are represented as a fraction of the ultimate tensile strength of the material, S_U.

For many structural materials the upper stress above which the material behavior must be evaluated by low-cycle fatigue analysis is given by 0.9 (S_U) at 10^3 cycles where S_U is the ultimate tensile strength of the material.

The endurance limit of an actual part (rather than that of the standard specimen) can be considerably less than S'_e. This difference can be accounted for by considering a number of empirical modifying factors—each of which corrects a specific effect;[5]

$$S_e = k_a k_b k_c k_d k_e k_f \left(S'_e \right)$$

where S_e = endurance limit of part
$\quad\quad S'_e$ = endurance limit of standard specimen
$\quad\quad k_a$ = surface finish factor
$\quad\quad k_b$ = size factor
$\quad\quad k_c$ = reliability factor
$\quad\quad k_d$ = temperature factor (=1 at room temperature)
$\quad\quad k_e$ = shape factor
$\quad\quad k_f$ = miscellaneous effects (environment, plating, residual stress, etc.)

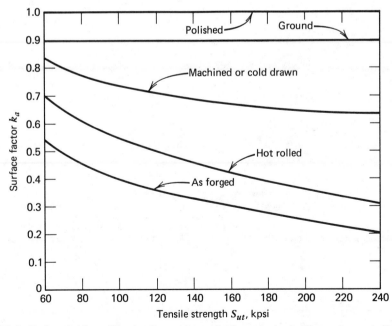

Fig. 4.5-6 Surface-finish modification factors for steel. These are the k_a factors for use in Eq. (4.1). (From Ref. 5.)

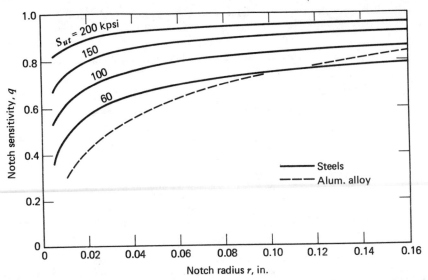

Fig. 4.5-7 Notch-sensitivity charts for steels and 2024-T wrought aluminum alloys subjected to reversed bending or reversed axial loads. Use the values for $r = 0.16$ in. for notch radii larger than 0.16 in. (From Ref. 5.)

Where possible, the various modifying factors should be obtained for the actual operating conditions of the component. However, a number of these factors are readily available for certain materials and are shown in Figs. 4.5-6 and 4.5-7, and Tables 4.5-1 and 4.5-2.[5] The shape factor, k_e, is actually a function of the theoretical stress concentration k_t and a notch sensitivity factor q as given by

$$k_e = \left[1 + q(k_t - 1)\right]^{-1}$$

Values of q are presented in Fig. 4.5-7 for various strength steels and an aluminum alloy.

TABLE 4.5-1 FATIGUE MODIFICATION FACTOR, k_b, TO ACCOUNT FOR COMPONENT SIZE

k_b	Section Size, d
1	$d \le 0.3$ in.
0.85	$0.3 < d \le 2$ in.
0.75	$d > 2$ in.

TABLE 4.5-2 RELIABILITY FACTORS k_c CORRESPONDING TO AN 8% STANDARD DEVIATION OF THE ENDURANCE LIMIT[a]

Reliability R	Standardized Variable z_R	Reliability Factor k_c
50	0	1.000
90	1.288	0.897
95	1.645	0.868
99	2.300	0.814
99.9	3.095	0.752
99.99	3.700	0.704

[a] From Reference 5.

Effect of Mean Stress on Fatigue Life

In addition to the applied cyclic stress, the mean stress (the average of the maximum and minimum applied stresses) can have a significant effect on the fatigue life of a component. The effect of mean stress is best described by a "failure diagram" or modified Goodman diagram shown in Fig. 4.5-8. In this diagram the mean stress is plotted on the abscissa and the alternating stress amplitude is plotted on the ordinate. Using this type of analysis, both failure by fatigue and failure by static loading is considered. The actual failure modes used can be tailored to meet the needs of the design. For instance, if designing for infinite life, the maximum value of the failure envelope on the ordinate would be limited to the endurance limit of the component. If, however, the number of cycles will be limited such that a safe design can be ensured for N cycles, then the maximum value of the ordinate would be the stress level, S_N, which corresponds to a fatigue life of N cycles. This value can be obtained from an S–N curve for the material, or from the empirical relationship developed for this regime of loading. Figure 4.5-8 illustrates this point where the static failure model is considered to be the yield strength of the material in both cases.

For pure torsional loading, where the stresses are expressed as shear stresses, a "shear failure diagram" can similarly be developed. The alternating and mean stresses are replaced by the corresponding values of alternating shear and mean shear stress. The material properties must also be modified to account for the shear behavior by making the following changes:

$$S_y \rightarrow S_{sy} = 0.58S_y$$

$$S_u \rightarrow S_{su} = 0.75S_u$$

$$S_e \rightarrow S_{se} = 0.58S_e \quad \text{(for steels)}$$

For the case of combined loading, where torsion, bending, and tension loading fluctuate, the failure criterion becomes complicated and the reader is referred to the literature for proposed design techniques. One widely accepted method is to use the "effective stress" criterion by defining

$$S_{\text{effective}} = \sqrt{S_1^2 + S_2^2 - S_1 S_2}$$

where S_1 and S_2 are the principal stresses.

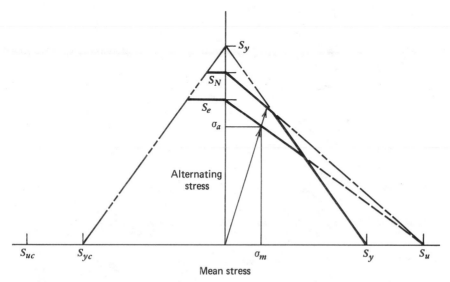

Fig. 4.5-8 "Failure diagram" or modified Goodman diagram showing the effect of mean stress on fatigue failure. Two failure criteria are shown plotted on the ordinate; S_e and S_N where S_e would correspond to infinite life and S_N is for a life of N cycles. Note that for finite life, a higher mean and alternating stress can be applied.

Low-Cycle Fatigue

Due to the tension compression nature of low-cycle fatigue testing, the specimen, test apparatus, and test procedure are more rigidly controlled compared to the high-cycle fatigue test requirements. The recommended specimens[8] have a solid circular cross section with a minimum diameter of 6.35 mm (0.25 in.) as shown in Fig. 4.5-9. Since specimens are subjected to compressive loads on half the loading cycle, special care must be taken to eliminate system backlash, and to align the specimen in

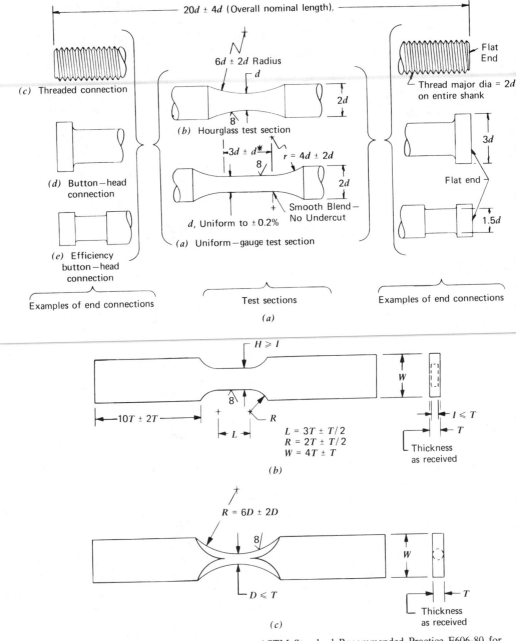

Fig. 4.5-9 Low-cycle fatigue specimens per ASTM Standard Recommended Practice E606-80 for constant-amplitude low-cycle fatigue testing.[8] (a) Dimension d is recommended to be 6.35 mm; (b) 2.54 mm (0.1 in.) $\leq I \leq T$; (c) 2.54 mm (0.1 in.) $\leq D$. (Reprinted with permission from the *Annual Book of ASTM Standards*. Copyright 1983, ASTM, 1916 Race St., Philadelphia, PA 19103.)

the test apparatus such that bending strains are minimized. The ASTM Standard Practice E606-80 places limits on the magnitude of bending-induced strains to < 5% of the minimum axial strain range imposed in the test program. A number of commonly used fixturing techniques are discussed in this ASTM practice[8] and are shown in Fig. 4.5-10. Tests may be run in either stress or strain control; however, total axial strain is the most commonly utilized control variable in a low-cycle fatigue test. The recorded data must include the initial series of hysteresis loops of axial stress (or load) versus the total plastic strain with subsequent recordings of single hysteresis loops at appropriate times throughout the test.

Results of these low-cycle tests can be employed to evaluate a materials cyclic stress–strain behavior and to determine a strain-life relationship. For many metals the following empirical relationships have been used for a convenient description of low-cycle fatigue data.[9]

For cyclic stress–strain behavior,

$$\frac{\Delta\sigma}{2} = K'\left(\frac{\Delta\varepsilon_p}{2}\right)^{n'}$$

where $\Delta\sigma$ = true stress range
$\Delta\varepsilon$ = true total strain range
$\Delta\varepsilon_p$ = true plastic strain range
K' = cyclic strength coefficient
n' = cyclic strain hardening exponent

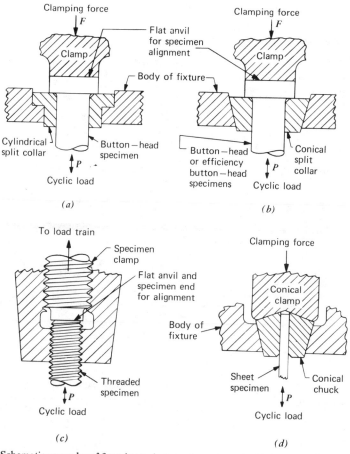

Fig. 4.5-10 Schematic examples of fixturing techniques for various specimen designs.[8] (a) Button-head fixture; (b) button-head or efficiency button-head fixture; (c) threaded specimen fixture; (d) sheet specimen fixture. The claming force should be greater than the cyclic load to avoid backlash within the specimen fixture. (Reprinted with permission from the *Annual Book of ASTM Standards*. Copyright 1983, ASTM, 1916 Race St., Philadelphia, PA 19103.)

Note, in the above relation, the plastic strain range can be replaced by total cyclic strain range by recognizing,

$$\Delta \varepsilon = \frac{\Delta \sigma}{E} + \Delta \varepsilon_p$$

For fatigue life evaluation N_f the following relationships are often employed:

$$\frac{\Delta \sigma}{2} = \sigma_f' (2 N_f)^b$$

$$\frac{\Delta \varepsilon_p}{2} = \varepsilon_f' (2 N_f)^c$$

$$\frac{\Delta \varepsilon}{2} = \left(\frac{\sigma_f'}{E} \right) (2 N_f)^b + \varepsilon_f' (2 N_f)^c$$

where b = fatigue strength exponent
          c = fatigue ductility exponent
          σ_f' = fatigue strength coefficient
          ε_f' = fatigue ductility coefficient
          E = modulus of elasticity

in addition to the variables defined above.

4.5-3   Fracture Mechanics Based Fatigue

The linear–elastic fracture mechanics (LEFM) based fatigue is commonly referred to as the "defect-tolerant approach" since it is based on the number of cycles required to propagate a flaw to a size in which failure occurs due to a static failure criteria. Using this approach, an initial flaw size is either assumed to exist or is observed, and the number of cycles, N_f, to propagate it to a size in which component failure occurs is calculated. Thus, while this method of analysis is more quantitative than the traditional approach, it requires that an initial flaw size be specified or that it be calculated. However, in many fatigue situations, the crack initiation phase can consume a major portion of the life of a component. For instance, in high-cycle fatigue, where the cycles to failure may exceed 10^6, the initiation phase can consume up to 90% of the life of the part. Regardless of this shortcoming, the damage-tolerant approach is useful in a number of important applications; for example, determining the initial flaw size which could lead to failure based on the anticipated service life, estimating the remaining life of a component if a flaw of a specified size is observed, and the calculation of the expected life of a component resulting from the maximum flaw size which remains in the component when it is put into service.

Fracture Mechanics Based Fatigue Models

The damage-tolerant approach utilizes the models developed for linear–elastic fracture mechanics to describe crack growth. Fatigue crack growth is plotted as da/dn versus $\Delta K'$ where, da/dn is the crack growth rate per cycle and $\Delta K'$ is a modified cyclic stress intensity factor. The modifications to the cyclic stress intensity factor can include effects of mean stress, plastic behavior, compressive loading, and so on. A schematic of the typical da/dn versus ΔK curve is shown in Fig. 4.5-11. Based on this typical model the fatigue crack growth behavior can be divided into three regions.[10] Region I, shown in Fig. 4.5-11, is referred to as the near threshold or threshold regime. In this regime the cyclic stress intensity is sufficiently low such that the crack growth per cycle is essentially zero. A value of ΔK_{th} is often specified below which cracks do not propagate under cyclic loading conditions and the prescribed test conditions (i.e., temperature, environment, rate, etc.). Based on experimental results, the fatigue crack threshold for a number of common metals was found to occur in the range

$$1.5 \times 10^{-4} \sqrt{\text{in.}} < \frac{\Delta K_{th}}{E} < 1.8 \times 10^{-4} \sqrt{\text{in.}}$$

where E is the elastic modulus in units of ksi and K_{th} is in ksi $\cdot \sqrt{\text{in.}}$[11] A second relationship that is often useful in threshold evaluation for steels is given by

$$\Delta K_{th} = 6.4(1 - 0.85R)$$

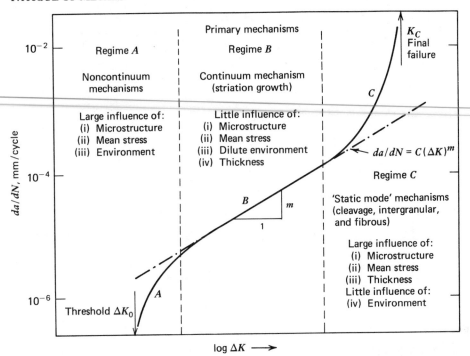

Fig. 4.5-11 Schematic variation of fatigue-crack growth rate da/dN with alternating stress intensity ΔK in steels, showing regimes of primary crack-growth mechanisms.[10]

where R is the load ratio, P_{min}/P_{max}, and ΔK_{th} is in ksi $\cdot \sqrt{in}$. This results in a generally conservative estimate of ΔK_{th} for ferritic-perlitic, martensitic, and austenitic steels under applied load ratios > 0.1.[12]

Region II is characterized by a linear relationship between $\log(da/dn)$ and $\log(\Delta K)$ which can be represented by the Paris relation[13]

$$\frac{da}{dn} = C(\Delta K')^n$$

where C and n are constants. In general, the exponent n is observed to range between 2 and 4 for most metals. Region II growth is characteristically modeled as a continuum growth and correlates with the observation of fatigue striations on the fatigue surface. Crack growth rates in this regime are typically from 10^{-6} to 10^{-3} mm/cycle and are relatively insensitive to microstructure, plane stress or plane strain constraint, load ratio, and in benign environments, the cyclic frequency.[10]

The models proposed for region II crack growth are generally applicable for constant ΔK loading. Under conditions of nonuniform loading, the models must be modified somewhat to account for the distribution of applied loads. When the loading is generally uniform, with occasional overloads, a crack retardation is observed. After the overload has occurred, a period of essentially no crack growth is followed by an increase in crack growth rate until the original growth rate has resumed. For variable amplitude loading that is random in nature, models similar to the Paris model can be employed where the cyclic stress intensity factor, ΔK, is replaced by a root-mean-square value of the cyclic stress intensity factor,

$$\Delta K_{rms} = \sqrt{\frac{\sum_{i=1}^{n} K_i^2}{n}}$$

Region III is generally referred to as the static failure regime since failure occurs within a relatively small number of cycles at stress intensities approaching the static failure values. In this region the crack growth rate predicted by the extrapolation of the region II behavior will underestimate the actual growth rate and will thus be nonconservative. This crack growth rate acceleration is believed to be due to the combined effects of fatigue (per the region II behavior) and static modes of failure such as ductile tear mechanisms.

In summary, quantitative fatigue crack growth predictions are limited to the evaluation of the threshold value, ΔK_{th}, and the model for growth in region II. When the component is subjected to loads which would place the behavior in the region III regime, static failure models are generally used.

Fracture Mechanics Based Fatigue Testing

Fatigue testing is generally employed to characterize two aspects of the LEFM-based fatigue models: the threshold, K_{th}, below which crack growth is not observed and the relationship between da/dn and

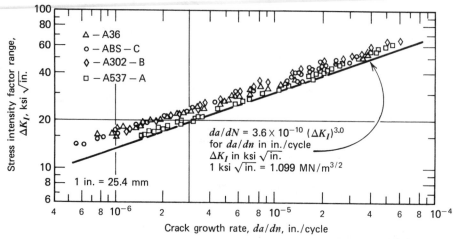

Fig. 4.5-12 Summary of fatigue crack growth data for ferrite-pearlite steels. (From Ref. 15.)

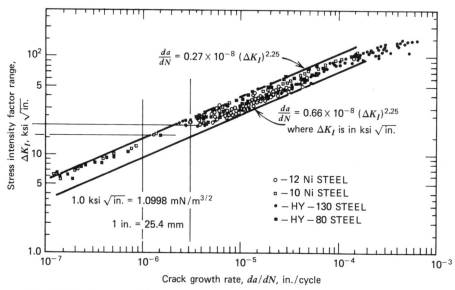

Fig. 4.5-13 Summary of fatigue crack propagation for martensitic steels. (From Ref. 15.)

ΔK in the regime of continuum crack growth, region II shown in Fig. 4.5-11. In either case the methods employed are similar to those used in fracture testing in that a crack is propagated in the material under a known load (or cyclic stress intensity in this case) and the crack length is recorded. Whereas in static fracture testing it is only important to know the crack length at the time of fracture, for a fatigue evaluation, the crack length must be monitored throughout the test in order to evaluate da/dn—the crack growth per cycle. A standard test method, ASTM E 647, has been developed to evaluate constant-load amplitude fatigue crack growth rates above 10^{-8} m/cycle.[14]

The crack length can be measured in a number of different ways. The three most commonly used methods are (i) direct observation of crack growth on the surface of the specimen,[14] (ii) measurement of the compliance of the specimen and in turn, relating this to crack length, and (iii) by use of the potential drop method. The potential drop method employs an external power supply to run a current through the specimen. The electric potential is measured from one side of the crack to the other, and as the crack length changes (which in turn changes the remaining ligament) the potential measured will change. Once the system is calibrated, this method provides a very efficient means of assessing small changes of crack growth. The crack length is used in two regards: to evaluate growth rate, da/dn, and also to calculate the stress intensity factor.

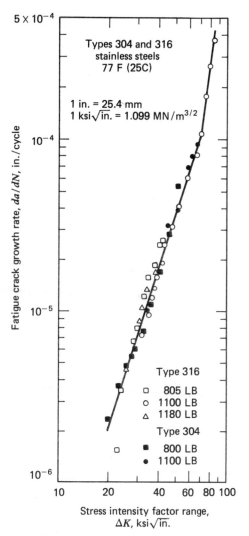

Fig. 4.5-14   Fatigue crack growth rates in type 304 and 316 stainless steels at room temperature as a function of range of stress intensity factor. (From Ref. 15.)

Measurement of the threshold stress intensity factor is very difficult to determine accurately due to the very precise measurement methods required to evaluate crack growth, the very low growth rates in this regime, and the length of time required to obtain the necessary data. Additionally, the threshold value is sensitive to the maximum value of the stress intensity and the load ratio R. To obtain the threshold value, one generally starts testing at a somewhat higher stress intensity than the expected threshold value and slowly decreases the load as crack growth is observed. Each time the load is decreased, generally less than 10% of its value, sufficient time must be allowed for continued crack growth. Sufficient run time at each load level is necessary since a false threshold may be observed due to the effect of crack growth retardation as the load is decreased. The effects of crack growth retardation can be minimized if the increment of load decrease is kept to a small fraction of the load level. This procedure of "load shedding" is continued until crack growth is no longer observed. Based on this load, measured crack length, and stress intensity solution for the specimen, the threshold value can be calculated.

Figures 4.5-12–4.5-16[15] show some typical da/dn versus ΔK curves for a number of common structural materials.

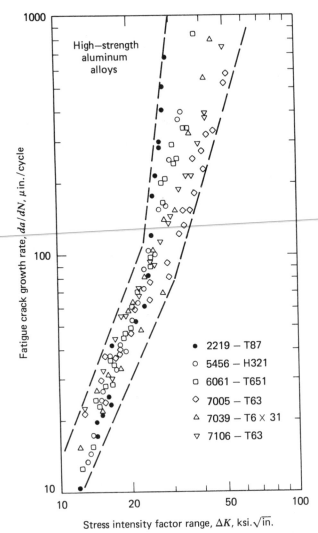

Fig. 4.5-15 Summary plot of da/dN versus ΔK for six aluminum alloys. The yield strengths of these alloys range from 34 to 55 ksi. (From Ref. 15.)

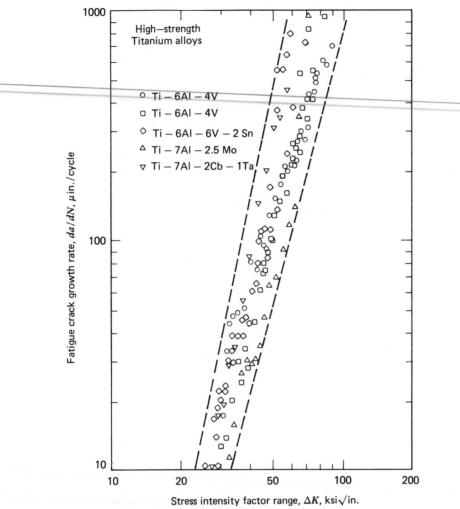

Fig. 4.5-16 Summary plot of da/dN versus ΔK data for five titanium alloys ranging in yield strength from 110 to 150 ksi. (From Ref. 15.)

REFERENCES

4.5-1 F. A. McClintock and A. S. Argon, "Fatigue," Ch. 18, *Mechanical Behavior of Materials*. Addison-Wesley, Reading, MA, 1966.

4.5-2 R. W. Hertzberg, *Deformation and Fracture Mechanics of Engineering Materials*. Wiley, New York, 1976, pp. 459–460.

4.5-3 ASM, *Metals Handbook, Failure Analysis and Prevention*, Vol. 10, 8th ed. American Society for Metals, 1975, pp. 17–21, pp. 95–125.

4.5-4 C. E. Richards and T. C. Lindley, *The Influence of Stress Intensity and Microstructure on Fatigue Crack Propagation in Ferretic Materials*, Vol. 4, Part 4, 1972, pp. 951–978.

4.5-5 J. E. Shigley, "Fatigue," *Mechanical Engineering Design*, 2nd ed. McGraw-Hill, New York, 1972, pp. 243–259.

4.5-6 American Society for Testing and Materials (ASTM), *Annual Book of ASTM Standards*, E466-82, Section 3, Vol. 03.01. ASTM, Philadelphia, 1983, pp. 566–571.

4.5-7 American Society for Testing and Materials (ASTM), *Annual Book of ASTM Standards*, E468-82, Section 3, Vol. 03.01. ASTM, Philadelphia, 1983, pp. 577–587.

4.5-8 American Society for Testing and Materials (ASTM), *Annual Book of ASTM Standards*, E606-80, Section 3, Vol. 03.01. ASTM, Philadelphia, 1983, pp. 652–669.

4.5-9 D. T. Raske and J. Morrow, "Mechanics of Materials in Low Cycle Fatigue Testing," Manual on Low Cycle Fatigue Testing, ASTM STP 465. American Society for Testing and Materials, Philadelphia, 1968, pp. 1–25.

4.5-10 R. D. Richie, "Near-Threshold Fatigue Crack Propagation in Steels," Review #245, *International Metals Reviews*, Nos. 5 and 6, (1979), pp. 205–230.

4.5-11 J. D. Harrison, "An Analysis of Data on Non-Propagating Fatigue Cracks on a Fracture Mechanics Basis," *British Welding Journal*, 2(3), (March 1970).

4.5-12 J. M. Barsom, "Fatigue Behavior of Pressure Vessel Steels," *WRC Bulletin* 194, Welding Research Council, New York, (May 1974).

4.5-13 P. C. Paris and F. Erdogan, "A Critical Analysis of Crack Propagation Laws," *J. Basic Engineering, Trans. ASME*, D, (1963), Vol. 85, p. 528.

4.5-14 American Society for Testing and Materials (ASTM), *Annual Book of ASTM Standards*, E647-83, Section 3, Vol. 03.01. ASTM, Philadelphia, 1983, pp. 710–730.

4.5-15 S. T. Rolfe and J. M. Barson, *Fracture and Fatigue Control in Structures*. Prentice-Hall, Englewood Cliffs, NJ, 1977, pp. 208–291.

4.6 FRACTURE OF METALS

R. D. Streit

4.6-1 Nature of Fracture

Fracture is the study of a material's resistance to crack propagation when it is subjected to a specified set of loading and environmental conditions. Due to the presence of flaws (or cracks) in a structure, failure may occur in a brittle manner, even though the material may be nominally ductile. Failures of this type can occur very rapidly, with little or no warning of imminent failure. Initially, the flaw may be a result of fabrication (e.g., weld cracking or machining), design (e.g., sharp corner or sharp contact points), or may be inherent in the material (e.g., forging lap or slag inclusions). From this initially flawed condition, the loading and environment can combine to extend the crack to its critical size—the crack size at which unstable crack growth occurs.

The field of fracture is divided into two major disciplines—that associated with structural analysis and that associated with the mechanical behavior of the material. The analysis discipline encompasses the evaluation of a component's fracture behavior under specified loading and environmental conditions due to the presence of one or more cracks in the component. It is assumed in such analyses that the material behavior is well characterized. Typical calculations deal with the evaluation of the stress intensity solution for different cracks in various component geometries; for example, the stress intensity solution for a circumferential crack in a pressure vessel. Other analysis efforts range from simple elastic stress calculations to detailed elastic-plastic crack growth and tearing analyses under complex loading conditions.

The evaluation of a material's resistance to crack growth and unstable crack propagation is concerned with the mechanical behavior of the material. The goals here are twofold; first, to characterize the true mechanical properties of the material in such a way that they can be used to determine the fracture behavior of any component of the same material under similar loading and environmental conditions. Fracture tests are carefully controlled to ensure that the properties being measured are not a function of the particular test configuration and test equipment. The ASTM has standardized specific tests to provide for a uniform code of fracture toughness evaluation. A number of these tests are discussed in Section 4.6-2.

The second goal of fracture evaluation is the characterization of specific properties that can be used as a means of comparing materials or screening materials. Notched impact tests are particularly useful toward this end since they tend to promote brittle behavior, are reasonably quick to complete and are relatively inexpensive compared to other fracture tests. Procedures for impact testing are also covered by ASTM standards. This aspect of fracture is discussed in Section 4.6-3.

Clearly, both structural analysis and material evaluation are critical to the field of fracture mechanics and must work together to assess the fracture response of a component. As such, the methods employed in fracture evaluation include both the required materials behavior and an analysis of the component in which the material is to be used.

Characteristics of Brittle Fracture

The toughness of a material is a measure of the energy required to initiate and propagate the fracture process. During brittle fracture very little energy is absorbed (i.e., low-fracture toughness) and, therefore, crack propagation can occur very rapidly, with little plastic deformation. For unstable fracture the elastic strain energy available for the fracture process is greater than that necessary to propagate a crack. In this context brittle fracture and unstable crack propagation are used interchangeably. Brittle fracture of most engineering structural materials can be described by linear-elastic fracture mechanics (LEFM) using the material's plane strain fracture toughness, the applied loading, the size and geometry of the flaw, and the geometry of the structure. The fracture surface is generally flat—exhibiting little deformation. However, the micromechanism of brittle fracture (or unstable crack propagation) may be any of a number of mechanisms ranging from low-strain cleavage to large-strain dimple rupture and tearing. In the former case it is clear that such behavior would be considered brittle. However, in the latter case, the material thickness, loading and environment can contribute to a mechanical behavior resulting in unstable crack growth, whereas, the microbehavior is clearly ductile.

Brittle Fracture of Ductile Materials

Materials that nominally behave in a ductile manner (i.e., with significant plastic deformation) can behave in a brittle manner in the presence of a crack under certain conditions. The conditions that tend to promote brittle behavior are (i) low temperature, (ii) high strain rate, and (iii) large section thicknesses (plane strain constraint). Under these conditions the crack may propagate in an unstable manner, providing little or no warning of imminent failure. The fracture surface may exhibit dimple rupture and shear—mechanisms generally associated with ductile behavior—however, the plastic deformation will be confined to a rather narrow region on either side of the crack plane.

The effect of reducing the temperature on lowering the fracture resistance of a material is clearly demonstrated by the temperature–transition curve observed on impact testing of most engineering materials. The characteristic shape of a transition–temperature curve for common structural steels is shown in Fig. 4.6-1. Under the impact test conditions (e.g., Charpy, dynamic tear, drop weight tear, etc.) some materials exhibit a very sharp transition–temperature regime between which a material may behave very ductile to the condition in which very little energy is absorbed in the fracture process. In addition, the thickness of the specimen and the nominal stress on the specimen play an important role in a given material's ductile-to-brittle transition–temperature (nil-ductility transition, NDT, temperature). Typically, operating temperatures greater than 60°F above the NDT temperature are often specified in codes for sections up to 1 in. thick.[1]

The rate of loading also plays an important role in a material's ductile-to-brittle transition behavior. As the loading rate is increased into the dynamic loading range, structural materials tend to behave in a more brittle manner. The effect of changing the loading rate on a typical steel is shown in Fig. 4.6-2. This effect of reducing the ductility with an increase in loading rate is analogous to the reduction of ductility with a decrease in temperature. For loading rates nominally in the slow or static regime, the material behavior is not strongly affected by changes in rate. Similarly, for very high loading rates such that the material's behavior is purely cleavage fracture, changes in loading rate will not change the fracture toughness. However, as with the transition–temperature region, over a "rate transition range" the fracture toughness can be dramatically influenced. In the study of rate effects on fracture, the loading rate is generally specified in terms of the rate of increase in the applied stress intensity since the strain rate around the crack tip of a crack is not uniform and is thus not clearly defined. The standard method for static fracture testing is valid for rates up to $2.75 \text{ MPa} \cdot \text{m}^{1/2} \cdot \text{s}^{-1}$.[2]

The third factor, the thickness of the material, influences the ductile-to-brittle behavior by changing the state of stress at the crack tip from one of biaxial tension (plane stress) to one of triaxial

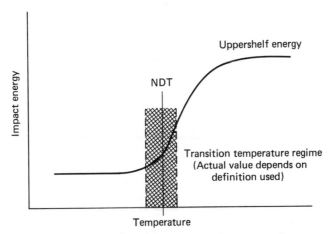

Fig. 4.6-1 Typical transition–temperature test results.

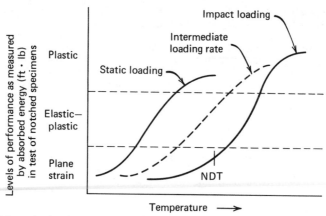

Fig. 4.6-2 Schematic showing relation between notch-toughness test results and levels of structural performance for various loading rates.[12]

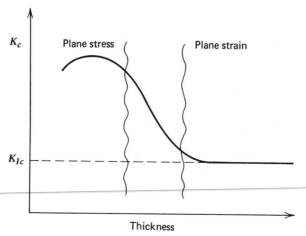

Fig. 4.6-3 Fracture toughness as a function of specimen thickness.

tension (plane strain). Under conditions of plane strain, the local deformation prior to fracture is significantly less than the plane stress case. This decrease in plastic zone size is due to the effective increase of the yield strength of the material at the crack tip. However, once the component thickness is sufficiently large, such that the complete crack front is in a state of plane strain, further increases in thickness will not reduce the fracture toughness of the material, Fig. 4.6-3. The required thickness for linear-elastic plane strain behavior has been established by the American Society for Testing and Materials (ASTM)[2] as $2.5(K_{Ic}/S_y)^2$. For design purposes, however, it is often desirable to have section thicknesses considerably less than this to ensure significant plastic deformation prior to fracture.

4.6-2 Fracture Mechanics

Fracture mechanics is a design tool used to predict the fracture of structures subjected to a prescribed loading and environmental conditions and that contain a crack. A number of methods can be employed to quantitatively evaluate a structure's safety against crack growth. These methods include linear-elastic fracture mechanics (LEFM), elastic-plastic fracture mechanics (EPFM), and fully plastic analysis. In LEFM unstable crack propagation is predicted based on the relationship between the material's plane strain fracture toughness, K_{Ic}, and the elastic stress field in the vicinity of the crack tip in a structure. The fracture toughness of a material is a unique materials property and cannot be deduced from other material tests (e.g., tensile tests). The EPFM approach is essentially an extension of the LEFM methods into the regime of nonlinear elastic behavior. Using the EPFM method, both the onset of fracture and the crack growth stability can be evaluated. These first two methods, LEFM and EPFM, are discussed in further detail in this section.

Fully plastic behavior is the extension of fracture mechanics to the regime where the fracture behavior of a structure can be fully described by the tensile properties of the material without regard to specific fracture properties. Under these conditions fracture analysis can be replaced by standard structural analysis in which the load-bearing area is reduced by the presence of the crack.

Linear–Elastic Fracture Mechanics

The most common form of analytical fracture mechanics is linear–elastic fracture mechanics (LEFM). LEFM is based on the principle that unstable crack propagation will occur when the stress intensity at the tip of a crack in the material equals or exceeds the fracture toughness of the material at the specified temperature, loading rate, and environmental conditions. Thus, by ensuring the applied stress intensity factor (based on an actual or assumed flaw size) is less than the fracture toughness, unstable crack propagation can be avoided. This design relationship is analogous to the association of a component's stress state with the material's tensile properties. LEFM enables the engineer to evaluate a material's resistance to crack propagation, and thus provide a quantitative means for certifying a material for fracture-safe applications.

LEFM Theory. Linear–elastic fracture mechanics is based on the elastic solution to the stress and strain field in the vicinity of the tip of a crack in a perfectly elastic material. While solutions for these stress and strain fields have been evaluated by various methods (e.g., boundary collocation,[3] conformal mapping,[4] polynomial stress functions,[5] etc.), the general form of the resulting equations for the stress components are given by

$$\sigma = \frac{K_I}{\sqrt{r}} F_1(\theta) + A_2 F_2(\theta) + \cdots \tag{4.6-1}$$

where r is the distance in front of the crack tip, K_I is a constant which characterizes the severity of the stress and strain field, A_2 is a second-order constant term in the series expansion, and the dimensionless functions $F_k(\theta)$ account for the angular position at which the stress is evaluated in the vicinity of the crack tip (Fig. 4.6-4). Equation (4.6-1) shows that at the crack tip, that is, as r goes to zero, the stresses become infinite. While this analysis is in reality only true for a purely elastic material, the parameter K_I, referred to as the stress intensity, uniquely characterizes the shape of the crack tip stress and strain field, and, as such, can relate to the structural behavior of a cracked configuration.

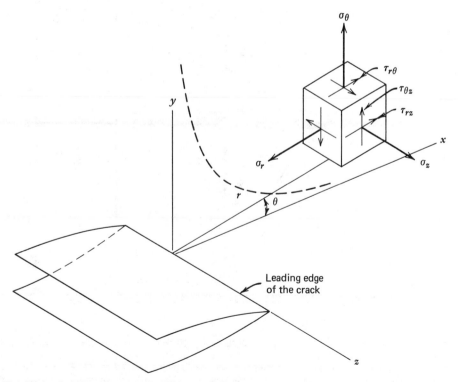

Fig. 4.6-4 Coordinates measured from the leading edge of a crack and the stress components in the crack tip stress field. The dashed line shows schematically the σ_θ stress along $\theta = 0$.

The limitations of LEFM lie in the fact that the analysis is strictly valid only for purely elastic materials. While some brittle materials may approach this behavior (e.g., glass, ceramics, certain high-strength steels), in reality, all materials exhibit some plastic deformation in the vicinity of the crack tip prior to fracture. In these cases LEFM can still be applicable providing that the zone of plastic deformation is small relative to the size of the region in which the stress intensity dominates the stress field solution. To meet this criterion, the characteristic dimensions (i.e., the crack length, the part thickness, and the uncracked ligament length) of the specimen or component must be greater than $2.5(K_{Ic}/S_y)^2$. This will ensure that the characteristic dimensions will be at least 50 times the plastic zone size. For cases in which this size requirement is not met, methods of testing and analysis that account for plastic deformation should be employed, for example, plastic zone correction[6] or elastic–plastic fracture mechanics.

LEFM Test Methods.   Fracture toughness values can be obtained, in principle, from any configuration for which a stress intensity solution (or equivalent) is available. Four standard specimen

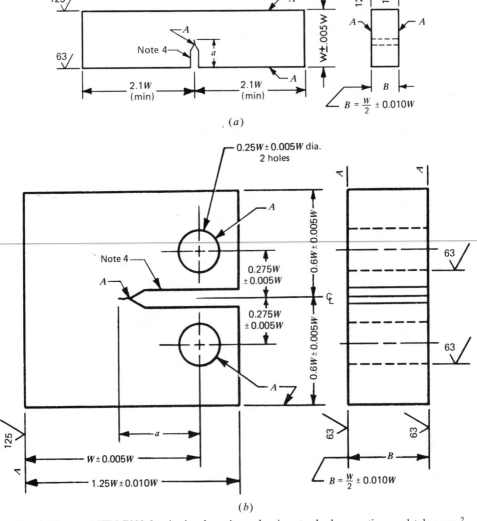

Fig. 4.6-5   (*a*) ASTM E399 3-point bend specimen showing standard proportions and tolerances.[2] (*b*) ASTM E399 compact tension specimen showing standard proportions and tolerances.[2] (*c*) ASTM E399 arc-shaped specimen showing standard proportions and tolerances.[2] (*d*) ASTM E399 disk-shaped compact specimen showing standard proportions and tolerances.[2] (Reprinted with permission from the *Annual Book of ASTM Standards*. Copyright 1983, ASTM, 1916 Race St., Philadelphia, PA 19103.)

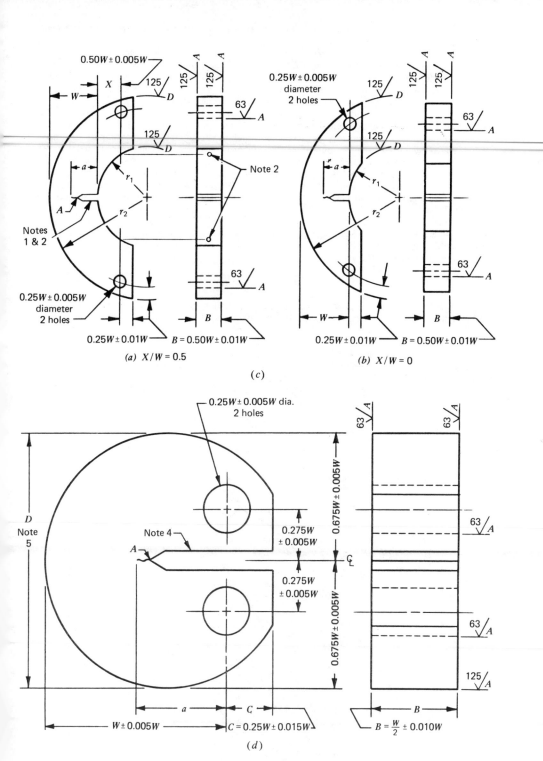

(a) $X/W = 0.5$

(b) $X/W = 0$

(c)

(d)

Fig. 4.6-5 (*Continued*)

geometries—the compact tension, three-point bend, arc-shaped, and the disk-shaped compact tension specimen—have been adopted by the ASTM[2] for obtaining plane strain fracture toughness (Fig. 4.6-5). The test method specifies specimen size requirements, loading fixture design, displacement measuring techniques, specimen preparation, and measurement requirements for valid K_{Ic} testing. To establish an initially sharp crack, careful fatigue precracking is also required.

Once a specimen has been prepared and fatigued according to the ASTM E399 specifications, the specimen is loaded to failure while recording the load–displacement curve (Fig. 4.6-6). For tests in which the fracture load, defined by P_Q, meets the standard's requirements, a preliminary toughness value, K_Q, is calculated from the appropriate stress intensity solution. Thus, having obtained K_Q, the specimen size requirements must be validated, and only then can K_Q be called K_{Ic}—the material's plane strain fracture toughness. Many sources of fracture toughness data are available in the literature. References 7–14 are a few of the sources available to obtain fracture toughness data. The range of toughness values for a number of common structural materials as obtained from these sources are given in Table 4.6-1.

LEFM Design Analysis. The design philosophy is simply to ensure that the applied stress intensity is less than the fracture toughness of the material. The stress intensity factor is related to the nominal stress, component geometry, and flaw size by

$$K_I = F(\text{geometry}) \, \sigma \sqrt{\pi a}$$

where F is a constant which depends only on the geometry of the part and crack, σ is the nominal stress, and a is the crack length. While the form of this equation may change due to the type of loading or geometry involved (e.g., using applied moment in place of stress or replacing the crack length a by the relative crack length $a/\text{thickness}$) the three basic components of the equation—loading, geometry, and crack size—are always incorporated. The stress intensity solutions for a number of cracked geometries have been evaluated and are readily available in a number of handbooks[15,16] as well as in the technical literature. A few of the more commonly employed geometries in design are shown in Fig. 4.6-7. Using the appropriate stress intensity solution, fracture will occur when the loading and/or crack size is increased such that

$$K_I > K_C$$

where K_C is the critical stress intensity factor for a particular material (including heat treatment,

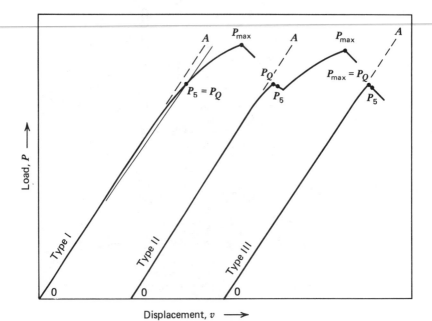

Fig. 4.6-6 Principal types of load displacement records obtained during ASTM E399 fracture toughness evaluation.[2] (Reprinted with permission from the *Annual Book of ASTM Standards.* Copyright 1983, ASTM, 1916 Race St., Philadelphia, PA 19103.)

TABLE 4.6-1   ROOM TEMPERATURE PLANE STRAIN FRACTURE TOUGHNESS (K_{IC}) OF METALS

Material	Condition	Form[a]	Yield Strength (ksi)	K_{Ic} (ksi · $\sqrt{\text{in.}}$) Min.*	Avg.	Max.*
Alloy Steels						
18Ni Maraging (200)	Aged 900°F, 6 hr	P	210	100	100	100
18Ni Maraging (250)	Aged 900°F, 6 hr	P	259	78	83	88
18Ni Maraging (300)	Aged 900°F	P	276	44	52	57
18Ni Maraging (300)	Aged 900°F, 6 hr	F	280	75	83	94
300M	Tempered at 600°F	F	259–262	43	49	52
300M	Tempered at 1000°F	F	230	64	—	66
4330V	Tempered at 525°F	F	203	77	82	84
4330V	Tempered at 800°F	F	191	93	96	100
4340	Tempered at 400°F	F	229–241	40	50	61
4340	Tempered at 500°F	P	217–238	45	50	56
4340	Tempered at 800°F	F	197–211	71	76	81
D6AC	Tempered at 1000°F[b]	F	213–221	64	76	84
Stainless Steels						
PH 13-8Mo	H1000	P	210–219	78	98	109
PH 13-8Mo	H950	F, 4 × 4 in.	210	70	79	92
PH 13-8Mo	H1000	F, 4 × 4 in.	212	60	100	130
PH 13-8Mo	H950	F, 8 in. dia.	210	47	54	58
PH 13-8Mo	H1000	F, 8 in. dia.	205	57	80	97
Aluminum Alloys						
2021	T81	P	61–66	26	28	36
2024	T851	P	59–66	18.5	24	28
2124	T851	P	64–66	21.5	26	32.5
7049	T73	F	61–68	29	32	34
7049	T73	E	73–75	27.5	30.5	35
7075	T651	P	70–76	24.5	27	30.5
7075	T7351	P	53–58	30.5	31.5	32.5
7075	T7651	P	63–65	26	27.5	28.5
Titanium Alloys						
Ti-6A1-4V	RA	—	118–121	74	88	98
Ti-6A1-4V	MA[c] (1300°F)	—	120–127	81	84	87
Ti-6A1-6V-2Sn	MA[c] (1300°F)	—	144–146	45	52	58
Ti-6A1-6V-2Sn	STA[d] (1050°F)	—	179	29	30	31

[a] E = extrusions, F = forgings, P = plate.
[b] 1000 to 1025°F.
[c] Mill annealed.
[d] Solution treated and aged.
*Typical minimum and maximum values are shown. This is not meant however that the material could not be better or worse under extreme processing, environmental or loading conditions.

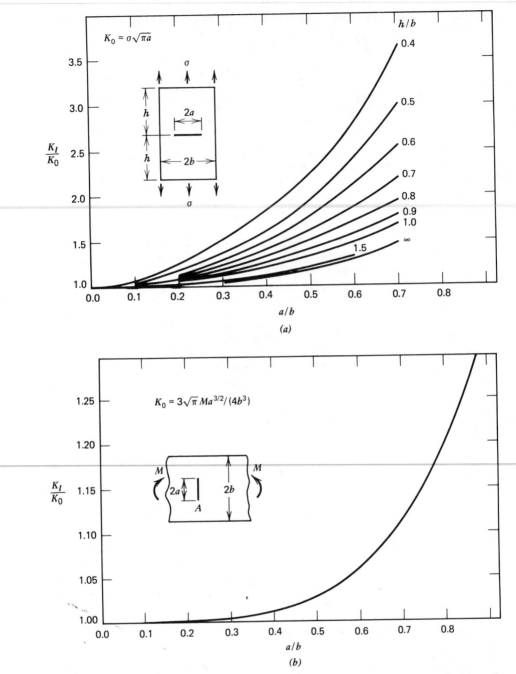

Fig. 4.6-7 (a) K_I for central crack in a rectangular sheet subjected to a uniform uniaxial tensile stress.[15] (b) K_I for tip A of a central crack in a finite width sheet subjected to a uniform bending moment.[15] (c) K_I for a central crack in a rectangular sheet with opposing forces at the center of the crack.[15] (d) K_I for an edge crack in rectangular sheet subjected to a uniform uniaxial tensile stress.[15] (e) K_I for an edge crack in a finite width sheet subjected to pure bending or three-point bending.[15] (f) K_I for a pressurized internal radial edge crack in a tube subjected to a uniform internal pressure.[15] (g) K_I for a semi-elliptical (thumbnail) crack in a large plate subjected to a uniform axial tension stress; above, the plate; below, dependence of the flaw shape parameter on the ratio of depth and crack surface length. $K_I = 1.12\sigma\sqrt{\pi a/Q}$.[12]

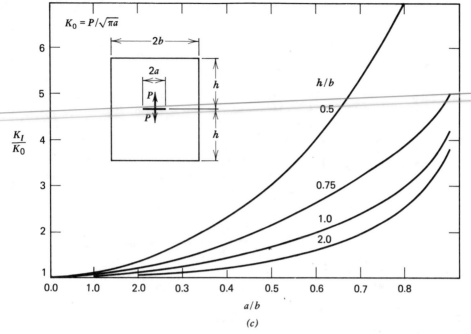

Fig. 4.6-7 (*Continued*)

weldments, etc.) at a given temperature, loading rate, and environment. For components under plane strain conditions, nominal loading rates and benign environments, K_C is equivalent to K_{Ic}.

Elastic–Plastic Fracture Mechanics

The applicability of LEFM is often limited by either the high toughness of many materials used in fracture-critical components or the requirement that specific components be capable of extensive plastic deformation prior to fracture. Such advances in materials development and design philosophy frequently require that the nonlinear material behavior, that is, plasticity, be embedded into the failure criterion. Methods employed for linear-elastic fracture toughness testing such as ASTM E399,[2] will often yield invalid data due to the specimen size requirements imposed for high-toughness materials. The problem is compounded since the material available for testing is usually limited by the engineering design under consideration. Further, a fracture criterion based on LEFM may give overly conservative estimates of the fracture strength of ductile fracture and can, therefore, restrict the use of perfectly adequate material in design. A fracture criterion that incorporates the basis of elastic–plastic material behavior must therefore be applied in cases in which linear elasticity can no longer be assumed and/or the specimens of the required size are not available for testing.

The two most widely accepted methods of elastic–plastic fracture mechanics (EPFM) are the crack opening displacement (COD) method and the J-integral approach. These concepts can be used to extend the range of fracture predictions beyond the linear–elastic regime. Both models can be employed in the evaluation of elastic–plastic fracture behavior (i.e., design analysis) and can be used as a measure of a material's fracture toughness and tearing resistance.

EPFM Theory. The crack opening displacement (COD) criterion to elastic–plastic fracture prediction has been accepted for a number of years in the European countries. Arguing that the fracture criterion should depend on local conditions at the crack tip, fracture will occur when the local strain or equivalently the crack opening displacement reaches a critical value.[17] Using Irwin's model of the plastic zone,[6] it is possible to show that the COD is related to the stress intensity K_I in the regime of LEFM by

$$\text{COD} = \frac{K_I^2}{S_y E}$$

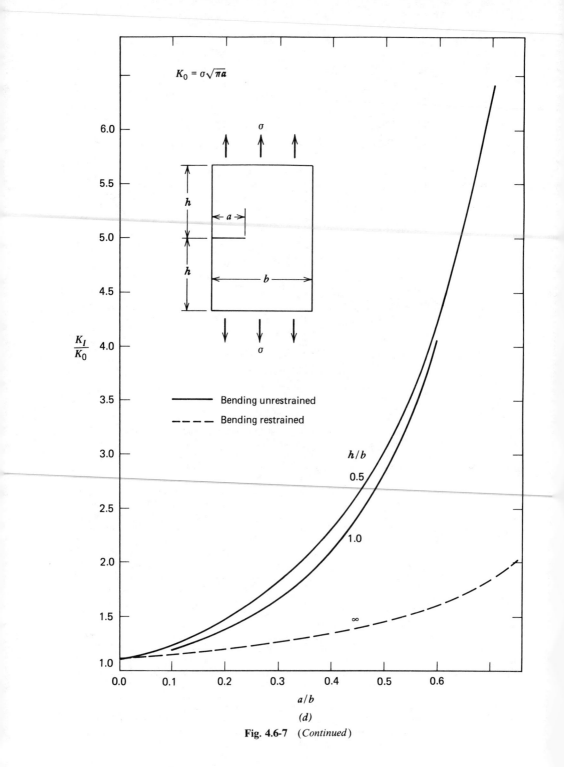

Fig. 4.6-7 (*Continued*)

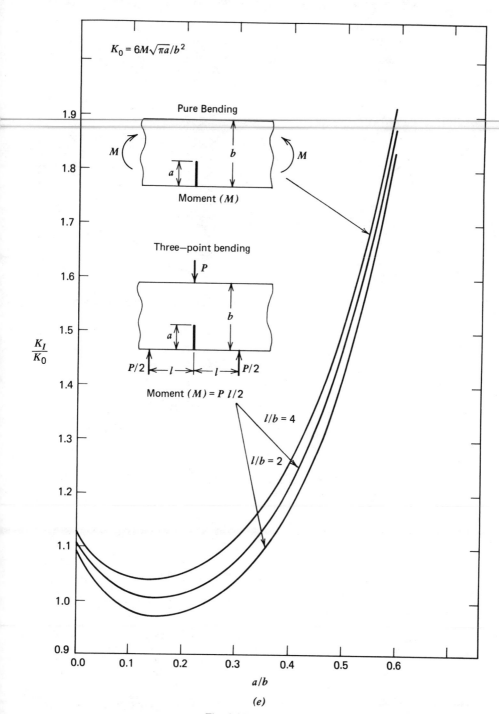

$K_0 = 6M\sqrt{\pi a}/b^2$

Pure Bending

M M

b

a

Moment (M)

Three—point bending

P

b

a

$P/2$ l l $P/2$

Moment $(M) = P\,l/2$

$l/b = 4$

$l/b = 2$

$\dfrac{K_I}{K_0}$

a/b

(e)

Fig. 4.6-7 (*Continued*)

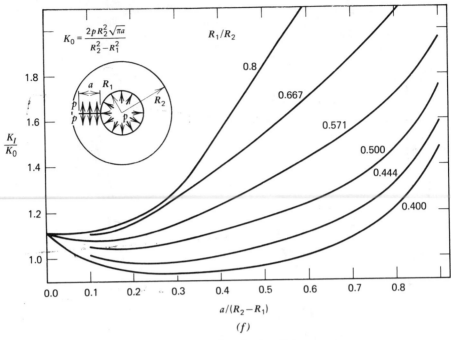

$$K_0 = \frac{2p R_2^2 \sqrt{\pi a}}{R_2^2 - R_1^2}$$

Fig. 4.6-7 (*Continued*)

The *J*-integral, originally defined by Rice,[18] is a path-independent energy line integral for two-dimensional problems. Using the model developed by Hutchinson,[19] and Rice and Rosengren[20] (HRR) to describe the stress–strain field in the vicinity of a crack, McClintock[21] concluded that *J* is a measure of the plastic stress and strain singularity near the tip of a crack. With such an interpretation, we may regard the field characterizing parameter *J*, for the plastic case, analogous to the stress intensity factor *K* in LEFM.

Rice has further shown that the *J* integral can be interpreted as the difference in potential energy between two identically loaded bodies having infinitesimally differing crack lengths. Thus, in the linear–elastic range, *J* is equal to the crack driving force *G*, and as a result J_{Ic} is related to the plane-strain fracture toughness K_{Ic} by

$$J_{Ic} = \frac{(1 - \nu^2)}{E} K_{Ic}^2 \tag{4.6-2}$$

where *E* is the elastic modulus, and ν is Poisson's ratio. Thus, using EPFM concepts, the plane strain fracture toughness K_{Ic} can be evaluated from specimens that meet the *J* thickness criterion. This thickness can be significantly less than that required for the standard K_{Ic} test as shown in the following section.

J_{Ic} **Test Methods.** A standard procedure for the determination of J_{Ic}, which can be used as a toughness value at the initiation of crack growth for metallic and certain nonmetallic materials, has been adopted by the ASTM. The ASTM E813[22] method involves using three-point bend or pin-loaded fatigue precracked specimens to determine *J* as a function of crack growth. Specimens are loaded such that the fatigue crack is extended some small amount, Δa_p, while load versus load-point displacement is obtained autographically on an *X-Y* recorder. The *J*-integral value associated with a given increment of crack extension is then determined from this load–displacement curve and plotted versus the physical crack growth, Δa_p. The curve of *J* versus *a* is referred to as the resistance curve or *R* curve. Four data points are required within the specified limits of crack growth for one complete test.

A blunting line, which approximates the artificial crack advance due to crack tip stretch, is then drawn. This estimate of crack tip blunting is calculated from material flow properties. The intersection of the blunting line and the *R* curve (determined by a least-square line of the *J* data points) then defines the initiation of actual crack growth, and, hence, J_{Ic}. This procedure is demonstrated in Fig.

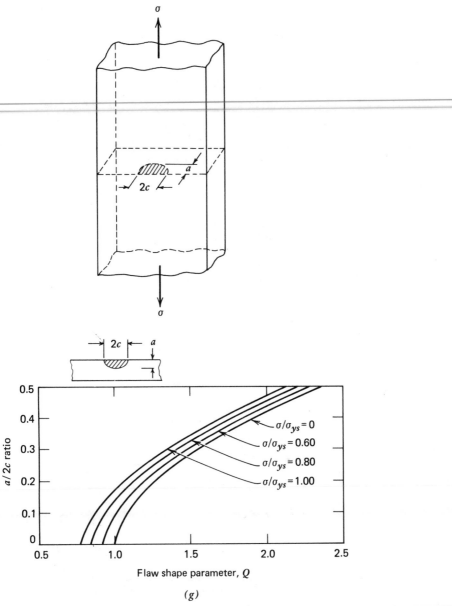

(g)

Fig. 4.6-7 (*Continued*)

4.6-8. The method requires that multiple specimens be used or that the crack length be monitored during the test using unloading compliance or another validated method.

Similar to the size requirement imposed on plane strain fracture toughness testing, certain specimen size constraints must be met for valid J_{Ic} testing. The remaining uncracked ligament (b) and the specimen thickness (B) must be large compared to the size dimension in which the stress and strain field is described by the J integral. Through an extensive analytical and experimental test program the ASTM has proposed that both b and B must be greater than $15(J/S_y)$ for points on the J resistance curve, and greater than $25(J_{Ic}/S_y)$ for the critical value of J_{Ic}. Further, since a nonlinear–elastic model is used to develop the HRR stress–strain field, the analysis is applicable to stationary cracks subjected to a monotonically increasing load. The maximum extent of crack growth

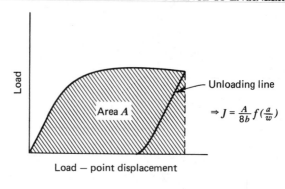

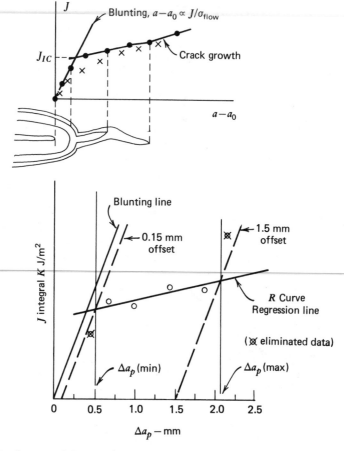

Fig. 4.6-8 Development of J versus Δa curve for the determination of J_{Ic} in ASTM E813 standard practice.[22]

must also be limited to maintain the linear J versus Δa relationship. This limit is set by a 1.5-mm (0.060 in.) offset to the blunting line (Fig. 4.6-8).

The results of a J_{Ic} fracture toughness test provide a fracture parameter that can be related to K_{Ic} using Equation (4.6-2), and, in turn, the critical flaw size and stress level. In addition, the slope of the J-Δa curve, dJ/da, can be used to calculate the tearing modulus T, as defined by Paris,[23]

$$T = \frac{E}{S_f^2} \frac{dJ}{da}$$

where S_f is the material flow stress (often defined as the average of the yield and ultimate strength) and E is the elastic modulus. The tearing modulus is a relative measure of a material's resistance to crack extension once a crack has been initiated. The two parameters, J_{Ic} and T, together describe the initiation and growth characteristics of a crack in the elastic–plastic regime.

EPFM Design Analysis. The fracture toughness J_{Ic} can be used to calculate the plane strain fracture toughness K_{Ic} [via Eq. (4.6-2)] regardless of the fact that the J specimen may not have met the size requirements for K_{Ic} plane strain testing. Once evaluated, the design principle of LEFM may be applied to the structure and a critical flaw size and/or stress level estimated (see LEFM Design Analysis). However, in cases in which the structure undergoes large plastic deformation prior to fracture this approach may be very conservative.

An elastic–plastic analysis incorporating the J integral may also be used directly to evaluate a given structure for crack growth initiation and unstable crack propagation. Crack growth will initiate when the applied loading is increased such that $J > J_{Ic}$. The stability of crack growth can be estimated based on the type of loading and the slope of the J resistance curve. While such calculations often require extensive computer analysis to determine the applied J and tearing modulus T, a number of the more common two-dimensional flaw geometries under various loading conditions have been evaluated and are presented in "Elastic–Plastic Fracture Handbook".[24]

Tables of Plane-Strain Fracture Toughness

The ASTM has standardized two procedures for obtaining valid plane-strain fracture toughness data.[2,22] These test methods, based on LEFM[2] and EPFM,[22] were discussed in previous sections. Numerous sources of fracture toughness data are available in the literature for particular materials, heat treatments, and processing. References 7–14 provide either a compilation of data or sources of data in which fracture toughness data can be found. Table 4.6-1 is a summary of data for some common structural materials as obtained from these references. The reader is cautioned, however, that data reported in the literature may not be strictly valid per the ASTM standards or may be limited to a particular heat treatment or mechanical processing.

4.6-3 Transition-Temperature Approach

Transition-temperature tests are used to evaluate the ductile-to-brittle failure transition as a function of the test temperature. These tests incorporate all three of the main factors that contribute to a material's ductile-to-brittle transition: low temperature, high loading rate, and notched geometry.

Transition-Temperature Test Methods

Various standard test methods have been adopted for impact loading of notched specimens. The most common tests employed are the Charpy and Izod tests (ASTM E23),[25] the dynamic tear test (ASTM E604),[26] and drop weight tear test (ASTM E436).[27] The Charpy test requires the smallest specimen of these impact tests and the V-notch version is probably the most common type of fracture test conducted. This Charpy test is a simple-beam impact test of a specimen 1×1 cm by 5.5 cm. long. A small 45° V-notch, 2 mm deep with an apex radius of 0.25 mm is machined into one side of the specimen. The standard Charpy specimens are shown in Fig. 4.6-9. The specimen is inserted into a pendulum-type impact machine and is fractured by the impact strike force of the pendulum (Fig. 4.6-10). The energy of fracture is measured by the energy loss in the pendulum, which, in turn, is measured by the difference in height it travels after fracturing the sample relative to its initial height. The pendulum shape, the pendulum head velocity, the height of drop, the system friction, as well as other particulars about the test are specified in the ASTM standard test method. Similar procedures exist for the other ASTM standard transition-temperature tests.

The results of these transition–temperature tests are generally interpreted in terms of (i) the temperature in which there is an abrupt increase in the energy required to propagate a crack—known as the nil-ductility transition (NDT) temperature and (ii) the "upper-shelf" energy for ductile fracture (Fig. 4.6-1). A number of typical Charpy–V notch transition curves for some common structural materials are shown in Fig. 4.6-11. Although these curves are referred to as "typical," the engineer is cautioned that such transition behavior is strongly dependent on material processing, chemistry, and heat treatment.

Transition–Temperature Design Approaches

Transition–temperature tests, while primarily used in material certification and screening, have also been incorporated in design. The methodologies are based on empirical relations between fracture mechanics properties (e.g., K_{Ic}) and transition-temperature test energy. A number of these correla-

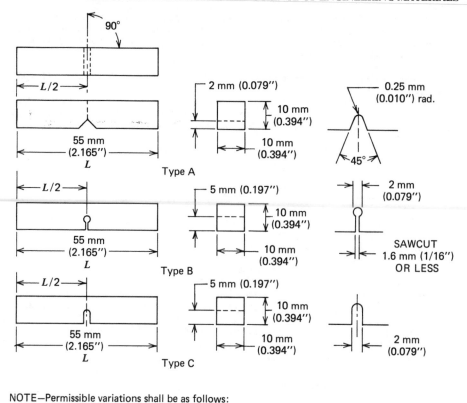

NOTE—Permissible variations shall be as follows:

Notch length to edge	±2°
Adjacent sides shall be at	90° ±10 min
Cross—section dimensions	±0.075 mm (±0.003 in.)
Length of specimen (L)	+0, −2.5 mm (+0, −0.100 in.)
Centering of notch (L/2)	±1 mm (±0.039 in.)
Angle of notch	±1°
Radius of notch	±0.025 mm (±0.001 in.)
Notch depth:	
Type A specimen	±0.025 mm (±0.001 in.)
Types B and C specimen	±0.075 mm (±0.003 in.)
Finish requirements	2 μm (63 μin.) on notched surface and opposite face; 4 μm (125 μin.) on other two surfaces

Fig. 4.6-9 Charpy (simple-beam) impact test specimens, Types A, B, and C. Type A is the commonly used Charpy–V notch specimen.[25]

tions are given in Reference 28. Among the more useful is the correlation between Charpy–V notch energy (CVN) and fracture toughness, K_{Ic}. For steels with yield strengths from 130–250 ksi and fracture toughnesses from 87–200 ksi · in.$^{1/2}$, in which the Charpy energy has reached its upper-shelf energy value in the range 16–60 ft · lb, the relation developed by Rolfe and Novak[29] is given by

$$\left(\frac{K_{Ic}}{S_y}\right)^2 = 5\left(\frac{CVN}{S_y} - 0.05\right)$$

where K_{Ic} is the fracture toughness (ksi · in.$^{1/2}$), S_y is the 0.2% yield strength (ksi) and CVN is the Charpy–V notch energy (ft · lb).

Another design approach, which employs the failure analysis diagram (FAD) developed by Pellini,[30] is based on three transition points defined by impact loading notched bend specimens. They include (i) FTP (fracture transition plastic) above which plastic deformation occurs without cracking, (ii) FTE (fracture transition elastic) below which a crack can propagate into material stressed below yield, and (iii) NDT (nil-ductility transition) which is defined as the temperature at which a small

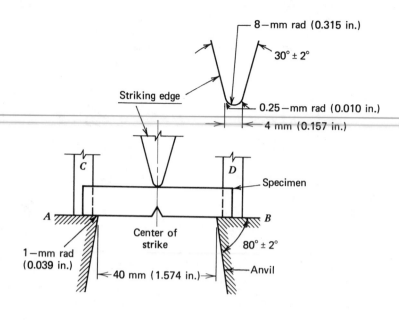

8—mm rad (0.315 in.)

30° ± 2°

Striking edge

0.25—mm rad (0.010 in.)

4 mm (0.157 in.)

C

D

Specimen

A

B

Center of
strike

80° ± 2°

1—mm rad
(0.039 in.)

40 mm (1.574 in.)

Anvil

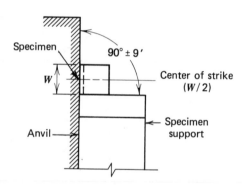

Specimen

90° ± 9′

W

Center of strike
(W / 2)

Specimen
support

Anvil

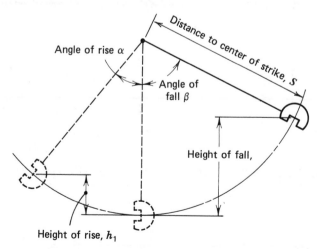

Distance to center of strike, S

Angle of rise α

Angle of
fall β

Height of fall,

Height of rise, h_1

Fig. 4.6-10 Schematic showing test setup for Charpy-type specimen in a pendulum-type impact test machine.[25]

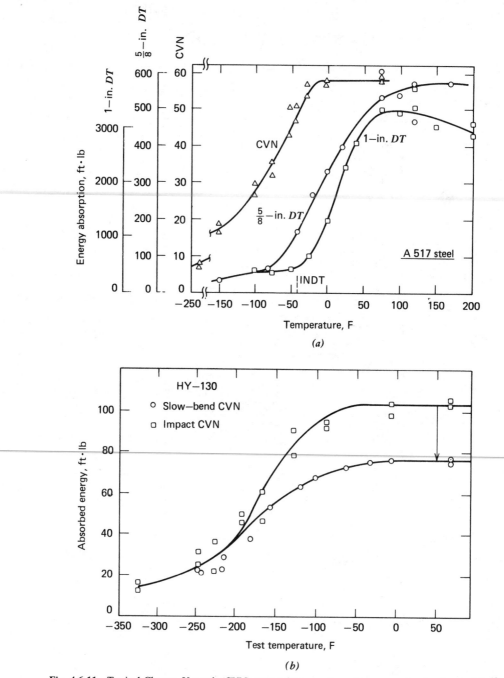

Fig. 4.6-11 Typical Charpy–V notch (CVN) curves for a number of common structural materials.[12]

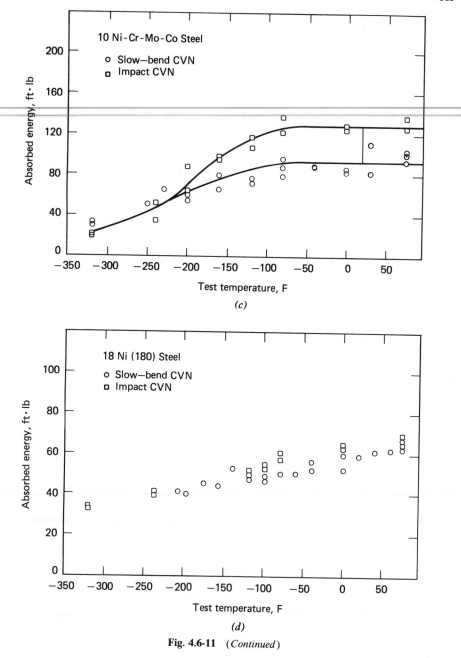

Fig. 4.6-11 (*Continued*)

crack will propagate at yield stress. As a rough rule of thumb, at temperatures greater than NDT + 30°F it is found that catastrophic crack propagation will not occur for nominal stresses below half the yield strength. At temperatures greater than NDT + 60°F (i.e., approximately FTE) even long cracks arrest if stresses are below the yield strength. For this reason, operating temperatures greater than and equal to the NDT + 60°F are often specified in codes for sections up to 1 in. thick. At greater thicknesses a larger shift in NDT is required. A large number of examples were cited by Pellini and Puzak to confirm these findings.[1]

One approach often used to combine traditional transition–temperature tests and fracture mechanics is through the ratio analysis diagram (RAD) as developed by Pellini.[31] The RAD presents dynamic

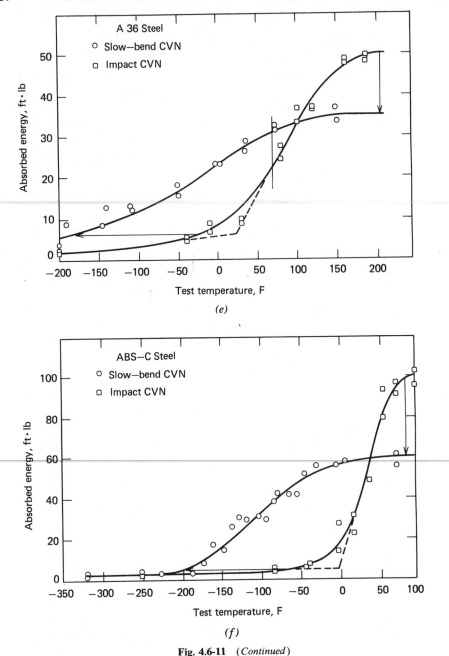

Fig. 4.6-11   (*Continued*)

tear energy and fracture toughness, K_{Ic}, as a function of yield strength and temperature (Fig. 4.6-12). Using the RAD, the designer can decide if the problem is in the regime of linear–elastic fracture mechanics, elastic–plastic fracture, or fully plastic behavior. While the RAD provides very valuable correlations, caution must be exercised in its application since specific materials can fall outside the bounds shown if material processing, chemistry, and heat treatment vary dramatically from the "accepted" practices.

4.6-4 Elements of Fracture Control

Various design philosophies may be employed in fracture-safe design. A few common practices include designing for an allowable "lifetime" based on fatigue crack growth predictions; employing

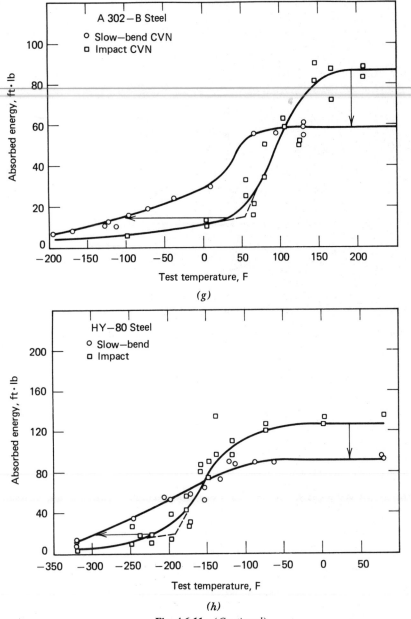

(g)

(h)

Fig. 4.6-11 (*Continued*)

failsafe concepts based on large plastic deformation, multiple-load paths, or redundant members; providing for leak-before-break; and/or developing suitable inspection procedures. Although differing in application, all methods are aimed at ensuring that the crack sizes remain less than the critical size for brittle fracture. Fracture control plans are a useful means of reducing the required design philosophy to practice.

A fracture control plan is a specific set of analyses, material and component tests, and recommendations developed for a particular structure to ensure its fracture integrity. As described by Rolfe and Barsom,[12] a fracture control plan should contain four basic elements:

1. *Identification* of the factors that may affect the fracture integrity of a structural member. This would include service conditions and loading (including static, dynamic, and cyclic events), geometric

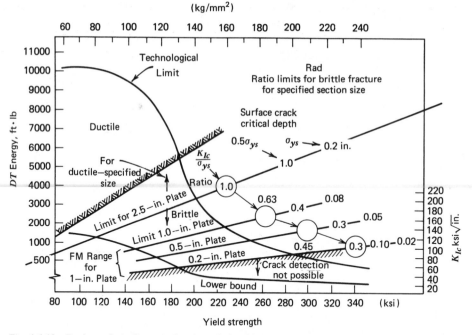

Fig. 4.6-12 Ratio analysis diagram showing relationship between yield strength and fracture parameters.

and weld particulars, material properties, heat treatment, processing and fabrication specifications, and service environment.

2. *Evaluation* of the relative importance of each of the factors identified—in particular, stress, toughness, and flaw size limitations. Assess the contribution of fatigue crack growth and corrosion leading to the critical flaw size, and evaluate the influence of residual stress and embrittlement on the design analysis.

3. *Determination* of tradeoffs and compromises relative to the various design alternatives. By considering the influence of tensile stress, flaw size, and material toughness with respect to cost, design life, load rating, and range of design applicability, an optimum design can be chosen. Include analysis of the factor of safety and fatigue life considerations. For example, for the same quality of fabrication and inspection, and the same fracture toughness, a reduction of the design stress may be regarded as an increase in the factor of safety or an extension of the fatigue life. These design alternatives must be optimized toward the design goal.

4. *Recommendations* of the specific design considerations to ensure the safety and reliability of the structure against failure. Such recommendations can include rated capacity or design stress level, material and/or fabrication specifications, inspection requirements and in-service surveillance, and allowable environments.

If employed in an organized and well-thought-out manner, the fracture control concept provides a very useful tool in ensuring the fracture integrity of an engineering structure. References 12, 32, and 33 include a number of fracture control plans as they have been applied to industry.

4.6-5 Analysis of Fracture Appearance

Analysis of the fracture surface can provide much information about the fracture process. This information can often be used to diagnose the cause of fracture and thus can be helpful in deciding on appropriate corrective action. While all fracture surfaces do not necessarily give complete information on the fracture process, analysis of the fracture surface can describe the nature of fracture (ductile or brittle), the type of loading and loading history, origin or point of crack initiation, and in some cases, the environment to which the component was subjected.

It should be noted at the outset, however, that in order to obtain the most information as well as the correct information about the fracture process, the fracture surface must be carefully handled from the time of the fracture through the time the analysis of the surface appearance is complete. Fracture surfaces should be kept clean and free of dirt, lint, oils, moisture, and so on. A plastic bag with appropriate desiccant provides a clean environment to store specimens. Avoid storing more than one specimen in a container in which rubbing may occur between the surfaces to be examined. If it is necessary to reduce the fracture surface size prior to examination, all cutting oils and loose debris should be cleaned from the surface prior to examination or storage. An alcohol or freon wash (using a squirt bottle) will generally provide adequate cleaning. Do not wipe the surfaces with a cloth or paper product as these will often leave unwanted debris on the fractured surface.

Fracture Mechanisms

The fracture process (or crack propagation) may progress as a result of a number of different fracture mechanisms. The mechanism is generally the result of the loading and load rate, the test temperature, and the environment as well as the characteristics of the material itself. Handbooks are available[34,35] that are dedicated to analysis of fracture appearance and fractography. In this section only a few of the more common fracture mechanisms will be discussed. Further, it should be noted that more than one mechanism may be active during the fracture process. Often the fracture surface depicts a series of events (e.g., fatigue crack initiation, slow stable ductile growth, interdispersed regions of cleavage fracture, ending in final fracture by tearing) that are as much a function of the loading history as the material characteristics.

Cleavage. Macroscopically, cleavage fracture is characterized by flat fracture surfaces exhibiting shiny angular flat facets. Words such as *glassy* and *rock-candy-like* have been used to describe the appearance of these fractured surfaces. Upon closer examination, the cleavage fracture surface typically features a "river pattern" texture (Fig. 4.6-13). The river pattern appearance is a result of the accommodation process in cleavage fracture as the crack passes through grains of different orientation in a polycrystalline material. Because of this process, the river pattern generally describes the direction of crack propagation as shown in Fig. 4.6-13. If the grain orientation is correct (low-angle boundaries) and the temperature and strain rate are appropriate, the cleavage fracture surface may also exhibit a "tongue" appearance on the cleavage facets. This results from a twining mode of deformation necessary to accommodate the different grain orientations (Fig. 4.6-14).

On the microscopic scale the cleavage process involves transcrystalline separation of the material along specific crystallographic planes. Because fracture proceeds along these specific planes, very little plastic deformation is required and therefore the material toughness is generally low. Cleavage fracture occurs in materials in which the crystal structure does not have sufficient slip systems to accommodate the applied deformation. Of the commonly used engineering materials the body-centered-cubic (BCC) and the hexagonal-close-packed (HCP) crystal structures are the most likely to exhibit cleavage fracture due to the limited slip systems in the crystal matrix. Face-centered-cubic (FCC) may also cleave; however, this generally occurs at very low temperatures when the major slip systems are inactive. Tables of the crystal structure of commonly used materials can be found elsewhere in this handbook.

Intergranular Fracture. A second mode of brittle fracture may arise due to intergranular cracking in which the crack follows along the grain boundaries rather than through the grains. This type of fracture results from a weakening of the grain boundaries in a polycrystalline material. This may be due to embrittlement, mechanical processing, or heat treating which causes a brittle phase to segregate at the grain boundary. The fracture surface has the appearance of the grain structure, often with many secondary cracks surrounding the embrittled grains. Visually, the fracture surface gives the effect of "mud cracking" as would be typical of a dried lake bed (Fig. 4.6-15).

Woody Fracture. As a consequence of anisotropic material structure resulting from mechanical processing (e.g., rolling, forging) a mechanical "fibering" can occur.[34] Such mechanical processing may result in alloy segregation, stringers, porosity, or inclusions (e.g., carbides) and second phases that align along specific orientations creating a weak path for crack growth. The fracture surface has the appearance of a "woody structure" resulting from the crack propagating along the degraded microstructure. Figure 4.6-16 shows a typical example of a fracture surface exhibiting a woody appearance.

Dimple Rupture. Dimple rupture is a ductile mode of material separation by which failure is a process of microvoid initiation, growth of the voids, and finally, coalescence. Microvoids are initiated at the interfaces between the matrix material and second-phase particles such as carbides and precipitates. Void growth is a result of a state of triaxial stress, which is characteristic of the region in front of a crack tip or at the center of a bar in tension. Once the voids grow to a critical size, the

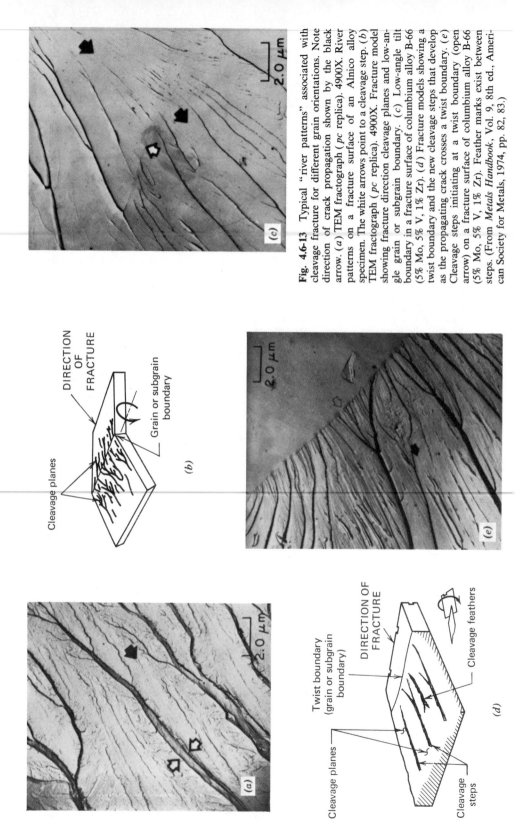

Fig. 4.6-13 Typical "river patterns" associated with cleavage fracture for different grain orientations. Note direction of crack propagation shown by the black arrow. (*a*) TEM fractograph (*pc* replica). 4900X. River patterns on a fracture surface of an Alnico alloy specimen. The white arrows point to a cleavage step. (*b*) TEM fractograph (*pc* replica). 4900X. Fracture model showing fracture direction cleavage planes and low-angle grain or subgrain boundary. (*c*) Low-angle tilt boundary in a fracture surface of columbium alloy B-66 (5% Mo, 5% V, 1% Zr). (*d*) Fracture models showing a twist boundary and the new cleavage steps that develop as the propagating crack crosses a twist boundary. (*e*) Cleavage steps initiating at a twist boundary (open arrow) on a fracture surface of columbium alloy B-66 (5% Mo, 5% V, 1% Zr). Feather marks exist between steps. (From *Metals Handbook*, Vol. 9, 8th ed., American Society for Metals, 1974, pp. 82, 83.)

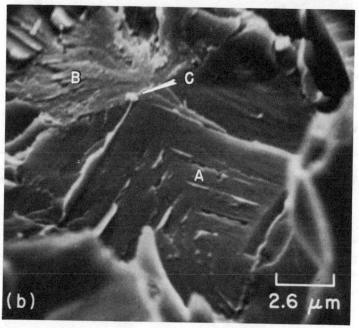

Fig. 4.6-14 Scanning electron micrographs of the "tongue" fracture in cleavage fracture of 1040 steel. Same area shown in both SEM fractographs but at different magnifications as noted. Grain *A* shows two well-defined sets of feathered tongues that are approximately orthogonal. A small carbide particle *C* at the boundary of grains *A* and *B* initiated the local cleavage crack through the surrounding grains. Secondary cracks are at *D* and *E*. Note succession of cleavage steps in grain *B*. (*a*) 1560X; (*b*) 3900X. (From *Metals Handbook*, Vol. 9, 8th ed., American Society for Metals, 1974, p. 65.)

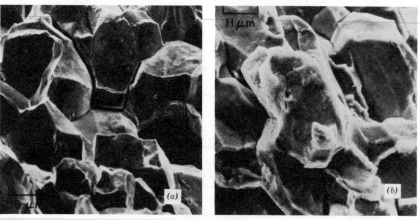

Fig. 4.6-15 Intergranular fracture surfaces of (*a*) sensitized 304 stainless steel component and (*b*) an alloy of nickel–copper. Both surfaces show the grain boundaries with essentially no deformation. (*a*) SEM fractograph, 180X. Intergranual corrosion fracture near a circumferential weld in a thick-wall tube made of type 304 stainless steel. The heat-affected zone of the weld was sensitized by the temperature of welding, and consequently it was susceptible to intergranual attack. Note the smooth grain facets and the secondary cracks at the grain boundaries. (*b*) SEM fractograph, 900X. Intergranular fracture in tubing made of nickel–copper alloy Monel 400, after exposure to high-temperature high-pressure steam. Secondary cracks are also intergranular. There is no grain-boundary deformation. The grainy appearance of grain facets suggests corrosion. (From *Metals Handbook*, Vol. 9, 8th ed., American Society for Metals, 1974, p. 77.)

Fig. 4.6-16 Typical "woody" fracture surface appearance shown at low magnification using light microscope. (*a*) Light fractograph, 10X. Mating surfaces of a fracture in a stud nut that was machined from 11L40 steel (0.15% Pb) heat treated to a hardness of Rockwell C31 to 32. Note the very woody nature of the fracture surfaces, which is a result of adding lead to improve the machinability of the steel. (*b*) Light fractograph, 6X. Surface of an impact fracture in a notched specimen of wrought iron. Here the longitudinal stringers of slag in the material are parallel to the direction of fracture, which gives the surface this typically "woody" appearance. (From *Metals Handbook*, Vol. 9, 8th ed., American Society for Metals, 1974, pp. 300, 396.)

(b)

Fig. 4.6-16 (*Continued*)

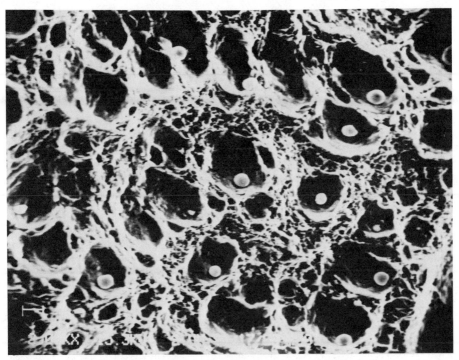

Fig. 4.6-17 Typical examples of fracture by microvoid initiation, growth, and coalescence. Note that some dimples contain carbide particles from which microvoids were initiated.

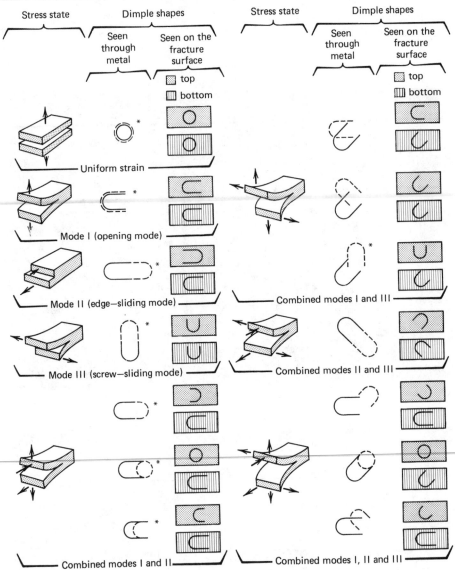

Fig. 4.6-18 Fourteen probable combinations of mating dimple shapes resulting from different stress states, which cause the crack tip to deform by various modes. (From *Metals Handbook*, Vol. 9, 8th ed., American Society for Metals, 1974, p. 105.)

material between the voids can no longer support the applied load and void coalescence and material failure results. Figure 4.6-17 shows a few typical examples of dimple rupture.

Because of this process of void growth to failure, the fracture surface can have the appearance of a fine array of cups—often with the initiating carbide or parcipitate at the center. The shape of these dimples (or cups) and its orientation to its mating half of the fracture surface can reveal information on the type of loading leading to fracture. Figure 4.6-18[34] shows a number of possible dimple shapes as a result of different applied stress states.

REFERENCES

4.6-1 W. S. Pellini and P. P. Puzak, "Fracture Analysis Diagram Procedures for the Fracture-Safe Engineering Design of Steel Structures," U.S. Naval Research Laboratory, Washington, D.C. (March 1963).

4.6-2 American Society for Testing and Materials (ASTM), *Annual Book of ASTM Standards*, E399-83, Section 3, Vol. 03.01. ASTM, Philadelphia, 1983, pp. 518–553.

4.6-3 B. Gross, J. E. Srawley, and W. F. Brown, "Stress Intensity Factors for a Single Edge Notch Tensile Specimen by Boundary Collocation of a Stress Function," *NASA* TN D-2395, (1964).

4.6-4 H. M. Westergaard, "Bearing Pressures and Cracks," *J. Appl. Mech.*, A (June 1939), pp. A-49 to A-53.

4.6-5 M. L. Williams, "On the Stress Distribution at the Base of a Stationary Crack," *J. Appl. Mech.*, **24** (1957), pp. 109–114.

4.6-6 G. Irwin, "Plastic Zone Near a Crack and Fracture Toughness," *Proc. 7th Sagamore Ordnance Materials Conference*, 1960. Syracuse University Research Institute, August 1960, p. IV-63.

4.6-7 *Aerospace Structural Metals Handbook*. Battelle Memorial Laboratory, Columbus, Ohio, (latest revision).

4.6-8 *Damage Tolerant Design Handbook, Part 1*. Battelle Memorial Laboratory, Columbus, Ohio, MCIC-HB-01, (latest revision).

4.6-9 M. C. Hudson and S. K. Seward, "A Compendium of Sources of Fracture Toughness and Fatigue Crack Growth Data for Metallic Alloys," *Int. J. of Fract.*, **14**(4), (1978), pp. R151–R184.

4.6-10 M. C. Hudson and S. K. Seward, "A Compendium of Sources of Fracture Toughness and Fatigue Crack Growth Data for Metallic Alloys, Part 2," *Int. J. of Fract.*, **20**(3), (1982), pp. R57–R117.

4.6-11 J. E. Campbell, "Plane-Strain Fracture Toughness Data for Selected Metals and Alloys," DMIC Report S-28, (June 1969).

4.6-12 S. T. Rolfe and J. M. Barson, *Fracture and Fatigue Control in Structures*. Prentice-Hall, Englewood Cliffs, NJ, 1977.

4.6-13 R. W. Hertzberg, *Deformation and Fracture Mechanics for Engineering*. Wiley, New York, 1976, pp. 369–372.

4.6-14 W. T. Mathews, *Plain Strain Fracture Toughness (K_{Ic}) Data Handbook for Metals*. Army Materials and Mechanics Research Center, 1974, AD-773-673.

4.6-15 D. P. Rooke and D. J. Cartwright, *Compendium of Stress Intensity Factors*. H. M. Stationary Office, London, 1976.

4.6-16 H. Tada, P. Paris, and G. Irwin, *The Stress Analysis of Cracks Handbook*. Del Research Corp., Hellertown, PA, 1973.

4.6-17 A. A. Wells, "Brittle Fracture Strength of Welded Steel Plates," *British Welding Journal* **8**, 259–277 (May 1961).

4.6-18 J. R. Rice "Path Independent Integral and the Approximate Analysis of Strain Concentration by Notches and Cracks," Transactions of the American Society of Mechanical Engineers, *J. Appl. Mech.*, **35** (June 1968), pp. 379–386.

4.6-19 J. W. Hutchinson, "Singular Behavior at the End of a Tensile Crack in a Hardening Material," *J. Mech. Phys. Solids*, **16** (1968), pp. 13–31.

4.6-20 J. R. Rice and G. F. Rosengren, "Plane Strain Deformation Near a Crack in a Power Law Hardening Material," *J. Mech. Phys. Solids*, **16** (1968), pp. 1–12.

4.6-21 F. A. McClintock, "Plasticity Aspects of Fracture," *Fracture*, Vol. 3, H. Liebowitz (ed.). Academic Press, New York, 1971, pp. 47–225.

4.6-22 American Society for Testing and Materials (ASTM), *Annual Book of ASTM Standards*, E813-81, Section 3, Vol. 03.01. ASTM, Philadelphia, 1983, pp. 762–780.

4.6-23 P. Paris, H. Tada, A. Zahoor and H. Ernst, "The Theory of Instability of the Tearing Mode of Elastic Plastic Crack Growth," *ASTM STP 668*. American Society for Testing and Materials, Philadelphia, 1979, pp. 5–36.

4.6-24 V. Kumar, M. D. German, and C. F. Shih, "An Engineering Approach for Elastic–Plastic Fracture Analysis," *EPRI* Report NP-1931 (July 1981).

4.6-25 American Society for Testing and Materials (ASTM), *Annual Book of ASTM Standards*, E23-82, Section 3, Vol. 03.01. ASTM, Philadelphia, 1983, pp. 198–221.

4.6-26 American Society for Testing and Materials (ASTM), *Annual Book of ASTM Standards*, E604-83, Section 3, Vol. 0301. ASTM, Philadelphia, 1983, pp. 640–651.

4.6-27 American Society for Testing and Materials (ASTM), *Annual Book of ASTM Standards*, E436-74, Section 3, Vol. 03.01. ASTM, Philadelphia, 1983, pp. 554–559.

4.6-28 "Rapid Inexpensive Test for Determining Fracture Toughness," Nat'l. Mat'l. Advisory Board, Nat'l. Academy of Sciences, NMAB-328, 1976.

4.6-29 S. T. Rolfe and S. R. Novak, "Slow Bend K_{Ic} Testing of Medium Strength, High Toughness Steels," *ASTM STP-463*. American Society for Testing and Materials, Philadelphia, 1970, p. 124.

4.6-30 W. S. Pellini, "Evolution of Engineering Principles for Fracture-Safe Design of Steel Structures," *NRL Report* 6957 (1969).

4.6-31 W. S. Pellini, "Criteria for Fracture Control Plans," *NRL Report* 7406 (1972).

4.6-32 P. Stanley, ed., *Fracture Mechanics in Engineering Practice*. Applied Science Publishers, London, 1977.

4.6-33 H. McHenry and S. T. Rolfe, "Fracture Control Practices for Metal Structures," *NBSIR* 79-1623, National Bureau of Standards (January 1980).

4.6-34 *Metals Handbook, Fractography and Atlas of Fractographs*, Vol. 9, 8th ed. American Society for Metals, Philadelphia, 1974.

4.6-35 *Failure Analysis of Metallic Materials by Scanning Electron Microscopy*, IITRI Fracture Handbook, Bhattacharyya, Johnson, Agarwal, & Howes, Eds. Metals Research Division, IIT Research Institute, Chicago, 1979.

4.7 CHARACTERISTICS OF CREEP

A. Goldberg

4.7-1 Introduction

When a material is subjected to a stress, it will deform continuously with time. This time-dependent deformation is known as creep. The process of creep occurs at all temperatures, even at stresses as low as a few pounds per square inch, but, at low temperature it is negligible at stresses below the yield strength and may require many lifetimes to be detectable. At high temperatures creep is important and will often determine the design stress that should not be exceeded in the structural component.

Creep regimes are usually classified according to either low $[(0-0.35)T_M]$, intermediate $[(0.35-0.65)T_M]$, or high $(> 0.65T_M)$ temperature regions, where T_M refers to the melting temperature in degrees Kelvin. Under normal strain rates $(10^{-4}-10^{-1}/s)$ these three regimes are governed by different creep laws. Below, we will describe a typical creep curve, the type of data generally obtained, and illustrations of the three principal temperature regimes.

4.7-2 Typical Creep Curve

A typical creep curve is shown in Fig. 4.7-1 depicting the three stages of creep. After an initial strain upon loading, ε_0, stage I begins; this stage shows the creep rate as a diminishing function of creep strain and is a result of strain hardening, which dominates deformation in this stage. Stage II corresponds to the steady-state creep-rate range, where there is a balance between the strain hardening and softening processes. The steady-state creep rate is given the symbol ε_s; it is also often designated as the minimum creep rate. The third stage, indicated as III in the figure, is also known as the tertiary

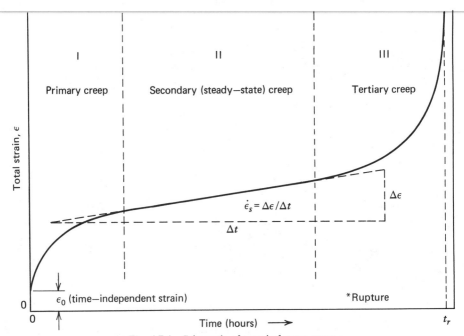

Fig. 4.7-1 Schematic of a typical creep curve.

stage; it shows the creep rate accelerating with strain, which is culminated by ultimate fracture of the material at t_r, the time to rupture.

4.7-3 Factors Influencing the Creep Curve

The nature of the creep curve is strongly influenced by the applied stress, by temperature, and by the microstructural state of a given material. Examples of the influence of stress and of temperature on the creep curve are shown in Figs. 4.7-2 and 4.7-3, respectively, for aluminum of commercial purity. An increase in stress or in temperature accelerates the creep process; the primary creep strain and steady-state creep rate are both increased. The predictive aspects of these variables are now well understood.

Materials in a metastable condition, however, may exhibit periods of accelerated creep, which are usually not accountable for in creep laws. For example, recrystallization of cold-worked material, concurrent precipitation from supersaturated phases in age hardening systems, diffusion in nonuniform (segregated) materials, or stress-induced phase changes can readily accelerate creep due to the increase in atomic mobility associated with these phenomena. In addition, one must consider the possibility of degradation due to high-temperature environmental attack resulting in a reduction in load-bearing capacity of the component. Furthermore, such degradation is likely to affect properties at ambient temperature.

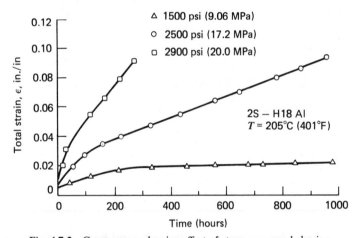

Fig. 4.7-2 Creep curves showing effect of stress on creep behavior.

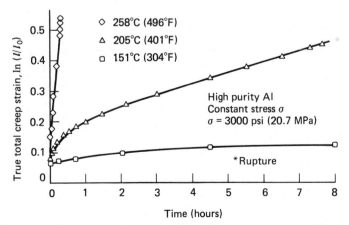

Fig. 4.7-3 Creep curves showing the effect of temperature on creep behavior.

4.7-4 Design Criteria

Design in structural materials for use at high temperatures, where creep is important, is often based on one of the following three considerations:

1. The stress to achieve rupture at a *fixed time*, for example, $t_r = 100,000$ hours (~ 11 years).
2. The stress where the total amount of *permanent strain* is not to exceed some small amount, for example, 0.5%.
3. The stress to produce a given *steady-state creep rate*, for example, $\dot{\varepsilon}_s = 10^{-10} \text{s}^{-1}$, a creep rate which represents 1% creep strain in 3.2 years.

For all three cases the stresses are usually presented in tabular form for a number of temperatures. The best source for such information is to contact the supplier for data on any specific material. Some data are also available in the references given here.

4.7-5 Stress–Rupture Tests

Stress–rupture data are commonly used to determine the design stress for a specific structural material to be used for elevated temperature service. Such data are obtained from creep machines where various constant loads are applied to tensile samples and the corresponding rupture times are recorded. Tests are performed at several temperatures while stresses are chosen to cause rupture at convenient times ranging from a few minutes (~ 0.2 hours) to a maximum of about 3 months (~ 2000 hours). The test results are plotted as stress as a function of time to rupture on logarithmic scale coordinates. An example of such a graph is shown in Fig. 4.7-4 for AISI 310 stainless steel at eight different temperatures ranging from 538°C (1000°F) to 983°C (1800°F). From such information accurate rupture life predictions for this 310 stainless steel, under conditions of very long time application (e.g., $t_r = 200,000$ hours), are possible. This is done through the use of time–temperature parameters. A number of different time–temperature parameters have been proposed. The parameters bring together all stress–rupture data at different temperatures onto a single common curve.

Time–Temperature Parameters

Examples are given in Fig. 4.7-5 illustrating superposition of the 25Cr–20Ni stainless steel rupture data (Fig. 4.7-4) with the use of four different time–temperature parameters. The most widely used parameter is the Larson–Miller parameter (Fig. 4.7-5a). The parameter most closely associated with physical processes occurring during creep is the Barrett–Ardell–Sherby parameter (Fig. 4.7-5c). The usefulness of these parameters is best described by means of an example. A 25Cr–20Ni stainless steel component is to be used at 800°C and is not to fracture in 200,000 hours of service. The question is to determine the maximum stress that can be applied to the component within the given requirements. The answer lies in evaluating the value of the parameter selected and utilizing the master curve

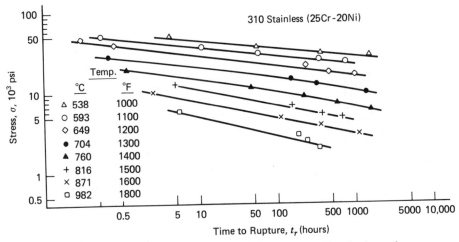

Fig. 4.7-4 Example of stress versus time-to-rupture plots. Note log-log scale.

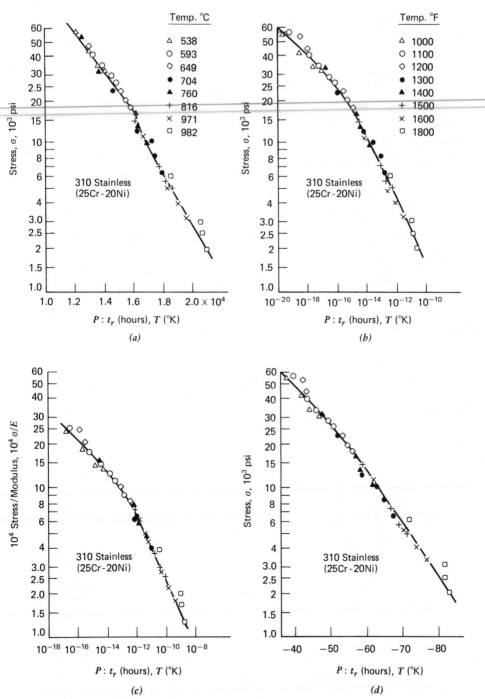

Fig. 4.7-5 Examples of correlations obtained for four different rupture time–temperature parameters. (*a*) Larson–Miller correlation. $P = T(c + \log_{10} t_r)$, $C = 14$. (*b*) Orr–Sherby–Dorn correlation. $P = t_r \exp(-Q/RT)$, $Q = 75.5$ kcal/mole. (*c*) Barrett–Ardell–Sherby–correlation. $P = t_r \exp(-Q/RT)$, $Q = 65.4$ kcal/mole. (*d*) Manson–Haferd correlation. $P = (T - T_a)/(\log t_r/t_{r_a})$, $T_a = 310°$K; $\log t_{r_a} = 14$.

developed for that parameter. Thus, in the Larson–Miller parameter, $P = T(C + \log_{10} t_r)$, the value of $P = (800 + 273)(14 + \log_{10} 200,000) = 1.88 \times 10^4$. From Fig. 4.7-5a the master curve indicates that the maximum stress σ is 4100 psi for this value of the parameter. Nearly the same value of stress would be obtained with the use of the other three parameters shown in Fig. 4.7-5.

Extrapolation of Stress–Rupture Data

Difficulties arise in the application of time–temperature parameters when predictions are made outside of the stress range where rupture data are available. For example, the value of P in the Larson–Miller parameter is uncertain at 1000 psi because the functional relation between σ and P is not known. Some judicious extrapolation is possible, however, with the Barrett–Ardell–Sherby parameter. The slope, in this case, is associated with a specific mechanism of creep, and it is the creep process that ultimately causes rupture. In the Barrett–Ardell–Sherby relation the stress is plotted as σ/E, where E is the dynamic elastic modulus of the material at the temperature of testing. Here, the parameter $P = t_r e^{-Q/RT}$ where Q is the activation energy for creep equal to that for lattice diffusion and R is the gas constant (8.32 J/mole/ °C). In Fig. 4.7-5c the slope of the σ/E versus P curve for the 25Cr–20Ni stainless steel is equal to -0.2 at low stresses. This covers the range of power-law creep where crystallographic slip within the grains is the predominant mechanism of deformation. At stresses below those investigated for this stainless steel, grain boundary sliding becomes the dominant deformation mechanism. The slope for this mechanism is -0.5. At yet lower stresses another mechanism of creep involving only stress-directed migration of atoms (diffusional creep) dominates deformation and the slope becomes -1.0. Figure 4.7-6 illustrates the most likely predicted curve for time-to-rupture behavior as a function of σ/E when extrapolations are made to low stresses for polycrystalline 25Cr–20Ni stainless steel based on these various deformation mechanisms. The predictions are from well-established constitutive equations for grain boundary sliding and for diffusional creep calculated for a typical grain size of 0.3 mm. As can be seen in Fig. 4.7-6, by taking into account the specific deformation mechanism, the predicted rupture life at very low stresses is much less than a linear extrapolation of the stress–rupture data would predict.

4.7-6 Deformation Maps

Creep regimes are also being depicted in deformation maps where the competing regimes are mapped out in terms of stress, temperature, grain size, and/or strain rate. Figures 4.7-7 and 4.7-8 show two such maps in which various mechanisms predominate over either a range of stress and temperature or stress and strain rate. The stress and temperature are normalized by the modulus and melting temperature, respectively. These maps are still in the developmental stage; they are receiving

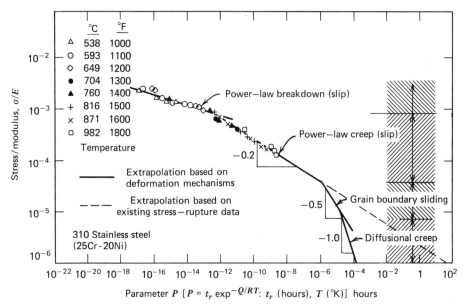

Fig. 4.7-6 Prediction of rupture life at very low stresses based on different criteria.

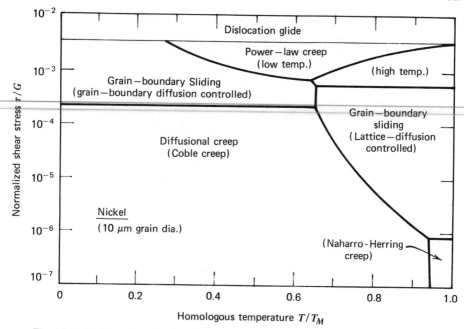

Fig. 4.7-7 Deformation map showing creep mechanism over a temperature and stress field.

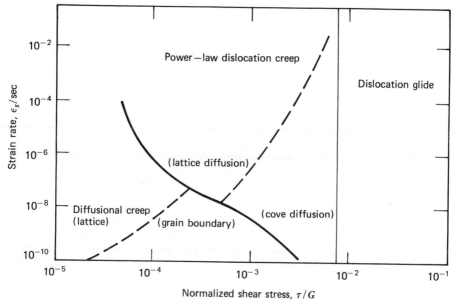

Fig. 4.7-8 Deformation map showing creep mechanism over a stress and strain rate.

continuously more attention by creep researchers. Undoubtedly with time, they will become part of the criteria for creep design.

4.7-7 Testing

Creep tests generally are performed by uniaxial loading through a 10- or 15-to-1 lever-arm system typically using a bank of some 10–20 units. Very high-temperature tests or very low-strength materials can use more economical direct-loading systems of simple design. Beam configurations with appropriate loading fixtures are generally used for brittle materials. In contrast to tension–compression testers, which are usually obtained from commercial sources, creep units are generally of in-house design. Recommended practices on conducting creep, creep–rupture, and stress–rupture tests of metallic materials are described in ASTM Standard E139-79 (Part 10).

BIBLIOGRAPHY

Aerospace Structural Metals Handbook. Metals and Ceramics Information Center, Battelle's Columbus Laboratories, Columbus, OH, 1983.

Ashby, Michael F., and David R. H. Jones. *Engineering Materials—An Introduction to Their Properties and Applications.* Pergamon Press, New York, 1980.

Bernasconi, G., and G. Piatti, (Eds.) "Creep of Engineering Materials and Structures," *Proceedings of the Joint Research Centre of the Commission of the European Communities, Ispra, Italy, 1978.* Applied Science, London, England, 1978.

Conway, J. B. *Creep-Rupture Data for the Refractory Metals at High Temperatures.* Gordon and Breach, New York, 1971.

Hitzl, L. C., and O. D. Sherby, "A Fundamental Study of Creep-Rupture Parameters as Applied to Several Heat Resistant Steels," *ASM* Publ. No. D8-100, (1968).

Sherby, O. D. , and A. K. Miller. "Combining Phenomenology and Physics in Describing the High Temperature Mechanical Behavior of Crystalline Solids," *J. Eng. Mat. Tech.*, **101**, 387, (1979).

Sherby, O. D., and P. M. Burke. "Mechanical Behaviour of Crystalline Solids at Elevated Temperatures," *Progress in Materials Sci.*, **13**, 325 (1968).

Structural Alloys Handbook. Metals and Ceramics Information Center, Battelle's Columbus Laboratories, Columbus, OH, 1983.

4.8 FUNDAMENTALS OF METALLOGRAPHIC PRACTICE

A. Goldberg

4.8-1 Introduction

By varying the composition and performing various thermomechanical treatments, a wide range of properties can be developed in most engineering metals and alloys. For metals of nominal purity, variations in mechanical properties are obtained primarily through control of the degree of cold work (deformed grains) and recrystallization (strain-free grains). For alloys the desired mechanical properties are usually obtained through specific heat treatments. The major types of heat treatments involve control of either transformation or precipitation mechanisms. Changes in properties of alloys can also be affected by cold or hot working, alone or in conjunction with a heat treatment. Deviations in composition from specifications may limit the ability to achieve optimum properties when using treatments specified for a given material. Impurities generally precipitate preferentially at grain or interphase boundaries. Excessive amounts of such impurities may form networks and if brittle or of low strength, they can greatly reduce the ductility of a material; if they have a low melting point, they will also limit the ability of the material to be hot worked. Insoluble impurities such as oxides, silicates, and sulfides are frequently present as stringers.

Changes in properties resulting from thermomechanical treatments are invariably reflected by corresponding changes in microstructures. Changes in crystal and defect structures may also be present, and these changes may also be reflected in microstructural changes. Thus, an understanding of microstructural features is extremely important in the development and control of acceptable mechanical properties. In addition, when establishing the origin (cause and location) of failures, macrostructural as well as microstructural features must be evaluated. The use of metallography is indispensable for evaluation of materials and material failures. A number of metallographic techniques are available for both macroscopic and microscopic examination.

4.8-2 Metallographic Techniques

Sample selection for macrostructural and microstructural examinations are discussed below.

   1. Evaluation of castings. Select samples from different regions to reflect different cooling rates and variations in composition from surface to center.

   2. Evaluation of weldments. Where practical, obtain cross sections of the complete weld including heat-affected zones adjacent to weld.

   3. Evaluation of segregation. Select samples with surfaces representing different orientations. For rolled shapes, select surfaces containing the longitudinal, short-transverse, and long-transverse directions.

4. Evaluation for quality assurance. Compare samples with photographs or metallographic samples representing accepted structures.

5. Evaluation of failures. Fracture surfaces must be protected at all times. If examining a crack, the crack surface can be exposed by completing the break through the section, but without damaging this surface. For materials with a ductile-to-brittle transition temperature, the break should be made below this temperature; this will permit identifying the termination of the in-service fracture. The surface should be dried immediately after it has been broken. Obtain sections remote from the crack and compare these with the corresponding cross section at the fracture or crack but only after complete evaluation of the fracture or crack surface.

4.8-3 Fractography of Failure Surfaces

The failure surface should be carefully examined for evidence of corrosion, fatigue, overload, or any combination of these. The various surface features associated with these different failure mechanisms are evaluated at low magnifications of up to about 50X, preferably using a stereomicroscope. Detection of fine surface features usually requires the use of the scanning electron microscope (SEM). Features such as fatigue striations, fracture paths, and fracture morphology (e.g., ductile dimples, cleavage, intergranular, transgranular, shear bands, corrosion product, etc.) provide information on the probable origin and history of the failure. Excessive corrosion (in stress corrosion, corrosion fatigue, or following crack development) may often obscure those features due to mechanical failure. Here, such information may be available only at the root (termination) of the failure. Removal of the corrosion product by cathodic cleaning may reveal some surface features. However, this should not be done until examination of the as-failed surface is exhausted. Features, that may not be detected using SEM, even after such cleaning, are often revealed by transmission electron microscopy of replicas taken of the cleaned surfaces.

Scanning electron microscopes are usually provided with a system for X-ray elemental analysis. This combination can give important information on the corrosion morphology and the corrosive environment. The energy-dispersive system will allow detection of sodium and heavier elements. The wave-dispersive system will allow detection of elements down to beryllium; in addition, the latter system provides better resolution of near-overlapping X-ray lines, and it can detect smaller elemental quantities than is possible using the energy-dispersive system.

Other techniques, in conjunction with fractography, often must be used to evaluate corrosion failures. Additional elemental information can be obtained using a variety of elemental analytical surface-spectroscopy tools. Those dependent on ion sputtering (e.g., sputter-ion mass spectroscopy) also provide in-depth surface measurements.

Examination of cross sections can provide important additional information on the sequence of events and the specific type (corrosive, mechanical, mechanical/corrosive) and path (intergranular, transgranular, branching) of attack. This can be supplemented with electron microprobe elemental analysis taken along the cross section of both corrosion product and the contiguous base material. The corrosion product can also be scraped off and further analyzed using both chemical analysis and X-ray diffraction.

4.8-4 Sample Preparation

Sectioning, cutting, mounting, grinding, polishing, and etching usually are required to evaluate microstructural features. To avoid modification of the original structure, deformation or heating of a sample should be kept at a minimum during its preparation.

Sectioning/Cutting

Sections to be metallographically examined can be obtained either by saw cutting, abrasive cutting, wire cutting, or electric discharge machining. Excessive distortion, excessive heating, material hardness, section size, minimum surface damage, and metallographic objectives are factors that enter in selecting the proper cutting methods. A variety of abrasive wheels are available, their use depending on materials and sizes to be cut. Coolants must be used to protect both the wheels and the material from overheating. For most purposes the entire wheel consists of the abrasive medium. Special thin "wafer" blades, in which the abrasive material is bonded onto a metal core, are available for producing low-distortion thin sections.

Wire cutting and electric discharge machining (EDM) are two methods that will yield minimum surface damage, eliminate burrs, and yield high-precision cuts. In wire cutting the cut may be effected either by a chemical solvent carried by the wire or by an abrasive wire itself. It can be used for cutting fragile materials, fine honeycomb structures, thin-walled tubing, single crystals and thin sections. It can give fine finishes with high tolerances. EDM is used in machining operations having special requirements involving unusual shaped cuts, removal of localized material in particular configurations, and especially for hard materials. It is also used for fragile or soft materials that might be damaged by the tool contact. In metallographic work it is particularly useful for producing strain-free thin sections. Both wire cutting and EDM are slow, tedious processes.

Mounts

Reasons for using mounts are handling convenience, uniformity for automatic preparation equipment, avoidance of sharp edges, protection of edges between externally cut and original surfaces, examining edges of thin sections, and preparation of flat surfaces. The mounts should resist both mechanical and chemical degradation during preparation and storage of the mounted sample. A variety of mounts are available. *Mechanical mounts* involve clamping the sample in some type of holder that can be easily handled. *Pressure mounts* require both heat and pressure for curing and use either thermosetting or thermoplastic materials. Temperatures and pressures can reach up to about 165°C (329°F) and 29 MPa (4.2 ksi), respectively. *Castable mounts* use cold-mounting plastics and consist of a resin and hardener.

Mechanical mounts are simple to use, avoid heat, and usually require minimum sample preparation; but, they are unsuitable for preparation or storage of large quantities of samples. However, a number of samples can readily be stacked and clamped in a single mount. The simplest clamp is ring shaped, cut from a pipe, and threaded with a clamping screw. Clamping between two flat plates is also commonly used. The clamping material should have similar polishing characteristics as the sample but be relatively inert to the sample etchants.

A large range of uncured plastic powders are available as the starting material for pressure mounts. Standard mold sizes are 1 in., $1\frac{1}{4}$ in., and $1\frac{1}{2}$ in. diameter. The sample should be undersized so that there is sufficient mount material to resist any spalling off. Bakelite and diallyl pthalate are the two most commonly used thermosetting resins (cured and hardened under pressure while hot). A number of resins are available for thermoplastics (fluid at molding temperature and harden as cooled under pressure), and these include polystyrene, polyvinyl chloride (PVC), polyvinyl formal (formvar), and methyl methacrylate (Lucite), which is transparent. Both thermosetting and thermoplastic resins have relatively high coefficients of expansion, and both have low adherence, especially to metals. Thus distortion of thin sections and shrinkage gaps may develop in the cooling cycle.

The castable mounts (cold-mounting plastics) normally consist of two components; the resin and hardener: liquid–liquid, liquid–solid, or melted solid–solid. The resins include polyesters, polystyrenes, acrylics, and epoxides. Molds can be made of a variety of materials: metals, Teflon, silicone rubber, and so on. Depending on the combination of the mount and mold materials, a mold-release agent may be required (e.g., silicone oil). Castable mounts offer a number of advantages over the pressure mounts: a wide choice of materials, additives can easily be added for increased abrasive resistance, allows use of vacuum impregnation to eliminate voids and thereby reduce bleeding of etchants, low curing stresses for delicate specimens, adaptable to hot cells, and a large number of samples can rapidly be serviced. A disadvantage may be the toxicity of some of the castable vapors.

Special techniques may have to be used to avoid distortion or damage to thin sections or to edges. This may be overcome by "double mounting," by adding abrasive-resistant fillers, by protecting with a hard enclosure (e.g., a metal ring), or by plating the surface. Mounts may require electrical conductivity where electrolytic polishing or etching is used. The simplest technique is to insert a self-tapping screw into a hole drilled to contact the bottom of the sample. A common method is to add conducting powder to the resins. A probe can be used on the metal surface, but this usually causes uneven attack.

Grinding

Flat, smooth surfaces are required for microscopic observations. This is accomplished through a sequence of grinding and polishing operations. These operations can be either manual or automatic or a combination of both. Grinding is performed on papers impregnated with either SiC or Al_2O_3 ranging from 40 to 800 grit mesh; a more common range is from 60 to 600 grit. Manual operations involve first the use of belt grinders (principally for rough grinding) and then paper laid on flat surfaces (for fine grinding). To avoid tearing of papers (and later of polishing cloths), edges and corners are leveled during rough grinding. In "grinding" a series of parallel "scratches" are introduced which are successively removed and replaced by finer scratches formed by each successive finer paper. The final scratches are subsequently eliminated during polishing. In grinding care must be taken to use light pressure, maintain the specimen surface flat, and eliminate all scratches introduced by the previous paper. The use of a lubricant such as kerosene facilitates hand grinding. To avoid transferring abrasives, the specimen should be rinsed between papers, and especially before the polishing operations.

Automatic grinding machines can handle a number of samples simultaneously and complete the operations (from coarse to fine) in several minutes. However, the machine torque may cause slight tilting of the mount resulting in removal of an excessive amount of surface from the sample. This does not seem to be a problem for samples with large surface areas (e.g., several inches diameter).

Polishing

The final stages of surface preparation involve removal of fine scratches introduced in the grinding operations (typically to 600 grit paper) and ultimately producing a scratch-free, highly polished, flat surface. A lack of flatness will limit the ability to use high magnifications. Polishing (buffing) is performed in either two or three stages on polishing wheels. A wide selection of polishing cloths and abrasives is available; the appropriate combinations depend on the relative hardness of the individual microconstituents as well as on the hardness of the overall sample. For very soft materials it may be necessary to go through several cycles of etching and final polishing in order to eliminate flowed metal introduced during the grinding and polishing operations. Polishing pressure on the sample can be applied either manually or with fitted weights when using automatic polishing machines.

Improper preparation techniques can result in artifacts. Insufficient polishing may show up as banded markings. The wrong choice of cloth and abrasive can cause extraction of microconstituents, producing voids. Abrasives can be imbedded and may be misinterpreted as impurities. Care must be taken to avoid any carryover of abrasives between wheels. Care must also be taken in mounting the polishing cloth and in maintaining its proper wetness. Distilled water is most commonly used. Care must be taken that the polishing ingredients do not preferentially attack the metal as this can result in undesirable relief as well as possible artifacts.

Cloths include nylon, silk, canvas, felt, velvet, with a variety of trade names. Differences are mainly in the nap, hardness, and wearing properties. Diamond pastes (with different size particles) are probably the most widely used abrasives. Other common abrasives are alundum, alumina, magnesium oxide, cerium oxide, chromium oxide, and silicon carbide. Most abrasives are available as pastes, powders, or suspensions. Diamond paste abrasives provide the quickest polishing rate with minimum relief between hard and soft components and can be used for all polishing stages. However, they may be unsuitable for very soft materials. Diamond pastes should be used with tight-weave, short-nap cloths. For soft materials (e.g., aluminum, lead, solders, etc.) other abrasives such as cerium oxide are preferred.

Electropolishing is an alternative to mechanical polishing for materials that are difficult to polish by conventional mechanical methods. It is particularly useful for many soft metals and alloys that work harden rapidly such as austenitic steels; the flowed or work-hardened surface structures introduced during grinding can be completely removed. Once the polishing parameters are established, the procedure is simple and fast. Disadvantages lie in the uneven rate of attack of different microconstituents resulting in possible pitting, phases showing in relief, and exaggeration of constituent sizes. Undulations also frequently develop, imposing an upper limit on usable magnification. Many of the electrolytes are either poisonous, explosive, and/or flammable. Also, the mounting material may be reactive to the electrolytes. New alloys or modifications of an existing alloy may require considerable time to establish a new set of polishing conditions. Information on electrolytes and polishing parameters for a variety of metals and alloys is readily available (e.g., ASTM E3).

Etching

Microstructural details are revealed by preferential attack of the microconstituent, either by dissolution or staining through the use of one or more etchants. Care must be taken that artifacts or mechanically disturbed surface metal are not mistaken as part of the microstructure. Artifacts may be introduced either during polishing or from the "after-effects" of etching due to bleeding from cavities and superficial staining.

A large number of etchants, together with appropriate procedures, are available and listed in various metallographic treatises and handbooks. Usually, the reagents are inorganic or organic acids or alkalies, often with other additives and dissolved in an appropriate solvent such as water, glycerine, glycol or alcohol, or mixtures of these. The purity of the reagents as well as the solvents is important. Different etchants may be used on the same metal to reveal different features; at times this can be done sequentially; but, usually repolishing is required between the two etchings. Most often the etching is completed within seconds, although it can extend up to many minutes. Even for a given material, depending on its thermomechanical history, the etching time can vary considerably. Care must be taken not to overetch, otherwise polishing must be repeated. In underetching, important features may be underdeveloped. Overetching can exaggerate certain microconstituents and obscure others. Etching can be performed by swabbing with cotton, immersion, or repeated dipping. To stop the etching reaction, the sample must then be immediately washed, preferably with distilled water, rinsed with ethyl alcohol, and dried in a stream of warm air. As an alternative to using alcohol, water can be removed with a blast of pressurized dry air.

Electroetching may be preferred for severely cold-worked metals and for alloys that exhibit surface passivity to conventional etchants, for example, heat- and corrosion-resistant, and some single-phase alloys. Many of the advantages and disadvantages related to electropolishing are applicable to electroetching. Heat etching (selective vaporization) and heat tinting (differential oxidation) can also be used to reveal microstructural features. These two techniques are especially suitable for materials with high porosity, such as castings, where excessive bleeding may occur following normal etching.

4.8-5   Evaluation of Microstructures

Grain size, inclusion content (type, size, and distribution), particle or precipitate content (shape, size, and distribution), constituent phases, segregation, flow lines, deformation bands, tears, corrosion, erosion, weld patterns, and hardened zones are among the many microstructural variables that are subject to metallographic evaluations. Quantitative measurements are usually obtained for grain sizes, dispersions (precipitates or inert particles), depth of corrosion, and volume percent phases. Analysis of inclusions (in steels) is made with reference to published charts (ASTM E45). Most microstructural evaluations, however, are qualitative or semiquantitative.

Macroscopic Evaluation

Features that can be revealed with the naked eye or up to magnifications of about 50X can best be evaluated using a stereomicroscope. Segregation cavities and flow and growth patterns in castings, deformation flow lines in forgings, surface defects, fractures, fatigue markings, duplex grain structures, and high inclusion contents are common features observed at a macroscopic level. Where etching is necessary, macroetching reagents are used; attack is more vigorous, usually requiring longer times compared to microetching, and frequently involving some heating.

Segregation of some steel impurities can be identified by contact printing techniques, for example, sulfur, phosphorus, and oxide printings. The three methods are similar and essentially involve placing a photographic paper, previously soaked in the appropriate solution, in contact with the polished and cleaned metal surface. A moderate pressure is applied for several minutes and the paper is then rinsed and subsequently treated as a photographic print, for example, fixing solution, and so on. In the case of sulfur identification, the aqueous solution is 2% H_2SO_4. The sulfuric acid reacts with iron sulfide or manganese sulfide giving off H_2S, which then attacks the AgBr to produce Ag_2S. The darkly colored areas of Ag_2S will correspond to the sulfur inclusions in the surface of the metal sample. The reactions and procedures are somewhat more complex for revealing either phosphorus or oxide segregation.

Quantitative Metallography

To obtain quantitative values, two-dimensional measurements are made and are either referred to as such or converted to a volumetric equivalent. Reference charts (ASTM E112) are most commonly used as a basis for grain size measurements. These are used in conjunction with grain size grids (eyepiece) at specific magnifications. For example, *with steels*, the *grain size number N* is defined as

$$N = \left(\frac{\log n}{\log 2} \right) + 1 \quad \text{or} \quad n = 2^{N-1}$$

where n is the number of grains per square inch at 100X. The larger the grain size the smaller N is. The grain size can be obtained either directly with eyepiece grids or by comparing the photographed grain size with the chart grain size. With *brasses* the grain size is identified in terms of grain diameter with reference charts at 75X. For *aluminum alloys* the average grain diameter (mm) is usually given. Unless some method has been specified for an alloy system, the average two-dimensional grain size diameter is specified.

Various techniques have been developed for quantitative measurements of particles, phases, surfaces, and so on. For example, the *point-count method* depends upon the number of points that fall on a given phase divided by the total number of test points. The *point-intersection method* depends on intersection of phases or grains by random lines. One can also count the number of points (phases or grains) within an area giving the number per area. Areal (or converted to volume) fractions can now be obtained quickly using a quantimet microscope-screen system with computer output that interprets various phases through their contrast differences.

Light Microscopy

Useful magnifications ranging from 30X to 1500X (with oil immersion) can be obtained with light microscopes. The clearness of the image and the resolution of fine detail depend primarily upon the effective numerical aperture (NA) of the objective lens. The fineness of detail F is given by

$$F = \frac{\lambda}{2NA}$$

where λ is the wavelength. The eyepiece (ocular) magnifies the image for the eye to resolve the details separated by F. Excessive magnification essentially provides a blurred image (empty magnification). The higher the NA the smaller both the field of view and the depth of focus (sharpness). Typically, a sample is examined at several magnifications. Several different types of objectives and eyepieces are available that attempt to minimize optical problems such as a curved image field and spherical aberration. To minimize the reflected light intensity or to enhance structural contrasts, filters are usually used.

The conventional form of viewing is by bright-field illumination. Here the objective lens serves first as a condensing system to the incident light beam, illuminating a small area of the sample and then forming an image of the reflected light. In dark-field illumination the incident light passes along the outside of the objective lens being reflected onto the sample by a concave reflector skirt enclosing the lens. Highly specular areas (ideal mirrorlike) not reflecting light back into the objective will appear dark. The less specular areas will appear bright. Thus the complementary image to that obtained by using bright-field illumination is observed. Oblique illumination may be used to increase contrast between structural components. This is obtained by off-centering the aperture diaphragm from the optical axis. Polarized light is often used to detect microstructural features such as nonmetallics, voids, cracks, and grain size especially in metals that are difficult to etch.

Electron Microscopy

Because of the sophisticated instrumentation, the operation of electron microscopes requires considerable training and expertise. The purpose here is only to point out some of the tools available. The use of electron wavelengths in contrast to visible-light wavelengths allows for several orders of magnitude increase in resolution and therefore in useful magnifications up to several hundred thousand. Reference to the scanning electron microscope (SEM) was made in Section 4.8-3. The transmission electron microscope (TEM) is valuable for obtaining in-depth microstructural details and crystallographic relations between various microstructural components (selected-area electron diffraction patterns); however, it is limited to the examination of thin foils and to replicas of surface features. Frequently, transmission studies of replicas will reveal finer details than are observed on the same surface using SEM. Methods of obtaining foils or replicas for TEM are available in a number of references. The scanning electron transmission microscope (STEM) combines the advantages of both systems; three-dimensional (high depth of field) and X-ray diffraction. In the latest SEMs facilities for electron microprobe analysis are also incorporated in the system, giving quantitative elemental linear traces.

Systems using ion beams in place of electron beams are also available for elemental analysis and for surface scanning of microstructural features. Ion probes, which work through sputtering, can provide in-depth analysis and "mapping."

BIBLIOGRAPHY

ASM Metals Handbook. Vols. 8 and 9, 8th ed., American Society for Metals, Metals Park, OH, 1973.

Kehl, George L. *The Principles of Metallographic Laboratory Practice*. McGraw-Hill, New York, 1949.

McCall, James L., and P. M. French (Eds.) *Metallography as a Quality Control Tool*. Plenum, New York, 1980.

McCall, James L., and P. M. French (Eds.) *Interpretive Techniques for Microstructural Analysis*. Plenum, New York, 1978.

McCall, James L., and P. M. French (Eds.) *Metallographic Specimen Preparation—Optical and Electron Microscopy*. Plenum, New York, 1974.

Murr, Lawrence E. *Electron and Ion Microscopy and Microanalysis—Principles and Applications*. Marcel Dekker, New York, 1982.

4.9   PRINCIPLES OF PHASE DIAGRAMS

A. Goldberg

4.9-1   Introduction

A phase diagram depicts the phases present in a system as a function of the variable parameters of the system. It provides information on the solubility of the components in the various phases as a function of temperature and/or pressure and what phases (or states for a single-component system) are present or in equilibrium with each other at any given temperature and/or pressure.

For a single-component system the variable parameters are typically pressure and temperature (T). The corresponding phase diagram shows the regions over which the different states are present, for example, solid, liquid, and gas, as a function of T and pressure. If a second component were added, the representation would require three dimensions. For metallic systems the pressure usually is held constant (atmospheric), with T and components being the variables. A binary system (i.e., two components, e.g., lead and tin) is thus represented in two dimensions (T versus composition). A ternary system (i.e., three components) would require a three-dimensional representation, namely, an equilateral triangular prism with the vertical edges representing T and a horizontal triangular section representing composition. Two-dimensional representations are made by taking either horizontal sections at a series of constant temperatures or vertical sections at a series of fixed composition ratios between any two of the components while varying the concentration of the third component. Having four components, together with T, gives four independent variables requiring a four-dimensional representation, and so forth for additional components.

Phase diagrams provide valuable information for the understanding of the behavior of materials, for the development and specifications of heat treatment, and for the development of new alloys. For example, *hot shortness* during hot working of steels can be related to the presence of a low melting phase; *cold shortness* in cold forming can be attributed to the presence of insoluble embrittling constituents. Certain combinations of phase changes suggest the potential for the development of desirable microstructures, for example, control of the eutectoid reaction in steels and precipitation from solid solution in many aluminum-base alloys can given rise to a large variety of properties.

The phase diagram assumes equilibrium. The actual structures (both micro and macro) that are obtained in practice usually arise from nonequilibrium cooling, which can lead either to detrimental or beneficial mechanical properties; for example, excessive segregation can lead to embrittlement, while the presence of dispersed phases can develop high strengths. Thus, a knowledge of both equilibrium and nonequilibrium conditions is important in the understanding of allowable composition ranges (specifications) and the production, fabrication, and heat treatment of metals and alloys.

4.9-2   Possible Types of Phase-Change Reactions

The types of phase-change reactions that can occur are independent of the number of components present for systems with two or more components. These reactions will be described with reference to binary (two-component) systems. The principal difference between binary and higher-order systems is that for each additional component the maximum number of phases allowed in a given type of reaction is increased by one. The maximum number of phases that can coexist in a binary system is three. The temperature at which this occurs is called the *invariant temperature*.

There are six possible invariant reactions involving the liquid and solid states: eutectic, peritectic, monotectic, syntectic, eutectoid, and peritectoid. These are illustrated in Fig. 4.9-1. The horizontal line indicates the invariant temperature, the invariant reaction, and the compositions of the three phases in equilibrium (extremities and an intermediate point). The remaining lines represent the temperature–composition limits over which a given phase can exist. The symbols L and S refer to liquid and solid, respectively. The solubility boundary limits of L are called *liquidus lines*, of S in contact with L *solidus lines*, and of S in contact with other solid phases *solvus lines*. With the exception of the syntectic reaction, these reactions appear in various combinations in a large number of alloy systems. The U–Pb system is one of the few systems that contains a syntectic reaction.

The invariant line represents zero degrees of freedom with respect to T and composition. For example, consider the eutectic reaction, $L \rightarrow S_1 + S_2$, for the eutectic composition E (Fig. 4.9-1a). At T_E, L of composition E, S_1 of composition x, and S_2 of composition y are in equilibrium. If the temperature is either raised or lowered, at least one of these phases must disappear. Similarly, a change in the composition of any one of these phases also requires the disappearance of one or more of the other phases, that is, the temperature and the composition of the three phases are fixed (zero degrees of freedom). In a two-phase region, for example, region $S_1 + S_2$ at T_S, S_1 and S_2 have compositions d and f. If the temperature is changed, the compositions of S_1 and S_2 must change accordingly along their respective solvus lines; if either S_1 or S_2 is changed, then the remaining S as well as T must be changed; that is, there is only one degree of freedom. In a one-phase region, such as in either L, S_1, or S_2, both T and composition can be changed simultaneously, that is, a one-phase region in a two-component system has two degrees of freedom (pressure constant). The same arguments can be applied to any of the other invariant reactions. The degrees of freedom corresponding to the number of phases present in a system is given by the *Gibbs phase rule*:

$$N = C - P + 1 \quad \text{(pressure constant)}$$

where N, C, and P refer to the degrees of freedom, components, and phases present. At the invariant reaction $N = 0$ and thus $P = C + 1$; for example, in a three-component system a *ternary eutectic* involves four phases ($L \rightarrow S_1 + S_2 + S_3$). A three-component system may also have any of the binary reactions. The same argument can be extended to four and higher-order systems.

Phase diagrams of real alloy systems consist of various combinations of the above reactions, the simplest case being one in which complete miscibility exists in both liquid and solid regions as shown in Fig. 4.9-2a. Both atoms A and B must have the same crystal structure, for example, Ag and Au, which shows complete miscibility in liquid and solid. Here the only reaction on cooling is $L \rightarrow S$. By contrast Fig. 4.9-2b illustrates a hypothetical system containing a number of invariant reactions (eutectic, peritectic, monotectic, eutectoid, and peritectoid). It also exhibits a composition X (δ phase)

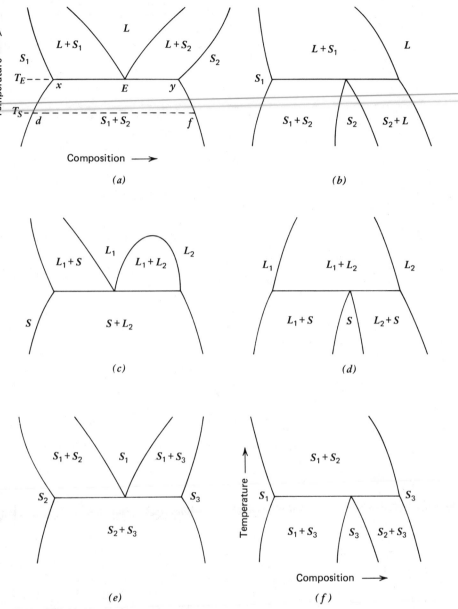

Fig. 4.9-1 The six possible binary invariant reactions. (a) Eutectic: $L \rightarrow S_1 + S_2$. (b) Peritectic: $L + S_1 \rightarrow S_2$. (c) Monotectic: $L_1 \rightarrow L_2 + S$. (d) Syntectic: $L_1 + L_2 \rightarrow S$. (e) Eutectoid: $S_1 \rightarrow S_2 + S_3$. ($f$) Peritectoid: $S_1 + S_2 \rightarrow S_3$.

having a *congruent melting point*. The κ phase melts *incongruently*. Composition X also shows a melting-point maximum. The solid phases are typically designated by the Greek alphabet α, β, γ, and so on.

Superimposed on a phase diagram may be other reactions such as magnetic changes, which are second-order reactions, and order–disorder reactions, which involve the formation of ordered atomic sites (superlattices). Table 4.9-1 contains examples of invariant reactions found in some of the more important binary alloy systems.

The reader should refer to the references listed for specific phase diagrams and for more detailed discussions.

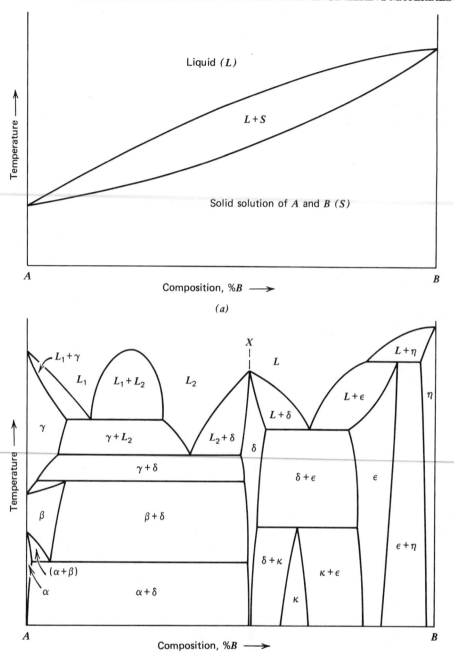

Fig. 4.9-2 Examples of (a) simple and (b) complex hypothetical, but typical, phase diagrams.

4.9-3 Interpretation of the Equilibrium Phase Diagram

Guidance in the use of the binary phase diagram under equilibrium conditions is given with reference to Fig. 4.9-3, which contains an eutectic and an eutectoid reaction. The α and β phases here consist of solid solutions of element B in the low- and high-temperature crystalline structures, respectively, of element A. Similarly, γ is a solid solution of A in the crystalline structure of B. Consider the eutectic composition X; at T_E it will solidify into an intimate mixture of β and γ particles of composition a and c which have successively (β, γ, β, γ, β, etc.) precipitated out from the liquid. The relative

TABLE 4.9-1 INVARIANT REACTIONS PRESENT IN SOME BINARY SYSTEMS

System	Eutectic	Eutectoid	Peritectic	Peritectoid	Monotectic
Fe–C	X	X	X		
Cu–Sn	X	X	X	X	
Cu–Zn		X			
Pb–Sn	X				
Ag–Cu	X				
Ag–Hg	X		X	X	
Ag–Ni	X				X
Al–Cu	X	X	X	X	
Al–Zn	X	X			
Be–Cu	X	X	X		
C–W	X	X	X		
Hg–Pb	X		X		
Pb–Zn	X				X
U–Zr		X		X	
W–Co	X		X	X	

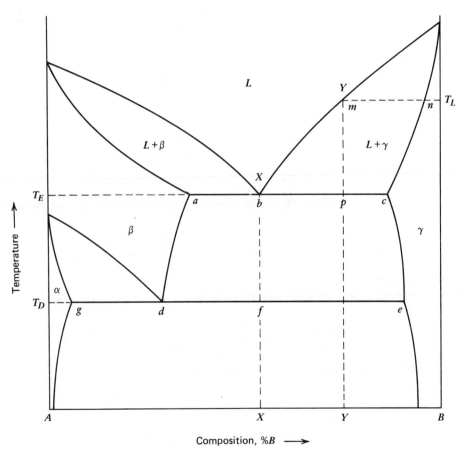

Fig. 4.9-3 Hypothetical phase diagram for elements A and B used in text for illustrating formation of "equilibrium" structures and lever law.

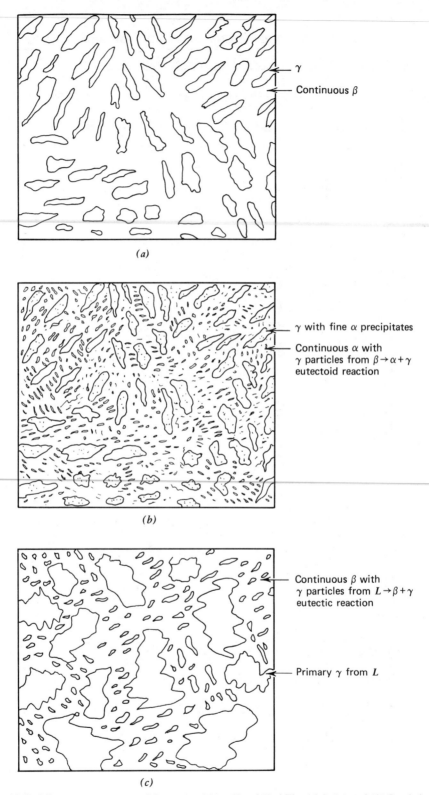

Fig. 4.9-4 Microstructures expected for compositions X and Y of Fig. 4.9-3. (a) and (b) Just below eutectic temperature composition X. (c) Just below eutectic temperature composition Y.

amounts of β and γ formed at T_E are given by the lever law: wt. % $\beta = (100)(c - b)/(c - a)$; wt. % $\gamma = (100)(b - a)/(c - a)$. The ratio of β to γ is then $(c - b)/(b - a)$. On cooling to T_D (eutectoid T), the solubility of element B in β changes along $a - d$ and the solubility of A in γ along $c - e$, with β particles precipitating out of γ and γ particles out of β. On reaching T_D the proportion of β to γ is now $(e - f)/(f - d)$. On passing through T_D all of β transforms to the $\alpha + \gamma$ eutectoid structure. The proportion of eutectoid to proeutectoid γ is also given by $(e - f)/(f - d)$. The relative amount of α to γ in the eutectoid structure is given by $(e - d)/(d - g)$.

The corresponding structures following the eutectic and eutectoid reactions for composition X (Fig. 4.9-3) are shown schematically in Figs. 4.9-4a and b, respectively. Depending on the energetics and kinetics of the system, a range of shapes (from spherical to rodlike to irregular) may be obtained. Because of the large differences in atomic diffusivity between solids and liquids, the ratio of the rate of nucleation to growth is much larger in the solid state; thus, structures resulting from solid-state reactions are usually much finer than those obtained involving a liquid phase. The predominant phase is usually continuous especially in the latter case. The further a reaction is depressed below the invariant T, the finer will be the corresponding structure. Theoretically, under equilibrium conditions invariant reactions must occur only at the invariant temperatures.

Consider next the solidification of composition Y, shown in Fig. 4.9-3. At T_L liquid of composition m is in equilibrium with γ of composition n, which is the first solid to precipitate out from this liquid. As the liquid continues to cool down to T_E, its composition changes from m to b along the liquidus boundary. Simultaneously, the γ phase in equilibrium with the liquid must change along the solidus from n to c. To maintain this equilibrium, interdiffusion of A and B atoms must take place between the γ core and the instantaneous γ-L interface. This requires very slow cooling, which usually causes the formation of excessively coarse, cored dendritic structures. In practice, such highly cored structures frequently exist in as-cast materials. For wrought products these structures can be eliminated by various thermomechanical and thermal treatments. (This problem is discussed in the next section.) On cooling from T_L to T_E, the proportion of primary γ to L changes from a trace to $(p - b)/(c - p)$. The eutectic liquid remaining then transforms to the eutectic structure, forming between the primary grains. This is illustrated schematically in Fig. 4.9-4c.

Since the eutectic is the liquid that solidifies last, it tends to form a continuous network. If the eutectic contains a high proportion of a weak constituent, then the solidified material will tend to be weak. For example, Bi forms a low-melting eutectic with Cu, with the eutectic being virtually all Bi. The Bi then forms thin films around the primary Cu grains rendering the Cu brittle; this requires only a trace of Bi. Another example is Cu in steels. Copper will introduce a low-melting eutectic that can cause hot shortness. Up to a few percent of Cu is commonly present in structural steels to improve their resistance to atmospheric corrosion. Nickel reverses the effect of Cu on the eutectic temperature; usually specifications call for a minimum ratio of 2:1 for Ni:Cu.

The basic scheme shown above for the two- and three-phase reactions, as well as for following any changes in solubility, is applicable to any binary system. It is also applicable with modifications to higher-order systems; however, the reader is advised to refer to texts on the subject when interpreting sections of ternary and higher-order phase diagrams, where the sections may not show the equilibrium solubility boundaries. Here, the relative proportion of phases present generally cannot be determined from the horizontal sections of these phase diagrams.

Table 4.9-2 contains a number of terms commonly used with reference to phase diagrams. A number of these terms are further described below in the various subsections of Section 4.9.

4.9-4 Nonequilibrium Conditions

Basis for Nonequilibrium Behavior

A reaction occurs because it results in a reduction in total energy of the system. At equilibrium, the driving force, which is given by the change in free energy, is zero; thus, the reaction can theoretically proceed in either direction depending on either the extraction or addition of heat. In practice, the new phase must first form stable critical-size nuclei, which then grow into the parent phase. The nucleation and growth are generally both temperature and time dependent. The critical size is obtained when the decrease in the (volume) free energy is greater than the increase in surface and strain energies associated with the formation of the new phase.

As one departs from equilibrium, for example by a temperature depression below an invariant temperature, the larger will be the driving force. This results in an increase in the number of stable nuclei and the formation of a finer microstructure. However, with a drop in temperature the diffusivity also drops, and since most phase and state changes, including precipitation phenomena, involve compositional changes, then at some lower temperature a reversal in the reaction rate may occur. (Theoretical discussions and corresponding nucleation equations illustrating the reversal in kinetics are described in texts concerned with nucleation and growth phenomena.) However, because of the large increase in instability of the high-temperature phase as the temperature is increasingly lowered, it is possible that this phase may be induced to transform into some lower energy state without any

TABLE 4.9-2 TERMS COMMONLY USED WITH REFERENCE TO PHASE DIAGRAMS, BOTH EQUILIBRIUM AND NONEQUILIBRIUM CONDITIONS[a]

Eutectic reaction	$L \rightarrow S_1 + S_2$
Peritectic reaction	$L + S_1 \rightarrow S_2$
Monotectic reaction	$L_1 \rightarrow L_2 + S$
Syntectic reaction	$L_1 + L_2 \rightarrow S$
Eutectoid reaction	$S_1 \rightarrow S_2 + S_3$
Peritectoid reaction	$S_1 + S_2 \rightarrow S_3$
Gibbs phase rule	$F = C - P + 1$ (for constant pressure)
Invariant temperature	Coexistence of phases with zero degrees of freedom, $F = 0$
Liquidus	Boundary between L and $L + S$
Solidus	Boundary between S and $L + S$
Solvus	Boundary between S_n and $S_n + S_{n+1}$
Primary	Formed directly from liquid
Pro	Prior to, for example, proeutectoid S
Hypo	Less than invariant composition
Hyper	More than invariant composition, for example, hypoeutectoid and hypereutectoid
Lever law	Gives relative proportion of phases
Tie line	Isotherm between phases for lever law
Solid solution	Atoms dissolved in crystal structure of the host phase
Substitutional	Soluble atoms occupying lattice sites
Interstitial	Soluble atoms between lattice sites
Superlattice	Substitutional preferred lattice sites
Terminal phases	Phases of elemental crystal structures
Intermediate phase	All phase between terminal phases
Intermetallic	Occurs over narrow composition range, usually at some simple stoichiometric ratio and generally having a complex crystal structure, for example, Fe_3C
Melting point maximum	Relative to adjacent compositions
Melting point minimum	Relative to adjacent compositions
Incongruent intermetallic	Phase decomposes before melting
Congruent intermetallic	Phase stable to melting point
Coring	Reflects compositional gradients across individual grains or dendrites
Dendrite	Jagged, skeletonlike solidified grain reflecting the preferred crystallographic growth directions of the crystal
Microsegregation	Segregation on a microscopic scale, usually across individual grains
Macrosegregation	Large-scale segregation, usually across a section
Supersaturated	In excess of equilibrium saturation
Precipitation	Formation of precipitates from a supersaturated phase
Martensitic	A reaction in which a metastable phase forms by shearing of the parent phase

[a] Reference is generally made to binary systems; but all definitions are applicable also to higher-order systems, but some with modifications, for example, $L \rightarrow S_1 + S_2$ changes to $L \rightarrow S_1 + S_2 + S_3$ and $L + S_1 \rightarrow S_2$ changes to $L + S_1 + S_2 \rightarrow S_3$ for ternary systems. An invariant line in a binary system becomes an invariant plane in a ternary system.

compositional change, such as by a shear process (martensitic). This product phase is metastable and may decompose by reheating up to some intermediate temperature to a more thermodynamically stable structure; such decomposition occurs in the tempering of a quenched steel. The shear process can be aided by an external stress, such that if this high-temperature nonequilibrium phase is retained at ambient temperature it may transform without a compositional change to the metastable phase when the material is strained (e.g., some austenitic steels transform to martensite when deformed, especially at cryogenic temperature). The products that form are either stress-assisted or strain-induced martensites. Similarly, precipitation systems (e.g., Al–Cu alloys), in which the solid-solution phase is supersaturated (i.e., its composition exceeds that of the equilibrium solvus line), can exhibit accelerated creep due to precipitation under load over some range of temperatures. In turn, the creep accelerates the precipitation. In general, concurrent stress and plastic strain will accelerate reactions in the solid state.

Segregation on Solidification

We will describe segregation and coring with reference to a hypothetical binary system AB, in which a eutectic reaction forms a solid solution of α and an intermetallic of β in the A-rich end of the phase diagram shown in Fig. 4.9-5. (This could be representative of many systems, e.g., Cu–Al, Al–Au, Al–Mg, Cr–C, Mg–Pb, etc.) Consider the solidification of composition X. On cooling the liquid, the first solid precipitates at T_1, where the composition line intersects the liquidus line at a. Here, a trace of solid of composition b is in equilibrium with liquid of composition a (essentially X). On cooling from T_1 to T_2 the compositions of the liquid and solid phases in equilibrium with each other follow along ac and bd, respectively.

In order for the two phases to remain uniform (homogeneous), a continuous change in composition of the liquid from a to c and of the solid from b to d is required, that is, A atoms both from the cores of the solid α grains (dendrites) and from the bulk of the liquid phase must diffuse to the L-α (equilibrium) interface while B atoms diffuse in the opposite directions. Uniformity is rapidly approached in the liquid phase. By contrast, in the solid, except under extremely slow cooling (equilibrium), concentration gradients develop and are largely retained down to ambient temperature; both microscopic (cored structures) and macroscopic (across a section) gradients are obtained.

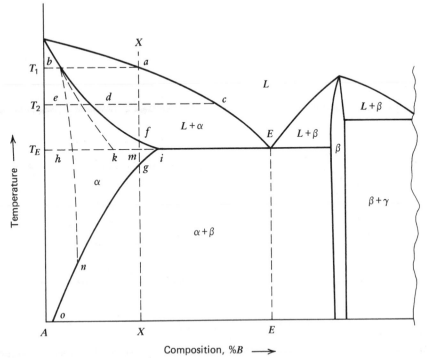

Fig. 4.9-5 Illustration of nonequilibrium solidification and cooling for a hypoeutectic composition in a schematic phase diagram representative of a number of systems.

In Fig. 4.9-5, under equilibrium conditions, composition X would be completely solidified at point f. At g, particles of intermetallic β will start to precipitate out within the α dendrites, yielding a structure illustrated in Fig. 4.9-6a. In an actual (nonequilibrium) case, while the surface of the dendrites may follow the solidus line, the center of a dendrite would follow along a curve such as beh (Fig. 4.9-5) with the average solid composition following along bk. Thus, the relative amount of eutectic liquid that would be present at T_E is equal to $(m - k)/(j - m)$. On final solidification the average solid composition has shifted from k to m corresponding to X, the composition of the original liquid. This eutectic that formed from the remaining eutectic liquid E will consist predominantly of the β intermetallic and will likely result in a continuous brittle network as shown in Fig. 4.9-6b. At T_E the α-phase composition ranges from h to i. On further cooling, as β precipitates out in the α grains, the extremes in α composition change along hn and in, converging at n and then following along no (if cooling is sufficiently slow for the necessary diffusion and precipitation to take place).

In this situation the segregation is manifested largely by a concentration gradient of the β precipitates from the core (low in β) to the surface (high in β) of the α dendrites, together with the continuous B network. (This only considers microsegregation; long-range segregation is discussed below.) Using the same logic, the nature of segregation and the potential for forming a brittle network can be determined for any type of phase diagram; however, interpretation of ternary and higher-order systems may be much more complex than that of binary systems.

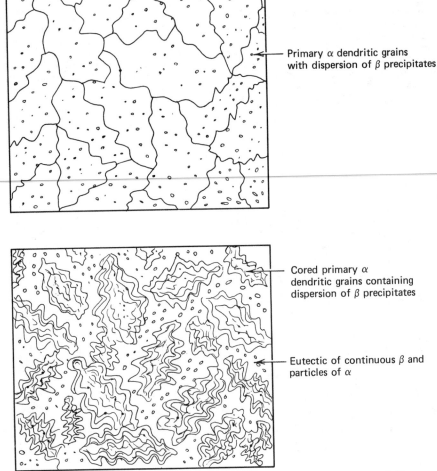

Primary α dendritic grains with dispersion of β precipitates

Cored primary α dendritic grains containing dispersion of β precipitates

Eutectic of continuous β and particles of α

Fig. 4.9-6 Microstructures expected for composition X under (a) equilibrium and (b) nonequilibrium conditions. (a) Very slow cooling approaching equilibrium. (b) Typical normal rate of solidification/cooling.

Segregation within individual dendrites (microsegregation) usually can readily be eliminated by simple homogenization treatments; however, temperature gradients exist in the casting between the surface and the interior of the melt, resulting in a corresponding time differential when solidification is initiated and completed across any given section of the casting (ingot). Thus, surface regions will be relatively rich in one constituent while the center will be relatively rich in some other constituent. This is referred to as macrosegregation. For example, with reference to composition X in Fig. 4.9-5, the surface regions of a casting may consist entirely of α while the last region to solidify would consist primarily of brittle eutectic. It is difficult to eliminate macrosegregation; its elimination can be considered only for ingots (for wrought materials), and by using a combination of long soaking periods and hot working. Segregation of relatively stable insoluble constituents (oxides, nonmetallics, intermetallics) usually cannot all be eliminated. At most, hot working may "break up" some of the continuous brittle phases. Rapid solidification or other methods, such as hot tops, used to minimize temperature gradients reduce the extent of macrosegregation; but, an increase in cooling rate will increase the degree of microsegregation.

Volume Change on Solidification

Most metals and alloys contract by about 4% on solidifying. Table 4.9-3 gives the change in volume during solidification of a number of metals. Shrinkage cavities form due to this decrease in volume. The most common are interdendritic cavities resulting from the impingement of growing dendrites. The liquid, which is entrapped between the interlocking dendritic arms, develops such cavities on solidification; however, the largest part of the volume change usually shows up as a large shrinkage cavity in the center of the ingot, known as a *shrinkage pipe*. This pipe is also associated with a high concentration of impurity constituents, the final liquid to solidify. In practice, these pipes are cut off and can account for scrapping as much as 30% of the ingot. The use of ingot hot tops and, more recently, continuous casting has virtually eliminated this problem. In cast components this shrinkage is largely eliminated by the use of risers. Rapid solidification of ingots will also minimize pipe formation.

In addition to shrinkage cavities, the volume change can also cause the solid to pull away from the mold wall. When this occurs, liquid from the interior of the mold then flows between the growing columnar dendrites (growth perpendicular to the mold walls) and solidifies on the mold wall resulting in an *inverse segregation*, in contrast to *normal segregation*. *Gravity segregation* can also occur where there are significant differences in density between the solid dendrites and liquid, for example, during solidification Si and Cu grains may rise to the surface in Cu–Si and Cu–Pb systems, respectively.

Homogenization and Precipitation

With reference to Figs. 4.9-5 and 4.9-6, the microsegregation developed on normal solidification of composition X can be eliminated by a homogenization treatment, heating the cast structure into the temperature range between g and f. Care must be taken to avoid melting by heating initially below m (T_E). Once β is completely eliminated, then one can proceed gradually somewhat above m being careful not to overshoot the solidus line at f. In practice, impurities would also be present, which will affect these temperature limits. Homogenization will usually develop excessively coarse structures; but the structure can be refined by various thermomechanical treatments, providing the part is not in its final shape.

Having developed a fine, chemically homogeneous grain structure, one can strengthen the structure by precipitation of fine precipitates (age hardening); these can be either coherent (early stages of aging and dependent on the ability to form a coherent interface), semicoherent, or incoherent precipitates. In practice, this is done by heating into the single-phase region within a specified temperature range for some given time followed by a rapid quench. The quench prevents precipitation, retaining the parent phase in a *supersaturated* state. Subsequent heating to some intermediate temperature allows

TABLE 4.9-3   VOLUME CHANGES ON SOLIDIFICATION

Element	Percent Change	Element	Percent Change	Element	Percent Change
Li	-1.6	Au	-4.85	Te	-3.1
Na	-2.46	Mg	-3.95	Si	$+10.7$
K	-2.5	Zn	-4.05	Ge	$+10.7$
Rb	-2.45	Cd	-4.50	Sn	-2.7
Cs	-2.55	Hg	-3.55	Pb	-3.4
Cu	-4.0	Al	-5.65	Sb	$+0.95$
Ag	-3.65	Ga	$+3.1$	Bi	$+3.25$

the precipitation to take place. Aging can occur at ambient temperature for some alloys; these quenched materials must be kept at subzero temperatures to avoid premature aging, for example, some Al–Cu alloys. Age hardening or precipitation hardening is an important strengthening mechanism for a large variety of alloys: Al-base, Ni-base, Cu-base, Ti-base, and so on.

4.9-5 Summary of Basic Microstructures and Properties

Single Phase. Consists of a network of grains revealed through differences in their relative crystallographic orientations and by the resulting grain boundaries.

1. Pure metals. fcc and bcc crystal structures generally have good formability; hcp, bct, and rhombohedral crystal structures generally have poor formability and tend to be brittle.

2. Solid solutions

(a) Terminal (primary) phases—substitutional or interstitial—formability similar to solvent atoms but with increased strength and hardness.

(b) Complete miscibility—substitutional atoms of the same crystal structure with many similarities such as valence, atomic size, periodic position, and so on; mechanical properties are optimum and electrical and thermal conductivity are a minimum at some intermediate, although different, composition.

(c) Intermediate (secondary) phases with moderate to significant solubility range; might have acceptable formability, but generally poorer than terminal phases; strength greater but ductility is less than that of terminal phases.

(d) Intermetallics (intermediate phases)–usually very limited solubility with complex crystal structures; usually brittle, hard with low conductivities; harmful if not properly controlled; usually beneficial if small amounts dispersed in a ductile continuous phase.

Mixtures. Properties are related to the ductility/formability of the continuous phase and the strengthening effect of the discontinuous phase.

1. Intimate mixtures

(a) Monotectic—usually a coarse globular structure.

(b) Eutectic—coarse to fine with a range of shapes.

(c) Eutectoid—various degrees of fineness, usually lamellar.

2. Surroundings

(a) Peritectic—product phase of $L + S_1 \rightarrow S_2$; may form a continuous network depending on the departure from equilibrium; it will have a significant influence on properties. Under equilibrium only the product phase should remain.

(b) Peritectoid—similar comments as for peritectic, except that structure is finer and the probability of retained parent phase S_1 ($S_1 + S_2 \rightarrow S_3$) is greater here. If the parent structure was predominantly S_1, it is possible that the final structure will consist of S_3 grains (containing remnants of S_2) in a continuous network of S_1 (containing S_3 particles).

3. Dispersions

(a) Precipitation from supersaturated solid-solution equilibrium phase; properties dependent on uniformity, size, shape, and aging stage.

(b) Precipitation from decomposition of metastable phase; properties dependent on degree of decomposition and growth of precipitates; usually fine globular type of precipitates.

BIBLIOGRAPHY

ASM Metals Handbook, Vol. 8, 8th ed. American Society for Metals, Metals Park, OH, 1973.

Brick, Robert M., Alan W. Pense, and Robert B. Gordon. *Structure and Properties of Engineering Materials.* McGraw-Hill, New York, 1977.

Bulletin of Alloy Phase Diagrams (started in 1980). American Society for Metals, Metals Park, OH.

Hansen, Max. *Constitution of Binary Alloys.* New York, 1958.

Moffatt, William G. *The Handbook of Binary Phase Diagrams* (4 volumes, started in 1978). General Electric Company, Schenectady, NY.

Reed-Hill, Robert E. *Physical Metallurgy Principles.* Van Nostrand, Princeton, NJ, 1973.

Rhines, Frederick N. *Phase Diagrams in Metallurgy.* Maple Press, York, PA, 1956.

Shunk, Francis A. *Constitution of Binary Alloys.* McGraw-Hill, New York, 1958.

4.10 PRINCIPLES OF HEAT TREATMENT

A. Goldberg

4.10-1 Introduction

Heat treatment of metals and alloys is performed for a number of purposes: homogenization, grain refinement, solutionizing, precipitation hardening, order hardening, recrystallization (softening), transformation (martensitic) hardening, tempering (softening), stress relieving, and various heat treatments to control the shape, size, and dispersion of phases associated with eutectoid transformations. Some of these heat treatments are briefly described in previous sections, where examples of nonequilibrium phenomena are presented. Some specific heat treatments are discussed also in sections dealing with commercial alloys. Here, the principles are presented for each generic type of heat treatment.

4.10-2 Homogenization

All alloys, in which the solid formed directly from the melt (in contrast to electroforming, vapor deposition processes, powder metallurgy, etc.) will develop some degree of chemical segregation on solidification. There are two types of segregation: *microsegregation*, which refers to local variations in composition, usually within individual cored grains or dendrites and *macrosegregation*, which refers to (large-scale) variations in composition across a section. Both types of segregation result from the continuous change in composition of the liquid (liquidus line) as solidification progresses. As heat is extracted from the surface of a casting, temperature gradients develop such that solidification progresses from the surface to the center of the casting (or ingot for wrought materials). Consequently, the components that lower the liquidus temperature continuously diffuse in the liquid toward the center of the casting (final liquid to solidify). Two main problems arise with respect to composition: (i) in effect, a range of alloy compositions is obtained largely with substandard properties and (ii) an excess of impurities, which often combine with other alloying elements to form intermetallic and nonmetallic compounds, usually segregate in the final liquid that solidifies. These compounds may appear as brittle networks in castings or as planes of weakness in the finished wrought materials.

Macrosegregation can be significantly decreased by rapid solidification (e.g., chill castings, continuous-strand castings); but, a rapid solidification rate may actually increase microsegregation by decreasing the time allowable for solid-state diffusion, even at the relatively high temperatures near the solidus line. However, microsegregation can be more readily eliminated than macrosegregation. Various techniques have been devised to minimize temperature gradients across an ingot section—such as a "hot top"—not only to reduce the concentration of impurities but also to minimize the formation of shrinkage cavities (pipes) near the top center of the ingot.

In large castings heat treatment is usually limited to stress relieving or to some modification of the microstructure, for example, through a phase transformation or precipitation; homogenization to eliminate segregation is rarely attempted. In ingots formed into wrought products, homogenization is greatly facilitated by the simultaneous reduction and heating (hot rolling, forging, and/or swaging) operations. Steel ingots are usually first heated to very high temperatures in soaking pits for several days before hot working. The hot deformation greatly accelerates atomic diffusivity; in addition, the large reduction in cross-sectional thickness reduces the migration path necessary to bring about homogenization.

Incomplete homogenization in hot-worked structures can show up in the form of *banded microstructures*; for example, in steels these bands are usually associated with phosphorus segregation, which then influences the local solubility of carbon and, in turn, the steel microstructure. By estimating the carbon distribution from the microstructure, measuring the distance between these bands, and using diffusion equations, an estimate can be made of the temperature–time combination required to virtually eliminate such segregation. Variations in grain size, precipitate density, and phase distribution may also indicate incomplete homogenization. Usually such variations are revealed through light microscopy. At times, segregation can only be detected by the use of other analytical techniques such as electron microprobe or X-ray diffraction (line width).

The temperature–time combinations required to obtain a chemically homogeneous microstructure is generally obtained by trial and error; frequently, the actual heat treatment may require soaking for days at just below the solidus temperature. In the initial stages of homogenization, care must be taken to avoid incipient melting of any (low-melting) constituents present in a cored structure. Liquid will form in the interdendritic regions, which on solidifying could be retained as weak or brittle grain boundary films in the final structure.

4.10-3 Grain Refinement

The solidification process and high-temperature heat treatments, such as homogenization, usually result in coarse structures. In wrought materials the initial coarse ingot structure is refined during the various stages of hot working. The degree of fineness depends largely on the finishing temperature. Subsequent cold work and recrystallization may further refine this grain structure. The dendritic configuration of the as-cast structure is usually eliminated during hot working; however, if some segregation remains, "ghosts" of the original structure may still be seen although the original dendrites now consist of many new grains.

Grains refinement can also be accomplished without deformation for alloys in which different solid phases are present above and below some elevated temperature. Heating above this temperature causes nucleation and growth of the high-temperature phase(s); on cooling the low-temperature phase(s) nucleates and grows usually forming a different grain size than was originally present for this phase. By appropriate choice of heating rate, temperature, time, and cooling rate parameters, which may involve more than one cycle, a large range of grain sizes may be obtained. Methods for measuring and designating grain sizes are given in Section 4.8. In some alloys (e.g., commercial purity aluminum, certain steels) a temperature region may exist where a rapid growth in grain size occurs, resulting in a duplex (small and large) grain structure. Such structures are undesirable in the final product.

4.10-4   Solution Heat Treatment and Precipitation Hardening

Solution heat treatment refers to heating an alloy into a temperature range corresponding to a single-phase region prior to precipitation hardening. Temperature limits are specified for each particular alloy. Overshooting the temperature may cause excessive grain growth and even melting. Too low a temperature may result in incomplete solution of existing precipitates, providing sites for premature nucleation during subsequent cooling.

Rapid quenching (typically in water) from the solutionizing temperature results in a supersaturated solid solution at room temperature. On heating to some intermediate temperature most alloys "age harden" by going through various stages from the supersaturated state to the formation of discrete precipitates. In the initial aging states, nuclei may form that are coherent or semicoherent with the parent phase, resulting in lattice strains accompanied by large increases in hardness; however, the degree of coherency is quite sensitive to small variations in time–temperature–composition parameters. With time, as these nuclei grow, coherency is lost and the nuclei form into discrete precipitates. A maximum in hardness occurs at an aging time that corresponds to this transition stage where a combination of coherent, and/or semicoherent, and discrete precipitates are present. Further loss in coherency and growth of precipitates results in a decrease in hardness.

If precipitation occurs during the quench, it is likely to form preferentially along the grain boundaries, and this should be avoided. Generally, the lower the temperature where precipitation occurs, the finer and more random will be the precipitate distribution. Aging temperatures may range from room temperature (for some aluminum alloys) to relatively high temperatures for cobalt- and nickel-base alloys. The hardness–temperature–time relationship for aging is shown schematically in Fig. 4.10-1. The time to reach peak hardness, as well as the magnitude of the peak hardness, decreases

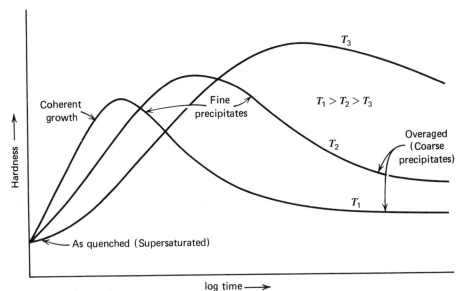

Fig. 4.10-1   Schematic illustrating effect of temperature (T) on aging.

TABLE 4.10-1 EXAMPLES OF PRECIPITATION HARDENING SYSTEMS

Alloy or System	Aging Temperature Range
Al–Cu 20xx alloys	70–350°F (21–177°C)
7075 (Al–Zn–Mg base)	240–260°F (116–127°C)
Cu–Be alloys	600–950°F (316–510°C)
Maraging steels (Fe–Ni base)	850–950°F (482–621°C)
17-4 PH (stainless steel)	900–1150°F (482–621°C)
Hastelloy R-235 (Ni-base)	1450–1800°F (788–982°C)
Ti-base α, α-β, and β alloys	1400–1800°F (760–982°C)

with an increase in aging temperature. If aging occurs at ambient temperature, then the supersaturated quenched material must be kept at subzero temperature until aging is desired. For example, aluminum–copper alloy rivets are kept in dry ice (CO_2) prior to being used; they will then develop near maximum hardness in about 5 to 7 days after reaching ambient temperature. Examples of precipitation (age) hardening alloys are given in Table 4.10-1. Aging temperature and time are determined by the specific composition, size, shape, and strength/ductility requirements. In some cases a sequence of several aging temperatures is used.

Solution annealing may be required to dissolve precipitates for purposes other than for subsequent age hardening, for example, dissolution of chromium carbides in stainless steels to regain their corrosion resistance qualities or dissolution of various phases or constituents for the development of specific properties resulting from transformation-related heat treatments. Reference to such treatments is given in other sections.

4.10-5 Order Hardening

A large number of systems contain superlattices whereby above a certain temperature range a solid solution exists in which the atoms occupy random sites. Below this temperature range, where the contribution of entropy is overcome by the decrease in lattice bonding energy, the atoms of one species preferentially segregate to one set of atomic sites, with the remaining sites occupied by the atoms of the other species. This results in an ordered solid solution lattice or superlattice. Examples of systems that exhibit superlattices are Cu–Zn, Au–Cu, Pt–Cu, Fe–Ni, Mn–Ni, Cu–Be, Fe–Co, Au–Ni, and many more. For example, in the Au–Cu system, complete ordering can occur at compositions corresponding to $AuCu_3$ and AuCu. In the $AuCu_3$ superlattice, the Au and Cu atoms occupy the corner and face positions, respectively. In the AuCu superlattice the two atomic species occupy alternate (002) lattice planes.

The disordered state can be retained in most systems by quenching from the disordered temperature region. Then, on reheating to some intermediate temperature, a change in mechanical properties is realized as ordering ensues, analogous to the behavior exhibited by precipitation hardening systems. Hardness, tensile strength, and elastic limit generally increase with ordering; however, ordering obtained by slowly cooling from the disordered region will not show as large an increase in mechanical properties as those obtained by reheating from the quenched state.

Order hardening is generally of minor practical importance with respect to mechanical properties; however the understanding of order–disorder behavior has received considerable attention with respect to mechanical properties as well as other properties; for example, electrical resistivity may decrease by an order of magnitude on ordering, and some alloys (Heusler alloys) are ordered in the ferromagnetic state. Numerous articles have been written on interpretations of specific-heat measurements and on various aspects of superlattice theories.

4.10-6 Recrystallization and Annealing of Cold-Worked Structures

Softening of cold-worked structures may be performed to allow a material to be further deformed or shaped without cracking, to develop a specified annealed state (temper) in a product, or to obtain some specific grain structure. Combinations of cold work and annealing may also be used to develop certain preferred crystallographic orientations relative to the principal deformation direction.

In cold working, changes in grain shape must reflect changes in external shape. Thus, in a rolling operation, the grains elongate in the rolling direction and compress in the short transverse direction. Heating to slightly elevated temperatures causes the elimination of point defects with a corresponding recovery in some physical properties, for example, electrical resistivity approaches that of the annealed state; however, no obvious microstructural change occurs in this *recovery* stage. At some higher temperature *recrystallization* occurs, whereby new strain-free equiaxed grains nucleate and grow,

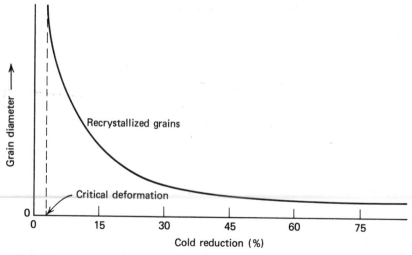

Fig. 4.10-2 Effect of cold work on recrystallized grain size shown schematically. Temperature: constant. Time: to complete recrystallization.

consuming the deformed elongated grains; the mechanical properties approach the original values of the annealed material.

The minimum recrystallization temperature depends on degree of cold work and time. The recrystallized grain size depends also on temperature. Small amounts of cold work, for example, 2 or 3% reduction in thickness may result in excessively large grain size due to very few nucleation sites. The minimum cold-worked strain required to cause recrystallization is referred to as the *critical strain*. The recrystallized grain size decreases rapidly with increased prior deformations of up to about 30% reduction. Figure 4.10-2 illustrates schematically the effect of degree of cold work on the recrystallized grain size. Higher temperatures require less time for recrystallization but result in larger grain sizes. Typically, a temperature is chosen where recrystallization is completed in about 10 minutes at temperature. Additional grain growth will occur following complete recrystallization. At low recrystallization temperatures this growth is usually small compared to the growth that occurs during the recrystallization process itself, but it becomes more significant at the higher temperatures. The extent of such aftergrowth varies considerably for different alloys and, in part, is affected by the trapping of grain boundaries by impurity atoms and precipitates.

The effect of temperature on the progress of recrystallization, as monitored by changes in hardness, is depicted schematically in Fig. 4.10-3. We note a slight drop in hardness during the recovery and grain-growth stages. In practice, where complete recrystallization is sought, 30–60 minutes at the recrystallization temperature is used.

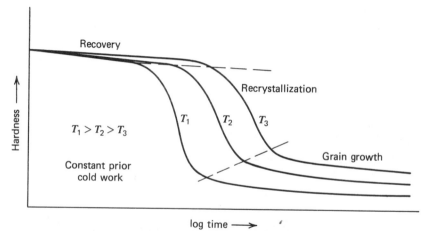

Fig. 4.10-3 Effect of temperature (T) on recrystallization kinetics shown schematically.

In *hot working*, deformation and recrystallization occur simultaneously. The resultant grain size depends on the finishing temperature. In *warm working*, "polygonization" often occurs, whereby fine low-angle-boundary strain-free grains form within the deforming grains. The high-angle boundaries between the original grains are retained, although the original grain shape changes. This polygonized grain structure usually develops ductility approaching that of the annealed condition but with greater strength. Polygonization can also occur by long-time annealing periods (months) below the recrystallization temperature. Concurrent straining will accelerate this process.

4.10-7 Transformation Hardening

There are a number of commercial alloys (e.g., Fe-, Ti-, Cu-, U-, Pu-base alloys) where a wide range of properties can be obtained by selecting the appropriate temperature–time–composition parameters for a given solid-state reaction. The eutectoid transformation is the most common of such reactions. On slowly cooling from some elevated temperature, a high-temperature phase, S_1, normally decomposes into an intimate mixture of the two low-temperature phases, S_2 and S_3, whose compositions differ from that of S_1. Sufficient time is required for atomic diffusion to accommodate these compositional changes before nucleation and growth of the product phases can occur. On continuous cooling, the nucleation and growth occurs over a range of temperatures, with the product phases becoming continuously finer as the transformation takes place at lower and lower temperatures. A corresponding increase in hardness is usually obtained. Furthermore, at some temperature the prevailing nucleation and growth mechanisms may change, with a corresponding change in the product. A rapid cooling rate may prevent transformation within the higher temperature range so that only the lower temperature product forms. Finally, if the cooling rate is sufficiently rapid to bypass both of these reactions, a temperature may be reached where the thermodynamic driving force for the transformation is so large that it can occur without a compositional change. This diffusionless transformation from the parent phase to the product phase may occur by a cooperative shearing process between groups of atoms, each of which move less than an interatomic distance. The resulting product is usually either plate-, lathe-, or needle-like in shape; this is a martensitic reaction and the product is called martensite. The product phase is supersaturated, with its crystal lattice usually distorted from its equilibrium state and generally having a density different than that of the parent phase. The internal strains resulting from the combination of a shape change and a volume change between the product and parent phases and the lattice distortion of the product phase can result in extensive hardening, although frequently with a large corresponding loss in ductility. The martensitic reaction is usually only one of several alternative eutectoid reactions that can occur.

The kinetics and formation of the transformation products associated with a eutectoid reaction are usually depicted in the form of a temperature–time–transformation (TTT) diagram. Figure 4.10-4 shows such a diagram for a 0.8 carbon steel (AISI 1080). This diagram is obtained by quenching a number of samples from the high-temperature parent phase to various temperatures and monitoring the changes that occur at each temperature, for example, along *xy*. The diagram is therefore an isothermal TTT diagram. The start and finish of the reactions occurring are indicated by the subscripts *s* and *f*, respectively, and they are plotted as a function of temperature and time. Similar diagrams are available for other alloy systems, and many are more complex, but they are interpreted in the same manner as described below.

In practice we are primarily interested in the continuous-cooling diagram. It is similar to the one shown in Fig. 4.10-4 except that the *C* curves are shifted to longer times and displaced to somewhat lower temperatures relative to the isothermal counterpart. We will refer primarily to the isothermal TTT diagram. Isothermal TTT diagrams are available for most steels from the steel companies. Methods have been developed for conversion of the isothermal to the continuous cooling TTT diagram.

In Fig. 4.10-4, the symbols S_1, S_2, and S_3 are replaced by γ (austenite), α (ferrite), and Fe_3C (cementite), respectively, consistent with terms used in steel terminology. T_E refers to the eutectoid temperature; T_t refers to a transition temperature region of the transformation. Above T_t, the eutectoid product is called pearlite (*P*) and consists of alternate lamellae of α and Fe_3C; it is nucleated by the Fe_3C phase. Below T_t the eutectoid product is called bainite (*B*) and consists of fine Fe_3C particles outlining fingerlike formations of α; it is nucleated by the α phase which then precipitates out the Fe_3C. *M* refers to the martensite phase; as shown here it forms only by a drop in the temperature. P_s, B_s, M_s and P_f, B_f, M_f refer to the start and finish boundaries of the respective phases or structures.

In order to obtain 100% martensite, the cooling rate must be such as to prevent the formation of any high-temperature phases, for example, along *wxk*; the part must then be cooled to below M_f. Cooling along *wxy* gives 100% bainite. Cooling along *wz* results in the formation of fine pearlite, bainite, and possibly some martensite. Cooling along *wl* produces a relatively coarse pearlitic structure. A number of quenching media are used: oil, water, saline solutions, and various liquid polymers, the latter increasing in popularity due to their excellent quenching characteristics.

Extremely high-tensile surface stresses can be developed in a martensitic product formed by a fast quench. This is due to the 4–6% transformation volume change (increase) and the large temperature

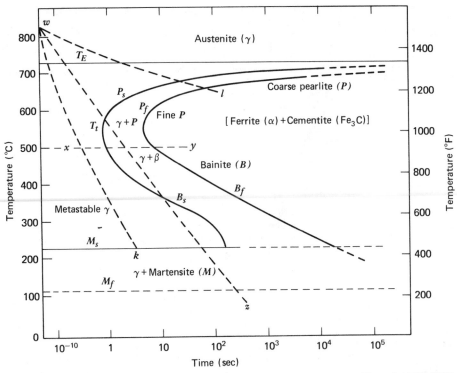

Fig. 4.10-4 Temperature–time–transformation (TTT) diagram for a eutectoid steel (AISI 1080). (Based on United States Steel I-T diagrams.)

gradient developed in a part on rapid quenching. When the surface transforms, the interior is still relatively soft γ, and it plastically accommodates the surface transformations; however, when the interior subsequently transforms and expands, high-tensile stresses are developed in the surface martensite; because of its low ductility, quench cracks may develop at the surface. A number of methods are available to overcome this problem.

The transformation curves may be shifted to longer times by the use of alloying elements, for example, Ni, Mo, Cr, V, Nb, and so on and thus allowing much slower cooling rates in order to bypass the high-temperature products. Quenching into a bath (liquid metal or molten salt) at a temperature just above M_s and holding until the part is uniform before finally cooling through the M region will also minimize the propensity toward surface cracks. This is known as *martempering*. A modification of this technique allows for some bainite to form following the temperature equalization; it is known as *austempering*.

Nonuniform quenching may also result in distortion; this is particularly a problem for nonsymmetrical parts. The susceptibility toward surface cracks and distortions increases with higher quench rates, higher quench temperature, and higher carbon contents. The minimum quench rate, or *critical cooling rate*, that is necessary to prevent the formation of the high-temperature transformation products gives a measure of the *hardenability of the steel*.

4.10-8 Tempering

The terms *temper* and *tempering* should not be confused with each other. *Temper* refers to a degree of cold work following recrystallization; it is used for a number of nonferrous alloys. *Tempering* (at times called *drawing*) consists of reheating a quenched steel sufficiently to decompose the brittle martensite into softer, less brittle products producing a quenched-and-tempered steel. The higher the tempering temperature (TT) up to the eutectoid temperature, the tougher the product. Where resistance to shock is important, TT in excess of 1000°F (538°C) should be used. The process involves the precipitation and growth of carbides as the excess carbon diffuses out of the metastable martensite. It *does not* involve softening through *recrystallization*, which in steels is referred to as *process annealing*. Tempering has also been used for heating air-cooled steels (not initially martensitic) up to, but below, the eutectoid temperature. Changes in a number of mechanical properties with TT is shown schematically in Fig. 4.10-5.

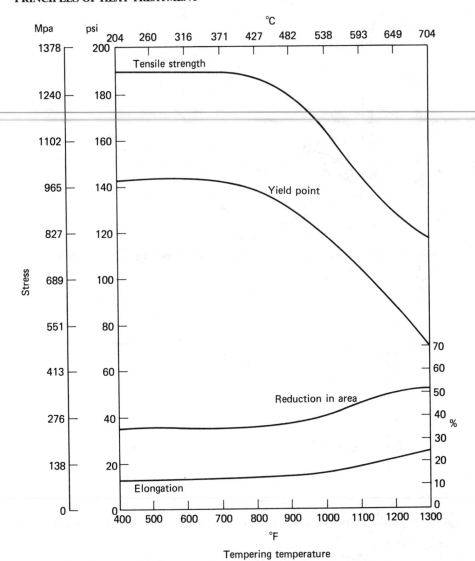

Fig. 4.10-5 Changes in room temperature tensile properties following tempering a eutectoid (AISI 1080) steel quenched from 1500°F (816°C) in oil. (Based on Bethlehem Steel, *Modern Steels and Their Properties*.)

The variation in toughness with TT is usually obtained by Charpy impact-energy measurements. Fig. 4.10-6 shows schematically the change in impact energy as a function of TT. Except for one and in some cases two TT regions, the toughness increases with an increase in TT. For virtually all steels, a minimum in toughness is obtained at about 500 to 600°F (260 to 316°C). This is referred to as the *"blue brittle" range* corresponding to the blue oxide film developed on the steel surface. The magnitude and position of the trough depends on composition and tempering time. It is believed to be due to precipitation of embrittling constituents such as nitrides; it may also be due to a metastable distorted carbide phase that starts to form somewhat below the blue brittle range. The third possibility is the formation, on subsequent cooling, of additional martensite from retained austenite. A fall-off in toughness, indicated by the dashed line, may be obtained for some steels slowly cooled from TT, in contrast to the solid line, which is representative of the same steels quenched from TT. A steel exhibiting this second behavior is said to be subject to *temper embrittlement*. The embrittlement is obtained on slowly cooling through the range 850–1100°F (454–593°C); it is associated with the formation of extremely fine precipitates. Steels that exhibit this phenomenon when slowly cooled through this range show a shift in the ductile-to-brittle transition temperature to higher temperatures

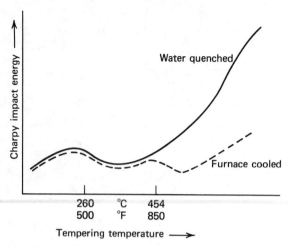

Fig. 4.10-6 Effect of cooling rate from tempering temperature on room temperature toughness typical of a steel susceptible to temper embrittlement.

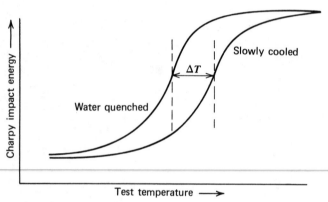

Fig. 4.10-7 Shift in ductile-to-brittle transition temperature, ΔT, typical of a steel susceptible to temper embrittlement.

and an apparent drop in the upper shelf values. For steels that have a low transition temperature, the embrittlement effect may not be revealed by ambient temperature Charpy tests as the "shelf" values may have converged (see Fig. 4.10-7).

The embrittling effect is very sensitive to composition and, to some extent, on prior manufacturing history. The Cr, Cr–Ni, and high-Mn low-alloy steels appear to be most susceptible. Molybdenum decreases the tendency. It appears to occur less frequently in plain-carbon and Ni steels. A higher degree of deoxidation during steel making and lower finishing forging temperatures also decrease the effect.

As a precaution, steels suspect of being susceptible to temper embrittlement should not be tempered in the range 850–1100°F (454–593°C) and should be quenched when tempered above this range. Caution must be exercised in avoiding quench cracks where sharp changes in section occur (e.g., threads, holes, ribs, etc), especially in water quenching of previously oil-hardened steels from tempering temperature below 850°F (454°C). Other phenomena (e.g., secondary hardening) occur on tempering of steels and these are discussed in the reaction dealing specifically with steels.

4.10-9 Stress Relieving

Residual stresses are introduced by cold working, hot working, shaping, forming, machining, grinding, welding, hardening heat treatments, contractions from mold constraints on solidification, and in general, any operation that results in uneven temperature distributions. Quenching operations may also introduce quench cracks where high residual surface tensile stresses develop. Residual stresses may cause cracking during subsequent hardening heat treatments. Cracking. especially edge cracks, may also occur on subsequent cold-forming operations. Residual stresses may develop warpage in machined parts as the machining upsets the balance between the residual stresses. Residual stresses can also develop during (rough) machining operations as thermal gradients result from the localized heating in the part. Most tools should be stress relieved after rough machining.

Large castings are especially susceptible to developing high residual stresses. Materials that are hardened by a transformation involving a volume increase on cooling, will contain surface tensile stresses, for example, steels. If no transformation is involved, and the material is rapidly cooled, compressive stresses are developed on the surface as the internal (softer) material completes its (cooling) contraction. Subsequent machining may expose tensile regions. Residual tensile stresses may significantly increase the susceptibility of a material to stress corrosion cracking (e.g., Cu-base alloys, especially Cu containing over 20% Zn). Stresses may also have to be reduced to permit further mechanical fabrication. Alloys having a low modulus can develop high distortions from relatively low residual stresses.

For materials that are hardened by solutionizing, quenching, and finally aging, a stress-relief anneal above the aging temperature will affect the as-aged properties, especially if aging is normally performed at a relatively low temperature. Here a compromise between a reduction in residual quenching stresses and a loss of age-hardened properties must be considered (e.g., for Al-base alloys). For non-age-hardening materials that have developed stresses in some cold-forming operation, the stress-relief anneal temperature must be kept below the recrystallization temperature if it is desired to retain the mechanical properties introduced by cold working. If the soft condition is specified, then care must still be taken not to exceed a temperature that can result in excessive grain growth or oxidation. The part should be slowly cooled from the annealing temperature to avoid distortion. Heating into a temperature range that could sensitize the material to corrosion must be avoided (e.g., stainless steels).

Most heat-resistant alloys are placed in service without stress relieving, especially castings; however, if the shape is relatively complex, it may crack on initial heating; distortions may be unacceptable because of stringent dimensional tolerances. Such components may require a prior stress-relief anneal.

Parameters for stress-relief annealing are dependent on the specific alloy, its shape, service requirements, and past history. Several examples follow. Tool steels are heated, after rough machining and before hardening, to within a temperature range from 1150 to 1250°F (621–677°C). Unalloyed nodular cast irons are stress relieved at 950–1050°F (510–566°C); austenitic nodular cast irons at 1150–1250°F (621–677°C); both are furnace cooled to 550°F (288°C) and then air cooled. Martensitic stainless steels are stress relieved at 300–700°F (149–371°C) to remove transformation-hardening stresses. Austenitic stainless steels have to be stress relieved at temperatures approaching 1850°F (1000°C) in order to avoid the sensitization temperature range. Aluminum alloys (non-age hardening) are stress relieved at 560–750°F (343–399°C) depending on alloy and degree of softening desired (e.g., for further cold reduction). Ni-base alloys are stress relieved within the range 800–1600°F (427–871°C). For most alloys, usually an anneal time of one hour plus one hour per inch of thickness is used. For titanium alloys periods of 15 minutes to 8 hours are recommended covering a range from 900 to 1300°F (482 to 704°C) depending mostly on the composition. Magnesium-base alloys are stress relieved 300–800°F (149 to 427°C) for 15–60 minutes.

Ideally, stress relief is an annealing treatment to eliminate residual stresses without producing recrystallization (in cold-worked structures), overaging (in age-hardening systems), and/or decomposition and growth (in transformation-hardening systems). In practice, some recrystallization, some overaging, and/or some phase decomposition/growth occurs either intentionally to introduce some softening or as a compromise between acceptable residual stresses and required properties.

BIBLIOGRAPHY

Brick, Robert M., Alan W. Pense, and Robert B. Gordon. *Structure and Properties of Engineering Materials*. McGraw-Hill, New York, 1977.

"Heat Treating," *Metals Handbook*, Vol. 4, 9th ed. American Society for Metals, Metals Park, OH, 1981.

"Properties and Selection: Iron and Steels," *Metals Handbook*, Vol. 1, 9th ed. American Society for Metals, Metals Park, OH, 1978.

Reed-Hill, Robert E., *Physical Metallurgy Principles*. Van Nostrand, Princeton, NJ, 1973.

CHAPTER 5
EXPERIMENTAL STRESS ANALYSIS

W. F. KIRKWOOD

Consultant, Experimental Stress Analysis
Livermore, California

5.1 INTRODUCTION

For over the past 45 years experimental stress analysis has been a factor in engineering design. The air–space industry has relied upon the fundamental techniques of load and strain measurement for experimental stress analysis of aircraft and missiles. Nuclear pressure vessel reactors have been assessed using three-dimensional photoelastic models for preliminary design. Whenever it is necessary to determine the structural integrity of a scaled-down model or a full-scale prototype, the basic techniques of experimental stress analysis can aid in the design.

In this chapter the effort has been concentrated on three stress analysis experimental methods that are most commonly used. These are:

1. The bonded electrical strain gauge.
2. Photoelasticity using two- and three-dimensional models and birefringent coatings.
3. Brittle coating (stresscoat) techniques.

The reader should also become familiar with the engineering journals related to experimental stress analysis and also the manufacturers of stress analysis equipment. The engineering journals present the latest research and applications papers and the manufacturers have a wealth of technical bulletins and "how to do it techniques."

5.2 BONDED STRAIN GAUGE

In order to perform an experimental structural analysis of a loaded structure, the investigator requires an accurate method to measure deformation, which may then be converted to unit strain. Simple uniaxial specimens with mechanical extensometers were the principal means for characterizing mechanical properties of materials in the pioneer days of materials testing. These direct measuring mechanical systems are heavy, bulky, have a low-frequency response and are not easily attached to free-formed surfaces. In 1938 E. E. Simmons, Jr., at the California Institute of Technology and Arthur Ruge at Massachusetts Institute of Technology independently developed techniques for bonding fine wire to structures. When these structures were loaded, the displaced surface transmitted all strains directly to the wire. This led to the bonded wire strain gauge and later to the bonded metal foil gauge.

The versatility of these bonded wire and metal strain gauges has been a boon to stress analysts because patterns of bonded gauges may be located over large structures and along selected planes of interest for monitoring strains during structural tests. The strain data may then be substituted into the appropriate elasticity equations to determine the uniaxial or biaxial stress values dependent on the number and direction of strain values measured at that point.

5.3 STRAIN GAUGE CONFIGURATIONS

Resistance strain gauges in general use today are usually packaged and sold under the title of "foil gauges" although sources of resistance wire gauges are still available. For general engineering practice a great number of geometric gauge designs in uniaxial or multiaxial patterns have been developed for applications to satisfy the needs for structural testing and transducer manufacturers. This type of

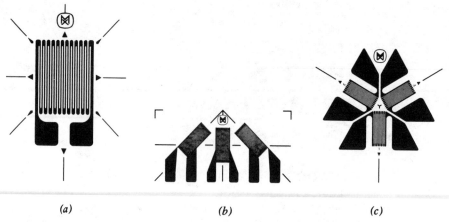

Fig. 5.3-1 Foil strain gauge designs. (a) Uniaxial; (b) rectangular three-element rosette; (c) delta rosette.

gauge is produced by a printed circuit process from thin foils of selected alloys. Foil thicknesses may range between 2.54×10^{-6} and 12.7×10^{-6} m (approximately 0.0001–0.0005 in.). With these very thin foils, the surface area of the conductor is large, resulting in low shear stresses in the insulating matrix (the gauge backing and adhesive) when the strain is transferred from the specimen to the grid.

Figure 5.3-1 is typical of modern foil strain gauge design. The printed circuit techniques enable the manufacturers to produce gauge lengths ranging from 0.20 to 12.70 mm. Most structural applications will use gauge lengths between 6.35 and 12.7 mm.

5.4 AXIAL RESPONSE AND STRAIN SENSITIVITY

When a straight wire is elastically stretched, the cross-sectional area will be reduced by Poisson's effect and a change in resistance will occur. This resistance is given by

$$R = p\frac{L}{A} \tag{5.4-1}$$

where R is the resistance in ohms, L is the length of the conductor, p is the resistivity constant, and A is the cross-sectional area of the conductor. Examination of this formula indicates the changes in length and area are additive in causing the wire resistance to increase as long as the material resistivity remains constant. This change in resistance of the conductor due to the change in length is defined by the term *strain sensitivity*:

$$\text{Strain sensitivity} = \frac{\Delta R/R}{\Delta L/L} \tag{5.4-2}$$

5.5 GAUGE FACTOR AND TRANSVERSE SENSITIVITY

The sensitivity or output characteristic of a bonded strain gauge may be defined by the term *gauge factor*:

$$F = \frac{\Delta R/R}{\Delta L/L} \tag{5.5-1}$$

This definition corresponds to that for strain sensitivity. However, in practice, the gauge factor for a bonded gauge is lower than that for the corresponding strain sensitivity of the grid material. The lower value is caused by "end-loops" in a foil or wire grid pattern resulting in a larger ΔR. This also causes an undesirable characteristic of bonded gauges termed *transverse sensitivity*. This is the ratio of gauge response perpendicular to the grid versus gauge response parallel to the grid. This is defined as

$$\frac{F_T}{F_L} = K_T \tag{5.5-2}$$

usually expressed as a percentage. The manufacturer has special equipment for determining gauge factor and cross sensitivity, and these values are included with each gauge package for that particular lot number.

5.6 TEMPERATURE EFFECTS

The basic function of the bonded gauge is to measure the change in resistance due to deformation of the structure to which it is bonded. When steady-state testing is being conducted, any changes in temperature of the structure will induce apparent strain due to change in resistance of the gauge material and differences in the expansion coefficients of the gauge and structure. The mismatch due to differences in thermal coefficient of expansion may be compensated by a dummy gauge applied to similar material as the test structure and placed in the adjacent leg of the Wheatstone bridge as shown in Fig. 5.6-1. Perry and Lissner discuss temperature compensation in detail for an assortment of structural and transducer applications.[1]

In order to produce strain gauges suitable to a large variety of tasks, the manufacturers have expended much effort to produce gauges compensated for temperature effects and designed for application on various materials. Gauges employing Constantan or modified Karma alloys have been proven to provide a reasonable match to structural materials. Figure 5.6-2 shows the typical apparent strain characteristics for the most common alloys employed in self-temperature-compensated strain gauges. When the test environment has a varying temperature and the apparent strain curve has a steep slope, significant static strain errors may be encountered.

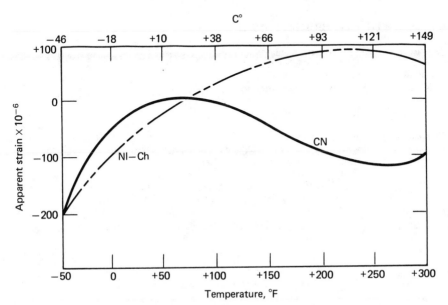

Fig. 5.6-1 Wheatstone bridge circuit compensated for temperature.

Fig. 5.6-2 Typical characteristics of constantan and nickel-chrome alloy self temperature-compensated strain gauges.

REFERENCES

5.6-1 C. C. Perry and H. R. Lissner, *The Strain Gauge Primer*, McGraw-Hill, New York, 1955, pp. 62–73.

5.7 BASIC INSTRUMENTATION

The bonded resistance strain gauge will sense small changes in resistance when the metal grid is strained. In order to measure this resistance change, a suitable circuit must be employed and the resistance change will relate to strain by the relationship

$$\varepsilon = \frac{\Delta L}{L} = \frac{\Delta R / R}{F} \qquad (5.7\text{-}1)$$

The value of ΔR will be very small and the accepted electrical circuit for measuring it is the Wheatstone bridge shown in Figure 5.8-2.

This very small change in ΔR is illustrated in the following example: The specimen is a steel bar:

$$E = 207 \times 10^9 \text{ Pa}$$

A 350-Ω strain gauge is bonded to the bar, $F = 2.04$, and uniaxial stress is 7×10^6 Pa. From Hookes law

$$\text{Uniaxial strain } (\varepsilon) = \frac{\sigma}{E} \qquad (5.7\text{-}2)$$

which gives us a solution for

$$\varepsilon = \frac{7 \times 10^6}{207 \times 10^9} = 0.0000338 \text{ m/m}$$

From Equation (5.7-1)

$$\Delta R = \varepsilon \times F \times R = 0.0000338 \times 2.04 \times 350$$

$$\Delta R = 0.02414 \ \Omega$$

The ability to measure this very small resistance change is beyond the resolution capabilities of conventional ohm-meters and therefore the Wheatstone bridge circuit is employed.

5.8 WHEATSTONE BRIDGE CIRCUITS

This circuit provides a convenient method for measuring changes in bridge output and is applicable to both static and dynamic testing. When the bridge is used as a direct readout device, the output voltage is read and related to the strain at the active gauge. A second and very popular mode is to employ the bridge as a null balance system where the output voltage E_0 is adjusted to a zero value.

5.8-1 Balanced Wheatstone Bridge

Wheatstone bridge theory is covered in the texts on elementary physics. However, detailed explanations and theoretical work relating to the equations and practical applications of strain gauges to engineering experiments are thoroughly covered by References 1–3.

The Wheatstone bridge circuit will be properly balanced by proportioning the four resistances R_1, R_2, R_3, and R_4. When this condition is attained, the system is stable and no current will flow in the galvanometer branch. It is this null balance feature that enables it to be used for static strain measurements.

From Fig. 5.8-1, the Wheatstone bridge is balanced when

$$\frac{R_1}{R_4} = \frac{R_2}{R_3} \qquad (5.8\text{-}1)$$

When the bridge has been balanced, the value of R_1 may be determined.

$$R_1 = \frac{R_2}{R_3} \times R_4 \qquad (5.8\text{-}2)$$

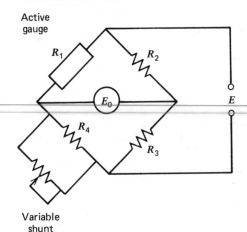

Fig. 5.8-1 Balanced Wheatstone bridge.

In this equation R_1 will be the resistance of an active strain gauge attached to the structural member being tested. R_2 and R_3 are considered fixed resistances, and R_4 will be a variable resistance. When the structure is loaded, R_1 will sense a change in resistance causing an unbalanced readout at the galvanometer. This change is ΔR_1 and the strain recorded is from Equation (5.7-1):

$$\varepsilon = \frac{\Delta R_1}{R_1 F}$$

For the convenience of the test engineer, modern instrumentation used in recording strains from static tests are calibrated in terms of unit strain $\times 10^{-6}$ and is dimensionless.

5.8-2 Unbalanced Wheatstone Bridge

When the bridge is employed as a direct reading instrument (Fig. 5.8-2), there will be no high-resistance shunt available in the fourth leg of the bridge for balancing purposes. In this case the output is the voltage change E_0 and the data is then converted to strain.

In practical engineering measurement applications, it is convenient to employ strain gauges or fixed resistances of equal value in the four legs of the bridge. If one of these resistances R is the active strain gauge, the output voltage will be

$$E_0 = \frac{E F \varepsilon R_G}{4(R + R_G)} \tag{5.8-3}$$

In practice it is convenient to use the open circuit voltage for $R \to \infty$ and to express the output in the dimensionless form

$$\frac{E_0}{E} = \frac{F \varepsilon}{4} \tag{5.8-4}$$

where $E_0/E =$ millivolt per volt
$\qquad \varepsilon =$ microstrain $= $ m/m or in./in. $\times 10^{-6}$
$\qquad F =$ gauge factor

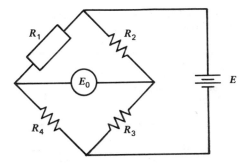

Fig. 5.8-2 Unbalanced Wheatstone bridge.

Table 5.8-1 shows a series of Wheatstone bridge configurations that are commonly employed in the solution of strain measurement problems and transducer designs.

Two assumptions are made when employing these unbalanced bridge configurations. They are:

1. $(R_1/R_4)_{nom} = 1 = (R_2/R_3)_{nom}$ when two or less active arms are used.
2. The supply voltage E = constant.

From Table 5.8-1, Cases 1, 2, 4, and 5 will have a noticeable nonlinear output at very high strains. This error will ordinarily be small. In most cases of structural testing and laboratory experiments on metals, if the part is not stressed beyond the elastic limit, this error may be ignored. Even if a tensile member is strained in tension to 17 000 microstrain, the error is only 300 microstrain which is added to the indicated strain making the true strain 17 300 microstrain.

There is an excellent technical note by The Measurements Group, that discusses the errors due to Wheatstone bridge nonlinearity in detail for those wishing to pursue this subject.

TABLE 5.8-1   WHEATSTONE BRIDGE CONFIGURATIONS AND OUTPUT EQUATIONS

Case No.	Bridge Configuration	Application	Output E_0/E (MV/V)
1	(diagram) ϵ, E_0, E	Single active strain gauge.	$\dfrac{E_0}{E} = \dfrac{F\epsilon \times 10^{-3}}{4}$
2	(diagram) ϵ, E_0, E, $-\nu\epsilon$	Two active gauges measuring the two principal strains on a uniaxially stressed member.	$\dfrac{E_0}{E} = \dfrac{F\epsilon(1+\nu) \times 10^{-3}}{4}$
3	(diagram) ϵ, E_0, E, $-\epsilon$	Two active gauges with equal and opposite strains—typical of bending-beam arrangement.	$\dfrac{E_0}{E} = \dfrac{F\epsilon}{2} \times 10^{-3}$
4	(diagram) ϵ, E_0, E, ϵ	Two active gauges with equal strains of same sign—used on opposite sides of column with low temperature gradient (bending cancellation, for instance).	$\dfrac{E_0}{E} = \dfrac{F\epsilon \times 10^{-3}}{2}$
5	(diagram) ϵ, $-\nu\epsilon$, E_0, E, $-\nu\epsilon$, ϵ	Four active gauges on a uniaxially stressed member. Two aligned with maximum principal strain, two aligned with the minimum principal strain. Doubles the output of Case 2.	$\dfrac{E_0}{E} = \dfrac{F\epsilon(1+\nu) \times 10^{-3}}{2}$
6	(diagram) ϵ, $-\nu\epsilon$, E_0, E, $-\epsilon$, $\nu\epsilon$	Four active gauges in uniaxial stress field—two aligned with maximum principal strain, two "Poisson" gauges (beam).	$\dfrac{E_0}{E} = \dfrac{F\epsilon(1+\nu) \times 10^{-3}}{2}$
7	(diagram) ϵ, $-\epsilon$, E_0, E, $-\epsilon$, ϵ	Four active gauges with pairs subjected to equal and opposite strains (See Figs. 5.10-1 and 5.10-2).	$\dfrac{E_0}{E} = F\epsilon \times 10^{-3}$

REFERENCES

5.8-1 C. C. Perry and H. R. Lissner, *The Strain Gauge Primer*, McGraw-Hill, New York, 1955, pp. 55–59.

5.8-2 J. W. Dally and W. F. Riley, *Experimental Stress Analysis*, McGraw-Hill, New York, 1965, pp. 460–467.

5.8-3 R. C. Dove and P. H. Adams, *Experimental Stress Analysis and Motion Measurement*, Merrill, 1964, pp. 77–91.

5.8-4 Measurements Group, Inc., Micromeasurements Division, Raleigh, N.C.

5.9 ROSETTE ANALYSIS

In a uniaxial stress problem the state of stress may be quite simply solved from the measured strain by

$$\sigma = E\varepsilon \tag{5.9-1}$$

In the biaxial stress field if the direction of the principal stress is known, the state of stress from the two principal strains ε_1 and ε_2 is

$$\sigma_1 = \frac{E}{1 - \mu^2}(\varepsilon_1 + \mu\varepsilon_2) \tag{5.9-2}$$

$$\sigma_2 = \frac{E}{1 - \mu^2}(\varepsilon_2 + \mu\varepsilon_1) \tag{5.9-3}$$

When the problem involves conditions where the principal stress directions are not readily apparent, strain gauges may be employed in rosette configurations to solve the principal strain values and directions. Table 5.9-1 shows the configurations and related equations for determining the magnitudes and directions for the principal stresses. When solving the equations for $\tan 2\phi$ in each of the rosette configurations, ϕ will have two values, ϕ_1 and ϕ_2. If gauge a is always located on the xx axis, then the direction of ε_1, from ε_2 may be determined by reference to the last column in Table 5.9-1.

A standard Wheatstone bridge color-coded wiring diagram is shown in Fig. 5.9-1. This diagram has been used in the present configuration for over 20 years and simplifies lead wire identification and reduces installation error.

5.10 TRANSDUCER APPLICATIONS

In addition to the direct determination of strain, bonded gauges are employed in many applications for force and motion measurements. When strain gauges are designed into these types of devices, they are referred to as transducers and play a large part in industrial and laboratory applications. Such transducers are employed in weighing scales, displacement devices, torque meters, dynamometers, and pressure measuring equipment. The following sections discuss examples of elementary strain sensing devices and how specific bridge circuits are employed to obtain optimum output and temperature compensation.

5.10-1 Ring-Type Load Transducer

Figure 5.10-1 shows a compact circular ring employed by many test organizations. This particular design was produced in several diameters and employed in the static test laboratory at Lawrence Livermore National Laboratory to measure forces on regions of the test structures where it was impractical to locate larger commercial-load cells.

A two-dimensional photoelastic model confirmed the proper place to locate the strain gauges and is discussed in Section 5.17 on two-dimensional photoelastic models. The ring-type transducer has maximum bending occurring at the position where the strain gauges are bonded. Under a tensile load, gauges R_1 and R_3 are strained in compression and gauges R_2 and R_4 are strained in tension. This Wheatstone bridge configuration has an output of four times of a single strain gauge response. This would be read out on a strain indicator as unit strain:

$$\varepsilon_{\text{total}} = -\varepsilon_1 - (+\varepsilon_2) - \varepsilon_3 - (+\varepsilon_4)$$

where the plus or minus sign before each unit strain denotes tension or compression, respectively. Due to the ring design and position of each strain gauge, each will sense an equal numerical strain. Also, the temperature effects in this bridge configuration are nullified.

In order for this design to perform undamaged and to have a linear output, the maximum ring stress should be below 70% of yield.

TABLE 5.9-1 EQUATIONS FOR DETERMINING PRINCIPAL STRAINS, PRINCIPAL STRESSES, AND THE MAXIMUM PRINCIPAL STRESS ANGLE ϕ_1, FROM THREE-ELEMENT STRAIN GAUGE ROSETTES

Type of Rosette	Gauge Configuration	Principal Strain and Stress Magnitude	Principal Angle ϕ_1 Measured from A	ϕ, Lies Between 0 and 90° When
Rectangular Three element		$\varepsilon_{1,2} = \dfrac{\varepsilon_A + \varepsilon_C}{2} \pm \dfrac{1}{2}\sqrt{(\varepsilon_A - \varepsilon_C)^2 + 2(\varepsilon_B - \varepsilon_A - \varepsilon_C)^2}$ $\sigma_{1,2} = \dfrac{E}{2}\left[\dfrac{\varepsilon_A + \varepsilon_C}{1 - \mu} \pm \dfrac{1}{1 + \mu}\sqrt{(\varepsilon_A - \varepsilon_C)^2 + (2\varepsilon_B - \varepsilon_A - \varepsilon_C)^2} \right]$	$\tan 2\phi_1 = \dfrac{2\varepsilon_B - \varepsilon_A - \varepsilon_C}{\varepsilon_A - \varepsilon_C}$	$\varepsilon_B > \dfrac{\varepsilon_A + \varepsilon_C}{2}$
Delta three element		$\varepsilon_{1,2} = \dfrac{\varepsilon_A + \varepsilon_B + \varepsilon_C}{3} \pm \dfrac{\sqrt{2}}{3}\sqrt{(\varepsilon_A - \varepsilon_B)^2 + (\varepsilon_B - \varepsilon_C)^2 + (\varepsilon_C - \varepsilon_A)^2}$ $\sigma_{1,2} = E\left[\dfrac{\varepsilon_A + \varepsilon_B + \varepsilon_C}{3(1 - \mu)} \pm \dfrac{\sqrt{2}}{3(1 + \mu)}\sqrt{(\varepsilon_A - \varepsilon_B)^2 + (\varepsilon_B - \varepsilon_C)^2 + (\varepsilon_C - \varepsilon_A)^2} \right]$	$\tan 2\phi_1 = \dfrac{\sqrt{3}(\varepsilon_C - \varepsilon_B)}{2\varepsilon_A - (\varepsilon_B + \varepsilon_C)}$	$\varepsilon_C > \varepsilon_B$

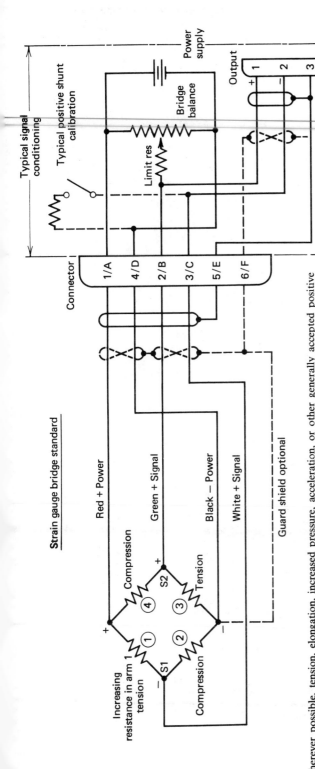

Strain gauge bridge standard

1. Wherever possible, tension, elongation, increased pressure, acceleration, or other generally accepted positive quantities shall produce positive output signals as indicated.

2. The bridge elements shall be arranged so that functions producing positive output will cause increasing resistance in arms 1 and/or 3 and decreasing resistance in arms 2 and/or 4 of the bridge.

3. The auxiliary wiring for 6 or 8 wire system will have same color but as a tracer on white wire or imsa colors.

4. Quarter bridge—when only one bridge element is active use arm 1 (arms 2, 3, and 4 as dummy elements).

5. Half bridge—when a tension and compression component is to be measured use arms 1 and 2 (arms 3 and 4 as dummy elements).

6. The direction or position of the function producing a positive output signal shall be indicated on transducers. Shunt calibration resistor shown will produce a positive output signal. The following markings are suggested.

   + ↕ tension load cells, universal load cells, micrometers, etc. (− ↕ compression load cells)
   + → accelerometers and flow meters
   ↑ + ↓ torque transducers
   + differential pressure cells at the port where the higher pressure causes positive output signals

7. For shielded-type bridge systems pins 5/E, 7/G, 9/I shall be shield terminals for 4, 6 and 8 wire systems.

Fig. 5.9-1 Western Regional Strain Gauge Committee Wiring Standards.

405

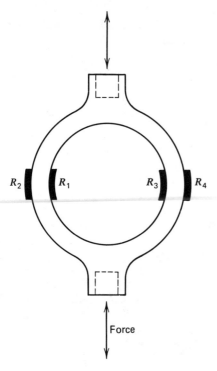

Fig. 5.10-1 Strain-gauged load ring.

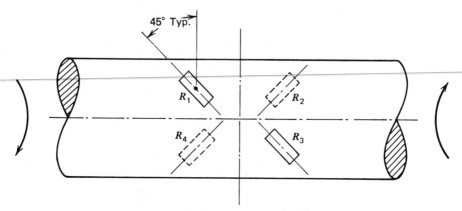

Fig. 5.10-2 Strain-gauged torque bar.

5.10-2 Torque Bar

Figure 5.10-2 shows the typical strain gauge layout on a round bar or tube for measuring torque. The location of the strain gauges on the 45° planes to the centerline of the body is in agreement with the theoretical position of the principal stress planes. The use of four strain gauges will again give the four times response from the bridge configuration of Fig. 5.10-1.

5.10-3 Pressure Sensors

A popular and reliable pressure transducer may be constructed using the principal of a thin flat plate with edges clamped. Timoshenko[1] (Part 11) shows the maximum principal stress in the plate is σ_R at the clamped boundary and for the same surface at the center of the plate, the radial stress is one-half of σ_R.

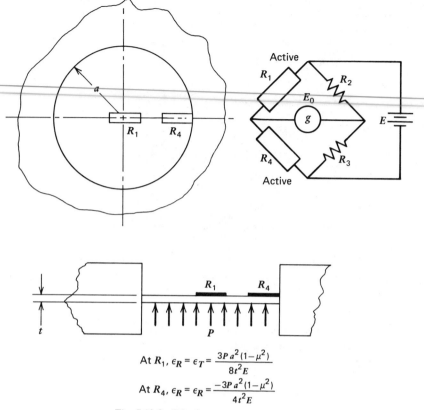

$$\text{At } R_1, \; \epsilon_R = \epsilon_T = \frac{3P\,a^2(1-\mu^2)}{8t^2E}$$

$$\text{At } R_4, \; \epsilon_R = \epsilon_R = \frac{-3P\,a^2(1-\mu^2)}{4t^2E}$$

Fig. 5.10-3 Diaphragm pressure sensor.

In actual transducer design the diaphragm plate should be an inherent part of the structure.

Figure 5.10-3 shows two strain gauges on the diaphragm located to record the radial strain at the center and as close as possible to the fixed edge to record the maximum radial strain. For this Wheatstone bridge configuration the output equation is

$$\frac{E_0}{E} = F(\varepsilon_1 + \varepsilon_4) \times 10^{-3}$$

where ε is strain $\times 10^{-6}$. If it is possible to locate two more gauges opposite R_1 and R_4, this output would be doubled.

There are now commercially available strain gauges specifically designed for applications to pressure sensor configurations. The commercial strain gauge literature has excellent examples of these configurations (Fig. 5.10-4) and the Measurements Group of Vishay Intertechnology has a very complete technical note on diaphragm design.[2]

REFERENCES

5.10-1 S. P. Timoshenko, *Strength of Materials*, 3rd ed., D. Van Nostrand, New York, 1968.
5.10-2 *Design Considerations for Diaphram Pressure Transducers*, TN-129-3, Measurements Group, Inc., P.O. Box 27777, Raleigh, N.C. 27611.

5.11 DYNAMIC STRAIN MEASURING

When a specimen or structure in motion is to be tested, the strains are most easily monitored with a potentiometer circuit. Such a circuit is shown in Fig. 5.11-1 with one active strain gauge. In this application the output is capacitance coupled to the indicator and only changes in potential across the gauge (ΔE_G) are recorded.

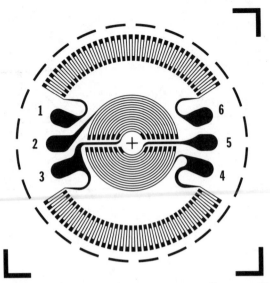

Fig. 5.10-4 Strain gauge designed for use with pressure sensors. (Courtesy of Micro-Measurements Division, Measurements Group, Inc., Raleigh, North Carolina.)

The power source may be either alternating or direct current. This text will assume direct current so the equations may be written in terms of resistance (R) instead of total impedance (Z). This circuit may be analyzed similar to the Wheatstone bridge, and the resulting sensitivity equation is

$$\frac{dE_G}{\varepsilon} = \frac{R_b R_G F E}{(R_b + R_G)^2} \tag{5.11-1}$$

where R_G = strain gauge resistance
R_b = a ballast resistance
E_g = potential difference across R_G
F = gauge factor

If it is desired to employ this equation in terms of current I_g, then

$$\frac{dE_G}{\varepsilon} = \frac{R_b R_G F I_G}{R_b + R_G} \tag{5.11-2}$$

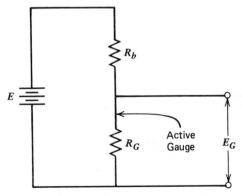

Fig. 5.11-1 Potentiometer circuit for dynamic strain measurements.

These two equations provide the circuit sensitivity for small changes in R_G and for the no-load condition across the gauge element. For calibration purposes these equations may be used to compute the potentiometer circuit output.

5.12 PHOTOELASTIC STRESS AND STRAIN ANALYSIS

Photoelasticity is the term employed since the early 1930s to describe the techniques for the application of plastic materials to experimental stress analysis. Over a hundred years earlier, Sir David Brewster[1] observed that glass exhibited color patterns when viewed under stress in polarized light. This same birefringent phenomenon observed in photoelastic models is explained in Maxwell's electromagnetic theory.

Photoelastic techniques have become very popular in the industrial and academic laboratories for investigating structures of complicated geometry where the existing mathematical methods do not lend themselves to ready solutions. Models of selected plastics when loaded and viewed in polarized light show a whole field display of the state of stress throughout the structure. From this information and with the proper application of the laws of similitude, the designer may perform a quantitative analysis and initiate any desired geometry changes.

The material presented in this handbook is generally in elementary form and more rigorous material may be studied by referring to the many excellent texts and the proceedings of technical and scientific journals.

REFERENCES

5.12-1 D. Brewster, *Phil. Trans. Roy. Soc.*, London, 156 (1816).

5.13 POLARISCOPE

Maxwell's electromagnetic theory may be simplified in this discussion by eliminating the electric disturbance and imagine the light as a simple transverse wave motion in space. These vibrating waves will vary in length according to the color of the light. A simple analogy would be a rope tied to a wall with a transverse vibration applied at the other end. These vibrations may be random in magnitude and direction and is the manner in which ordinary light is produced.

The photoelastic effect is achieved by placing a filter in the field of transverse vibrating light waves and confining them to parallel planes (see Fig. 5.13-1). The material employed in these filters is Polaroid, a commercially available product that polarizes light very efficiently.

A basic plane polariscope is a simple assembly consisting of a light source and two polarizing elements (Fig. 5.13-2). The element nearest the light is called the polarizer and the second element, the analyzer. When the planes of both elements are parallel, light in the plane of the page will pass through to the camera or viewer. This arrangement is called a parallel polariscope.

By slowly rotating the analyzer, the observer will note a gradual diminishing of the light intensity until the element is rotated 90° at which time the light emission is extinguished and a crossed polariscope is achieved.

5.14 STRESSED TRANSPARENT MODELS

Photoelastic stress analysis is based on the property of some transparent materials to become doubly refracting when stressed. When a model fabricated from a material manufactured for photoelastic stress analysis is placed in the plane of the vibrating light field between the polarizer and analyzer and

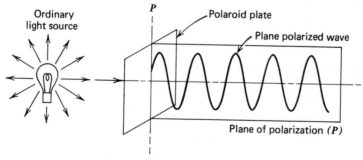

Fig. 5.13-1 Plane polarized light.

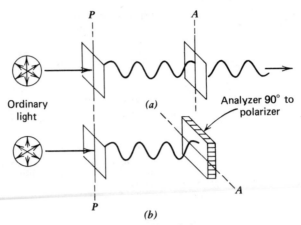

Fig. 5.13-2   Basic polariscope.

then loaded, the ray of light is split into two components vibrating at right angles to each other, which are also the planes of principal stresses in the model. In addition to this phenomenon, the light travels along these perpendicular paths at two velocities dependent on the principal stress magnitudes.

This lag of one principal stress component behind the other is called the relative retardation and is measured as a length. It is assumed that the third principal stress acting perpendicular to the plane of the model will have equal influence and not disturb the relative retardation of the two light components. For two-dimensional plane stress models, this third principal stress is considered to be zero.

5.15   STRESS OPTICAL COEFFICIENT

When a two-dimensional photoelastic model is stressed in its plane, at each point on the model surface there is a maximum and minimum principal stress termed σ_1 and σ_2. These two stresses are assumed constant throughout the model thickness, and the phase difference between them depends on the principal stress magnitudes. Also, each fringe is a locus of points of equal shear stress where

$$\tau = \frac{\sigma_1 - \sigma_2}{2} \qquad (5.15\text{-}1)$$

When this plane stress model is viewed in the circular polariscope, a series of light and dark bands appear covering the model surface. The bands are called fringes and are a result of the retardation difference of the two principal stresses. This fringe pattern is related to the retardation by equations developed by Brewster through experiments, which include the following model thickness:

$$R = Ct(\sigma_1 - \sigma_2) \qquad (5.15\text{-}2)$$

where R = relative retardation mm
   C = a constant named the stress optical coefficient, mm^2/N
   t = model thickness, mm
 σ_1, σ_2 = maximum and minimum principal stresses, Pa

In order to employ the observed fringe pattern for an experimental analysis of the model stresses, it is necessary to express R in terms of the fringes which are proportional to the phase difference. With a known wavelength λ for the monochromatic light source, then

$$N\lambda = R \qquad (5.15\text{-}3)$$

By combining Equations (5.15-2) and (5.15-3),

$$N\lambda = Ct(\sigma_1 - \sigma_2)$$

let

$$\frac{\lambda}{C} = f$$

then

$$\sigma_1 - \sigma_2 = \frac{Nf}{t}$$

(5.15-4)

where N = number of fringes
t = model thickness, mm
f = fringe stress coefficient, N/mm fringe

This equation is the basic solution employed in the photoelastic method of stress analysis.

5.16 CIRCULAR POLARISCOPE

For general photoelastic work it is necessary to install two quarter wave plates in the light path, one on each side of the photoelastic model. Figure 5.16-1 shows the behavior of light waves when polarized in the circular polariscope. These plastic plates are prestressed uniformly during manufacture to produce a one-quarter wave retardation throughout the plate area.

The principal planes of these plates (designated by $\lambda/4$) are located at 45° to the axis of the polarizer. In this orientation the polariscope is set up to remove the directional sensitivity or the observance of isoclinics to be discussed later. Isoclinics are removed to prevent them from interfering with the general fringe pattern.

Figure 5.16-2 is a schematic diagram of how simply a practical polariscope may be constructed. This design by Leven[1] was modified and Fig. 5.16-3 shows the polariscope that was employed for many years at Lawrence Livermore National Laboratory. All the rotating elements are remote controlled from the console at the camera position. The long wooden bellows contains a 610-mm (24-in.) focal length camera that permits the image on the frosted glass film holder to be enlarged by a factor of 5.

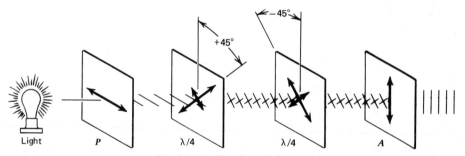

Fig. 5.16-1 Circular polariscope.

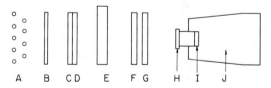

A – Light Source – Two Staggered Rows of Green and White
      Fluorescent Lamps.
B – Flashed Opal Glass Diffusion Screen.
C & G – Polaroid Disks in Rotatable Mount.
D & F – Quarter Wave Plate Disks in Rotatable Mount and
      Removable for Conversion to Plane Polariscope.
E – "Frozen Stress" Slice
H – Camera Lens – Long Focal Length
I – Filter Wratten 77A
      or Corning I–60, 4–96, and 3–68 Combined.
J – Camera – Long Bellows Extension.

Fig. 5.16-2 Diffused light circular polariscope design.

Fig. 5.16-3   Practical diffused light polariscope. (Courtesy Lawrence Livermore National Laboratory.)

REFERENCES

5.16-1   M. M. Leven, *Epoxy Resins for Photoelastic Use*, Macmillan, New York, 1963, pp. 145–165.

5.17   TWO-DIMENSIONAL MODEL ANALYSIS

This type of model is most beneficial in dealing with cases where quick results are required for solving a plane stress problem. The stresses in this system will be parallel to the plane of the model, and all other stress components are zero. Frocht,[1] Dalley and Riley,[2] and Heywood[3] show excellent examples of two- and three-dimensional analysis.

5.17-1   Materials for Two-Dimensional Analysis

The plastic employed in photoelastic models should be carefully selected from available commercial materials. These will usually be in sheet form approximately 254 mm (10 in.) square with optional thicknesses from 3.18 to 12.7 mm. During the past half-century many materials for two-dimensional photoelastic analysis have been developed, but only a few plastics have been found that provide the necessary characteristics for practical work. Table 5.17-1 lists some commercially available plastics presently employed for two-dimensional models. For the individual wishing to try casting sheets of epoxy material, Agarwal and Teufel[4] adequately address this problem. However, excellent shop facilities and some experience in handling epoxy formulations are a definite prerequisite for producing quality plastics for two- or three-dimensional photoelastic models.

5.17-2   Calibration of the Model Material

The material employed in either two- or three-dimensional photoelastic models may show some variation in optical and mechanical properties. When purchasing material in sheet form from the manufacturer, the material fringe value f may be identified on the sheet, and a close approximation of the modulus of elasticity should be available in the manufacturer's literature. In order for the experimenter to obtain the most accurate results, the exact value of f should be determined from a calibration specimen taken from the same sheet from which the model was fabricated. The solution of f may be obtained from Equation (5.15-4)

$$f = (\sigma_1 - \sigma_2)\frac{t}{N}$$

TABLE 5.17-1   MATERIAL PROPERTIES FOR TWO-DIMENSIONAL MODELS

Material	Young's Modulus (mPa)	Stress Optical Coefficient (kN/m)	Time Edge Effects	Remarks
CR-39	1725	16	Poor	Inexpensive and suitable for approximate solutions or demonstrations
Cast epoxy	3270	10.2–11.4	Good	Properties depend on casting procedure and type of hardener used. Good model material
Polycarbonate	2350–2480	7.0	Very good	Properties depend on manufacturer
Polyurethane sheet	—	0.525–0.875	—	Used for dynamic studies or demonstration models
PSM-1	3100	10.50	Very good	Commercial material in sheet form Measurements Group Inc. P.O. Box 27777 Raleigh, N.C. 27611

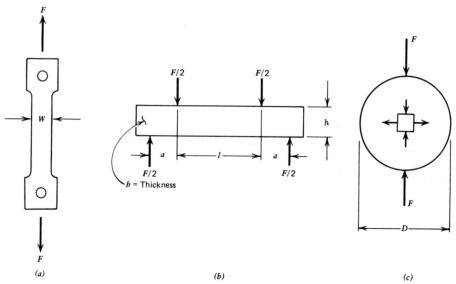

Fig. 5.17-1   Calibration specimens. (*a*) Tension specimen; (*b*) beam in pure bending (*c*) diametrically loaded disk.

In lieu of f, some photoelasticians use the term C, which is in Pa/mm N. Simple arithmetic will show this to be equal to f.

Figure 5.17-1 shows the three calibration models that are practical for determining the fringe value f.

5.17-3   Tensile Specimen Calibration

Install the specimen in a suitable load frame located between 2 two-quarter wave plates and with the polarizers crossed to give a dark field. All photoelastic work may be pursued using white or

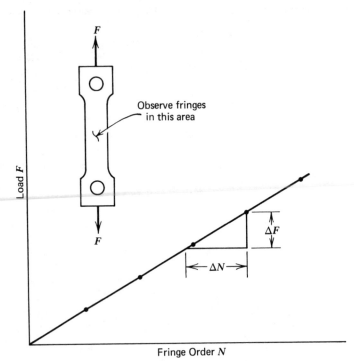

Fig. 5.17-2   Tensile specimen calibration curve.

monochromatic light; most experimenters use white light for locating the black fringe that usually identifies the zero-order fringe, for watching the tint of passage to identify the first-order and higher fringes as the load is increased. For general evaluations in both two- and three-dimensional model work, monochromatic light is the usual source for stress analysis and photography. As the tension specimen is loaded, the light intensity in the nominal section will alternate from minimums to maximum to minimums again. Each time the minimum light intensity occurs with increasing load, the next higher fringe order has been passed. Figure 5.17-2 shows an example of such a load versus fringe order (N) plot. Knowing the unit stress in the specimen, the value of f may be determined from equation (5.15-4):

$$f = \frac{F}{WN} \tag{5.17-1}$$

In this case there is no need to determine the stress, just take the slope $\Delta F/\Delta N$ and divide by the nominal width W. The material fringe value may be determined without knowing the model thickness.

5.17-4   Beam in Pure Bending

The beam in pure bending is a very practical method for calibrating the photoelastic model material. A four-point loading fixture applies the load to produce a section of pure bending along the length of the beam. This produces the type of fringe pattern shown in Fig. 5.17-3a where the fringes are uniformly spaced from a value of $N = 0$ at the neutral axis to N maximum at the upper and lower boundaries. The beam in this picture is a frozen stress specimen made from a Jones and Dabney epoxy. From Fig. 5.17-1b the dimensions are

$$h = 25.4 \text{ mm (1 in.)}$$

$$a = 25.4 \text{ mm (1 in.)}$$

$$b = 6.375 \text{ mm (0.251 in.) thickness}$$

$$F = 15.41 \text{ N (3.465 lb)}$$

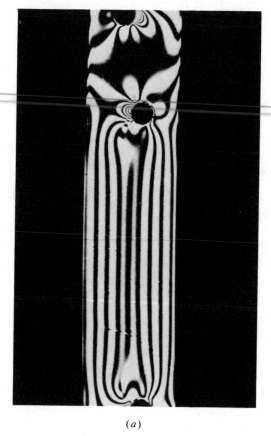

(a)

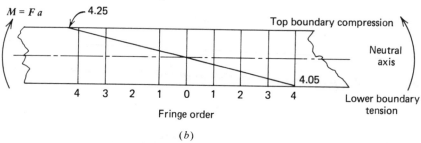

(b)

Fig. 5.17-3 Pure bending calibration specimen.

From the load and beam dimensions the maximum stress at the upper and lower boundaries may be determined from the flexure equation:

$$\sigma_{max} = \frac{MY}{I} = \frac{12F/2aY}{bh^3} = \frac{6Fah/2}{bh^3}$$

$$\sigma_{max} = \frac{3Fa}{bh^2} \tag{5.17-2}$$

where σ_{max} = stress at the upper and lower boundary
$M = (F/2)$
a = constant bending moment along l
$I = bh^3/12$

Let the maximum principal stress at the lower and upper boundaries equal $\pm \sigma_1$ and let $\sigma_2 = 0$.
By combining Equations (5.17-2) and (5.15-4), the material fringe value may be determined:

$$f = \frac{3Fa}{Nh^2} \tag{5.17-3}$$

From Fig. 5.17-3a the fringe orders at the upper and lower boundaries may be graphically extrapolated as shown in Fig. 5.17-3b. The values of N at the boundaries may be different due to a small time edge effect that is inherently characteristic of some photoelastic materials. Therefore, the average value of these fringe orders should be used in Equation (5.17-3). This gives a value of

$$f = \frac{3Fa}{Nh^2} = \frac{3 \times 15.41(25.4 \times 10^{-3})}{4.15(25.4 \times 10^{-3})^2}$$

$$f = 0.4386 \text{ kN/m } (2.55 \text{ lb/in. fringe})$$

5.17-5 Diametrically Loaded Disk

When a solid circular disk is loaded by two concentrated loads 180° apart, the fringe pattern will appear as shown in Fig. 5.17-4. The theoretical solution to this problem is furnished in the theory of elasticity.[5] At the center of the disk, the difference of the two principal stresses is

$$\sigma_1 - \sigma_2 = \frac{8F}{\pi t D} = \frac{Nf}{t}$$

where F = force
       t = thickness
       D = diameter
       N = fringe order or number
       f = material fringe value

and

$$f = \frac{8F}{\pi DN} \tag{5.17-4}$$

This calibration technique is suitable for use in determining the material fringe values of either two- or three-dimensional model materials. The numerical value of N at the center of the disk is easily

Fig. 5.17-4 Diametrically loaded disk.

determined by making a load versus N plot for each one-half- or full-order fringe as it occurs. If the disk has the stresses frozen in when calibrating for a three-dimensional frozen stress model, it will be necessary to employ either a compensation or the Tardy method to determine the exact fringe value.

5.17-6 Tension Link Analysis

Figure 5.17-5 is a light-field photograph of a loaded tension link. The dark bands are one-half-order fringes, the light bands represent the full orders and the fringe values are noted in the field of the photograph. The maximum stress occurs at the 4.75 fringe order position.

Figure 5.17-6 shows a plot of the tangential fringe order distribution around one quadrant of the hole. Notice how the sign of the stresses changes from a maximum negative stress at the top centerline, passes through zero, and becomes a maximum in tension at the horizontal centerline.

This example clearly shows the value of photoelasticity in providing the complete boundary stress distribution for a plane stress problem.

5.17-7 Load Ring Analysis

This example is another plane stress model employed to examine the stress distribution in a circular ring. These load or force rings are frequently used in structural test setups as load measuring transducers. The photoelastic model (Fig. 5.17-7) has been loaded in compression across the vertical diameter imposing a combined state of axial and flexural stress throughout the structure.

The maximum tensile fringe order of 5.25 is shown in the field of the photograph at the change of wall section near the loading boss. This criteria will determine the limiting loads for the actual metal prototype. For the installation of strain gauges, this is most easily accomplished on the inner and outer diameters at the horizontal centerline.

Fig. 5.17-5   Tension link model.

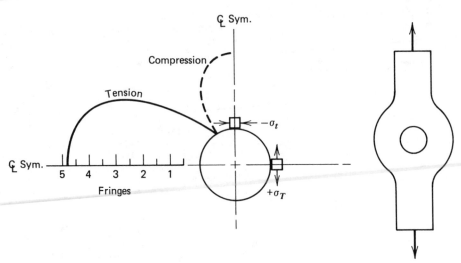

Fig. 5.17-6 Stress distribution in fringes around one quadrant of the hole boundary.

Fig. 5.17-7 Light-field photograph of a load ring under axial compression.

At these positions the fringes are approximately $4\frac{1}{2}$ inside compression and about $3\frac{1}{2}$ tension on the outside. The numerical difference is due to the negative axial stress component that adds to the compressive bending stress component and subtracts from the tensile.

REFERENCES

5.17-1 M. M. Frocht, *Photoelasticity*, Vol. 1, Wiley, New York, 1957, pp. 231–243, 297–322.

5.17-2 J. W. Dally and W. F. Riley, *Experimental Stress Analysis*, McGraw-Hill, New York, 1965, pp. 308–323.

5.17-3 R. B. Heywood, *Designing by Photoelasticity*, Chapman and Hall, London, 1952, pp. 109–124, 314–372.

5.17-4 R. B. Agarwal and L. W. Teufel, "Epon 828 Epoxy: A New Photoelastic Model Material," *Exp. Mech.* **23**(1), 30–35, March, 1983.

5.17-5 S. P. Timoshenko, *Strength of Materials*, 3rd ed., D. Van Nostrand, New York, 1968, pp. 98–99.

5.18 BABINET COMPENSATOR

The Babinet compensator is a device designed for accurate determination of the fringe order in the field or boundary of a photoelastic model when loaded in a fixture or to evaluate slices removed from three-dimensional frozen stress models. When such a device is employed, retardations to a hundredth of a wavelength may be measured.

The Babinet compensator employs two quartz wedges as shown in Fig. 5.18-1.

The wedges are machined to right triangle geometry with a 2.5° sliding plane. Triangle ABC has an optic axis lying in the plane of the paper or triangle, and the wedge DEF has its optical axis at right angles to ABC.

In the instrument one wedge is fixed and the other wedge is displaced along the slide plane by a fine micrometer adjustment. When the micrometer screw is brought to a position where the thickness of the wedges are equal to each other, the compensator is in a neutral position.

To employ the compensator, place it between the quarter wave plate and the model in the polariscope. Select the point on the model where the fringe order is to be determined and align the compensator axis with one of the principal stress directions. This direction is determined from the isoclinic passing through the point of interest. The compensator lead screw is now turned until extinction occurs. The reading of the compensator is proportional to the fringe order in the model.

Referring to Fig. 5.18-1, let d_0 be the wedge thickness at the datum point (where the engraved pair of crossed lines intersect). With the origin O as the datum point, let x denote the distance from O to O_1 on the moving wedge where the thickness is again d_0.

The equation for the total retardation R_p produced in the basic Babinet compensator is

$$R_p = \frac{K}{\lambda}(d - d_0) = Cx \qquad (5.18\text{-}1)$$

where K = a constant

C = a constant

λ = wavelength of the light

and d, d_0, and x are the distances shown in Fig. 5.18-1.

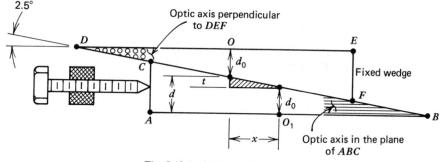

Fig. 5.18-1 Babinet compensator.

The latest commercial compensation equipment makes the use of the compensator a simpler task. The quartz wedges have been replaced by juxtaposed frozen stress epoxy strips. The birefringence is linearly distributed along their length and proportional to the displacement of one strip to another. When extinction is achieved in the field of the compensator, a number appears in a small window of the instrument. The fringe value is then determined from a graph or table supplied by the manufacturer. Another commercial form of the improved compensator is one that converts the strip displacement into a digital voltmeter display of the difference in principal stresses.

5.19 TARDY COMPENSATION

This very popular method of determining fractional fringe orders requires no extraneous apparatus and employs only the optical elements of the polariscope.

In a standard circular polariscope employing white light, remove the quarter wave plates. Apply a light load to the two-dimensional photoelastic model and slowly rotate the polarizer and analyzer while observing the model. A black band moving across the field relative to this rotation represents a locus of all points along which the principal stresses σ_1 and σ_2 are perpendicular and parallel to the polarizing axis. These black lines are called isoclinics and will be described in Section 5.21.

With the polarizer and analyzer crossed, rotate these elements until the isoclinic rests on the exact spot in the model where the fringe number is to be determined. This will usually be at some low fringe number where the fringe spacing is wide or at a free boundary. Replace the quarter wave plates and the isoclinics will disappear. Switch to monochromatic light and perform the following operations.

Rotate the analyzer through an angle ϕ until extinction occurs at the point being investigated. The fringe order at this point is

$$n = r + \frac{\phi}{180} \tag{5.19-1}$$

where r is the order of the last visible fringe.

When the extinction occurs by rotating the analyzer in the opposite direction, then

$$n = (r + 1) - \frac{\phi}{180} \tag{5.19-2}$$

The application of either equation depends on knowing the value of r before bringing the point of interest to extinction. Generally the fractional order the investigator of a two- or three-dimensional model is most interested in occurs at a free or loaded boundary, as illustrated in Fig. 5.19-1.

Here again, the polarizer and analyzer are crossed and aligned either perpendicular or tangent to point 0 because the principal stresses are perpendicular and parallel, respectively, to a free boundary. The procedure is as follows:

1. The angle of the analyzer $\phi_a = 0$.
2. Rotate the analyzer until the fringe r just touches the boundary at point 0. This is ϕ_b.
3. Continue to rotate the analyzer until r vanishes from the field of view. This is angle ϕ_c.

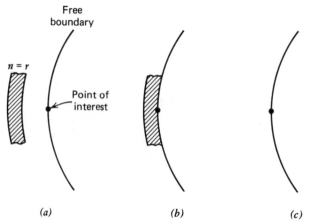

Fig. 5.19-1 Illustration of the "Tardy" method of fringe-order determination: (a) $\phi_a = 0$; (b) ϕ_b; (c) ϕ_c; $n_0 = r + (\phi_b + \phi_c)/360$.

The fringe order at the position 0 of the free boundary is

$$n_0 = r + \frac{\phi_b + \phi_c}{360}$$

(5.19-3)

This method is commonly referred to as the Tardy in–out method, and with a little practice accuracies of less than 1/30th of a fringe can be achieved.

5.20 COLOR MATCHING

The English scientists Coker and Filon performed most of their photoelastic experiments with white light. The determination of the principal stresses with white light requires a reliable scaling chart showing the relationship of fringe order to colors of the spectrum. The discussion of photoelastic coatings uses examples where, on most occasions, the experiment will employ white light so the introduction to color matching is appropriate here.

If white light is used in an experiment, the loaded model will be covered by a series of colored bands called isochromatics. Any continuous color along these bands represents a locus of all points along which the difference of the two principal stresses are equal, just as the black fringes represent the same phenomenon in monochromatic light. In white light these colored bands occur at intervals between 5750 Å to nearly 5800 Å according to the light source. Table 5.20-1 lists the relationship of colors and fringe orders for the first four fringes. Color matching beyond the fourth-order fringe becomes impractical.

TABLE 5.20-1 COLOR TO FRINGE ORDER RELATIONSHIP FOR A STRESSED MODEL VIEWED IN A DARK FIELD USING WHITE LIGHT

Color	Retardation (Å)	Fringe Order
Black	0	0
Gray	1,600	0.28
White	2,600	0.45
Yellow	3,500	0.60
Orange	4,600	0.79
Red	5,200	0.90
Tint of passage 1[a]	5,770	1.00
Blue	6,200	1.06
Blue-green	7,000	1.20
Green-yellow	8,000	1.38
Orange	9,400	1.62
Red	10,500	1.81
Tint of passage 2[a]	11,500	2.00
Green	13,500	2.33
Green-yellow	14,500	2.50
Pink	15,500	2.67
Tint of passage 3[a]	17,300	3.00
Green	18,000	3.10
Pink	21,000	3.60
Tint of passage 4[a]	23,000	4.00
Green	24,000	4.13

[a] The tint of passage is a sharp dividing zone occurring between red and blue in the first-order fringe, red and green in the second-order fringe, and pink and green in the third-, fourth-, and fifth-order fringes. Beyond five fringes, white-light analysis is not adequate. Note that the first-order fringe is the only one with a blue component. This is helpful in identifying the fringe order.

5.21 ISOCLINICS

When the directions of the principal stresses in a plane polariscope coincide with the plane of polarization of the analyzer, then, while passing through the model, the light will vibrate in the plane of the polarizer and be absorbed by the analyzer. At that point in the model the fringe representing the principal stress difference is covered by the black band of the isoclinic and will interfere with the analysis. These isoclinics are removed by inserting the two quarter wave plates to create circularly polarized light.

However, the isoclinic phenomenon is necessary in tracing the principal stress directions and for employing the Babinet compensator and the Tardy method (see Fig. 5.21-1).

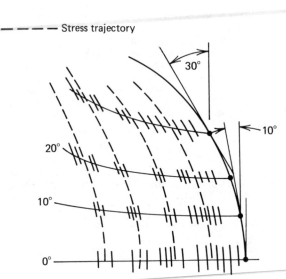

Fig. 5.21-1 Stress trajectories and isoclinics.

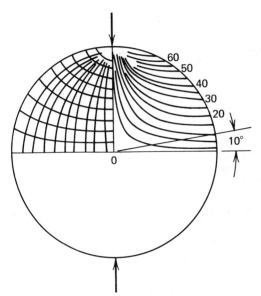

Fig. 5.21-2 Isoclinics and stress trajectories traced from a disk in diametric compression. See Fig. 5.17-4.

Frocht has several illustrations of plotting isoclinics from photoelastic models and the development of stress trajectories. Figure 5.21-2 shows isoclinics and stress trajectories traced from the diametrically loaded disk (Fig. 5.17-4).

5.22 THREE-DIMENSIONAL PHOTOELASTIC MODELS

Three-dimensional models provide a method for the structural analyst to duplicate a full- or modified-scale version of a structure in plastic form and by employing the frozen stress technique to determine the state of stress at any position of interest. The fabrication of these models requires care and attention to detail and should not be attempted without some experience in machining plastics or else with the support of proper shop craftsmen.

The "frozen stress" method for three-dimensional photoelasticity has been in use for over 45 years after Oppel's initial work in 1936. Today the epoxy resins are the principal plastic, formulated with phthalic and hexahydrophthalic anhydrides to produce stress-free castings suitable for machining models for photoelastic analysis.

5.22-1 Equipment

1. Standard circular polariscope.
2. Bandsaw.
3. Oven—capable of 180°C max. temperature.
4. Temperature programmer.
5. Temperature controller—capable of $\pm \frac{1}{2}°$ C/hr.
6. Metal lathe.
7. Milling machine.

5.22-2 Casting Materials and Procedures

The ability to mix the epoxy formulations and obtain quality castings is extremely dependent on the proper mixing and temperature control during these operations. The selection of the epoxy resin will have a definite impact on the material quality relative to clarity, machining, material fringe value, and figure of merit. M. M. Leven, while on the staff at Westinghouse Research Laboratories, conducted considerable experiments to develop a suitable three-dimensional photoelastic model material. A compilation of these results is published in References 1 and 2 and has been the basic formulations for much of the three-dimensional photoelastic work over the past 25 years. Although Leven based his final formulation on ERL-2774 which produces a quality product, the experimenter will also achieve excellent results with Jones and Dabney 510, which goes under the trade name Epi-Rez or Araldite 6020. Other resins should be selected at the discretion of the user. Table 5.22-1 lists some of the properties of two epoxy formulations with the phthalic and Hex hardeners.

The two epoxies in Table 5.22-1 have the same weight formulations: 100 parts by weight epoxy, 42 parts by weight phthalic-anhydride, and 20 parts by weight hexahydrophthalic anhydride.

Casting Procedure

1. Clean all molds with an OSHA-approved solvent and coat with a good quality mold release compound.
2. Heat the epoxy to about 120°C and the Hex anhydride to about 100°C in separate containers.
3. Pour the Hex anhydride into the epoxy. Use a cheese cloth filter.
4. Add the phthalic anhydride flakes to the mixture and then stir until all ingredients have dissolved.
5. Pour the mixture into the previously prepared molds and allow to gel for a 1- to 3-day period at a minimum of 92°C

TABLE 5.22-1 PROPERTIES OF EPOXIES AT CRITICAL TEMPERATURE

Material	T_{cr} (°C)	f_{eFF} (N/fringe/m)	E_{eFF} (mPa)	Q (1/mm)
ERL 2774[a]	162–175	411–539	37–45	362–422
J & D 510	160–170	415–437	35–38	340–380

[a] Reference 2.

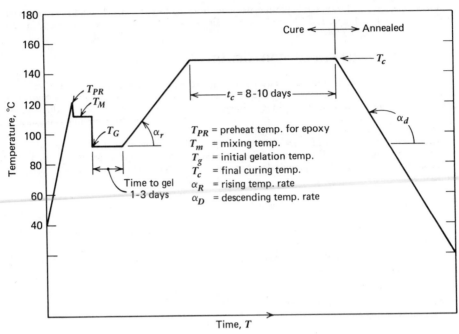

Fig. 5.22-1   Temperature versus time for curing hexahydropthalic anhydride formulations.

6. Raise temperature to 150°C at 0.5–1°C per hour using the lower value for castings greater than 300 mm. Hold for 8–10 days.
7. Slowly reduce the temperature at 0.25–0.50°C per hour using the lower value for larger castings.

An alternative to this procedure is to slowly raise the temperature to 100°C after gelation and then slowly decrease to room temperature. Remove the casting or castings (do not clean them); replace them in the oven and raise to 100°C. Some economy may be gained this way if several castings are to be made from a single mold.

Figure 5.22-1 graphically displays the temperature–time profile for the casting and curing procedure.

REFERENCES

5.22-1 M. M. Leven, "Practical Three-Dimensional Photoelasticity as an Aid to Design," Scientific Paper 6-94459-2-P6, Westinghouse Research Laboratories, Jan. 1957.
5.22-2 M. M. Leven, *Epoxy Resins for Photoelastic Use*, Macmillan, New York, 1963, pp. 145–165.

5.23   MACHINING AND CEMENTING

After casting, the epoxy resins are readily machined into most desired shapes with the proper technique and equipment. Figure 5.23-1 shows a large hemishell with a removable top containing ports. This model was cast as a solid hemisphere and machined to final dimensions. The machining operation should employ carbide-tipped tools using high cutting speeds and low feeds to prevent edge chipping. When taps or dies are used, the operation should be done in several steps. Threads cut on a machine should be chased.

After stress freezing of the model, slices may be removed along the planes to be investigated by a band saw. Finish cutting of these slices will be accomplished most satisfactorily by fly-cutting on a milling machine.

Photoelastic models may be easily cemented at all clean flat surfaces and joints. A mixture of 100 parts by weight of epoxy and 10 parts by weight Hardener 951 manufactured by Furane Plastics has proven to make a structurally sound joint. Roughen the surface lightly with emory cloth and load the joint with a light pressure. Allow to dry 24 hr at room temperature, then cure at 50°–60°C.

Fig. 5.23-1 Hemishell of 24 in. diameter.

5.24 STRESS FREEZING

The locking of stresses into models is based on the diphase behavior of polymeric materials. A practical analogy would be as discussed below.

Figure 5.24-1 is a sketch of the popular "Spring-Ice" analogy used in explaining stress freezing. This four-part diagram has:

1. A spring frozen in a can of ice.
2. The same spring axially loaded. Note the small spring deflection.
3. A spring in a can of water. The spring is displaced by a load P.
4. The can of water containing the spring displaced by load P is now frozen as a block of ice.

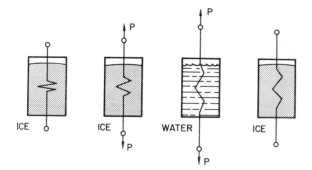

SPRING-ICE ANALOGY SHOWS HOW STRAINS AND OPTICAL

EFFECTS ARE FROZEN IN

Fig. 5.24-1 Spring Ice analogy for frozen-in strains and optical effects.

Now at this point if the block of ice is sawed perpendicular to the load P, the ice and spring will come out in sections, but the composite assembly of each spring and ice component will still keep the spring sections with the frozen displacement unchanged. Of course, when the ice sections melt, the displacements will disappear. A wet rope frozen under load will display the same effect when sliced.

In three-dimensional photoelastic models a similar phenomenon takes place. Above the 150°C critical temperature, the model material experiences a 100 times reduction in modulus of elasticity. When this occurs, the very soft model reacts by displacing as a function of applied loads. When the temperature is slowly reduced to ambient, all displacements and stresses are permanently locked into the model (similar to when the block of ice was sliced, but the spring was permanently displaced until the ice melts). This model may now be sliced and viewed in the polariscope for stress analysis. The only procedure to return the model to an unloaded state would be to repeat the stress freezing procedure with the model unloaded.

5.24-1 Testing Three-Dimensional Models

After assembly the three-dimensional photoelastic model is loaded into a fixture that will duplicate similar boundary loading to be expected in the prototype. A calibration specimen made from one of the billets used in fabricating the model should be located in the oven, loaded to a safe stress, and subjected to the same thermal-time history as the model. A stress freezing cycle may be accomplished by following the suggested temperature–time graph as shown in Fig. 5.24-2. Caution should be used in choosing temperature rates for both the ascending and descending temperature change rates. A thick-wall shell subjected to a rapid temperature change may develop thermal stresses that will not be "locked in" during the descending portion of the stress freezing cycle and will cause serious errors in the analysis. This could be readily detected when the loads or tractions are removed because the displacements would cause the model to move in the fixture. This would also be apparent in a pressure vessel during removal of heads or bandsawing across a principal plane when, as the final cut through the section is made, the saw kerf suddenly widens. Suggested rates of change are tabulated in the field of Fig. 5.24-1. These numbers are conservative, but when a valuable experiment is involved, may save a lot of grief.

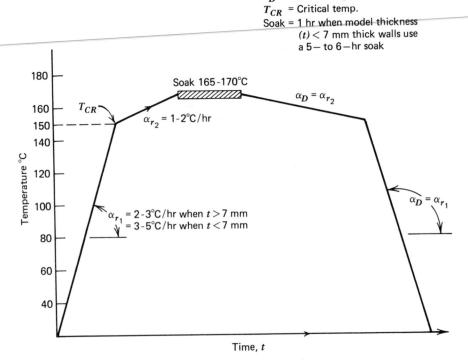

Fig. 5.24-2 Stress freezing cycle.

5.24-2   Slicing "Frozen Stress" Models

Upon completion of the stress freezing cycle, the designer or principal investigator will have the model and calibration specimen removed. It is very important to plan the locations on the model from which slices are to be removed. Such a plan is usually made during the model design so it may be sent to the machine shop and be immediately available for implementation.

Slices selected parallel to a principal stress plane, such as a plane of symmetry in a pressure vessel or perpendicular to a free boundary, provide a section of the model that may be observed in normal incidence. These same slices may be subsequently subsliced for viewing with normal incidence in the plane perpendicular to the original slice. Some investigators employ the oblique incidence technique instead of subslicing in order to evaluate interior stresses. Although this is a time saver, the possibility of error is enhanced and the saving in time is not justified.

5.25   THREE-DIMENSIONAL MODEL ANALYSIS

Figure 5.25-1 is a full-scale three-dimensional model of a thick-wall cylindrical pressure vessel used in equation of state studies by Lawrence Livermore National Laboratory physicists. The prototype operates at pressures to 1 GPa, and the main structural concern was the stresses around the four tapered sapphire window openings. The photoelastic model incorporated epoxy plugs in lieu of sapphire windows and the ends of the cylinder were modified for convenience of pressurizing. These modifications were distant enough from the windows so they did not affect the stress patterns. The epoxy windows were sealed along the tapered walls with vacuum grease. The slicing plan is indicated in Fig. 5.25-1 by the circled numbers.

From Fig. 5.25-2 the maximum hoop stresses at the inner wall may be determined. This dark-field photograph has concentric circles of fringes increasing to a maximum of 4.12 at inner walls. The slice is 6.43 mm thick and the material fringe value is 420 N/m fringe.

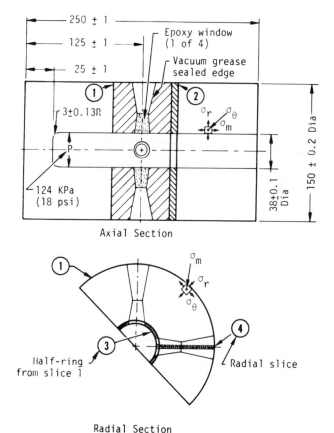

Axial Section

Radial Section

Fig. 5.25-1 Pressure vessel model for photoelastic analysis.

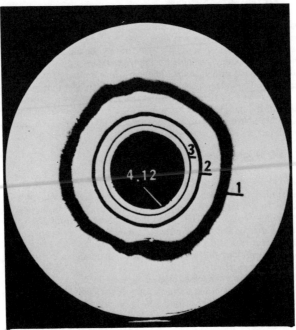

Fig. 5.25-2 Dark-field photograph of slice 2 from Fig. 5.25-1.

The principal stresses for the cylinder will be

$$\sigma_\theta = \text{hoop stresses}$$

$$\sigma_M = \text{meridional or axial stress}$$

$$\sigma_r = \text{radial stress}$$

(at the inner cylinder wall, σ_r = pressure and is denoted by a minus sign).
From Equation (5.15-4):

$$\sigma_\theta - (-\sigma_r) = \frac{Nf}{t}$$

$$\sigma_\theta = \frac{4.12 \times 420}{6.43 \times 10^{-3}} - 124 \times 10^3$$

$$\sigma_\theta = 145 \text{ kPa}$$

This is the nominal hoop stress in the cylinder and agrees reasonably well with the 141-kPa theoretical value.

To determine the maximum stress at the window opening on the inside wall, section 1 was removed and a surface (circular sector) slice 3 was machined from it. Figure 5.25-3 shows this surface slice after final machining to 2.74 mm.

At the inside window opening $\sigma_m = -P$, and the maximum fringe order is 2.78.

Equation (5.15-4) is again used to solve the maximum stress. In this case the model stresses are solved and plotted around the boundary of the hole in dimensionless form σ/P.

$$\frac{\sigma_\theta}{P} = \frac{Nf}{Pt} - \frac{\sigma_m}{P} = \frac{Nf}{Pt} - 1$$

$$\frac{\sigma_\theta}{P} = 2.4365$$

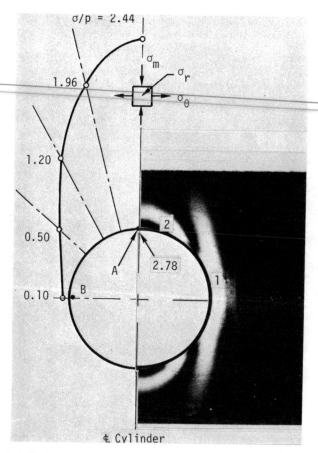

Fig. 5.25-3 Dark-field photograph of slice 3 showing stresses around the window hole at the inside wall.

This proved to be the highest stressed region of the pressure vessel with the maximum stress concentration at the top and bottom of the window. At this point the stress concentration factor is

$$K = \frac{(\sigma/P)_{\text{max}}}{(\sigma/P)_{\text{nom}}} = \frac{2.4365}{145/124} = 2.08$$

The dimensionless σ/P ratio solutions may be directly related to solving the prototype stresses in pressure vessel photoelastic models.

5.25-1 Rotating Disk

The next example is a model of a rotating slotted disk. This is a plane stress problem but employs the frozen stress technique again. This model was mounted and keyed to a motor shaft extending through the wall of the stress freezing oven. Figure 5.25-4 is a light-field photograph showing the one-half-order fringes after freezing. Disk dimensions are 140 mm outside diameter and the shaft hole is 25.4 mm. The slots are 12.7 mm long × 3.3 mm wide. The disk is 6.35 mm thick and the rotation speed was 2670 rpm.

Figure 5.25-5 shows the isoclinics and stress trajectories in the region of the slots. The letters P and Q in the field of the sketch denote the principal stresses.

5.26 MODEL TO PROTOTYPE RELATIONS

Photoelastic models are fabricated in either a two- or three-dimensional configuration based on the type of analysis required to evaluate the design parameters. Two-dimensional models are fabricated

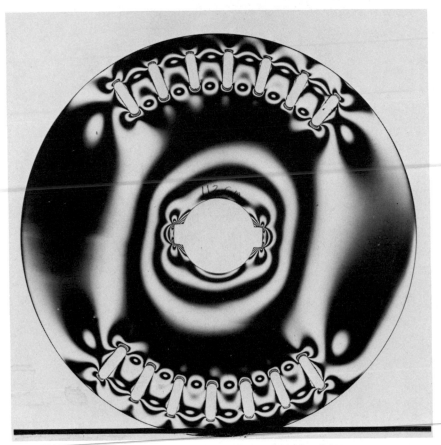

Fig. 5.25-4 Stress fringes in a slotted rotating disk.

from thin sheets of plastic material where the stresses across the plate thickness are considered constant and the model is deemed to be in a "plane stress" state. Confirmation of this plane stress state in two-dimensional models was made by Frocht and Leven.[1]

In the analysis of a photoelastic model the stresses must be related to a prototype design made from another material and possessing different elastic constants. However, the stress distribution obtained from plane stress or strain models is considered independent of elastic constants, and the photoelastic results may be applied to the prototype after applying the dimensional values to the similitude equations. This condition will hold true as long as Poisson's ratio is not effected by:

1. Body forces or mass of the model is very small and does not effect fringes.
2. Static testing only.

Similarity between the model and prototype will hold true in the case of two-dimensional models as long as they are simply connected. Multiply connected bodies containing holes with the resultant forces on the boundary reducing to zero (along a centerline of symmetry) will not be influenced by Poisson's ratio. Even if these forces around holes or discontinuities do not reduce to zero, the influence on the maximum principal stress will not exceed 7%.

5.26-1 Similitude Equations

The similarity between model and prototype is based on geometry, scale, and applied loads. For practical applications the use of dimensionless ratios between the model and prototype will provide the necessary equations for engineering analysis. For an elementary two-dimensional model, the stress at a point is proportional to the quantity F/A, which in terms of the model is F/tl. The stress in the

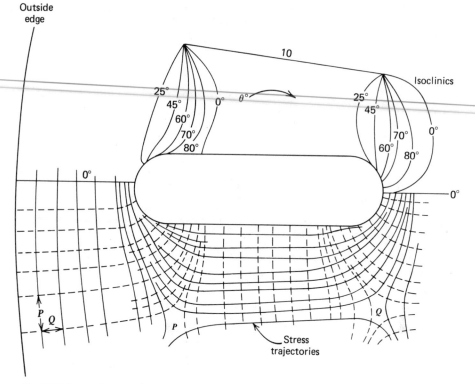

Fig. 5.25-5 Isoclinics and stress trajectories around a slot in the rotating disk of Fig. 5.25-4.

prototype or actual engineering part can be obtained from the relationship

$$\sigma_p = \sigma_m \times \frac{F_p}{F_m} \times \frac{t_m}{t_p} \times \frac{l_m}{l_p} \qquad (5.26\text{-}1)$$

where σ = the stress at a given point
F = applied load
t = thickness
l = a typical length

and the subscripts p and m refer to prototype and model, respectively.

The displacement in a two-dimensional model is proportional to the quantity F/E_t and the derived displacement equation is

$$d_p = d_m \times \frac{F_p}{F_m} \times \frac{t_m}{t_p} \times \frac{E_m}{E_p} \qquad (5.26\text{-}2)$$

where E is the Young's modulus. For a three-dimensional model geometrically similar to the prototype, the relationships are

$$\sigma_p = \sigma_m \times \frac{F_p}{F_m} \times \left(\frac{l_m}{l_p}\right)^2 \cdots \qquad (5.26\text{-}3)$$

and

$$d_p = d_m \times \frac{F_p}{F_m} \times \frac{l_m}{l_p} \times \frac{E_m}{E_p} \qquad (5.26\text{-}4)$$

Pressure vessel models lend themselves to very simplified analysis because the stress to internal pressure ratios of geometrically similar bodies are equal:

$$\frac{\sigma_p}{P_p} = \frac{\sigma_m}{P_m}$$

REFERENCES

5.26-1 M. M. Frocht and M. M. Leven, "On the State of Stress in Thick Bars," *J. Appl. Phys.*, 13, May 1942.

5.27 PHOTOELASTIC COATINGS

This is a special and very practical area of photoelasticity that has gained much popularity in the structural test field over the past 30 years. Photoelastic or birefringent coatings are flat or contoured sheets of epoxy formulations that are bonded to structures to be tested. When the specimen is loaded, the photoelastic coating exhibits the characteristic fringe pattern representative of the difference in principal strains on the surface of the part. This resulting fringe pattern presents the state of strain across the complete area covered by the coating, showing the critical areas by the magnitude of the fringe pattern. These higher fringe orders will usually show up around fillets, holes, and near loading points. If the structure yields during testing, the coating will show the residual strain pattern upon release of the load.

Figure 5.27-1 is a schematic illustrating the basic reflection polariscope employed in photoelastic coating evaluations. The camera is a removable item usually in position for documenting the fringe pattern after the part has been analyzed by the test engineer. Figure 5.27-2 is a commercially available reflecting polariscope.

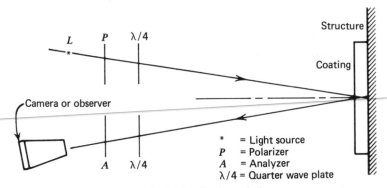

Fig. 5.27-1 Basic reflection polariscope.

Fig. 5.27-2 Commercial reflecting polariscope.

5.27-1 Selecting Photoelastic Coatings

The ability to develop an in-house epoxy formulation suitable for birefringent coating applications is cost and timewise impractical for general applications. Commercially available materials and instrumentation advertised in the technical journals for experimental mechanics and strain instrumentation provide both the guidance and selection methods for testing with birefringent coatings.
The following parameters will effect coating selection:

1. Desired sensitivity.
2. Severe contour effects.
3. Reinforcing effect.
4. Strain range.
5. Test environments.

All the above circumstances may not always be satisfied and prudent compromises will sometimes be necessary.

5.27-2 Coating Fringe Value and Calibration

Photoelastic coatings on loaded structures will exhibit the characteristic fringe patterns similar to stressed photoelastic models when observed through a reflecting polariscope. The difference of the principal stresses at the surface of the loaded structure is

$$\sigma_1^S - \sigma_2^S = \frac{Nf_\sigma}{2t} \frac{E^S(1 + \mu^C)}{E^C(1 + \mu^S)} \tag{5.27-1}$$

where σ_1, σ_2 = principal stresses
E = Young's modulus
μ = Poisson's ratio
t = coating thickness
f_σ = material fringe value of the coating N/m per fringe
N = the observed fringe order

The superscripts S and C refer to the structure and coating, respectively.
An alternative and very practical way to perform the experimental stress analysis is in terms of the principal strains. The equation in this case is

$$\varepsilon_1^C - \varepsilon_2^C = \frac{Nf_e}{2t} \tag{5.27-2}$$

where the superscript C refers to the coating, ε_1 and ε_2 are the principal strains and f_e is the coating fringe value in terms of strain per fringe.

5.27-3 Coating Calibration

The value of f_e may be determined from suitable calibration specimens such as those employed in two- or three-dimensional photoelastic models. An aluminum cantilever beam shown in Fig. 5.27-3 with a coating on the top surface is very convenient for determining f_e. At any position L measured from the loaded free end of the beam, the theoretical difference in the principal strains is

$$\varepsilon_1^S - \varepsilon_2^S = 6(1 + \mu_S)\frac{LF}{b_S t_S^2 E_S} \tag{5.27-3}$$

Now this represents the true strain in the bent cantilever beam at the top fibers but does not account for the coating thickness. The strains vary linearly through the coating thickness and a correction factor, C_2, for a plate in bending[1] will provide the correct strain difference on the surface of the structure (see Fig. 5.27-4):

$$\left(\varepsilon_1^S - \varepsilon_2^S\right) = C_2\left(\varepsilon_1^C - \varepsilon_2^C\right)_{avg} \tag{5.27-4}$$

This calculation corrects the observed strain in the coating to the true strain in the top fibers of the beam.

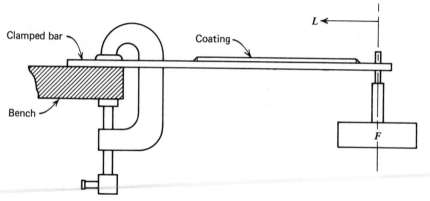

Fig. 5.27-3 Calibration bar for strain f_e.

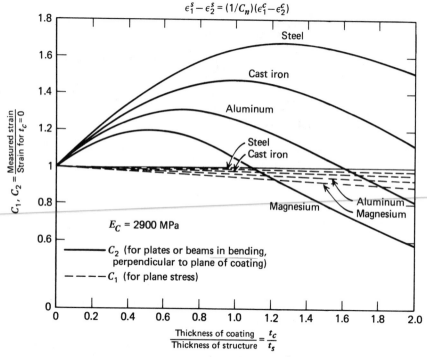

Fig. 5.27-4 Correction factors for bending (C_2) and plane stress (C_1).

Substituting Equations (5.27-3) and (5.27-4) into the strain optic Equation (5.27-2) gives the corrected material fringe value in terms of the beam properties loads and photoelastic coating properties:

$$f_e = \frac{12(1 + \mu_S)}{Nbt_S^2 E_S} LFt_C C_2 \tag{5.27-5}$$

5.27-4 Reinforcing Effects and Correction Factors

Birefringent coatings on structural members such as beams or pressure vessel walls add little to the load-carrying capacity of the part. In the case of thin beams or plastic low-modulus materials, the reinforcing effect must be corrected. For example, a 25 mm × 25 mm cross section of an aluminum tension link with a 3.18 mm thick epoxy coating will be used to show any reinforcing effect. The bar is loaded and the reinforcing effect is evaluated by determining the load ratio between the bar and coating. Assume the following:

Area of the aluminum bar = 1 (unity)
Area of the coating = 0.125 units
Elastic modulus: coating = 0.45 units
aluminum = 10 units

Establish the following ratio:

$$\frac{\text{Load in bar}}{\text{Load in coat}} = \frac{F_b}{F_c} = \frac{\sigma_b A_b}{\sigma_c A_c} = \frac{E_b \varepsilon_b A_b}{E_c \varepsilon_c A_c}$$

where $\varepsilon_b = \varepsilon_c$, then

$$\frac{F_b}{F_c} = \frac{E_b A_b}{E_c A_c} = \frac{10 \times 1}{0.45 \times 0.125} = 1422$$

which means the aluminum is carrying over 1400 times the load in the coating and reinforcing may be considered negligible.

In most cases corrections are going to be made for bending problems and for axial plane stress problems where the material thickness is low or the modulus is low. Figure 5.27-4 presents correction factors to cover an array of materials for specimens either axially loaded or in bending from Zandman.[1]

REFERENCES

5.27-1 F. Zandman, S. Redner, and J. W. Dally, *Photoelastic Coatings*, Iowa State, SESA, 1977, p. 88.

5.28 SEPARATION OF PRINCIPAL STRAINS OR STRESSES

The fundamental assumption in the practical use of photoelastic coatings is that the fringe pattern represents the difference of principal strains or stresses on the surface of the loaded structure. If there is a hole or edge representing a free boundary (where the coating terminates), the single principal strain may be determined with reasonable accuracy from the fringe pattern because there is only one principal stress at a free boundary. If the structure is uniaxially stressed where one principal stress is numerically zero, Equations (5.27-1) and (5.27-2) will provide the complete solution. In cases where it is known, one principal stress is small and may be considered negligible, Equation (5.27-1) may be used to perform the analysis. Sometimes it may be necessary to determine the values of the principal strains and stresses. There are several methods to accomplish this, and for photoelastic coatings the oblique incidence method is probably the simplest because the commercially available reflecting polariscope usually has the necessary accessories.

5.28-1 Oblique Incidence Method

Drucker[1] derived this method for examining two-dimensional photoelastic models or slices from three-dimensional models. Zandman, Redner, and Dally[2] give the derivation of the equations for employing the oblique incidence method to separating the principal strains in photoelastic coatings. The procedure for solving the principal strain values is as follows:

1. Determine the direction of principal strains at the point to be investigated. This is determined by positioning the isoclinic over that point.
2. Take a normal incidence reading of the fringe value N_0.
3. Take an oblique incidence reading (Fig. 5.28-1) N_θ. It is advisable to establish a three-dimensional coordinate system such as shown in Fig. 5.28-1
4. The following equations will provide the numerical values of the principal strains:

$$\varepsilon_x = \frac{f_e}{2t_C} \frac{1}{(1 + \mu_C)\sin^2\theta} \left[N_\theta(1 - \mu_C)\cos\theta - N_0(\cos^2\theta - \mu_C) \right] \qquad (5.28\text{-}1)$$

$$\varepsilon_y = \frac{f_e}{2t_C} \frac{1}{(1 + \mu_C)\sin^2\theta} \left[N_\theta(1 - \mu_C)\cos\theta - N_0(1 - \mu_C\cos^2\theta) \right] \qquad (5.28\text{-}2)$$

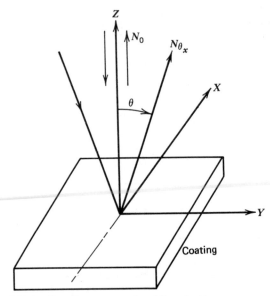

Fig. 5.28-1   Coordinate system for oblique incidence measurements.

The derivation of these equations assumes the fringe orders N_0 and N_θ are positive when $\varepsilon_x - \varepsilon_y$ are positive. Prior to performing the strain solutions, determine the sign (tension or compression) of the fringe value by employing a compensator.

   After solving the principal strains, these values are substituted into Equations (5.9-2) and (5.9-3) for solving the principal stresses.

REFERENCES

5.28-1   D. C. Drucker, "Photoelastic Separation of Principal Stresses by Oblique Incidence," *J. Appl. Mech. Trans. ASME* **10**(3), pp. A156–160, Sept. 1943.

5.28-2   F. Zandman, S. Redner, and J. W. Dally, *Photoelastic Coatings*, Iowa State, SESA, 1977, p. 88.

5.29   BRITTLE COATINGS—INTRODUCTION

The brittle coating method of experimental stress analysis has been in use since DeForest, Ellis, and Stern[1] made it practical for experimenters in 1941. Prior to this, the technique was tried on structural members where the mill scale formed after hot rolling and oxides formed on heated surfaces. When these members were assembled into structures and loaded, if excessive strains were produced, it would cause the scale to flake off when the material yielded. Thus the designer had an approximate idea of the integrity of his structure.

   In 1941 Greer Ellis, working with the Magnaflux Corporation, perfected a resin-based material that was marketed under the name Stresscoat. This brittle coating was expressly developed to simplify and reduce the time and cost of experimental stress analysis and also to improve product designs. It was also an improvement over the mill scale technique.

   The original Stresscoat formulation was a simple combination of zinc resinate base, carbon disulfide solvent, and dibutyl phthalate for a plasticizer. The plasticizer permitted the performance of the brittle coating to be varied in strain sensitivity to suit threshold strain requirements.

   Surfaces with biaxial stress are difficult to analyze. By weighing the advantages and disadvantages of brittle lacquers for experimental stress analysis, the test engineer may elect a qualitative or quantitative evaluation. Even the qualitative approach will provide the necessary positions of high stresses and isoclinics that are an important aid for accurate orientation of strain gauges. Although the following sections describe methods for complete quantitative analysis, the practical way to employ Stresscoat is to observe the crack patterns for high-stress points and stress directions and use the bonded strain gauges for obtaining accurate strain measurements for engineering analysis.

   This variation permits the experimenter to select a strain sensitivity of the coating to match the lowest strain value on the structure in which he or she is interested.

5.29-1 Advantages of Brittle Coating Method

1. Like the photoelastic coatings, it provides a picture of the principal strain directions and approximate strain values.
2. It may be employed in a static or dynamic condition where the prototype structure is under actual service loads.
3. The prototype structure may be used and this avoids expensive models and similitude equations.
4. The data are gathered by visual means and converted to stress without a rigorous mathematical analysis.

5.29-2 Disadvantages of Brittle Coating Method

1. Threshold strains are time, temperature, and humidity dependent.
2. The position of threshold strains must be rapidly determined, requiring a trained eye.
3. Coating thickness and biaxiality may effect results.

5.29-3 Specimen and Structure Preparation

The brittle lacquers may be used on all types of metals, glass, wood, and plastics. After careful cleaning of the surface to be coated with a suitable OSHA-approved solvent, it should be dried and then given a coating of quick-drying aluminum lacquer. This is called an undercoat, and the reflective qualities are helpful in crack detection. This uniformity in the undercoating color makes it easier to judge the thickness of the brittle lacquer during spraying.

5.29-4 Strain-Indicating Coating

Throughout the industrial nations of the world, brittle lacquers are available for applications to structures. Within the United States several companies market this material, and their staffs provide excellent guidance for practical applications. These strain-indicating coatings are available in the standard formulations for ambient temperature analysis in aerosol cans, and the more recent formulations function independently of humidity. For high-temperature testing ceramic-base brittle coatings may be employed. It provides the means to determine approximate values of high-temperature operating stresses where temperatures may run from 700 to 1000°F. The ceramic brittle lacquer will operate successfully in a water or oil environment.

5.29-5 Calibrating the Brittle Coating

Brittle lacquer coatings should be sprayed on the test structure and on suitable aluminum bars of approximately 25 mm width, 250 mm length, and 6.25 mm thickness during the same spraying operation. These bars may then be clamped to a table and loaded in a cantilever mode to cause cracks

Fig. 5.29-1 Cam loading fixture and calibration scale for determining Stresscoat threshold sensitivity.

to appear along the top tensile-stressed fibers. With all the loading geometry known, the flexure equation for a cantilever beam will provide the solution to plot the strain versus distance from the load point along the beam. For the convenience of conducting these types of tests, it is advisable to purchase a complete set of apparatus and materials from a local supplier. This apparatus will include a fixture to load and locate the sealed threshold crack values along the top fibers as they appear on the calibration bar. This will be the strain value that causes the first cracks to appear during loading.

Figure 5.29-1 shows the standard cam loading fixture for calibration beams.

5.29-6   Determining Threshold Strain

Tensile Strain

1. Scrape off last inch of coating from each end of the bar.
2. Apply load in 1 second and immediately locate the crack closest to the loading end. Easiest way to locate cracks is to shine a flashlight on the surface normal to the crack direction and at a 30°–60° angle.
3. For a cantilever beam mounted on a bench, determine the distance from the load point to the crack and determine the theoretical strain. This is the threshold strain value for this coating. If the bar has been stressed in a commercial cam-loaded fixture, remove the bar and place it in the strain scale that comes with the kit and read off the strain sensitivity from the crack that is parallel to the lowest strain sensitivity on the numbered scale along the side.

Compression Strain

It should be evident to the reader by now that brittle lacquers only sense tensile strains on the components they are covering. However, if some of the procedures for determining the threshold tensile strain are reversed, then the compressive strains may be evaluated by the following steps:

1. Spray calibration bar and load it in the cantilever beam mode or calibration fixture with the coating down.
2. Allow it to dry the prescribed time.
3. Release the load and determine the strain threshold value using the same procedure for tensile bars.

5.29-7   Creep Correction

Brittle lacquer coatings under load will creep as a function of time. This will relieve the stress in the coating and if the load is maintained for a long period, it will have considerable relaxation. The coating manufacturers provide creep correction charts that should be requested by the investigator when the material is purchased (see Fig. 5.29-2).

A very positive way for the experimenter to maintain an awareness of the changing threshold strain value due to creep is to subject the calibration bar and specimen to simultaneous time and load histories. The creep effect is similar for both parts and the calibration bar will always indicate the value of the threshold strain.

5.29-8   Measuring Strain Under Static Loading

The brittle coating cracks that occur when the specimen is loaded will first occur at the position of maximum tensile strain. As each increment of load is applied, and the extent of the cracks is detected, if a marking pen is used to connect the ends of all cracks, this will give a record of strain versus load across the surface. As these patterns form and are marked, the threshold strain value from the calibration bar should be noted with each load. The calibration bar will always be loaded and unloaded in the same time frame as the specimen.

If the investigator is interested in a complete stress analysis on a structure where the surface is in a state of biaxial stress, then both principal strains must be determined. This may be accomplished by holding the final load for 2 hours after recording the tensile strains. The coating should now be in a relaxed unstrained state. Release the load in similar increments as before and the compression strains as they relax will cause the coating to crack perpendicular to them, thus providing the second set of principal strain values. The relaxation chart (Fig. 5.29-3) illustrates how the strain threshold is affected by hold time and release time on the part.

The routines that have been described here are possible with skilled technicians and experimenters who are aware of the limitations of brittle coatings.

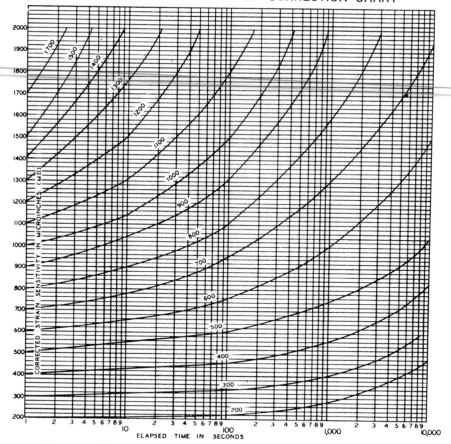

Fig. 5.29-2 Stresscoat creep chart. (Courtesy Magnaflux Corp.—Stresscoat Inc.)

5.29-9 Dynamic Strain Measuring

When the need arises to test machinery in motion, the brittle coatings may be employed for a qualitative evaluation of maximum stress locations and stress trajectories. These tests may require the use of coatings of varying strain thresholds ranging from 600 to 1200 microstrain in order to cover the complete strain profiles from the high to lower stressed areas while operating at a single speed.

A final note on dynamic testing is to use brittle lacquers in a qualitative mode and proceed with the experimental stress analysis using strain gauges or photoelastic coatings.

5.29-10 Stresscoat Examples

Figure 5.29-4 is a plane stress model of a double cantilever beam fracture mechanics specimen. The material is a piece of 6061-T6 aluminum 3.175 mm (0.125 in.) thick. After loading, the crack pattern was enhanced for viewing by spraying the surface with stati-flux powder. A complete picture of the principal stress directions in the region of the notch and across the specimen is clearly visible.

Figure 5.29-5 shows the strain pattern in a tupp loaded statically to determine strain directions for locating strain gauges.

REFERENCE

5.29-1 A. V. DeForest, G. Ellis, and F. B. Stern, "Brittle Coating for Quantitative Strain Measurements," *J. Appl. Mech.*, **9**(4), pp. A184–188, 1942.

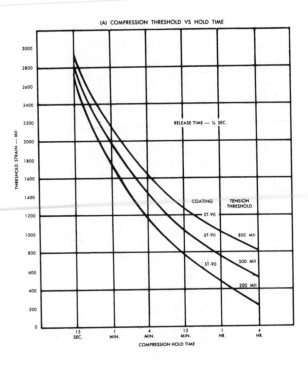

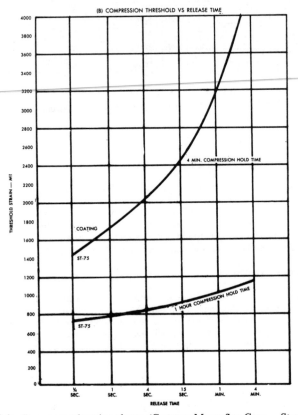

Fig. 5.29-3 Stresscoat relaxation charts. (Courtesy Magnaflux Corp.—Stresscoat Inc.)

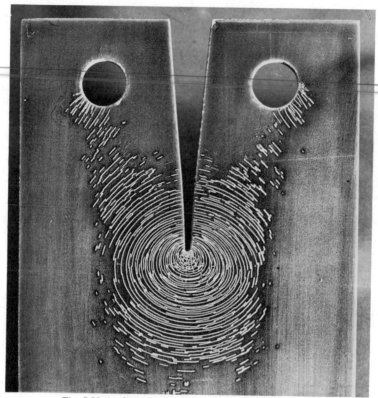

Fig. 5.29-4 Crack pattern in a double cantilever beam.

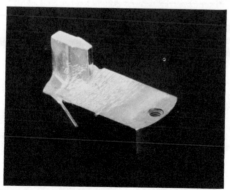

Fig. 5.29-5 Crack pattern in a tupp used in impact testing.

5.30 DISPLACEMENT GAUGES AND EXTENSOMETERS

Displacement measuring instruments employing electric signals to indicate static or dynamic motion have been commercially available for many years. They are used fundamentally for determining small motions on structures or specimens that may be of interest for engineering analysis.

5.30-1 Linear Position Transducers

Normally, these devices are designed for measuring displacements encountered in mechanical and thermal physical tests of U.S. military equipment and for applications to laboratory-type tests. A few examples would be:

1. Displacements during static structural tests.
2. Displacements of specimens subjected to low-frequency vibration tests.
3. Displacement of hydraulic rams in universal test machines.
4. Cross-head motion in screw-driven universal test machines.
5. Any displacement where it is convenient to attach a small lightweight transducer.

Figure 5.30-1 shows a photograph and specifications of a potentiometric linear position transducer. This is an electromechanical device that converts straight-line displacement into an equivalent

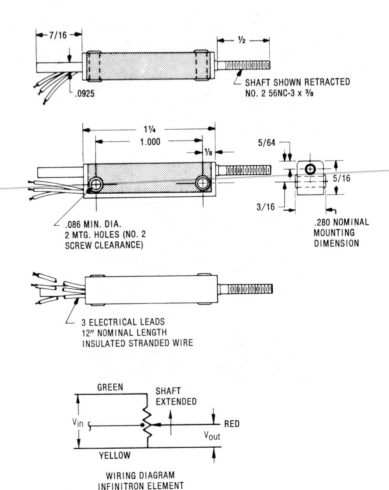

Fig. 5.30-1 Linear displacement transducer. (Courtesy of Bourns Instruments, Inc., Riverside, California.)

electrical output signal. A precision contact attached to the movable shaft wipes a resistive element oriented in a plane parallel to the linear motion of the shaft. Three electrical terminals—two are connected to the ends of the resistive element and one is connected to the end of the movable contact —allow an input voltage to be divided as a function of the shaft position. The transducer case will usually have either tapped or through holes for mounting. The movable shaft may be threaded or smooth depending on the model selected.

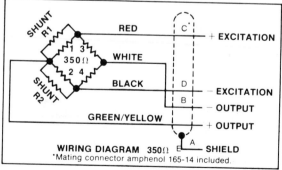

The excitation required is 12 volts DC maximum and 8 volts DC nominal. Gage factors are approximately 3mv/volt.

Fig. 5.30-2 Extensometer for axial testing. (Courtesy of MTS Systems Corporation.)

5.30-2   Axial Extensometers

Extensometers or clip-on gauges are designed to measure strains on mechanical test specimens. These specimens are usually tested in a universal structural test machine and the thermal environment may impose severe requirements on the extensometer.

The basic geometry of the extensometer is shown in Fig. 5.30-2. Two arms attached to the specimen by either springs or elastic bands determine the gauge length by the spacing between them. During testing the specimen will either elongate or contract depending on the type of axial load. Any displacement in the specimen is transmitted through the arms and strains a beam gauged in a four-arm Wheatstone bridge configuration.

A multiaxial clip-on gauge is shown in Fig. 5.30-3. This dual transducer will simultaneously measure both the axial and transverse strain in a uniaxial test specimen and provides the means to generate stress–strain curves and Poisson's ratio simultaneously.

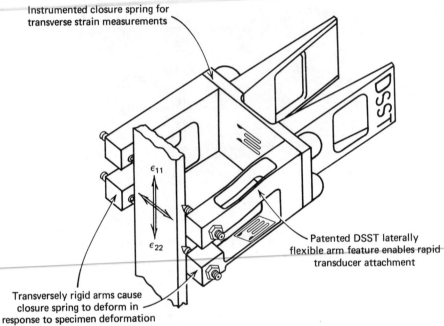

Instrumented closure spring for transverse strain measurements

ϵ_{11}

ϵ_{22}

Patented DSST laterally flexible arm feature enables rapid transducer attachment

Transversely rigid arms cause closure spring to deform in response to specimen deformation

Fig. 5.30-3 Biaxial clip-on extensometer designed to measure axial and transverse strains on mechanical test specimens. (Courtesy of Measurements Technology, Inc.)

CHAPTER 6

INTRODUCTION TO ELASTICITY, PLASTICITY, AND ELASTIC STABILITY

JOE W. McKINLEY

President, Joe W. McKinley, Inc.
Laguna Beach, California

6.1 THEORY OF ELASTICITY

6.1-1 Equilibrium in a Two-Dimensional Stress System

Many real-life stress situations are biaxial. When this occurs, it is very important to see how the stresses combine to affect the structure. Any structure or portion of structure in a general load situation such as shown in Fig. 6.1-1 will experience σ_x, σ_y, and τ_{xy} as functions of x and y which vary in a (usually) continuous fashion throughout the region.

Some examples of regions where these kinds of stress situations might occur are the surface of a pressure vessel, the skin of an aircraft or missile, and the floor of a building. The recognition of the biaxial stress situation is important. Serious errors will result if the uniaxial stress assumptions are used when the stress distribution is biaxial. The biaxial stress problem has been addressed by exact mathematical treatment and by numerical methods such as numerical relaxation.[1-5] The general problem is also widely addressed by the finite-element method.[6-9]

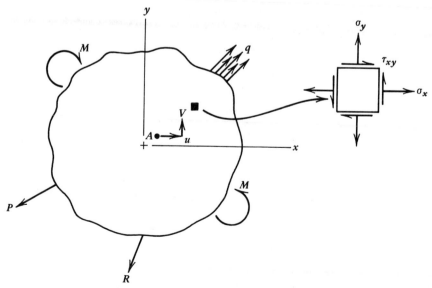

Fig. 6.1-1 Continuous stress variations in a given region.

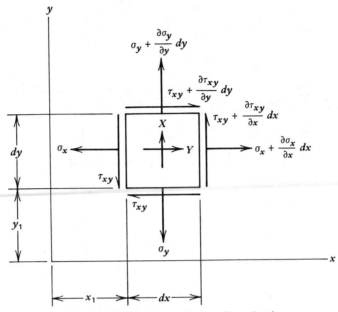

Fig. 6.1-2   Equilibrium element for two-dimensional stress.

The problem of the variation of stresses at one point in a structure where the reference axis is rotated was discussed in Chapter 3. The solution of this problem yields the *principal stresses*. In this chapter we discuss the problem finding the stresses throughout the structure for a given geometry and loading. Here the stresses can be defined as follows:

$$\sigma_x = f_1(x, y)$$

$$\sigma_y = f_2(x, y)$$

$$\tau_{xy} = f_3(x, y)$$

We shall restrict ourselves to *plane stress*. By this we understand that the stresses such as those in Fig. 6.1-1 vary with x and y, but that all z direction stresses are zero.

A typical two-dimensional stress element is shown in Fig. 6.1-2. Forces X and Y are body forces (gravity, electrical, or inertial fields, etc.).

In the x direction (consider the z direction thickness to be unity)

$$-\sigma_x \, dy + \left(\sigma_x + \frac{\partial \sigma_x}{\partial x} dx\right) dy + \left(\tau_{xy} + \frac{\partial \tau_{xy}}{\partial y} dy\right) dx - \tau_{xy} \, dx + X \, dx \, dy = 0$$

$$\frac{\partial \sigma_x}{\partial x} + \frac{\partial \tau_{xy}}{\partial y} + X = 0 \qquad (6.1\text{-}1)$$

In the y direction

$$\frac{\partial \sigma_y}{\partial y} + \frac{\partial \tau_{xy}}{\partial x} + Y = 0 \qquad (6.1\text{-}2)$$

These are the two-dimensional, plane-stress *equations of equilibrium*.

6.1-2   Polar Coordinates and Axial Symmetry

Many problems are more naturally expressed in polar coordinates. References 1–5 treat this problem mathematically and numerically. Consider the element of Fig. 6.1-3. If R and T denote the various body forces per unit volume, and the element is subjected to the radial and tangential stresses as

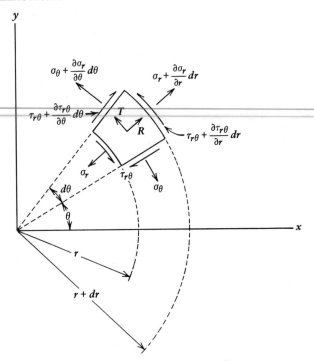

Fig. 6.1-3   Elasticity in polar coordinates.

shown, equilibrium in the radial and tangential directions yields

$$\frac{\partial \sigma_r}{\partial r} + \frac{1}{r}\frac{\partial \tau_{r\theta}}{\partial \theta} + \frac{\sigma_r - \sigma_\theta}{r} + R = 0$$

$$\frac{1}{r}\frac{\partial \sigma_\theta}{\partial \theta} + \frac{\partial \tau_{\rho\theta}}{\partial r} + \frac{2\tau_{r\theta}}{r} + T = 0 \qquad (6.1\text{-}3)$$

These are the *equations of equilibrium in polar coordinates.*
In an axially symmetric problem, the equilibrium Equations (6.1-3) reduce to

$$\frac{d\sigma_r}{dr} + \frac{1}{r}(\sigma_r - \sigma_\theta) + R = 0 \qquad T = 0 \qquad (6.1\text{-}4)$$

Both the structure and the loading must be axially symmetrical.
An example of a practical solution to an axially symmetric problem is that of the thick-walled cylinder under internal pressure, p. See Fig. 6.1-4. The solutions are

$$\sigma_r = -\frac{a^2 b^2 p}{(b^2 - a^2)r^2} + \frac{a^2 p}{b^2 - a^2}$$

$$\sigma_\theta = \frac{a^2 b^2 p}{(b^2 - a^2)r^2} + \frac{a^2 p}{b^2 - a^2} \qquad (6.1\text{-}5)$$

Another example of a polar coordinate problem is that of contact stress shown in Fig. 6.1-5. The solution is

$$\sigma_r = -\frac{2w}{\pi r}\sin\theta$$

$$\sigma_\theta = 0$$

$$\tau_{r\theta} = 0 \qquad (6.1\text{-}6)$$

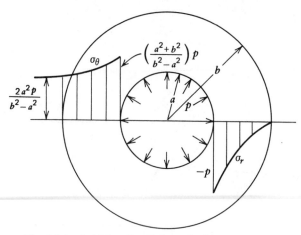

Fig. 6.1-4 A thick-wall cylinder under internal pressure.

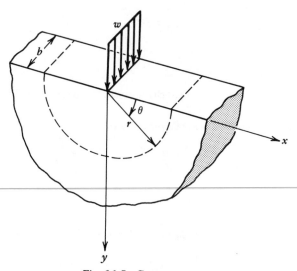

Fig. 6.1-5 Contact stresses.

See Reference 6. See also Section 6.4.

Additional applications in polar coordinates will be given after we discuss the equations of compatibility and the stress function.

6.1-3 Equations of Compatibility

Consider the displacements of particles in a two-dimensional stress field such as that illustrated in Fig. 6.1-6. Particle A will actually move when the structure is loaded. We may think of its motion, or displacement, as being composed of "horizontal" displacement u and "vertical" displacement v. If ABC and D form the corners of the element shown in Fig. 6.1-6, then the deformation of the element is able to be calculated from linear geometry. Strains are shown to be

$$\varepsilon_x = \frac{u + (\partial u/\partial x)\, dx - u}{dx}$$

$$\varepsilon_x = \frac{\partial u}{\partial x} \tag{6.17}$$

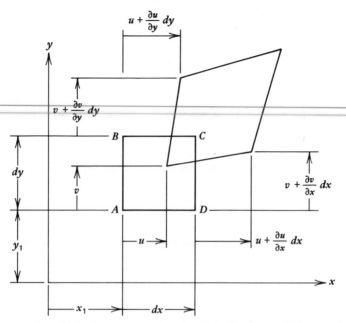

Fig. 6.1-6   Displacements in a two-dimensional stress field.

Similarly

$$\varepsilon_y = \frac{\partial v}{\partial y} \tag{6.1-8}$$

The change in angle from original right angle is

$$\gamma_{xy} = \frac{u + (\partial u/\partial y)\,dy - u}{dy} + \frac{v + (\partial v/\partial x)\,dx - v}{dx}$$

$$\gamma_{xy} = \frac{\partial u}{\partial y} + \frac{\partial v}{\partial x} \tag{6.1-9}$$

Equations (6.1-7), (6.1-8), and (6.1-9) are called the strain–displacement equations. In this case three strain components are expressed in terms of two displacement components. In order that the strains are ultimately related to the displacements in such a way as to form a *continuously* deformed structure, an additional relationship is needed. It is obtained in the following way.

Differentiate Equation (6.1-9) with respect to x and y

$$\frac{\partial^2 \gamma_{xy}}{\partial x\,\partial y} = \frac{\partial^2}{\partial x\,\partial y}\frac{\partial u}{\partial y} + \frac{\partial^2}{\partial x\,\partial y}\frac{\partial v}{\partial x} \tag{6.1-10}$$

u and v are continuous functions of x and y. The order of differentiation is interchangeable and Equation (6.1-10) can be written

$$\frac{\partial^2 \gamma_{xy}}{\partial x\,\partial y} = \frac{\partial^2}{\partial y^2}\frac{\partial u}{\partial x} + \frac{\partial^2}{\partial x^2}\frac{\partial v}{\partial y} \tag{6.1-11}$$

or

$$\frac{\partial^2 \varepsilon_x}{\partial y^2} + \frac{\partial^2 \varepsilon_y}{\partial x^2} = \frac{\partial^2 \gamma_{xy}}{\partial x\,\partial y} \tag{6.1-12}$$

Equation (6.1-12) is called the equation of compatibility. The two-dimensional stress–strain relationships for plane stress may be written as

$$\varepsilon_x = \frac{1}{E}\left(\sigma_x - \nu\sigma_y\right)$$

$$\varepsilon_y = \frac{1}{E}\left(\sigma_y - \nu\sigma_x\right)$$

$$\gamma_{xy} = \frac{2(1 + \nu)}{E}\tau_{xy}$$

To tie it all together, the concept of a stress function is introduced such that

$$\sigma_x = \frac{\partial^2\psi}{\partial y^2} \tag{6.1-13}$$

$$\sigma_y = \frac{\partial^2\psi}{\partial x^2} \tag{6.1-14}$$

$$\tau_{xy} = \frac{-\partial^2\psi}{\partial x\,\partial y} \quad \text{(if body forces are omitted)} \tag{6.1-15}$$

Mathematical combination gives

$$\frac{\partial^4\psi}{\partial x^4} + 2\frac{\partial^4\psi}{\partial x^2\partial y^2} + \frac{\partial^4\psi}{\partial y^4} = 0 \tag{6.1-16}$$

The stress function ψ, which satisfies Equation (6.1-16) will satisfy equilibrium, compatibility and stress–strain. Equation (6.1-16) is called the biharmonic equation.

6.1-4 Application to a Simple Beam

As an illustration consider the beam of Fig. 6.1-7. This beam has been treated in Chapter 3 by elementary beam theory. Now we shall use the results of the two-dimensional theory of elasticity.

The knowledge we have of beam theory will help to select a stress function ψ. The bending moment, and therefore the stress, increases linearly with x. At any station x, the stress is a linear function of y.

Therefore, it is reasonable to assume

$$\frac{\partial^2\psi}{\partial y^2} = \sigma_x = C_1 xy \tag{6.1-17}$$

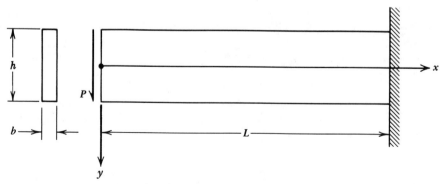

Fig. 6.1-7 Application of elasticity to a beam.

We integrate Equation (6.1-17) twice. Thus

$$\psi = \frac{C_1}{6} xy^3 + yf_1(x) + f_2(x) \tag{6.1-18}$$

Then

$$\nabla^4 \psi = 0$$

yields

$$\frac{y \, df_1^4(x)}{dx^4} + \frac{d^4 f_2(x)}{dx^4} = 0 \tag{6.1-19}$$

The only way Equation (6.1-19) can be satisfied for all x and y values is

$$\frac{df_1^4(x)}{dx^4} = 0 \tag{6.1-20}$$

and

$$\frac{d^4 f_2(x)}{dx^4} = 0 \tag{6.1-21}$$

These equations can be solved by conventional methods for $f_1(x)$ and $f_2(x)$.

$$f_1(x) = c_2 x^3 + c_3 x^2 + c_4 x + c_5 \tag{6.1-22}$$

and

$$f_2(x) = c_6 x^3 + c_7 x^2 + c_8 x + c_9 \tag{6.1-23}$$

Then ψ becomes

$$\psi = \frac{C_1}{6} xy^3 + y\left(c_2 x^3 + c_3 x^2 + c_4 x + c_5 \right) + c_6 x^3 + c_7 x^2 + c_8 x + c_9 \tag{6.1-24}$$

Ignore the weight of the beam and assume the body forces are zero.

$$\sigma_y = \frac{\partial^2 \psi}{\partial x^2} = 6c_2 yx + 6c_6 x + 2c_3 y + 2c_7 \tag{6.1-25}$$

$$\tau_{xy} = \frac{-\partial^2 \psi}{\partial x \, \partial y} = \frac{-c_1}{2} y^2 - 3c_2 x^2 - 2c_3 x - c_4 \tag{6.1-26}$$

Apply the known *boundary conditions*, at $y = \pm h/2$, $\sigma_y = 0$, because these surfaces are not loaded.
Therefore,

$$6\left(c_2 \frac{h}{2} + c_6 \right) x + 2\left(c_3 \frac{h}{2} + c_7 \right) = 0$$

$$6\left(-c_2 \frac{h}{2} + c_6 \right) x + 2\left(-c_3 \frac{h}{2} + c_7 \right) = 0 \tag{6.1-27}$$

In order that Equations (6.1-27) be zero for all values of x

$$c_2 \frac{h}{2} + c_6 = 0 \qquad c_3 \frac{h}{2} + c_7 = 0$$

$$-c_2 \frac{h}{2} + c_6 = 0 \qquad -c_3 \frac{h}{2} + c_7 = 0$$

From these equations

$$c_2 = c_3 = c_6 = c_7 = 0$$

We now have, from Equation (6.1-26)

$$\tau_{xy} = \frac{-c_1 y^2}{2} - c_4 \tag{6.1-28}$$

The surfaces are free of shear stress for all $y = \pm h/2$, thus,

$$\frac{-c_1}{8} h^2 - c_4 = 0$$

or

$$c_4 = \frac{-c_1 h^2}{8}$$

The integration of the shear stress times area over the end of the beam must add up to P.

$$-\int_{-h/2}^{h/2} \tau_{xy} b \, dy = \int_{-h/2}^{h/2} \frac{c_1}{8} b \left(4y^2 - h^2\right) dy = P \tag{6.1-29}$$

or

$$c_1 = \frac{-12 P}{h^3 b} \tag{6.1-30}$$

Since $I = bh^3/12$, we finally obtain

$$\sigma_x = \frac{-Pxy}{I}$$

$$\sigma_y = 0$$

$$\tau_{xy} = \frac{-P}{2I}\left(\frac{h^2}{4} - y^2\right) \tag{6.1-31}$$

Equations (6.1-31) correspond exactly with elementary beam theory. To obtain the displacements, use the stress–strain relations, also

$$\varepsilon_x = \frac{\partial u}{\partial x} = \frac{\sigma_x}{E} = \frac{-Pxy}{EI}$$

$$\varepsilon_y = \frac{\partial v}{\partial y} = \frac{-\nu \sigma_x}{EI} = \frac{\nu Pxy}{EI}$$

$$\gamma_{xy} = \frac{\partial u}{\partial y} + \frac{\partial v}{\partial x} = \frac{2(1+\nu)}{E} \tau_{xy} = \frac{-(1+\nu)P}{EI}\left(\frac{h^2}{4} - y^2\right)$$

Integrating:

$$u = \frac{-P}{2EI} x^2 y + g_1(y)$$

Integrating:

$$v = \frac{\nu P}{2EI} xy^2 + g_2(x)$$

and then note that

$$\frac{dg_1}{dy} - \frac{P}{EI}\left(1 + \frac{\nu}{2}\right)y^2 = \frac{-dg_2}{dx} + \frac{P}{2EI}x^2 - \frac{(1+\nu)P}{4EI}h^2 \tag{6.1-32}$$

Setting the left and right sides of Equation (6.1-32) equal to a_1 and integrating, we have

$$g_1(y) = \frac{P}{3EI}\left(1 + \frac{\nu}{2}\right)y^3 + a_1 y + a_2$$

$$g_2(x) = \frac{P}{6EI}x^3 - \frac{(1+\nu)P}{4EI}h^2 x - a_1 x + a_3 \qquad (6.1\text{-}33)$$

substituting

$$u = \frac{-P}{2EI}x^2 y + \frac{P}{3EI}\left(1 + \frac{\nu}{2}\right)y^3 + a_1 y + a_2$$

$$v = \frac{\nu P}{2EI}xy^2 + \frac{P}{6EI}x^3 - \frac{(1+\nu)P}{4EI}h^2 x - a_1 x + a_3$$

The boundary conditions can be written

$$u, v\big|_{\substack{x=L \\ y=0}} = 0$$

$$\frac{\partial v}{\partial x}\bigg|_{\substack{x=L \\ y=0}} = 0$$

Substituting the boundary conditions results in

$$u = \frac{-P}{2EI}x^2 y + \frac{P}{3EI}\left(1 + \frac{\nu}{2}\right)y^3 + \frac{P}{2EI}\left[L^2 - (1+\nu)\frac{h^2}{2}\right]y$$

$$v = \frac{\nu P}{2EI}xy^2 + \frac{P}{6EI}x^3 - \frac{PL^2}{2EI}x + \frac{PL^3}{3EI}$$

The equation of the elastic axis is v evaluated at $y = 0$.

$$v = \frac{Px^3}{6EI} - \frac{PL^2 x}{2EI} + \frac{PL^3}{3EI} \qquad (6.1\text{-}34)$$

Equation (6.1-34) checks with elementary beam theory.

6.1-5 Application to a Plate

In polar coordinates the stress function ψ is defined so as to satisfy the relations

$$\sigma_r = \frac{1}{r}\frac{\partial \psi}{\partial r} + \frac{1}{r^2}\frac{\partial^2 \psi}{\partial \theta^2}$$

$$\sigma_\theta = \frac{\partial^2 \psi}{\partial r^2}$$

$$\tau_{r\theta} = \frac{1}{r^2}\frac{\partial \psi}{\partial \theta} - \frac{1}{r}\frac{\partial^2 \psi}{\partial r \partial \theta} = -\frac{\partial}{\partial r}\left(\frac{1}{r}\frac{\partial \psi}{\partial \theta}\right) \qquad (6.1\text{-}35)$$

As an example, consider a flat plate[5] subjected to uniform loading as shown in Fig. 6.1-8. A small circular hole causes significantly higher stresses in the region around the hole. This is the phenomenon of *stress concentration*, which is discussed in Section 3.8.

In the absence of any hole, we have obviously

$$\sigma_x = S \qquad \sigma_y = \tau_{xy} = 0$$

which may be derived from the stress function

$$\psi_1 = \tfrac{1}{2}Sy^2$$

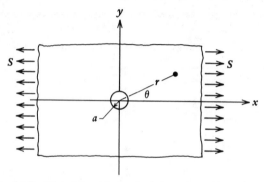

Fig. 6.1-8 Flat plate with circular hole under uniform loading.

Note that ψ satisfies the biharmonic equation and is therefore the exact solution. In terms of cylindrical coordinates, since $y = r\sin\theta$, we have

$$\psi_1 = \tfrac{1}{2}Sr^2\sin^2\theta = \tfrac{1}{4}Sr^2(1 - \cos 2\theta)$$

from which it follows

$$\sigma_{r_1} = \frac{1}{r}\frac{\partial\psi_1}{\partial r} + \frac{1}{r^2}\frac{\partial^2\psi_1}{\partial\theta^2} = \frac{1}{2}S(1 + \cos 2\theta)$$

$$\sigma_{\theta_1} = \frac{\partial^2\psi_1}{\partial r^2} = \frac{1}{2}S(1 - \cos 2\theta)$$

$$\tau_{r\theta_1} = -\frac{\partial}{\partial r}\left(\frac{1}{r}\frac{\partial\psi_1}{\partial\theta}\right) = -\frac{1}{2}S\sin 2\theta$$

When a hole of radius a is drilled through the plate, the boundary conditions become

$$\sigma_r = \tau_{r\theta} = 0 \qquad \text{at } r = a$$

and

$$\sigma_r = \sigma_{r_1}, \ \sigma_\theta = \sigma_{\theta_1}, \ \tau_{r\theta} = \tau_{r\theta_1} \text{ at } r = \infty$$

Guided by the expression of ψ_1, we shall assume a trial stress function of the form

$$\psi = f_1(r) + f_2(r)\cos 2\theta$$

where $f_1(r)$ and $f_2(r)$ are unknown functions of r. Substituting into the biharmonic equation for polar coordinates

$$\left(\frac{\partial^2}{\partial r^2} + \frac{1}{r}\frac{\partial}{\partial r} + \frac{1}{r^2}\frac{\partial^2}{\partial\theta^2}\right)\left(\frac{\partial^2\psi}{\partial r^2} + \frac{1}{r}\frac{\partial\psi}{\partial r} + \frac{1}{r^2}\frac{\partial^2\psi}{\partial\theta^2}\right) = 0$$

The solution is shown to be

$$\sigma_r = \frac{S}{2}\left(1 - \frac{a^2}{r^2}\right) + \frac{S}{2}\left(1 + \frac{3a^4}{r^4} - \frac{4a^2}{r^2}\right)\cos 2\theta$$

$$\sigma_\theta = \frac{S}{2}\left(1 + \frac{a^2}{r^2}\right) - \frac{S}{2}\left(1 + \frac{3a^4}{r^4}\right)\cos 2\theta$$

$$\tau_{r\theta} = -\frac{S}{2}\left(1 - \frac{3a^4}{r^4} + \frac{2a^2}{r^2}\right)\sin 2\theta$$

The stresses in an infinite plate with an elliptical hole were first discussed by Inglis[10] and among others by Muskhelishvili.[11] If one of the principal axes of the elliptical hole coincides with the direction of the tension, the stresses at the ends of the axis of the hole perpendicular to the direction of the tension are

$$\sigma = S\left(1 + 2\frac{a}{b}\right)$$

where $2a$ is the axis of ellipse perpendicular to the tension and $2b$ is the other axis. When $a = b$, $\sigma = 3S$ when the hole is circular. If the ratio a/b is very large, the maximum stress at the edge of the hole becomes very large and there is a high stress concentration. This explains why cracks perpendicular to the directions of the applied forces tend to spread.

6.1-6 Other Applications

Wang[5] has presented some special applications. The other elasticity references do much the same. The methods for the thick-wall cylinder with internal pressure can be used to develop results for stresses in the case of *shrink* or *force fits*. There are practical cases where it is desirable to force or shrink an external member on an internal one. It is often required to find the pressure that will be entailed by a given difference in diameter, or interference.

Suppose that two cylinders are engaged by shrink fit or force fit so that, after assembly, the inner cylinder has radii a and b and the outer cylinder has radii b and c (Fig. 6.1-9). When $a = 0$, this gives us the case of a cylinder shrunk on a solid shaft.

Now let p be the radial pressure between the two cylinders. If these cylinders were disengaged, the inner cylinder would evidently expand and the external cylinder would contract. By the principle of superposition, the removal of the pressure p is equivalent to the imposition of a negative p on the outer surface of the inner cylinder and on the inner surface of the outer cylinder. Thus, if we let $p_i = 0$, $p_0 = -p$, and $r = b$ we obtain the increase in the external radius of the inner cylinder, which is

$$u_1 = \frac{bp}{E_i(b^2 - a^2)}\left[(1 + \nu_1)a^2 + (1 - \nu_1)b^2\right]$$

where E_1, ν_1 pertain to the material of the inner cylinder. Similarly, the radial displacement of the inner surface of the outer cylinder can be found by substituting $p_i = -p$, $p_0 = 0$, $r = C$ and changing the symbols a and b to b and c, respectively. Hence

$$u_2 = -\frac{bp}{E_2(c^2 - b^2)}\left[(1 + \nu_2)c^2 + (1 - \nu_2)b^2\right]$$

where E_2, ν_2 pertain to the material of the outer cylinder.

The inner cylinder, after disengagement, will therefore have an external radius $b + u_1$, while the outer cylinder will have an internal radius $b + u_2$. The difference in diameter, corresponding to a radial pressure p, after disengagement is given by

$$\delta = 2(u_1 - u_2)$$

$$= 2bp\left[\frac{(1 + \nu_1)a^2 + (1 - \nu_1)b^2}{E_1(b^2 - a^2)} + \frac{(1 + \nu_2)c^2 + (1 - \nu_2)b^2}{E_2(c^2 - b^2)}\right]$$

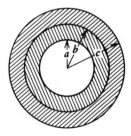

Fig. 6.1-9 Cylinders in shrink or force fit.

When both cylinders are made of the same material, we have

$$\delta = \frac{4b^3(c^2 - a^2)}{(b^2 - a^2)(c^2 - b^2)} \frac{p}{E}$$

For a cylinder shrunk on a solid shaft, $a = 0$, and the above formula reduces to

$$\delta = \frac{4bc^2}{c^2 - b^2} \frac{p}{E}$$

If the magnitude of σ_r or σ_θ in the cylinders after engagement is specified, the value of p can be determined from the appropriate thick-cylinder formulas. The application of the formulas to the design of big guns has been discussed in detail by Southwell.[12]

The stress produced in a disk rotating at high speed is important in many practical instances, including the design of disk wheels in steam and gas turbines. Wang[5] presents the following points: the stresses due to tangential forces being transmitted are usually small in these cases, and the large stresses are due to the centrifugal forces of the rotating disk. Let us first consider the case of a thin disk with constant thickness. The body force is now the centrifugal force, which is

$$F_r = \rho \omega^2 r$$

where ρ is the mass density of the material of the disk and ω is the angular velocity. It is evident that the stress distribution in the disk must be symmetrical with respect to the axis of rotation. The equilibrium equation is therefore

$$\frac{d\sigma_r}{dr} + \frac{\sigma_r - \sigma_\theta}{r} + \rho \omega^2 r = 0$$

The corresponding stress components are shown to be

$$\sigma_r = \frac{\psi}{r} = -\frac{3 + \nu}{8} \rho \omega^2 r^2 + \frac{c_1}{2} + \frac{c_2}{r^2}$$

$$\sigma_\theta = \frac{d\psi}{dr} + \rho \omega^2 r^2 = -\frac{1 + 3\nu}{8} \rho \omega^2 r^2 + \frac{c_1}{2} - \frac{c_2}{r^2}$$

For a solid disk of radius b with no external forces applied at the boundary, we have $\sigma_r = 0$ at $r = b$. The stress components are

$$\sigma_r = \frac{3 + \nu}{8} \rho \omega^2 (b^2 - r^2) \qquad \sigma_\theta = \tfrac{1}{8} \rho \omega^2 [(3 + \nu)b^2 - (1 + 3\nu)r^2]$$

The maximum stress occurs at the center of the disk and is

$$\sigma_r = \sigma_\theta = \frac{3 + \nu}{8} \rho \omega^2 b^2$$

If the disk has a circular hole of radius a at the center, the condition that no external forces are applied at the boundaries requires $\sigma_r = 0$ at $r = b$ and $r = a$. The stress components are

$$\sigma_r = \frac{3 + \nu}{8} \rho \omega^2 \left(b^2 + a^2 - \frac{a^2 b^2}{r^2} - r^2 \right)$$

$$\sigma_\theta = \frac{3 + \nu}{8} \rho \omega^2 \left(b^2 + a^2 + \frac{a^2 b^2}{r^2} - \frac{1 + 3\nu}{3 + \nu} r^2 \right)$$

The maximum stress occurs at the inner boundary and is

$$\sigma_\theta = \frac{3 + \nu}{4} \rho \omega^2 b^2 \left(1 + \frac{1 - \nu}{3 + \nu} \frac{a^2}{b^2} \right)$$

If the circular hole is very small, $(a/b)^2$ is negligible compared with 1 and we find the maximum value of the stress is now twice that for a solid disk. That is, by making a small circular hole a rotating disk, we shall double the maximum stress in the disk.

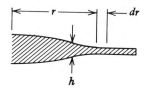

Fig. 6.1-10 Rotating disk with thickness as a function of r.

6.1-7 Rotating Disk of Variable Thickness

The method may be used to treat the problem of a rotating disk the thickness of which is a function of the distance r from the axis (Fig. 6.1-10). If we let σ_r and σ_θ denote the mean radial and tangential stresses at r, and h the variable thickness, the equation of equilibrium of such an element is

$$\frac{d}{dr}(hr\sigma_r) - h\sigma_\theta + \rho\omega^2 hr^2 = 0$$

Suppose the thickness of the disk varies according to the equation

$$h = cr^{-\beta}$$

where c is a constant and β is any number. The stresses are shown to be

$$\sigma_r = \frac{3+\nu}{8-(3+\nu)\beta}\rho\omega^2 b^2\left[\left(\frac{r}{b}\right)^{q_1+\beta-1} - \left(\frac{r}{b}\right)^2\right]$$

$$\sigma_\theta = \frac{3+\nu}{8-(3+\nu)\beta}\rho\omega^2 b^2\left[q_1\left(\frac{r}{b}\right)^{q_1+\beta-1} - \frac{1+3\nu}{3+\nu}\left(\frac{r}{b}\right)^2\right]$$

For a disk with uniform thickness, we have $b = 0$.

6.1-8 Summary

This section has dealt with the application of the principles of equilibrium, compatibility, and stress–strain to the solution for stresses and deflections in elastic continua. The principles are combined mathematically to form a governing equation—the biharmonic equation. Examples of the application of this procedure have been given. The problems solved by this technique form a vast literature. The classical references are given at the end of this section.

REFERENCES

6.1-1 S. P. Timoshenko, *History of the Strength of Materials*, McGraw-Hill, New York, 1953.
6.1-2 Joseph E. Shigley, *Mechanical Engineering Design*, McGraw-Hill, New York, 1963.
6.1-3 A. G. Durelli, E. A. Phillips, and C. H. Tsao, *Analysis of Stress and Strain*, McGraw-Hill, New York, 1958.
6.1-4 S. P. Timoshenko and J. N. Goodier, *Theory of Elasticity*, McGraw-Hill, New York, 1951.
6.1-5 C. T. Wang, *Applied Elasticity*, McGraw-Hill, New York, 1953.
6.1-6 Richard G. Budynas, *Advanced Strength and Applied Stress Analysis*, McGraw-Hill, New York, 1977.
6.1-7 Richard H. Gallagher, *Finite Element Analysis Fundamentals*, Prentice-Hall, Englewood Cliffs, NJ, 1975.
6.1-8 Larry J. Segerlind, *Applied Finite Element Analysis*, Wiley, New York, 1976.
6.1-9 Hayrettin Kardestuncer, *Elementary Matrix Analysis of Structures*, McGraw-Hill, New York, 1974.
6.1-10 C. E. Inglis, *Stresses in a Plate Due to the Presence of Cracks and Sharp Corners*, Trans. Inst. Naval Arch., London, 1913.
6.1-11 N. Muskhelishvili, *Bull. Acad. Sci. Russ.*, Vol. xiii, p. 663, 1919.
6.1-12 R. V. Southwell, *Theory of Elasticity*, Oxford University Press, London and New York, 1941, pp. 408–423.

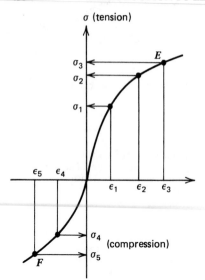

Fig. 6.2-1 Stress–strain curve into the plastic zone (after Reference 1).

6.2 PLASTICITY*

6.2-1 Stress–Strain Curve

The stress–strain curve for materials that are typically plastic is shown in Fig. 6.2-1. See also Section 3.1-1 for the definitions of uniaxial stress and strain.

6.2-2 Plastic Bending of Beams

The development of bending stresses in a beam will result in a linear distribution of stress as long as the material remains elastic. When the material becomes inelastic, as in Fig. 6.2-1 (at approximately ϵ_1), the stress distribution becomes nonlinear as shown in Fig. 6.2-2. The material may have a different stress–strain curve in compression and tension. The assumption that plane sections remain plane (see Section 3.3) is retained and thus stress σ_3, for example, may be obtained from the linear strain distribution of Fig. 6.2-1 through Fig. 6.2-3. The equations of statics still apply

$$\int_A \sigma \, dA = 0 \qquad \int_A \sigma y \, dA = 0$$

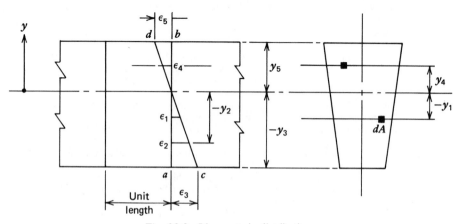

Fig. 6.2-2 Linear strain distribution.

*This discussion follows Popov[1], pp. 135–142, 440–447, and 70–73.

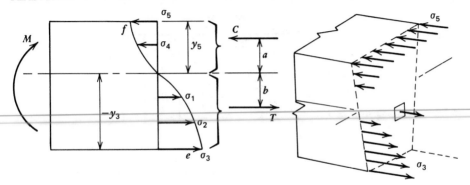

Fig. 6.2-3   Example of non-linear stress distribution.

The solution to the equations of equilibrium is a trial and error process. The development of this process for a beam of rectangular cross section is shown in Fig. 6.2-4.

As another example of inelastic bending, consider a rectangular beam of elastic–plastic material (Fig. 6.2-5). A sharp separation of the member into distinct elastic and plastic zones is possible. If the material has a distinct *yield point*, then no fiber of the beam will sustain a stress higher than σ_{yp} until the distribution of Fig. 6.2-5c is reached.

Example 6.2-1.   Determine the plastic or the ultimate capacity in flexure of a mild steel beam of rectangular cross section. Consider the material to be ideally elastic–plastic.[2]

Solution.   The idealized stress–strain diagram is shown in Fig. 6.2-6a. It is assumed that the material has the same properties in tension and compression. The strains that can take place during yielding are much greater than the maximum elastic strain (15 to 20 times the latter quantity). Therefore, since unacceptably large deformations of the beam occur along with very large strains, the plastic moment may be taken as the ultimate moment.

The stress distribution shown in Fig. 6.2-6b applies after a large amount of deformation takes place. In computing the resisting moment, the stresses corresponding to the triangular areas *abc* and *bde* may be neglected without unduly impairing the accuracy. They contribute little resistance to the

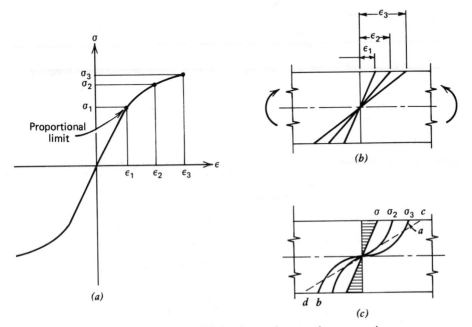

Fig. 6.2-4   Inelastic bending in a beam of rectangular cross section.

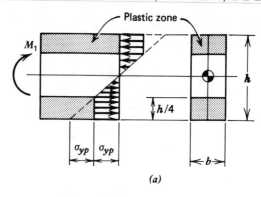

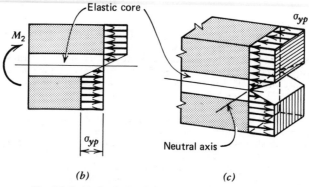

Fig. 6.2-5   Inelastic bending of an elastic-plastic beam.

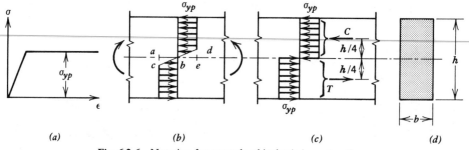

Fig. 6.2-6   Notation for example of inelastic beam bending.

applied bending moment because of their short moment arms. Hence the idealization of the stress distribution to that shown in Fig. 6.2-6c is permissible and has a simple physical meaning. The whole upper half of the beam is subjected to a uniform compressive stress σ_{yp}, while the lower half is all under a uniform tension σ_{yp}. That the beam is divided evenly into a tension and a compression zone follows from symmetry. Numerically

$$C = T = \sigma_{yp}\left(\frac{bh}{2}\right) \quad \text{(i.e., stress} \times \text{area)}$$

Each one of these forces acts at a distance $h/4$ from the neutral axis. Hence the plastic or ultimate resisting moment of the beam is

$$M_p \equiv M_{\text{ult}} = C\left(\frac{h}{4} + \frac{h}{4}\right) = \sigma_{yp}\frac{bh^2}{4}$$

where b is the breadth of the beam and h is its height.

For static loads such as occur in buildings, ultimate capacities can be determined using plastic moments. The procedures based on such concepts are referred to as the plastic method of analysis or design. For such work plastic section modulus Z is defined as

$$M_p = \sigma_{yp} Z$$

For the rectangular beam analyzed above $Z = bh^2/4$.

~~The Steel Construction Manual~~[3] provides a table of plastic section moduli for many common steel shapes. For a given M_p and σ_{yp} the solution for Z is very simple.

The method of limit or plastic analysis is unacceptable in situations where fatigue properties of the material are important.

Example 6.2-2. Find the residual stresses in a rectangular beam upon removal of the ultimate bending moment.

Solution. The stress distribution associated with an ultimate moment is shown in Fig. 6.2-7a. The magnitude of this moment has been determined in the preceding example and is $M_p = \sigma_{yp} bh^2/4$. Upon release of this plastic moment M_p every fiber in the beam can rebound elastically. The elastic range during the unloading is double that which could take place initially. Therefore, since $M_{yp} = \sigma_{yp} bh^2/6$ and the moment being released is $\sigma_{yp}(bh^2/4)$ or $1.5 M_{yp}$, the maximum stress calculated on the basis of elastic action is $\frac{3}{2}\sigma_{yp}$ as shown in Fig. 6.2-7b. Superimposing the initial stresses at M_p with the elastic rebound stresses due to the release of M_p, one finds the residual stresses (Fig. 6.2-7c). Note that both tensile and compressive longitudinal microresidual stresses remain in the beam. The tensile zones are shaded in the figure. If such a beam were machined by gradually reducing its depth, the release of the residual stresses would cause undesirable deformations of the bar.

Example 6.2-3. Determine the moment resisting capacity of an elastic–plastic rectangular beam.

Solution. To make the problem more definite consider a cantilever loaded as in Fig. 6.2-8a. If the beam is made of ideal elastic–plastic material and the applied force P is large enough to cause yielding, plastic zones will be formed (shown shaded in the figure). At an arbitrary section aa the corresponding stress distribution will be as shown in Fig. 6.2-8c. The elastic zone extends over the depth of $2 y_0$. Noting that within the elastic zone the stresses vary linearly and that everywhere in the plastic zone the axial stress is σ_{yp}, one finds that the resisting moment M is

$$M = -2\int_0^{y_0}\left(-\frac{y}{y_0}\sigma_{yp}\right)(b\,dy)\,y - 2\int_{y_0}^{h/2}(-\sigma_{yp})(b\,dy)\,y$$

$$= \sigma_{yp}\frac{bh^2}{4} - \sigma_{yp}\frac{by_0^2}{3} = M_p - \sigma_{yp}\frac{by_0^2}{3}$$

It is interesting to note that, in this general equation, if $y_0 = 0$, the moment capacity becomes equal to the plastic or ultimate moment. On the other hand, if $y_0 = h/2$, the moment reverts back to the limiting elastic case where $M = \sigma_{yp} bh^2/6$. When the applied bending moment along the span is known, the elastic–plastic boundary can be determined by solving for y_0. As long as an elastic zone or core remains, the plastic deformations cannot progress without a limit. This is a case of contained plastic flow.

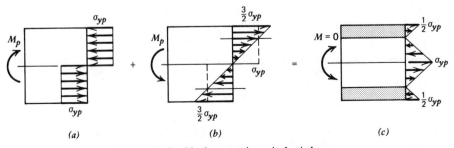

(a)   (b)   (c)

Fig. 6.2-7   Residual stresses in an inelastic beam.

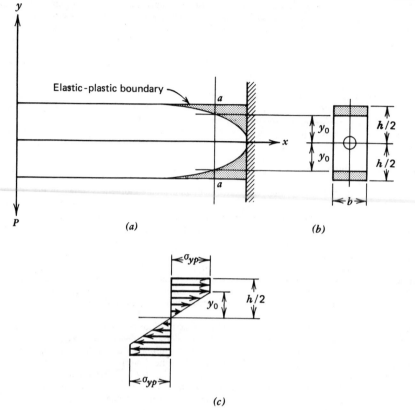

Fig. 6.2-8 Example of inelastic cantilever beam.

6.2-3 Limit Analysis of Beams

It is important to note that in elastic–plastic bending there are three stages of loading. First, there is the range of linear elastic response. Then a portion of a structure yields as the remainder continues to deform elastically. This is the range of contained plastic flow. Finally, the structure continues to yield at no increase in load. At this stage the plastic deformation of the structure becomes unbounded. This condition corresponds to the limit load for the structure.

Since the same general behavior is exhibited by elastic–plastic beams, the objective now is to develop a procedure for determining the limit loads for them. By bypassing the earlier stages of loading and going directly to the determination of the limit load, the procedure becomes relatively simple. For background, some of the results previously established will be reexamined.

Typical moment–curvature relationships of elastic–plastic beams having several different cross sections are shown in Fig. 6.2-9. Note especially the rapid ascent of the curves toward their respective asymptotes as the cross sections plastify. This means that very soon after exhausting the elastic capacity of a beam, a rather constant moment is both achieved and maintained. This condition is likened to a plastic hinge. In contrast to a frictionless hinge capable of permitting large rotations at no moment, the plastic hinge allows large rotations to occur at a constant moment. This constant moment is approximately M_p, the ultimate or plastic moment for a cross section. The ratio of moments defined in Fig. 6.2-9 is the shape factor k.

Using plastic hinges, a sufficient number can be inserted into a structure at the points of maximum moments to create a kinematically admissible collapse mechanism. Such a mechanism, permitting unbounded moment of a system, enables one to determine the ultimate or limit-carrying capacity of a beam or of a frame. This approach will now be illustrated by an example.

Example 6.2-4. A force P is applied at the middle of a simply supported beam, Fig. 6.2-10a. If the beam is made of a ductile material, what is the limit load P_{ult}? Neglect the weight of the beam.

Solution. The shape of the moment diagram is the same regardless of the load magnitude. For any value of P, the maximum moment $M = PL/4$, and if $M \leq M_{yp}$, the beam behaves elastically.

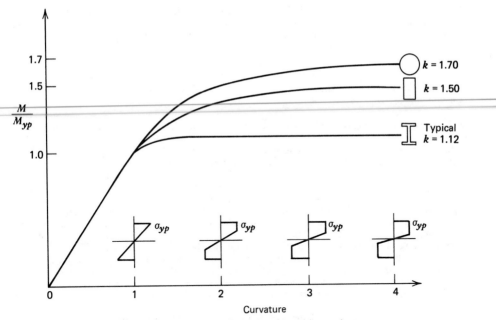

Fig. 6.2-9 Inelastic bending for various cross sections.

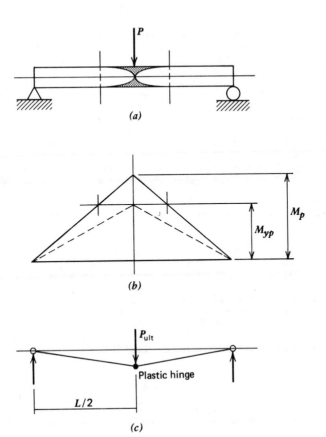

(a)

(b)

(c)

Fig. 6.2-10 Example of beam with plastic hinge.

Once M_{yp} is exceeded, contained yielding of the beam commences and continues until the maximum plastic moment M_p is reached. At that instant a plastic hinge is formed in the middle of the span forming the collapse mechanism shown. By setting the plastic moment M_p equal to $PL/4$ with $P = P_{ult}$, one obtains the result sought:

$$P_{ult} = \frac{4M_P}{L}$$

Note that consideration of the actual plastic zone, shaded in Fig. 6.2-10a, is unnecessary in this calculation.

6.2-4   Shearing Stresses and Deformations in Circular Shafts in the Inelastic Range[1]

The torsion formula for circular sections is based on Hooke's law. Therefore, it applies only up to the point where the proportional limit of a material in shear is reached in the outer annulus of a shaft.

The solution can be extended to include inelastic behavior of a material. The equilibrium requirements at a section must be met. The deformation assumption of linear strain variation from the axis remains applicable. Only the difference in material properties affects the solution.

A section through a shaft is shown in Fig. 6.2-11a. The linear strain variation is shown schematically on the same figure. Some possible mechanical properties of materials in shear, obtained, for example, in experiments with thin tubes in torsion, are as shown in Fig. 6.2-11b, c, and d. The corresponding shearing-stress distribution is shown to the right in each case. The stresses are determined from the strain. For example, if the strain is a at an interior point, Fig. 6.2-11, the corresponding stress is found from the stress–strain diagram. This procedure is applicable to solid shafts as well as to integral shafts made of concentric tubes of different materials, providing the corresponding stress–strain diagrams are used. The derivation for a linearly elastic material is simply a special case of this approach.

After the stress distribution is known, the torque T carried by these stresses is found as before, that is,

$$T = \int_A [\tau(dA)]\rho$$

Although the shearing-stress distribution after the elastic limit is exceeded is nonlinear and the elastic torsion formula does not apply, it is sometimes used to calculate a fictitious stress for the ultimate torque. The computed stress is called the modulus of rupture; see the largest ordinates of the dashed lines on Fig. 6.2-11f and g. It serves as a rough index of the ultimate strength of a material in torsion. For a thin-walled tube the stress distribution is very nearly the same, regardless of the mechanical properties of the material (Fig. 6.2-12). For this reason experiments with thin-walled tubes are widely used in establishing the shearing stress–strain (τ–γ) diagrams.

If a shaft is strained into the plastic range and the applied torque is then removed, every "imaginary" annulus rebounds elastically. Because of the differences in the strain paths, which cause permanent set in the material, residual stresses develop. This process will be illustrated in one of the examples that follow.

For determining the rate of twist of a circular shaft or tube, consider the following:

$$\frac{d\phi}{dx} = \frac{\gamma_{max}}{c} = \frac{\gamma_a}{\rho_a}$$

Here either the maximum shearing strain or the strain at ρ_a determined from the stress–strain diagram must be used.

Example 6.2-5.   A solid steel shaft of 24 mm diameter is so severely twisted that only an 8 mm diameter elastic core remains on the inside (Fig. 6.2-13a). If the material properties can be idealized as shown in Fig. 6.2-13b, what residual stresses and residual rotation will remain upon release of the applied torque?

Solution.   To begin, the magnitude of the initially applied torque and the corresponding angle of twist must be determined. The stress distribution corresponding to the given condition is shown in Fig. 6.2-13c. The stresses vary linearly from 0 to 160 MN/m^2 when $0 \le \rho \le 4$ mm; the stress is constant 160 MN/m^2 for $\rho > 4$ mm. The release of the torque T causes elastic stresses (Fig. 6.2-13d). The difference between the two stress distributions, corresponding to no external torque, gives the residual stresses.

$$T = \int_A \tau\rho \, dA = \int_0^c 2\pi\tau\rho^2 \, d\rho = \int_0^{0.004} \left[\frac{\rho}{0.004} 160\right] 2\pi\rho^2 \, d\rho$$

$$+ \int_{0.004}^{0.012} (160) 2\pi\rho^2 \, d\rho = (16 + 557)10^{-6} MN \cdot m$$

$$= 573 \times 10^{-6} MN \cdot m = 573 N \cdot m$$

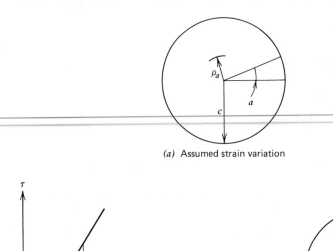

(a) Assumed strain variation

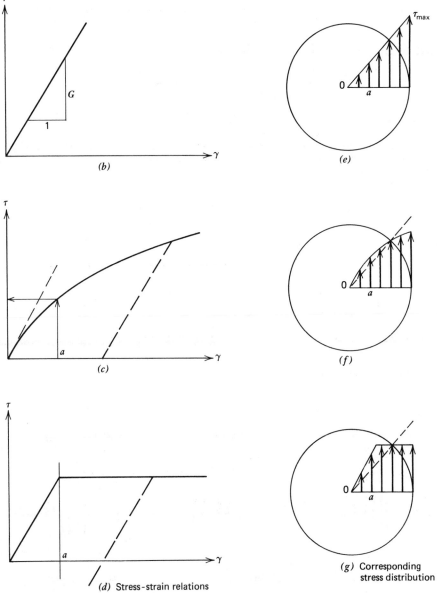

(b)

(c)

(d) Stress-strain relations

(e)

(f)

(g) Corresponding
stress distribution

Fig. 6.2-11   Linear strains and nonlinear stresses in a circular shaft.

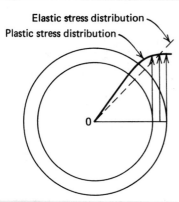

Elastic stress distribution
Plastic stress distribution

Fig. 6.2-12   Plastic stress distribution in thin-wall tubes.

(Note the smallness of the contribution of the first integral.)

$$\tau_{max} = \frac{T_c}{J} = \frac{573 \times 0.012}{(\pi/32)(0.024)^4} = 211 \times 10^4 \text{ N/m}^2 = 211 \text{ MN/m}^2$$

At $\rho = 12$mm, $\tau_{residual} = 211 - 160 = 51$ MN/m^2.

Two diagrams of the residual stresses are shown in Fig. 6.2-13e. For clarity the initial results are replotted from the horizontal line. In the center shaded portion of the diagram, the residual torque is clockwise; and exactly equal residual torque acts in the opposite direction in the inner portion of the shaft.

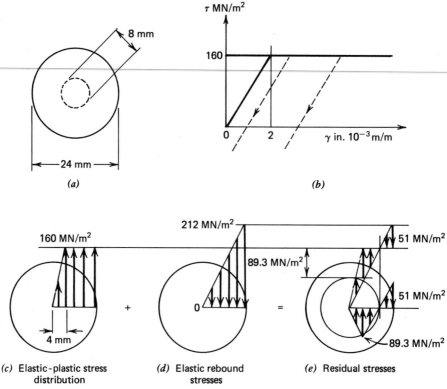

(a)

τ MN/m^2

160

0   2 γ in. 10^{-3} m/m

(b)

160 MN/m^2
4 mm

(c) Elastic-plastic stress distribution

212 MN/m^2
89.3 MN/m^2

(d) Elastic rebound stresses

51 MN/m^2
51 MN/m^2
89.3 MN/m^2

(e) Residual stresses

Fig. 6.2-13   Residual stresses in a circular shaft.

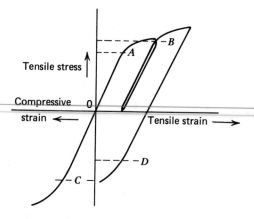

Fig. 6.2-14 Bauschinger effect.

The initial rotation is best determined by calculating twist of the elastic core. At $\rho = 4$ mm, $\gamma = 2 \times 10^{-3}$. The difference between the inelastic and the elastic twists gives the residual rotation per unit length of shaft. If the initial torque is reapplied in the same direction, the shaft responds elastically.

Inelastic $\qquad\qquad \dfrac{d\phi}{dx} = \dfrac{\gamma_a}{\rho_a} = \dfrac{2 \times 10^{-3}}{0.004} = 0.5$ rad/m

Elastic $\qquad\qquad \dfrac{d\phi}{dx} = \dfrac{T}{JG} = \dfrac{573}{(\pi/32)(0.024)^4(80) \times 10^9} = 0.22$ rad/m

Residual $\qquad\qquad \dfrac{d\phi}{dx} = 0.5 - 0.22 = 0.28$ rad/m

6.2-5 Strain Hardening[4]

Toward the end of the nineteenth century, Bauschinger discussed the phenomenon of strain hardening. Refer to Fig. 6.2-14. Bauschinger discovered: (1) that if a part is subjected to static stress beyond the elastic limit and then the load is removed, a *subsequent* reapplication of load in the same direction (tensile or compressive) will result in the observation of a higher elastic limit; (2) if the subsequent stress is applied in the opposite direction to the first stress, there will be a reduction in the elastic limit; and (3) toughness and ductility will be reduced by stressing above the elastic limit. For a better understanding, consider Fig. 6.2-14. The tensile elastic limit, initially, is at A; the compressive elastic limit is at C. After removing and reapplying the tensile load, the tensile elastic limit is at B and the compressive elastic limit is at D.

Strain hardening improves hardness and decreases toughness and ductility. Cold deformation or cold work are conditions in which the Bauschinger effect is operable. Such deformation in metals at room temperature is what is meant by cold working. The effects of cold work can be removed. The working of metals at elevated temperatures, hot working, will not result in the introduction of the Bauschinger effect or the associated alteration of mechanical properties. An example of this is forging.

REFERENCES

6.2-1 E. P. Popov, *Mechanics of Materials*, 2nd ed., Prentice-Hall, Englewood Cliffs, NJ, 1976, 1952.

6.2-2 A. Nadai, *Plasticity*, Engineering Societies Monograph, McGraw-Hill, New York, 1931.

6.2-3 *Manual of Steel Construction* (AISC Handbook), American Institute of Steel Construction, 7th ed., New York, 1970.

6.2-4 Carl A. Keyser, *Materials of Engineering*, Prentice-Hall, Englewood Cliffs, NJ, 1956, p. 651.

6.3 ELASTIC STABILITY AND BUCKLING

6.3-1 Conditions of Equilibrium and Euler's Theory

The consideration of elastic stability, or buckling, arises frequently in the stress analysis of structures in all fields of engineering. This topic is often the governing criterion in aircraft and missiles, building frames, ships, oil refinery storage tanks, and a large assortment of civil and mechanical structures.

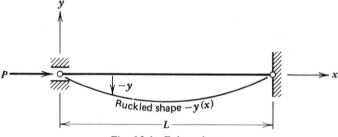

Fig. 6.3-1 Euler column.

Elastic buckling is a phenomenon that defines a failure mode in which the structure "collapses" without *necessarily* having any portion of the structure exceeding the yield strength of the material from which the structure is made. A familiar example is the "spring out" of a long slender object such as a fishing rod or wooden yardstick when a force is applied on the end. When the load is released, the structure returns to its unstressed condition and is undamaged. Nevertheless it has failed as a structural element when it buckles.

A column is a member that carries compression along its longitudinal axis. The elastic stability of columns was first studied by Euler in approximately 1757.[1] The basic equation for column buckling is still called Euler's equation and the basic buckling load is often referred to as the Euler load for the column. Euler's equation is for the *pinned, uniform, elastic, initially straight* column as shown in Fig. 6.3-1. The product of modulus of elasticity E and minimum moment of inertia I is bending rigidity EI, which is constant with length. The differential equation governing such an elastic curve is

$$\frac{d^2y}{dx^2} = \frac{M}{EI}$$

The bending moment for the column is given by $M = P(-y)$. So that

$$\frac{d^2y}{dx^2} = \frac{-Py}{EI}$$

The result is shown to be

$$P_{cr} = P_E = \frac{\pi^2 EI}{L^2}$$

Euler's equation gives the lowest, or fundamental buckling load. Theoretically the column will buckle at the loads corresponding to $n = \pm 2, \ldots$ but in actuality will buckle at the load P_E unless artificially forced to a higher load.

The equation can be rearranged to give the "buckling stress." We divide by A and use area property called radius of gyration, ρ, where $I = \rho^2 A$. Then

$$\sigma_E = \frac{\pi^2 E}{(L/\rho)^2}$$

This equation plots as a parabola with σ_E versus L/ρ as shown in Fig. 6.3-2.

6.3-2 End Conditions Other Than Pinned

When the end conditions are different than the basic Euler column (pinned), the boundary conditions of the solution for the equation are different. This problem has been treated extensively in the literature.[2,3,7,8] The customary way to present the results is to define an equivalent length, L_e, for columns with other end conditions with the Euler column as the reference. This is illustrated in Fig. 6.3-3.

Euler's equation may then be written

$$\sigma_{cr} = \frac{\pi^2 E}{(L_e/\rho)^2} \tag{6.3-1}$$

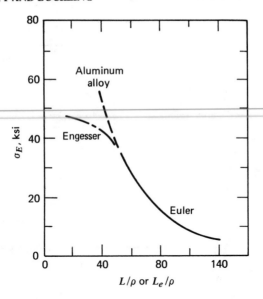

Fig. 6.3-2 Euler and Engesser theories, σ_E or σ_{cr} versus L/ρ and L_e/ρ.

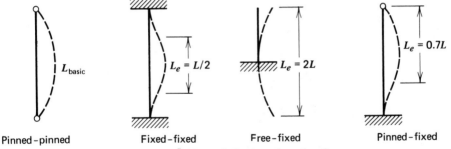

Fig. 6.3-3 Definitions of effective column lengths.

And Fig. 6.3-2 will plot the same, but the abscissa is (L_e/ρ). Thus the concept of equivalent length takes care of different end conditions. All aluminum columns now fall on the curve of Fig. 6.3-2. The equivalent lengths, L_e, are tabulated for many column end conditions. See Reference 6.3-4 for an excellent compilation.

Sometimes the end *fixity coefficient*, C_E, is tabulated where

$$C_E \equiv \frac{L_e}{L}$$

Samples of these curves for steel, aluminum, and magnesium are shown in Figs. 6.3-4–6.3-6.[9] Material that has been welded is subject to the cutoffs in allowable compressive stress shown by the dashed lines.

6.3-3 Effects of Inelastic Buckling

Many columns are designed to yield and buckle at about the same load. This can be shown to give an efficient column.[1] There is some degree of uncertainty regarding inelastic column behavior. One way to handle this problem is to use the tangent modulus, E_t.

$$\sigma_{cr} = \frac{\pi^2 E_t}{\left(L_e/\rho\right)^2} \tag{6.3-2}$$

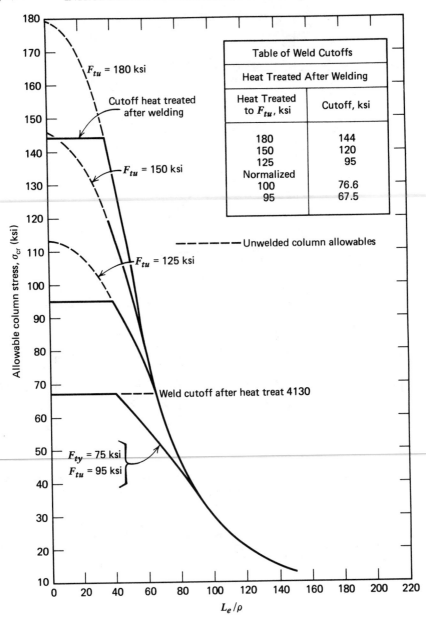

Fig. 6.3-4   Column stresses for steels.

Equation (6.3-2) is *Engesser's equation*, developed about 1898.[10] The effect of inelastic considerations on the aluminum alloy of Fig. 6.3-2 is shown by the flattening of the curve at the higher stresses. Examination of Figs. 6.3-4–6.3-6 reveal that the curves are sometimes "chopped" or drawn horizontally at some upper limit stress.

6.3-4   Effects of Eccentricity

In actuality, columns are not initially straight. References 1, 3, and 7 develop the mathematics of the column with initial imperfection, the column which is not initially straight, and the column subjected simultaneously to lateral loading (the *beam column*). The column with initial lack of straightness—intentional or accidental—is shown in Fig. 6.3-7.

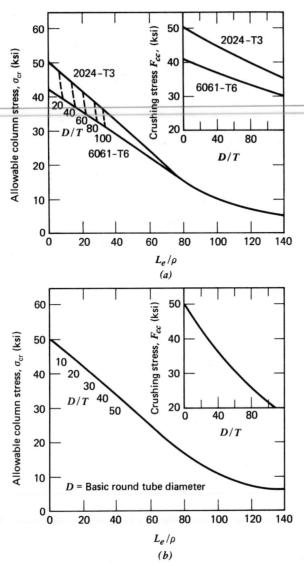

Fig. 6.3-5 Column stresses for aluminums. (*a*) Round 2024 and 6061 tubing; (*b*) streamline 2024-T3 tubing.

The initial maximum imperfection is denoted by e. We define the column *eccentricity ratio* as e/r_c where r_c is the core radius $= \rho^2/c$ or Z/A; another cross section property. The critical stress for such a column is given implicitly by

$$F_{ty} = \sigma_{cr}\left(1 + \frac{ec}{\rho^2}\sec\frac{L}{2\rho}\sqrt{\frac{\sigma_{cr}}{E}}\right) \tag{6.3-3}$$

A plot of the solution to this equation for a steel alloy is given in Fig. 6.3-8.[1,3,7]

Note that the concept of plotting σ_{cr} versus L_e/ρ allows all of the column conditions: end fixity, inelasticity, imperfections to be plotted on the same chart.

Many handbooks for civil engineering structures will show empirical data.

6.3-5 Local Crushing of Columns

As hollow circular columns become thinner (the D/t ratio becomes higher), they may be subjected to local crushing or crippling. This is illustrated in Fig. 6.3-9.

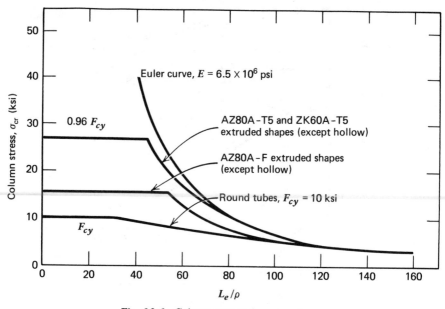

Fig. 6.3-6 Column stresses for magnesiums.

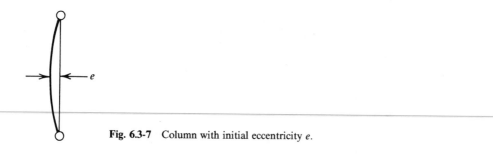

Fig. 6.3-7 Column with initial eccentricity e.

Figure 6.3-5 shows the curves for aluminum alloy local crushing stress versus D/t ratio. These columns must be checked for Euler buckling and local crushing. The cutoff values of D/t are shown for these tubes on the σ_{cr} versus L_e/ρ curves.

Figure 6.3-5 can also be used to determine local crushing for circular tubes in bending since the compressive bending stress has essentially the same effect as the direct compressive stress. Figure 6.3-10 gives the local buckling stresses for round tubes in torsion.[12]

6.3-6 Stability of Flat Plates

A thin, flat panel may be compressed along opposing edges while all four edges of the panel are supported in some fashion[5] (Fig. 6.3-11). The plate, which is free to rotate along the edges but is restrained against z direction displacements, is said to be *simply supported* or *clamped*. If the edges of the plate are built-in, the plate is *fixed*. Various combinations of these conditions are also possible. Flat plates buckle by "springing out" in the center, or in multiple sine waves m as shown in Fig. 6.3-11.

The equation for critical stress of such a panel is

$$\sigma_{cr} = \frac{\pi^2 K_c E}{12(1 - \nu^2)} \left(\frac{t}{b} \right)^2 \tag{6.3-4}$$

Figure 6.3-12 shows the buckling coefficient, K_c, for various edge conditions.[3,6] Many other kinds of loading and edge conditions may be considered.[4]

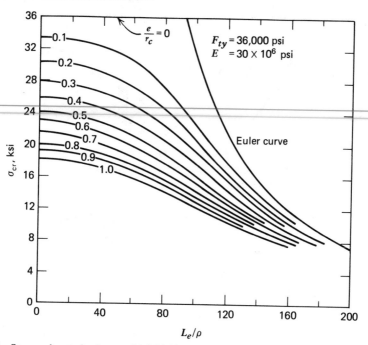

Fig. 6.3-8 Stresses for steel columns with initial imperfections. e/r_c gives the imperfection parameter.

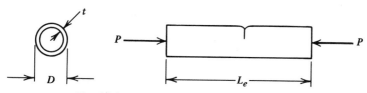

Fig. 6.3-9 Local crushing of a tubular column.

A similar equation governs the buckling of panels in shear,[3]

$$\tau_{cr} = \frac{\pi^2 K_s E}{12(1 - \nu^2)} \left(\frac{t}{b}\right)^2 \tag{6.3-5}$$

Values of K_s are shown in Fig. 6.3-13. This figure is for all four edges simply supported or all four edges clamped.

6.3-7 Lateral Buckling of Beams

A relatively long-span beam can fail in an instability mode by rotating and springing out sideways. Several cases are illustrated below.[3,20] (See Fig. 6.3-14.)

$$M_{cr} = \frac{\pi\sqrt{EI_y GJ}}{L}$$

where

$$I_y = \frac{b^3 h}{12} \qquad J = \frac{b^3 h}{3}\left(L - 0.630\frac{b}{h}\right)$$

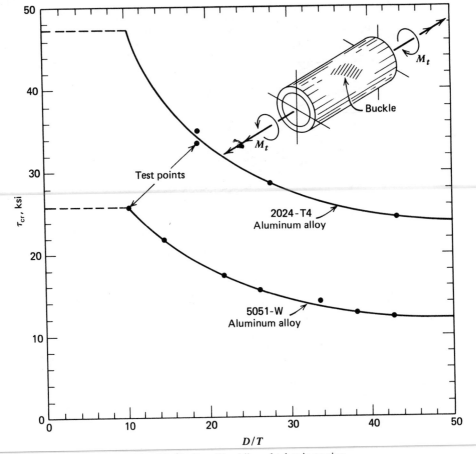

Fig. 6.3-10 Local buckling of tubes in torsion.

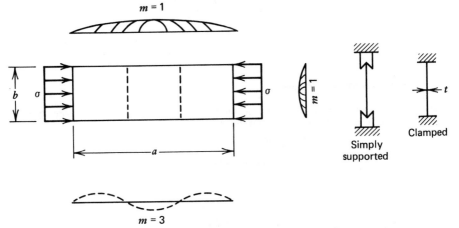

Fig. 6.3-11 Buckling modes of plates in compression.

474

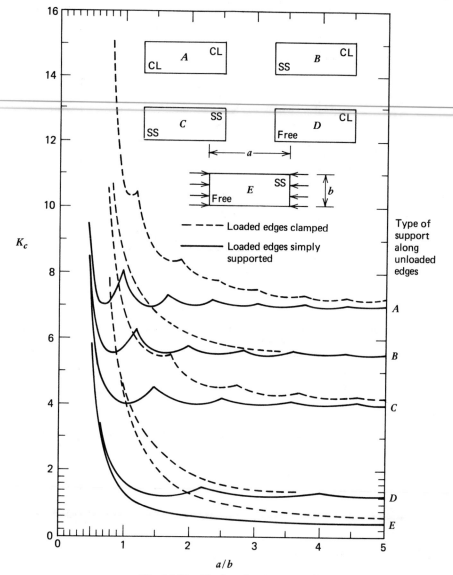

Fig. 6.3-12 K_c chart for plates.

See Fig. 6.3-15. And

$$P_{cr} = \frac{k\sqrt{EI_yGJ}}{L^2}$$

$$I_y = \frac{hb^3}{12} \qquad J = \frac{hb^3}{3}\left(L - 0.630\frac{b}{h}\right)$$

See Fig. 6.3-16.

When the load P is applied at distance g above the centroid of cross section, the parameter k and the critical load are calculated by the following equations

$$k = 4.01\left(1 - \frac{g}{L}\sqrt{\frac{EI_y}{GJ}}\right)$$

$$P_{cr} = \frac{k\sqrt{EI_yGJ}}{L^2}$$

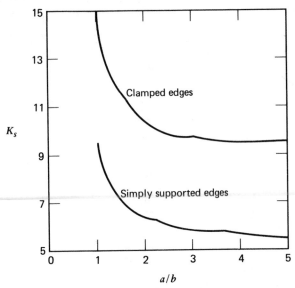

Fig. 6.3-13 K_s chart for plates.

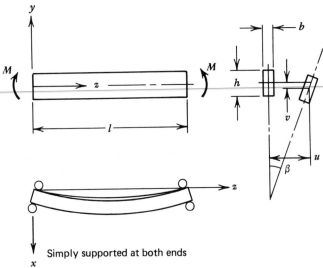

Elastic lateral buckling of
narrow beam with
rectangular cross section

Fig. 6.3-14 Elastic lateral buckling of narrow beam with rectangular cross section under applied moment M.

6.3-8 Buckling of Curved Panels

A sample of the critical shear stress relationships for curved panels is shown in Figs. 6.3-17 and 6.3-18.[1]

6.3-9 Stability of Cylindrical Shells

Cylindrical shells are closed, thin-walled structures that develop their strength from their shape. They are structurally similar to a beer can. Such structures, or slight modifications thereof, are extensively

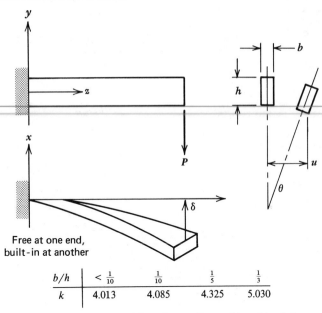

b/h	$< \frac{1}{10}$	$\frac{1}{10}$	$\frac{1}{5}$	$\frac{1}{3}$
k	4.013	4.085	4.325	5.030

Fig. 6.3-15 Lateral buckling of a cantilever with end load P.

Values of k

$\dfrac{L^2 GJ}{EC_w}$	0.1	1	2	3	4	6	8	10
k	44.3	15.7	12.2	10.7	9.76	8.69	8.03	7.58

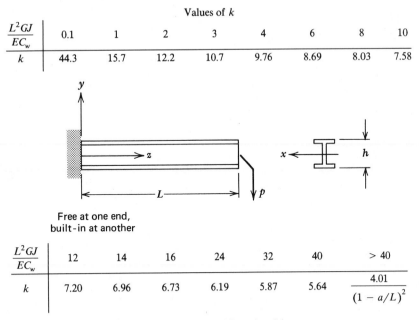

Free at one end,
built-in at another

$\dfrac{L^2 GJ}{EC_w}$	12	14	16	24	32	40	> 40
k	7.20	6.96	6.73	6.19	5.87	5.64	$\dfrac{4.01}{(1 - a/L)^2}$

Fig. 6.3-16 Lateral buckling of an I beam.

used in missiles and aircraft. There is also a wide application of shell structures, usually slightly thicker in civil, architectural, and mechanical fields.

The buckling response of thin shells is a local in-and-out pattern. The diamond pattern frequently observed in circular cylinders under axial compression is illustrated in Fig. 6.3-19.

The classical theory of shells is generally outside the scope of this handbook. However, certain application results are given below. The expressions for the cylindrical shell in axial compression are obtained from References 4 and 13.

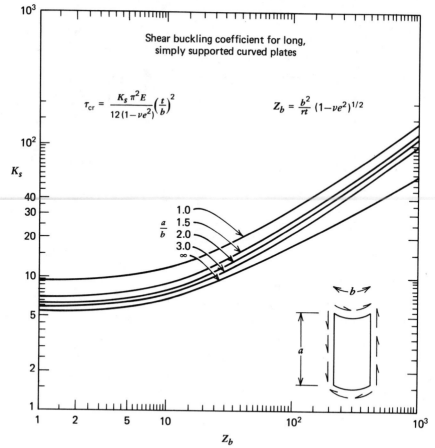

Fig. 6.3-17 K_s chart for curved long panels.

The first basic equation is

$$\sigma_{cr} = K_c \frac{\pi^2 D}{L^2 t} \tag{6.3-6}$$

where

$$D = \frac{Et^3}{12(1 - v^2)}$$

K_c is plotted versus Z in Fig. 6.3-20.

Some of the correlation of experimental and theoretical results is shown in Fig. 6.3-21.

The cylindrical shell in bending is illustrated in Fig. 6.3-22. The critical bending moment[9,14,15] is

$$M_{cr} = \alpha \frac{E}{(1 - v^2)} Rt^2$$

where for a long tube

$$\alpha = \frac{2\sqrt{2}\,\pi}{9} = 0.99$$

The value of $\alpha = 0.99$ is theoretical. The average value of test results is reported as $\alpha = 1.14$, with a minimum (conservative) value being $\alpha = 0.72$.

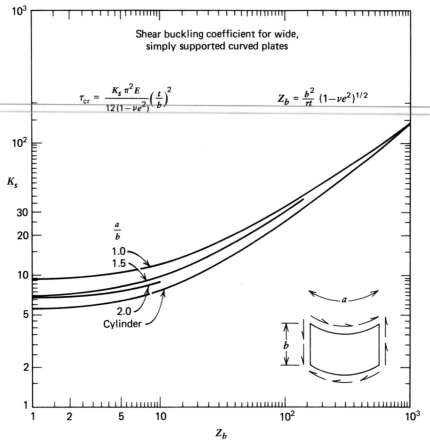

Fig. 6.3-18 K_s chart for curved wide panels.

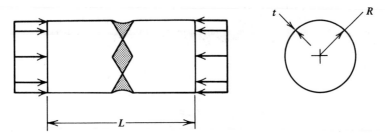

Fig. 6.3-19 Typical buckling pattern for thin-wall cylinder in axial compression.

The cylindrical shell in torsion[9,16] is depicted in Fig. 6.3-23.
The critical shear stress is given by

$$\tau_{cr} = K_s \frac{\pi^2 D}{L^2 t}$$

where

$$D = \frac{Et^3}{12(1 - \nu^2)}$$

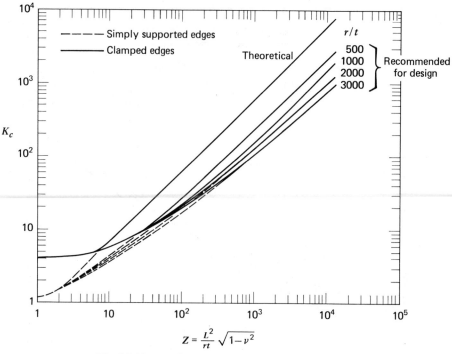

Fig. 6.3-20 K_c chart for cylinders in axial compression.

Some results for K_s versus Z are shown in Fig. 6.3-24. A comparison with experimental results is shown in Fig. 6.3-25.

6.3-10 Beam Columns

Columns that are acted on by lateral loads, such as a beam carries are called beam columns.[17] In this case the bending moment along the length of the beam is appreciably affected by the compressive forces. The deflections are not linearly proportional to load. Figure 6.3-26 shows a beam that is acted on by lateral loads and end moments. The addition of the axial compressive forces makes it a beam column. The primary moment, M', is the bending moment of the structure as a beam (no axial force). This is incremented by the moment of the axial load, which is $-Py$. At a cross section located by coordinate x it is $M = M' - Py$. Then

$$\frac{d^2M}{dx^2} = \frac{d^2M'}{dx^2} - P\frac{d^2y}{dx^2}$$

Since

$$\frac{d^2y}{dx^2} = \frac{M}{EI}$$

We have

$$\frac{d^2M}{dx^2} + \frac{PM}{EI} = w$$

for which the general solution is

$$M = C_1\sin\frac{x}{j} + C_2\cos\frac{x}{j} + wj^2$$

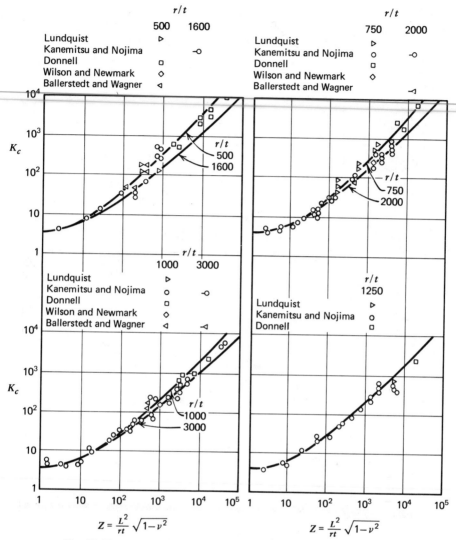

Fig. 6.3-21 Experimental K_c values for cylinders in axial compression.

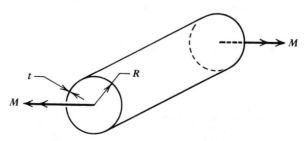

Fig. 6.3-22 Thin-wall cylinder in bending.

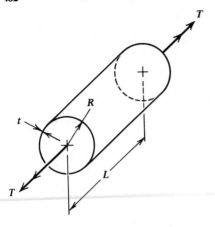

Fig. 6.3-23   Thin-wall cylinder in torsion.

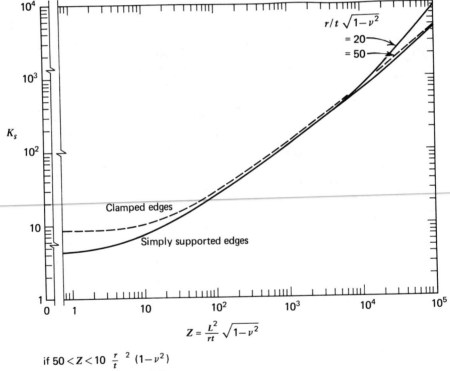

$$Z = \frac{L^2}{rt}\sqrt{1-\nu^2}$$

if $50 < Z < 10\ \frac{r}{t}^2\ (1-\nu^2)$

$$K_s = 0.85\ Z^{3/4}$$

Fig. 6.3-24 K_s chart for cylinders in torsion.

The term j is defined as

$$j^2 = \frac{EI}{P}$$

The constants C_1 and C_2 are evaluated from the boundary conditions for various problems. Some of the solutions are shown in Table 6.3-1. Beam columns are tabulated in Reference 4. An approximate method for beam columns can be used.[17] If the bending moment in a beam column with a

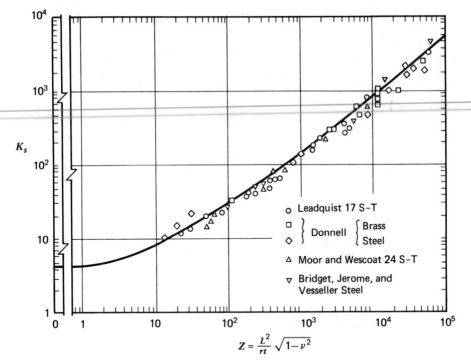

Fig. 6.3-25 Experimental K_s values for cylinders in torsion.

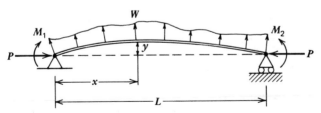

Fig. 6.3-26 Typical beam-column.

compressive load P is compared to a beam with no axial load, the following results.

$$M = \frac{M'}{1 - (P/P_{cr})}$$

6.3-11 Summary

The body of literature on the elastic stability of columns, tubes, beams, plates, stiffened plates, shells and stiffened shells is quite staggering and the material in this section only represents a sample of the available formulas. Similarly, the references at the end of this section represent a beginning.

Reference 4 is to be especially recommended for handbook type of information. It is extremely thorough, comprehensive and practical, and it contains an extensive bibliography. For those interested in a study of derivations and theory, the classic text by Timoshenko and Gere[7] is still hard to beat.

REFERENCES

6.3-1 F. R. Shanley, *Mechanics of Materials*, McGraw-Hill, New York, 1967.
6.3-2 E. F. Bruhn, *Analysis and Design of Flight Vehicle Structures*, Tri-State Offset Co., 1973.
6.3-3 G. Gerard, *Introduction to Structural Stability Theory*, McGraw-Hill, New York, 1962.

6.3-4   Column Research Committee of Japan, Ed., *Handbook of Structural Stability*, Corona Publishing, Tokyo, 1971.

6.3-5   P. S. Bulson, *The Stability of Flat Plates*, American Elsevier, New York, 1969.

6.3-6   R. J. Roark, *Formulas for Stress and Strain*, McGraw-Hill, New York, 1965.

6.3-7   S. P. Timoshenko and J. M. Gere, *Theory of Elastic Stability*, McGraw-Hill, New York, 1961.

6.3-8   F. Bleich, *Buckling Strength of Metal Structures*, McGraw-Hill, New York, 1952.

6.3-9   *Metallic Materials and Elements for Aerospace Vehicle Structures*, Dept. of Defense, Washington, D.C., 1966.

TABLE 6.3-1   VALUES OF C_1 AND C_2 FOR BEAM COLUMNS

$$M = C_1\sin\frac{x}{j} + C_2\cos\frac{x}{j} + \omega j^2; \quad j^2 = \frac{EI}{P}$$

Loading		C_1	C_2
End moments, $\omega = 0$			
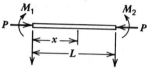		$\dfrac{M_2 - M_1\cos\dfrac{L}{j}}{\sin\dfrac{L}{j}}$	M_1
Uniform Load, $\omega = \omega_0$			
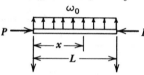		$\dfrac{\omega_0 j^2\left(\cos\dfrac{L}{j} - 1\right)}{\sin\dfrac{L}{j}}$	$-\omega_0 j^2$
Triangular loading, $\omega = \dfrac{x}{L}\omega_0$			
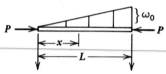		$\dfrac{-\omega_0 j^2}{\sin\dfrac{L}{j}}$	0
Concentrated load,			
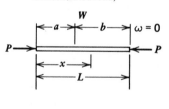	$x < a$	$-\dfrac{W_j\sin\dfrac{b}{j}}{\sin\dfrac{L}{j}}$	0
	$x > a$	$\dfrac{W_j\sin\dfrac{a}{j}}{\tan\dfrac{L}{j}}$	$-W_j\sin\dfrac{a}{j}$
Couple, $\omega = 0$			
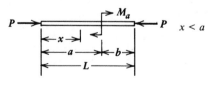	$x < a$	$-\dfrac{M_a\cos\dfrac{b}{j}}{\sin\dfrac{L}{j}}$	0
	$x > a$	$-\dfrac{M_a\cos\dfrac{a}{j}}{\tan\dfrac{L}{j}}$	$M_a\cos\dfrac{a}{j}$

6.3-10 S. P. Timoshenko, *History of the Strength of Materials*, McGraw-Hill, New York, 1953.

6.3-11 G. Gerard and H. Becker, *Handbook of Structural Stability. Part III. Buckling of Curved Plates and Shells*, NACA TN 3883.

6.3-12 R. L. Moore, *Torsional Strength of Aluminum Alloy Round Tubing*, NACA TN 879, 1943.

6.3-13 S. B. Batdorf, M. Shildcrout, and M. Stein, *Critical Stress of Thin-Wall Cylinders in Axial Compression*, NACA Report 887, 1947.

6.3-14 L. G. Brasier, "On the Flexure of Thin Cylindrical Shells and Other Thin Sections," *Proc. Roy. Soc. London, Ser. A.* **116**, 104–114, 1927.

6.3-15 H. S. Suer, L. A. Harris, W. T. Skene, and R. J. Benjamin, "The Bending Stability of Thin-Walled Unstiffened Circular Cylinders Including the Effect of Internal Pressure," *J. Aero Sci.* **25**(5), 281–287, May 1958.

6.3-16 S. B. Batdorf, *A Simplified Method of Elastic-Stability Analysis for Thin Cylindrical Shells*, NACA Report 874, 1947.

6.3-17 D. J. Peery, *Aircraft Structures*, McGraw-Hill, New York, 1950.

6.4 CONTACT STRESS PROBLEMS

6.4-1 Pressure Between Two Spherical Bodies in Contact

Timoshenko[1] has discussed the problem solved by Hertz in 1881. Our discussion follows Shigley.[2] When two bodies with curved surfaces are in contact, compressive stresses develop. This is seen in wheels and rails, cams and followers, ball bearings and mating gear teeth. Consider two solid spheres pressed together (see Fig. 6.4-1). A circular area of contact is developed. The radius of this circular area is

$$a = \sqrt[3]{\frac{3F}{8} \frac{(1 - \mu^2)[(1/E_1) + (1/E_2)]}{(1/d_1) + (1/d_2)}}$$

The pressure within the spheres has a hemispherical distribution as shown in Fig. 6.4-1. The maximum pressure (stress) is at the center of the area of contact and is

$$p_{max} = \frac{3F}{2\pi a^2}$$

The equations are general and apply to the contact of a sphere and plane surface ($d = \infty$) or to an internal spherical surface (d is negative). The stress at varying depths below the surface is given in Fig.

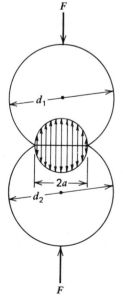

Fig. 6.4-1 Two spheres in contact.

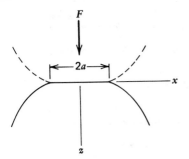

Fig. 6.4-2   Coordinate system for two spheres in contact.

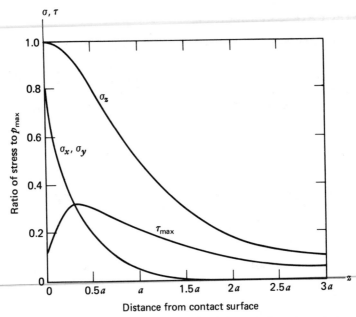

Distance from contact surface

Fig. 6.4-3   Stress distributions versus depth for two spheres in contact.

6.4-3. The formulas for the principal stresses are[2]

$$\sigma_x = \sigma_y = \frac{4a[(1/d_1) + (1/d_2)]}{\pi(1 - \mu^2)[(1/E_1) + (1/E_2)]}\left[(1 + \mu)\left(\frac{z}{a}\cot^{-1}\frac{z}{a} - 1\right) + \frac{1}{2}\frac{a^2}{a^2 + z^2}\right]$$

$$\sigma_z = -\frac{4a[(1/d_1) + (1/d_2)]}{\pi(1 - \mu^2)[(1/E_1) + (1/E_2)]}\left(\frac{a^2}{a^2 + z^2}\right)$$

Figure 6.4-2 defines the coordinate system. Figure 6.4-3 shows the principal stresses for a depth up to $3a$ in terms of the maximum pressure.

6.4-2   Cylindrical Surfaces

Two cylinders in contact will develop a rectangular area of contact, Fig. 6.4-4. The length of the cylinders is taken as 1. Then the half width of the contact area is

$$b = \sqrt{\frac{2F}{\pi l}\frac{(1 - \mu^2)[(1/E_1) + (1/E_2)]}{(1/d_1) + (1/d_2)}}$$

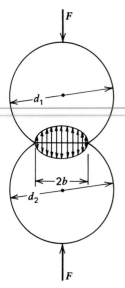

Fig. 6.4-4 Two cylinders in contact.

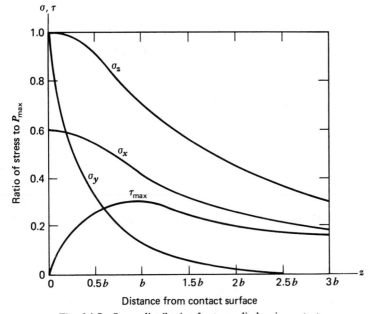

Fig. 6.4-5 Stress distribution for two cylinders in contact.

The maximum pressure (stress) is

$$P_{max} = \frac{2F}{\pi bl}$$

The above will apply to a cylinder and a plane surface if d is taken as infinite. Plots of the state of stress for this case are shown in Fig. 6.4-5.

6.4-3 Contact Stresses in Gears

Buckingham[3] defined a load–stress factor, also called a wear factor by modifying the Hertz equations. The modified equation is

$$b^2 = \frac{4F}{\pi l}(1 - \mu^2)\frac{(1/E_1) + (1/E_2)}{(1/r_1) + (1/r_2)}$$

where b = width of the rectangular contact area
  F = contact force, lb
  l = length of the cylinder, in.
  μ = Poisson's ratio
  E = modulus of elasticity, psi
  r = cylinder radius, in.

The maximum pressure is related to the surface endurance limit, S_{fe}

$$S_{fe} = \frac{2F}{\pi bl}$$

By using an average μ of 0.3

$$2.857 S_{fe}^2 \left(\frac{1}{E_1} + \frac{1}{E_2} \right) = \frac{F}{l} \left(\frac{1}{r_1} + \frac{1}{r_2} \right)$$

The left-hand side of the equation is defined as a constant K_1. We see

$$K_1 = \frac{F}{l} \left(\frac{1}{r_1} + \frac{1}{r_2} \right)$$

Table 6.4-1 lists some of the load–stress factors, K_1, as found by Reference 4. A conservative value of the surface endurance limit for steels can be found by

$$S_{fe} = 400(\text{BHN}) - 10,000$$

TABLE 6.4-1   LOAD–STRESS FACTORS AS FOUND BY TALBOURDET[a,b]

Material 1	Hardness BHN	R_c	Material 2	Hardness BHN	R_c	K_1, No Sliding	K_1, 9% Sliding[c]
ASTM 20 C.I.	150		ASTM 20 C.I.	150		1,300	1,050
G.M. Meehanite	220		G.M. Meehanite	220		1,950	1,500
Nodular C.I.	225		Nodular C.I.	225		3,400	1,850
ASTM 30 C.I.	280		ASTM 30 C.I.	280		4,200	3,400
1.05C tool steel[d]		61	ASTM 20 C.I.	150		1,000	900
1.05C tool steel[d]		61	Nodular C.I.	225		2,000	1,750
1.05C tool steel[d]		61	ASTM 35 C.I.	240		2,300	2,100
1.05C tool steel[d]		61	ASTM 30 C.I.	280		3,100	2,500
1.05C tool steel[d]		61	AISI 1020 steel	140		1,700	1,350
1.05C tool steel[d]		61	AISI 4150 steel	285		9,000	6,700
1.05C tool steel[d]		61	AISI 6150 steel	285		1,850	
1.05C tool steel[d]		61	AISI 1020 steel[e]		54	13,000	8,500
1.05C tool steel[d]		61	AISI 1340 steel[e]		50	10,000	8,000
1.05C tool steel[d]		61	AISI 4340 steel[e]		50	13,000	9,000
1.05C tool steel[d]		61	Phosphor bronze	72		1,000	
SAE 39 cast alum	65		ASTM 30 C.I.	350		300	

[a] Charles Lipson and L. V. Colwell (eds.), "Engineering Approach to Surface Damage," University of Michigan Summer Session, p. 71, Ann Arbor, Mich., 1958.
[b] These factors were obtained under ideal conditions, and so their full values should never be used in design.
[c] The percent sliding is the ratio of the difference in peripheral velocities to the peripheral velocity of the driven roll.
[d] Through-hardened.
[e] Case-hardened.

This value can be substituted into

$$K_1 = 2.857 S_{fe}^2 \left(\frac{1}{E_1} + \frac{1}{E_2} \right)$$

to obtain K_1. If the two materials in contact have different Brinell hardnesses (BHN), then the lesser value should be used.

6.4-4 Cams

When cams fail or wear, they do so by a surface fatigue action. An adaptation of the above equations gives

$$\frac{F}{l} = \frac{K_1}{(1/r_R) + (1/r_C)}$$

where K_1 = load–stress factor from Table 6.4-1
 F = contact force
 l = face width of cam or follower, whichever is narrower
 r_R = roller radius, in.
 r_C = cam curvature at point considered, in.

For a flat face follower, r_R is infinite, and

$$\frac{F}{l} = K_1 r_C$$

6.4-5 Tables

Roark[5] has presented extensive tabulated matter in this area (see Table 6.4-2).

REFERENCES

6.4-1 S. Timoshenko and J. N. Goodier, *Theory of Elasticity*, 2nd ed. McGraw-Hill, New York, 1951.
6.4-2 Joseph Edward Shigley, *Mechanical Engineering Design*, McGraw-Hill, New York, 1963.
6.4-3 Earl Buckingham, *Analytical Mechanics of Gears*, McGraw-Hill, New York, 1949, Chap. 23.
6.4-4 Charles Lipson and L. V. Colwell (Eds.), *Engineering Approach to Surface Damage*, University of Michigan Summer Session, Ann Arbor, Mich., 1958.
6.4-5 Raymond J. Roark, *Formulas for Stress and Strain*, 4th ed. McGraw-Hill, New York, 1965.

6.5 FUNDAMENTALS OF THE FINITE-ELEMENT METHOD

6.5-1 An Overview of the Finite-Element Method

Occasionally a revolutionary method will come along in a certain engineering field. The finite-element method has revolutionized conventional stress analysis[6,7] and is currently still having its impact felt, not only in stress analysis, but in areas such as aerodynamics and heat transfer.[1] The finite-element method brings together three areas: (1) matrix algebra, (2) the digital computer and accompanying numerical analysis, and (3) stress analysis. Desai and Abel[2] give a good historical development of this method.

There are several reasons for the revolutionary effect of the finite-element method:

1. It utilizes a digital computer and thus handles large amounts of processing in a short time. This includes automation of input and graphic display of output.
2. It has the feature of permitting the user to assemble the properties of a complex, large structure from a number of elements with simple properties.
3. It is capable of handling static, dynamic, stability, and nonlinear problems for a very broad range of structures.

The truss is assembled as a collection of axial (two-force) members. Each of the truss members is referred to as an element. Since the elements are finite in size, as opposed to the differentially small element in the "classical" method, we get the name finite element. The usual name for the axial force element is the rod element. Other types of elements will be discussed near the end of this section.

TABLE 6.4-2 STRESSES BETWEEN SURFACES IN CONTACT[a]

Conditions and Case No.	Formulas for Dimensions of Contact Area and for a Maximum Stress
1. Sphere on a flat plate. Axes parallel. P = load	$a = 0.721\sqrt[3]{PD\left[\dfrac{1-\nu_1^2}{E_1} + \dfrac{1-\nu_2^2}{E_2}\right]}$ $\text{Max } s_c = 0.918\sqrt[3]{\dfrac{P}{D^2\left[\dfrac{1-\nu_1^2}{E_1} + \dfrac{1-\nu_2^2}{E_2}\right]^2}}$ If $E_1 = E_2 = E$ and $\nu_1 = \nu_2 = 0.3$, $a = 0.881\sqrt[3]{\dfrac{PD}{E}}$, $\text{Max } s_c = 0.616\sqrt[3]{\dfrac{PE^2}{D^2}}$, $\text{Max } s_t = 0.133\,(\text{Max } s_c)$, $y = 1.55\sqrt[3]{\dfrac{P^2}{E^2 D}}$ $\text{Max } S_s = \tfrac{1}{3}(\text{Max } S_c)$, at depth $\tfrac{1}{4}a$ below surface of plate (approximate values)
2. Sphere on a sphere. P = total load	$a = 0.721\sqrt[3]{P\left(\dfrac{D_1 D_2}{D_1 + D_2}\right)\left[\dfrac{1-\nu_1^2}{E_1} + \dfrac{1-\nu_2^2}{E_2}\right]}$ $\text{Max } s_c = 0.918\sqrt[3]{\dfrac{P\left(\dfrac{D_1+D_2}{D_1 D_2}\right)^2}{\left[\dfrac{1-\nu_1^2}{E_1} + \dfrac{1-\nu_2^2}{E_2}\right]^2}}$, $y = 1.04\sqrt[3]{\dfrac{P^2(D_1+D_2)}{D_1 D_2}\left(\dfrac{1-\nu_1^2}{E_1} + \dfrac{1-\nu_2^2}{E_2}\right)^2}$ If $E_1 = E_2 = E$ and $\nu_1 = \nu_2 = 0.3$, $a = 0.881\sqrt[3]{\dfrac{P}{E}\dfrac{D_1 D_2}{D_1 + D_2}}$, $\text{Max } s_c = 0.616\sqrt[3]{PE^2\left(\dfrac{D_1+D_2}{D_1 D_2}\right)^2}$, $\text{Max } S_s = \tfrac{1}{3}(\text{Max } s_c)$, $\text{Max } s_t = 0.133\,(\text{Max } s_c)$, $y = 1.55\sqrt[3]{\dfrac{P^2}{E^2}\dfrac{(D_1 - D_2)}{D_1 D_2}}$

490

3. Sphere in spherical socket. P = total load

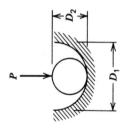

$$a = 0.721 \sqrt[3]{P \frac{D_1 D_2}{D_1 - D_2}\left[\frac{1 - \nu_1^2}{E_1} + \frac{1 - \nu_2^2}{E_2}\right]} \qquad \text{Max } s_c = 0.918 \sqrt[3]{\frac{\left(\dfrac{D_1 - D_2}{D_1 D_2}\right)^2}{P\left[\dfrac{1 - \nu_1^2}{E_1} + \dfrac{1 - \nu_2^2}{E_2}\right]^2}}$$

If $E_1 = E_2 = E$ and $\nu_1 = \nu_2 0.3$, $a = 0.881 \sqrt[3]{\frac{P}{E}\frac{D_1 D_2}{D_1 - D_2}}$, $\text{Max } s_c = 0.616 \sqrt[3]{PE^2\left(\frac{D_1 - D_2}{D_1 D_2}\right)^2}$, $\text{Max } S_s = \frac{1}{3}(\text{Max } s_c)$,

$$\text{Max } s_t = 0.133 \,(\text{Max } s_c), \quad y = 1.55 \sqrt[3]{\frac{P^2}{E^2}\frac{(D_1 - D_2)}{D_1 D_2}}$$

4. Cylinder between flat Plates p = load per linear in. = P/L

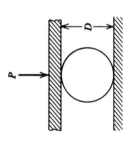

$$b = 1.6 \sqrt{pD\left[\frac{1 - \nu_1^2}{E_1} + \frac{1 - \nu_2^2}{E_2}\right]} \qquad \text{Max } s_c = 0.798 \sqrt{\frac{p}{D\left[\dfrac{1 - \nu_1^2}{E_1} + \dfrac{1 - \nu_2^2}{E_2}\right]}}$$

Total compression of cylinder between two plates is:

$$\Delta D = 4p\left(\frac{1 - \nu^2}{\pi E}\right)\left(\frac{1}{3} + \log_e \frac{2D}{b}\right)$$

If $E_1 = E_2 = E$ and $\nu_1 = \nu_2 = 0.3$, $b = 2.15 \sqrt{\frac{pD}{E}}$, $\text{Max } s_c = 0.591 \sqrt{\frac{pE}{D}}$

For $E = 30{,}000{,}000$, $\nu_1 = \nu_2 = 0.25$, $b = 0.0004\sqrt{pD}$, $\text{Max } S_c = 3190 \sqrt{\frac{p}{D}}$, $\text{Max } S_s = 958 \sqrt{\frac{p}{D}}$ at depth $0.393b$ below surface of plane

Mutual approach of remote points in two plates $= 4p\frac{1 - \nu^2}{\pi E}\log_e\frac{\pi EL}{p(1 - \nu^2)}$

TABLE 6.4-2 *(Continued)*

Conditions and Case No.	Formulas for Dimensions of Contact Area and for a Maximum Stress

5. Cylinder on cylinder. Axes parallel. p = load per linear in.

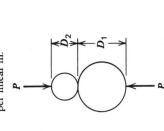

$$b = 1.6\sqrt{p\frac{D_1 D_2}{D_1+D_2}\left[\frac{1-\nu_1^2}{E_1}+\frac{1-\nu_2^2}{E_2}\right]} \qquad \text{Max } s_c = 0.798\sqrt{\frac{p\dfrac{D_1+D_2}{D_1 D_2}}{\left[\dfrac{1-\nu_1^2}{E_1}+\dfrac{1-\nu_2^2}{E_2}\right]}}$$

If $E_1 = E_2 = E$ and $\nu_1 = \nu_2 = 0.3$, $b = 2.15\sqrt{\dfrac{p}{E}\dfrac{D_1 D_2}{D_1+D_2}}$, Max $s_c = 0.591\sqrt{pE\dfrac{D_1+D_2}{D_1 D_2}}$,

$$y = \frac{2(1-\nu^2)}{E}\frac{p}{\pi}\left(\frac{2}{3}+\log_e\frac{2D_1}{b}+\log_e\frac{2D_2}{b}\right)$$

6. Cylinder in circular groove, p = load per linear in.

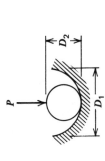

$$b = 1.6\sqrt{p\frac{D_1 D_2}{D_1-D_2}\left[\frac{1-\nu_1^2}{E_1}+\frac{1-\nu_2^2}{E_2}\right]} \qquad \text{Max } s_c = 0.798\sqrt{\frac{p\dfrac{D_1-D_2}{D_1 D_2}}{\left[\dfrac{1-\nu_1^2}{E_1}+\dfrac{1-\nu_2^2}{E_2}\right]}}$$

If $E_1 = E_2 = E$ and $\nu_1 = \nu_2 = 0.3$, $b = 2.15\sqrt{\dfrac{p}{E}\dfrac{D_1 D_2}{D_1-D_2}}$, Max $s_c = 0.591\sqrt{pE\dfrac{D_1-D_2}{D_1 D_2}}$

7. Cylinder on cylinder. Axes at right angles. P = total load

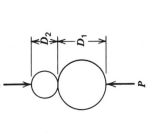

$$c = \alpha\sqrt[3]{P\frac{D_1 D_2}{D_1+D_2}\left[\frac{1-\nu_1^2}{E_1}+\frac{1-\nu_2^2}{E_2}\right]}, \quad d = \beta c, \quad \text{Max } s_c = \frac{1.5P}{\pi c d}, \quad y = \lambda\sqrt[3]{\frac{P^2\left(\dfrac{D_1+D_2}{D_1 D_2}\right)}{\left(\dfrac{E_1}{1-\nu_1^2}+\dfrac{E_2}{1-\nu_2^2}\right)^2}}$$

where α and β and λ depend on ratio $\dfrac{D_1}{D_2}$ and have values as follows:

$\dfrac{D_1}{D_2}$	1	$1\frac{1}{2}$	2	3	4	6	10
α	0.908	1.045	1.158	1.350	1.505	1.767	2.175
β	1	0.765	0.632	0.482	0.400	0.308	0.221
λ	2.080	2.060	2.025	1.950	1.875	1.770	1.613

If $E_1 = E_2 = 30{,}000{,}000$, $\nu_1 = \nu_2 = 0.25$, $c = 0.00397\alpha \sqrt[3]{P\dfrac{D_1 D_2}{D_1 + D_2}}$

For these values of E and ν and for values of $\dfrac{D_1}{D_2}$ between 1 and 8, Max $S_s = \dfrac{11{,}750}{\left(\dfrac{R_1}{R_2}\right)^{0.271}} \sqrt[3]{\dfrac{P}{R_2^{\,2}}}$, where $R_1 = \dfrac{1}{2}D_1$, $R_2 = \dfrac{1}{2}D_2$

(Approximate formula)

8. General case of two bodies in contact. $P = $ total pressure

At point of contact minimum and maximum radii of curvature are R_1 and $R_1{}'$ for Body 1, R_2 and $R_2{}'$ for Body 2.

Then $\dfrac{1}{R_1}$ and $\dfrac{1}{R_1{}'}$ are *principal curvatures* of Body 1. and $\dfrac{1}{R_2}$ and $\dfrac{1}{R_2{}'}$ of Body 2, and in each body the principal curvatures are mutually perpendicular.

The plane containing curvature $\dfrac{1}{R_1}$ in Body 1 makes with the plane containing curvature $\dfrac{1}{R_2}$ in Body 2 the angle ϕ. Then:

Max $s_c = \dfrac{1.5P}{\pi c d}$, $c = \alpha \sqrt[3]{\dfrac{P\delta}{K}}$, $d = \beta \sqrt[3]{\dfrac{P\delta}{K}}$, and $y = \lambda \sqrt[3]{\dfrac{P^2}{K^2\delta}}$, where $\delta = \dfrac{4}{\dfrac{1}{R_1} + \dfrac{1}{R_2} + \dfrac{1}{R_1{}'} + \dfrac{1}{R_2{}'}}$

and $K = \dfrac{8}{3}\dfrac{E_1 E_2}{E_2(1 - \nu_1^2) + E_1(1 - \nu_2^2)}$

α and β are given by the following table, where $\theta = \operatorname{arc\,cos}\dfrac{1}{4}\delta \sqrt{\left(\dfrac{1}{R_1} - \dfrac{1}{R_1{}'}\right)^2 + \left(\dfrac{1}{R_2} - \dfrac{1}{R_2{}'}\right)^2 + 2\left(\dfrac{1}{R_1} - \dfrac{1}{R_1{}'}\right)\left(\dfrac{1}{R_2} - \dfrac{1}{R_2{}'}\right)\cos 2\phi}$

θ	0°	10°	20°	30°	35°	40°	45°	50°	55°	60°	65°	70°	75°	80°	85°	90°
α	∞	6.612	3.778	2.731	2.397	2.136	1.926	1.754	1.611	1.486	1.378	1.284	1.202	1.128	1.061	1.00
β	0	0.319	0.408	0.493	0.530	0.567	0.604	0.641	0.678	0.717	0.759	0.802	0.846	0.893	0.944	1.00
λ	—	0.851	1.220	1.453	1.550	1.637	1.709	1.772	1.828	1.875	1.912	1.944	1.967	1.985	1.996	2.00

TABLE 6.4-2 (Continued)

Conditions and Case No.	Formulas for Dimensions of Contact Area and for a Maximum Stress
9. Rigid knife-edge across edge of semi-infinite plate. Load p lb. per linear in. 	At any point Q, $s_c = \dfrac{2\,p\cos\theta}{\pi r}$
10. Rigid block of width $2b$ across edge of semi-infinite plate. Load p lb. per linear in. 	At any point Q on surface of contact, $s_c = \dfrac{p}{\pi\sqrt{b^2 - x^2}}$
11. Uniform pressure p lb. per sq. in. over length L across edge of semi-infinite plate 	At any point O_1 outside loaded area, $y = \dfrac{2p}{\pi E}\left[(L + x_1)\log_e \dfrac{d}{L + x_1} - x_1\log_e \dfrac{d}{x_1}\right] + pL\left(\dfrac{1-v}{\pi E}\right)$ At any point O_2 inside loaded area, $y = \dfrac{2p}{\pi E}\left[(L - x_2)\log_e \dfrac{d}{L - x_2} + x_2\log_e \dfrac{d}{x_2}\right] + pL\left(\dfrac{1-v}{\pi E}\right)$ Where y = deflection relative to a remote point A distant d from edge of loaded area At any point Q, $S_c = 0.318\,p(\alpha + \sin\alpha)$ $S = 0.318\,p\sin\alpha$

494

12. Rigid cylindrical die of radius r on syrface of semi-infinite body, total load P lb.

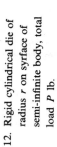

$$y = \frac{P(1 - v^2)}{2RE}$$

At any point Q on surface of contact $s_c = \frac{P}{2\pi R\sqrt{R^2 - r^2}}$

Max $s_c = \infty$ at edge

Min $s_c = \frac{P}{2\pi R^2}$ at center

13. Uniform pressure p lb. per sq. in. over circular area of radius R on surface of semi-infinite body

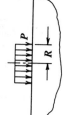

Max $y = \frac{2pR(1 - v^2)}{E}$ at center

y at edge $= \frac{4pR(1 - v^2)}{\pi E}$

Max $S_s = 0.33\, p$ at point $0.638R$ below center of loaded area

14. Uniform pressure p lb. per sq. in. over square area of sides $2b$ on surface of semi-infinite body

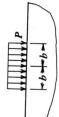

Max $y = \frac{2.24pb(1 - v^2)}{E}$ at center

$y = \frac{1.12pb(1 - v^2)}{E}$ at corners

Average $y = \frac{1.90pb(1 - v^2)}{E}$

[a]From Reference 5.

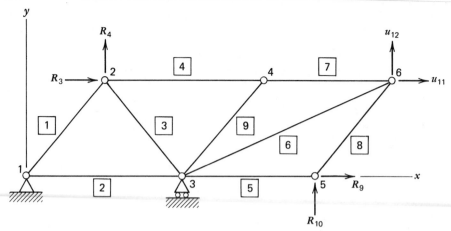

Fig. 6.5-1   Typical plane truss with node numbers, element numbers, loads, and deflection notation.

A point where two elements intersect or a point where the structure is supported is called a node. In Fig. 6.5-1 the nodes are numbered 1 through 6. The elements are also numbered $\boxed{1}$, $\boxed{2}$, and so on.

This is a particular subtopic of the finite-element method known as the stiffness method. It is also commonly called the displacement method. In this method the displacements of the nodes are the unknowns and the general solution of the structural equations centers on finding the nodal displacements. Other information, such as the element forces and stresses, can be calculated once the structural displacements are known. In the truss of Fig. 6.5-1, the significant displacements are the vertical and horizontal displacements of each of the nodes. The displacements of node 6 (u_{11} and u_{12}) are shown for illustration. A node can also be loaded and R_3, R_4, R_9, R_{10} are shown in Fig. 6.5-1 as examples.

The displacements and loads are positive if they are in the plus x and y directions of the user-defined coordinate axes as shown; otherwise they are negative. The location of each node is specified by giving its x and y coordinates in the system shown.

An important point is that the structure is mathematically modeled by examining the load–deflection behavior of its nodes. This is true of this truss and it is true of all structures, in general, which are analyzed by the finite-element method. This means that all the loading and displacement occurs at the nodes and that all the distributed stiffness properties of the structure must be "lumped" into the nodes.

Lumping the structural stiffness properties at the nodes is accomplished by assembling the structural stiffness matrix. The general entry of a stiffness matrix K_{ij} is, by definition, the force at the ith node, which is present when a unit displacement is given to the jth node and the other nodes are held fixed.

The overall structural stiffness matrix is actually affected by several factors:

1. The location of the nodes.
2. The size of the elements and the element material properties.
3. The way the elements are connected (to which nodes).

It is extremely fortunate that all of the factors can be dealt with by considering the elements one by one, and that the relatively complex job of assembling the structural stiffness matrix consists of taking each element's stiffness matrix, transforming the stiffness matrix to structural coordinates, and adding the result to the proper location of the structural stiffness matrix.

The elements are also characterized by their load–displacement relationships. Figure 6.5-2 shows a rod element. Such elements make up the example truss of Fig. 6.5-1. The element extremities are called nodes. The nodal loads P_1 and P_2 must be related to the nodal displacements δ_1 and δ_2. This is done by writing the element stiffness matrix. The element stiffness matrix is

$$\{P\}_{2\times1} = [k]_{2\times2}\{\delta\}_{2\times1}$$

or

$$\begin{Bmatrix} P_1 \\ P_2 \end{Bmatrix} = \frac{AE}{L}\begin{bmatrix} 1 & -1 \\ -1 & 1 \end{bmatrix}\begin{Bmatrix} \delta_1 \\ \delta_2 \end{Bmatrix}$$

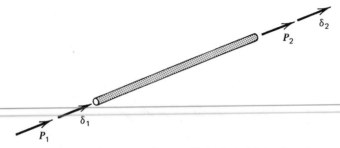

Fig. 6.5-2 Rod element coordinates with loads and deflections shown.

The only difficult part is to transform the element stiffness matrix, which is written in element coordinates, to the xy structural coordinate system. It is the element-by-element assembly process that gives the finite-element method its power. If a library of element types is available to us, we can assemble more complicated structures—much like building it up from a set of tinker toys. The elements are each described completely by their (relatively) simple stiffness matrices.

We shall now discuss some of the more common types of elements. Most large-scale programs will have the following element types (all are generally three dimensional):

1. Beam element (bar). Takes bending in two planes, shear in two planes and torsion (Fig. 6.5-3).
2. Rod element. Takes axial forces. May take torsion (Fig. 6.5-4).
3. Shear panels. In-plane shear (Fig. 6.5-5).
4. Membrane elements. In-plane shear and normal stresses (also Fig. 6.5-5).
5. Plate elements. In-plane shear and normal stress plus bending stress and plate shear (Fig. 6.5-6).
6. Axisymmetric elements. Represent the cross section of axisymmetric structures such as shells (Fig. 6.5-7).
7. Solid polyhedral elements. For building up "bulky" structures that call for three-dimensional elasticity solutions (Fig. 6.5-8).

6.5-2 Concepts of Stiffness and the Rod Element Stiffness Matrix

The definition of the ijth term of the stiffness matrix is given as follows: k_{ij} = force necessary at the ith node to cause the structure to have a unit displacement at the jth node while the other nodes are held at zero displacement.

The stiffness matrix relationships are usually a little difficult to see clearly.

From the definition of stiffness and a little reflection, we can observe a general truth. If we give a unit displacement to the jth node of an element (or structure) and hold other displacements equal zero, we can write down the jth column of the stiffness matrix as the nodal forces required to do this. For the rod we can do as shown below (see Fig. 6.5-2).

6.5-3 Matrix Coordinate Transformations and the Rod Element Transformation Matrix

Consider the forces and displacements expressed in two coordinate systems

System 1	System 2	
$\{F\}$	$(Q\}$	Forces
$\{X\}$	$\{q\}$	Displacements

We can generally find a scheme for transferring the displacements from one coordinate system to another, say

$$\{q\} = [T]\{X\}$$

Then the work expressed in the two coordinate systems is an invariant, and this work can be expressed by

$$\{X\}^T\{F\} = \{q\}^T\{Q\}$$

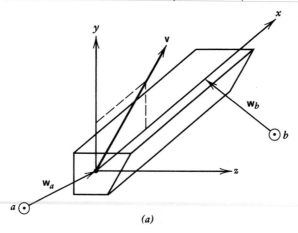

(a)

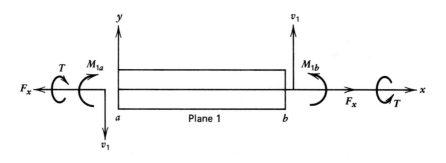

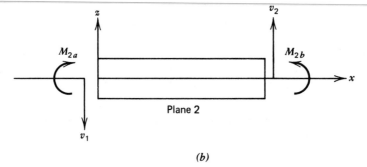

(b)

Fig. 6.5-3 Three-dimensional beam element.[3] (a) Element coordinate system; (b) element forces.

or

$$\{X\}^T\{F\} = \{X\}^T[T]^T\{Q\}$$

and, since the $\{X\}$ are arbitrary displacements, we have

$$\{F\} = [T]^T\{Q\}$$

and we see that the force transformation from $\{Q\}$ to $\{F\}$ is given. This type of transformation is called a *contragradient* transformation.

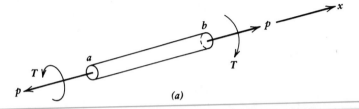

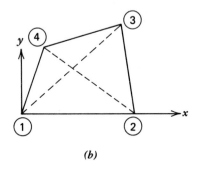

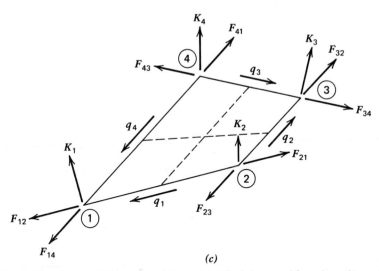

Fig. 6.5-4 Axial and torsional forces on rod element. (*a*) Rod element with torsion; (*b*) quadrilateral panel element; (*c*) quadrilateral shear panel element.[3]

Consider how the stiffness matrix and load–deflection relationship will transform under a general transformation of coordinates. We will assume that the transformation is contragradient.

We wish to have the transformation from the element coordinate system where

$$\{P\} = [K_e]\{\delta\}$$

where $[K_e]$ is the element stiffness matrix to the structure coordinate system where

$$\{R\} = [K_{M_e}]\{u\}$$

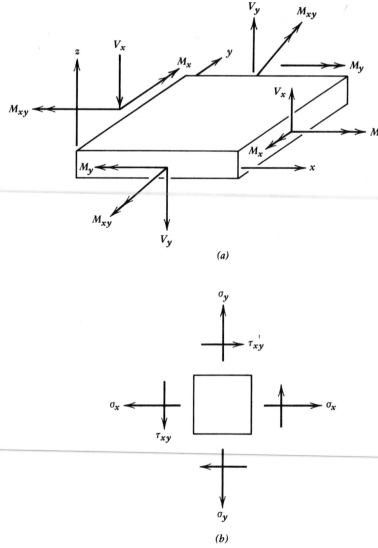

Fig. 6.5-5 Three-dimensional plate element.[3]

where $[K_{M_e}]$ is the element stiffness matrix transformed to master, or structure, coordinates. We assume that we can find a displacement transformation

$$\{\delta\} = [T]\{u\}$$

The principle of contragradience says

$$\{R\} = [T]^T\{P\}$$

Substitute

$$\{R\} = [T]^T[K_e]\{\delta\}$$

Substitute

$$\{R\} = [T]^T[K_e][T]\{u\}$$

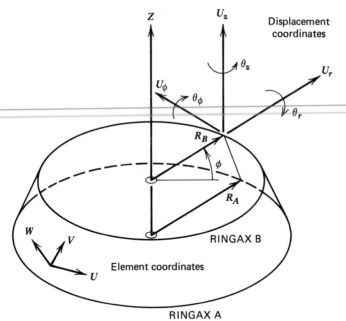

Fig. 6.5-6 Conical ring element.[3]

Compare

$$\{R\} = [K_{M_e}]\{u\}$$

We see that

$$[K_{M_e}] = [T]^T[K_e][T]$$

This will provide the transformation of the element stiffness matrix in element coordinates to the element stiffness matrix in master coordinates provided that we can write the displacement transformation $[T]$. We shall illustrate this procedure for the rod element.

We wish to derive the relation

$$\{\delta\} = [T]\{u\}$$

for the rod element. This means that we will transform from element displacements δ_1 and δ_2 to structural displacements u_1, u_2, u_3, and u_4 as shown in Fig. 6.5-8.

The element is oriented in the xy coordinate system at angles θ and β. The direction cosines of the angles may be expressed as

$$l = \cos\theta$$

$$m = \cos\beta$$

We may then write

$$\delta_1 = l\mu_1 + mu_2$$

$$\delta_2 = lu_3 + mu_4$$

or

$$\begin{Bmatrix} \delta_1 \\ \delta_2 \end{Bmatrix}_{2\times1} = \begin{bmatrix} l & m & 0 & 0 \\ 0 & 0 & l & m \end{bmatrix}_{2\times4} \begin{Bmatrix} u_1 \\ u_2 \\ u_3 \\ u_4 \end{Bmatrix}_{4\times1}$$

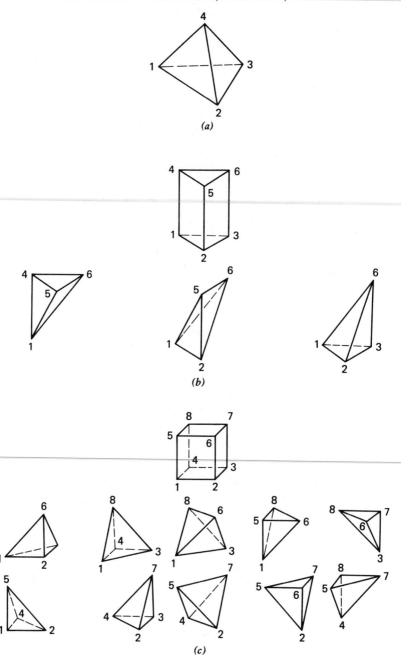

Fig. 6.5-7 Various three-dimensional solid elements.[3] (a) Tetrahedron; (b) wedge and one of its six decompositions; (c) hexahedron and its two decompositions.

Now we have

$$[K_{e_M}]_{4\times4} = \begin{bmatrix} l & 0 \\ m & 0 \\ 0 & l \\ 0 & m \end{bmatrix}_{4\times2} \times \frac{AE}{L}\begin{bmatrix} 1 & -1 \\ -1 & 1 \end{bmatrix}_{2\times2}\begin{bmatrix} l & m & 0 & 0 \\ 0 & 0 & l & m \end{bmatrix}_{2\times4}$$

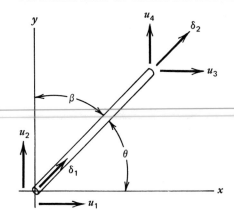

Fig. 6.5-8 Structure and element coordinates for the rod element.[9]

This expression shows how the rod element, which "stiffens" coordinates δ_1 and δ_2 in the *element* system, is transformed to "stiffen" coordinates u_1, u_2, u_3, and u_4 in the structure system.

At this point we can begin to visualize the assembly of the structure. Each rod element is attached to two structure nodes. The element stiffness matrix $[K_e]$ is transformed to element stiffness in master coordinates $[K_{ME}]$. In this way the element stiffness *adds to* the structure stiffness for the four structure freedoms associated with the two nodes to which the element is attached.

If the element stiffness matrix in master coordinates is multiplied out, it will give

$$[K_{M_e}] = \frac{AE}{L} \begin{bmatrix} l^2 & lm_2 & -l^2 & -lm \\ lm & m^2 & -lm & -m^2 \\ -l^2 & -lm & l^2 & lm \\ -lm & -m^2 & lm & m^2 \end{bmatrix}_{4\times4}$$

This is the final result for the rod element.

Next, consider the triangular membrane element, Fig. 6.5-9. This element is widely used to model structures for two-dimensional, in-plane stresses. The triangular element can readily be used to "polygonize" any two-dimensional shape.

The thickness of the triangle is t and the definitions of dimensions a, b, and c are given in Fig. 6.5-9. Then the element stiffness matrix is

$$K_e =$$

$$
\begin{array}{l}
1) \\ 2) \\ 3) \\ 4) \\ 5) \\ 6)
\end{array}
\begin{bmatrix}
b^2 + \lambda_1(c-a)^2 & & & & & \\
-b^2 - \lambda_1 c(c-a) & b^2 + \lambda_1 c^2 & & \text{Symmetric} & & \\
\lambda_1 a(c-a) & -\lambda_1 ac & \lambda_1 a^2 & & & \\
-\lambda_2 b(c-a) & \lambda_1 cb + vb(c-a) & -\lambda_1 ab & \lambda_1 b^2 + (c-a)^2 & & \\
\lambda_1 b(c-a) + vcb & -\lambda_2 bc & \lambda_1 ab & -\lambda_1 b^2 - c(c-a) & \lambda_1 b^2 + c^2 & \\
-vab & vab & 0 & a(c-a) & -ac & a^2
\end{bmatrix}
$$

where

$$\lambda_1 = \tfrac{1}{2}(1-v) \qquad \lambda_2 = \tfrac{1}{2}(1+v)$$

If θ is defined according to Fig. 6.5-10, then the transformation matrix $[T]$ for this element is

$$[T] = \begin{bmatrix}
\cos\theta & 0 & 0 & -\sin\theta & 0 & 0 \\
0 & \cos\theta & 0 & 0 & -\sin\theta & 0 \\
0 & 0 & \cos\theta & 0 & 0 & -\sin\theta \\
\sin\theta & 0 & 0 & \cos\theta & 0 & 0 \\
0 & \sin\theta & 0 & 0 & \cos\theta & 0 \\
0 & 0 & \sin\theta & 0 & 0 & \cos\theta
\end{bmatrix}_{6\times6}$$

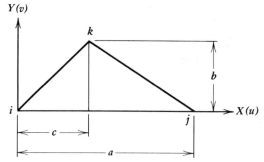

Fig. 6.5-9 Coordinates for the triangular membrane element.[9]

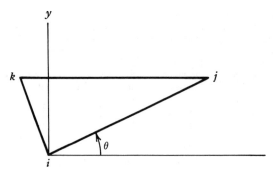

Fig. 6.5-10 Transformation of coordinates for the triangular membrane element.[9]

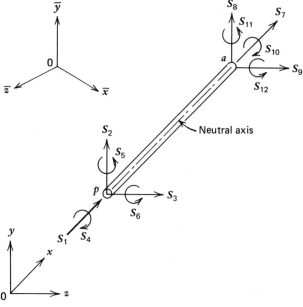

Fig. 6.5-11 Element and structure coordinates for the three-dimensional beam.[5]

$$
\begin{bmatrix}
\dfrac{EA}{l} \\[4pt]
0 & \dfrac{12EI_z}{l^3(1+\Phi_y)} \\[4pt]
0 & 0 & \dfrac{12EI_y}{l^3(1+\Phi_z)} \\[4pt]
0 & 0 & 0 & \dfrac{GJ}{l} \\[4pt]
0 & 0 & \dfrac{-6EI_y}{l^2(1+\Phi_z)} & 0 & \dfrac{(4+\Phi_z)EI_y}{l(1+\Phi_z)} \\[4pt]
0 & \dfrac{6EI_z}{l^2(1+\Phi_y)} & 0 & 0 & 0 & \dfrac{(4+\Phi_y)EI_z}{l(1+\Phi_y)} \\[4pt]
\dfrac{-EA}{l} & 0 & 0 & 0 & 0 & 0 & \dfrac{AE}{l} \\[4pt]
0 & \dfrac{-12EI_z}{l^3(1+\Phi_y)} & 0 & 0 & 0 & \dfrac{-6EI_z}{l^2(1+\Phi_y)} & 0 & \dfrac{12EI_z}{l^3(1+\Phi_y)} \\[4pt]
0 & 0 & \dfrac{-12EI_y}{l^3(1+\Phi_z)} & 0 & \dfrac{6EI_y}{l^2(1+\Phi_z)} & 0 & 0 & 0 & \dfrac{12EI_y}{l^3(1+\Phi_z)} \\[4pt]
0 & 0 & 0 & \dfrac{-GJ}{l} & 0 & 0 & 0 & 0 & 0 & \dfrac{GJ}{l} \\[4pt]
0 & 0 & \dfrac{-6EI_y}{l^2(1+\Phi_z)} & 0 & \dfrac{(2-\Phi_z)EI_y}{l(1+\Phi_z)} & 0 & 0 & 0 & \dfrac{6EI_y}{l^2(1+\Phi_z)} & 0 & \dfrac{(4+\Phi_z)EI_y}{l(1+\Phi_z)} \\[4pt]
0 & \dfrac{6EI_z}{l^2(1+\Phi_y)} & 0 & 0 & 0 & \dfrac{(2-\Phi_y)EI_z}{l(1+\Phi_y)} & 0 & \dfrac{-6EI_z}{l^2(1+\Phi_y)} & 0 & 0 & 0 & \dfrac{(4+\Phi_y)EI_z}{l(1+\Phi_y)}
\end{bmatrix}
$$

Symmetric

Fig. 6.5-12 Three-dimensional beam element stiffness matrix[5]

6.5-4 Catalog of Element Stiffness and Transformation Matrices

The following element stiffness matrices are taken from Przemieniecki,[5] although they are widely developed elsewhere. The references and the literature may be consulted for the development of more advanced element stiffness matrices. The market today contains a wide choice of programs and excellent support is widely available.[4,8]

The three-dimensional beam element stiffness is given by Przemieniecki[5] (p. 68).

Figure 6.5-11 shows the element, element coordinate system, and master coordinate system. Figure 6.5-12 shows the element stiffness matrix k. Figure 6.5-13 is for the general two-dimensional beam, and Fig. 6.5-14 is for the two-dimensional beam with the shear displacements neglected. The beam is

$$
k = \begin{bmatrix}
\dfrac{EA}{l} & & & & & \\[2ex]
0 & \dfrac{12EI_z}{l^3(1+\Phi_y)} & & \text{Symmetric} & & \\[2ex]
0 & \dfrac{6EI_z}{l^2(1+\Phi_y)} & \dfrac{(4+\Phi_y)EI_z}{l(1+\Phi_y)} & & & \\[2ex]
-\dfrac{EA}{l} & 0 & 0 & \dfrac{EA}{l} & & \\[2ex]
0 & \dfrac{-12EI_z}{l^3(1+\Phi_y)} & \dfrac{-6EI_z}{l^2(1+\Phi_y)} & 0 & \dfrac{12EI_z}{l^3(1+\Phi_y)} & \\[2ex]
0 & \dfrac{6EI_z}{l^2(1+\Phi_y)} & \dfrac{(2-\Phi_y)EI_z}{l(1+\Phi_y)} & 0 & \dfrac{-6EI_z}{l^2(1+\Phi_y)} & \dfrac{(4+\Phi_y)EI_z}{l(1+\Phi_y)}
\end{bmatrix}
$$

Fig. 6.5-13　Two-dimensional beam element general stiffness matrix.[5]

$$
k = \dfrac{EI_z}{l^3} \begin{bmatrix}
\dfrac{Al^2}{I_z} & & \text{Symmetric} & & & \\[2ex]
0 & 12 & & & & \\[1ex]
0 & 6l & 4l^2 & & & \\[1ex]
-\dfrac{Al^2}{I_z} & 0 & 0 & \dfrac{Al^2}{I_z} & & \\[2ex]
0 & -12 & -6l & 0 & 12 & \\[1ex]
0 & 6l & 2l^2 & 0 & -6l & 4l^2
\end{bmatrix}
$$

Fig. 6.5-14　Two-dimensional beam element, shear stiffness deleted.[5]

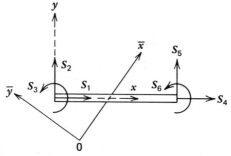

Fig. 6.5-15　Element and coordinate systems for two-dimensional beam in bending.

shown in Fig. 6.5-15. The transformation matrix $[T]$ is given in Fig. 6.5-16; the l_{ox}, and so on, are the direction cosines. Figure 6.5-17 gives the two-dimensional $[T]$ matrix.

The rectangular plate with in-plane forces is shown in Fig. 6.5-18. Figure 6.5-19 gives the stiffness matrix. Figure 6.5-20 gives the $[T]$ matrix, where the λ_{qr} and so on, are the matrices of direction cosines,

$$\lambda_{qr} = [l_{qr} m_{qr} n_{qr}]$$

$$\left[\begin{array}{c|c|c|c}
\begin{matrix}\lambda_{ox}\\ \lambda_{oy}\\ \lambda_{oz}\end{matrix} & 0 & 0 & 0 \\ \hline
0 & \begin{matrix}\lambda_{ox}\\ \lambda_{oy}\\ \lambda_{oz}\end{matrix} & 0 & 0 \\ \hline
0 & 0 & \begin{matrix}\lambda_{ox}\\ \lambda_{oy}\\ \lambda_{oz}\end{matrix} & 0 \\ \hline
0 & 0 & 0 & \begin{matrix}\lambda_{ox}\\ \lambda_{oy}\\ \lambda_{oz}\end{matrix}
\end{array}\right]$$

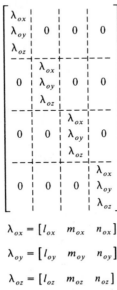

$$\lambda_{ox} = [l_{ox} \quad m_{ox} \quad n_{ox}]$$

$$\lambda_{oy} = [l_{oy} \quad m_{oy} \quad n_{oy}]$$

$$\lambda_{oz} = [l_{oz} \quad m_{oz} \quad n_{oz}]$$

Fig. 6.5-16 Transformation matrix for three-dimensional beam.[5]

$$[\mathrm{T}] = \begin{bmatrix}
l_{ox} & m_{oz} & 0 & 0 & 0 & 0 \\
l_{oy} & m_{oy} & 0 & 0 & 0 & 0 \\
0 & 0 & 1 & 0 & 0 & 0 \\
0 & 0 & 0 & l_{ox} & m_{ox} & 0 \\
0 & 0 & 0 & l_{oy} & m_{oy} & 0 \\
0 & 0 & 0 & 0 & 0 & 1
\end{bmatrix}$$

Fig. 6.5-17 Transformation matrix for two dimensional beam.[5]

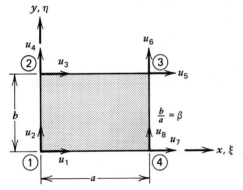

Fig. 6.5-18 Quadrilateral panel.[5]

$$k = \frac{Et}{12(1 - \nu^2)}$$

$$\beta = \frac{b}{a}$$

	1	2	3	4	5	6	7	8
1	$4\beta + 2(1 - \nu)\beta^{-1}$							
2	$\frac{3}{2}(1 + \nu)$	$4\beta^{-1} + 2(1 - \nu)\beta$						
3	$2\beta - 2(1 - \nu)\beta^{-1}$	$-\frac{3}{2}(1 - 3\nu)$	$4\beta + 2(1 - \nu)\beta^{-1}$					
4	$\frac{3}{2}(1 - 3\nu)$	$-4\beta^{-1} + (1 - \nu)\beta$	$-\frac{3}{2}(1 + \nu)$	$4\beta^{-1} + 2(1 - \nu)\beta$				
5	$-2\beta - (1 - \nu)\beta^{-1}$	$-\frac{3}{2}(1 + \nu)$	$-4\beta + (1 - \nu)\beta^{-1}$	$-\frac{3}{2}(1 - 3\nu)$	$4\beta + 2(1 - \nu)\beta^{-1}$			
6	$-\frac{3}{2}(1 + \nu)$	$-2\beta^{-1} - (1 - \nu)\beta$	$\frac{3}{2}(1 - 3\nu)$	$2\beta^{-1} - 2(1 - \nu)\beta$	$\frac{3}{2}(1 + \nu)$	$4\beta^{-1} + 2(1 - \nu)\beta$		
7	$-4\beta + (1 - \nu)\beta^{-1}$	$\frac{3}{2}(1 - 3\nu)$	$-2\beta - (1 - \nu)\beta^{-1}$	$\frac{3}{2}(1 + \nu)$	$2\beta - 2(1 - \nu)\beta^{-1}$	$-\frac{3}{2}(1 - 3\nu)$	$4\beta + 2(1 - \nu)\beta^{-1}$	
8	$-\frac{3}{2}(1 - 3\nu)$	$2\beta^{-1} - 2(1 - \nu)\beta$	$\frac{3}{2}(1 + \nu)$	$-2\beta^{-1} - (1 - \nu)\beta$	$\frac{3}{2}(1 - 3\nu)$	$-4\beta^{-1} + (1 - \nu)\beta$	$-\frac{3}{2}(1 + \nu)$	$4\beta^{-1} + 2(1 - \nu)\beta$

Symmetric

Fig. 6.5-19 Stiffness matrix for quadrilateral panel.[5]

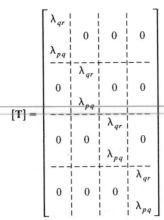

$$[\mathbf{T}] = \begin{bmatrix} \lambda_{qr} & & & \\ & 0 & 0 & 0 \\ \lambda_{pq} & & & \\ \hline & \lambda_{qr} & & \\ 0 & & 0 & 0 \\ & \lambda_{pq} & & \\ \hline & & \lambda_{qr} & \\ 0 & 0 & & 0 \\ & & \lambda_{pq} & \\ \hline & & & \lambda_{qr} \\ 0 & 0 & 0 & \\ & & & \lambda_{pq} \end{bmatrix}$$

Fig. 6.5-20 Transformation matrix for quadrilateral panel.[5]

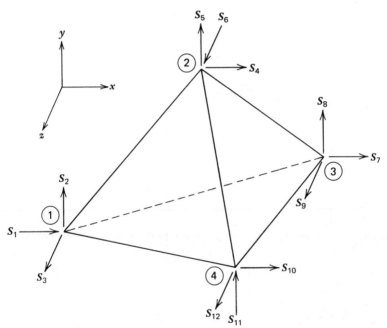

Fig. 6.5-21 Tetrahedral three-dimensional solid element.[5]

For the tetrahedral element, Fig. 6.5-21, the element stiffnesses are calculated directly in the master coordinates and are shown in Figs. 6.5-22 and 6.5-23. Note that typical term A_{pqr}^{ij} represents the area projection of triangle pqr on the ij coordinate plane.

6.5-5 Advanced Applications

The solution to more advanced problems is dependent on finding suitable advanced elements with which to build a model. There are a large number of choices available, and these are widely cataloged in the literature. The choice of element can affect the results significantly. See Figs. 6.5-24–6.5-41.

$$k_n = \frac{E}{9(1+\nu)(1-2\nu)V} \times$$

	1	2	3	4	5	6
1	$(1-\nu)A_{432}^{zx}A_{432}^{zx}$					
2	$\nu A_{432}^{zx}A_{432}^{yx}$	$(1-\nu)A_{432}^{yx}A_{432}^{yx}$				
3	$\nu A_{432}^{zx}A_{432}^{xz}$	$\nu A_{432}^{yx}A_{432}^{xz}$	$(1-\nu)A_{432}^{xy}A_{432}^{xy}$			
4	$-(1-\nu)A_{431}^{yz}A_{432}^{zx}$	$-\nu A_{431}^{yz}A_{432}^{yx}$	$-\nu A_{431}^{yz}A_{432}^{xz}$	$(1-\nu)A_{431}^{yz}A_{431}^{yz}$		
5	$-\nu A_{431}^{xz}A_{432}^{zx}$	$-(1-\nu)A_{431}^{xz}A_{432}^{yx}$	$-\nu A_{431}^{xz}A_{432}^{xz}$	$\nu A_{431}^{xz}A_{431}^{yz}$	$(1-\nu)A_{431}^{xz}A_{431}^{xz}$	
6	$-\nu A_{431}^{xy}A_{432}^{zx}$	$-\nu A_{431}^{xy}A_{432}^{yx}$	$-(1-\nu)A_{431}^{xy}A_{432}^{xy}$	$\nu A_{431}^{xy}A_{431}^{yz}$	$\nu A_{431}^{xy}A_{431}^{xz}$	$(1-\nu)A_{431}^{xy}A_{431}^{xy}$
7	$(1-\nu)A_{421}^{zx}A_{432}^{zx}$	$\nu A_{421}^{zx}A_{432}^{yx}$	$\nu A_{421}^{zx}A_{432}^{xy}$	$-1(1-\nu)A_{421}^{zx}A_{431}^{yz}$	$-\nu A_{421}^{zx}A_{431}^{xz}$	$-\nu A_{421}^{zx}A_{431}^{xy}$
8	$\nu A_{421}^{yx}A_{432}^{zx}$	$(1-\nu)A_{421}^{yx}A_{432}^{yx}$	$\nu A_{421}^{yx}A_{432}^{xy}$	$-\nu A_{421}^{yx}A_{431}^{yz}$	$-(1-\nu)A_{421}^{yx}A_{431}^{xz}$	$-\nu A_{421}^{yx}A_{431}^{xy}$
9	$\nu A_{421}^{xy}A_{432}^{zx}$	$\nu A_{421}^{xy}A_{432}^{yx}$	$(1-\nu)A_{421}^{xy}A_{432}^{xy}$	$-\nu A_{421}^{xy}A_{431}^{yz}$	$-\nu A_{421}^{xy}A_{431}^{xz}$	$-(1-\nu)A_{421}^{xy}A_{431}^{xy}$
ν 10	$-(1-\nu)A_{321}^{zx}A_{432}^{zx}$	$-\nu A_{321}^{zx}A_{432}^{yx}$	$-\nu A_{321}^{zx}A_{432}^{zy}$	$(1-\nu)A_{321}^{zx}A_{431}^{yz}$	$\nu A_{321}^{zx}A_{431}^{xz}$	$\nu A_{321}^{zx}A_{431}^{xy}$
11	$-\nu A_{321}^{yx}A_{432}^{zx}$	$-(1-\nu)A_{321}^{yx}A_{432}^{yx}$	$-\nu A_{321}^{yx}A_{432}^{zy}$	$\nu A_{321}^{yx}A_{431}^{yz}$	$(1-\nu)A_{321}^{yx}A_{431}^{xz}$	$\nu A_{321}^{yx}A_{431}^{xy}$
12	$-\nu A_{321}^{xy}A_{432}^{zx}$	$-\nu A_{321}^{xy}A_{432}^{yx}$	$-(1-\nu)A_{321}^{xy}A_{432}^{zy}$	$\nu A_{321}^{xy}A_{431}^{yz}$	$\nu A_{321}^{xy}A_{431}^{xz}$	$(1-\nu)A_{321}^{xy}A_{431}^{xy}$
	1	2	3	4	5	6

Fig. 6.5-22 Stiffness matrix for k_n of the tetrahedral three-dimensional solid element.[5] This element

$$k_s = \frac{E}{18(1+\nu)V} \times$$

	1	2	3	4	5	6
1	$(A_{432}^{zx})^2 + (A_{432}^{zy})^2$					
2	$A_{432}^{yz}A_{432}^{zx}$	$(A_{432}^{yz})^2 + (A_{432}^{zy})^2$				
3	$A_{432}^{yz}A_{432}^{zx}$	$A_{432}^{yz}A_{432}^{zy}$	$(A_{432}^{zx})^2 + (A_{432}^{zy})^2$			
4	$-A_{431}^{zx}A_{432}^{zx} - A_{431}^{zy}A_{432}^{zy}$	$-A_{431}^{zx}A_{432}^{yz}$	$-A_{431}^{zx}A_{432}^{zx}$	$(A_{431}^{zx})^2 + (A_{431}^{zy})^2$		
5	$-A_{431}^{yz}A_{432}^{zx}$	$-A_{431}^{yz}A_{432}^{yz} - A_{431}^{zy}A_{432}^{zy}$	$-A_{431}^{yz}A_{432}^{zx}$	$A_{431}^{yz}A_{431}^{zx}$	$(A_{431}^{yz})^2 + (A_{431}^{zy})^2$	
6	$-A_{431}^{xy}A_{432}^{zx}$	$-A_{421}^{xy}A_{432}^{zy}$	$-A_{431}^{zx}A_{432}^{zx} - A_{431}^{zy}A_{432}^{zy}$	$A_{431}^{xy}A_{431}^{zx}$	$A_{431}^{zx}A_{431}^{zy}$	$(A_{431}^{zx})^2 + (A_{431}^{zy})^2$
7	$A_{421}^{zx}A_{432}^{zx} + A_{421}^{zy}A_{432}^{zy}$	$A_{421}^{zx}A_{432}^{yz}$	$A_{421}^{xy}A_{432}^{yz}$	$-A_{421}^{zx}A_{431}^{zx} - A_{421}^{zy}A_{431}^{zy}$	$-A_{421}^{zx}A_{431}^{yz}$	$-A_{421}^{xy}A_{431}^{zy}$
8	$A_{421}^{yz}A_{432}^{zx}$	$A_{421}^{yz}A_{432}^{yz} + A_{421}^{zy}A_{432}^{zy}$	$A_{421}^{xy}A_{432}^{zy}$	$-A_{421}^{yz}A_{431}^{zx}$	$-A_{421}^{yz}A_{431}^{yz} - A_{421}^{zy}A_{431}^{zy}$	$-A_{421}^{xy}A_{431}^{zy}$
9	$A_{421}^{xy}A_{432}^{zx}$	$A_{421}^{xy}A_{432}^{zy}$	$A_{421}^{zx}A_{432}^{zx} + A_{421}^{zy}A_{432}^{zy}$	$-A_{421}^{xy}A_{431}^{zx}$	$-A_{421}^{xy}A_{431}^{zy}$	$-A_{421}^{zx}A_{431}^{zx} - A_{421}^{zy}A_{431}^{zy}$
10	$-A_{321}^{zx}A_{432}^{zx} - A_{321}^{zy}A_{432}^{zy}$	$-A_{321}^{zx}A_{432}^{zy}$	$-A_{321}^{xy}A_{432}^{zy}$	$A_{321}^{zx}A_{431}^{zx} + A_{321}^{zy}A_{431}^{zy}$	$A_{321}^{zx}A_{431}^{zy}$	$A_{321}^{xy}A_{431}^{zy}$
11	$-A_{321}^{yz}A_{432}^{zx}$	$-A_{321}^{yz}A_{432}^{yz} - A_{321}^{zy}A_{432}^{zy}$	$-A_{321}^{xy}A_{432}^{zy}$	$A_{321}^{yz}A_{431}^{zx}$	$A_{321}^{yz}A_{431}^{zy} + A_{321}^{zy}A_{431}^{zy}$	$A_{321}^{xy}A_{431}^{zy}$
12	$-A_{321}^{xy}A_{432}^{zx}$	$-A_{321}^{xy}A_{432}^{zy}$	$-A_{321}^{zx}A_{432}^{zx} - A_{321}^{zy}A_{432}^{zy}$	$A_{321}^{xy}A_{431}^{zx}$	$A_{321}^{xy}A_{431}^{zy}$	$A_{321}^{zx}A_{431}^{zx} + A_{321}^{yz}A_{431}^{zy}$
	1	2	3	4	5	6

Fig. 6.5-23 Stiffness matrix for k_s of the tetrahedral three-dimensional solid element.[5] Note:

7	8	9	10	11	12
$(1-\nu)A_{421}^{yz}A_{421}^{yz}$					
$\nu A_{421}^{zx}A_{421}^{yz}$	$(1-\nu)A_{421}^{zx}A_{421}^{zx}$				
$\nu A_{421}^{xy}A_{421}^{yz}$	$\nu A_{421}^{xy}A_{421}^{zx}$	$(1-\nu)A_{421}^{xy}A_{421}^{xy}$			
$-(1-\nu)A_{321}^{zx}A_{421}^{yz}$	$-\nu A_{321}^{yz}A_{421}^{zx}$	$-\nu A_{321}^{yz}A_{421}^{xy}$	$(1-\nu)A_{321}^{yz}A_{431}^{yz}$		
$-\nu A_{321}^{zx}A_{421}^{yz}$	$-(1-\nu)A_{321}^{zx}A_{421}^{zx}$	$-\nu A_{321}^{zx}A_{421}^{xy}$	$\nu A_{321}^{zx}A_{321}^{yz}$	$(1-\nu)A_{321}^{zx}A_{321}^{zx}$	
$-\nu A_{321}^{xy}A_{421}^{yz}$	$-\nu A_{321}^{xy}A_{421}^{zx}$	$-(1-\nu)A_{321}^{xy}A_{421}^{xy}$	$\nu A_{321}^{xy}A_{321}^{yz}$	$\nu A_{321}^{xy}A_{321}^{zx}$	$(1-\nu)A_{321}^{xy}A_{321}^{xy}$

is formed directly from master coordinates. Note: $k = k_n + k_s$. See Fig. 6.5-23.

7	8	9	10	11	12
$(A_{421}^{zx})^2 + (A_{421}^{xy})^2$					
$A_{421}^{yz}A_{421}^{zx}$	$(A_{421}^{yz})^2 + (A_{421}^{xy})^2$				
$A_{421}^{yz}A_{421}^{xy}$	$A_{421}^{zx}A_{421}^{xy}$	$(A_{421}^{zx})^2 + (A_{421}^{yz})^2$			
$-A_{321}^{zx}A_{421}^{zx}$ $-A_{321}^{xy}A_{421}^{xy}$	$-A_{321}^{yz}A_{421}^{zx}$	$-A_{321}^{xy}A_{421}^{xy}$	$(A_{321}^{zx})^2 + (A_{321}^{xy})^2$		
$-A_{321}^{yz}A_{421}^{zx}$	$-A_{321}^{yz}A_{421}^{zx}$ $-A_{321}^{xy}A_{421}^{xy}$	$-A_{321}^{xy}A_{421}^{zx}$	$A_{321}^{yz}A_{321}^{zx}$	$(A_{321}^{yz})^2 + (A_{321}^{xy})^2$	
$-A_{321}^{yz}A_{421}^{xy}$	$-A_{321}^{zx}A_{421}^{xy}$	$-A_{321}^{xy}A_{421}^{zx}$ $-A_{321}^{xy}A_{421}^{xy}$	$A_{321}^{yz}A_{321}^{xy}$	$A_{321}^{zx}A_{321}^{xy}$	$(A_{321}^{zx})^2 + (A_{321}^{yz})^2$

$k = k_n + k_s$. See Fig. 6.5-22.

Fig. 6.5-24 Portions of Boeing 747 analyzed by finite-element methods.[5]

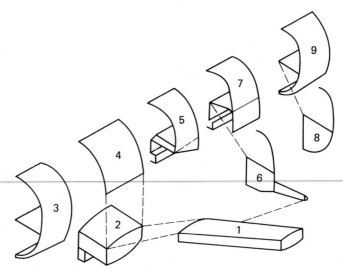

Fig. 6.5-25 Numbering of substructures for Boeing 747.[5]

Substructure	Description	Nodes	Unit[a] Loadcases	Beams	Plates	Interact[b] Freedoms	Total Freedoms
1	Wing	262	14	355	363	104	796
2	Wing Center	267	8	414	295	198	880
3	Body	291	7	502	223	91	1026
4	Body	213	5	377	185	145	820
5	Body	292	7	415	241	200	936
6	Bulkhead	170	10	221	103	126	686
7	Body	285	6	392	249	233	909
8	Bulkhead	129	10	201	93	148	503
9	Body	286	7	497	227	92	1038
Total		2195	63	3374	1979	555	7594

[a]Some unit loadcases involve more than one substructure.
[b]Several freedoms of substructure interface contribute to a single freedom.

Fig. 6.5-26 Modeling details for substructures for Boeing 747.[9]

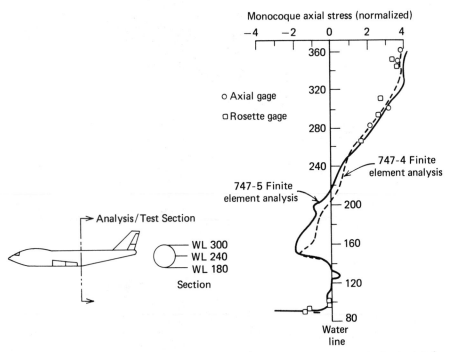

Fig. 6.5-27 A comparison of finite-element analysis and stress test results for the Boeing 747.[5]

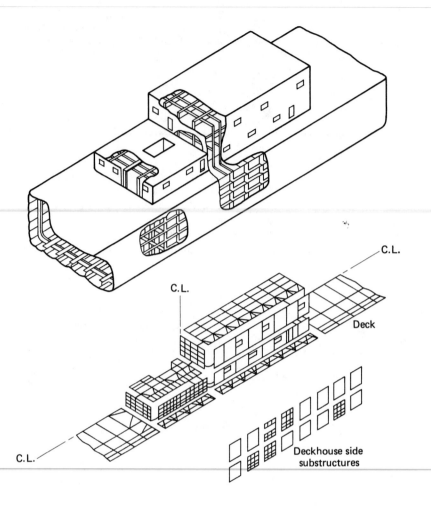

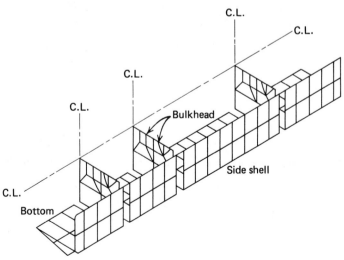

Fig. 6.5-28 Ship stress analysis by the finite-element method.[5,10]

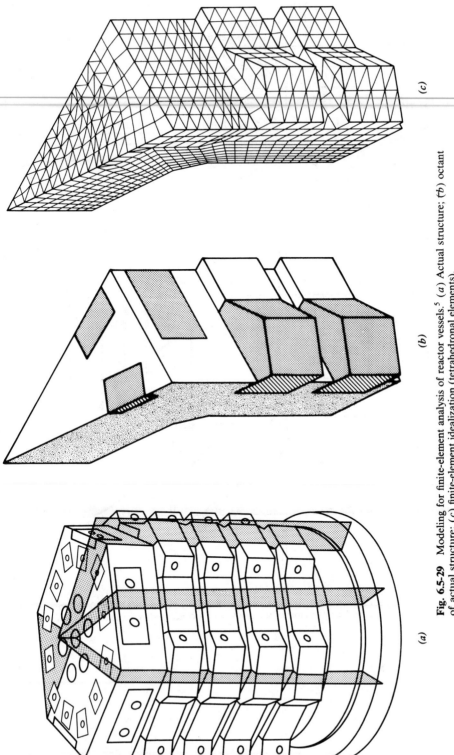

Fig. 6.5-29 Modeling for finite-element analysis of reactor vessels.[5] (*a*) Actual structure; (*b*) octant of actual structure; (*c*) finite-element idealization (tetrahedronal elements).

(*a*)

(*b*)

(*c*)

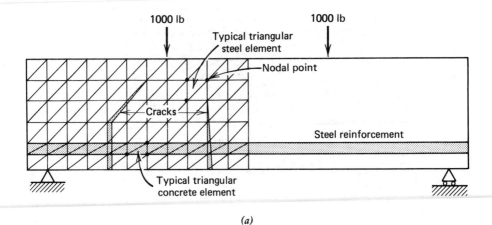

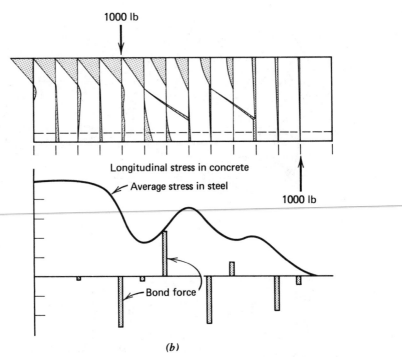

Fig. 6.5-30 (*a*) Finite-element analysis of a reinforced concrete beam.[5] (*b*) Calculated stress distribution. For an interesting report of the analysis of a concrete dam structure, see Reference 2.

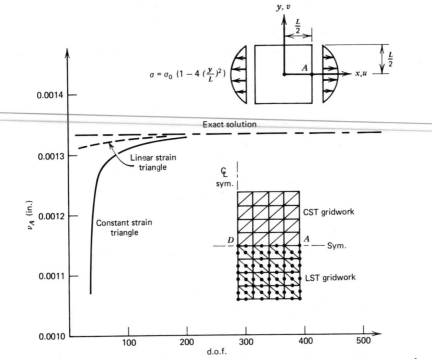

Fig. 6.5-31 Comparison of the accuracy of two different elements for the problem shown.[5]

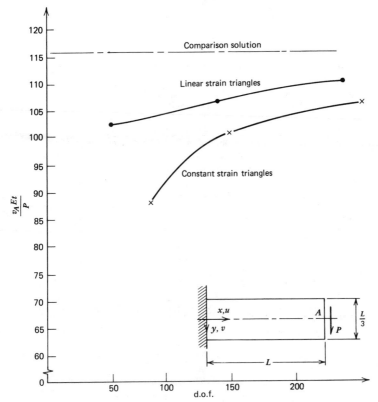

Fig. 6.5-32 Comparison of the accuracy of two different elements for the problem shown.[5]

517

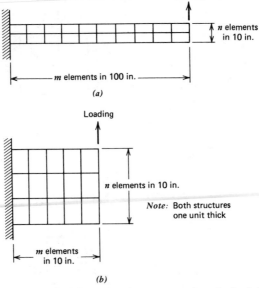

Loading

n elements in 10 in.

— *m* elements in 100 in. —

(a)

Loading

n elements in 10 in.

Note: Both structures one unit thick

m elements in 10 in.

(b)

Fig. 6.5-33 Three-dimensional modeling to test for accuracy of results for (a) slender and (b) deep cantilever beams.[5] See Fig. 6.5-34.

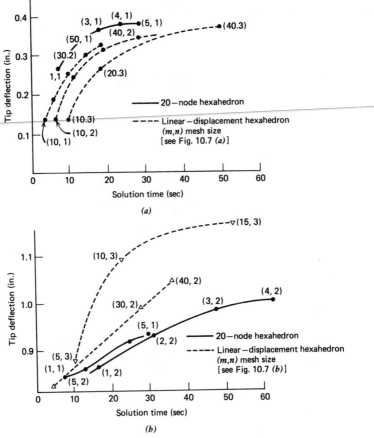

(a)

(b)

Fig. 6.5-34 Comparison of efficiency of solution for the problem of Fig. 6.5-33.

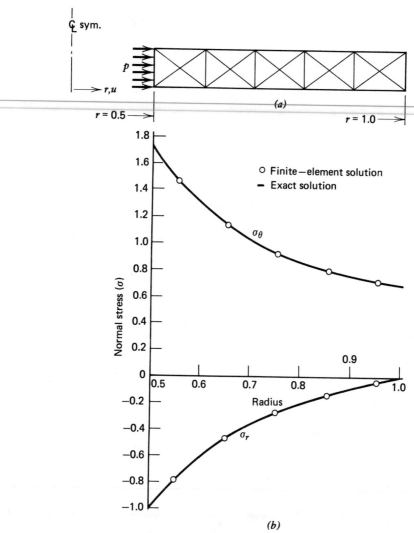

Fig. 6.5-35 Analysis of thick-walled cylinder under internal pressure (axisymmetric element application).[5] (*a*) Finite-element idealization; (*b*) calculated stresses.

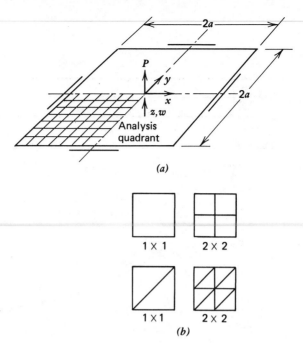

(a)

1 × 1   2 × 2

1 × 1   2 × 2

(b)

Fig. 6.5-36 Element comparison for plate in bending.[5] See Fig. 6.5-37. Grids for (*a*) rectangular and (*b*) triangular elements.

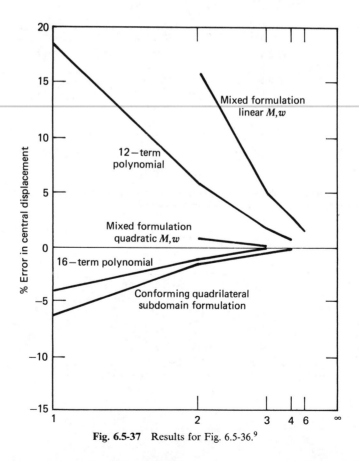

Fig. 6.5-37 Results for Fig. 6.5-36.[9]

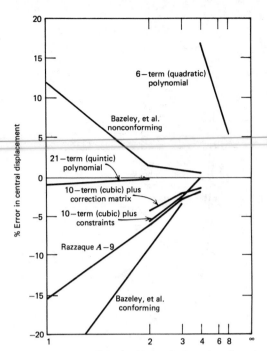

Fig. 6.5-38 A comparison of results for different kinds of triangular plate elements applied to the same problem.[5]

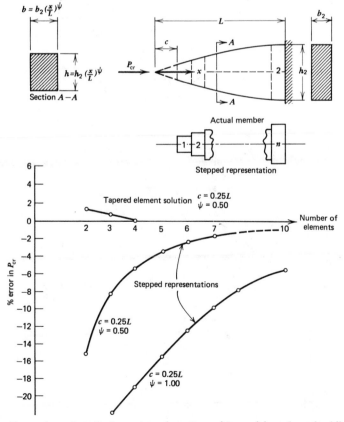

Fig. 6.5-39 Comparison of results for various elements used to model a column buckling problem.[5]

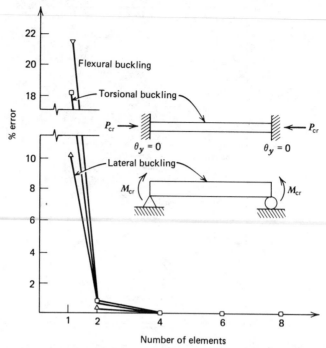

Fig. 6.5-40 Accuracy of various representations for finite-element application to elastic stability.[5]

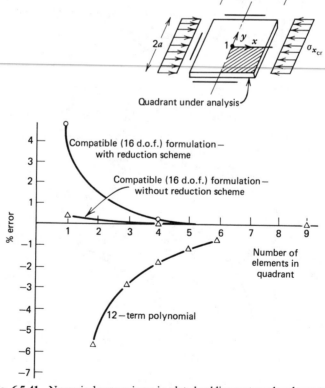

Fig. 6.5-41 Numerical comparisons in plate buckling rectangular elements.[5]

REFERENCES

6.5-1 O. C. Zienkiewicz, *The Finite Element Method in Engineering and Science*, McGraw-Hill, London, 1971.

6.5-2 C. S. Desai and J. F. Abel, *Introduction to the Finite Element Method*, Van Nostrand Reinhold, New York, 1972.

6.5-3 Richard H. MacNeal, ed., *NASTRAN Theoretical Manual*, NASA SP-221, Washington, D.C., 1970.

6.5-4 Z. Ziedans, *Survey of Advanced Structural Design Techniques*, ASME Paper 69-DE-13, New York, 1969.

6.5-5 J. S. Przemieniecki, *Theory of Matrix Structural Analysis*, McGraw-Hill, New York, 1968.

6.5-6 R. J. Roark and W. C. Young, *Formulas for Stress and Strain*, 5th ed. McGraw-Hill, New York, 1968.

6.5-7 T. Baumeister, ed., *Mark's Mechanical Handbook*, 6th ed. McGraw-Hill, New York, 1967.

6.5-8 B. M. Irons, *Matrix and Digital Computer Methods in Structural Analysis*, Int. J. Num. Meth. Engr., **2**(1), Jan–Mar 1970.

6.5-9 J. S. Przemieniecki, *Theory of Matrix Structural Analysis*, McGraw-Hill, New York, 1968.

6.5-10 J. W. McKinley, *Fundamentals of Stress Analysis*, Matrix Publishers, Portland, OR, 1979.

6.5-11 H. A. Kamel, W. Birchler, D. Liu, J. W. McKinley, and W. R. Reid, "Computer Analysis of Large Tankers," Presented at Tanker Conference, American Petroleum Institute, Pocono, PA, May 12, 1969.

CHAPTER 7
STRAIGHT MEMBERS

RAYMOND D. CIATTO

Manager, Design
Teledyne Engineering Services
Waltham, Massachusetts

7.1 INTRODUCTION

Straight members may be used alone in simple structures such as pipe supports consisting of simple beams or tension members; or thousands of straight members may be combined to construct a large building frame in which the individual members interact with one another. Piping systems are composed of both straight and curved members; the straight members in this case are pressurized circular tubes. Both building frames and piping systems are analyzed using the same structural engineering theory of frame analysis.

Using computerized structural analysis, these straight members, which are beams or columns, can be isolated as finite elements in a complex frame. The structure can be discretized into a number of finite elements interconnected at "nodal" points. Nodal displacements and rotations (or forces and moments) comprise the member degree of freedom. An individual straight member may be assigned as many as 12 degrees of freedom, 6 at each end. Such a member is illustrated in Fig. 7.1-1. For convenience, the forces and deflections are located at the ends of the member. If the member has loads applied somewhere between the ends, say at midspan for instance, it is simply divided into a number of elements such that the loads are at the ends of the element. More discussion is presented in Section 7.11 on computer analysis.

Beams are those straight members that resist loads applied laterally to the member axis. In Fig. 7.1-1 all degrees of freedom may be involved in a beam, except 1 and 7, the axial degree of freedom.

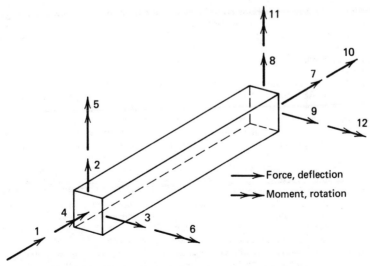

Fig. 7.1-1 Twelve degrees of freedom of a straight member.

All of the others, except torsional degrees of freedom 4 and 10, tend to impose internal bending or *flexure* within the member, an important consideration for beams. Beam flexure is discussed in Section 7.3.

Columns are those vertical straight members whose loads impose axial compression along the member axis. Although this involves only 2 degrees of freedom, 1 and 7, it deserves special attention because of the complexity of analysis of column buckling and the sudden, often catastrophic, failure that can occur when a column is overloaded. Both horizontal and vertical straight members in compression are analyzed for column buckling (Section 7.9).

Beam columns may have all of the loads or deflections of Fig. 7.1-1 except the torsional degrees of freedom, 4 and 10. Of course the axial loads must impose compression, not tension, in beam columns as they do in ordinary columns. The compressive axial loads tend to reduce bending stiffness, whereas tensile axial loads tend to increase bending stiffness. Also somewhat complex in analysis as discussed in Section 7.10, beam columns cannot support full axial or bending loads, which they can resist when one or the other load is absent. An exception is reinforced-concrete columns that exhibit an increase in moment capacity with applied axial compressive load.

Straight members, also called linear members by some analysts, are characterized by their small depth-to-length ratio. Although it is possible to analyze them as three-dimensional solids, they usually are considered to have a unidirection stress field. And, of course, plane sections are assumed to remain plane after bending or axial loads are applied.

Some of the common materials used to construct straight members are discussed in this chapter. For beams the analysis procedures are not affected by the choice of material. But, for columns, the member response is very sensitive to material selection. Of course, materials for linear (straight) members may be used alone or in combination with one another as in composite construction (Section 7.6).

7.2 DESIGN CRITERIA

After a straight member is designed (to determine dimensions) and analyzed (to determine stresses and deflection), it is evaluated to determine if it meets design criteria. For beams, bending stress usually governs the acceptability and, for columns, compressive axial stress determines the member acceptability. These quantities are compared to allowable values, which are usually based on some fraction of yield stress. For short, deep beams shear stress may be more dominant than bending stress and it may govern the design. Even if all members in a structure meet the stress criteria, excessive deflection at the point of load application may be cause for redesign. This is especially true for pipe support structures where low stiffness can result in high stress in the piping. In buildings, excessive deflection can damage plaster walls and ceilings, and so the building codes impose deflection limits. A minimum fundamental natural frequency is often a requirement given in the design criteria for a structure. The members are required to be stiff enough to avoid resonance with vibrating machinery, or they must avoid excitation due to seismic base motion.

There are several construction codes[1-7], used in the United States, that give stress and deflection criteria for linear structures. The construction codes require that members be designed with a *factor of safety*, which is the ratio of member strength to applied member load. When using elastic stress analysis methods, also called "working stress" methods, the factor of safety is applied by comparing the computed member stress to an allowable value, which is some fraction of yield or ultimate stress. Since many materials exhibit large amounts of strain hardening, the member's factor of safety may range from approximately 1.5 to 3.0, or even higher.

When the ultimate strength design method is used, the factor of safety may also be in this range of 1.5–3.0. But instead of using allowable stresses as criteria, the loads are multiplied by "load factors" and compared to ultimate values. This method is used in Part 2 of the AISC Construction Specification,[1] in the ACI Building Code[2] for concrete structures and in Appendix XVII-4000 of the ASME Boiler and Pressure Vessel Code,[3] Section III, Division 1. All of these codes give acceptance criteria for straight members as well as other structural elements.

It is important to note the classification of member stresses into primary and secondary categories because stress criteria may differ for these two categories. A *primary* stress is one which maintains equilibrium of the member. Its deflection is not limited, and, when primary stress reaches some high value, the member can be expected to fail. On the other hand, *secondary* stresses result from self-constraint of the structure, and deflections or rotations associated with these stresses are self-limiting. The tensile stress in a tension rod is primary, whereas the bending stress at the ends of a bolted truss member is secondary. Often, the latter stress is ignored by designers and analysts, but, nevertheless, it does exist, and it can affect the integrity of a structure. Design criteria such as the ASME Code[3] allow higher limits for secondary stress but, because of the difficulty in categorizing stress, the analysts often lump the primary and secondary stresses together, and they conservatively compare the combined stresses to primary stress limits. In any event the secondary stresses in structures composed of straight members will reduce the capacity of any beam column member and, hence, the secondary stress should be computed and evaluated.

REFERENCES

7.2-1 *AISC Specification for the Design, Fabrication, and Erection of Structural Steel for Buildings* (with commentary), American Institute of Steel Construction, Chicago, November 1978.

7.2-2 *ACI Building Code Requirements for Reinforced Concrete (ACI 318-77)*, American Concrete Institute, Detroit, 1977.

7.2-3 *ASME Boiler and Pressure Vessel Code, Section III, Nuclear Power Plant Components, Division 1, Subsection NF, Component Supports*, American Society of Mechanical Engineers, New York, 1980.

7.2-4 *Cold-Formed Steel Design Manual*, American Iron and Steel Institute, Washington, D.C., March 1977.

7.2-5 *Stainless Steel Cold-Formed Structural Design Manual*, American Iron and Steel Institute, Washington, D.C., 1974.

7.2-6 *Specifications for Aluminum Structures*, The Aluminum Association, Inc., Washington, D.C., April 1976.

7.2-7 *Timber Construction Manual*, American Institute of Timber Construction, Wiley, New York, 1974.

7.3 BEAM FLEXURE

Beam bending, or flexure, may be analyzed by working stress theory using the following assumptions:

1. Under beam loading, all stresses are elastic.
2. Deflections are small.
3. The straight members are slender, that is, the ratio of depth to length is small ($\leq \frac{1}{10}$).
4. Plane sections, normal to the member axis, remain plane after loading.

Many strength of materials texts give elastic beam theory, including References 1 and 2. Using differential equations, a variety of solutions for beam problems have been developed for many boundary and loading conditions. The deformed shape of an elastic beam appears in Fig. 7.3-1. It may be shown[1] that the curvature of a beam differential element is

$$\frac{1}{\rho} = \frac{M}{EI} \tag{7.3-1}$$

where ρ = radius of curvature
M = bending moment
E = material modulus of elasticity
I = section moment of inertia

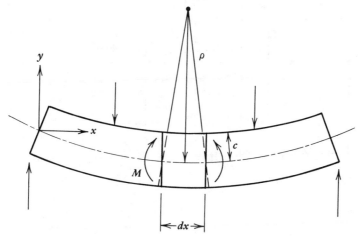

Fig. 7.3-1 Beam differential element.

TABLE 7.3-1 FORMULAS FOR BEAMS

Structure	Shear ↑↓	Moment	Slope	Deflection ↓
Simply Supported Beam				
	$S_A = -\dfrac{M_0}{L}$	M_0	$\theta_A = \dfrac{M_0 L}{3EI}$ $\theta_B = -\dfrac{M_0 L}{6EI}$	$y_{max} = 0.062\,\dfrac{M_0 L^2}{EI}$ at $x = 0.442L$
	$S_A = \dfrac{W}{2}$	$M_C = \dfrac{WL}{4}$	$\theta_A = -\theta_B = \dfrac{WL^2}{16EI}$	$y_C = \dfrac{WL^3}{48EI}$
	$S_A = \dfrac{Wb}{L}$ $S_B = -\dfrac{Wa}{L}$	$M_a = \dfrac{Wab}{L}$	$\theta_A = \dfrac{Wab}{6EIL}(L+b)$ $\theta_B = -\dfrac{Wab}{6EIL}(L+a)$	$y_a = \dfrac{Wa^2 b^2}{3EIL}$
	$S_A = \dfrac{wL}{2}$	$M_C = \dfrac{wL^2}{8}$	$\theta_A = -\theta_B = \dfrac{wL^3}{24EI}$	$y_c = \dfrac{5wL^4}{384EI}$

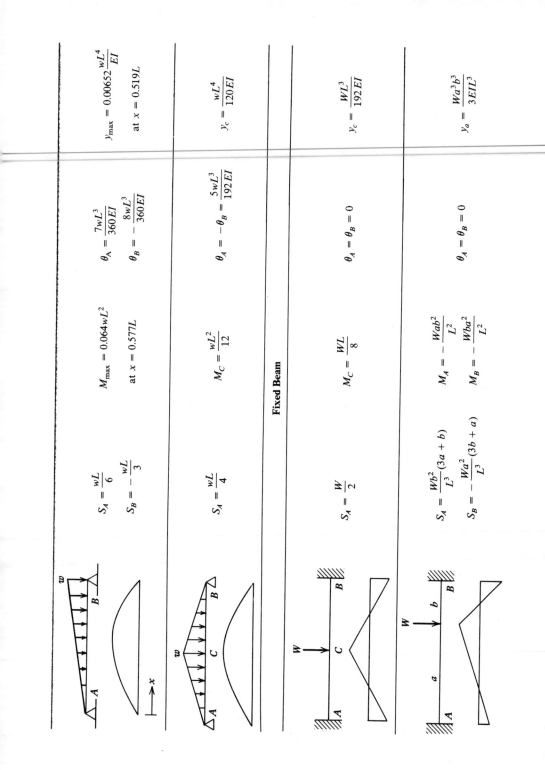

$$M_{max} = 0.064wL^2$$
$$\text{at } x = 0.577L$$

$$y_{max} = 0.00652\frac{wL^4}{EI}$$
$$\text{at } x = 0.519L$$

$$S_A = \frac{wL}{6}$$
$$S_B = -\frac{wL}{3}$$

$$\theta_A = \frac{7wL^3}{360EI}$$
$$\theta_B = -\frac{8wL^3}{360EI}$$

$$M_C = \frac{wL^2}{12}$$

$$S_A = \frac{wL}{4}$$

$$\theta_A = -\theta_B = \frac{5wL^3}{192EI}$$

$$y_c = \frac{wL^4}{120EI}$$

Fixed Beam

$$M_C = \frac{WL}{8}$$

$$S_A = \frac{W}{2}$$

$$\theta_A = \theta_B = 0$$

$$y_c = \frac{WL^3}{192EI}$$

$$M_A = -\frac{Wab^2}{L^2}$$
$$M_B = -\frac{Wba^2}{L^2}$$

$$S_A = \frac{Wb^2}{L^3}(3a+b)$$
$$S_B = -\frac{Wa^2}{L^3}(3b+a)$$

$$\theta_A = \theta_B = 0$$

$$y_a = \frac{Wa^3b^3}{3EIL^3}$$

TABLE 7.3-1 (Continued)

Structure	Shear $\uparrow\downarrow$	Moment	Slope	Deflection $\downarrow$
		Fixed Beam (Continued)		
	$S_A = \dfrac{wL}{2}$	$M_A = M_B = -\dfrac{wL^2}{12}$	$\theta_A = \theta_B = 0$	$y_C = \dfrac{wL^4}{384EI}$
	$S_A = \dfrac{3wL}{20}$ $S_B = -\dfrac{7wL}{20}$	$M_A = -\dfrac{wL^2}{30}$ $M_B = -\dfrac{wL^2}{20}$	$\theta_A = \theta_B = 0$	$y_{max} = 0.00131\dfrac{wL^4}{EI}$ at $x = 0.525L$
	$S_A = \dfrac{wL}{4}$	$M_A = M_B = -\dfrac{5wL^2}{96}$	$\theta_A = \theta_B = 0$	$y_C = \dfrac{0.7wL^4}{384EI}$
		Cantilever Beam		
	0	M_0	$\theta_A = \dfrac{M_0 L}{EI}$	$y_A = -\dfrac{M_0 L^2}{2EI}$

$$W$$

$$M_B = -WL$$

$$\theta_A = -\frac{WL^2}{2EI}$$

$$y_A = \frac{WL^3}{3EI}$$

$$S_B = -wL$$

$$M_B = -\frac{wL^2}{2}$$

$$\theta_A = -\frac{wL^3}{6EI}$$

$$y_A = \frac{wL^4}{8EI}$$

$$S_B = -\frac{wL}{2}$$

$$M_B = -\frac{wL^2}{6}$$

$$\theta_A = -\frac{wL^3}{24EI}$$

$$y_A = \frac{wL^4}{30EI}$$

$$S_B = -\frac{wL}{2}$$

$$M_B = -\frac{wL^2}{2}$$

$$\theta_A = -\frac{wL^3}{8EI}$$

$$y_A = \frac{11wL^4}{120EI}$$

TABLE 7.3-1 (*Continued*)

Structure	Shear $\uparrow\downarrow$	Moment	Slope	Deflection $\downarrow$
		Propped Cantilever		

$$S_A = -\frac{3M_0}{2L}$$

$$M_B = -\frac{M_0}{2}$$

$$\theta_A = \frac{M_0 L}{4EI}$$

$$y_{max} = \frac{M_0 L^2}{27EI}$$
$$\text{at } x = \frac{L}{3}$$

$$S_A = \frac{5W}{16}$$

$$M_B = -\frac{3WL}{16}$$
$$M_C = \frac{5WL}{32}$$

$$\theta_A = \frac{WL^2}{32EI}$$

$$y_{max} = 0.00932\frac{WL^3}{EI}$$
$$\text{at } x = 0.447L$$

$$S_A = \frac{Wb^2}{2L^3}(a + 2L)$$
$$S_B = -\frac{Wa}{2L^3}(3L^2 - a^2)$$

$$M_B = -\frac{Wab}{L^2}\left(a + \frac{b}{2}\right)$$

$$\theta_A = \frac{Wab^2}{4EIL}$$

$$y_a = \frac{Wa^2 b^3}{12EIL^3}(3L + a)$$

$$S_A = \frac{3wL}{8}$$

$$M_B = -\frac{wL^2}{8}$$

$$\theta_A = \frac{wL^3}{48EI}$$

$$y_{max} = 0.0054\frac{wL^4}{EI}$$
$$\text{at } x = 0.422L$$

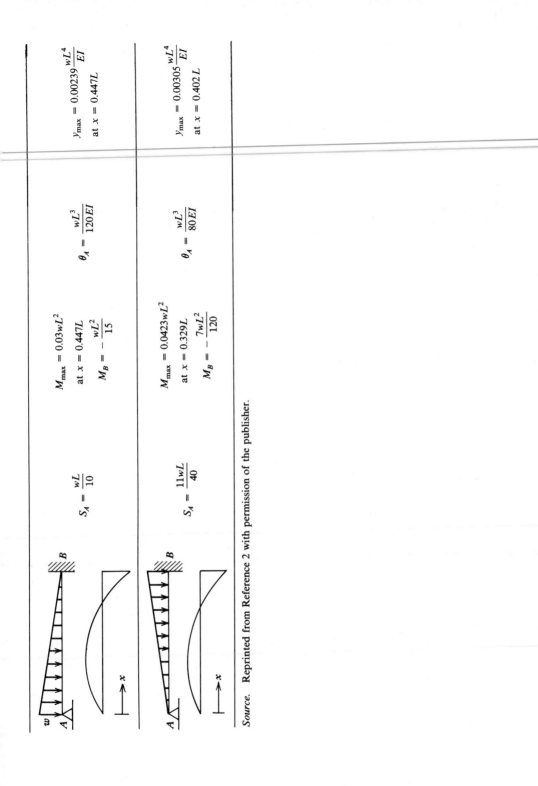

$M_{max} = 0.03wL^2$
at $x = 0.447L$
$M_B = -\dfrac{wL^2}{15}$

$S_A = \dfrac{wL}{10}$

$\theta_A = \dfrac{wL^3}{120EI}$

$y_{max} = 0.00239\dfrac{wL^4}{EI}$
at $x = 0.447L$

$M_{max} = 0.0423wL^2$
at $x = 0.329L$
$M_B = -\dfrac{7wL^2}{120}$

$S_A = \dfrac{11wL}{40}$

$\theta_A = \dfrac{wL^3}{80EI}$

$y_{max} = 0.00305\dfrac{wL^4}{EI}$
at $x = 0.402L$

Source. Reprinted from Reference 2 with permission of the publisher.

Also, for small deflections it may be shown that

$$\frac{1}{\rho} = \frac{d^2y}{dx^2}$$

(7.3-2)

where x = coordinate along beam axis
$\quad\quad y$ = beam deflection, normal to axis

and, hence,

$$\frac{d^2y}{dx^2} = \frac{M}{EI}$$

(7.3-3)

which is the fundamental differential equation of elastic beam theory. Differentiating Equation (7.3-3) provides expressions for internal shear force V and applied load w.

$$\frac{d^3y}{dx^3} = \frac{V}{EI}$$

(7.3-4)

$$\frac{d^4y}{dx^4} = \frac{w}{EI}$$

(7.3-5)

Boundary conditions are necessary to determine the expressions for beams after the general solutions to the differential equation are found. Three typical sets of boundary conditions are

1. Simple beam (pin connections at ends)

$$y = \frac{d^2y}{dx^2} = 0 \quad \text{at } x = 0, L$$

(7.3-6)

2. Fixed-ended beam

$$y = \frac{dy}{dx} = 0 \quad \text{at } x = 0, L$$

(7.3-7)

3. Cantilever beam

$$y = \frac{dy}{dx} = 0 \quad \text{at } x = 0$$

$$\frac{d^2y}{dx^2} = \frac{M}{EI} \quad\quad \text{at } x = L$$

$$\frac{d^3y}{dx^3} = \frac{V}{EI} \quad\quad \text{at } x = L$$

(7.3-8)

A set of solutions to typical problems of isolated beams is given in Table 7.3-1,[2] which gives solutions for shear, moment, rotation, and deflection.

After the analyst determines the internal bending moment in a beam, the bending stress may be determined. This stress is normal to the plane section under consideration and, hence, it is a direct (normal) stress parallel to the member axis. Figure 7.3-2 illustrates the linear distribution of bending stress at a member section. For a straight member in pure flexure, the *neutral axis* is coincident with the geometric centroidal axis of the member cross section. From Fig. 7.3-2 it can be seen that stress on one side of the neutral axis is tensile (positive), while on the other side it is compressive (negative) but the bending stress is zero at the neutral axis. The following expression gives the bending stress

$$\sigma = \frac{My}{I}$$

(7.3-9)

where y is the distance from the neutral axis. Since the bending stress is maximum at the extreme fiber of the section, that is, one surface of the beam, it follows that

$$\sigma_{max} = \frac{M}{S}$$

(7.3-10)

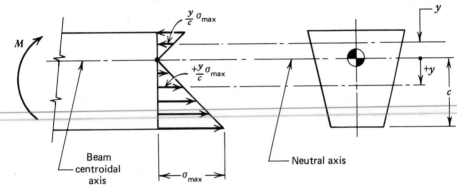

Fig. 7.3-2 Beam in pure flexure.

where S = section modulus = I/c
$\quad\quad c$ = distance from the neutral axis to the outermost fiber

If the beam section is symmetric with respect to the neutral axis, as it would be with a rectangular or circular cross section, then the maximum tensile bending stress is equal to the maximum compressive bending stress. Table 7.3-2 gives section moduli of some typical cross sections.

TABLE 7.3-2   SECTION MODULI

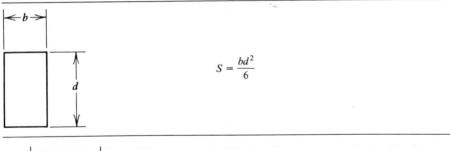

$$S = \frac{bd^2}{6}$$

$$S = \frac{1}{6d}\left[bd^3 - (b - t_w)(d - 2t_f)^3 \right]$$

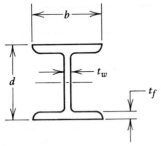

TABLE 7.3-2 (*Continued*)

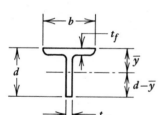

$$\bar{y} = \frac{\frac{1}{2}\left[t_f^2 b + \left(d^2 - t_f^2\right)t_w\right]}{t_f b + \left(d - t_f\right)t_w}$$

$$S_{\text{top}} = \frac{1}{\bar{y}}\left\{\frac{bt_f^3}{12} + \frac{t_w\left(d - t_f\right)^3}{12} + bt_f\left(\bar{y} - \frac{t_f}{2}\right)^2 + \left(d - t_f\right)t_w\left[\left(d - \bar{y}\right) - \left(\frac{d - t_f}{2}\right)\right]^2\right\}$$

$$S_{\text{bottom}} = \frac{1}{d - \bar{y}}\left\{\frac{bt_f^3}{12} + \frac{t_w\left(d - t_f\right)^3}{12} + bt_f\left(\bar{y} - \frac{t_f}{2}\right)^2 + \left(d - t_f\right)t_w\left[\left(d - \bar{y}\right) - \left(\frac{d - t_f}{2}\right)\right]^2\right\}$$

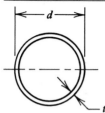

$$\bar{y} = \frac{2d^2 t_w + \left(b - 2t_w\right)t_f^2}{2\left[2dt_w + \left(b - 2t_w\right)t_f\right]}$$

$$S_{\text{top}} = \frac{1}{\bar{y}}\left[\frac{2t_w d^3}{12} + \frac{\left(b - 2t_w\right)t_f^3}{12} + 2dt_w\left(\frac{d}{2} - \bar{y}\right) + t_f\left(b - 2t_w\right)\left(\bar{y} - \frac{t_f}{2}\right)^2\right]$$

$$S_{\text{bottom}} = \frac{1}{d - \bar{y}}\left[\frac{2t_w d^3}{12} + \frac{\left(b - 2t_w\right)t_f^3}{12} + 2dt_w\left(\frac{d}{\bar{y}} - \bar{y}\right)^2 + t_f\left(b - 2t_w\right)\left(\bar{y} - \frac{t_f}{2}\right)^2\right]$$

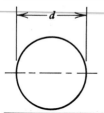

$$S = \frac{\pi d^3}{32}$$

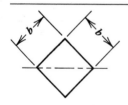

$$S = \frac{\pi d^2 t}{4}$$

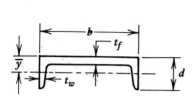

$$S = \frac{b^3}{6\sqrt{2}}$$

TABLE 7.3-2 (*Continued*)

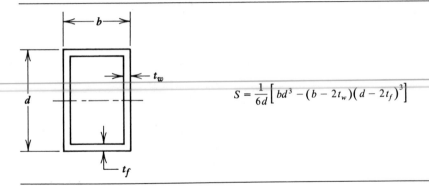

$$S = \frac{1}{6d}\left[bd^3 - (b - 2t_w)(d - 2t_f)^3\right]$$

In some cases a beam will have no axis of symmetry in its cross section. This is true for an angle with unequal legs or a Z section. In this case the principal axes of the section should be determined and the bending moments should then be resolved into components about the principal axes. After determining section properties for each axis, the member stresses may be found as shown before. If the member does have an axis of symmetry, but the load is askew to that axis, then the resulting bending moment can be resolved into components about the principal axes that are the same as the axes of symmetry. A full explanation of this class of problem, known as unsymmetrical bending, may be found in Reference 3. See also Chapter 3.

REFERENCES

7.3-1 E. P. Popov, *Introduction to Mechanics of Solids*, Prentice-Hall, Englewood Cliffs, NJ, 1968.
7.3-2 E. H. Gaylord and C. N. Gaylord, *Structural Engineering Handbook*, 2nd ed., McGraw-Hill, New York, 1979.
7.3-3 F. B. Seely and J. O. Smith, *Advanced Mechanics of Materials*, 2nd ed., Wiley, New York, 1952.

7.4 SHEAR STRESS AND DEFORMATION

While beam bending stress may be the most dominant in a straight member with a transverse load, the internal shear forces can result in stress and strain that affect the integrity and response of the member. Of course, for a beam in pure flexure, there is no shear force and, hence, no shearing stress nor strain. But this load condition exists only for a small minority of structural beams and, in fact, bending and shear usually occur simultaneously. Often, the shear force is maximum at the member ends, and the designer must use care in designing connections to resist the maximum shear forces in the beam.

From Equations (7.3-3) and (7.3-4) the shear force at any section may be determined from

$$V = \frac{dM}{dx} \tag{7.4-1}$$

A rigorous analysis[1] of the beam internal stress gives an expression for shear "flow," which is a shear force per unit distance,

$$q = \frac{VQ}{I} \tag{7.4-2}$$

where q = shear flow
   V = shear force over entire section
   Q = first moment of area beyond point under consideration about neutral axis (Fig. 7.4-1)

From equilibrium, it may be shown that shear stress in the plane of the cross section is equal to shear stress normal to the cross section. Dividing Equation (7.4-2) by thickness t gives the shear stress formula

$$\tau = \frac{VQ}{It} \tag{7.4-3}$$

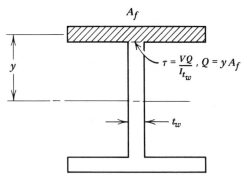

Fig. 7.4-1   Shear stress at flange/web connection.

where t is the cross-section thickness at the point under consideration. The shear stress is maximum at the neutral axis whereas the bending stress, as shown in Section 7.3, is maximum at the extreme fibers of the beam. It may be shown that the shear stress distribution in a rectangular section is parabolic. The maximum value is

$$\tau_{max} = \frac{3}{2}\frac{V}{A} \qquad (7.4\text{-}4)$$

where A is the rectangular section area. For wide-flange sections the applicable codes usually require the average shear stress in the beam web be compared to an allowable value. For this evaluation it is generally assumed that all shear stress in the flanges can be neglected; hence the computed beam shear stress is simply the shear force divided by the web cross-section area.

A common point of interest is the shear stress or shear flow at the web-to-flange connection in a wide-flange beam (Fig. 7.4-1). In Equation (7.4-2) or (7.4-3), Q is the flange area multiplied by the distance between the flange centroid and the beam neutral axis. The web thickness, t_w, is used in Equation (7.4-3). If the beam consists of plates welded together, then t_w is the effective thickness of the weld. Or if the beam is built of wood with the flange screwed to the web, then the shear flow can be computed and the spacing can be determined given the size of the fasteners.

Shear strain is related to shear stress by the following expression:

$$\gamma = \frac{\tau}{G} \qquad (7.4\text{-}5)$$

where γ = shear strain
$\quad\quad\; G$ = material shear modulus

Without derivation, the shear deformation over a segment of beam with a constant internal shear force, V, is

$$y_s = \frac{VL}{A_s G} \qquad (7.4\text{-}6)$$

where L = segment length
$\quad\quad\; A_s$ = equivalent shear area = $\frac{5}{6}A$ for a rectangular section

This is illustrated in Fig. 7.4-2 where both flexural deformation, y_f, and shear deformation, y_s, are shown. For most slender members the shear deflection given by Equation (7.4-6) is negligible compared to bending deflection. Hence, the deflections given by the expressions in Table 7.3-1 neglect shear deformation. However, some beams resist large loads over a short span. Such a beam is proportioned so that it is usually deep compared to its length. Its bending deflection will be small compared to its shear deflection. A concrete shear wall in a building structure is an example in which shear deflection is many times greater than bending deflection. In general, shear deformation can be neglected when the ratio of beam length to depth is ≥ 10. Several cases of shear deformation are given by Blake.[2]

Shear deformation, along with rotary inertia, is an important secondary effect in Timoshenko theory of beam dynamics. Cowper[3] has shown that the shear shape factor, K, applied to beam deflection varies somewhat according to the solution of the equations of beam motion.

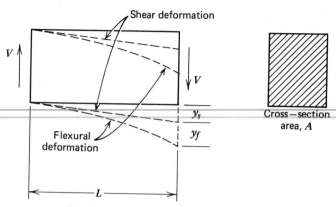

Fig. 7.4-2   Shear and flexural deformation.

REFERENCES

7.4-1   E. P. Popov, *Introduction to Mechanics of Solids*, Prentice-Hall, Englewood Cliffs, NJ, 1968.
7.4-2   A. Blake, *Practical Stress Analysis in Engineering Design*, Dekker, New York, 1982.
7.4-3   G. R. Cowper, "The Shear Coefficient in Timoshenko's Beam Theory," *ASME J. Appl. Mech.*, June 1966.

7.5   TORSIONAL RESPONSE OF STRAIGHT MEMBERS

In Fig. 7.1-1 degrees of freedom 4 and 10 are the torsional degrees of freedom of a straight member. A rigorous analysis of the torsional response of a member may be performed using elasticity theory.[1] It can be shown that the following expression gives the rotation over a member segment with a constant value of torque

$$\theta = \frac{T}{GJ} \tag{7.5-1}$$

where θ = rotation per unit length of the member
T = torque (torsional moment)
G = shear modulus
J = torsional constant dependent on the member cross-section geometry

The torsional constant, J, for several cross-section shapes is given in Table 7.5-1. For a circular section the torsional constant is equal to the polar moment of inertia, but this is not so for other shapes. While J may be derived from elasticity theory for a limited number of shapes, the membrane analogy may be used to derive J for irregular shapes, including structural steel shapes.

TABLE 7.5-1   TORSIONAL CONSTANTS

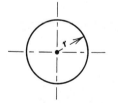

$$\theta = \frac{T}{GJ}, \; \tau = \text{maximum shear stress}$$

$$J = \frac{\pi r^4}{2} \qquad \tau = \frac{Tr}{J}$$

$$J = 2\pi r^3 t \qquad \tau = \frac{Tr}{J}$$

TABLE 7.5-1 (*Continued*)

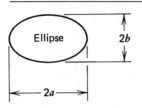

$$J = \frac{\pi a^3 b^3}{a^2 + b^2} \qquad \tau = \frac{2T}{\pi ab^2}$$

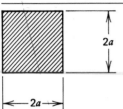

$$J = 2.25a^4 \qquad \tau = \frac{0.6T}{a^3}$$

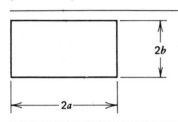

$$J = ab^3 \left[\frac{16}{3} - 3.36\frac{b}{a}\left(1 - \frac{b^4}{12a^4}\right) \right]$$

$$\tau = \frac{T(3a + 1.8b)}{8a^2 b^2}$$

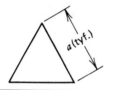

$$J = \frac{a^4\sqrt{3}}{80} \qquad \tau = \frac{20T}{a^3}$$

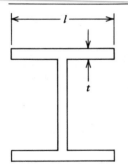

$$J \cong \sum_{i=1}^{n} \frac{l_i t_i^3}{3}$$

where n = number of plates

$$\tau \cong \frac{Tt_i}{J} \text{ for plate } i$$

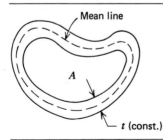

$$J = \frac{4A^2 t}{S} \qquad \tau = \frac{T}{2At}$$

where A = area enclosed by the mean line of boundaries
S = mean line length

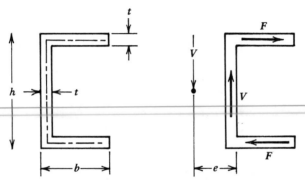

Fig. 7.5-1 Shear center for a channel. F is the transverse shear force induced by applied shear force, V; e is the shear center $= b^2h^2t/4I$; I is the cross-section moment of inertia.

Shear stress is the dominant stress resulting from torsion. In a circular section the shear stress is proportional to the distance from the center of the member.[2]

$$\tau = \frac{T\rho}{J} \quad \text{(circular sections only)} \tag{7.5-2}$$

where τ = torsional shear stress

ρ = distance from center (member axis) to the point of interest

$J = \pi R^4/2$ for a solid, circular shaft

R = member radius

Equation (7.5-2) does not apply to noncircular cross sections, although as a rule the torsional shear stress is greater at the perimeter of a section than at the centroid of the member.

Table 7.5-1 gives torsional constants for a variety of cross sections. These can be used to compute torsional rotation by Equation (7.5-1), and the table also gives expressions for maximum torsional shear stress.

The designer should be aware of the existence of warping in members with noncircular cross sections under torsional loads. Plane sections tend not to remain plane for rectangular sections, wide flanges, channels, and so on. Hence, warping stress, normal to the cross section, is induced when the cross section is constrained to remain plane. The American Institute of Steel Construction has published a handbook[3] which gives formulas for computing warping stress.

In order to avoid torsional shear stress in a member which is loaded in shear or bending, the loads must be applied through the member's shear center.[2] When the load is in the plane of symmetry, then it is on the shear center. But sometimes the load cannot be applied in a plane of symmetry. For instance, when the load is parallel to the web of a channel, the shear center is some distance away from the cross section, as shown in Fig. 7.5-1. For an angle or tee section the shear center coincides with the intersections of the legs of the angle or the web and flange of the tee. When the load cannot be applied to the shear center, then the torsional stresses. shear, and warping stress must be evaluated and added to other stresses. Other details of torsional analysis can be found in Chapter 3.

REFERENCES

7.5-1 C. Wang, *Applied Elasticity*, McGraw-Hill, New York, 1953.

7.5-2 E. P. Popov, *Introduction to Mechanics of Solids*, Prentice-Hall, Englewood Cliffs, NJ, 1968.

7.5-3 *Torsion Analysis of Rolled Steel Sections*, American Institute of Steel Construction, Chicago, IL, 1983.

7.6 COMPOSITE BEAMS

Different materials can be used to construct a straight beam. When these materials are bonded together to form a single beam, the member is called a composite beam. When forces are applied to this member, the different materials act together, as if they were one material, that is, plane sections remain plane and there is strain compatibility between the different materials. One of the most common forms of composite construction is reinforced concrete. Another example is a timber beam with steel cover plates attached.

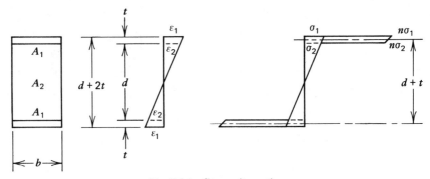

Fig. 7.6-1 Composite section.

A simple elastic stress analysis of a composite beam can be performed using a *transformed cross section*. Noting that if the materials remain elastic, then the only significant material difference is the modulus of elasticity. All materials can be transformed to one single material by adjusting the moduli of elasticity.[1] This concept is illustrated in Fig. 7.6-1 for a beam composed of two materials. A soft material such as wood is sandwiched between two cover plates such as steel. Because plane sections remain plane, the strain distribution is linear. But the stresses in materials 1 and 2 at the interface between the materials are

$$\sigma_1 = E_1 \varepsilon \tag{7.6-1}$$

$$\sigma_2 = E_2 \varepsilon \tag{7.6-2}$$

where σ_1, σ_2 = bending stresses in the two materials
$\quad E_1, E_2$ = material moduli of elasticity
$\quad\quad \varepsilon$ = strain

Dividing Equation (7.6-1) by (7.6-2) and rearranging gives an expression for the cover plate material stress

$$\sigma_1 = n\sigma_2 \tag{7.6-3}$$

where n is the modular ratio = E_1/E_2. Now, in order to use the flexure formula [Eq. (7.3-9) or (7.3-10)], the cover plate material can be transformed to an equivalent section of material 1 as illustrated in Fig. 7.6-2. The section properties (I, S) can be computed using this transformed section. After computing material 2 bending stress, based on the transformed section, the actual material 1 stress can be computed using Equation (7.6-3). The transformed section can also be used to compute shear flow and shear stress per Equations (7.4-2) and (7.4-3), respectively, and the shear flow can be used to design shear connectors between the two materials.

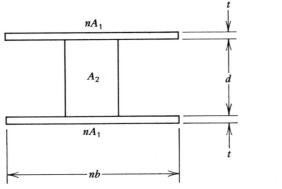

Fig. 7.6-2 Transformed section.

REFERENCE

7.6-1 G. A. Olsen, *Elements of Mechanics of Materials*, Prentice-Hall, Englewood Cliffs, NJ, 1958.

7.7 ULTIMATE BENDING STRENGTH OF BEAMS

It is a well-known fact that structures can exhibit some plastic deformation under load and still perform their intended functions with a large margin of safety. When a framed structure is loaded, it has the ability to *redistribute* its internal forces and moments if there is some yielding of the material at a few locations. In order for the redistribution to take place, the material must have adequate *ductility* so that there can be deformation at the yielded location without fracture. These phenomena were observed by researchers, and the ultimate strength method, or plastic design method, as applied to steel structures, was developed. Their method is well documented in ASCE Manual 41, *Plastic Design in Steel, A Guide and Commentary*.[1] As a result of laboratory and theoretical research, the AISC Specification,[2] Part 2, now provides rules for the plastic design of structures.

The steel alloys which AISC allows to be used in plastic design are

Structural steel, ASTM A36

High-strength low-alloy structural steel, ASTM A242

High-strength low-alloy structural manganese vanadium steel, ASTM A441

Structural steel with 42,000 psi minimum yield point, ASTM A529

High-strength low-alloy columbium-vanadium steels of structural quality, ASTM A572

High-strength low-alloy structural steel with 50,000 psi minimum yield point to 4 in. thick, ASTM A588

The method can also be applied to other ductile materials. In fact, the ultimate strength method is the preferred method used by the ACI Building Code[3] for reinforced-concrete structures that exhibit substantial ductility when the correct amount of reinforcing steel is used. Other considerations of plastic design are included in Chapter 6.

The materials listed above can exhibit nonductile behavior under certain conditions. Material imperfections and manufacturing process can be the source of this problem. Also, when structural steel, including A36, is subjected to cold service temperatures it can become brittle. The Charpy impact test is used to determine if a heat of steel may exhibit nonductile behavior. It measures the absorbed energy under impact load. Construction codes such as the AASHTO Code[4] for bridges require Charpy impact tests under certain conditions.

7.7-1 Ultimate Moment of a Steel Beam

As discussed in Section 7.3, the main principle of beam flexure is "plane sections remain plane." This principle is utilized to determine the ultimate bending strength of a beam cross section. Figure 7.7-1 shows that when the bending moment is applied in successive increments the stress distribution is linear until the extreme fiber yields load in increment 2. As the moment increases, due to an increase in external load, the yielding propagates into the internal fibers of the beam, and the stress distribution becomes nonlinear in load increments 3 and 4 even though the strain distribution is linear because "plane sections remain plane." Finally, in load increment 5, the material is fully yielded; it is in compression on one side of the neutral axis and in tension on the other side. Of course, with the increase in bending moment, the beam section has also developed significant curvature, or rotation, and the moment/rotation diagram for each corresponding load increment is shown in Fig. 7.7-2, which shows why material ductility is a necessary prerequisite for plastic behavior, When the section is fully plastic, as in load increment 5, a *plastic hinge* has developed and the moment cannot be

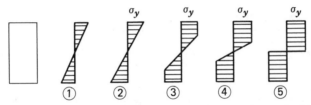

Fig. 7.7-1 Successive stages of stress distribution. (Reprinted from Reference 1 with the permission of the American Society of Civil Engineers.)

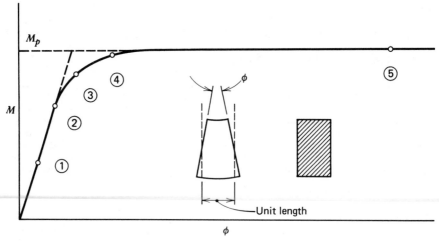

Fig. 7.7-2 Moment–curvature relationship for beam of rectangular section in bending. (Reprinted from Reference 1 with permission of the American Society of Civil Engineers.)

increased as illustrated in Fig. 7.7-2. The corresponding plastic or ultimate moment is expressed by

$$M_P = \int_A \sigma_y \, y \, dA \qquad (7.7\text{-}1)$$

where σ_y = yield stress level
y = distance from the neutral axis
A = cross-section area

Instead of using Equation (7.7-1), the plastic moment is more conveniently computed from

$$M_P = \sigma_y Z \qquad (7.7\text{-}2)$$

where Z is the plastic modulus of a section. For sections that are irregular, the plastic modulus can be expressed as a function of the elastic section modulus, S, discussed in Section 7.3

$$Z = KS \qquad (7.7\text{-}3)$$

where K is the ratio of plastic to elastic moduli given for some sections in Table 7.7-1. Tables of plastic moduli for rolled steel sections are given in the AISC Manual of Steel Construction.[2] The designer must use care when computing the ultimate moment by Equation (7.7-1) or (7.7-2). It is conceivable that the compression flange of the member could buckle prior to developing the full plastic moment. The ultimate moment, M_u, is then less than the plastic moment, M_P

$$M_u < M_P \qquad (7.7\text{-}4)$$

Section 7.8 discusses flange buckling and gives safeguards to avoid this condition.

TABLE 7.7-1   SECTION SHAPE FACTORS

Flanges Only				Solid Round	
$K = 1$	$K = 1$ to 1.5		$K = 1.5$	$K = 1.7$	$K = 2.0$

7.7-2 Ultimate Moment of a Reinforced-Concrete Beam

Equation (7.7-1) applies to a homogeneous beam, but a similar expression for a composite member such as a reinforced-concrete beam can be derived.[5] The bending strength (ultimate moment) of such a straight member may be computed by utilizing equilibrium and strain compatibility. Figure 7.7-3a illustrates a singly reinforced rectangular concrete beam. The cross-section strain is shown in Fig. 7.7-3b. Research has shown that the maximum useful concrete compressive strain is

$$\varepsilon_{c,max} = 0.003 \tag{7.7-5}$$

Also, the steel tensile force is maximum when the steel strain is equal to or greater than the yield strain

$$\varepsilon_s \geq \varepsilon_y \tag{7.7-6}$$

Since reinforced concrete generally uses carbon steel reinforcing, it is reasonable to assume an elastic–perfectly-plastic stress–strain relationship. Hence, when Equation (7.7-6) is satisfied, the steel tension force T, shown in Fig. 7.7-3c, is

$$T = A_s f_y \tag{7.7-7}$$

where A_s = cross-section area of tensile steel
f_y = yield strength of tensile steel

Noting that concrete is weak in tension and therefore assumed to develop no tensile stress, the stress distribution is shown in Fig. 7.7-3c, but the concrete compressive stress can be approximated by the rectangular stress block shown in Fig. 7.7-3d. An average stress of $0.85f_c'$ where f_c' is the concrete compressive strength, is used for this stress distribution and the block has a depth of $\beta_1 x$. For most concrete strengths, $\beta_1 = 0.85$, but it shall not be less than 0.65 as required by Section 10.2.7 of the ACI Building Code.[3]

In order to maintain equilibrium in the cross section, the concrete compressive force must equal the steel tensile force ($C = T$). The flexural strength is then easily determined from statics (see Eq. 3.35 of Reference 6)

$$M_n = A_s f_y \left(d - 0.59 \frac{A_s f_y}{f_c' b} \right) \tag{7.7-8}$$

where M_n = bending moment capacity of the cross section
d = beam depth, distance from extreme compression fiber to centroid of reinforcement steel
f_c' = concrete compressive strength
b = beam width

Equation (7.7-8) assumes that, as the flexural strength of the member is developed, the reinforcement steel will yield before the concrete crushing strain ($\varepsilon_c = 0.003$) is attained. This is desirable since a concrete compression failure may be sudden and without warning. On the other hand, tensile yielding of the steel is slow and gradual as the member capacity develops, and visible deflections, accompanied by concrete cracking, will warn of an impending failure. The concrete beam failure mode is a function of the *reinforcement ratio* ρ

$$\rho = \frac{A_s}{bd} \tag{7.7-9}$$

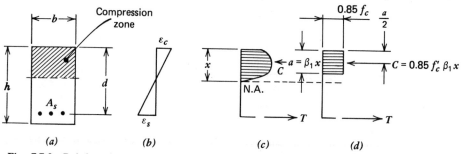

Fig. 7.7-3 Reinforced-concrete beam. (*a*) Beam; (*b*) strain; (*c*) actual stress distribution; (*d*) rectangular stress block.

where terms are defined above. For low values of ρ, failure will be by tensile yielding of the steel whereas high values of ρ will result in a sudden concrete compression failure. A *balanced condition* is said to exist when the reinforcement ratio is such that the two failure modes occur simultaneously, that is, the concrete strain reaches 0.003 and the steel strain reaches the yield strain

$$\varepsilon_s = \varepsilon_y = \frac{f_y}{E} \tag{7.7-10}$$

where f_y = yield strength of steel
E = steel modulus of elasticity

The ACI Building Code[3] requires that the reinforcement ratio not exceed 75% of the balanced condition ratio in order to ensure beam ductility. Any concrete design text such as Section 3.5 of Reference 6 provides a method for computing the balanced condition ratio, ρ_b. If it is shown that the

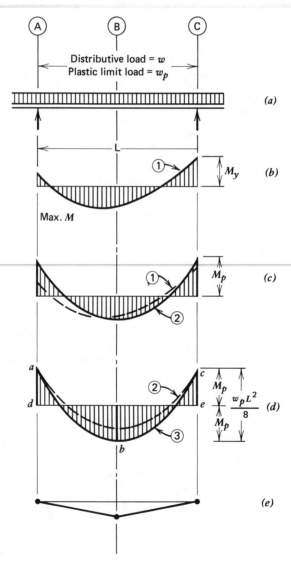

Fig. 7.7-4 Statical method of analysis applied to a continuous beam. (Reprinted from Reference 1 with permission of the American Society of Civil Engineers.)

reinforcement ratio, ρ, is less than ρ_b, then Equation (7.7-8) is valid for computing the flexural strength.

7.7-3 Limit Load

In order to make use of the ultimate moment, the designer must determine the "limit load," which is the maximum applied load(s) the structure can withstand when enough plastic hinges have formed to develop a collapse mechanism. Energy principles[1,7] can be used to determine the limit loads through the mechanism method or the statical method. The latter method is illustrated in Fig. 7.7-4, which shows a beam on multiple supports with a distributed load. The moment, M_y, shown in Fig. 7.7-4b, is the moment at first yield (corresponding to increment 2 in Fig. 7.7-1). The collapse mechanism is shown in 7.7-4e. From statics of a simple beam, the limit load is

$$\frac{w_P L^2}{8} = M_P + M_P \tag{7.7-11}$$

Rearranging gives

$$w_P = \frac{16 M_P}{L^2} \tag{7.7-12}$$

where w_P = distributed limit load
L = span length

Once the limit analysis has determined the collapse (limit) load, it can be compared to the applied design load. To ensure safe design margins, load factors (LF) are applied to the design loads before they are compared to the limit or ultimate loads, viz.,

$$(\text{LF}) P \le P_{\text{ult}} \tag{7.7-13}$$

where (LF) = load factor
P = applied design load
P_{ult} = limit load

In Equation (7.7-12), w_P is an example of the applied design load. The AISC specification requires the following load factors.

LF = 1.7 for dead load plus live load

LF = 1.3 for dead load plus live load plus wind or seismic forces

For concrete structures, the ACI Code[3] has similar load factors. For nuclear component and pipe supports, the ASME Boiler and Pressure Vessel Code[8] imposes load factors that range from 1.1 to 1.7 depending on the support service requirements for each load category, that is, under certain circumstances large deformations of the support are tolerable and, hence, the load factor is low, while for "normal" loads the factor is high.

REFERENCES

7.7-1 *Plastic Design in Steel—A Guide and Commentary*, Welding Research Council and American Society of Civil Engineers, ASCE, New York, 1971.

7.7-2 *AISC Specification for the Design, Fabrication, and Erection of Structural Steel for Buildings* (with commentary), American Institute of Steel Construction, Chicago, November 1978.

7.7-3 *ACI Building Code Requirements for Reinforced Concrete (ACI 318-77)*, American Concrete Institute, Detroit, 1977.

7.7-4 *AASHTO Standard Specifications for Highway Bridges*, 12th ed., with annual supplements, American Association of State Highway and Transportation Officials, Washington, D.C., 1977.

7.7-5 C. S. Whitney and E. Cohen, *Guide for Ultimate Strength Design of Reinforced Concrete*, *Am. Concrete Inst. J.*, **53**, Nov. 1956.

7.7-6 C. Wang and C. G. Salmon, *Reinforced Concrete Design*, 3rd ed., Harper & Row, New York, 1979.

7.7-7 M. R. Horne, *Plastic Theory of Structures*, M.I.T. Press, Cambridge, MA, 1971.

7.7-8 *Boiler and Pressure Vessel Code*, Section III, Nuclear Power Plant Components, Division 1, Appendix XVII-4000, Subsection NF, Component Supports, American Society of Mechanical Engineers, New York, 1980.

7.8 BUCKLING OF BEAM COMPRESSION ELEMENTS

The internal bending moment in a straight beam applies normal tensile stress to one side of the neutral axis and normal compressive stress to the other. If the beam is a wide-flange or I shape, the entire flange on one side of the neutral axis will be in direct compression. If the flange is unsupported, it may be susceptible to buckling just as if it were a column under compressive load.

A beam flange may buckle in one of two ways: local buckling and lateral torsional buckling. Figure 7.8-1 illustrates local buckling modes. Flange buckling due to beam bending is shown in Fig. 7.8-1*a*. In the vicinity of the applied load, or the end sections, the web may buckle as well as the flange, as illustrated in Fig. 7.8-1*b*. Web buckling in a steel wide-flange section is avoided by keeping the web compressive stress within the code limits allowed by the AISC Specification.[1] Designers can also avoid local buckling of both webs and flanges by welding stiffeners to thin compression elements as provided by the various design codes.

Strain hardening plays a role in avoiding local buckling. The effect of strain hardening of a carbon steel material on the moment–curvature relationship is illustrated in Fig. 7.8-2,

where M_p = plastic moment
$\quad EI$ = beam rigidity (elastic)
$\quad EI_{st}$ = strain hardening beam rigidity
$\quad \phi_p$ = curvature corresponding to M_p
$\quad \phi_{st}$ = curvature at onset of strain hardening
$\quad \varepsilon_y$ = yield strain
$\quad \varepsilon_{st}$ = strain at onset of strain hardening

The design codes control the maximum width–thickness ratios of flanges so that local buckling will not occur before the onset of strain hardening; ϕ_{st} in Fig. 7.8-2 is the curvature at the onset of strain hardening. For instance, the AISC specification requires beam flanges to be proportioned such that

$$\frac{b}{t} \leq \frac{95}{\sqrt{F_y}} \tag{7.8-1}$$

where b = one-half the nominal width
$\quad t$ = flange thickness
$\quad F_y$ = yield stress, ksi

The *Specifications for Aluminum Structures* also has width–thickness ratio limits in order to avoid local buckling.

Lateral torsional buckling is another possible buckling mode of the beam compression flange. This buckling mode is illustrated in Fig. 7.8-3. The compression flange buckles as a column. There are analytical solutions available[3,4] to determine the critical moment that can cause a flange to buckle. The member design can avoid this form of buckling by providing support for the compression flange and the design codes provide guidance on lateral support spacing. Also, they limit the allowable bending stress when adequate lateral support is not provided.

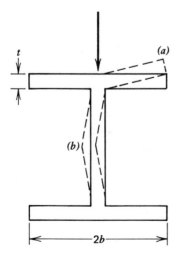

Fig. 7.8-1 Local buckling modes. (*a*) Flange buckling and (*b*) web crippling.

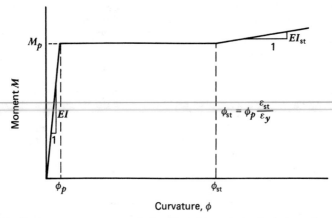

Fig. 7.8-2 Idealized moment–curvature relationship (with strain hardening). ε_{st} is strain hardening strain; ε_y is yield strain.

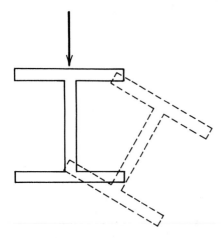

Fig. 7.8-3 Lateral torsional buckling.

REFERENCES

7.8-1 *AISC Specification for the Design, Fabrication, and Erection of Structural Steel for Buildings* (with commentary), American Institute of Steel Construction, Chicago, November 1978.

7.8-2 *Specifications for Aluminum Structures*, The Aluminum Association, Washington, D.C., April 1976.

7.8-3 B. G. Johnston, *Guide to Stability Design Criteria for Metal Structures*, 3rd ed., Wiley, New York, 1976.

7.8-4 M. Ojalvo and R. R. Weaver, "Unbraced Length Requirements for Steel I-Beams," *J. Structural Div.*, American Society of Civil Engineers, New York, March 1978.

7.9 CENTRALLY LOADED COLUMNS

The centrally loaded column is a highly idealized structural element because placing a compressive load exactly on the member's centerline is impossible. Nevertheless, it does warrant attention by structural designers and analysts. Theoretical solutions to centrally loaded column problems do have practical application when the column is fairly straight, with a maximum camber of 1/1000 of its length and a load eccentricity that is about the same order. When the column is not straight or when the distance between the axial compressive load and the member's centerline is large, then the member, which is subjected to combined axial compression and bending, is called a *beam column*. Mathematically, beam columns are treated very differently from centrally loaded columns (see Section 7.10 on beam columns).

Unlike the elastic or plastic deformation of beams, the response of centrally loaded columns is very sensitive to the material compressive stress–strain diagram, whose shape has a pronounced effect on columns used in practical design. Hence, the following sections discuss the response of columns fabricated from different materials.

The mode of failure associated with a long centrally loaded column is *buckling*. For a long column this can be a sudden catastrophic collapse of the straight member. There may be no apparent deformation of the member as load is increased; then suddenly the member displaces laterally from its centerline and the stress pattern switches from a uniform compressive stress to combined bending and compression that cannot sustain the applied load because the stresses are high. The *stability* of the member is lost. This failure mode is very different from that of beam columns that deform gradually in a flexural or bending mode as load is increased, thus providing a visible warning mechanism of impending failure. Of course, beams or beam columns may deform suddenly if they fail in local or lateral torsional buckling as discussed in Section 7.8.

7.9-1 Elastic Buckling of a Long Column

Mathematically, elastic column buckling is a characteristic value problem. As the compressive force, P, on a column is increased, it reaches a point at which a *bifurcation*[1] exists; that is, there are two possible equilibrium positions, one undeflected, and in the other the columns deflected from its long axis. The corresponding load, P_{cr}, is called the critical buckling load. This is easy to visualize using the rigid link mechanism in Fig. 7.9-1. For low values of axial force, P, the spring will easily keep the two links in a straight line. But, at some higher value of load, say P_{cr}, the center connection may displace by an amount, Δ, without violating the principal of virtual work. From equilibrium

$$P_{cr}\Delta = k\Delta\frac{L}{2} \tag{7.9-1}$$

where P_{cr} = critical axial load
Δ = unknown deflection
k = spring rate
L = length

From Equation (7.9-1) it follows that

$$P_{cr} = k\frac{L}{2} \tag{7.9-2}$$

Notice that we have not been able to determine the actual value of the deflection, Δ.

Elastic columns exhibit a bifurcation similar to that of the rigid link mechanism. Consider the initially straight, centrally loaded long column with pin ends in Fig. 7.9-2. The column stability is lost when the axial compressive load is of such a magnitude, P_{cr}, that the column may be either straight or deflected. By equating internal and external bending moments and utilizing the relationship for beam

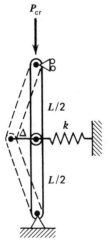

Fig. 7.9-1 Rigid link mechanism.

Fig. 7.9-2   Elastic column.

curvature [Eq. (7.3-2)], we obtain the following differential equations.[2]

$$\frac{d^2y}{dx^2} + k^2y = 0 \qquad (7.9\text{-}3)$$

where

$$k^2 = \frac{P}{EI} \qquad (7.9\text{-}4)$$

The general solution of Equation (7.9-3) is

$$y = A \sin kx + B \cos kx \qquad (7.9\text{-}5)$$

where A and B are constants. It may be shown that for nontrivial solutions of Equation (7.9-5),

$$k = \frac{n\pi}{L} \qquad (7.9\text{-}6)$$

where $n = 1, 2, 3$. Therefore, from Equation (7.9-4)

$$P = \frac{n^2\pi^2}{L^2} EI \qquad (7.9\text{-}7)$$

The critical buckling load for the pin-ended column is the smallest load possible such that $n = 1$.

$$P_{cr} = \frac{\pi^2 EI}{L^2} \qquad (7.9\text{-}8)$$

This expression, first derived by Euler,[3] can be used for other boundary conditions by determining an equivalent or *effective length*, KL, where K is called an effective length factor. Note again that the actual value of deflection, y, has not been determined for the bifurcation condition.

The effective length factor, K, is shown for different end conditions[4] in Fig. 7.9-3a. The nomograph[4] in Fig. 7.9-3b gives effective length factors for semirigid end conditions in which beams or girders that frame into the ends of columns provide partial moment restraint.

It is convenient to express the critical column load in terms of axial compressive stress

$$\sigma_{cr} = \frac{P_{cr}}{A} = \frac{\pi^2 E}{(KL/r)^2} \qquad (7.9\text{-}9)$$

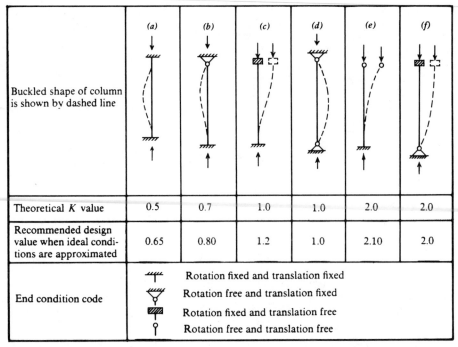

	(a)	(b)	(c)	(d)	(e)	(f)
Buckled shape of column is shown by dashed line						
Theoretical K value	0.5	0.7	1.0	1.0	2.0	2.0
Recommended design value when ideal conditions are approximated	0.65	0.80	1.2	1.0	2.10	2.0
End condition code	Rotation fixed and translation fixed					
	Rotation free and translation fixed					
	Rotation fixed and translation free					
	Rotation free and translation free					

Fig. 7.9-3(a) Effective length factor K for various end conditions. (Reprinted with permission of American Institute of Steel Construction.)

where σ_{cr} = critical stress
A = cross-section area
E = elastic modulus
KL = effective length
r = minimum radius of gyration = $(I/A)^{1/2}$

Most columns, such as wide-flange sections, have different moments of inertia about different cross-section axes. If the effective length is the same for both axes, the column will buckle about the axis that has the smallest moment of inertia and, hence, the smallest radius of gyration.

7.9-2 Inelastic Column Buckling

For a very short column the critical buckling stress should coincide with the compressive yield stress while the Euler stress, given by Equation (7.9-9), governs for long slender columns. The solid line in Fig. 7.9-4 shows the theoretical relationship between critical stress, σ_{cr}, and effective length, KL. But column tests confirm that while very long columns behave according to Equation (7.9-9) and very short columns collapse at F_y, the intermediate range of slenderness ratios gives failures somewhat below these values, as shown by the dotted line in Fig. 7.9-4. The effects of residual stress, particularly due to rolling or welding of wide-flange carbon steel sections,[5] play an important role for columns in the intermediate range of effective length or slenderness ratio. The tangent modulus theory, which is valid for a *rectangular* cross section, can account for residual stress. Figure 7.9-5 illustrates the use of the method in developing a relationship or "column curve" for the intermediate range of effective length. Notice in Fig. 7.9-5a that the compressive stress–strain curve* is rounded in the vicinity of yielding because the residual stress, when added to the applied compressive stress, causes the column section to lose stiffness at the proportional limit. The dotted line in the figure is drawn for an ideal coupon, without residual stress. An auxiliary curve of tangent modulus versus stress is shown in Fig. 7.9-5b and the column curve is shown in Fig. 7.9-5c by solving for L/r for known values of σ_{cr} and E_t from

$$\sigma_{cr} = \frac{\pi^2 E_t}{(L/r)^2} \tag{7.9-10}$$

*Developed from a stub column test in accordance with the requirements of the Structural Stability Research Council.[6]

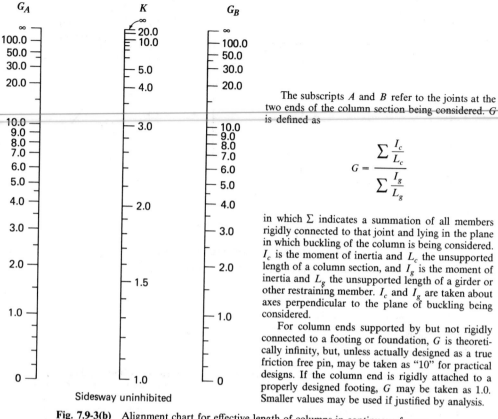

$$G = \frac{\sum \dfrac{I_c}{L_c}}{\sum \dfrac{I_g}{L_g}}$$

The subscripts A and B refer to the joints at the two ends of the column section being considered. G is defined as

in which $\sum$ indicates a summation of all members rigidly connected to that joint and lying in the plane in which buckling of the column is being considered. I_c is the moment of inertia and L_c the unsupported length of a column section, and I_g is the moment of inertia and L_g the unsupported length of a girder or other restraining member. I_c and I_g are taken about axes perpendicular to the plane of buckling being considered.

For column ends supported by but not rigidly connected to a footing or foundation, G is theoretically infinity, but, unless actually designed as a true friction free pin, may be taken as "10" for practical designs. If the column end is rigidly attached to a properly designed footing, G may be taken as 1.0. Smaller values may be used if justified by analysis.

Fig. 7.9-3(b) Alignment chart for effective length of columns in continuous frames.

This formula is only valid for a rectangular section bent about its weak axis. But it could also be used for a wide-flange section bent about its strong axis or a wide-flange section having no residual stress bent about its weak axis.

Studies by Shanley have shown that the tangent modulus theory gives the lower bound to critical stress and the actual critical stress may be slightly higher.[2,7]

The tangent modulus formula, Equation (7.9-10), has limited application, and it is somewhat difficult to use since it depends on a number of material parameters. Other approximate column

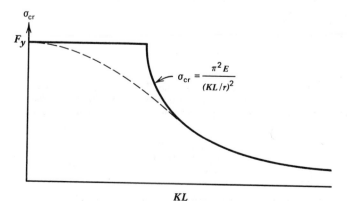

Fig. 7.9-4 Theoretical column buckling (solid line) versus actual buckling curve (dotted).

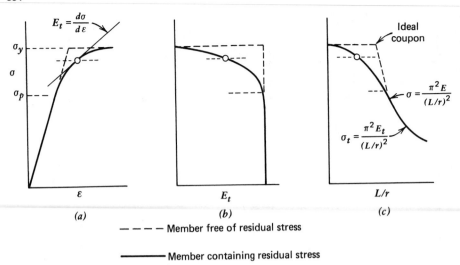

Fig. 7.9-5 Influence of residual stress on column curve.

formulas for various materials have been developed, and they are given in Sections 7.9-3 to 7.9-5 where appropriate factors of safety are also discussed.

7.9-3 Carbon Steel Columns

The residual compressive stress, caused by cooling after a hot wide-flange member is rolled at the mill, can be as high as 20 ksi.[8] Cold straightening of carbon steel members is also done at the mill to limit the out-of-straightness to 1/1000 of the length. This, too, imposes residual stress in the member and affects the critical load. Because the residual stress pattern varies so much, the tangent modulus formula, besides being difficult to use and limited to certain shapes, is not considered reliable. Therefore, much research, including testing, has been performed to develop a column curve that is reliable in the intermediate range of slenderness ratio, where most practical steel columns lie. The accepted column curve in the United States* was developed by the Column Research Council, now called the Structural Stability Research Council, and it is given in two parts. The first part is a parabolic curve for short and intermediate columns that have low values of slenderness ratio and the second part is the Euler curve for long columns. The AISC specification[4] gives these two formulas with the factors of safety (FS) as follows.

$$F_a = \frac{F_y\left\{1 - \left[(KL/r)^2/2C_c^2\right]\right\}}{\text{FS}} \tag{7.9-11}$$

where F_a = allowable compressive stress of centrally loaded column
KL/r = slenderness ratio
$C_c = (2\pi^2 E/F_y)^{1/2}$
E = elastic modulus
F_y = yield stress
$\text{FS} = \frac{5}{3} + 3(KL/r)/8C_c - (KL/r)^3/8C_c^3$

Equation (7.9-11) is valid for slenderness ratios less than C_c, which is the slenderness ratio corresponding to $F_y/2$. For slenderness ratios that exceed C_c, the Euler formula, modified to include the factor of safety is used.

$$F_a = \frac{12\pi^2 E}{23(KL/r)^2} \tag{7.9-12}$$

*Other column curves for countries around the world are given in Reference 9.

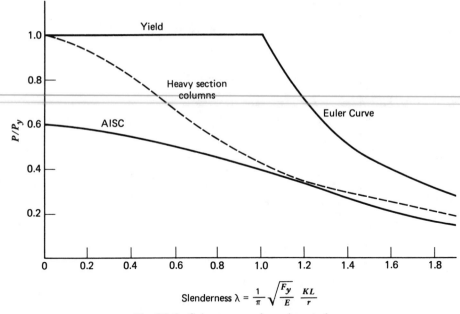

Fig. 7.9-6 Column curves for carbon steel.

Hence, the factor of safety for steel columns varies from 1.67 to 1.92. The CRC curve used by AISC is plotted in Fig. 7.9-6, where it is compared to the Euler curve and the straight-line yield curve. In Fig. 7.9-6, the critical load, normalized with respect to the compressive force at yield, P_y, is plotted as a function of slenderness ratio parameter λ, where λ is defined in the figure. In tests of actual columns the critical buckling loads lie between the two curves. For very short columns these AISC requirements are shown to be conservative. But, for most intermediate columns, these requirements have some reasonable safety factor, except for heavy sections for which they may be unconservative as shown by the dashed line* in Fig. 7.9-6.

7.9-4 Aluminum Columns

The behavior of aluminum columns is reasonably well defined by the tangent modulus theory since the compressive stress–strain curve is nonlinear (Fig. 7.9-7). This has been approximated by a straight-line formula[6] in the short to intermediate range.

The critical stress, σ_c, in ksi is

$$\frac{KL}{r} \leq C_c$$

$$\sigma_c = B_c - D_c\left(\frac{KL}{r}\right) \qquad (7.9\text{-}13)$$

For long columns the Euler formula is used.

$$\frac{KL}{r} > C_c$$

$$\sigma_c = \frac{\pi^2 E}{(KL/r)^2} \qquad (7.9\text{-}14)$$

*See Fig. 3.32 of Reference 6. This is the lower bound computed column strength curve.

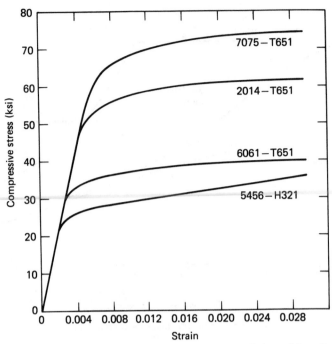

Fig. 7.9-7 Minimum compressive stress–strain curve for sheet and plate of four aluminum alloys. (Reprinted from Reference 6 with permission of the publisher.)

In Equations (7.9-13) and (7.9-14), the constants for alloys that are not artificially aged are

$$B_c = \sigma_{cy}\left[1 + \left(\frac{\sigma_{cy}}{1000}\right)^{1/2}\right]$$

$$D_c = \frac{B_c}{20}\left(\frac{6B_c}{E}\right)^{1/2}$$

$$C_c = \frac{2B_c}{3D_c}$$

and for alloys that are artificially aged

$$B_c = \sigma_{cy}\left[1 + \left(\frac{\sigma_{cy}}{2250}\right)^{1/2}\right]$$

$$D_c = \frac{B_c}{10}\left(\frac{B_c}{E}\right)^{1/2}$$

$$C_c = 0.41\left(\frac{B_c}{D_c}\right)$$

where σ_{cy} = compressive yield strength, 0.2% offset, ksi
       E = elastic modulus, ksi

For buildings a factor of safety of 1.95 is used while for bridges a factor of safety of 2.2 is used.

7.9-5 Austenitic Stainless Steel Columns

The stress–strain curve for austenitic stainless steel is nonlinear and subject to changes by cold working. Also, as shown in Fig. 7.9-8, typical stainless steels are highly anisotropic, and it would not

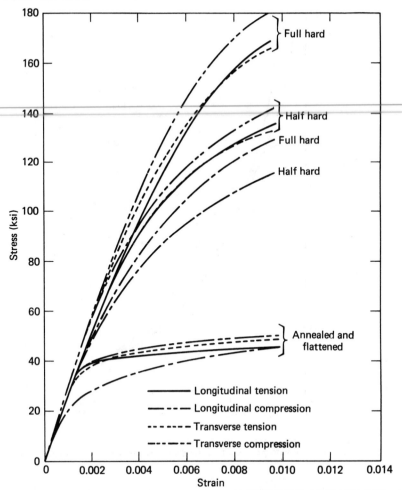

Fig. 7.9-8 Representative stress–strain curves, types 302 and 304 stainless steel. (Copyright, ASTM, Philadelphia. Reprinted from ASTM Special Technical Publication 454 with permission.)

be suitable to approximate the compressive stress–strain curve with the tensile curve. It has been shown that the tangent modulus formula is suitable for computing column strength for this material. Tests by Johnson and Winter[10] show that when the stub column compressive stress–strain curve is used to determine E_t, the tangent modulus, the results agree well with the tangent modulus formula but the CRC curve (used for carbon steel design) overestimates the column strength, as shown in Fig. 7.9-9.

Because research for stainless steel members has been more limited than that for carbon steel columns, the AISI *Stainless Steel Cold Formed Structural Design Manual*[11] uses a factor of safety of 2.15 on the tangent modulus formula [Eq. (7.9-10)]. The allowable compressive stress for *fully effective* sections (those sections not subject to local buckling) is

$$F_a = 4.60 \frac{E_t}{(KL/r)^2} \leq \frac{F_y}{2.15} \qquad (7.9\text{-}15)$$

where E_t = tangent modulus in compression

$\dfrac{KL}{r}$ = slenderness ratio

F_y = yield strength in compression in direction of applied load

Reference 11 provides design rules for columns with cross sections that are not fully effective.

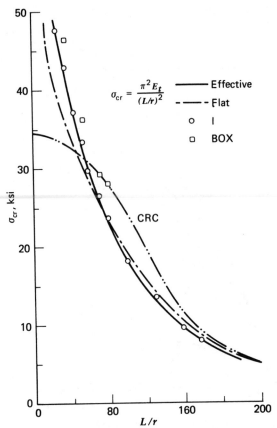

Fig. 7.9-9 Critical column stress versus slenderness ratio for type 304 stainless steel, annealed. (Reprinted from Reference 10 with permission of the American Society of Civil Engineers.)

A straight-line approximation to the tangent modulus formula has been proposed by some researchers in unpublished reports. For short and intermediate columns

$$\frac{KL}{r} \leq 120$$

$$F_a = \frac{F_y}{2.15} - \left(\frac{\dfrac{F_y}{2.15} - 6}{120} \right) \frac{KL}{r} \tag{7.9-16}$$

For long columns

$$\frac{KL}{r} > 120$$

$$F_a = 12 - \frac{1}{20}\left(\frac{KL}{r} \right) \tag{7.9-17}$$

where F_a is in ksi.

7.9-6 Reinforced-Concrete Columns

Structures that are supported by reinforced-concrete columns use either spiral reinforced circular columns or tied rectangular columns. Both types contain longitudinal steel reinforcement that is enclosed by either the spiral reinforcement of the circular columns or the transverse ties of the rectangular reinforcement. Although they may be designed such that their ultimate compressive loads are equivalent, the circular columns exhibit greater ductility than the rectangular sections, which reach their ultimate compressive loads when the longitudinal reinforcement buckles in between the points restrained by the lateral ties.[12] Only rectangular columns will be discussed further in this section.

It is rare that a concrete column is loaded concentrically. The load conditions in a building usually result in bending combined with axial load. Hence, most concrete columns of practical interest are beam columns.[13] These are discussed in Section 7.10-2. However, if a *short column*, whose slenderness ratio (KL/r) is less than 22,* is loaded concentrically in compression, the ultimate compressive force is

$$P_0 = 0.85f_c'\left(A_g - A_{st}\right) + f_y A_{st} \tag{7.9-18}$$

where P_0 = ultimate compressive force
f_c' = concrete ultimate compressive strength
A_g = gross area of column cross section
A_{st} = area of steel reinforcement
f_y = yield strength of steel reinforcement

REFERENCES

7.9-1 F. Bleich, *Buckling Strength of Metal Structures*, McGraw-Hill, New York, 1952.

7.9-2 L. Tall, *Structural Steel Design*, 2nd ed., Ronald Press, New York, 1974.

7.9-3 S. P. Timoshenko, *Theory of Elastic Stability*, McGraw-Hill, New York, 1961.

7.9-4 *AISC Specification for the Design, Fabrication, and Erection of Structural Steel for Buildings* (with commentary), American Institute of Steel Construction, Chicago, November 1978.

7.9-5 L. S. Beedle and L. Tall, *Basic Column Strength*, Transactions of the American Society of Civil Engineers, 1962.

7.9-6 B. G. Johnston, *Guide to Stability Design Criteria for Metal Structures*, 3rd ed., Wiley, New York, 1976.

7.9-7 R. H. Batterman and B. G. Johnston, "Behavior and Maximum Strength of Metal Columns," *J. Structural Div.*, American Society of Civil Engineers, New York, April 1967.

7.9-8 L. Tall, "Recent Developments in the Study of Column Behavior," *J. Institution of Engineers, Australia*, December 1964.

7.9-9 Structural Stability Research Council, et al., "Stability of Metal Structures—A World View," *Eng. J.*, American Institute of Steel Construction, Chicago, 3rd quarter, 1981.

7.9-10 A. L. Johnson and G. Winter, "Behavior of Stainless Steel Columns and Beams," *J. Structural Div.*, American Society of Civil Engineers, New York, October 1966.

7.9-11 *Stainless Steel Cold-Formed Structural Design Manual*, American Iron and Steel Institute, Washington, D.C., 1974.

7.9-12 C. Wang and C. G. Salmon, *Reinforced Concrete Design*, 3rd ed., Harper & Row, New York, 1979.

7.9-13 *ACI Building Code Requirements for Reinforced Concrete (ACI 318-77)*, American Concrete Institute, Detroit, 1977.

7.10 BEAM COLUMNS

Those straight members that are subjected to simultaneous bending and compressive axial loads are called beam columns. The response of these members is nonlinear, that is, deflections are not a linear function of axial force in the elastic and inelastic ranges of material behavior. Elastic response theory of beam columns[1] shows that the following fourth-order differential equation governs the response

$$EI\frac{d^4y}{dx^4} + P\frac{d^2y}{dx^2} = w \tag{7.10-1}$$

This is similar to the beam flexure formula, Equation (7.3-5), and terms are the same except that the axial load P has been added. Several solutions are given by Timoshenko[1] for various load and boundary conditions. Although Equation (7.10-1) is linear, the solutions are nonlinear. One solution, for a column with axial load P and eccentricity e at both ends, leads to the so-called secant formula.[2]

$$\sigma_{max} = \frac{P}{A}\left[1 + \left(\frac{ec}{r^2} + \frac{e_0 c}{r^2}\right)\sec\frac{L}{2r}\left(\frac{P}{AE}\right)^{1/2}\right] \tag{7.10-2}$$

---

*The complete rule for determining if a column is short is given in Section 7.10 and Reference 3.

where σ_{max} = maximum stress at the midspan
  P = axial compressive load
  A = cross-section area
  e = eccentricity of applied load
  e_0 = assumed equivalent eccentricity accounting for defects, etc.
  c = distance from neutral axis to extreme fiber
  r = radius of gyration
  L = member length
  E = material modulus of elasticity

It has been shown[2] that the lateral deflection or bending stress computed from simple beam theory (Section 7.3) is increased approximately under axial load by an *amplification factor* (AF)

$$AF = \frac{1}{(1 - P/P_e)} \qquad (7.10\text{-}3)$$

where P_e = Euler buckling load = $\pi^2 E/(KL/r)^2$
  KL/r = slenderness ratio

Hence, the theoretical stress in Equation (7.10-2) can be computed by first calculating the nominal bending stress imposed by the moment, $M = P \times e$, and then multiplying by the factor, AF.

7.10-1 Steel Beam Columns

In the inelastic range beam column response can be computed numerically[3,4] using the finite-difference method. As in centrally loaded steel columns, the residual stress distribution affects the member response and the finite-difference solution can take this into account. Testing[5,6] has confirmed the accuracy of the finite-difference procedure for computing beam column response and load capacity. But, in spite of the existence of digital computers to handle the numerical integration, an approximate interaction for computing beam column load capacity is widely accepted for carbon steel members and may be applied to other homogeneous straight members in combined bending and compression. The amplification factor, Equation (7.10-3), is the basis for the well-known Column Research Council Interaction Formula which appears in Part 2 of the AISC code[7] and Appendix XVII-4000 of the ASME code[8]

$$\frac{P}{P_{CR}} + \frac{C_m M}{M_p (1 - P/P_e)} \le 1.0 \qquad (7.10\text{-}4)$$

where M = applied moment
  M_p = moment capacity [Equation (7.7-2)]
  P_e = Euler buckling load [computed for both short and long columns in Equation (7.10-4)]
  P = applied axial compressive force
  P_{CR} = compressive force capacity if member were a centrally loaded column, Equation (7.9-11) or (7.9-12).
  C_m = coefficient whose value shall be taken as follows:

1. For compression members in frames subject to joint translation (side sway), $C_m = 0.85$.
2. For restrained compression members in frames braced against joint translation and not subject to transverse loading between their supports in the plane of bending,

$$C_m = 0.6 - 0.4\frac{M_1}{M_2}$$

but not less than 0.4, where M_1/M_2 is the ratio of the smaller to larger moments at the ends of that portion of the member unbraced in the plane of bending under consideration. M_1/M_2 is positive when the member is bent in double curvature, negative when bent in single curvature.

3. For compression members in frames braced against joint translation in the plane of loading and subjected to transverse loading between their supports, the value of C_m may be determined by rational analysis. However, in lieu of such analysis, the following values may be used:

   (a) For members whose ends are restrained

$$C_m = 0.85$$

(b)   For members whose ends are unrestrained

$$C_m = 1.0$$

This empirical formula [Equation (7.10-4)] gives good agreement with numerical solutions of beam columns. In the event that Equation (7.10-4) shows the member to be stable, another criterion is applied to determine if plastic hinges will form at the ends of the member. For a wide-flange section with bending about the strong axis, this criterion is

$$\frac{P}{P_y} + \frac{M}{1.18M_p} \le 1.0 \qquad (7.10\text{-}5)$$

where P_y is the compressive force at yield.

The stability and hinge criteria are illustrated in Fig. 7.10-1. These two criteria are given in AISC,[7] Part 2, and the ASME code,[8] Appendix XVII-4000, but the moment capacity, M_p, in Equation (7.10-4) is replaced by M_m, which accounts for lateral torsional buckling, a possibility that the designer must take into account when the major-axis and minor-axis slenderness ratios are significantly different (Section 7.8).

$$M_m = 1.07 - \frac{\dfrac{L}{r_y}\sqrt{F_y}}{3160} M_p \le M_p \qquad (7.10\text{-}6)$$

where L = member length
r_y = weak axis radius of gyration
L/r_y = slenderness ratio

Factors of safety are inherent in the load factors (Section 7.7) applied to the axial compressive force P and moment M, which are typically 1.7 and 1.3 for dead load plus live load and 1.3 for dead load plus live load plus wind or seismic forces, respectively.[7]

The above formulas have been converted to stress criteria and another term has been added for weak axis bending. The stress criteria for beam columns are

$$\frac{f_a}{F_a} + \frac{C_{mx}f_{bx}}{\left(1 - f_a/F'_{ex}\right)F_{bx}} + \frac{C_{my}f_{by}}{\left(1 - f_a/F'_{ey}\right)F_{by}} \le 1.0 \qquad (7.10\text{-}7)$$

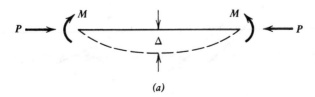

(a)

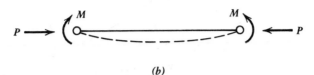

(b)

Fig. 7.10-1   Stability and yield criteria. (*a*) beam column stability:

$$\frac{P}{P_{CR}} + \frac{C_m M}{M_p\left(1 - \dfrac{P}{P_e}\right)} \le 1.0$$

(*b*) plastic hinge formation:

$$\frac{P}{P_y} + \frac{M}{1.18M_p} \le 1.0$$

and

$$\frac{f_a}{0.60S_y} + \frac{f_{bx}}{F_{bx}} + \frac{f_{bx}}{F_{by}} \le 1.0 \qquad (7.10\text{-}8)$$

where f_a = compressive stress, P/A
$\qquad A$ = cross-section area
$\qquad F_a$ = critical axial load stress for member without applied moment
$\qquad f_{bx}, f_{by}$ = applied bending stresses about x and y axes
$\qquad F_{bx}, F_{by}$ = allowable bending stresses for x and y axes
$\qquad F_{ex}', F_{ey}'$ = Euler buckling stress for x and y axes = $\frac{12}{23}\pi^2 E/(KL/r)^2$

These appear in Part 1 of the AISC code and in Appendix XVII-2000 of the ASME code. Factors of safety are included in the allowable stress terms and the applicable code should be consulted for the value of allowable stress corresponding to the material used. Several texts, including References 7.10-2 and 7.10-9, give a complete background on the development of these formulas. They were written specifically for wide-flange sections, but they also work well for other sections such as thick-walled tubes. The designer should keep in mind, however, that the moment multiplier, 1.18, in the denominator of the second term of Equation (7.10-5), was derived specifically for strong-axis bending of wide-flange members and other factors would be appropriate for other sections.

It is clear that when both the axial compressive load and end moments are *primary*, the beam column design should adhere to both stability and plastic hinge criteria. When it is certain that the end moment is secondary and the axial force is primary, then the stability criteria, Equation (7.10-4) or (7.10-7) should be applied. But in this case the formation of a plastic hinge at the end of the member will not result in collapse, so the limits of Equation (7.10-5) or (7.10-8) can be exceeded without violating the integrity of the structure. The ASME code limits the right-hand side of Equation (7.10-5) or (7.10-8) to 1.5 to maintain linearity of the support stiffness in those cases in which the bending

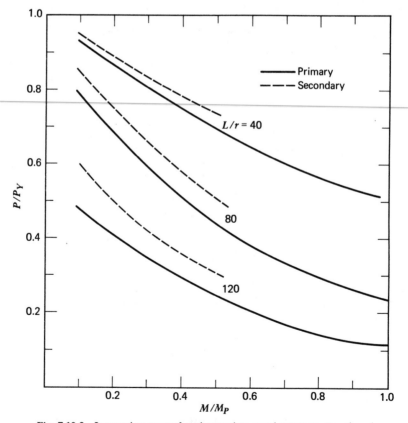

Fig. 7.10-2 Interaction curves for pipe section, equal moments at each end.

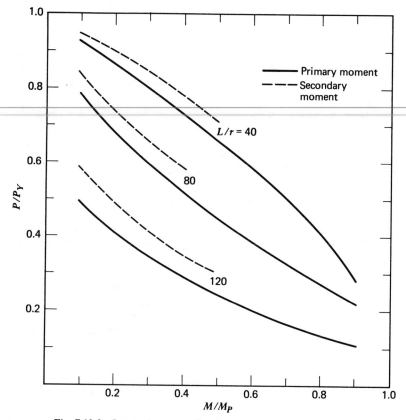

Fig. 7.10-3 Interaction curves for pipe section, moment at one end.

moment is secondary. It must be emphasized, however, that the stability criteria must be evaluated for both primary and secondary loads.

When the secondary moment is imposed by a member that is very heavy and stiff, compared to the beam column, the beam or girder that frames into the ends of the beam column will restrain the member against lateral deflection. Hence, a higher axial load capacity exists when the secondary moment is imposed by a heavy restraining member whose rotation at the end of the beam column is limited. But even secondary moment in this case reduces axial load capacity because of the curvature imposed into the beam column. Interaction charts, computed numerically for both primary and secondary moments (imposed by heavy restraining beams or girders) are given in Figs. 7.10-2–7.10-5 for a few cases.

7.10-2 Reinforced-Concrete Beam Columns

Building columns that are constructed of reinforced concrete almost always have bending moments applied simultaneously with the axial compressive forces. Hence, the subject of reinforced-concrete columns is presented in this Section, 7.10, even though they are referred to as "columns," not "beam columns."

Most reinforced-concrete columns used in building construction are *short*, and the methods used to evaluate column buckling (Sections 7.9-1 and 7.9-2) need not apply. However, a simple check of the slenderness ratio should be performed to determine if the column is indeed short. Section 10.11.4 of the ACI Building Code[10] provides the following rules to determine if slenderness effects need be evaluated.

1. For compression members braced against side sway, effects of slenderness may be neglected when

$$\frac{KL}{r} < 34 - 12\left(\frac{M_1}{M_2}\right) \qquad (7.10\text{-}9)$$

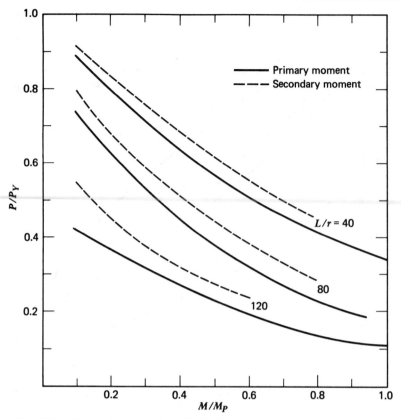

Fig. 7.10-4   Interaction curves for welded wide-flange, equal moments at each end.

2. For compression members not braced against side sway effects of slenderness may be neglected when

$$\frac{KL}{r} < 22 \qquad\qquad (7.10\text{-}10)$$

where M_1 = value of smaller end moment on compression member calculated by conventional elastic frame analysis; positive if member is bent in single curvature, negative if bent in double curvature

   M_2 = value of larger end moment on compression member calculated by conventional elastic frame analysis, always positive

In the event that the column is shown to be slender, rather than short, the ACI Building Code[10] provides a method for accounting for the amplified bending moment similar to that given by Equation (7.10-3).

   In the remainder of this section, statics and strain compatibility will be utilized to determine the load capacity of *short rectangular* concrete columns. First, the compressive force and eccentricity that give the balanced condition[11] will be determined. As shown in Fig. 7.10-6a, an axial force, P_b, is applied to the column with an eccentricity, e_b, simultaneously producing the concrete ultimate strain, 0.003, and the tensile steel yield strain, ε_y, shown in Fig. 7.10-6b. The corresponding stress diagram, shown in Fig. 7.10-6c, gives the assumed rectangular stress block for concrete compression as discussed in Section 7.7-2. Strain compatibility gives the location of the neutral axis

$$\frac{x_b}{d} = \frac{0.003}{(f_y/E_s) + 0.003} \qquad\qquad (7.10\text{-}11)$$

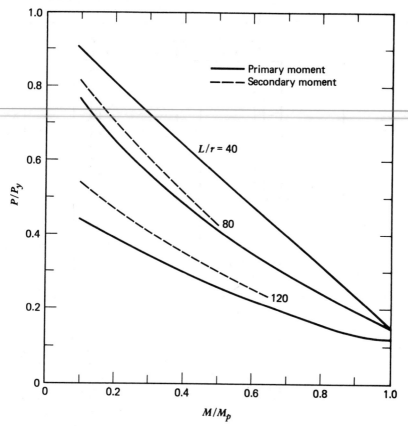

Fig. 7.10-5 Interaction curves for welded wide-flange moment at one end.

Terms are shown in Fig. 7.10-6. The concrete compressive force is

$$C_c = 0.85f_c'\beta_1 x_b b - 0.85f_c'A_s'$$ (7.10-12)

Assuming the compression steel yields, the compressive force in the steel is

$$C_s = A_s'f_y$$ (7.10-13)

and the tensile steel force is

$$T = A_s f_y$$ (7.10-14)

Force and moment equilibrium can be used to obtain the ultimate balanced compressive force, P_b, and moment, M_b.

$$P_b = C_c + C_s - T$$

$$= 0.85f_c'\beta_1 x_b b + A_s'\left(f_y - 0.85f_c'\right) - A_s f_y$$ (7.10-15)

and

$$M_b = P_b e_b = C_c\left(d - \frac{a}{2} - d''\right) + C_s(d - d' - d'') + Td''$$ (7.10-16)

Each column analyzed should be checked to determine if the compressive steel actually does yield per Equation (7.10-13). Chances are good that it will since it should be near the compression face where the strain is 0.003.

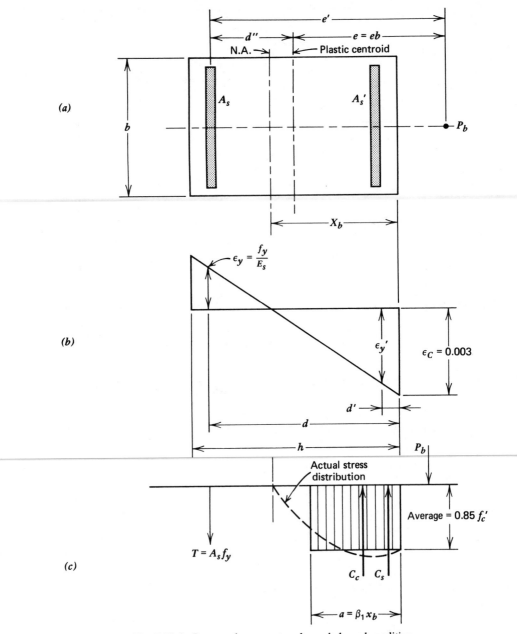

Fig. 7.10-6 Rectangular concrete column, balanced condition.

The equations above can be used to give one point on an interaction diagram for a cross section. Equilibrium and strain compatibility can be used again to determine other points, when either tension controls, as in the case of concrete beams, or when compression controls. The interaction diagram can be drawn by selecting values of eccentricity, e, such that

$$e < e_b \quad \text{or} \quad e > e_b$$

and solving for the corresponding force, P_u, and moment, $P_u e$. Such a diagram is shown in Fig. 7.10-7 where it is observed that axial force $P < P_b$ enhances the moment capacity of a member. The maximum ultimate force, allowed by the ACI Building Code,[10] is shown by the straight line at $P_{u(\max)}$. Many such interaction curves are available in nondimensionalized form[12,13] to facilitate analysis of concrete columns without using the cumbersome statics approach. The Portland Cement

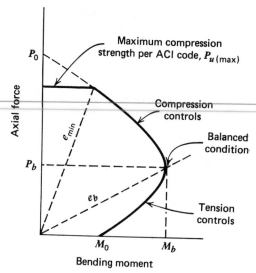

Fig. 7.10-7 Interaction diagram for a reinforced concrete column.

Association (Skokie, Illinois) and the American Concrete Institute (Detroit, Michigan) publish design aids for concrete columns.

REFERENCES

7.10-1 S. P. Timoshenko, *Theory of Elastic Stability*, McGraw-Hill, New York, 1961.

7.10-2 B. G. Johnston, *Guide to Stability Design Criteria for Metal Structures*, 3rd ed., Wiley, New York, 1976.

7.10-3 R. L. Ketter, E. L. Kaminsky, and L. S. Beedle, *Plastic Deformation of Wide—Flange Beam Columns*, Transactions of the American Society of Civil Engineers, 1955.

7.10-4 T. V. Galambos and R. L. Ketter, "Columns Under Combined Bending and Thrust," *J. Eng. Mechanics Div.*, American Society of Civil Engineers, April 1959.

7.10-5 R. C. Van Koren and T. V. Galambos, "Beam Column Experiments," *J. Structural Div.*, American Society of Civil Engineers, April 1964.

7.10-6 T. J. Dwyer and T. V. Galambos, "Plastic Behavior of Tubular Beam Columns," *J. Structural Div.*, American Society of Civil Engineers, August 1965.

7.10-7 *AISC Specification for the Design, Fabrication, and Erection of Structural Steel for Buildings* (with commentary), American Institute of Steel Construction, Chicago, November 1978.

7.10-8 *ASME Boiler and Pressure Vessel Code*, Section III, Nuclear Power Plant Components, Division 1, Subsection NF, Component Supports, American Society of Mechanical Engineers, New York, 1980.

7.10-9 L. Tall, *Structural Steel Design*, 2nd ed., Ronald Press, New York, 1974.

7.10-10 *ACI Building Code Requirements for Reinforced Concrete (ACI 318-77)*, American Concrete Institute, Detroit, 1977.

7.10-11 C. Wang and C. G. Salmon, *Reinforced Concrete Design*, 3rd ed., Harper & Row, New York, 1979.

7.10-12 E. H. Gaylord and C. N. Gaylord, *Structural Engineering Handbook*, 2nd ed., McGraw-Hill, New York, 1979.

7.10-13 Special Publication No. 17A, *Ultimate Strength Design Handbook*, American Concrete Institute, Detroit, 1970.

7.11 COMPUTER ANALYSIS OF BEAM AND COLUMN STRUCTURES

Fundamentals of the finite-element method are given in Chapter 6. The mathematical steps of matrix analysis can be found in Chapter 1. This section indicates those aspects of the finite-element theory that directly apply to beam and column structures.

A complete structure, composed of many straight members, may be analyzed by the *stiffness* method, also called the *displacement* method. The digital computer may be programmed to analyze structures by this method, and there are several software packages available for this purpose. References 1–3 are a few of the programs that use this method. *Finite-element* theory, as it is applied to structures, is a generalization of the stiffness method, and these programs include finite-element analysis of plates, shells, and solids. Straight members can be treated as linear finite elements, and most programs are capable of combining linear elements with other finite elements in the same analysis. An example is a roof (plate or shell elements) supported by columns (linear elements).

7.11-1 Linear Analysis

The stiffness method theory is given in detail in References 4 and 5. Deflections, Δ_i, at the nodal points (points at which straight members are joined) in the structure shown in Fig. 7.11-1 define the overall response of the structure. One to six degrees of freedom (DOF) can be selected for analysis at each node; for the plane frame structure composed of beam and column members, it would be appropriate to analyze three degrees of freedom per node as shown in the figure. For *linear elastic* response the forces at each degree of freedom can be expressed by a set of simultaneous equations of the form

$$F_1 = K_{11}\Delta_1 + K_{12}\Delta_2 + \cdots + K_{1j}\Delta_j + \cdots + K_{1n}\Delta_n$$

$$F_2 = K_{21}\Delta_1 + K_{22}\Delta_2 + \cdots + K_{2j}\Delta_j + \cdots + K_{2n}\Delta_n$$

$$\vdots \quad \vdots \quad \vdots \qquad \vdots \qquad \vdots$$

$$F_j = K_{j1}\Delta_1 + K_{j2}\Delta_2 + \cdots + K_{jj}\Delta_j + \cdots + K_{jn}\Delta_n \qquad (7.11\text{-}1)$$

$$\vdots \quad \vdots \quad \vdots \qquad \vdots \qquad \vdots$$

$$F_n = K_{n1}\Delta_1 + K_{n2}\Delta_2 + \cdots + K_{nj}\Delta_j + \cdots + K_{nn}\Delta_n$$

where F_i = nodal force or moment at ith DOF

K_{ij} = stiffness coefficient, force or moment in ith DOF corresponding to a unit displacement or rotation in the jth DOF with all other degrees of freedom restrained, including i. For K_{ii} only the ith DOF is moved a unit displacement

Δ_j = deflection or rotation at jth DOF

or in matrix notation Equation (7.11-1) becomes

$$\mathbf{F} = \mathbf{k}\Delta \qquad (7.11\text{-}2)$$

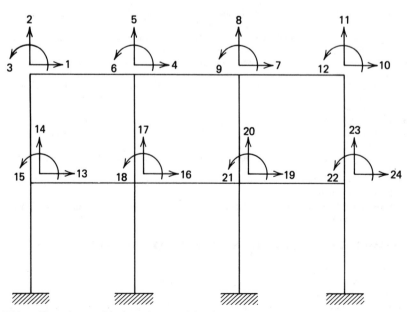

Fig. 7.11-1 Plane frame with three degrees of freedom per node. (Reprinted from Reference 1 with permission of the publisher.)

where $\mathbf{F}$ = force vector

$$= F_1, F_2, \ldots, F_n$$

$\mathbf{K}$ = stiffness matrix

$$= \begin{bmatrix} K_{11}K_{12} \cdots K_{1n} \\ \vdots \\ K_{n1}K_{n2} \cdots K_{nn} \end{bmatrix}$$

Δ = displacement vector

$$= \{\Delta_1, \Delta_2, \ldots, \Delta_n\}$$

For known forces and stiffness coefficient, Equation (7.11-2) can be solved for the displacements

$$\Delta = \mathbf{K}^{-1}\mathbf{F} \tag{7.11-3}$$

where $\mathbf{K}^{-1}$ is the inverse of stiffness matrix.

Often, there are no applied forces for many structure degrees of freedom. Hence, some elements of the force vector may equal zero. This creates no special problem and the method is still valid. Or sometimes the loads are applied between the nodes, as they are with distributed loads. In that case equivalent joint loads are computed and applied at the structure degrees of freedom.

Assembling the coefficients of the stiffness matrix is accomplished by first computing the individual stiffness of the straight members. Then, at each structure degree of freedom, the stiffnesses are summed. But care must be taken because usually the structure (global) coordinate system and the member coordinate system do not coincide. A rotation matrix R may be used to transform the member stiffness into the structure coordinate system

$$\mathbf{k} = \mathbf{R}\mathbf{k}'\mathbf{R}^{-1} \tag{7.11-4}$$

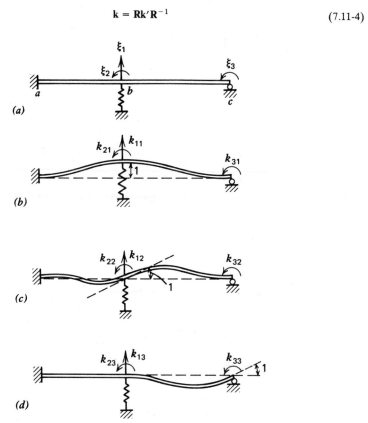

Fig. 7.11-2 Structure degrees of freedom. (a) Degrees of freedom ξ_1, ξ_2, and ξ_3; (b) forces in ξ_1, ξ_2, and ξ_3 due to a unit displacement in ξ_1; (c) forces in ξ_1, ξ_2, and ξ_3 due to a unit displacement in ξ_2; (d) forces in ξ_1, ξ_2, and ξ_3 due to unit displacement in ξ_3. (Reprinted from Reference 1 with permission of the publisher.)

where $\mathbf{R}$ = rotation matrix = $\begin{bmatrix} \lambda_{1m} \cdots \lambda_{1m} \\ \vdots \\ \lambda_{m1} \cdots \lambda_{mm} \end{bmatrix}$

λ_{ij} = direction cosines between ith structure coordinate (DOF) and jth member coordinate (DOF)

$\mathbf{k}, \mathbf{k}'$ = member stiffness matrix in structure (global) and member coordinates (DOF), respectively

The corresponding elements of the small member stiffness matrices (1 to 12 DOF) may be combined with one another to form the larger structure stiffness matrix after the member matrices are transformed to the structure system by Equation (7.11-4) where necessary.

The simple structure in Fig. 7.11-2 illustrates the method. In this structure there are two members and only three degrees of freedom. The fixity at a and the rigid support at c eliminate these

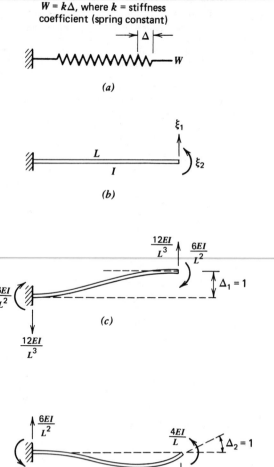

$W = k\Delta$, where k = stiffness coefficient (spring constant)

(a)

(b)

(c)

(d)

Fig. 7.11-3 Stiffness coefficients. (a) One degree of freedom spring system; (b) cantilever beam as a system with two degrees of freedom ξ_1 and ξ_2; (c) forces in the degrees of freedom to maintain a unit displacement pattern $\Delta_1 = 1$, $\Delta_2 = 0$; (d) forces in the degrees of freedom to maintain a unit displacement pattern $\Delta_1 = 0$, $\Delta_2 = 1$. (Reprinted from Reference 1 with permission of the publisher.)

TABLE 7.11-1 SPACE FRAME MEMBER STIFFNESS MATRIX

	1	2	3	4	5	6	7	8	9	10	11	12
1	$\dfrac{EA_X}{L}$	0	0	0	0	0	$-\dfrac{EA_X}{L}$	0	0	0	0	0
2	0	$\dfrac{12EI_Z}{L^3}$	0	0	0	$\dfrac{6EI_Z}{L^2}$	0	$-\dfrac{12EI_Z}{L^3}$	0	0	0	$\dfrac{6EI_Z}{L^2}$
3	0	0	$\dfrac{12EI_Y}{L^3}$	0	$-\dfrac{6EI_Y}{L^2}$	0	0	0	$-\dfrac{12EI_Y}{L^3}$	0	$-\dfrac{6EI_Y}{L^2}$	0
4	0	0	0	$\dfrac{GI_X}{L}$	0	0	0	0	0	$-\dfrac{GI_X}{L}$	0	0
5	0	0	$-\dfrac{6EI_Y}{L^2}$	0	$\dfrac{4EI_Y}{L}$	0	0	0	$\dfrac{6EI_Y}{L^2}$	0	$\dfrac{2EI_Y}{L}$	0
6	0	$\dfrac{6EI_Z}{L^2}$	0	0	0	$\dfrac{4EI_Z}{L}$	0	$-\dfrac{6EI_Z}{L^2}$	0	0	0	$\dfrac{2EI_Z}{L}$
7	$-\dfrac{EA_X}{L}$	0	0	0	0	0	$\dfrac{EA_X}{L}$	0	0	0	0	0
8	0	$-\dfrac{12EI_Z}{L^3}$	0	0	0	$-\dfrac{6EI_Z}{L^2}$	0	$\dfrac{12EI_Z}{L^3}$	0	0	0	$-\dfrac{6EI_Z}{L^2}$
9	0	0	$-\dfrac{12EI_Y}{L^3}$	0	$\dfrac{6EI_Y}{L^2}$	0	0	0	$\dfrac{12EI_Y}{L^3}$	0	$\dfrac{6EI_Y}{L^2}$	0
10	0	0	0	$-\dfrac{GI_X}{L}$	0	0	0	0	0	$\dfrac{GI_X}{L}$	0	0
11	0	0	$-\dfrac{6EI_Y}{L^2}$	0	$\dfrac{2EI_Y}{L}$	0	0	0	$\dfrac{6EI_Y}{L^2}$	0	$\dfrac{4EI_Y}{L}$	0
12	0	$\dfrac{6EI_Z}{L^2}$	0	0	0	$\dfrac{2EI_Z}{L}$	0	$-\dfrac{6EI_Z}{L^2}$	0	0	0	$\dfrac{4EI_Z}{L}$

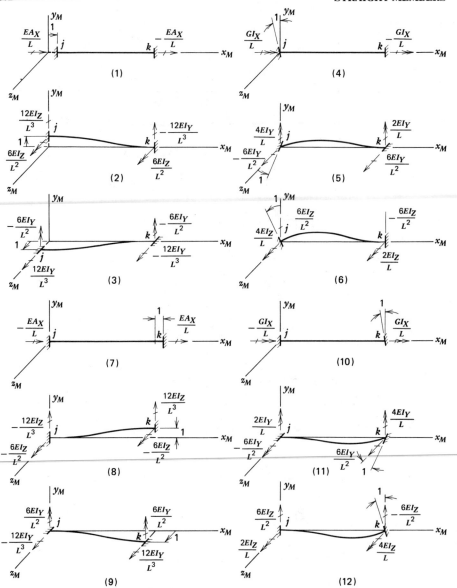

Fig. 7.11-4 Member stiffnesses—obtained by imposing unit translation and rotations in each degree of freedom.

displacement degrees of freedom. The spring support at b may represent another portion of the linear structure or a foundation stiffness. Individual member stiffnesses are shown in Figs. 7.11-1–7.11-3. The member displacements shown should not be confused with the structure degrees of freedom. Global stiffnesses at the center support are computed by adding the appropriate member and spring stiffnesses together, that is, in the first structure DOF, the stiffnesses, K_{11}, K_{12}, and K_{13} are computed by adding corresponding values from the two beam members and the spring. In this example rotation of the member stiffness matrix is not necessary.

It should be noted that both member and structure stiffness matrices are symmetric, viz.,

$$k_{ij} = k_{ji}$$

and

$$K_{ij} = K_{ji}$$

Also, the inverse of the rotation matrix in Equation (7.11-4) is equal to the transpose.

$$\mathbf{R}^{-1} = \mathbf{R}^T \tag{7.11-5}$$

The elements of $\mathbf{R}^{-1}$ may be established simply by interchanging the direction cosines in the rows and columns of $\mathbf{R}$.

The general stiffness matrix for a member with 12 DOF is given in Table 7.11-1. The elements of the matrix are computed from the unit displacements shown in Fig. 7.11-4. Shear deformation (Section 7.4) has been neglected.

Dynamic analysis of framed or piping structures is performed by considering mass and time at the nodal points of the structure. The programs cited above[1-3] are capable of dynamic analysis and the complete theoretical treatment is given in Reference 6. The method is used extensively for dynamic analysis of building structures and piping systems, including analysis of steady-state vibrations, and transient response, including earthquake and blast loadings.

7.11-2 Nonlinear Analysis

The details of nonlinear analysis of frameworks and piping networks composed of straight members are complex. However, certain situations, such as elevated temperature that reduces the yield stress of the material, require nonlinear analysis. References 7 and 8 give complete descriptions of the techniques used to analyze structures that are loaded into the inelastic material range and that experience large deflections. The simultaneous equations (7.11-1) [or (7.11-2) in matrix form] are the basic stiffness equations used in both linear and nonlinear analysis. It is obvious that, upon application of small loads or small deflections, the linear form of these displacement equations remains valid. But as load is increased, the stiffness coefficients become dependent on the member forces. When yielding or large deflection of a member takes place, the reduced or modified stiffness can be computed by an iterative procedure such as the Newton–Raphson method.

In most nonlinear structural analyses, the load is increased in small increments from the linear range into the nonlinear range. In each nonlinear increment, it may be necessary to perform several interative solutions to converge to the correct solution.

Nonlinear analysis of beam and column structures can be dynamic as well as static, although dynamic analyses can be expensive for large analytical models. Some programs have "gap" and "hook" elements that conveniently account for clearances in slotted joints or lift-off of foundations which are not tied to ground. Programs such as ANSYS,[3] ABAQUS,[9] and NONSAP[10] are capable of performing nonlinear analysis including creep analysis.

REFERENCES

7.11-1 *SAP 6/7 Structural Analysis Program for Static and Dynamic Analysis*, Structural Mechanics Computer Laboratory, University of Southern California, Los Angeles, 1982.

7.11-2 *STARDYNE User Information Manual*, Control Data Corp., Minneapolis, Minnesota, 1980.

7.11-3 *ANSYS Engineering Analysis System User's Manual*, Swanson Analysis Systems, Elizabeth, PA, 1979.

7.11-4 J. M. Gere and W. Weaver, *Analysis of Framed Structures*, Van Nostrand, Princeton, 1965.

7.11-5 J. S. Przemieniecki, *Theory of Matrix Structural Analysis*, McGraw-Hill, New York, 1968.

7.11-6 R. W. Clough and J. Penzien, *Dynamics of Structures*, McGraw-Hill, New York, 1975.

7.11-7 O. C. Zienkiewicz, *The Finite Element Method*, 3rd ed., McGraw-Hill, New York, 1977.

7.11-8 K. J. Bathe, *Finite Element Procedures in Engineering Analysis*, Prentice-Hall, Englewood Cliffs, NJ, 1982.

7.11-9 *ABAQUS—Structural and Heat Transfer Analysis Program*, Hibbitt, Karlsson & Sorensen, Providence, RI, 1982.

7.11-10 K. J. Bathe, E. L. Wilson, and R. H. Idding, *NONSAP—Structural Analysis Program for Static and Dynamic Response of Nonlinear Systems*, SESM Report No. 74-3, University of California, Berkeley, Feb. 1974.

CHAPTER **8**

CURVED MEMBERS

FREDERICK C. NELSON

Dean of Engineering
Tufts University
Medford, Massachusetts

8.1 ASSUMPTIONS

Curvature strongly affects the way a structure carries external loads. For example, a simply supported horizontal beam carries a uniformly distributed vertical load by a linear distribution of vertical shear and a parabolic distribution of bending moment; there is no axial force unless the displacement is large. On the other hand if this beam has the form of a parabolic arch, a uniformly distributed vertical load is carried completely by axial forces; there is no shear or bending moment. The general case is a combination of these two systems of internal forces; that is, the lateral loading of a curved structure will, in general, induce a combination of bending, transverse shear, and direct stress.

A common example of a curved structure is a thin-walled shell. The loading of thin shells of revolution by pressure can cause both membrane (direct) stress and bending stress.

This chapter will consider only beamlike curved members, that is, a curved member whose cross-sectional dimensions are comparable in size to each other but much less than the member's developed length. Such structures display the characteristic coupling between in-plane bending and membrane forces due to curvature and, in addition, can display coupling between out-of-plane bending and torsion.

8.2 IN-PLANE DEFORMATION OF THIN, CURVED BEAMS

Figure 8.2-1 shows an infinitesimal segment of a curved beam. The plane of the paper is the initial plane of curvature of the beam, and its deformation is confined to this plane. This latter requirement will be met if the beam cross section is doubly symmetric and one axis of symmetry lies in the initial plane of curvature. It is also assumed that the curved beam is so thin that it is not necessary to distinguish between the radius to the inner fiber and the radius to the outer fiber. In addition, if $R/h \geq 10$, there is close agreement between the bending stresses in the curved beam and those in the corresponding straight beam; if the comparison is based on deflection instead of stress, the criterion of agreement can be relaxed to $R/h \geq 5$. Application of Newton's law for moment equilibrium, force equilibrium in the radial direction, and force equilibrium in the tangential direction gives[1]

$$\frac{dM}{ds} - R\frac{dF}{ds} = 0 \tag{8.2-1}$$

$$\frac{F}{R} + \frac{dV}{ds} + q = 0 \tag{8.2-2}$$

$$\frac{dF}{ds} - \frac{V}{R} = 0 \tag{8.2-3}$$

Combining Equations (8.2-1) and (8.2-3) gives

$$\frac{dM}{ds} = V \tag{8.2-4}$$

which is the same result as that obtained for a shearing force V in straight beams. It is consistent with

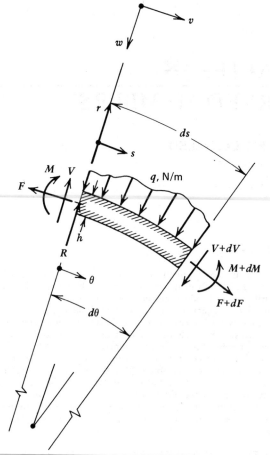

Fig. 8.2-1 In-plane loading and deformation of a curved-beam element.

the above assumptions to assume that the straight-beam expression for elastic energy

$$U = \frac{1}{2} \int_0^l \frac{M^2}{EI_1} \, ds \qquad (8.2\text{-}5)$$

where I_1 is the area moment of inertia for in-plane bending and l is the developed length sufficiently accurate for thin, curved beams.

It can be shown[1] that M is related to coordinates w and v by

$$M = -EI_1 \frac{d}{ds}\left(\frac{dw}{ds} + \frac{v}{R} \right) \qquad (8.2\text{-}6)$$

and that

$$F = EA\left(\frac{dv}{ds} - \frac{w}{R} \right) \qquad (8.2\text{-}7)$$

Note that if $R \to \infty$, these equations reduce to the straight-beam results and that bending and membrane behavior become uncoupled.

If it is reasonable to assume that curved-beam elements bend without extension, this implies

$$\frac{dv}{ds} = \frac{w}{R} \qquad (8.2\text{-}8)$$

This simplification was first proposed by Rayleigh.[2]

For the special case of a thin circular beam of constant cross section that bends without extension, the above assumptions lead to

$$\frac{d^4 w}{d\theta^4} + 2\frac{d^2 w}{d\theta^2} + w = \frac{qR^4}{EI_1}$$

(8.2-9)

for equilibrium and

$$U = \frac{EI_1}{2R^3} \int_0^\alpha \left(\frac{d^2 w}{d\theta^2} + w \right)^2 d\theta$$

(8.2-10)

for energy where α is the opening angle of the ring segment.

REFERENCES

8.2-1 J. Prescott, *Applied Elasticity*, Dover Publications, New York, 1961, Sec. 195.
8.2-2 Rayleigh, *The Theory of Sound*, Vol. I, 2nd ed., Dover Publications, New York, 1945, p. 384.

8.3 OUT-OF-PLANE DEFORMATION OF THIN, CURVED BEAMS

Figure 8.3-1 shows an element of a curved beam subjected to out-of-plane loading. Application of Newton's law for out-of-plane force equilibrium, bending moment equilibrium, and torque equi-

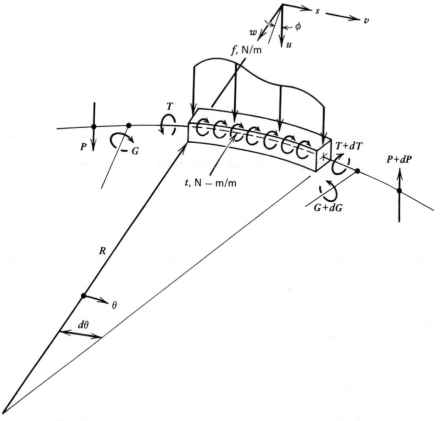

Fig. 8.3-1 Out-of-plane loading and deformation of a curved-beam element.

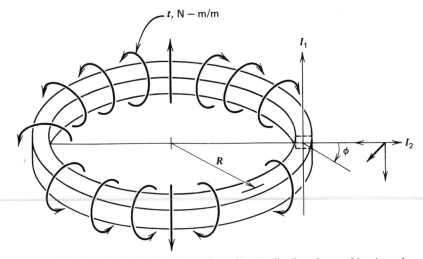

Fig. 8.3-2 Twisting of a circular ring by couples uniformly distributed around its circumference.

librium gives[1]

$$\frac{dP}{ds} - f = 0 \tag{8.3-1}$$

$$\frac{dG}{ds} - \frac{T}{R} + P = 0 \tag{8.3-2}$$

$$\frac{dT}{ds} + \frac{G}{R} + t = 0 \tag{8.3-3}$$

The force–displacement relations are[1]

$$\frac{G}{EI_2} + \frac{\phi}{R} + \frac{d^2u}{ds^2} = 0 \tag{8.3-4}$$

where I_2 is the area moment of inertia for out-of-plane bending, and

$$\frac{T}{C} - \frac{d\phi}{ds} + \frac{1}{R}\frac{du}{ds} = 0 \tag{8.3-5}$$

where C is the torsional rigidity. Again, if $R \to \infty$, the straight-beam results are recovered and the bending and twisting behavior become uncoupled.

For the special case of a thin circular ring of constant cross section that is uniformly twisted in its own plane (see Fig. 8.3-2), the above equations reduce to[2] $G = -Rt$ and $\phi = -GR/EI_2 = tR^2/EI_2$. This case is useful for the design of flanges on pipes and pressure vessels.[3]

One must be careful about situations that violate the above assumptions. For example, if the curved beam is initially planar but the principal axes of inertia are skewed with respect to this plane, the in-plane and out-of-plane deformation are coupled.[4] In addition, there are problems of practical interest, for example, helical springs, in which the curved beam is not initially planar.[5] Also the equations given above assume that the beam is linear with respect to both material and kinematic behavior. The analysis of nonlinear situations is best carried out using numerically based procedures such as the finite-element method.[6,7]

REFERENCES

8.3-1 J. Prescott, *Applied Elasticity*, Dover Publications, 1961, Sec. 195.

8.3-2 S. Timoshenko, *Strength of Materials, Part II*, 3rd ed., Van Nostrand, New York, 1956, p. 138 ff.

8.3-3 J. F. Harvey, *Theory and Design of Modern Pressure Vessels*, 2nd ed., Van Nostrand Reinhold, New York, 1974, Sec. 4.9.

8.3-4 H. R. Meck, "Three-Dimensional Deformation and Buckling of a Circular Ring of Arbitrary Section," *J. Engr. Industry, Trans. ASME*, 266–272 (1969).

8.3-5 A. J. Durelli, V. J. Parks, and H. M. Hasseem, "Helices Under Load," *J. Engr. Industry, Trans. ASME*, **97**, 853–858 (1975).

8.3-6 A. K. Noor, W. H. Greene, and S. J. Hartley, "Nonlinear Finite Element Analysis of Curved Beams," *Comput. Meth. Appl. Mech. Engr.* **12**, 289–307 (1977).

8.3-7 O. A. Fettahlioglu, and T. G. Toridis, "Elastic-Plastic Analysis of Curved Structures Subjected to Static and Dynamic Loads," *J. Press. Vessel Tech., Trans. ASME* **98**, 126–134 (1976).

8.4 DEFLECTIONS OF THIN, CURVED BEAMS

The previous sections have discussed thin curved beams from a Newtonian point of view; if, instead, a point of view based on energy is adopted, it is possible to calculate deflections directly. The two most common procedures are the dummy load method[1] and Castigliano's second theorem.[2]

Consider three sample cases. First, the in-plane deflection of the circular arc cantilever shown in Fig. 8.4-1. The distribution of bending moment is $M(\theta) = -PR\sin\theta$ where positive moment puts the inner fibers in tension. Using Equation (8.2-5),

$$U = \frac{P^2 R^3}{2 E I_1} \int_0^\pi \sin^2\theta \, d\theta = \frac{P^2 R^3 \pi}{4 E I_1}$$

and from Castigliano's second theorem,

$$w(180°) = \frac{dU}{dP} = \frac{\pi P R^3}{2 E I_1}$$

Second, consider the out-of-plane deflection caused by the torsional loading shown in Fig. 8.4-2. The out-of-plane bending moment due to T_0 is $T_0\sin\theta$; the distribution of torque is $T_0\cos\theta$. The associated change in curvature and twist are $T_0\sin\theta/EI_2$ and $T_0\cos\theta/C$. The moment and torque distributions

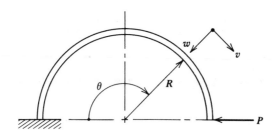

Fig. 8.4-1 Circular arc clamped at $\theta = 0°$ and loaded by a radial load at $\theta = 180°$.

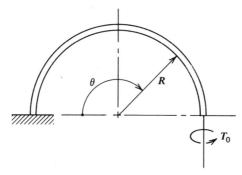

Fig. 8.4-2 Circular arc clamped at $\theta = 0°$ and loaded by a torque T_0 at $\theta = 180°$.

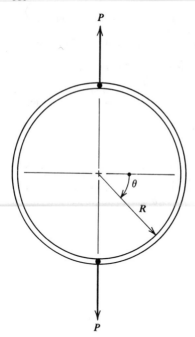

Fig. 8.4-3   Circular ring loaded by two radial loads.

due to a positive unit load at $\theta = 180°$, applied normal to the plane of curvature, are, respectively, $-R \sin \theta$ and $-R(1 + \cos \theta)$. Application of the dummy load method gives

$$u(180°) = \int_0^\pi \left(\frac{T_0 \sin \theta}{EI_2} \right)(-R \sin \theta)\, R\, d\theta + \int_0^\pi \left(\frac{T_0 \cos \theta}{C} \right)[-R(1 + \cos \theta)]\, R\, d\theta$$

$$= -\frac{T_0 R^2}{EI_2} \frac{\pi}{2}(1 + \lambda)$$

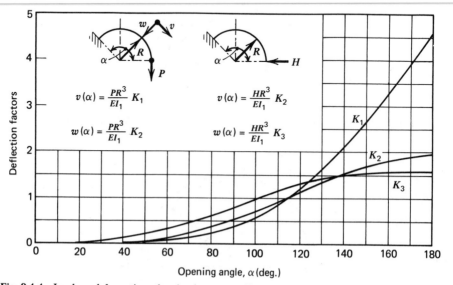

Fig. 8.4-4   In-plane deformation of a circular arc cantilever under a tip load. $K_1 = (6\alpha + \sin 2\alpha - 8 \sin \alpha)/4$; $K_2 = (\cos 2\alpha - 4 \cos \alpha + 3)/4$; $K_3 = (2\alpha - \sin 2\alpha)/4$, where α is in radians. (Used by permission from A. Blake, *Design of Curved Members for Machines* Krieger, 1979.)

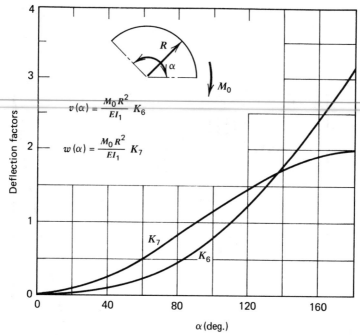

Fig. 8.4-5 In-plane deformation of a circular arc cantilever under a tip moment. $K_6 = \alpha - \sin \alpha$; $K_7 = 1 - \cos \alpha$. (Used by permission from A. Blake, *Design of Curved Members for Machines*, Krieger, 1979.)

where λ is the ratio of flexural to torsional rigidity. For metal structures of circular cross section the rigidity ratio is equal to $1 + \nu$ where ν is Poisson's ratio.

Thirdly, consider a complete circular ring with diametrically opposed loads as in Fig. 8.4-3. The distribution of in-plane bending moment in the quadrant between $\theta = 0$ and $\theta = \pi/2$ is $M(\theta) = M_0 - (PR/2)(1 - \cos \theta)$. M_0 is the statically indeterminate moment at $\theta = 0$; the transverse shear at $\theta = 0$ is zero due to symmetry. The value of M_0 can be found by using the principle of minimum complementary energy.[3] For linear elastic material and in the absence of prescribed displacements, the complementary energy is equal to the stress energy. Then M_0 must minimize

$$U = 4 \left\{ \frac{R}{2EI_1} \int_0^{\pi/2} \left[M_0 - \frac{PR}{2}(1 - \cos \theta) \right]^2 d\theta \right\}$$

which gives the result $M_0 = PR(\frac{1}{2} - 1/\pi)$. Castigliano's theorem can then be used to find the radial displacement at the point $\theta = 90°$; namely,

$$w(90°) = -\frac{PR^3}{EI_1} \left(\frac{\pi}{8} - \frac{1}{\pi} \right)$$

These techniques have been used to accumulate a large catalog of results for circular arc beams[4-6]; more recent results are also available.[7-10] A selection of results from References 5 and 10 is given in Figs. 8.4-4–8.4-26.* Centrally loaded circular arches are considered in Figs. 8.4-23–8.4-26[5]; design information is also available for circular arches with more general loading conditions.[11,12]

---

*Note that the design charts in Figs. 8.4-4, 8.4-5, and 8.4-6 give vertical and horizontal displacements of the free end of the circular arc beam.

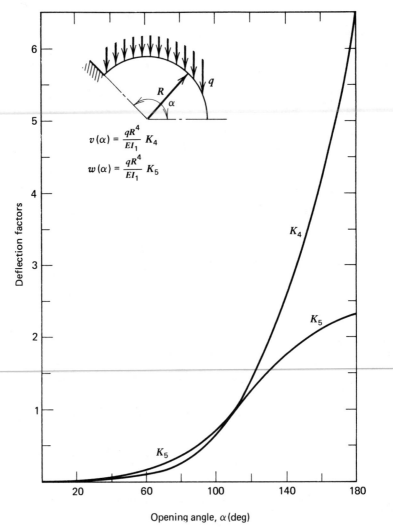

Fig. 8.4-6 In-plane deformation of a circular arc cantilever with a uniformly distributed vertical load. $K_4 = (2\alpha \sin 2\alpha - 8\alpha \sin \alpha + 3 \cos 2\alpha - 16 \cos \alpha + 2\alpha^2 + 13)/8$; $K_5 = (4\alpha + 2\alpha \cos 2\alpha - 3 \sin 2\alpha)/8$. (Used by permission from A. Blake, *Design of Curved Members for Machines*, Krieger, 1979.)

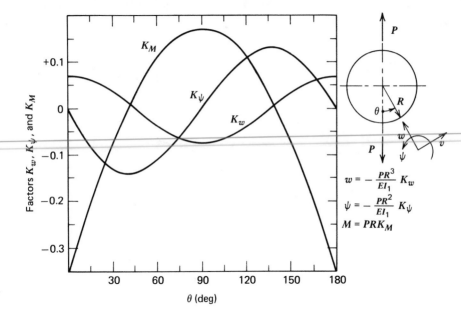

Fig. 8.4-7   Circular ring loaded by two radial loads.

Symbol	Function	Range of Application
K_w	$0.2500 \sin \theta + (0.3927 - 0.2500\theta)\cos \theta - 0.3183$	$0-\pi$
K_ψ	$(0.2500\theta - 0.3927)\sin \theta$	$0-\pi$
K_M	$0.5000 \sin \theta - 0.3183$	$0-\pi$

(Used by permission from A. Blake, *Design of Curved Members for Machines*, Krieger, 1979.)

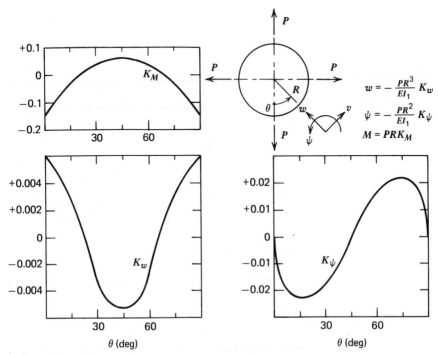

Fig. 8.4-8   Circular ring loaded by two pairs of radially outward loads. Formulas for K_w, K_ψ, and K_M can be obtained by a superposition of the formulas in the caption to Fig. 8.4-7. (Used by permission from A. Blake, *Design of Curved Members for Machines*, Krieger, 1979.)

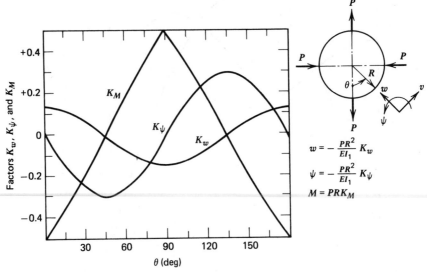

Fig. 8.4-9 Circular ring loaded by two pairs of radial loads; one pair radially outward and one pair radially inward. Formulas for K_w, K_ψ, and K_M can be obtained by a superposition of the formulas in the caption to Fig. 8.4-7. (Used by permission from A. Blake, *Design of Curved Members for Machines*, Krieger, 1979.)

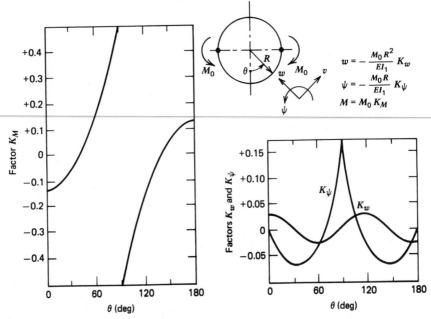

Fig. 8.4-10 Circular ring loaded by two moments.

Symbol	Function	Range of Application
K_w	$0.5000 - 0.3183\theta \sin\theta - 0.4775 \cos\theta$	$0-\pi/2$
K_ψ	$0.1592 \sin\theta - 0.3183\theta \cos\theta$	$0-\pi/2$
K_M	$0.5000 - 0.6366 \cos\theta$ $-0.5000 - 0.6366 \cos\theta$	$0-\pi/2$ $\pi/2-\pi$

(Used by permission from A. Blake, *Design of Curved Members for Machines*, Krieger, 1979.)

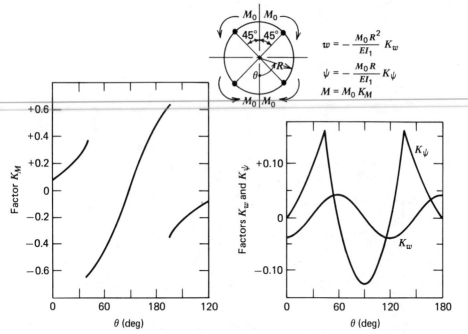

Fig. 8.4-11 Circular ring loaded by two pairs of moments. Formulas for K_w, K_ψ, and K_M can be obtained by a superposition of the formulas in the caption to Fig. 8.4-10. (Used by permission from A. Blake, *Design of Curved Members for Machines*, Krieger, 1979.)

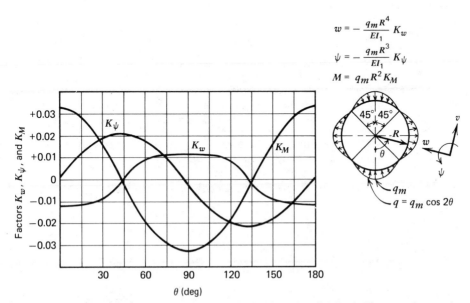

Fig. 8.4-12 Circular ring loaded by a cosine distribution of radial load. These results can be compared to those of Fig. 8.4-9 to assess the effect of spreading of concentrated loads on stress and deflection. (Used by permission from A. Blake, *Design of Curved Members for Machines*, Krieger, 1979; see also H. M. Haydl and S. P. Yip, "Influence of Spreading of Concentrated Loads on Stress Resultants in Circular Rings," *Proc. Inst. Civil Engr.* **61**, 401–409 (1976).)

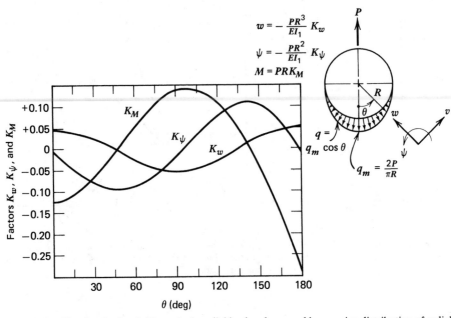

$$w = -\frac{PR^3}{EI_1} K_w$$

$$\psi = -\frac{PR^2}{EI_1} K_\psi$$

$$M = PRK_M$$

$$q = q_m \cos\theta$$

$$q_m = \frac{2P}{\pi R}$$

Fig. 8.4-13 Circular ring loaded by a single radial load and opposed by a cosine distribution of radial load (e.g., single-point lifting of a short cylindrical vessel half full of fluid).

Symbol	Function	Range of Application
K_w	$0.1989\theta \sin\theta + (0.4081 - 0.0796\theta^2)\cos\theta - 0.3618$	$0 - \pi/2$
	$(0.3750 - 0.0398\theta)\sin\theta + (0.3658 - 0.2500\theta)\cos\theta - 0.3618$	$\pi/2 - \pi$
K_ψ	$(0.0796\theta^2 - 0.2092)\sin\theta + 0.0398\theta \cos\theta$	$0 - \pi/2$
	$(0.2500\theta - 0.4055)\sin\theta + (0.1250 - 0.0398\theta)\cos\theta$	$\pi/2 - \pi$
K_M	$0.3183\theta \sin\theta + 0.2387\cos\theta - 0.3618$	$0 - \pi/2$
	$0.5000 \sin\theta - 0.0796\cos\theta - 0.3618$	$\pi/2 - \pi$

(Used by permission from A. Blake, *Design of Curved Members for Machines*, Krieger, 1979.)

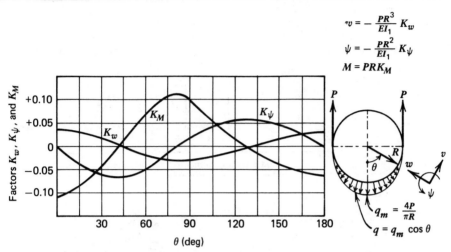

$$v = -\frac{PR^3}{EI_1} K_w$$

$$\psi = -\frac{PR^2}{EI_1} K_\psi$$

$$M = PRK_M$$

Fig. 8.4-14 Circular ring loaded by two tangential loads and opposed by a cosine distribution of radial load.

Symbol	Function	Range of Application
K_w	$0.5570\theta \sin\theta + (0.9382 - 0.1592\theta^2)\cos\theta - 0.9053$	$0-\pi/2$
	$(0.0796\theta - 0.2500)\sin\theta + 0.0681\cos\theta + 0.0947$	$\pi/2-\pi$
K_ψ	$(0.1592\theta^2 - 0.3812)\sin\theta + 0.2387\theta\cos\theta$	$0-\pi/2$
	$0.0115\sin\theta + (0.0796\theta - 0.2500)\cos\theta$	$\pi/2-\pi$
K_M	$0.6366\theta \sin\theta + 0.7958\cos\theta - 0.9053$	$0-\pi/2$
	$0.1592\cos\theta + 0.0947$	$\pi/2-\pi$

(Used by permission from A. Blake, *Design of Curved Members for Machines*, Krieger, 1979.)

Concentrated End Load

$$u(0) = \frac{PR^3}{EI_2}(B_1 + \lambda B_2)$$

$$\psi(0) = -\frac{PR^2}{EI_2}(\lambda B_3 + B_4)$$

$$\varphi(0) = -\frac{PR^2}{EI_2}(\lambda B_3 - B_1)$$

$$G = -PR\sin\beta$$

$$T = -PR(1 - \cos\beta)$$

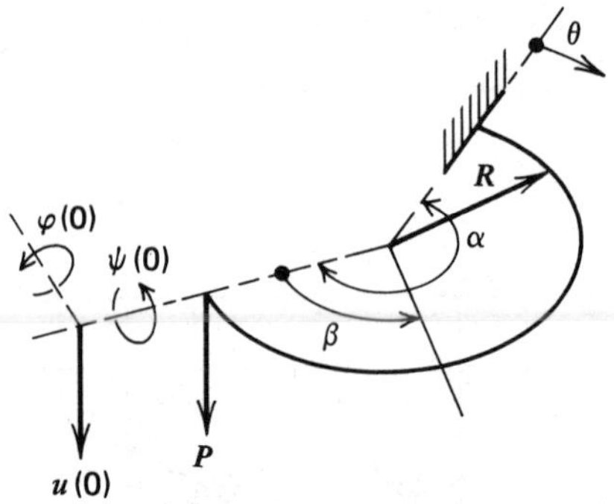

End Twisting Moment

$$u(0) = \frac{T_0 R^2}{EI_2}(\lambda B_5 - B_1)$$

$$\psi(0) = -\frac{T_0 R}{EI_2}(\lambda - 1)B_4$$

$$\varphi(0) = -\frac{T_0 R}{EI_2}(\lambda B_6 + B_1)$$

$$G = T_0 \sin\beta$$

$$T = -T_0 \cos\beta$$

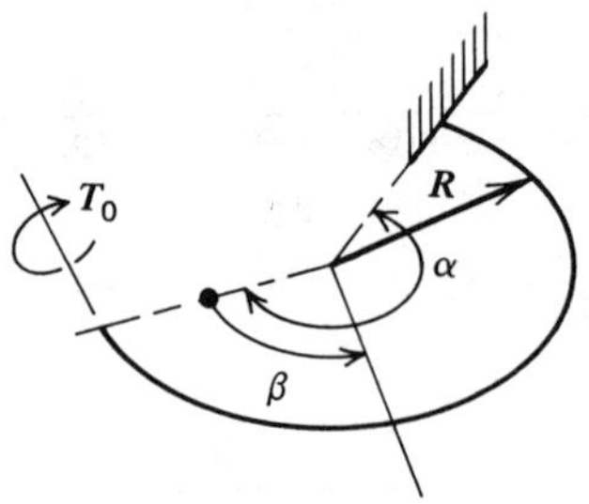

Uniform Loading

$$u(0) = \frac{qR^4}{EI_2}(\lambda B_7 - B_3)$$

$$\psi(0) = -\frac{qR^3}{EI_2}(\lambda B_8 + B_5)$$

$$\varphi(0) = -\frac{qR^3}{EI_2}(\lambda B_9 + B_3)$$

$$G = -qR^2(1 - \cos\beta)$$

$$T = -qR^2(\beta - \sin\beta)$$

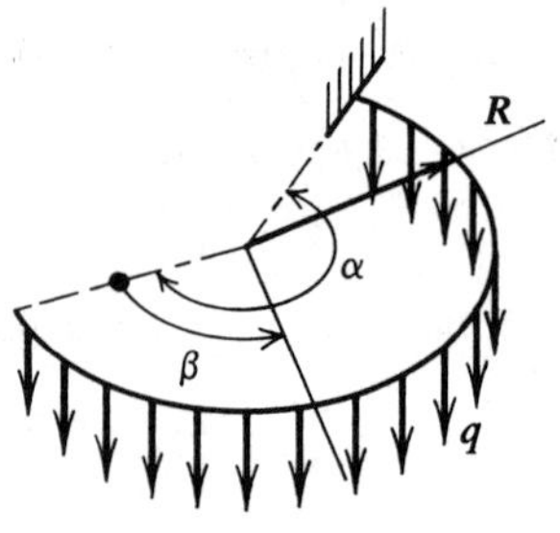

Fig. 8.4-15 Out-of-plane loading of circular arc cantilevers. See Figs. 8.4-16 and 8.4-17 for factors B_1 through B_9. Deflection and rotations are at $\beta = 0$. λ is the ratio of flexural to torsional rigidity. (Used by permission from A. Blake, *Design of Curved Members for Machines*, Krieger, 1979.)

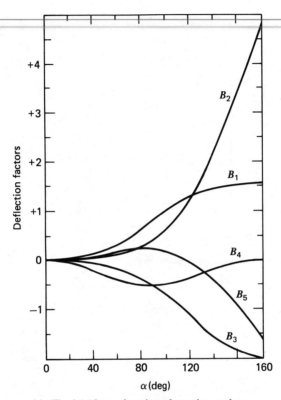

Fig. 8.4-16 Factors used in Fig. 8.4-15 as a function of opening angle α

$$B_1 = (2\alpha - \sin 2\alpha)/4$$

$$B_2 = (6\alpha + \sin 2\alpha - 8\sin\alpha)/4$$

$$B_3 = (4\cos\alpha - \cos 2\alpha - 3)/4$$

$$B_4 = (\cos 2\alpha - 1)/4$$

$$B_5 = (4\sin\alpha - \sin 2\alpha - 2\alpha)/4$$

where α must be in radians. (Used by permission from A. Blake, *Design of Curved Members for Machines*, Krieger, 1979.)

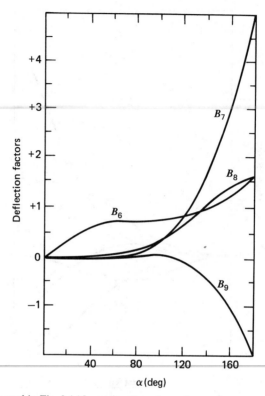

Fig. 8.4-17 Factors used in Fig. 8.4-15 as a function of opening angle α

$$B_6 = (2\alpha + \sin 2\alpha)/4$$

$$B_7 = (2\alpha^2 - \cos 2\alpha - 4\alpha \sin \alpha + 1)/4$$

$$B_8 = (4 \sin \alpha + \sin 2\alpha - 4\alpha \cos \alpha - 2\alpha)/4$$

$$B_9 = (4 \cos \alpha + \cos 2\alpha + 4\alpha \sin \alpha - 5)/4$$

where α must be in radians. (Used by permission from A. Blake, *Design of Curved Members for Machines*, Krieger, 1979.)

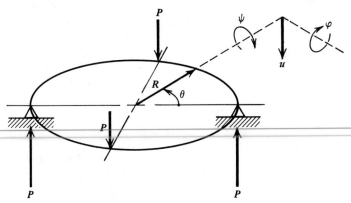

Fig. 8.4-18 Out-of-plane loading of a two-point supported ring by two concentrated loads. See Fig. 8.4-21 for a plot of the factors A_1, A_2, A_4, A_5, A_{10}, A_{11}, and A_{12}. For $0 \le \theta \le 90°$

$$u = \frac{PR^3}{EI_2}(A_1 + \lambda A_2)$$

$$\psi = -\frac{PR^2}{EI_2}(A_4 + \lambda A_5)$$

$$\varphi = -\frac{PR^2}{EI_2}(1 + \lambda)A_7$$

$$G = PR(A_{10} - A_{11})$$

$$T = PR(A_{11} - A_{12})$$

where λ is the ratio of flexural to torsional rigidity.

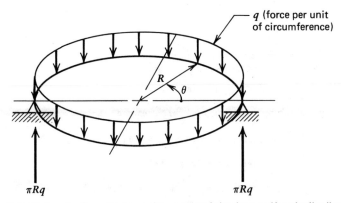

Fig. 8.4-19 Out-of-plane loading of a two-point supported ring by a uniformly distributed load. See Fig. 8.4-22 for a plot of the factors B_1 through B_8. For $0 \le \theta \le 90°$,

$$u = \frac{qR^4}{EI_2}[(1 + \lambda)B_1 - \lambda B_2]$$

$$\psi = -\frac{qR^3}{EI_2}[(1 + \lambda)B_3 - \lambda B_4]$$

$$\varphi = -\frac{qR^3}{EI_2}(1 + \lambda)[B_5 + B_6 - 1]$$

$$G = -2qR^2B_5 + qR^2$$

$$T = -qR^2B_7 + qR^2B_8.$$

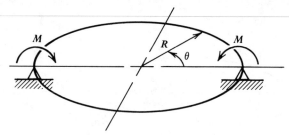

Fig. 8.4-20 Out-of-plane loading of a two-point supported ring by concentrated moments. See Fig. 8.4-21 for a plot of the factors A_3, A_6, A_8, A_9, A_{10}, and A_{11}. This case may be combined with those in Figs. 8.4-18 and 8.4-19 to provide solutions to ring trunnions or gimbals, see Reference 10; additional cases of out-of-plane loading of rings may be found in Reference 8. For $0 \le \theta \le 90°$,

$$u = \frac{MR^2}{EI_2}(1 + \lambda)A_3$$

$$\psi = -\frac{MR}{EI_2}(1 + \lambda)A_6$$

$$\varphi = -\frac{MR}{EI_2}[(1 - \lambda)A_8 + (1 + \lambda)A_9]$$

$$G = -MA_{10}$$

$$T = -MA_{11}.$$

Fig. 8.4-21 Factors used in Figs. 8.4-18 and 8.4-20 as a function of location angle θ

$$A_1 = [2(\theta - 1)\sin\theta + (2\theta + 2 - \pi)\cos\theta + \pi - 2]/8$$

$$A_2 = [(6 + 2\theta - \pi)\cos\theta - 2(3 - \theta)\sin\theta + \pi + 4\theta - 6]/8$$

$$A_3 = [2\theta\cos\theta + \pi(1 - \cos\theta) - 2\sin\theta]/8$$

$$A_4 = [2\theta\cos\theta + (\pi - 2\theta)\sin\theta]/8$$

$$A_5 = [(\pi - 2\theta - 4)\sin\theta + 2(\theta - 2)\cos\theta + 4]/8$$

$$A_6 = [(\pi - 2\theta)\sin\theta]/8$$

$$A_7 = [2(1 - \theta)\sin\theta + (\pi - 2 - 2\theta)\cos\theta]/8$$

$$A_8 = \sin\theta/4$$

$$A_9 = [(\pi - 2\theta)\cos\theta]/8$$

$$A_{10} = \sin\theta/2$$

$$A_{11} = \cos\theta/2$$

$$A_{12} = (1 - \sin\theta)/2$$

where θ is in radians. (Used by permission from Reference 10.)

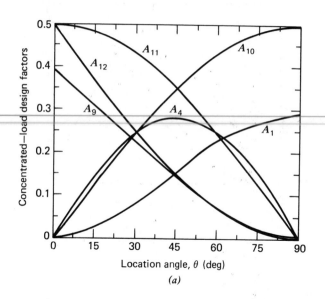

(a)

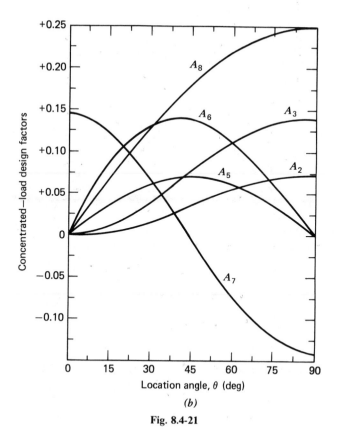

(b)

Fig. 8.4-21

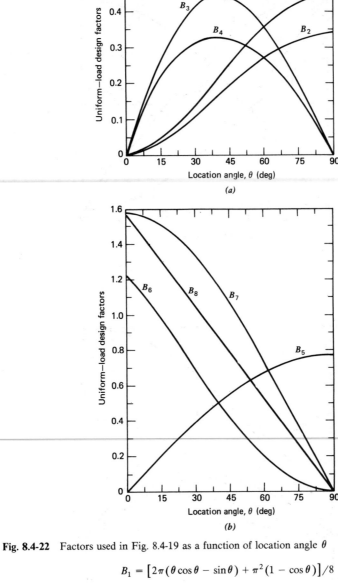

Fig. 8.4-22 Factors used in Fig. 8.4-19 as a function of location angle θ

$$B_1 = \left[2\pi(\theta\cos\theta - \sin\theta) + \pi^2(1 - \cos\theta)\right]/8$$

$$B_2 = \left[4\theta^2 - 4\pi(\theta - \sin\theta)\right]/8$$

$$B_3 = \left[(\pi^2 - 2\pi\theta)\sin\theta\right]/8$$

$$B_4 = \theta - \pi(1 - \cos\theta)/2$$

$$B_5 = \pi\sin\theta/4$$

$$B_6 = \left[(\pi^2 - 2\pi\theta)\cos\theta\right]/8$$

$$B_7 = \pi\cos\theta/2$$

$$B_8 = \pi/2 - \theta$$

where θ is in radians. (Used by permission from Reference 10.)

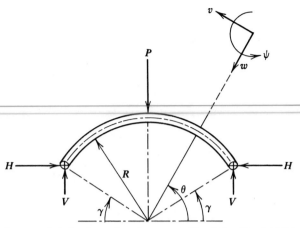

Fig. 8.4-23 Simply supported circular arch under a central load. See Fig. 8.4-24 for factors G_1 and G_2.

$$w(\theta = 90°) = \frac{PR^3}{EI_1} G_1$$

$$H = PG_2$$

$$V = \tfrac{1}{2}P$$

(Used by permission from Reference 5.)

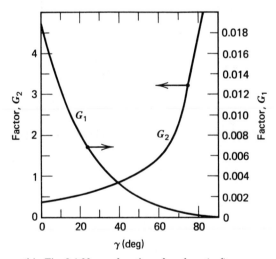

Fig. 8.4-24 Factors used in Fig. 8.4-23 as a function of angle γ (rad)

$$G_1 = \tfrac{1}{8}\left[(\pi - 2\gamma)(1 + 2\cos^2\gamma) - 8\cos\gamma + 3\sin 2\gamma\right]$$

$$- \frac{[4\sin\gamma + 3\cos 2\gamma - (\pi - 2\gamma)\sin 2\gamma - 1]^2}{8(\pi - 2\gamma)(1 + 2\sin^2\gamma) - 24\sin 2\gamma}$$

$$G_2 = \frac{4\sin\gamma + 3\cos 2\gamma - (\pi - 2\gamma)\sin 2\gamma - 1}{2(\pi - 2\gamma)(1 + 2\sin^2\gamma) - 6\sin 2\gamma}$$

(Used by permission from Reference 5.)

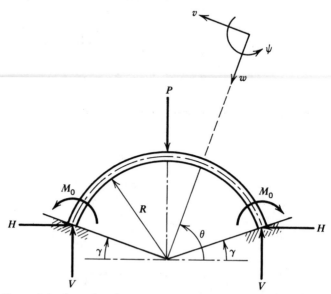

Fig. 8.4-25 Clamped circular arch under a central load. See Fig. 8.4-26 for the factors J_1, J_2, and J_3.

$$w(\theta = 90°) = \frac{PR^3}{EI_1}J_1$$

$$H = PJ_2$$

$$M_0 = PRJ_3$$

$$V = \tfrac{1}{2}P$$

(Used by permission from Reference 5.)

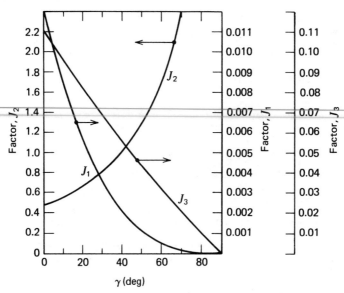

Fig. 8.4-26 Factors used in Fig. 8.4-25 as a function of angle γ (rad)

$$J_1 = \tfrac{1}{8}\left[(\pi - 2\gamma)(1 + 2\cos^2\gamma) - 8\cos\gamma + 3\sin 2\gamma\right]$$

$$-\frac{4A_2A_4^2 - 8A_1A_3A_4 + 0.5(\pi - 2\gamma)A_1^2}{4\left[(\pi - 2\gamma)A_2 - 8A_3^2\right]}$$

$$J_2 = \frac{(\pi - 2\gamma)A_1 - 8A_3A_4}{2(\pi - 2\gamma)A_2 - 16A_3^2}$$

$$J_3 = \frac{A_1A_3 - A_2A_4}{(\pi - 2\gamma)A_2 - 8A_3^2}$$

$$A_1 = 4\sin\gamma + 3\cos 2\gamma - (\pi - 2\gamma)\sin 2\gamma - 1$$

$$A_2 = (\pi - 2\gamma)(1 + 2\sin^2\gamma) - 3\sin 2\gamma$$

$$A_3 = 0.5(\pi - 2\gamma)\sin\gamma - \cos\gamma$$

$$A_4 = 1 - 0.5(\pi - 2\gamma)\cos\gamma - \sin\gamma$$

(Used by permission from Reference 5.)

REFERENCES

8.4-1 N. J. Hoff, *The Analysis of Structures*, Wiley, New York, 1956, p. 63ff.

8.4-2 C. H. Norris and J. B. Wilbur, *Elementary Structural Analysis*, 2nd ed., McGraw-Hill, New York, 1960, p. 380ff.

8.4-3 N. J. Hoff, ref. 1, p. 342.

8.4-4 R. J. Roark and W. C. Young, *Formulas for Stress and Strain*, 5th ed., McGraw-Hill, New York, 1975, chap. 8.

8.4-5 A. Blake, *Practical Stress Analysis in Engineering Design*, Dekker, New York, 1982, Part IV.

8.4-6 W. Griffel, *Handbook of Formulas for Stress and Strain*, Ungar, New York, 1966, pp. 130–152.

8.4-7 J. Y. Liu and Y. P. Chiu, "Analysis of a Thin Elastic Ring under Arbitrary Load," *J. Engr. Industry, Trans. ASME*, **96**, 870–876 (1974).

8.4-8 J. L. Houtman, "Design Coefficients for Rings Loaded Out-of-Plane," *J. Engr. Industry, Trans. ASME*, **98**, 1976, 369–374.

8.4-9 J. R. Barber, "Force and Displacement Influence Functions for the Circular Ring," *Jour. Strain Analysis*, **13**(2), 1978, 77–81.

8.4-10 A. Blake, "Gimbal Ring Design," *Machine Design*, pp. 92–93, April 8, 1976.

8.4-11 R. J. Roark and W. C. Young, ref. 4, Table 18.

8.4-12 G. S. Glushkov, I. R. Egorov, and V. V. Ermolov, *Handbook of Formulas for the Analysis of Complex Frames and Arches*, Israel Program for Scientific Translations, 1967, Chap. III.

8.5 THICK, CURVED BEAMS

The previous sections assumed that the curved beam was thin ($R/h \geq 10$) and that only the bending energy contributed to deflection. When the curved beam is thick ($R/h < 10$), the implications of these assumptions must be modified. The nature of the required modifications can be seen by recourse to the theory of plane stress in linear elasticity.

For example, consider the elasticity solution to the problem shown in Fig. 8.5-1.[1] The exact bending stress σ_θ is plotted in Fig. 8.5-2 for inside radius a variable and outside radius $b = a + 1$; that is, for $a = 0.5$, $R/h = a/1 = 0.5$; for $a = 1$, $R/h = 1$; for $a = 2$, $R/h = 2$; for $a \to \infty$, $R/h \to \infty$, which is the straight-beam result and the thin, curved-beam limit. Figure 8.5-2 shows that when R/h becomes small, the bending stress variation across the thickness deviates significantly from the linear result of a straight beam; in fact, the stress variation becomes hyperbolic. Also the neutral axis ($\sigma_\theta = 0$) no longer coincides with the centroidal axis ($h = 0.5$). The elasticity solution also shows the existence of a radial stress, but this stress is not shown in Fig. 8.5-2. The thick, curved-beam problem has also been considered within the context of an engineering theory; this theory, which is due to Winkler, retains the hypothesis that plane cross sections remain plane during bending but allows for the fact that, due to the initial curvature, the longitudinal fibers between adjacent cross sections have unequal initial lengths.

In terms of the notation of Fig. 8.5-1, it can be shown[2] that this approach leads to

$$\sigma_\theta = -\frac{M}{AR}\left(1 + \frac{1}{z}\frac{\rho}{R + \rho}\right) \tag{8.5-1}$$

where

$$z = \frac{1}{AR}\int \frac{\rho^2 \, dA}{R + \rho} \tag{8.5-2}$$

If e is the distance from the centroidal axis to the neutral axis, it can be further shown that

$$e = \frac{zR}{z + 1} \tag{8.5-3}$$

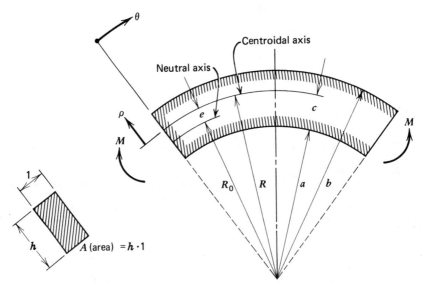

Fig. 8.5-1 Thick, circular beam under pure bending.

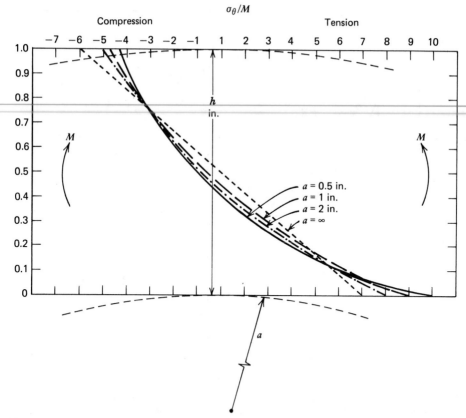

Fig. 8.5-2 Tangential (bending) stress distribution in circular arc beams under pure bending. The cross section is rectangular with unit thickness. (Used by permission from Reference 1.)

which implies

$$R_0 = R - e = \frac{R}{z + 1} \tag{8.5-4}$$

When R becomes large, Equation (8.5-1) becomes equal to $-M\rho/I$ and $e \to 0$ so that $R \to R_0$. Defining a stress correction factor K at $\rho = -c$ so that

$$(\sigma_\theta)_{max} = K \frac{Mc}{I} \tag{8.5-5}$$

allows the bending stress at the concave extreme fiber of a thick, curved beam to be more easily determined. It can be shown that this is the maximum bending stress across the cross section; in particular, it is larger than the bending stress at the convex extreme fiber. Values of K for some compact cross sections are shown in Fig. 8.5-3. More extensive data are available in Roark and Young.[3] Inspection of Fig. 8.5-3 shows that reasonably wide variation in cross-sectional shape results in only a modest variation in K. This led Blake[4] to suggest an analytical approximation that is sufficiently accurate for the design of thick, curved beams of compact cross section:

$$K = 0.35 \left[\frac{3\left(\frac{R}{c}\right)^2 - \left(\frac{R}{c}\right) - 1}{(R/c)^2 - R/c} \right] \tag{8.5-6}$$

If values of z in Equation (8.5-2) are needed, they can be found from Reference 2 or computed by numerical integration of Equation (8.5-2). Since z is likely to be quite small, care is required in order

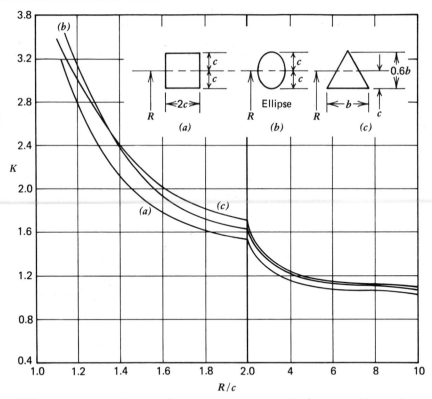

Fig. 8.5-3 Stress correction factor at the concave extreme fiber of a thick, curved beam of compact cross section. (Used by permission from Peery and Azar, Aircraft Structures, 2nd ed., McGraw-Hill, New York, 1982.)

to compute z accurately. For compact cross sections the Winkler theory gives good agreement with the elasticity solutions. For example, for a rectangular, curved beam of unit width, unit thickness, and an inner radius of $a = 2.0$, Fig. 8.5-2 implies $(\sigma_\theta)_{max} = 7M$; using Fig. 8.5-3, $K = 1.15$ and Equation (8.5-5) implies $(\sigma_\theta)_{max} = 6.9M$. However, for thin-walled, open sections, such as I beams or T beams, the Winkler theory can lead to unacceptable error.[5] There seems to be little design information available for thick curved beams with thin-walled open sections. It is recommended that thin-beam theory[6] be used together with conservative choices of K from Fig. 8.5-3; for an accurate calculation of deflection, it may be necessary to include deflection due to shear.[7]

If the thick curved beam is subjected to a direct load F at the centroid as well as a bending moment M, Equation (8.5-5) must be augmented. For a thick curved member the stress due to F is not F/A because the initial lengths of the fibers vary over the cross section. In fact, the stress due to F at the concave extreme fiber will be higher than F/A. Seely and Smith[8] suggest using the same correction factor for axial stress as for bending stress; then

$$(\sigma_\theta)_{max} = K\left(\frac{F}{A} + \frac{Mc}{I}\right) \tag{8.5-7}$$

In using energy methods to calculate the deflection of thick curved beams, the stress energies due to axial load and transverse shear must be added to the stress energy due to bending moment. The total stress energy is[9]

$$U_\sigma = \frac{1}{2}\int \frac{F^2}{AE}\,ds + \frac{1}{2}K_s\int \frac{V^2}{GA}\,ds + \frac{1}{2}\int \frac{M^2}{EAeR}\,ds - \int \frac{MF}{EAR}\,ds \tag{8.5-8}$$

The factor K_s corrects for the nonuniform distribution of the shear stress, and its value depends on the shape of the cross section; Roark[10] recommends $K_s = \frac{6}{5}$ for a rectangular cross section and $K_s = \frac{10}{9}$ for a solid circular cross section.

REFERENCES

8.5-1 E. E. Sechler, *Elasticity in Engineering*, Wiley, New York, 1952, Sec. 8.6.
8.5-2 F. B. Seely, and J. O. Smith, *Advanced Mechanics of Materials*, 2nd ed., Wiley, New York, 1957, Sec. 48.
8.5-3 R. J. Roark and W. C. Young, *Formulas for Stress and Strain*, 5th ed., McGraw-Hill, New York, 1975, Chap. 8, Table 16.
8.5-4 A. Blake, *Design of Curved Members for Machines*, Krieger, Melbourne, FL, 1979, p. 206.
8.5-5 F. B. Seely and J. O. Smith, ref. 2, Sec. 51.
8.5-6 A. Gjelsvik, *The Theory of Thin Walled Bars*, Wiley-Interscience, New York, 1981.
8.5-7 L. H. Donnell, *Beams, Plates, and Shells*, McGraw-Hill, New York, 1976, pp. 147–158.
8.5-8 F. B. Seely and J. O. Smith, ref. 2, p. 152.
8.5-9 F. B. Seely and J. O. Smith, ref. 2, Sec. 53.
8.5-10 R. J. Roark and W. C. Young, ref. 3, p. 185.

8.6 FINITE-ELEMENT METHOD FOR CURVED MEMBERS

The analytical methods described in Section 8.5, the exact formulation of elasticity, or the engineering formulation due to Winkler can be used to provide approximate solutions for thick, curved members of practical interest. The first application of the Winkler theory was to chain links; subsequently, it has been used for lifting hooks, bearing races and housings, proving rings, eye bars, clevises, machine frames (e.g., cutting machines and punch presses), C-clamps, and pipe elbows.

Current practice uses analytical methods for preliminary design. For final design it is more likely that the degree of approximation between the analytical model and the actual structure will be reduced by using a finite-element model. Such a model is shown in Fig. 8.6-1.

In the case of a thin curved ring, the finite-element stiffness matrix can be written explicitly.[1]

REFERENCE

8.6-1 H. R. Meck, "An Accurate Polynomial Displacement Function for Finite Ring Elements," *Comput. & Structures* **11**, 265–269 (1980).

8.7 THERMAL STRESSES IN CIRCULAR BEAMS

The analysis of the free thermal stresses in a thin, circular beam can be simplified by assuming a one-dimensional stress distribution. Using the notation of Fig. 8.2-1, it is assumed that $\sigma_\theta \gg \sigma_r, \sigma_z$. Of course as R/h becomes small, a circular beam becomes more like a sector of a circular plate with a hole in the center and the stress distribution becomes more two-dimensional, that is, $\sigma_\theta, \sigma_r \gg \sigma_z$. Similarly, if the axial length of a circular beam becomes large compared to its radial thickness, the structure becomes more like a panel of a cylindrical shell and the stress distribution again becomes

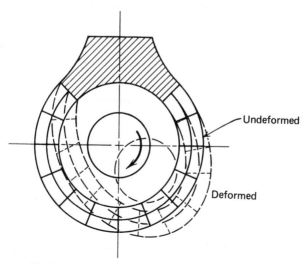

Fig. 8.6-1 Finite-element model of a curved member.

two-dimensional, that is, $\sigma_\theta, \sigma_z \gg \sigma_r$. Solutions for both the one-dimensional[1] and two-dimensional[2] situations are available. In this section only the one-dimensional approximation is discussed.

In addition, only free thermal stresses are considered; that is, the thermal stresses in an unrestrained structure. Boundary restraints induce isothermal stresses that must be added to the free thermal stresses. In general, thermal stress analysis consists of a two-step process: determination of the free thermal stresses and deformations followed by a calculation of the discontinuity stresses required to achieve compatibility with the attached structures.

8.7-1   Circular Ring with Axial Temperature Variation

Let the axial coordinate z be measured from a plane centrally located along the axial width of the ring. The ring will remain plane during free thermal deformation if the axial temperature distribution is symmetric, that is, $T(z) = T(-z)$. If this symmetry were not present, the ring would twist out of its initial plane of curvature. For the symmetric situation[1]

$$\sigma_\theta = E\alpha\big[T_{avg} - T(z)\big] \tag{8.7-1}$$

where T_{avg} is average temperature across the ring width. There is also a radial shear stress in the ring, but this is much smaller than σ_θ provided the ring width is much smaller than R.

8.7-2   Circular Ring with Radial Temperature Variation

A circular ring with a radial temperature variation $T(r)$ can be viewed as a set of concentric rings composed of the same material, with each ring at a different uniform temperature. This viewpoint leads to[1]

$$\sigma_\theta = E\alpha\big[T_{avg} - T(r)\big] \tag{8.7-2}$$

where T_{avg} is the average temperature across the ring thickness. There is also a radial stress σ_r but for R/h large, σ_r is small compared to σ_θ. This viewpoint allows a straightforward extrapolation to the case of concentric rings made of different materials.[1]

8.7-3   Circular Ring with Temperature Variation in r and z

The cases of Sections 8.7-1 and 8.7-2 can be combined to give[1]

$$\sigma_\theta = E\alpha\big[T_{avg} - T(r, z)\big] \tag{8.7-3}$$

where T_{avg} is $T(r, z)$ averaged over both r and z. The variation of temperature in z must be symmetric about the midplane of the ring's axial length.

REFERENCES

8.7-1   D. Burgreen, *Elements of Thermal Stress Analysis*, C. P. Press, 1971, Chap. 2.
8.7-2   D. Burgreen, ref. 1, Chap. 4.

CHAPTER **9**

PLATES

FREDERICK C. NELSON

Dean of Engineering
Tufts University
Medford, Massachusetts

9.1 BACKGROUND AND ASSUMPTIONS

Flat plates and membranes are planar structural elements whose thickness is much smaller than their in-plane dimensions. They are used to carry surface loads, usually pressure loads.

Flat plates carry pressure loads by out-of-plane shear and, as for beams, this shear is associated with a distribution of bending moment; for plates the bending moment distribution is two dimensional and usually accompanied by a distribution of torsional moment.

Flat membranes carry pressure loads by deflecting so that in-plane loads acquire out-of-plane components and, as for strings, linear response is obtained only if the membrane is pretensioned; for membranes the pretension is usually two dimensional.

Curved membranes can react pressure loads in a linear manner without pretension. Indeed, the study of this situation constitutes the membrane theory of thin shells. Compilations of membrane stresses in shells are available, see Reference 1 for a catalog of results.

The analogy of a plate as a two-dimensional lattice of beams or a membrane as a two-dimensional network of strings was used by many early investigators. In the eighteenth century Euler used the latter analogy to predict the vibration behavior of flat rectangular membranes, and James Bernoulli used the former analogy to predict the vibration behavior of flat rectangular plates; however, Bernoulli's plate equation was wrong because it neglected torsional resistance. The contributions of many nineteenth century investigators, principally Germain, Navier, Kirchhoff, and Love, were required before a satisfactory theory was established for isotropic plates with straight and curved edges.

This section considers only the static theory of plates. The classical (linear) theory of elastic isotropic plates is based on the following assumptions (adapted by permission from Reference 2, p. 28):

1. The material of the plate is elastic, homogeneous, and isotropic.
2. The plate is initially flat.
3. The thickness of the plate is small compared to its other dimensions. The smallest in-plane dimension of the plate is at least 10 times larger than its thickness.
4. The deflections are small compared to the plate thickness. A maximum deflection of one-fifth of the thickness is considered as the limit for small-deflection theory. This limitation can also be stated in terms of length; that is, the maximum deflection is less than one-fiftieth of the smaller span length.
5. The slopes of the deflected middle surface are small compared to unity.
6. The deformations are such that straight lines, initially normal to the middle surface, remain straight lines and normal to the middle surface (deformations due to transverse shear will be neglected).
7. The deflection of the plate is produced by displacement of points of the middle surface normal to its initial plane.
8. The stresses normal to the middle surface are of a negligible order of magnitude.
9. The strains in the middle surface can be neglected in comparison with strains due to bending.

The first eight assumptions are logical extensions of flat-beam theory. Assumption 9 is the counterpart of the assumption of inextensional bending in curved-beam theory. Assumption 6 is often referred to as the Kirchhoff–Love hypothesis.

REFERENCES

9.1-1 A. Pflüger, *Elementary Statics of Shells*, F. W. Dodge Corp., 1961.
9.1-2 R. Szilard, *Theory and Analysis of Plates*, Prentice-Hall, Englewood Cliffs, NJ, 1974.

9.2 ISOTROPIC, ELASTIC PLATES: EQUATION OF EQUILIBRIUM

The theory is formulated in terms of moments and forces per unit length of the middle surface of the plate. These quantities are shown in their positive directions for both cartesian and polar coordinates in Fig. 9.2-1. This figure also shows the positive direction of the deflection w in the thickness direction z.

The equation of equilibrium in terms of displacement is

$$D\nabla^2\nabla^2 w = p_z \tag{9.2-1}$$

where D is the bending stiffness of the plate given by

$$D = \frac{Eh^3}{12(1 - \nu^2)} \tag{9.2-2}$$

where p_z is the surface pressure loading, ∇^2 is Laplace's differential operator, and ν is Poisson's ratio.
In cartesian coordinates (x, y)

$$\nabla^2 = \frac{\partial^2}{\partial x^2} + \frac{\partial^2}{\partial y^2} \tag{9.2-3}$$

so that

$$\nabla^2\nabla^2 = \frac{\partial^4}{\partial x^4} + 2\frac{\partial^4}{\partial x^2 \partial y^2} + \frac{\partial^4}{\partial y^4} \tag{9.2-4}$$

In polar coordinates

$$\nabla^2 = \frac{\partial^2}{\partial r^2} + \frac{1}{r^2}\frac{\partial^2}{\partial \phi^2} + \frac{1}{r}\frac{\partial}{\partial r} \tag{9.2-5}$$

and if, as is often the case, the loading of a circular plate is symmetric about the z axis, $w = w(r)$ only and

$$\nabla^2 = \frac{\partial^2}{\partial r^2} + \frac{1}{r}\frac{\partial}{\partial r} = \frac{1}{r}\frac{\partial}{\partial r}\left(r\frac{\partial}{\partial r}\right) \tag{9.2-6}$$

The moments are related to deflection as follows:
Cartesian coordinates

$$m_x = -D\left(\frac{\partial^2 w}{\partial x^2} + \nu\frac{\partial^2 w}{\partial y^2}\right) \tag{9.2-7}$$

$$m_y = -D\left(\frac{\partial^2 w}{\partial y^2} + \nu\frac{\partial^2 w}{\partial x^2}\right) \tag{9.2-8}$$

$$m_{xy} = m_{yx} = -D(1 - \nu)\frac{\partial^2 w}{\partial x \partial y} \tag{9.2-9}$$

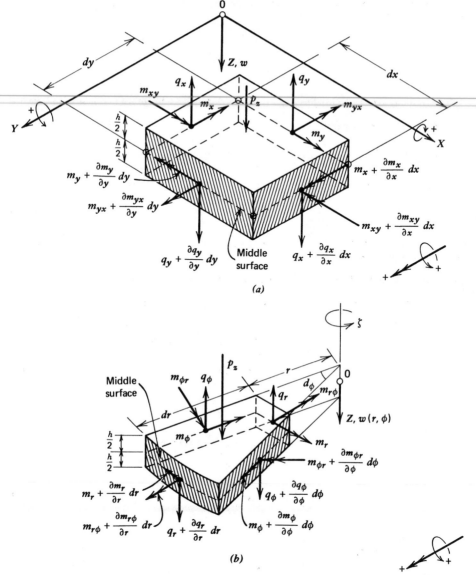

Fig. 9.2-1 Plate elements showing positive directions of forces and moments per unit length. (a) Element of a rectangular plate; (b) element of a circular plate. (Used by permission from Reference 2, pp. 30 and 91.)

Polar coordinates

$$m_r = -D\left[\frac{\partial^2 w}{\partial r^2} + \nu\left(\frac{1}{r^2}\frac{\partial^2 w}{\partial \phi^2} + \frac{1}{r}\frac{\partial w}{\partial r}\right)\right] \qquad (9.2\text{-}10)$$

$$m_\phi = -D\left(\frac{1}{r}\frac{\partial w}{\partial r} + \frac{1}{r^2}\frac{\partial^2 w}{\partial \phi^2} + \nu\frac{\partial^2 w}{\partial r^2}\right) \qquad (9.2\text{-}11)$$

$$m_{r\phi} = m_{\phi r} = -D(1-\nu)\frac{\partial}{\partial r}\left(\frac{1}{r}\frac{\partial w}{\partial \phi}\right) \qquad (9.2\text{-}12)$$

The maximum bending stress is found from the bending moments by using

$$[\sigma_\alpha]_{\text{max}} = \frac{6m_\alpha}{h^2}$$

$$[\sigma_\beta]_{\text{max}} = \frac{6m_\beta}{h^2} \tag{9.2-13}$$

These relations assume the bending stress is linearly distributed through the thickness with its neutral axis at the middle surface of the plate.

The maximum torsional stress is given by

$$[\tau_{\alpha\beta}]_{\text{max}} = \frac{6m_{\alpha\beta}}{h^2} \tag{9.2-14}$$

and the maximum transverse shear stress is given by

$$[\tau_{\alpha z}]_{\text{max}} = \frac{3}{2}\frac{q_\alpha}{h} \tag{9.2-15}$$

In Equation (9.2-15), the value of q_α must be found from equilibrium. For cartesian coordinates $\alpha \to x$; $\beta \to y$; for polar coordinates $\alpha \to r$, $\beta \to \phi$.

9.3 ISOTROPIC, ELASTIC PLATES: BOUNDARY CONDITIONS

The general solution to Equation (9.2-1) requires appropriate boundary conditions in order to be applicable to specific situations. The classical boundary conditions for a straight edge are shown in Fig. 9.3-1.

These boundary conditions have the following mathematical forms:

1. Fixed edge

$$w = 0 \quad \text{at } x = a \tag{9.3-1}$$

$$\frac{\partial w}{\partial y} = 0 \quad \text{at } x = a \tag{9.3-2}$$

2. Free edge

$$\frac{\partial^2 w}{\partial x^2} + \nu\frac{\partial^2 w}{\partial y^2} = 0 \quad \text{at } x = a \tag{9.3-3}$$

$$\frac{\partial^3 w}{\partial x^3} + (2 - \nu)\frac{\partial^3 w}{\partial x\,\partial y^2} = 0 \quad \text{at } x = a \tag{9.3-4}$$

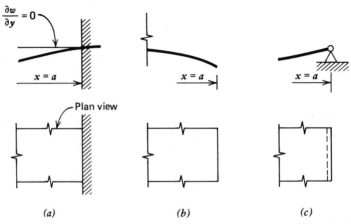

(a)   (b)   (c)

Fig. 9.3-1 Classical boundary conditions for a straight edge of a plate. (a) Clamped edge; (b) free edge; (c) simply supported edge. (Used by permission from Reference 2, p. 38.)

3. Simple support

$$w = 0 \quad \text{at } x = a \tag{9.3-5}$$

$$\frac{\partial^2 w}{\partial x^2} + \nu \frac{\partial^2 w}{\partial y^2} = 0 \quad \text{at } x = a \tag{9.3-6}$$

Conditions (9.3-3) and (9.3-6) require that the bending moment m_x given by Equation (9.2-7) vanishes at the edge $x = a$. Condition (9.3-4) requires that $q_x + \partial m_{xy}/\partial y$ vanishes at $x = a$; this boundary condition was first obtained by Kirchhoff and its physical justification given by Kelvin and Tait.[1]

Similar boundary conditions exist for circular edges,[2] and boundary conditions can also be formulated for elastically supported edges.[3]

It should be pointed out that for large deflections the coupling of bending and in-plane stretching requires a more comprehensive set of boundary conditions, for example, one must distinguish between a clamped edge with full restraint of in-plane displacement (clamped-fixed) and a clamped edge with no restraint of in-plane displacements (clamped-free).

REFERENCES

9.3-1 S. P. Timoshenko and S. Woinowsky-Krieger, *Theory of Plates and Shells*, 2nd ed., McGraw-Hill, New York, 1959, pp. 83–84 and 88–92.
9.3-2 S. P. Timoshenko and S. Woinowsky-Krieger, ref. 1, Chap. 3.
9.3-3 R. Szilard, *Theory and Analysis of Plates*, Prentice-Hall, Englewood Cliffs, NJ, 1974, pp. 40–42.

9.4 ISOTROPIC, ELASTIC PLATES: STRAIN ENERGY

The elastic energy written in terms of displacement is called the strain energy. The bending strain energy for a plate in cartesian coordinates is

$$U = \frac{1}{2} \int_A D \left\{ \left(\frac{\partial^2 w}{\partial x^2} + \frac{\partial^2 w}{\partial y^2} \right)^2 - 2(1 - \nu) \left[\frac{\partial^2 w}{\partial x^2} \frac{\partial^2 w}{\partial y^2} - \left(\frac{\partial^2 w}{\partial x \partial y} \right)^2 \right] \right\} dx\, dy \tag{9.4-1}$$

and in polar coordinates

$$U = \frac{1}{2} \int_A D \left\{ \left(\frac{\partial^2 w}{\partial r^2} + \frac{1}{r} \frac{\partial w}{\partial r} + \frac{1}{r^2} \frac{\partial^2 w}{\partial \phi^2} \right)^2 - 2(1 - \nu) \left[\frac{\partial^2 w}{\partial r^2} \left(\frac{1}{r} \frac{\partial w}{\partial r} + \frac{1}{r^2} \frac{\partial^2 w}{\partial \phi^2} \right) \right. \right.$$
$$\left. \left. - \left(\frac{\partial}{\partial r} \left\langle \frac{1}{r} \frac{\partial w}{\partial \phi} \right\rangle \right)^2 \right] \right\} r\, d\phi\, dr \tag{9.4-2}$$

The strain energy is combined with the potential energy of the external forces to give the total potential energy, Π. For cartesian coordinates and an external pressure loading p_z,

$$\Pi = U - \int_A p_z w(x, y)\, dx\, dy$$

where U is given by Equation (9.4-1).

9.5 ISOTROPIC, ELASTIC PLATES: SOLUTIONS

A large number of plate problems have been considered. A recent bibliography[1] contains more than 12,000 references. In addition to the classic book by Timoshenko,[2] several new monographs are available.[3–5] Design information is also widely available.[4,6–9]

Early analyses of plate problems featured analytical solutions; that is, solutions of the differential equations (9.2-1) subject to boundary conditions such as described in Section 9.3. The books by Timoshenko and Panc contain many examples of this approach. Analytical solutions were supplemented by approximate methods based on the differential equation (e.g., Galerkin's method) or on the total potential energy expression in Section 9.4 (e.g., the Rayleigh–Ritz method); these methods are discussed in books on energy and variational methods.[10,11]

The more recent tendency is to go directly to numerical methods. Chapter 2 of Szilard[4] and Chapter 5 of Ugural[5] give presentations of the finite-difference method and the finite-element method

Loading and Edge Conditions Case Number		Stress C	Deflection K	L
1	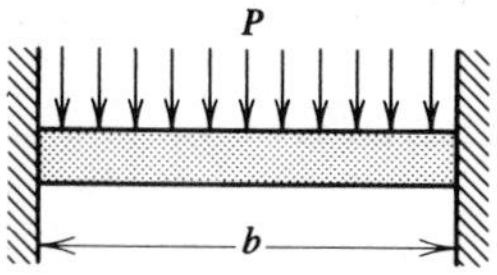 All edges clamped, uniform load over entire surface	C_1	K_1	b
2	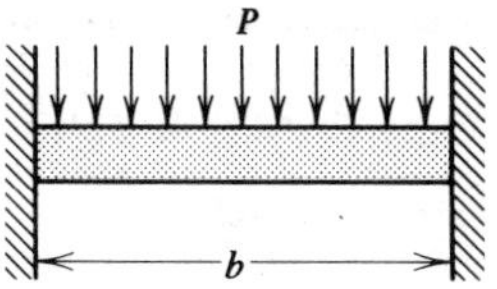 Long edges clamped, short edges supported. Uniform load over entire surface	C_2	K_2	b
3	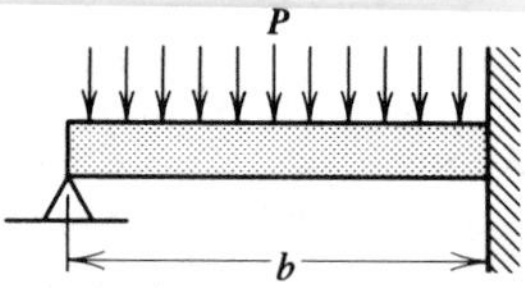 One long edge clamped, other three edges supported. Uniform load over entire surface	C_3	K_3	b
4	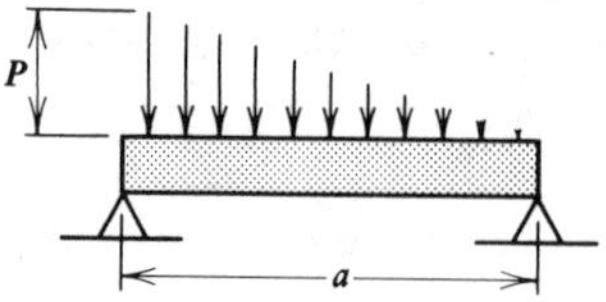 All edges supported. Hydrostatic pressure varying along length	C_4	K_4	b
5	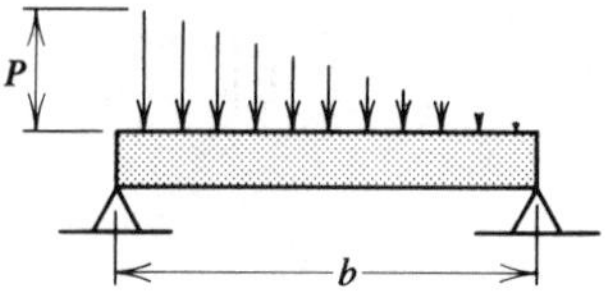 All edges supported. Hydrostatic pressure varying along breadth	C_5	K_5	b
6	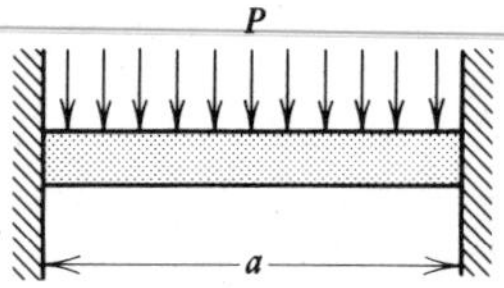 Short edges clamped, long edges supported. Uniform load over entire surface	C_6	K_6	b
7	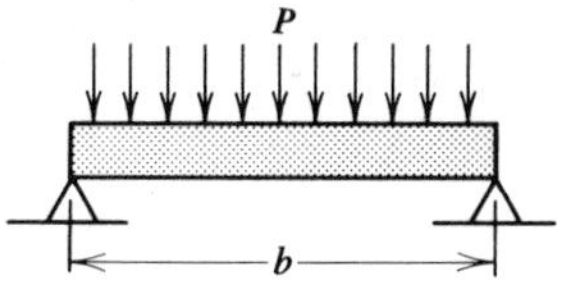 All edges supported. Uniform load over entire surface	C_7	K_7	b

Fig. 9.5-1 Maximum stress and deflection of rectangular, isotropic, elastic plates for $\nu = 0.3$.

$$\sigma_{max} = CpL^2/h^2 \quad \text{and} \quad w_{max} = KpL^4/Eh^3$$

Values of L are given in each case description. Values of C and K for each case are plotted in Charts 1 and 2; note that a and b are defined so that $a \geq b$. Case 14 is covered by Chart 3 ($a = 1.4b$), Chart 4 ($a = b$), and Chart 5 ($a = 2b$); case 15 is covered by Chart 6; case 16 by Chart 7. (Used by permission from Reference 13, pp. 183, 185, and 186.)

Loading and Edge Conditions Case Number	Stress C	Deflection K	L
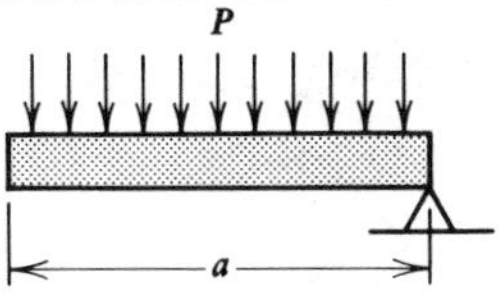 8 — One short edge free. Other three edges supported. Uniform load over entire surface.	C_8	K_8	b
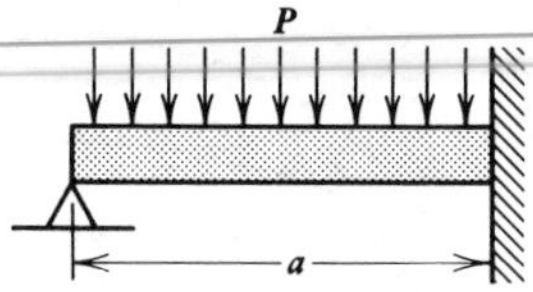 9 — One short edge clamped. Other three edges supported. Uniform load over entire surface.	C_9	K_9	b
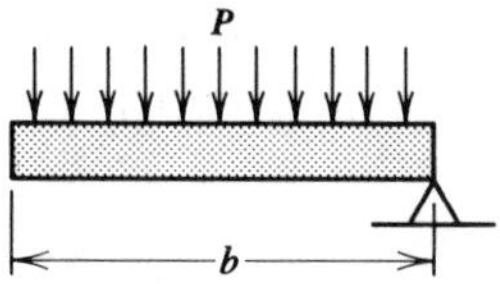 10 — One long edge free. Other three edges supported. Uniform load over entire surface.	C_{10}	K_{10}	a
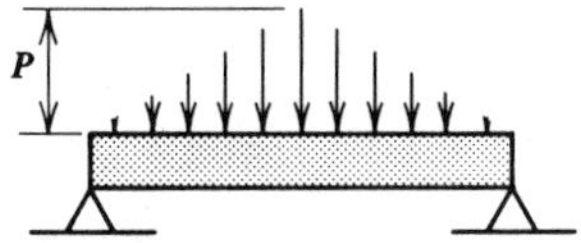 11 — All edges supported. Distributed load in form of a triangular prism.	C_{11}	K_{11}	b
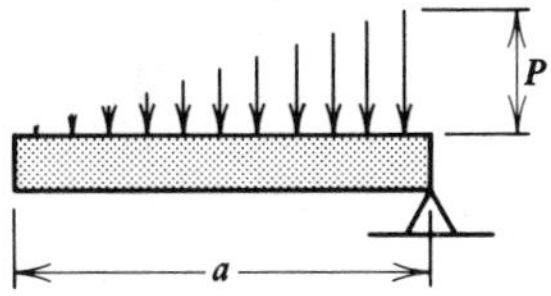 12 — One short edge free. Other three edges supported. Distributed load varying linearly along length.	C_{12}	K_{12}	b
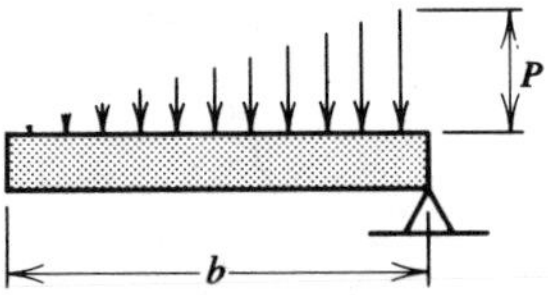 13 — One long edge free. Other three edges supported. Distributed load varying linearly along breadth.	C_{13}	K_{13}	a
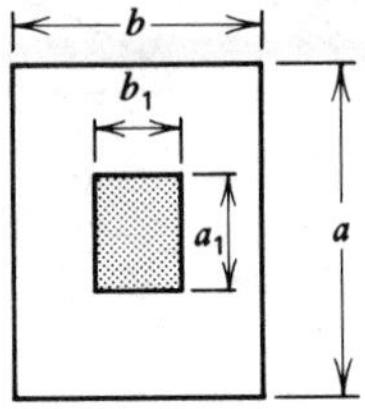 14 — All edges supported, uniformly distributed load over shaded portion.	$\sigma_{max} = \dfrac{C_{14}\, pa_1 b_1}{h^2}$ Charts 3, 4, 5		
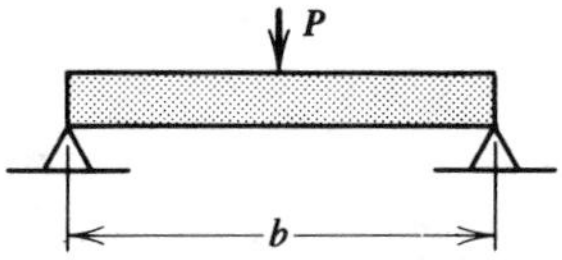 15 — All edges supported. Single concentrated load at center.	$w_{max} = K_{15} Pb^2 / Eh^3$ Chart 6		
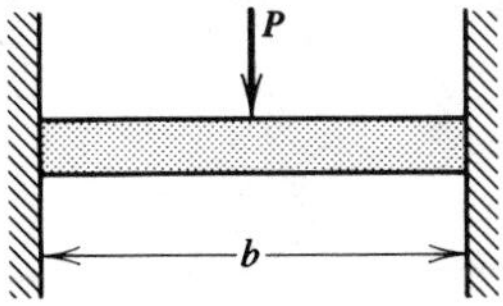 16 — All edges clamped. Single concentrated load at center.	$w_{max} = K_{16} Pb^2 / Eh^3$ Chart 7		

Fig. 9.5-1 (*Continued*)

609

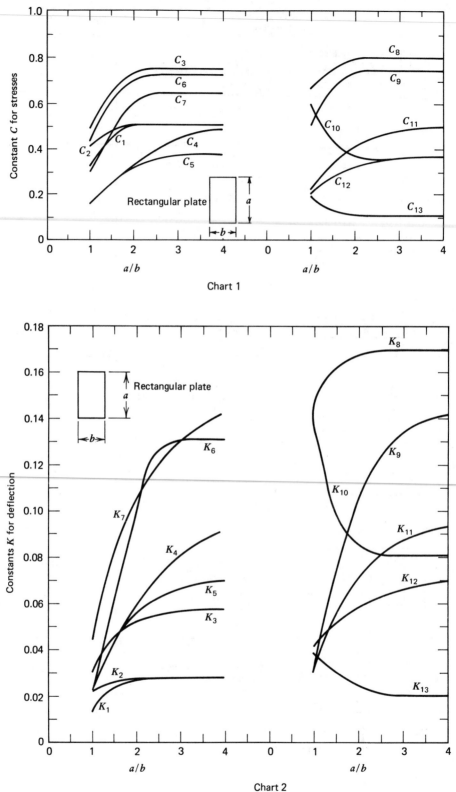

Fig. 9.5-1 (*Continued*)

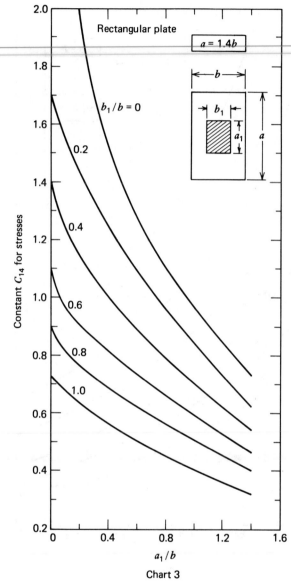

Chart 3

Fig. 9.5-1 (*Continued*)

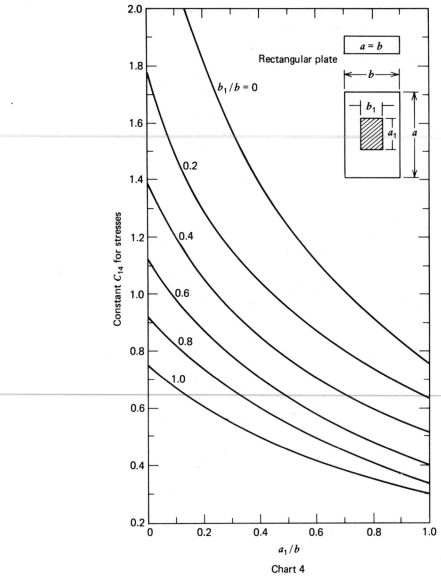

Chart 4

Fig. 9.5-1 (*Continued*)

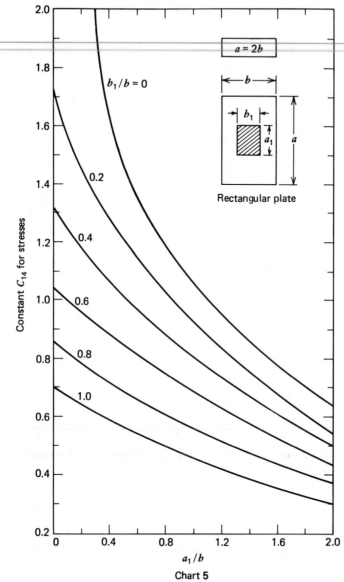

Chart 5

Fig. 9.5-1 (*Continued*)

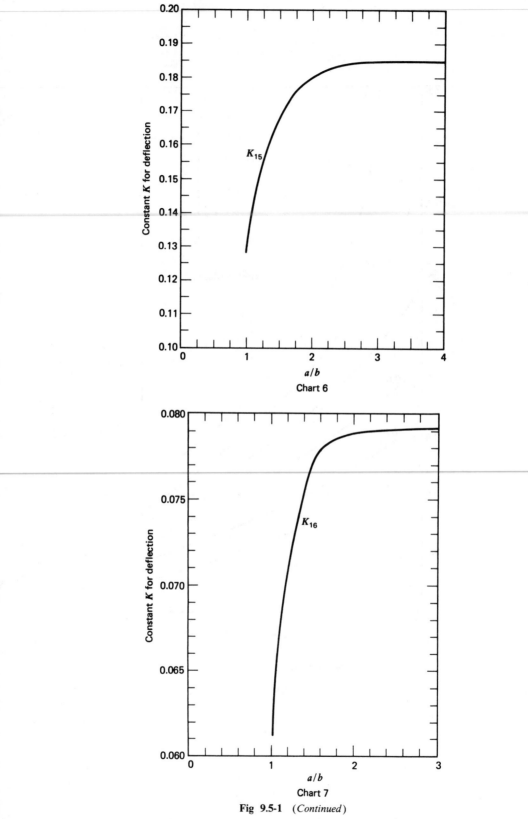

Chart 6

Chart 7

Fig 9.5-1 (*Continued*)

	Loading and Edge Conditions / Case Number	Total Applied Load $P(N)$	Deflection β	Stress λ
1	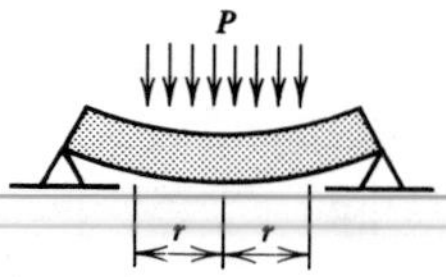 Outer edge supported. Uniform load over concentric circular area of radius r	$p\pi r^2$	β_6	λ_6
2	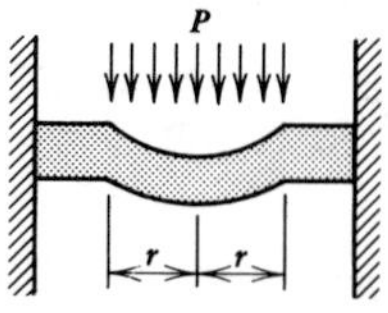 Outer edge clamped. Uniform load over concentric circular area of radius r.	$p\pi r^2$	β_2	λ_2
3	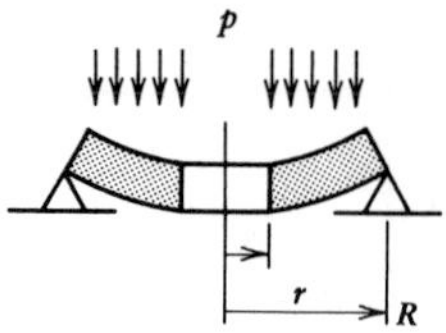 Outer edge supported. Inner edge free. Uniform load over entire actual surface.	$p\pi(R^2 - r^2)$	β_4	λ_4
4	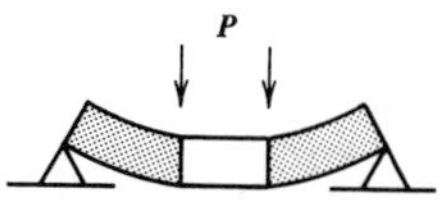 Outer edge supported. Inner edge free. Loaded along inner edge.	P	β_7	λ_7
5	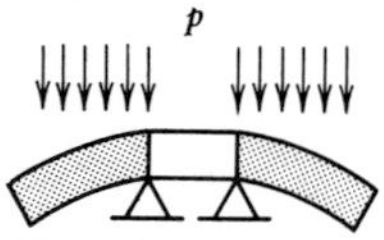 Outer edge free, inner edge supported. Uniform load over entire actual surface.	$p\pi(R^2 - r^2)$	β_1	λ_1
6	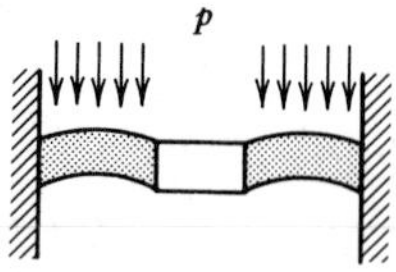 Outer edge clamped. Inner edge free. Uniform load over entire actual surface.	$p\pi(R^2 - r^2)$	β_3	λ_3
7	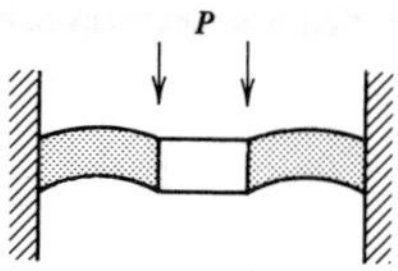 Outer edge clamped. Inner edge free. Loaded along inner edge.	P	β_5	λ_5
8	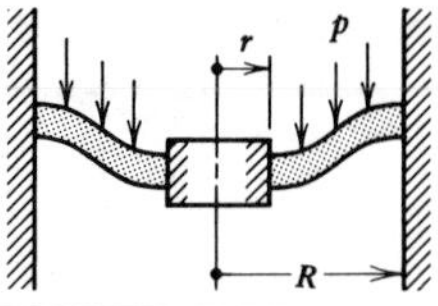 Outer edge clamped. Inner edge fixed against rotation. Uniform load over entire actual surface.	$p\pi(R^2 - r^2)$	β_8	λ_8

Fig. 9.5-2 Maximum stress and deflection of circular, isotropic, elastic plates for $\nu = 0.3$

$$\sigma_{max} = \lambda P/h^2 \quad \text{and} \quad w_{max} = \beta PR^2/Eh^3$$

where P is applied load, R is radius to outside edge, and r is radius to inside edge. Values of λ and β are given in Charts 8, 9, 10, and 11. (Used by permission from Reference 13, p. 187.)

	Loading and Edge Conditions Case Number	Total Applied Load $P(N)$	Deflection β	Stress λ
9	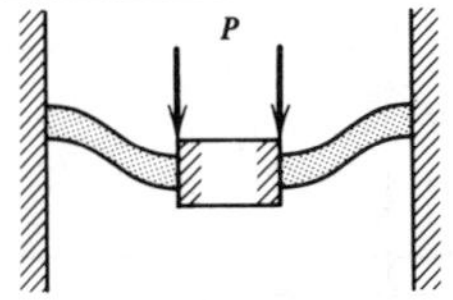 Outer edge clamped. Inner edge fixed against rotation. Loaded along inner edge.	P	β_{10}	λ_{10}
10	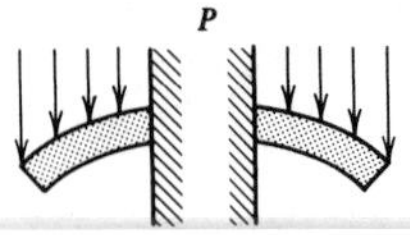 Outer edge free. Inner edge clamped. Uniform load over entire actual surface.	$p\pi(R^2 - r^2)$	β_{12}	λ_{12}
11	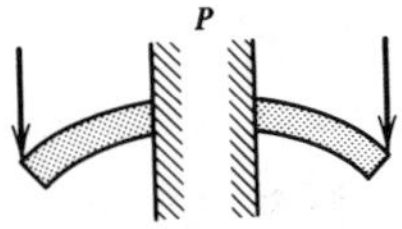 Outer edge free. Inner edge clamped. Loaded along outer edge.	P	β_{13}	λ_{13}
12	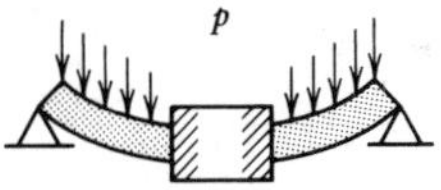 Outer edge supported. Inner edge fixed against rotation. Uniform load over entire actual surface.	$p\pi(R^2 - r^2)$	β_{11}	λ_{11}
13	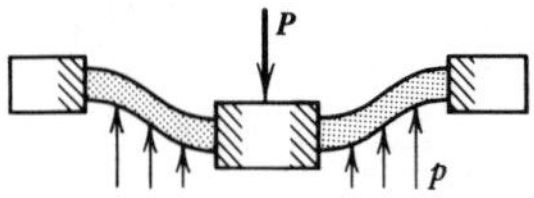 Both edges fixed against rotation. Balanced loading.	$p\pi(R^2 - r^2)$	β_9	λ_9
14	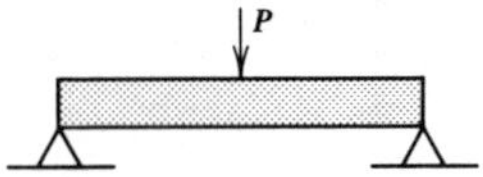 Outer edge supported. Single concentrated load at center.	P	$w_{\max} = 0.986\dfrac{PR^2}{Eh^3}$	
15	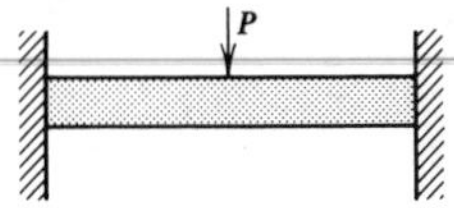 Outer edge clamped. Single concentrated load at center.	P	$w_{\max} = 0.217\dfrac{PR^2}{Eh^3}$	

Fig. 9.5-2 (*Continued*)

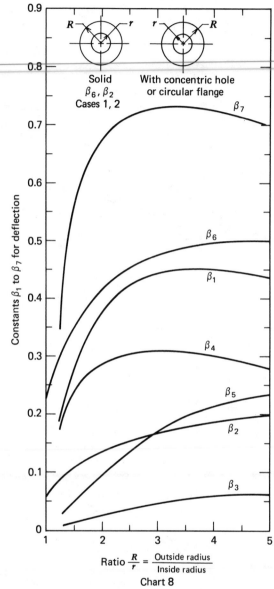

Fig. 9.5-2 (*Continued*)

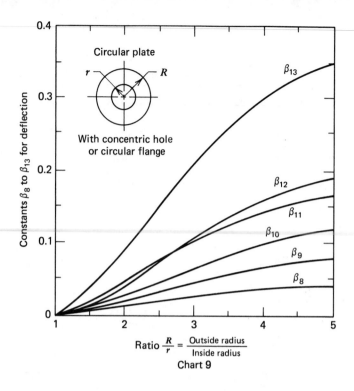

Chart 9

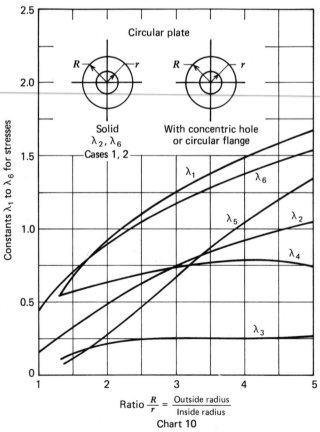

Chart 10

Fig. 9.5-2 (*Continued*)

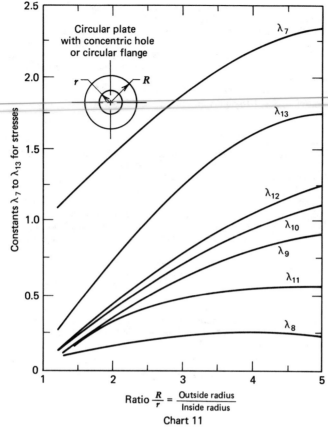

Chart 11

Fig. 9.5-2 (*Continued*)

applied to plates. The latter method is popular because a finite-element mesh can be easily varied in size from one portion of the structure to another without significant penalties in solution time or accuracy. This is convenient when the plate has an irregular boundary or internal holes and/or cutouts. Of equal importance is the fact that large-capacity computer programs based on the finite-element method[12] are widely available. An important aspect of finite-element analysis is the judgment as to whether the computer output is correct. Together with physical intuition, the above exact and approximate solutions are very useful in making this judgment.

For the preliminary design of rectangular or circular plates often only the maximum deflection or maximum stress is required. This information can be presented in a compact, graphical form,[13] see Figs. 9.5-1 and 9.5-2.

There is not much design information available for plates that are not either rectangular or circular. There are a few solutions available for elliptical plates, triangular plates, and sector-shaped plates.[14,15] These analytical solutions are complicated. Consequently, if such solutions are not readily available, the use of numerical or approximate methods is recommended.

REFERENCES

9.5-1 M. Naruoka, *Bibliography on Theory of Plates*, Gehodo Publishing, Tokyo, 1981.

9.5-2 S. P. Timoshenko and S. Woinowsky-Krieger, *Theory of Plates and Shells*, 2nd ed., McGraw-Hill, New York, pp. 83–84 and 88–92.

9.5-3 V. Panc, *Theories of Elastic Plates*, Noordhoff, Leiden, The Netherlands, 1975.

9.5-4 R. Szilard, *Theory and Analysis of Plates*, Prentice-Hall, Englewood Cliffs, NJ, 1974, pp. 40–42.

9.5-5 A. C. Ugural, *Stresses in Plates and Shells*, McGraw-Hill, New York, 1981.

9.5-6 A. Blake, *Practical Stress Analysis in Engineering Design*, Dekker, New York, 1982.

9.5-7 R. J. Roark, and W. C. Young, *Formulas for Stress and Strain*, 5th ed., McGraw-Hill, New York, 1975.

9.5-8 R. Bareš, *Tables for the Analysis of Plates, Slabs and Diaphragms Based on Elastic Theory*, McDonald & Evans, Plymouth, England, 1979.

9.5-9 *Astronautic Structures Manual*, Vol. II, Part I, NASA TM-73306, 1975.

9.5-10 S. H. Crandall, *Engineering Analysis*, McGraw-Hill, New York, 1956, Chap. 4.

9.5-11 H. L. Langhaar, *Energy Methods in Applied Mechanics*, Wiley, New York, 1962, Chap. 5.

9.5-12 W. Pilkey, K. Saczalski, and H. Schaeffer, *Structural Mechanics Computer Programs*, University of Virginia Press, Charlottesville, 1974.

9.5-13 W. Griffel, *Handbook of Formulas for Stress and Strain*, Ungar, New York, 1966, Section 51.

9.5-14 R. Szilard, ref. 4, Sec. 1.16 and Appendix A.

9.5-15 A. C. Ugural, ref. 5, Chap. 4.

9.6 COMBINED BENDING AND IN-PLANE LOADING OF ISOTROPIC, ELASTIC PLATES: BUCKLING AND LARGE-DEFLECTION (NONLINEAR) BEHAVIOR

The classical plate theory summarized in Sections 9.2–9.4 does not account for in-plane forces. When these forces are tensile, plate deflection can produce components that materially increase the stiffness of the plate to lateral loads (see Fig. 9.6-1); when they are compressive, they can induce plate buckling.

The formulation of the plate equation under combined bending and in-plane loading is due to Von Karman. When the in-plane forces per unit of middle surface length n_x, n_y, n_{xy} shown in Fig. 9.6-2 are independent of w, the generalization of Equation (9.2-1) for cartesian coordinates is

$$D\nabla^2\nabla^2 w(x,y) = p_z + n_x\frac{\partial^2 w}{\partial x^2} + n_y\frac{\partial^2 w}{\partial y^2} + 2n_{xy}\frac{\partial^2 w}{\partial x\,\partial y} \tag{9.6-1}$$

The values of n_x, n_y, and n_{xy} are obtained from solution of the in-plane loaded plate. For example, consider a rectangular plate that is compressed in its middle surface by forces uniformly distributed along the sides $x = 0$ and $x = a$. Let F_x be the compressive force and let the length of the side be b; then by a simple strength of materials analysis

$$n_x = -\frac{F_x}{b}$$

$$n_y = n_{xy} = 0$$

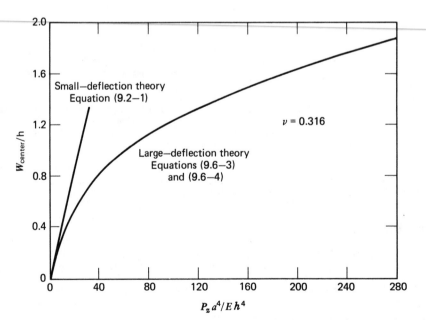

Fig. 9.6-1 Load (abscissa) versus deflection (ordinate) for a simply supported, square plate with uniform pressure load p_z and edge length a. Note that the small-deflection (linear) theory is accurate for $w_{center}/h \le 0.2$. (Used by permission from Reference 8, p. 341).

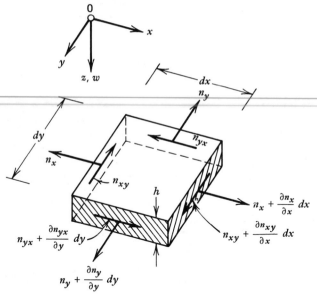

Fig. 9.6-2 Membrane forces per unit length on an element of a rectangular plate. (Used by permission from Reference 8, p. 152).

and for $p_z = 0$, Equation (9.6-1) becomes

$$D\nabla^2\nabla^2 w + n_x \frac{\partial^2 w}{\partial x^2} = 0 \qquad (9.6\text{-}2)$$

Values of n_x that give nontrivial solutions are the buckling loads. The theory of plate buckling is well established[1-3] and design information is available.[4-6] In particular, Gerard discusses the stability of plate-type structures such as multiweb beams, box beams, plates with various stiffener cross sections, and grid-stiffened plates. An extensive collection of results (293 cases) has been compiled by the Column Research Committee of Japan;[7] it includes plates, plate structures, and sandwich plates. Figure 9.6-3 is a synthesis of several of the design charts available for plate buckling.

When n_x, n_y, and n_{xy} are functions of w, the equations of equilibrium become geometrically nonlinear. For cartesian coordinates these equations are[8]

$$D\nabla^2\nabla^2 w(x, y) = h\left[\frac{\partial^2 w}{\partial x^2}\frac{\partial^2 \Phi}{\partial y^2} + \frac{\partial^2 w}{\partial y^2}\frac{\partial^2 \Phi}{\partial x^2} - 2\frac{\partial^2 w}{\partial x\,\partial y}\frac{\partial^2 \Phi}{\partial x\,\partial y}\right] + p_z \qquad (9.6\text{-}3)$$

$$\frac{1}{E}\nabla^2\nabla^2\Phi(x, y) = \left(\frac{\partial^2 w}{\partial x\,\partial y}\right)^2 - \frac{\partial^2 w}{\partial x^2}\frac{\partial^2 w}{\partial y^2} \qquad (9.6\text{-}4)$$

where Φ is the Airy stress function defined by

$$n_x = h\frac{\partial^2 \Phi}{\partial y^2} \qquad n_y = h\frac{\partial^2 \Phi}{\partial x^2} \qquad n_{xy} = -h\frac{\partial^2 \Phi}{\partial x\,\partial y} \qquad (9.6\text{-}5)$$

Exact solutions of Equations (9.6-3) and (9.6-4) are extremely cumbersome,[9] and use of an appropriate finite-element computer code[10] is recommended. However, design charts based on a perturbation method of solution are available[11] for the large deflection of circular plates under axisymmetric lateral load.

The above equations give insight into the postbuckling behavior of plates.[5,9] This is illustrated in Fig. 9.6-4. The ability of a buckled plate to carry additional load should be contrasted with the inability of a buckled column to support loads above those predicted by the linear theory. The buckled plate in Fig. 9.6-4 develops significant tensile stresses σ_y in its center region, and these in-plane tensile

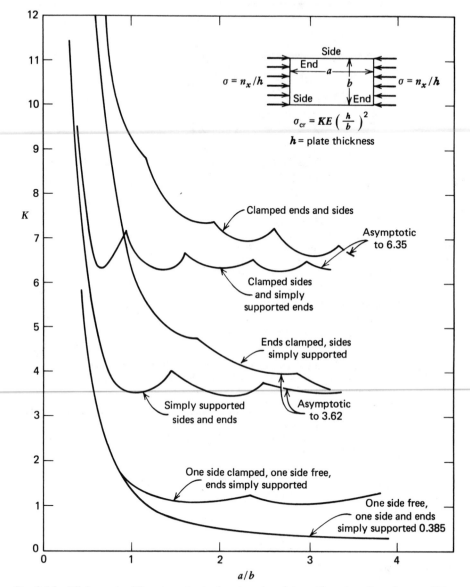

Fig. 9.6-3 Minimum buckling stress (σ_{cr}) of rectangular plates with various boundary conditions. (Used by permission from Peery & Azar, *Aircraft Structures*, McGraw-Hill, 1982, p. 348.)

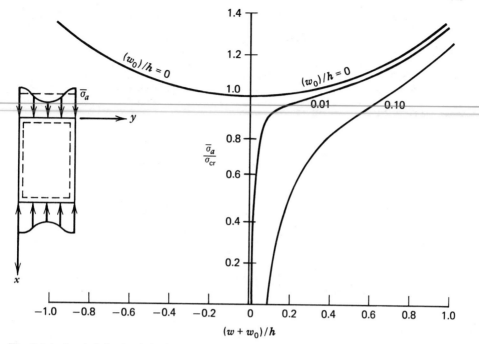

Fig. 9.6-4 Load–deflection behavior of a simply supported plate that is compressed in the axial direction:

$\bar{\sigma}_a$ = average stress of axial compression,

σ_{cr} = critical buckling stress predicted by Equation (9.6-2)

w_0 = a measure of initial geometric imperfection

(Used by permission from Reference 5, p. 46.)

stresses stiffen the plate against further lateral deflection and thus permit the plate to carry additional load; no such stresses exist for columns.

REFERENCES

9.6-1 S. Timoshenko and J. M. Gere, *Theory of Elastic Stability*, McGraw-Hill, New York, 1961.

9.6-2 F. Bleich, *Buckling Strength of Metal Structures*, McGraw-Hill, New York, 1952.

9.6-3 D. O. Brush and B. O. Almroth, *Buckling of Bars, Plates, and Shells*, McGraw-Hill, New York, 1975.

9.6-4 R. J. Roark and W. C. Young, *Formulas for Stress and Strain*, 5th ed., McGraw-Hill, New York, 1975, Chap. 14.

9.6-5 G. Gerard, *Introduction to Structural Stability Theory*, McGraw-Hill, New York, 1962.

9.6-6 *Astronautic Structures Manual*, Vol. II, Part II, NASA TM-73306.

9.6-7 Column Research Committee of Japan, *Handbook of Structural Stability*, Corona Publishing, Tokyo, 1971.

9.6-8 R. Szilard, *Theory and Analysis of Plates*, Prentice-Hall, Englewood Cliffs, NJ, 1974, pp. 340–344.

9.6-9 C-Y. Chia, *Nonlinear Analysis of Plates*, McGraw-Hill, New York, 1980.

9.6-10 W. Pilkey, K. Saczalski, and H. Schaeffer, *Structural Mechanics Computer Programs*, University of Virginia Press, Charlottesville, 1974, Chap. 1.

9.6-11 L. G. Watson and J. M. Chudobiack, "Design Charts for Finite Deflection Analysis of Circular Plates Subjected to Axisymmetric Loading," ASME paper 83-WA/DE-10.

9.7 ORTHOTROPIC PLATES

The plate theory of the previous sections assumes an isotropic plate; that is, at each point of the middle surface the elastic constants are the same in all directions. If the elastic constants vary with direction, the plate is anisotropic. The general theory of anisotropic plates[1] is used to predict the behavior of systems such as asymmetrically laminated plates made from multiple plies of composite material[2] and monolithic plates made from anisotropic crystals. Asymmetrically laminated plates exhibit coupling between bending and in-plane forces that considerably complicates their theory. This section will consider only the simpler, but still widely applicable, case of uncoupled behavior.

Many practical plates permit an additional simplification inherent in the fact that they have two orthogonal, in-plane directions of elastic symmetry. Such plates are said to be orthotropic. Orthotropic plates can include sandwich plates,[3] fiber-reinforced plates,[4] and plates reinforced with closely spaced ribs,[5] see Fig. 9.7-1.

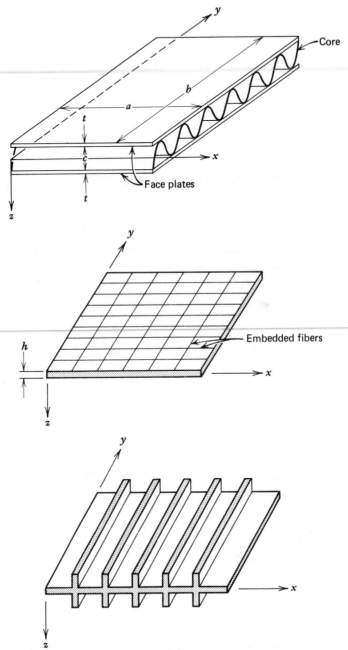

Fig. 9.7-1   Various plates that can be analyzed by uncoupled, orthotropic plate theory.

If the x and y axes are the directions of elastic symmetry, classical isotropic plate theory can be extended to the orthotropic case to give[5]

$$D_{xx}\frac{\partial^4 w}{\partial x^4} + 2H_{xy}\frac{\partial^4 w}{\partial x^2 \partial y^2} + D_{yy}\frac{\partial^4 w}{\partial y^4} = p_z \qquad (9.7\text{-}1)$$

where

$$D_{xx} = \frac{C_{xx}h^3}{12} \qquad D_{yy} = \frac{C_{yy}h^3}{12} \qquad D_{xy} = \frac{C_{xy}h^3}{12} \qquad (9.7\text{-}2)$$

$$G_{xy} = \frac{Gh^3}{12} \qquad H_{xy} = D_{xy} + 2G_{xy}$$

in which the constants $C_{\alpha\beta}$ and G are defined by

$$\left\{ \begin{array}{c} \sigma_x \\ \sigma_y \\ \tau_{xy} \end{array} \right\} = \left[\begin{array}{ccc} C_{xx} & C_{xy} & 0 \\ C_{xy} & C_{yy} & 0 \\ 0 & 0 & G \end{array} \right] \left\{ \begin{array}{c} \varepsilon_x \\ \varepsilon_y \\ \gamma_{xy} \end{array} \right\} \qquad (9.7\text{-}3)$$

The moments per unit length can be obtained from

$$m_x = -\left(D_{xx}\frac{\partial^2 w}{\partial x^2} + D_{xy}\frac{\partial^2 w}{\partial y^2} \right) \qquad (9.7\text{-}4)$$

$$m_y = -\left(D_{yy}\frac{\partial^2 w}{\partial y^2} + D_{xy}\frac{\partial^2 w}{\partial x^2} \right) \qquad (9.7\text{-}5)$$

$$m_{xy} = -2G_{xy}\frac{\partial^2 w}{\partial x \partial y} \qquad (9.7\text{-}6)$$

In the isotropic case

$$C_{xx} = C_{yy} = \frac{E}{1-\nu^2} \qquad C_{xy} = \frac{\nu E}{1-\nu^2} \qquad G = \frac{E}{2(1+\nu)}$$

and the above relations reduce to those of Section 9.2. It should be emphasized that the above theory does not include deformation due to transverse shear. Large-deflection theory can be obtained as in Section 9.6.[6]

The problem of analyzing orthotropic plates, such as those in Fig. 9.7-1, is reduced to finding values for D_{xx}, D_{yy}, D_{xy}, and G_{xy} or equivalently C_{xx}, C_{yy}, C_{xy}, and G. Timoshenko[5] gives ways to calculate these constants for reinforced-concrete slabs, plywood panels, corrugated sheets, and rib-stiffened plates, see Fig. 9.7-2. It is also possible to find the elastic constant by a series of simple experiments.[7]

If the stiffeners are only on one side of the plate, there is coupling between bending deflection and in-plane strains and Equation (9.7-1) is only approximate; but nevertheless it is often used. Coupled equations and formulas for calculating the associated elastic constants are available for plates integrally stiffened on one side.[8] If the spacing of the stiffening ribs is small, the rib-stiffened plate can be replaced by an orthotropic or antisotropic plate of constant thickness. If the rib spacing is large, the rib must be treated as a discrete element. For simple arrangements of ribs this can be done by analytical methods for either rectangular plates[9] or circular plates.[10] Finite-element methods can accept plates of arbitrary shape and arbitrary rib configurations.

In the case of orthotropic sandwich plates with soft cores, it is necessary to extend Equation (9.7-1) to include the effect of deformation due to transverse shear. The theory is given by Plantema[3] and various design considerations are described by Bishop.[11]

Geometry	Rigidities

A. Reinforced concrete slab with x and y directed reinforcement steel bars

$$D_{xx} = \frac{E_c}{1-v_c^2}\left[I_{cx} + \left(\frac{E_s}{E_c}-1\right)I_{sx}\right],$$

$$D_{yy} = \frac{E_c}{1-v_c^2}\left[I_{cy} + \left(\frac{E_s}{E_c}-1\right)I_{sy}\right]$$

$$G_{xy} = \frac{1-v_c}{2}\sqrt{D_{xx}D_{yy}}, \quad H_{xy} = \sqrt{D_{xx}D_{yy}}, \quad D_{xy} = v_c\sqrt{D_{xx}D_{yy}}$$

v_c: Poisson's ratio for concrete

E_c, E_s: Elastic modulus for concrete and steel, respectively

$I_{cx}(I_{sx}), I_{cy}(I_{sy})$: Moment of inertia of the slab (steel bars) about neutral axis in the section x = constant and y = constant, respectively

B. Plate reinforced by equidistant stiffeners

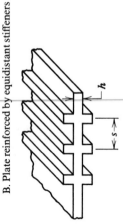

$$D_{xx} = H_{xy} = \frac{Eh^3}{12(1-v^2)}, \quad D_{yy} = \frac{Eh^3}{12(1-v^2)} + \frac{E'I}{s}$$

E, E': Elastic modulus of plate and stiffeners, respectively

v: Poisson's ratio of plating

s: Spacing between centerlines of stiffeners

I: Moment of inertia of the stiffener cross section with respect to midplane of plating

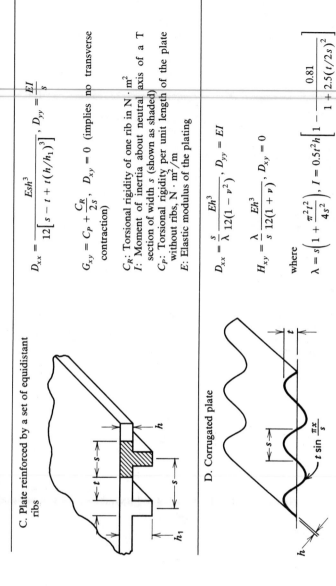

C. Plate reinforced by a set of equidistant ribs

$$D_{xx} = \frac{Esh^3}{12\left[s - t + t(h/h_1)^3\right]}, \quad D_{yy} = \frac{EI}{s}$$

$$G_{xy} = C_P + \frac{C_R}{2s}, \quad D_{xy} = 0 \text{ (implies no transverse contraction)}$$

C_R: Torsional rigidity of one rib in $N \cdot m^2$
I: Moment of inertia about neutral axis of a T section of width s (shown as shaded)
C_P: Torsional rigidity per unit length of the plate without ribs, $N \cdot m^2/m$
E: Elastic modulus of the plating

D. Corrugated plate

$$D_{xx} = \frac{s}{\lambda}\frac{Eh^3}{12(1 - \nu^2)}, \quad D_{yy} = EI$$

$$H_{xy} = \frac{\lambda}{s}\frac{Eh^3}{12(1 + \nu)}, \quad D_{xy} = 0$$

where

$$\lambda = s\left(1 + \frac{\pi^2 t^2}{4s^2}\right), \quad I = 0.5t^2 h\left[1 - \frac{0.81}{1 + 2.5(t/2s)^2}\right]$$

Fig. 9.7-2 Formulas for calculating the stiffness coefficients of various orthotropic plates. (Used by permission from A.C. Ugural, *Stresses in Plates and Shells*, McGraw-Hill, New York, 1981; p. 143.)

627

REFERENCES

9.7-1 S. G. Lekhnetskii, *Anisotropic Plates*, Gordon & Breach, New York, 1968.
9.7-2 J. E. Ashton and J. M. Whitney, *Theory of Laminated Plates*, Technomic Press, Stamford, CT, 1970.
9.7-3 F. J. Plantema, *Sandwich Construction*, Wiley, New York, 1966.
9.7-4 J. R. Vinson and T-W. Chow, *Composite Materials and Their Use in Structures*, Wiley, New York, 1975, Chap. 6.
9.7-5 S. P. Timoshenko and S. Woinowsky-Krieger, *Theory of Plates and Shells*, 2nd ed., McGraw-Hill, New York, 1959, Chap. 11.
9.7-6 R. Szilard, *Theory and Analysis of Plates*, Prentice-Hall, Englewood Cliffs, NJ, p. 377.
9.7-7 S. W. Tsai, "Experimental Determination of the Elastic Behavior of Orthotropic Plates," *J. Engr. Industry, ASME*, pp. 315–318, Aug. 1965.
9.7-8 N. F. Dow, C. Lebove, and R. F. Hubka, "Formulas for the Elastic Constant of Plates with Integral Waffle-Like Stiffening," NACA report 1195, 1954.
9.7-9 L. H. Donnel, *Beams, Plates, and Shells*, McGraw-Hill, New York, 1976. p. 204ff.
9.7-10 C. B. Biezeno and R. Grammel, *Engineering Dynamics, Vol II, Elastic Problems of Single Machine Elements*, Blackie & Sons, London, 1956, pp. 293–299.
9.7-11 W. Bishop, "Honeycomb Sandwich Structures," in *Thin Walled Steel Structures*, K. C. Rockey and H. V. Hill (eds.), Gordon & Breach, New York, 1968.

9.8 INELASTIC PLATES

9.8-1 Elastic-Plastic Behavior

The application of the theory of metal plasticity to plates leads to analytical problems of substantial difficulty[1] and few solutions exist. Current practice is to replace the analytical approach with a numerical approach based on the finite-element methods.[2]

As the plastic region engulfs more and more of the plate, one must establish whether the plate fails by excessive deformation or by the more catastrophic mechanisms of plastic collapse or inelastic buckling.

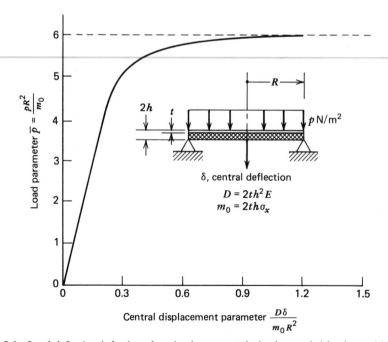

Fig. 9.8-1 Load-deflection behavior of a simply supported circular sandwich plate subjected to uniform pressure. The face plates are rigid perfectly plastic with yield stress σ_y; the core resists the transverse shear without deformation. (Used by permission from Reference 1, p. 376.)

9.8-2 Limit Load

Figure 9.8-1 shows the load-deflection behavior of a simply supported circular sandwich plate subjected to a monotonically increasing uniform pressure load.[1] When the load parameter $\bar{p} \rightarrow 6$, the plate exhibit large deflection increments for small-load increments. The load $\bar{p} = 6$ is called the limit load. The analytical determination of limit loads for plates can either proceed by elastic-plastic analysis, as in Fig. 9.8-1, or by application of the theorems of limit analysis.[3,4] Limit analysis gives only upper and lower bounds on the limit load, but it is much simpler to use.

If the edges of a thin circular plate are restrained from radial motion, membrane action will stiffen the plate and allow it to carry loads above the limit load, see Fig. 9.8-2.[5] Reinforced-concrete plates with clamped edges can behave very differently from the steel plates of Fig. 9.8-2. There is no plateau in the vicinity of the limit load and failure is by sudden snap-through to a large deflection determined by the membrane stiffness of the reinforcement.[3,5]

9.8-3 Inelastic Buckling

In plates as well as columns, it is possible for the buckling stress (σ_{cr}) to exceed the proportional limit. In this case Hooke's law must be replaced by a plastic stress–strain law. The effect of plasticity on buckling stress appears in the modulus and Poisson's ratio terms, and it is possible to unify elastic and inelastic buckling results by defining a plasticity reduction factor η

$$\eta = \frac{\sigma_{cr} \text{ (plastic)}}{\sigma_{cr} \text{ (elastic)}} \tag{9.8-1}$$

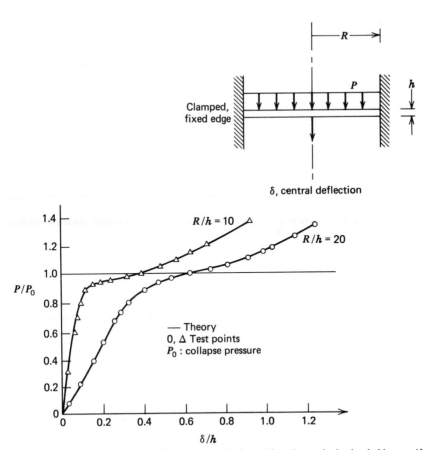

Fig. 9.8-2 Load-deflection behavior of a circular steel plate with a clamped edge loaded by a uniform pressure p. (Used by permission from Reference 5, p. 95.)

Values of η are available for a variety of plate shapes, boundary conditions, and materials.[6-8]

There is little design information available on postbuckling behavior of plates that buckle inelastically.

REFERENCES

9.8-1 J. B. Martin, *Plasticity: Fundamentals and General Results*, MIT Press, Cambridge, MA 1975, Chap. 8.

9.8-2 W. Pilkey, K. Saczalski, and H. Schaeffer, *Structural Mechanics Computer Programs*, University of Virginia Press, Charlottesville, 1974, Chap. 1.

9.8-3 M. A. Save and C. E. Massonnet, *Plastic Analysis and Design of Plates, Shells and Disks*, North-Holland, Amsterdam, 1972.

9.8-4 R. Szilard, *Theory and Analysis of Plates*, Prentice-Hall, Englewood Cliffs, NJ, 1974, Chap. 7.

9.8-5 R. C. Calladine, "Simple Ideas in the Large-Deflection Plastic Theory of Plates and Slabs," in *Engineering Plasticity*, J. Heyman and F. A. Leckie (eds.), Cambridge University Press, New York, 1968, p. 93.

9.8-6 G. Gerard, *Introduction to Structural Stability Theory*, McGraw-Hill, New York, 1962, p. 47.

9.8-7 *Astronautic Structures Manual*, Volume II, Part II, Section C2, NASA TM-73306.

9.8-8 Column Research Committee of Japan, *Handbook of Structural Stability*, Corona Publishing, Tokyo, 1971.

9.9 THERMAL AND CYCLIC LOADING OF PLATES

In the previous sections the independent variable is force, and it is assumed to be applied monotonically and sustained at its final value. Instead, many plates and plate structures are loaded by cyclic forces, prescribed deformation, or intense heating. The structural response of plates to these conditions is important in power plant structures and is covered by the associated design codes.[1]

9.9-1 Heated Plates

Consider an unloaded plate for which

Classical plate theory is valid, that is, the assumptions of Section 9.1 are satisfied.

The thermoelastic problem is uncoupled, that is, the strain field does not induce a temperature field.

The temperature field is steady state, that is, temperature $T = T(x, y, z)$ is found from

$$\nabla^2 T = 0 \tag{9.9-1}$$

and appropriate boundary conditions.[2]

The properties of the plate material E, ν, α (coefficient of thermal expansion) are independent of temperature or, alternatively, mean values are used.

In cartesian coordinates define

$$T_{n_T} = \frac{1}{h} \int_{-h/2}^{+h/2} T(x, y, z) \, dz \tag{9.9-2}$$

$$T_{m_T} = \frac{1}{h^2} \int_{-h/2}^{+h/2} T(x, y, z) z \, dz \tag{9.9-3}$$

where T_{n_T} is associated with thermoelastic in-plane forces (n_T) and T_{m_T} with thermoelastic bending moments (m_T). The small-deflection equations for plates are then[3]

$$D\nabla^2\nabla^2 w(x, y) = -\frac{E\alpha h^2}{1 - \nu} \nabla^2 T_{m_T} \tag{9.9-4}$$

$$\frac{1}{E}\nabla^2\nabla^2 \Phi(x, y) = -\alpha \nabla^2 T_{n_T} \tag{9.9-5}$$

Equation (9.9-4) is the analog of (9.2-1) and Equation (9.9-5) governs the in-plane problem.[4] For large deflection, Equations (9.9-4) and (9.9-5) are coupled and nonlinear.[3] The advantage of the above

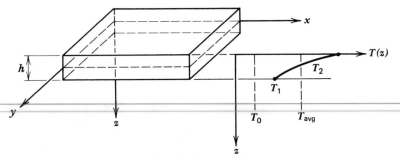

Fig. 9.9-1 Plate with temperature gradient through its thickness

$$T(h/2) = T_1$$

$$T(-h/2) = T_2$$

$$T_{avg} = T_{n_T} = \frac{1}{h} \int_{-h/2}^{h/2} T(z)\, dz$$

$T_0 = $ initial, uniform temperature of the plate.

formulation is that all the solution techniques applied to Equation (9.2-1) can be applied to Equation (9.9-4) with the substitution

$$p_z = -\frac{E\alpha h^2 \nabla^2 T_{m_T}}{1 - \nu}$$

Some finite-element computer codes can solve both the heat conduction equation for temperature and the equilibrium equations for stresses on a timestep-by-timestep basis; this allows thermal stresses to be evaluated as a function of time.

It is often possible to obtain simple relations for thermal stress, and these are helpful in the preliminary design of plates. For example, consider a plate clamped along its edges and subjected to a temperature distribution $T = T(z)$, see Fig. 9.9-1. Let $\Delta T = T(z) - T_{avg}$, then[5]

$$\sigma_z = \tau_{xz} = \tau_{yz} = \tau_{xy} = 0$$

$$\sigma_x = \sigma_y = -\frac{E\alpha \Delta T}{1 - \nu} \qquad (9.9\text{-}6)$$

This is valid whether T is the steady-state temperature gradient or a transient gradient. In particular, consider a transient gradient such that $T = T_0$ except at $z = h/2$ where $T = T_1$. Then $T_{avg} \approx T_0$ and $\sigma_z = \sigma_x = \sigma_y = 0$, except at $z = h/2$ where $\sigma_x = \sigma_y = -E\alpha(T_1 - T_0)$. This is an example of "skin" stress caused by a thermal shock to the surface $z = h/2$. Such skin stresses are often produced by the sudden application of a hot or cold fluid to the surface of a structure. In fluids that develop large thermal boundary layer film coefficients, such as water, the severity of the skin stresses is reduced by the temperature drop across the thermal boundary layer. Design charts are available to estimate the resulting reduction factor.[6]

If the temperature distribution is steady state, the gradient is linear, so that $T_{avg} = (T_1 + T_2)/2$ and at the surface $z = h/2$ Equation (9.9-6) implies

$$\sigma_x\left(\frac{h}{2}\right) = \sigma_y\left(\frac{h}{2}\right) = -\frac{E\alpha}{2(1 - \nu)}(T_1 - T_2) \qquad (9.9\text{-}7)$$

Conversely, if the plate edges are free, a linear gradient in z produces no thermal stress; however, the plate deforms into a spherical shape.

There are a number of in-plane analytical solutions for circular plates with either purely radial temperature distributions, $T = T(r)$, or axially symmetric temperature distributions which preclude bending, that is, $T = T(r, z) = T(r, -z)$.[7]

Alternatively, thermal stress analysis can be divided into a two-step process: first, the determination of the free thermal stresses, that is, the thermal stresses and deformations in the unloaded, unrestrained structure; and second, the calculation of the discontinuity stresses required to achieve geometric compatibility with the attached structures.

9.9-2   Cyclically Loaded Plates

Cyclic loading and heating cause fatigue. However, if the plate becomes plastic during part of the cyclic, more complicated responses can occur:

Shakedown. After an initial cycle of load or temperature which produces plastic strain, the structure responds elastically.

Alternating plasticity. Cyclic load or temperature establishes an elastic-plastic hysteresis loop which considerably accelerates the process of fatigue damage.

Incremental collapse. A sustained in-plane load combined with cyclic load or deformation can cause additive, incremental plastic strain and lead to unacceptable deformation and collapse.

These phenomena, like creep, are generally associated with high-temperature service and form a specialized area of plate analysis. Much of the design information available is for beams[8] and must be extrapolated for plates and plate structures. Some information is available on creep design of plates by analytical[9] or numerical[10] methods.

REFERENCES

9.9-1 *ASME Boiler and Pressure Code*, The American Society of Mechanical Engineers, New York, 1980, Section VIII, Division 2 and Section III, Division 1 (revised code issued every 3 years).

9.9-2 H. S. Carslaw and J. C. Jaeger, *Conduction of Heat in Solids*, 2nd ed., Clarendon Press, Oxford, 1967.

9.9-3 R. Szilard, *Theory and Analysis of Plates*, Prentice-Hall, Englewood Cliffs, NJ, 1974, Sec. 3.5.

9.9-4 D. Burgreen, *Elements of Thermal Stress Analysis*, C. P. Press, Jamaica, NY, 1971, Chap. 3.

9.9-5 D. Burgreen, ref. 4, Sect. 1.7.

9.9-6 D. Burgreen, ref. 4, p. 195.

9.9-7 D. Burgreen, ref. 4, Chap. 4.

9.9-8 D. Burgreen, *Design Methods for Power Plant Structures*, C. P. Press, Jamaica, NY, 1975.

9.9-9 R. K. Penny and D. L. Marriott, *Design for Creep*, McGraw-Hill, New York, 1971, p. 61ff.

9.9-10 H. Kraus, *Creep Analysis*, Wiley-Interscience, New York, 1980, pp. 203–205.

9.10   SPECIAL APPLICATIONS

There are a number of practical design problems for plates that deserve mention: continuous plates, perforated plates, variable-thickness plates, plates on elastic foundation, and plate brackets. Only references to the formulation of these special applications will be presented since these designs are now mainly effected with computer models. The analytical solutions are useful primarily for simple designs or to verify the computer models.

9.10-1   Continuous Plates

A plate that rests on a series of intermediate supports (elastic or rigid) is called a continuous plate, see Fig. 9.10-1. Between the panels of span l_i, l_{i+1}, the conditions of continuity of displacement and slope must be satisfied; also m_x must be continuous across a support that has zero torsional rigidity. The theory of continuous plates is well established[1] and an extensive design guide is available;[2] this design guide considers skew slabs, $0 < \phi < 90°$, as well as unskewed slabs, $\phi = 90°$.

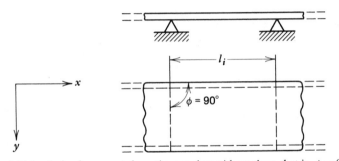

Fig. 9.10-1 A simply supported, continuous plate without skew, that is, $\phi = 90°$.

9.10-2 Perforated Plates

The bending of a plate with a hole can be formulated as a problem in linear elasticity and solved by the complex potential method of Muskhelishvili. Using this method, Savin[3] has solved the bending of a rectangular plate with holes of circular, elliptic, square, rectangular, or triangular shape. Savin's results have been summarized for design use by Griffel.[4] The case of a circular hole in a very large plate can also be solved using the classical plate theory.

However, these solutions can be in error if they neglect plate deformation due to transverse shear. If the hole diameter is less than three times the plate thickness, the size of the error can become unacceptable.[5] This error can be eliminated by using a plate theory that incorporates shear deformation;[6] alternatively, experimental data or an appropriate finite-element method can be used.

There is a large amount of design information on stress concentration around holes and notches in plates.[7,8] Figure 9.10-2 applies to a central elliptical hole in a plate subjected to bending, and Fig. 9.10-3 to a similar plate subjected to tension. When the elliptical holes become cracklike, that is, $b/a \gg 1$, the predicted elastic stresses become very large. The actual stresses will be limited by local plasticity.

A case of practical interest is a circular plate having a large number of holes arranged in a particular pattern and which is subjected to pressure loading: for example, plates used as heat exchanger tubesheets. Design rules for perforated tubesheets are available in the TEMA standards.[9] Design of perforated plates by analysis is also possible.[10] Stress concentration factors for square, rectangular, and diamond patterns of circular holes are given in Griffel.[4]

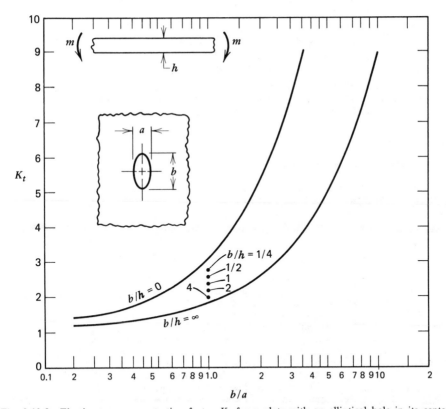

Fig. 9.10-2 Elastic stress concentration factor K_t for a plate with an elliptical hole in its center subjected to pure bending

$$\sigma_{max} = K_t \left(6m/h^2 \right)$$

Note that $b/a = 1$ corresponds to a circular hole. (Used by permission from Reference 7, p. 268.)

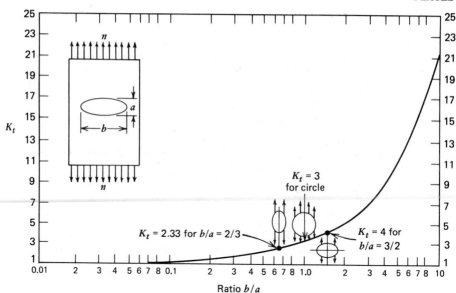

Fig. 9.10-3 Elastic stress concentration factor K_t for a plate with an elliptic hole subjected to inplane force per unit length, n.

$$\sigma_{max} = K_t n/h$$

where h is the plate thickness. This variation of K_t with b/a is for an infinite plate; it is sufficiently accurate for a finite width plate if the hole dimensions are less than about one-fifth the plate width. Note that as $b/a \to 0$, a crack aligned with n, $K_t \to 1$ while for $b/a \to \infty$, a crack transverse to n, $K_t \to \infty$. (Used by permission from Reference 4, p. 247.)

9.10-3   Plates with Variable Thickness

Finite-element methods allow the design of plates with practically arbitrary variation in plate thickness, h. Analytical methods are much more limited. For h slowly variable, the small-deflection equilibrium equation is[11]

$$\nabla^2(D\nabla^2 w) - (1 - \nu)\left(\frac{\partial^2 D}{\partial x^2}\frac{\partial^2 w}{\partial y^2}\right) - 2\frac{\partial^2 D}{\partial x \partial y}\frac{\partial^2 w}{\partial x \partial y} + \frac{\partial^2 D}{\partial y^2}\frac{\partial^2 w}{\partial x^2} = p_z \qquad (9.10\text{-}1)$$

where $D(x, y) = Eh^3/12(1 - \nu^2)$. Exact solutions are available[12] for rectangular plates with exponential or linear variations in D; for circular sector plates where D varies as r (radius), r^2, or r^3; and for circular plates with rotational symmetry for which D varies with ρ^k (where $\rho = r/r_1$) and $[1 - \rho^k]^n$ (if $k = 1$, $n = 3$, $r_1 =$ radius at which $h = 0$, this corresponds to a linearly tapered plate) and $\exp[-\rho^k]$. Circular plates with linearly varying thickness are discussed in Szilard and design tables are presented.[13] It should be emphasized that in most cases the middle surface must be plane. Also, if the variable rigidity plate is a rotating disk, for example, a turbine rotor, centrifugal body forces must be included in the design analysis.[14]

Alternatively, it is possible to approximate a circular plate of variable thickness with a series of annular plates of constant thickness, see Fig. 9.10-4. In such cases it is necessary to match displacement, slopes, and bending moments at the discontinuities. An example of this procedure is given by Ugural.[15]

9.10-4   Plates on Elastic Foundation

The motivation for this formulation is a series of practical design problems dealing with road pavements, building support slabs, and equipment base plates.

The method of approach follows the well-known theory of beams on elastic foundation. One assumes a locally reacting (Winkler-type) foundation such that the pressure $p(x, y)$ between founda-

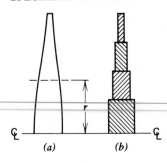

Fig. 9.10-4 (*a*) Profile of a circular plate with variable thickness; (*b*) Model of (*a*) with constant thickness plates. (Used by permission from Ugural Reference 15, p. 47.)

tion and the plate is linearly proportional to the local point displacement $w(x, y)$. The constant of proportionality, k, is the foundation modulus in force per unit of surface area per unit of deflection, for example, N/m^3 or Pa/m. The equation of equilibrium becomes

$$D\nabla^2\nabla^2 w + kw = p_z \qquad (9.10-2)$$

and its solution is straightforward for either circular or rectangular plates.[16]

The foundation can consist of a variety of materials (metal, concrete, elastic mats, soil), and the determination of k must be done with care, especially for soils since soil properties are usually nonlinear and site dependent. General guidance on soil modulus values is available[17] based on static and small-amplitude dynamic testing.

Table 9.10-1 gives the guidelines of Barkan for various soils in their elastic range. Various examples of specific design procedures are available[17-20] for the foundations of machine tools, reciprocating engines, forge hammers, motor generators, and turbine generators.

TABLE 9.10-1   RECOMMENDED DESIGN VALUES OF PERMISSIBLE BEARING LOAD AND FOUNDATION MODULUS FOR SOILS[a]

Soil Group Category	Soil Group	Permissible Load on Soil under Action of Static Load Only (kg/cm²)	Coefficient of Elastic Uniform Compression k (kg/cm³)
I	Weak soils (clays and silty clays with sand, in a plastic state; clayey and silty sands; also soils of categories II and III with laminae of organic silt and of peat)	Up to 1.5	Up to 3
II	Soils of medium strength (clays and silty clays with sand, close to the plastic limit; sand)	1.5–3.5	3–5
III	Strong soils (clays and silty clays with sand, of hard consistency; gravels and gravelly sands; loess and loessial soils)	3.5–5	5–10
IV	Rocks	Greater than 5	Greater than 10

[a] Used by permission from Reference 17.

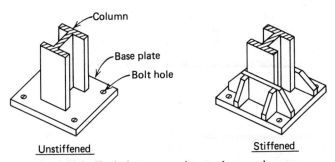

Fig. 9.10-5 Typical moment-resistant column anchorages.

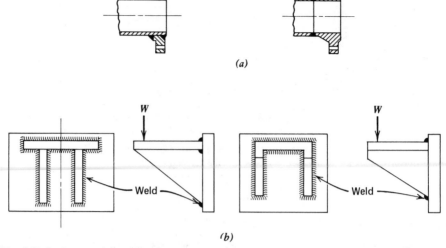

Fig. 9.10-6 Structural element made with plate segment. (*a*) Pipe flanges; (*b*) structural brackets. (Part (*a*) used by permission from Baumeister & Marks, *Mechanical Engineers Handbook*, 6th ed., McGraw-Hill, 1958, pp. 8–190; (*b*) used by permission from Blake, *Practical Stress Analysis in Engineering Design*, Dekker, 1982, p. 463.)

Pressure vessels, pumps, and heat exchangers are often supported on columns, one end of the column being anchored to a steel or concrete slab by means of bolts passing through a base plate, see Fig. 9.10-5. It is necessary to design the anchorage system so that the base plate, the bolts, and the slab can safely withstand the applied load. The base plates are usually comparable in size to the column end and therefore require special design procedures.[21]

9.10-5 Flanges and Plate Brackets

Sections of circular plates and rectangular plates are often welded together to form useful structural entities such as pipe flanges and structural brackets, see Fig. 9.10-6. The usual configurations are strongly three dimensional and an accurate stress analysis requires a finite-element model. Such an analysis may cost more than the structure it analyzes, and it is therefore useful to have conservative design methods that are simple and inexpensive. For example, flanges can be designed by code rule[22] or by approximate analyses based on plate and beam theory.[23]

There are no codes for structural brackets so simplified analysis[23] must depend on beam and plate theory and the standard methods for the design of weldments.[24]

REFERENCES

9.10-1 K. Marguerre and H-T. Woernle, *Elastic Plates*, Blaisdell, Waltham, MA, 1969, Chap. 8 and 9.

9.10-2 C. Schleicher and B. Wegener, *Continuous Skew Slabs*, 2nd ed., VEB Verlag für Bauwesen, Berlin, 1971.

9.10-3 G. N. Savin, *Stress Concentration Around Holes*, Pergamon Press, New York, 1961, Chap. VI.

9.10-4 W. Griffel, *Handbook of Formulas for Stress and Strain*, Ungar, New York, 1966, Chap. IX.

9.10-5 S. P. Timoshenko and S. Woinowsky-Kreiger, *Theory of Plates and Shells*, 2nd ed., McGraw-Hill, New York, Sec. 74.

9.10-6 S. P. Timoshenko and S. Woinowsky-Kreiger, ref. 5, Sect. 39.

9.10-7 C. Lipson and R. C. Juvinall, *Handbook of Stress and Strength*, Macmillan, New York, 1963, Chap. 20.

9.10-8 R. E. Peterson, *Stress Concentration Factors*, Wiley, New York, 1974, Chap. 2 (notches and grooves) and Chap. 4 (holes).

9.10-9 *Standards of Tubular Exchanger Manufactures Association*, 6th ed., Tubular Exchanger Manufacturers Association Inc., New York, 1978.

9.10-10 S. S. Gill (ed.), *The Stress Analysis of Pressure Vessels and Pressure Vessel Components*, Pergamon Press, New York, 1970, Chap. 11 (also see Reference 9.9-1).

9.10-11 E. H. Mansfield, *The Bending and Stretching of Plates*, Pergamon Press, New York, 1964, Chap. 1.

9.10-12 E. H. Mansfield, ref. 11, Chap. V.

9.10-13 R. Szilard, *Theory and Analysis of Plates*, Prentice-Hall, Englewood Cliffs, NJ, 1974, p. 130 and Appendix A.

9.10-14 J. P. Den Hartog, *Advanced Strength of Materials*, McGraw-Hill, New York, 1952, Chap. II.

9.10-15 A. C. Ugural, *Stresses in Plates and Shells*, McGraw-HIll, New York, 1981, pp. 47–55.

9.10-16 R. Szilard, ref. 13, pp. 136–143.

9.10-17 D. D. Barkan, *Dynamics of Bases and Foundations*, McGraw-Hill, New York, 1962.

9.10-18 F. E. Richart, R. D. Woods, and J. R. Hall, *Vibration of Soils and Foundations*, Prentice-Hall Englewood Cliffs, NJ, 1970, Chap. 10.

9.10-19 P. Srinivasulu and C. V. Vaidyanathan, *Handbook of Machine Foundations*, McGraw-Hill, New York, 1977.

9.10-20 A. Major, *Dynamics in Civil Engineering*, vol. III (*Foundations for High Speed Machinery, Steam and Nuclear Power Plants, Structural Details*), Collets, Budapest, Hungary, 1980.

9.10-21 L. J. DiLuna and J. A. Flaherty, "An Assessment of the Effect of Plate Flexibility on the Design of Moment Resistant Baseplates," ASME 79-PVP-50.

9.10-22 *ASME Boiler and Pressure Vessel Code*, Section VIII, Division 1, The American Society of Mechanical Engineers, New York, 1980.

9.10-23 A. Blake, *Practical Stresses Analysis in Engineering Design*, Dekker, New York, 1982, Chap. 33.

9.10-24 O. W. Blodgett, *Design of Welded Structures*, James F. Lincoln Arc Welding Foundation, Cleveland, Ohio, 1966.

CHAPTER 10

PRESSURE VESSEL AND PIPING DESIGN

DONALD F. LANDERS
Executive Vice President
Teledyne Engineering Services
Waltham, Massachusetts

10.1 INTRODUCTION

This section provides some of the background material for the rules contained in the pressure vessel and piping design subsections of Section III of the ASME *Boiler and Pressure Vessel Code*.[1]

The rules for vessels and piping design are discussed, particularly with respect to background information. A significant list of references is provided that allows the user of this chapter to follow the derivation of the current criteria and to use this chapter in conjunction with the equations, rules, and guides contained therein to more reasonably apply those requirements in developing a design.

REFERENCE

10.1-1 *Boiler and Pressure Vessel Code*, Section III, Nuclear Power Plant Components, Division 1, The American Society of Mechanical Engineers, New York, 1980.

10.2 DEVELOPMENT OF DESIGN CRITERIA

10.2-1 Prior to 1963

A detailed history of ASME code activities, including technical considerations through 1943 was prepared by Green.[1] Although it may seem strange to discuss nuclear technology with reference to such an early time period, the reader of that history can not help but be impressed with the tremendous influence of a basic engineering approach on health and safety and on component reliability. Had there been a problem previously? Consider the fact that in the period 1898–1903 there were some 1600 boiler explosions in the United States which killed approximately 1200 people. This should be contrasted with the more recent experience reported in WASH-1285,[2] that "no welded boiler drum or Section VIII pressure vessel has ever suffered a disruptive failure in central station service in the United States."

By the term *basic engineering approach* we mean a combination of elementary engineering principles, such as a simple stress analysis and the tensile test, and of intelligent application of experience. In essence, what we would identify today as primary membrane stresses were kept below one-fifth of the ultimate tensile strength of the material. As a contribution to conservation of material, the nominal factor of safety was reduced to 4 as a World War II emergency measure. Experience with such vessels was good, so the basic factor of safety was reduced to this value in the postwar years. As far as more complex stresses and their effects were concerned, the 1950 edition of the code recognized their existence, and stated, "It is recognized that high localized and secondary bending stresses may exist in vessels designed and fabricated in accordance with these rules. Insofar as practical, design rules for details have been written to hold such stresses at a safe level consistent with experience."

During the early 1950s three situations arose that caused a reconsideration of the code design criteria:

1. *Economics*, which were the result of two opposite influences:
   (a) Because so few vessels were failing, was the factor of safety still too high?

(b) Because a few vessels in severe service were failing, should more detailed criteria be adopted?

2. *Technology*, because of the rapid development in stress analysis techniques and in knowledge of material properties which greatly improved the ability to predict structural response to service loadings.

3. *Nuclear power*, as a potentially significant source of energy, but one that would be more sensitive to failures because of considerations of public health and cost and difficulty of repair.

The third of these was first approached by the Code Committee, through the 1954 appointment of a Task Group of the Subcommittee on Power Boilers, which was reorganized as a Special Committee of the Main Committee in March 1955. A similar coordinated effort was started by the Naval Reactors Program in November 1954, although the latter was clearly in recognition of all three situations. An early documentation of information related to this effort was presented by Cooper.[3] A major contribution in this area, although not specifically related to nuclear power, was the 1955 API paper by Murphy, Soderberg, and Rossheim.[4] ASME responded to the first two of the three itemized situations by the appointment in 1955 of the Special Committee to Review Code Stress Basis. The assignment of this committee was to work toward the establishment of a basic philosophy of pressure vessel design which balances the factors of material properties, design, fabrication, inspection, stress allowables, and so on, taking into account the considerations of service and safety with the general objective of providing a safe and economic solution to the problem. The next year of note was 1958. By this time the Special Committee on Nuclear Power had recommended that the rules of Section I, Power Boilers, and Section VIII, Unfired Pressure Vessels, were proper for use in nuclear power when modified by Code Cases 1224, 1225, 1226, and 1228. However, as a result of considerable interaction between the two special committees and the Naval Reactor effort it was realized that further consideration had to be given to this recommendation, and it was agreed that the Special Committee to Review Code Stress Basis should complete its efforts on the development of a code section specific to nuclear power. On December 1, 1958 the Navy's "Tentative Structural Design Basis for Reactor Pressure Vessels and Directly Associated Components (Pressurized, Water Cooled Systems)" (TSDB) was published as PB 151987. This TSDB contained the first comprehensive set of rules for the design of nuclear vessels and was applied to many of the early commercial nuclear power plants.

The closing report on this early period was a 1962 paper prepared by E. O. Bergman[5] who was chairman of the Special Committee on Nuclear Power. By this time the code cases modifying the rules of Sections I and VIII had been further developed and were collected into a nuclear series, as follows:

Case	Title
1270N	General Requirements for Nuclear Vessels
1271N	Safety Devices
1272N	Containment and Intermediate Containment Vessels
1273N	Nuclear Reactor Vessels and Primary Vessels
1274N	Special Material Requirements
1275N	Inspection Requirements
1276N	Special Equipment Requirements

The majority of these nuclear cases imposed requirements in addition to those applicable to nonnuclear vessels in order to set a higher level of construction requirements for vessels in nuclear service.

10.2-2   1963 through 1971

Section III, Nuclear Vessels, of the ASME Boiler and Pressure Vessel Code was first issued in 1963, with a second edition in 1965 and subsequent editions every three years since 1965. Although the design criteria contained in the new Section III was described in a large number of papers by various authors, ASME prepared a "Criteria" document in 1964 which described the criteria in specific detail. A revised version of this document was published in 1969.[6] A considerably revised and updated criteria document is expected in the mid-1980s.

It is descriptive to term the earlier design method as "design by rule" and the new methods incorporated in Section III as "design by analysis." Specifically, design by rule involves substitution into simple formulas, the use of geometric rules, and the consideration of a design pressure at a design temperature; whereas, design by analysis involves the performance of comprehensive stress analyses and the specific consideration of operating conditions in addition to the steps included in design by rule. The design by rule steps are retained in Section III to assure that maximum advantage is taken of past experience with the method, although the nominal factor of safety has been reduced from 4 to 3.

The remaining steps of the design by analysis procedure are intended to assure that the nominal factor of safety is achieved or exceeded through the explicit consideration of additional modes of failure.

The most important of these additional modes of failure considered in Section III is that of fatigue. Before 1950, the study of the fatigue of metals was limited largely to situations involving millions of cycles, and was used mostly in the design of rotating machinery and vibrating systems. The major loads on pressure-containing systems occur only a few thousand times in the service life of the system, and fatigue was usually ignored in design even though it was a common cause of failure. In the 1950s the pioneering work of Coffin[7] and Manson[8] showed that low-cycle fatigue is characterized by the strain range, not the stress range, of the cycle. This led the way to the application of quantitative fatigue analysis in vessel design, with associated criteria to assure that elastic methods of stress analysis properly predicted the effect, as developed by Langer.[9]

Perhaps the best summary of the background to the code rules as they existed during this period is the two-volume compilation published by ASME[10] in 1972. The first volume contains Part I, "Analysis for Design," and the second contains Part II, "Structural Components," and Part III, "Structural Dynamics." These volumes give excellent summaries of the state of the art, reprints of the most important papers, and extensive but selected bibliographies. For the reader interested in an earlier period, a similar book was published by ASME in 1960.[11]

The original period of code development ended with publication of the 1971 edition of Section III, which integrated rules prepared outside of the code committee, but on a similar basis, for nuclear piping, pumps and valves with those previously contained for vessels. Inclusion of these rules in Section III affected a significant addition to the criteria, in the form of a simplified elastic–plastic analysis method which permitted waiving the limit on primary-plus-secondary stresses. Although the specific method adopted and prepared by B. F. Langer was never published, it is a fairly direct modification of the method for nuclear piping developed by Tagart.[12]

10.2-3 After 1971

Increased attention to the definition and consequences of postulated accidents led to the addition in 1972 of rules for the evaluation of the response of nuclear components to such events. These rules were contained in Appendix F of Section III. The original title, "Rules for the Evaluation of Faulted Conditions," was changed in 1976 to "Level D Service Limits." The purpose of this change was to clarify the intent that the code provides various levels of service limits but does not define the system or component conditions that are to be evaluated with respect to a given level. Although Appendix F is a nonmandatory appendix, other requirements in the code make these rules mandatory unless alternative or supplementary criteria are defined by the owner.

REFERENCES

10.2-1 A. M. Green, Jr., *History of the Boiler Code*, The American Society of Mechanical Engineers, New York, 1955.

10.2-2 The Advisory Committee on Reactor Safeguards, WASH-1285, *Report on the Integrity of Reactor Vessels for Light-Water Power Reactors*, January 1974.

10.2-3 W. E. Cooper, *Proposed Structural Design Basis for Nuclear Reactor Pressure Vessels*; D. J. Hughes, S. McLain, and C. Williams, *Problems in Nuclear Engineering*, Pergamon Press, New York, 1955, pp. 258–287.

10.2-4 J. J. Murphy, C. R. Soderberg, Jr., and D. B. Rossheim, *Considerations Affecting More Economic but Equal Safe Pressure Vessel Construction Utilizing Either Present-Day Ductile Material or New High-Strength Less Ductile Materials*, API, Division of Refining, 35 (III), 1955, pp. 258–287.

10.2-5 E. O. Bergman, "The Basis and Content of a Nuclear Pressure Vessel Code," Paper No. 94, Nuclear Congress, 1962.

10.2-6 *Criteria of the ASME Boiler and Pressure Vessel Code for Design by Analysis in Sections III and VIII*, Criteria for Design by Analysis, Division 2, The American Society of Mechanical Engineers, New York, 1969.

10.2-7 L. F. Coffin, Jr., "A Study of the Effects of Cyclic Thermal Stress on a Ductile Metal," *Trans. Am. Soc. Mech. Eng.* **76**, 931 (1954).

10.2-8 S. S. Manson, "Thermal Stress in Design—Part 19, Cyclic Life of Ductile Materials," *Machine Design* 139 (July 7, 1960).

10.2-9 B. F. Langer, "Design of Pressure Vessels for Low Cycle Fatigue," The American Society of Mechanical Engineers, *J. Basic Engr.* **84**(3), 389 (Sept. 1966).

10.2-10 *Pressure Vessels and Piping: Design and Analysis*, 2 vols, The American Society of Mechanical Engineers, New York, 1972.

10.2-11 *Pressure Vessel and Piping Design*, The American Society of Mechanical Engineers, New York, 1960.

10.2-12 S. W. Tagart, Jr., "Plastic Fatigue Analysis of Pressure Components," The American Society of Mechanical Engineers Paper No. 68-PVP-3, 1968.

10.3   VESSEL DESIGN

10.3-1   Introduction

Section III, Division 1[1] provides rules for three classes of pressure vessels (classes 1, 2, and 3) and one class of containment vessels (class MC, where the M signifies metal).

Figure 10.3-1 summarizes the classes and identifies the applicable subarticle. For class 2 vessels alternative design subarticles exist. For containment vessels both subarticles must be applied. In order to provide unifying concepts, the design method is categorized as either design by rule or as design by analysis. Each of the subsections has a lead-in subarticle NX-3100 (the letter X is used here to indicate that the reference applies to NB, NC, ND, and NE—on a BPVC agenda it would also mean NF and NG) with common contents. Also, there are requirements in NX-3300 that are common to all subsections, such as the categorization of welds. These types of information are discussed in Section 10.3-3. The design by analysis approach for class 1 vessels is discussed in Section 10.3-4. The design by rule approach is presented in Section 10.3-5, in order to describe the design requirements that are based upon Section VIII, Division 1.

The basis for the alternative class 2 vessel rules is described by Fig. 10.3-1 as one which combines design by rule and design by analysis and is based upon the rules of Section VIII, Division 2.[2] These are discussed in Section 10.3-6 and are of interest to the class 1 vessel designer for purposes of initial design. For metal containment Fig. 10.3-1 indicates that both the design by analysis method, based on a modification of NB-3200, and the design by rule method, based on Section VIII, Division 1, are required.

10.3-2   Design Methods

The design procedures of the various sections of the ASME Boiler and Pressure Vessel Code range from substitution into simple equations to the performance of complex analyses at the limits of present knowledge in the field of solid mechanics. For purposes of this discussion the extremes of these procedures will be termed *design by rule* and *design by analysis*, respectively. This distinction is useful in discussing the apparent contrast between the design procedures of the earlier sections of the code (I, IV, and VIII, Division 1), and the newer sections (III, class 1, class 2 alternative rules, and VIII, Division 2). The word *apparent* is required in the previous sentence because the major difference between the procedures is in the number of failure modes considered rather than in the technical basis of the procedure.

Originally class 2, 3, and MC vessels were designed in accordance with rules that were identical to those of Section VIII of the code. In fact, the Section III design rules applicable to such components simply referenced Section VIII. The summer 1972 addenda contained a complete rewrite of Subsection NE, class MC components, which eliminated the need for reference to Section VIII by incorporating those Section VIII rules specifically applicable to class MC components. Experience in working with the summer 1972 addenda indicated such strong advantages for this approach that Subsection NC and ND were similarly rewritten for publication of the 1974 edition.

In general, the design by rule procedure combines a sound evaluation of the stress effects of a design value of internal pressure and a reasonably low allowable stress, selected at the design temperature, with additional rules for construction that are based upon experience. In contrast, the design by analysis procedure generally permits a higher allowable stress for an equivalent design

Class	Subarticle	Method	Basis
1	NB-3300	Analysis	NB-3200
2	NC-3300	Rule	VIII, Div. 1
	or		
	NC-3200	Rule and analysis	VIII, Div. 2
3	ND-3300	Rule	VIII, Div. 1
MC	NE-3200	Analysis	Modified NB-3200
and	and		
CC*	NE-3300	Rule	VIII, Div. 1

Fig. 10.3-1   Summary of vessel design rules. (*Metal portions not backed by concrete.)

pressure evaluation and then requires that additional evaluation be performed with respect to actual conditions of operation. In order to simplify this additional evaluation, more restrictions on materials, fabrication and examination are also required. If the vessel is to be subjected to extremely severe operational conditions, particularly thermal cycling, recognition of those effects in the design by analysis procedures will result in a more reliable design than do the design by rule procedures which are incomplete with respect to the thermal stresses. For moderate service conditions either vessel will probably perform satisfactorily. The failure criterion used with the design by rule procedures is the maximum normal stress theory, rather than the maximum shear stress theory, which is used in design by analysis.

Consideration of Design Pressure and Design Mechanical Loads

The development of the code equations used with the design by rule procedure included, wherever available, a detailed elastic analysis. General membrane stresses were limited to the tabulated S value, local membrane stresses limited to $1.5S$, and primary plus secondary stresses limited to $4S$. When such analyses were not available, judgment and experience were applied to the preparation of rules that hold pressure stresses at a safe level consistent with experience.

Evaluation for design pressure conditions, in combination with the design mechanics loads which are used in design by analysis rules must result in stresses that satisfy the primary stress intensity limits. For the general membrane and bending stress limits discontinuity effects are not included. Gross structural discontinuity effects are included when evaluating local membrane stresses. It is a consequence of this procedure that the required analyses are generally quite simple, involving only simple configurations.

Although rules that follow the design by rule method state that consideration should be given to "superimposed loads such as other vessels, operating equipment, insulation, corrosion resistant or erosion resistant linings, and piping" as well as to "reactions of supporting lugs, rings, saddles or other types of supports," the rules do not explicitly consider these loadings. Many such loads are beyond the control of the vessel designer, and are not made known to him by the purchaser. It is a consequence of this situation that superimposed internal and external loads are often not considered by the vessel designer, at least as a portion of his code responsibilities. Section VIII, Division 1, provides nonmandatory guidance with respect to these problems in Appendix D, "Suggested Good Practice Regarding Internal Structures," and Appendix G, "Suggested Good Practice Regarding Design of Supports."

Design by analysis methods provide for specific consideration of the primary stress effects of such superimposed internal and external loads only when such loads are identified as design mechanical loads by the owner's design specification. When so defined, the effects must be combined with design pressure loadings and included within the stress limits. If such loads are not defined as design mechanical loads, the primary stress effects of such loads need not be evaluated. Failure to define internal and external loads as design mechanical loads is one of the more common errors in the preparation of the owner's design specification.

The requirements that apply the design by rule procedure have adopted the maximum stress criterion, and the sections that apply the design by analysis procedure have adopted the maximum shear stress criterion. If radial stresses are small, so that the axial and circumferential stresses are the two significant principal stresses, the two criteria give identical results if both stresses are tensile or compressive; and, the maximum shear stress criterion is more conservative than the maximum normal stress criterion if one of the stresses is tensile and the other compressive. The maximum difference between the two criteria is a factor of 2 if the stresses are equal and opposite in sign.

If one is considering only membrane pressure stresses, as is the situation when the design by rule method is used, both stresses are almost always of the same sign. An important exception is in the knuckle region of a torispherical or ellipsoidal head, where the circumferential stress is compressive. This helps to explain why some difficulties have been experienced with very thin heads. External or internal loads may also cause compressive longitudinal stresses, but the membrane effects are generally sufficiently small compared to the pressure stresses that the error in using the maximum normal stress criterion is not usually significant. One must conclude from this discussion that the use of the simpler maximum normal stress criterion is generally justified when only membrane stresses are important. The same argument could be generated with respect to primary bending stresses such as those that occur in flat heads.

Consideration of Operating Cycles

In discussing operating conditions we should recognize that the operating pressure is usually lower than the design pressure, that superimposed internal and external loads are present, that thermal stresses may result from nonisothermal temperature conditions, and that vibratory effects may be present from flow-induced vibrations. The first two of these loadings have been considered in the preceding section with respect to static design conditions. Thermal stresses are generally neglected in that stage because, except for very brittle materials which are not used in vessels, steady-state thermal conditions do not cause failure. Therefore, a reason for giving further consideration to operating conditions is to evaluate the possibility of fatigue failures.

In addition to requiring consideration of pressure and superimposed internal and external loads, as previously discussed, requirements that follow the design by rule method require consideration of "impact loads, including rapidly fluctuating pressure," "wind loads, and earthquake loads where required," and "the effects of temperature gradients on maximum stresses."

Such requirements are completely silent as to the criteria that are to be used in evaluating these operating conditions, and such evaluation is seldom performed as part of the code design effort.

Even if one were to attempt to provide such an evaluation, it would seldom be meaningful because of certain types of design and fabrication details permitted and by the lack of detailed nondestructive examination of local areas significant to fatigue evaluation. For example, consider the problem of evaluating thermal stresses in the vicinity of a pad-type nozzle reinforcement. The most important factor in controlling the temperature distribution is the fit-up between the pad and the vessel, yet this cannot be controlled in a meaningful way. Other examples are the use of non-full-penetration attachment welds and many of the permitted nozzle weld details.

Therefore, we must conclude that a vessel designed to the design by rule procedure does not have a known margin for resistance to such operating conditions. This is not necessarily a major criticism of these rules, because experience has indicated that such vessels generally have an adequate service life. Instead, it is an indication that if a vessel has experienced fatigue failures in service, or if a contemplated vessel is going to be subjected to severe cyclic operating conditions, one should not expect the design by rule method to provide adequate consideration of such operating conditions.

In contrast, the design by analysis method provides explicit rules for the operating cycles defined by the design specification. The stresses, including interaction effects but not the effects of local structural discontinuities (stress concentrations), are evaluated for all conditions of operation and the maximum range of the stress differences is compared with an allowable range of $3S_m$. If the calculated range is less than $3S_m$, the elastic analysis technique is validated and one proceeds, after incorporating the effects of local structural discontinuities and local thermal stress, with the fatigue evaluation. If the calculated stress intensity range exceeds $3S_m$, the simplified elastic–plastic corrections are incorporated before performing the fatigue evaluation. The fatigue evaluation includes the stress cycles that result from all of the specified operating cycles and a linear damage summation technique is used to evaluate satisfaction of the fatigue conditions.

10.3-3   Common Requirements

Much of the content of NX-3100 is common to all of the four subsections considered. The more important common subjects will be discussed here. Figure 10.3-2 indicates the subjects considered by NX-3100.

Figure 10.3-3 identifies the loading conditions that must be considered for class MC vessels. The listing for other vessels is identical to this except that reactions to impingement and earthquake loads have been specifically identified for class MC. This is not to imply that such loadings may be ignored for other classes.

Although the loadings to be considered are essentially the same for all classes, the manner of treating such loadings differs from class to class. This is an important aspect that will receive further consideration in this section.

Responsibility for any effects of corrosion is assigned to the owner and not covered in the code by other than simple statements such as that in Fig. 10.3-4, which apply to class 1 vessels. This is a major responsibility assigned to the owner, and the large numbers of corrosion-type failures in nuclear power plants indicate the need for additional caution.

Dimensions of standard products are required to comply with listed standards and specifications, as illustrated by Fig. 10.3-5. However, compliance with these standards does not eliminate or replace the need to meet the remaining design requirements of the subsection—so special designs may be required. Note that the standards are designated by applicable dates.

External pressure design rules apply only to cylinders under external pressure and axial load or to spheres under external pressure.

The design curves for external pressure included in Section III, and implemented by NB-3133 and elsewhere, are based upon work conducted by R. G. Sturm at the University of Illinois in the 1930s. The construction of these charts is discussed by Bergman[3] in a paper prepared when these charts were introduced into the code in 1952. A typical chart, that for Type 304 stainless steel, is included herein as Fig. 10.3-6 where the geometry and material curves of the code have been superimposed in the manner in which these charts were published previously.

The theory behind the development of these charts utilizes a tangent modulus approach to the elastic–plastic buckling problem. The results are supported by a large series of tests on models performed by Sturm and by others. The lines on the chart running from the upper left toward the lower right are dependent solely on the geometry of the problem. The parameter on the abscissa axis

Subject Considered	−3100 Paragraph in			
	NB	NC	ND	NE
Loading Criteria	10	10	10	10
Loading conditions	11	11	11	11
Design loadings	12	12	12	12
Design pressure	12.1	12.1	12.1	12.1
Design temp.	12.2	12.2	12.2	12.2
Design mech. loads	12.3	12.3	12.3	12.3
Allowable stress	NA	12.4	12.4	12.4
Service Conditions	13	13	13	13
Levels A, B, C, D	in NCA-2145			NE
Testing conditions	in NCA-2142.2			14
Casting quality	NA	NA	15	NA
Special Considerations	20	20	20	20
Corrosion	21	21	21	21
Cladding	22	22	22	22
Primary stress	22.1	22.1	22.1	22.1
Design dimensions	22.2	22.2	22.2	22.2
Bearing stress	22.4	NA	NA	22.3
Q & F Stress	22.3	NA	NA	NA
Max. allowable stress	NA	NA	NA	22.4
Max. allowable temp	NA	NA	NA	22.5
Welding	23	NA	NA	NA
Dissimilar metals	23.1	23	23	23
Fillet weld att.	23.2	NA	NA	NA
Environmental eff.	24	NA	NA	NA
General Design Rules				
Scope	31	31	31	31
Design reports	NA	31.1	31.1	NA
Proof tests	NA	31.2	31.2	NA
Dimensional stds.	32	32	32	32
Ext. pressure rules	33	33	33	33
General	33.1	33.1	33.1	33.1
Nomenclature	33.2	33.2	33.2	33.2
Detailed rules	33.3	33.3	33.3	33.3
	−33.6	−33.8	−33.8	−33.7
Material properties	NA	NA	NA	34
Leak tightness	34	NA	NA	NA
Attachments	35	NA	NA	NA
Openings	36	NA	NA	NA

Fig. 10.3-2 Contents of NX-3100.

The loadings that shall be taken into account in designing a vessel shall include, but are not limited to, those in (A) through (I) below:

(A) Internal and external pressure
(B) Impact loads including rapidly fluctuating pressure
(C) Weight of the vessel and normal contents under all conditions, including additional pressure due to static and dynamic head of liquids
(D) Superimposed loads such as other components, operating equipment, insulation, corrosion resistant or erosion resistant linings, and piping
(E) Wind loads, snow loads, and vibration loads where specified
(F) Reactions of supporting lugs, rings, saddles, or other types of supports
(G) Temperature effects
(H) Reactions to steam and water jet impingement
(I) Earthquake loads

Fig. 10.3-3 Loading conditions.

Material subject to thinning by corrosion, erosion, mechanical abrasion, or other environmental effects shall have provision made for these effects during the design or specified life of the component by a suitable increase in or addition to the thickness of the base metal over that determined by the design formulas. Material added or included for these purposes need not be of the same thickness for all areas of the component if different rates of attack are expected for the various areas. It should be noted that the tests on which the design fatigue curves (Figs. I-9.0) are based did not include tests in the presence of corrosive environments which might accelerate fatigue failure.

Fig. 10.3-4 Corrosion.

A is strain. Parameter B on the right ordinate is stress, the curves plotted on this chart being stress–strain curves at various temperatures with the stress reduced by a factor of 4. The procedure is such that charts as drawn are consistent with the nominal factor of safety of 4 used in Section VIII of the code, where such charts were first published. Rather than redrawing these charts with a factor of safety of 3 for Section III class 1 vessels, the equations in NB-3100 contain a multiplying factor of $\frac{4}{3}$ for the evaluation of cylinders. A similar factor is not applied to spheres for the reasons discussed below.

A reading of Bergman's paper[3] will indicate that there was considerable controversy at the time these charts were first included within the code as to whether or not some specific correction for deviations from ideal geometry, consistent with the code fabrication requirements on tolerances, should be included. This correction factor was not included, and subsequent tests and applications have indicated that the true factor of safety with the Section VIII procedure is somewhat less than the nominal factor of 4, being approximately 3.7 for cylinders and about 3.0 for spheres depending on the actual deviations. This is the reason why the reduced factor of safety is not provided for spheres.

In order to facilitate the definition of design, fabrication, and examination requirements for vessels welds, they are categorized in the manner shown by Fig. 10.3-7.

When the design by rule method is used for design, the permissible geometries are illustrated. The following indicate a few examples and provide a comparison between the rules applicable to various classes.

1. **Unstayed flat heads and covers.** See Fig. 10.3-8. The capital letter under each figure indicates the subsection that includes that geometry. The letter C indicates those permitted for NC-3300 vessels; other rules may apply for NC-3200 vessels.

2. **Tapered transitions.** Figure 10.3-9a illustrates the requirements as shown in NC/ND-3366-1. Figure 10.3-9b is the similar figure for NB/NE. The important difference is that a 3:1 taper is required for NC/ND but the taper for NB/NE is established by analysis.

3. **Reinforcement of openings.** The general requirements for reinforcement of openings are given in NX-3331 and in NC-3231. For class 1 vessels, class 2 vessels to NC-3200, and class NE vessels the rules may be waived based on analysis. For other class 2 vessels and for class 3 vessels the rules must be satisfied. For class 1 vessels (in NB-3339) and class 2 vessels to NC-3200 (in NC-3239) alternative reinforcement rules are provided that may require less reinforcement. The basic method for class 2 vessels designed for NC-3300, class 3 vessels and class 4 containment vessels are described in NX-3332 through 3336.

Standard	Designation
Pipes and Tubes	
Welded and seamless wrought steel pipe	ANSI B36.10-1975
Stainless steel pipe	ANSI B36.19-1965 (Rev. 1971)
Fittings, Flanges, and Gaskets	
Steel pipe flanges and flanged fittings	ANSI B16.5-1977
Factory made wrought steel buttwelding fittings	ANSI B16.9-1978[1]
Forged steel fittings, socket-welding and threaded	ANSI B16.11-1973
Ring-joint gaskets and grooves for steel pipe flanges	ANSI B16.20-1973
Nonmetallic gaskets for pipe flanges	ANSI B16.21-1972
Buttwelding ends	ANSI B16.25-1972
Wrought steel buttwelding short radius elbows and returns	ANSI B16.28-1964 (Rev. 1972)
Refrigeration flare type fittings	ANSI B70.1-1969
Stainless steel buttwelding fittings	MSS SP-43-1971
Steel pipeline flanges	MSS SP-44-1975
Factory made buttwelding fittings for Class 1 Nuclear piping applications	MSS SP-87-1977
Large diameter carbon steel flanges	API 605-1978
Standard for steel pipe flanges	AWWA C207-1955
Bolting	
Square and hex bolts and screws, including askew head bolts, hex cap screws, and lag screws	ANSI B18.2.1-1972
Square and hex nuts	ANSI B18.2.2-1972
Socket cap, shoulder, and set screws	ANSI B18.3-1969
Threads	
Unified inch screw threads (UN and UNR thread form)	ANSI B1.1-1974
Pipe threads (except Dryseal)	ANSI B2.1-1968
Dryseal pipe threads	ANSI B2.2-1968
Valves	
Steel valves	ANSI B16.34-1977

Fig. 10.3-5 Dimensional standards.

In each case there is an exemption from reinforcement of small openings. For class 1 and NC-3200 vessels, no reinforcement is required if a single opening has a diameter not exceeding $0.2(Rt)^{1/2}$. The exemption applies to NC/ND-3300 vessels if the opening has a diameter equal to or less than 2 in. nominal pipe size. For class MC, the maximum size of an unreinforced opening is defined as equal to or less than $2\frac{1}{2}$ in.

For cylindrical shells the area required for reinforcement is specified for each plane through the nozzle centerline. The maximum required area of reinforcement is equal to the area removed by the penetration, using the required thickness of the vessel in the computation. The required area must be provided within stated dimensional limits as measured along the nozzle centerline and as measured along the outside of the vessel. Figure 10.3-10, which is Figure UA-280 of Section VIII, Division 1, illustrates the procedure.

10.3-4 Design by Analysis (NB-3300)

In accordance with NB-3311 the requirements for acceptability of a vessel design are:

1. "The design shall be such that the requirements of NB-3100 and NB-3200 shall be satisfied."
2. "The rules of this subarticle shall be met. In cases of conflict between NB-3200 and NB-3300 the requirements of NB-3300 shall govern."

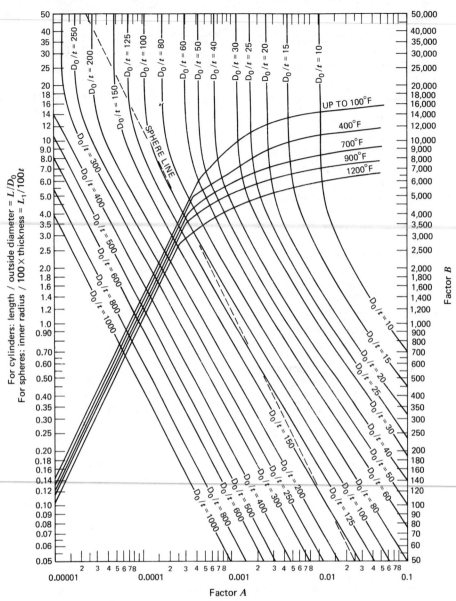

For cylinders: length / outside diameter = L/D_0
For spheres: inner radius / $100 \times$ thickness = $L_1/100t$

Factor A

Factor B

$D_0/t = 250$ $D_0/t = 200$ $D_0/t = 150$ $D_0/t = 125$ $D_0/t = 100$ $D_0/t = 80$ $D_0/t = 60$ $D_0/t = 50$ $D_0/t = 40$ $D_0/t = 30$ $D_0/t = 25$ $D_0/t = 20$ $D_0/t = 15$ $D_0/t = 10$

SPHERE LINE

$D_0/t = 300$ $D_0/t = 400$ $D_0/t = 500$ $D_0/t = 600$ $D_0/t = 800$ $D_0/t = 1000$

UP TO 100°F
400°F
700°F
900°F
1200°F

$D_0/t = 10$ $D_0/t = 15$ $D_0/t = 20$ $D_0/t = 25$ $D_0/t = 30$ $D_0/t = 40$ $D_0/t = 50$ $D_0/t = 60$ $D_0/t = 80$ $D_0/t = 100$ $D_0/t = 125$

$D_0/t = 150$ $D_0/t = 200$ $D_0/t = 250$ $D_0/t = 300$ $D_0/t = 400$ $D_0/t = 500$ $D_0/t = 600$ $D_0/t = 800$ $D_0/t = 1000$

Fig. 10.3-6 Chart for determining shell thickness of cylindrical and spherical vessels under external pressure when constructed of austenitic steel.

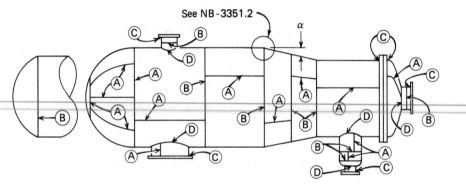

Fig. 10.3-7   Welded joint categories.

For class 1 vessels unique requirements are included for environmental effects, leak testing, attachments, and opening reinforcement. The latter two are quite clear and do not require discussion.

For environmental effects this paragraph relates to damage as a result of fast neutron irradiation. Because of this effect, nozzles or other structural discontinuities in ferritic vessels should preferably not be placed in regions of high neutron flux.

When a system leak tightness greater than that required or demonstrated by a hydrostatic test is required, the leak tightness requirements for each component shall be set forth in the design specification.

The requirements particular to class 1 vessel design may be summarized under the following series of topics:

1. Design considerations (NB-3320).
2. Opening and reinforcement (NB-3330).
3. Analysis of vessels (NB-3340).
4. Design of welded construction (NB-3350).
5. Special vessel requirements (NB-3360).

Several of these contain details of various aspects pertinent to the main topic of the subsubarticle. A brief outline of the major content of each of these subsubarticles is given in the following paragraphs.

Formulas are given in NB-3320 for the tentative thickness required for cylindrical and spherical shells for internal pressure. These are based on the maximum shear theory of failure and the code-allowable stress value S_m for primary membrane stress intensity.

NB-3330 covers the following four major aspects of the requirements for openings in a vessel:

1. Requirements on primary stresses (NB-3331).
2. Requirements on reinforcement (NB-3332–NB-3336) including a definition of when reinforcement is required and how much, the dimensional limits on material that may be considered as reinforcement, and the required strength of the reinforcing material.
3. The methods of weld attachment for nozzles, and other connections, to the shell (NB-3337).
4. Pressure stresses for fatigue evaluation at openings (NB-3338).

NB-3340 simply references the stress analysis that is required by NB-3214.

Four categories of welded joints (categories A–D) are defined in NB-3350 depending on the location of the joint in the vessel and are established for the purpose of specifying joint types and the degree of examination required.

The last subsubarticle of NB-3300, NB-3360, addresses the need for:

1. Providing adequate transitions at category A and B joints between sections that differ in thickness and evaluating the transition in accordance with the requirements of NB-3220.
2. Dimensioning bolted flange connections for external piping.
3. Provision of access openings, and their preferred configuration.
4. Vessel support design and evaluation must satisfy the stress limits of NB-3200 and the requirements of subsection NF.

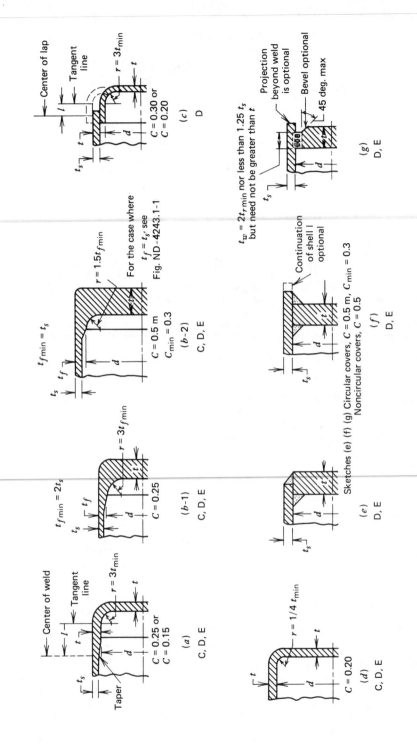

Center of lap

Tangent line

$r = 3t_{min}$

t

$C = 0.30$ or
$C = 0.20$

(c)

D

$t_w = 2t_{r\,min}$ nor less than $1.25\ t_s$
but need not be greater than t

Projection beyond weld is optional

Bevel optional

45 deg. max

(g)

D, E

$t_{f\,min} = t_s$

$r = 1.5t_{f\,min}$

For the case where
$t_f = t_s$, see
Fig. ND-4243.1-1

$C = 0.5$ m
$C_{min} = 0.3$

$(b\text{-}2)$

C, D, E

Continuation of shell l optional

$C = 0.5$ m, $C_{min} = 0.3$
$C = 0.5$

(f)

D, E

$t_{f\,min} = 2t_s$

$r = 3t_{f\,min}$

$C = 0.25$

$(b\text{-}1)$

C, D, E

(e)

D, E

Sketches (e) (f) (g) Circular covers, $C = 0.5$ m, $C_{min} = 0.3$
Noncircular covers, $C = 0.5$

Center of weld

Tangent line

$r = 3t_{min}$

t

Taper

$C = 0.25$ or
$C = 0.15$

(a)

C, D, E

$r = 1/4\ t_{min}$

t

$C = 0.20$

(d)

C, D, E

650

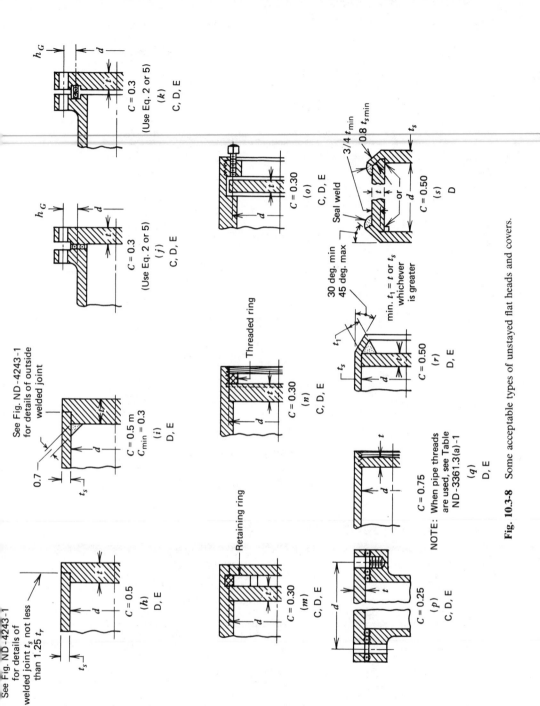

Fig. 10.3-8 Some acceptable types of unstayed flat heads and covers.

651

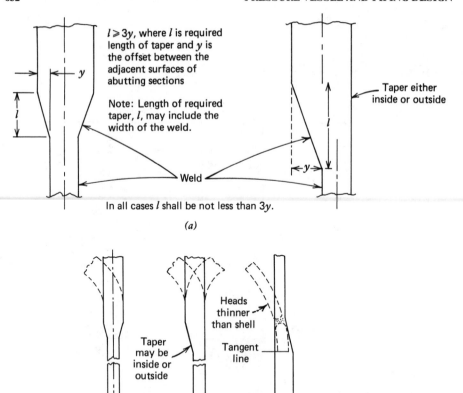

$l \geqslant 3y$, where l is required length of taper and y is the offset between the adjacent surfaces of abutting sections

Note: Length of required taper, l, may include the width of the weld.

Taper either inside or outside

Weld

In all cases l shall be not less than $3y$.

(a)

Taper may be inside or outside

Heads thinner than shell

Tangent line

NOTE: Length of taper may include the width of the weld.

(b)

Fig. 10.3-9 Transitions in thickness.

10.3-5 Design By Rule (NC-3300 and ND-3300)

NC-3310 and ND-3310 simply state that "vessel requirements as stipulated in the Design Specifications (NCA-3250) shall conform to the design requirements of this Article."

Although the basic design criteria and allowable stress values (i.e., factor of safety) are the same in NC-3300 and ND-3300, the overall class 2 vessel rules include restrictions on materials, fabrication details, and inspection requirements that ensure increased design integrity relative to class 3 rules. For example, the class 3 rules permit a variety of weld joints as well as limited or no radiographic inspection of welds, provided that the weld efficiency factor is accounted for in the minimum shell wall thickness calculation. The class 2 vessel rules restrict the types of weld joints permitted and require full radiographic examination for primary welded joints. These are discussed further with the aid of Fig. 10.3-11.

NC-2121 specifies that the materials permitted for use for class 2 components are limited to those ASME materials listed in Table I-7.0 in Appendix I of Section III. The requirement for class 3 components, ND-2121, permits the use of materials listed in Table I-7.0 and Table I-8.0. The class 3 materials list includes a wide variety of carbon, low-alloy and high-alloy steels, and nickel and other alloys generally as contained in Section VIII, Division 1. The class 2 rules are restrictive in that only certain materials from each group are permitted.

It should be noted here that the allowable stress values, S, are the same for common materials listed in Table I-7.0 and I-8.0 for class 2 and class 3 vessels designed in accordance with NC-3300 and ND-3300. The materials and S values in these subsections of Section III are those of Section VIII, Division 1. The factor of safety is 4 on ultimate strength or two-thirds of yield strength, whichever is more conservative.

NC-3300 and ND-3300 refer to several of the requirements contained in NC-3100 and ND-3100, respectively.

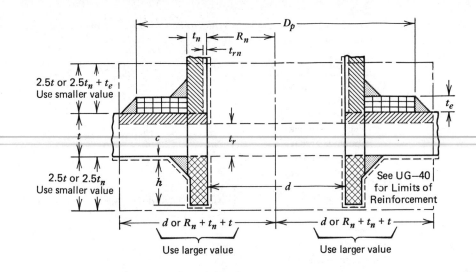

Without Reinforcing Element (Plate)

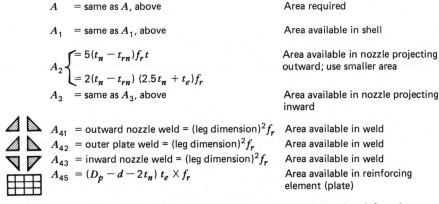

$\blacksquare$ $= A = d \times t_r \times F$ — Area required

$\diagdown\diagdown$ $= A_1$
$\begin{cases} = (E_1 t - F t_r)(d - R_n) \times 2 = (E_1 t - F t_r)d \\ = (E_1 t - F t_r)(R_n + t_n + t - R_n) \times 2 \\ = 2(E_1 t - F t_r)(t_n + t) \end{cases}$ — Area available in shell; use larger value

$\diagdown\diagdown$ $= A_2$
$\begin{cases} = (t_n - t_{rn}) f_r \times 2.5 t \times 2 = 5(t_n - t_{rn}) f_r t \\ = (t_n - t_{rn}) f_r \times 2.5 t_n \times 2 = 5(t_n - t_{rn}) f_r t_n \end{cases}$ — Area available in nozzle projecting outward; use smaller value

$\boxtimes$ $= A_3 = (t_n - c) f_r \times h \times 2 = 2(t_n - c) f_r \times h$ — Area available in nozzle projecting inward

$= A_{41}$ = outward nozzle weld = (leg dimension)$^2 f_r$ — Area available in weld

$= A_{43}$ = inward nozzle weld = (leg dimension)$^2 f_r$ — Area available in weld

If $A_1 + A_2 + A_3 + A_{41} + A_{43} \geqslant A$ — Opening is adequately reinforced

If $A_1 + A_2 + A_3 + A_{41} + A_{43} < A$ — Opening is not adequately reinforced so reinforcing element must be added or thicknesses increased

With Reinforcing Element (Plate Added)

A = same as A, above — Area required

A_1 = same as A_1, above — Area available in shell

A_2 $\begin{cases} = 5(t_n - t_{rn}) f_r t \\ = 2(t_n - t_{rn})(2.5 t_n + t_e) f_r \end{cases}$ — Area available in nozzle projecting outward; use smaller area

A_3 = same as A_3, above — Area available in nozzle projecting inward

A_{41} = outward nozzle weld = (leg dimension)$^2 f_r$ — Area available in weld

A_{42} = outer plate weld = (leg dimension)$^2 f_r$ — Area available in weld

A_{43} = inward nozzle weld = (leg dimension)$^2 f_r$ — Area available in weld

$A_{45} = (D_p - d - 2 t_n) t_e \times f_r$ — Area available in reinforcing element (plate)

If $A_1 + A_2 + A_3 + A_{41} + A_{42} + A_{43} + A_5 \geqslant A$, opening is adequately reinforced

Fig. 10.3-10 Reinforcement calculation.

Condition Considered	-33XX Paragraph in		
	NC	ND	NE
Ligaments (between holes)	29	29.6	NA
Category A, B, or C butt weld			
Welded from both sides and			
Fully radiographed	52	52.1 (a)	52
Spot radiography	NA	52.1 (b)	NA
No radiography	NA	52.1 (c)	NA
Single side with backing			
Fully radiographed	52	52.1 (a)	52
Spot radiography	NA	52.1 (b)	NA
No radiography	NA	52.1 (c)	NA
Other weld types	52	52.1 (c)	52
Fillet welds	56	56.1	56.
Nozzle attachment welds	59	59	59

Fig. 10.3-11 Structural efficiency factors.

The requirements of NC-3200 provide a completely independent set of rules from those in NC-3300. They cannot be interchanged in any way. There is no ND-3200.

Although the requirements are contained in -3112.4, the requirements for NC- and ND-3300 vessels are sufficiently different from those applicable to other vessels that they are discussed here.

Given the large list of loadings that are to be considered, which stresses are to be computed? NC/ND-3112.4(c) states that, using NC as an example:

"(c) The wall thickness of a component computed by the rules of this Subsection shall be determined so that the general membrane stress due to any combination of mechanical loadings listed in NC-3111 which are expected to occur simultaneously during a condition of loading for which service Level A is designated for the component does not exceed* the maximum allowable stress value permitted at the Design Temperature unless specifically permitted in other paragraphs of this subsection.

*It is recognized that high localized and secondary stresses may exist in components designed and fabricated in accordance with the rules of this Subsection; however, insofar as practical, design rules for details have been written to hold such stresses at a safe level consistent with experience."

NC- and ND-3112.4(c) are identical to UG-23(c) of Section VIII, Division 1, as it was prior to the Winter 1979 Addenda. The NC/ND paragraph in its present form has resulted in much consternation on the part of designers and therefore the footnote has been added in an attempt to qualify the rule. Unfortunately, definitions are not provided for membrane, secondary, and peak stresses, and the rule is open to interpretation unless, and perhaps even if, the designer is familiar with the rules for design by analysis.

Several interpretations of UG-23 have been employed in the past by vessel designers. One is that the rules require that the total stress, including discontinuity and thermal stresses, cannot exceed the allowable stress value S. Another is that since the term *membrane stress* is used, the code intent is to average stresses across the shell thickness and compare the average or membrane stress to the code-allowable stress. A third interpretation is that the intent of the rule is to include only mechanical loads and pressure and that the shell membrane stress shall not be greater than S at any location, including local areas adjacent to structural discontinuities.

The Winter 1979 Addenda to Section VIII, Division 1, revised UG-23(c) to read:

"(c) The wall thickness of a vessel computed by these rules shall be determined such that any combination of loadings listed in UG-22 that are expected to occur simultaneously during normal operation of the vessel will induce a maximum general primary membrane stress which does not exceed the maximum allowable stress value from the tables in Subsection C. Except where limited by special rules, such as those for cast iron in flanged joints, the above loads shall not induce a combined maximum primary membrane stress plus primary bending stress across the thickness which exceeds $1\frac{1}{2}$ times the maximum allowable stress value from the tables in Subsection C. It is recognized that high localized discontinuity stresses may exist in vessels designed and fabricated in accordance with these rules. Insofar as practical, design rules for details have been written to limit such stresses to a safe level consistent with experience.

Service Limit	Stress Limits
Design and Level A	$\sigma_m \leq 1.0S$ $(\sigma_m$ or $\sigma_L) + \sigma_b \leq 1.5S$
Level B	$\sigma_m \leq 1.10S$ $(\sigma_m$ or $\sigma_L) + \sigma_b \leq 1.65S$
Level C	$\sigma_m \leq 1.5S$ $(\sigma_m$ or $\sigma_L) + \sigma_b \leq 1.8S$
Level D	$\sigma_m \leq 2.0S$ $(\sigma_m$ or $\sigma_L) + \sigma_b \leq 2.4S$

Fig. 10.3-12 Stress limits for design and service loadings.

The maximum allowable stress values that are to be used in the thickness calculations are to be taken from the tables at the temperature which is expected to be maintained in the metal under the conditions of loading being considered. Maximum stress values may be interpolated for intermediate temperatures."

A similar revision should be considered for subsections NC/ND 3112.4(c) because the revised wording makes the intent very clear. In fact, the primary stress limits take on the same format for NC/ND-3300 vessels as for vessels that follow the design by analysis method.

The evidence that the intent of the present NC/ND-3112.4(c) is identical to that of the revised UG-23(c) is available in Table NC/ND 3321-1, which is included as Fig. 10.3-12.

The requirements therein for design conditions and service level A limits are as described by the revised UG-23(c). For service levels B, C, and D, the allowable values are multipliers of the limit applicable to service level A.

One of the characteristics of the design by rule method is the use of simple equations for the determination of the minimum required thickness of various structural elements. Figure 10.3-13 summarizes the structural elements for which equations are available for NC/ND-3300 and NE-3300 designs.

In some cases, because of fabrication details or geometries present, the usual allowable stress is multiplied by a structural efficiency factor, which is less than unity. Figure 10.3-11 identifies such factors in a manner similar to that used for Fig. 10.3-13.

Structural Element	Loading(s) Considered	-33XX Paragraph in		
		NC	ND	NE
Cylinder	Internal Pressure	24.3	24.3	24.3
Sphere	Internal Pressure	24.4	24.4	24.4
Formed heads	Internal			
Ellipsoidal	Pressure	24.6	24.6	24.6
Hemispherical		24.7	24.7	24.7
Torispherical		24.8	24.8	24.8
Conical		24.9	24.9	24.9
Toriconical		24.10	24.10	24.10
Reducer sections	Internal pressure	24.11	24.11	24.11
Flat heads	Int. press. & bolt load	24.11	24.11	24.11
Bolted-on dished heads	Int. press. & bolt load	26.2	26.2	26.2
Stayed surface	Internal pressure	Not allowed	29.1	Not allowed
Bolted flanges	Int. press. & bolt load	62	62	62
Plug welds	Various	NA	56.3	NA

Fig. 10.3-13 Minimum thickness formulas.

Structural Element Considered by the Rule	-33XX Paragraph in		
	NC	ND	NE
Nozzles	24.12	24.12	24.12
Nozzle—Piping transistion	24.13	24.13	NA
Quick actuating closures	27	27	27
Combination units	NA	28	28
Staybolts & welded stays	NA	29	NA
Openings	31	31	31
Reinforcement of openings			
General method	32–36	32–36	32–36
Alternative method	NA	NA	NA
Design of weldments	50–59	50–59	50–59
Tapered transistions	61	61.1	61
Staggered welds	NA	61.2	NA
Threaded connections	NA	61.3	NA
Access or inspection openings	63	63	63
Supports	64	64	64
Bellows expansion joints	65	65	65
Indices for fatigue	NA	NA	38
Evaluation of openings			
Closures on small penetrations	NA	NA	67

Fig. 10.3-14   Geometric rules.

When thickness equations are not given, geometric rules are used to communicate the design requirements. Figure 10.3-14 summarizes these in a manner similar to that used for Fig. 10.3-13.

10.3-6   Combined Methods of NC-3200

The rules of NC-3200 provide an alternative set of rules to those contained in NC-3300. The two sets of rules cannot be mixed or combined in any way—one is the alternative to the other. Although based on design by analysis procedures, design by rule concepts have been introduced in order to simplify applications. That is why we have used the term *combined methods*. The rules contained in NC-3200 are taken from Division 2 of Section VIII. They are based on design by analysis procedures in conjunction with the maximum shear failure criteria that result in a stress evaluation based on stress intensity rather than the maximum normal stress as employed in NC- and ND-3300. However, design procedures are provided in NC-3200 that often permit the designer to establish the vessel design on the basis of simplified rules, similar to those contained in NC- and ND-3300 and Section VIII, Division 1. Unfortunately, at the present time the NC-3200 rules cover only limited design configurations for the various pressure vessel parts. It is the intent of the code committee to include in the code additional design rules for other standard configurations as analytical and/or experimental design substantiation is acquired, thereby reducing the need for redundant stress analyses on common pressure parts. Subarticle NC-3200 also includes design by analysis criteria for configurations not covered by the simplified design rules. The design by analysis rules, contained in Appendix XIII and XIV of Section III, provide for the evaluation of design conditions and operating conditions, including fatigue evaluation. The rules are the same as those of Appendix 4 and 5 of Section VIII, Division 2, which are essentially those of NB-3200 for class 1 vessels. The NC-3200 rules are included for class 2 components to provide an added measure of design integrity and therefore should be considered applicable for vessels where more severe operating conditions or cyclic operating conditions prevail.

The rules of NC-3200 are more restrictive than those of NC/ND-3300 in that only materials acceptable for class 1 components are permitted. The use of Table I-1.0 materials and the S_m allowable stress intensity value are of particular significance because of the increased allowable stress values. Fabrication and weld details are more restrictive, full radiographic weld examination is required, and the designer is required to evaluate the cyclic nature of operational service of the vessel or vessel part.

NC-3211.1 states that:

1.   "These requirements provide specific design rules for some commonly used vessel shapes under pressure loadings and, within specified limits, rules for treatment of other loadings.

Simplified rules are also included for the approximate evaluation of design cyclic service life. Rules are not given which cover all details of design."

2. "When complete rules are not provided or when the vessel designer chooses, a complete stress analysis of the vessel or vessel region shall be performed considering all the loadings of NC-3212 and the Design Specifications. This analysis shall be done in accordance with Appendix XIII for all applicable stress categories. Alternatively, an experimental stress analysis shall be performed in accordance with Appendix II."

3. "When these alternative design rules are used, the special requirements of NC-4250, NC-5250, NC-6600, and NC-6700 shall be met."

4. "A Design Report shall be prepared by the Certificate Holder showing compliance with this Subarticle. This Design Report shall meet the requirements of NCA-3550 for a Design Report (Appendix C)."

NC-3211.1 defines the scope of the alternative rules. It should be noted that subparagraph (b) previously permitted the design by analysis rules only when design rules were not provided for a pressure part. A significant revision was made to this paragraph in the summer 1973 addenda that permits design by analysis rules to be employed at the discretion of the designer or user. The design by analysis methods of Appendix XIII or experimental methods of Appendix II may now be used in lieu of a given design rule to substantiate minimum thickness requirements.

NC-3211.2 states that:

1. "The design shall be such that the requirements of NC-3100 and this Subarticle are satisfied. In cases of conflict, the requirements of this Subarticle shall govern."

2. "The design shall be such that stress intensities do not exceed the limits given in NC-3216."

3. "For configurations where compressive stresses occur, the critical buckling stress shall be taken into account. For the special case of external pressure, the rules of NC-3133 shall be met."

The requirements of NC-3300 provide a completely independent set of rules from those in NC-3200. They cannot be interchanged in any way.

If the geometries and conditions are such that the design can be achieved using only the requirements of NC-3200 without reference to Appendices XIII and XIV, the stresses to be considered are the same as for NC-3300 vessels. The combinations of pressure, static head, and temperature to be considered are as given in Table NC-3215(a)(1) as shown by Fig. 10.3-15

If reference must be made to Appendices XIII and XIV, the stresses which must be considered are the same as those for class 1 (NB-3300) vessels. The previously cited quotation from NC-3211.1 gives some guidance in this respect. The fatigue exemption requirements of NC-3219 are summarized here. The requirements of condition B are identical to those applicable to class 1 vessels. The requirements of condition A are simplified and more conservative. Conditions AP and BP are similar but are more conservative and are for vessels containing pad-type nozzles.

A fatigue analysis is not required if condition A or condition B is satisfied:

Condition A

1. Material UTS $\leq$ 80,000 psi
2. $N_a + N_b + N_c + N_d \leq 1000$ cycles
   (a) Number of full-range design pressure cycles.
   (b) Number of pressure cycles where $\Delta P \geq 0.2 P_D$.
   (c) Number of changes in metal temperature between two adjacent points.
   (d) Number of temperature cycles when welds between materials having different coefficients of expansion are present.

Condition B. Same as for class 1 vessels.

Condition AP. For pad-type nozzles; same as condition A but limited to 400 cycles.

Condition BP. Same as condition B but with factors increased by $\frac{4}{3}$.

The materials permitted and basic allowable stress intensity values are those permitted for class 1 vessels. The allowable stress intensity for various types of operations is expressed as kS_m, where the k values are given by Table NC-3217-1 as shown in Fig 10.3-16. Although Section VIII, Division 2, also contains k values, they differ from those used for NC-3200 vessels.

The design rules include minimum shell thickness requirements for cylindrical, spherical, and conical shells and elliptical and hemispherical heads, as well as reducer sections and flat heads. The

1	2 Pressure at Top of Vessel	3 Pressure Due to Static Head	4 Temperature	5 Remarks
Condition (1)				
For vessel as a whole	Design pressure	None	Coincident metal	Pressure and temperature to be stamped on nameplate
At any point	Coincident pressure	Pressure to point under consideration due to static head of vessel contents	Design coincident temperature	Temperature at various points may vary, in which case the maximum for these conditions should be used for the vessel as a whole or coincident conditions for specific locations should be listed on the manufacturer's data report and stamping
Condition (2)				
At any point	Coincident pressure	Coincident pressure to point under consideration due to static head	Design temperature	Higher temperature and lower pressure combinations (than Condition 1) must be checked or a part may be designed for the maximum design pressure and the design temperature
Condition (3)				
For vessel as a whole	Test pressure	None	Test temperature	
At any point	Test pressure	Pressure at point under considera- tion due to static head	Test temperature	
Condition (4)				
For vessel as a whole or any part	Coincident pressure		Minimum permissible temperature	Minimum permissible temperature is used together with notch toughness tests or with low maximum stresses to determine suitability of material at service temperature
	Safety valve setting			Usually set above the operating pressure but not over the limits set in NC-7000

Fig. 10.3-15   Pressure and temperature relationships.

minimum shell thickness rules of NC-3200 are similar to those of NC and ND-3300 except the equations are based upon the maximum shear stress failure criteria.

The NC-3200 rules include more restrictions on geometry and weld details of the basic shell components than do the NC/ND-3300 rules. For example, the reducer rules require local reinforcement consideration. Flat head covers and blind flange rules are restrictive, and no design rules are provided for spherically dished covers with bolting flanges or ligaments. It should be noted that the elliptical head minimum thickness rule of NC-3224.6 results in a thicker shell than does the requirement of NC/ND-3300. The subgroup on design analysis performed a detailed study involving a series of parametric analyses involving inelastic behavior of elliptical and torispherical heads, which resulted in the development of the new design rules based on an improved definition of collapse load.

NC-3230 provides rules for openings and the required reinforcement. The rules are similar to those of NC/ND-3300 for fully radiographed welds. Local reinforcement equal to the cross-sectional areas of the opening must be added to the minimum required shell thickness. Area reinforcement limits are specified normal and parallel to the shell surface at the opening. The area reinforcement rules are intended to provide an adequate design for internal pressure loading only.

NC-3239 provides alternative reinforcement rules that were added in the summer 1973 addenda. These rules are based on experimental and finite-element studies from which the optimum placement

Service Limits[1]	k (note 2)
Design	1.0
Level A[3]	1.0
Level B[3]	1.1
Level C	1.2
Level D[4]	2.0
Test	1.25 for hydrostatic test and 1.15 for pneumatic test (see NC-3218.1 for special limits)

Notes:

1 For design limits, use design pressure at design metal temperature; for service limits, use service pressure at service metal temperature; for test limits, use test pressure at test metal temperature.

2 The condition of structural instability or buckling must be considered.

3 See NC-3219 and Appendix XIV.

4 When a complete analysis is performed in accordance with NC-3211.1 (c), the stress limits of Appendix F may be applied.

Fig. 10.3-16 Stress intensity k factors for design, service, and test load combinations.

of area reinforcement material was determined. NC-3239.7 also provides alternative stress index rules for nozzles designed in accordance with NC-3239. Conventional stress indices contained in Appendix XIII are used in conjunction with the area reinforcement rules in NC-3230.

When applied to the shell membrane stress, nozzle stress indices are factors that result in primary plus secondary plus peak stress values at the inside and outside corner of a nozzle. These values have been determined as a result of extensive tests sponsored by PVRC. The stress index values may only be used for fatigue evaluation of nozzles designed in accordance with code rules. The stress indices have been developed for internal pressure loading only; consequently, stress analysis is required in order to determine stress resulting from piping loads or thermal loads.

Nozzles with separate reinforcing pads are permitted by NC-3239.8; however, limitations are specified regarding materials and thickness and also the cyclic service requirements of NC-3219 must be satisfied.

REFERENCES

10.3-1 *Boiler and Pressure Vessel Code*, Section III, Nuclear Power Plant Components, Division 1, The American Society of Mechanical Engineers, New York, 1980.

10.3-2 *Boiler and Pressure Vessel Code*, Section VIII, Divisions 1 and 2, Rules for Unfired Pressure Vessels, The American Society of Mechanical Engineers, New York, 1980.

10.3-3 E. O. Bergman, "The Basis and Content of a Nuclear Pressure Vessel Code," Paper No. 94, Nuclear Congress, 1962.

10.4 NUCLEAR CLASS 1 CODE DESIGN MARGINS OF SAFETY—PIPING

The code margins of safety are a function of the class of the component being considered. Section III[1] provides three classes for all components excluding core support structures and containment vessels. Each class can be considered as a quality level, with class 1 being the highest and class 3 the lowest. These levels of quality exist because of the various requirements in materials, fabrication, erection, examination, and design. Design was placed last on the list because there is sufficient evidence to indicate that the other considerations listed are of equal or more importance than the design requirements. The following discussion will address design margins of safety only. Also, it will address the margins by class. The actual application of code rules is discussed in Section 10.6. Although the application of the rules discussed here and in Section 10.6 is only required for nuclear power piping they can and have been used by engineers to resolve a piping design or operating problem in many other industries. The particular areas of importance are related to collapse and fatigue protection.

Satisfaction of the code rules assures that violation of the pressure boundary will not occur. This assurance is provided in two ways:

1. Protection against catastrophic failure.
2. Protection against initiation and propagation of a crack through the pressure boundary.

Protection is provided as a function of the types of stresses being considered. Referring to the criteria documents,[2] the following stress categories are considered:

1. General primary membrane.*
2. Local primary membrane.*
3. Primary bending.*
4. Secondary.†
5. Peak.†

Definitions of these terms are given in NB-3200 of Section III[1] and justification for the chosen categories is given in Section II.[2]

It should be noted that for class 1 piping, no consideration of local primary membrane stress (P_L) is required. The basic reason for this is that bending is the major concern in piping systems, and control of the general primary membrane stress due to pressure by use of a minimum thickness equation provided sufficient control. Further, as indicated by the discussion in Section II,[2] there is some question as to whether the P_L stress is primary in nature. Certainly distortions of the pressure boundary are far more tolerable (in a geometric sense) in piping than in a reactor pressure vessel.

Having defined the stress categories that require protection, the choice of the basic stress limits must be examined. Reference 1 provides this explanation in sufficient detail. Those limits that are important to this discussion will be elaborated here as required.

10.4-1 Primary Bending Stress—Limit Load

This is the major area of concern for piping subjected to dynamic loading and controls the design of piping and piping support systems because of the magnitude of dynamic loading to be accommodated. The limit on this stress category is based on limit design theory and assumes the material is elastic–perfectly-plastic with no strain hardening, as can be seen in Fig. 1, Reference 2. Using this assumption, the stress limit was developed by providing a margin on the actual limit stress curve for combined tension and bending on a rectangular section. This is shown in Fig. 2, Reference 2. Quite often, the point is made that 1.5 is the "shape factor" for a particular section and should not be used for hollow circular sections. This point is usually made in reference to the allowable stress of $1.5 S_m$ for primary bending for class 1 components. It is important to recognize that the 1.5 factor used with S_m is not an attempt to provide a "shape factor" but is a factor which, when taken with the allowable stress (S_m), provides a margin on the theoretical limit load for any elastic–perfectly-plastic material. For ferritic materials, S_m can never be higher than $\frac{2}{3} S_y$ and therefore $1.5 S_m$ results in an allowable stress for primary bending which is never higher than S_y. Figure 2 indicates that the theoretical limit stress varies from $1.5 S_y$ with no membrane stress present to $1.0 S_y$ with only membrane stress present. It should be noted that the theoretical limit stress peaks at approximately 1.65 with a combination of bending and membrane stress when the membrane stress is approximately $\frac{1}{3} S_y$. The theoretical limit of 1.65 should not be used for design purposes since it would negate the margins of safety.

It is true that Fig. 2 does not represent the margin between code-allowable and theoretical limit load for a straight cylinder which has a shape factor of about 1.33 instead of 1.5. However, it should be recognized that for some elbows the addition of membrane stress due to pressure increases the capacity of the elbow to carry a bending movement. This is recognized in the winter 1981 addenda to NB-3600 of Section III, which provides an equation to determine the effect of pressure on collapse of an elbow. This effect varies from 0.0 to a value equal to that for a straight pipe. Generally, then, it can be said that for piping a minimum margin of approximately 1.3 between code stress limit and theoretical limit load for an elastic–perfectly-plastic material exists.

The above discussion addresses S_m for ferritic materials. For austenitic and some nonferrous materials, S_m at temperatures above 100°F may exceed $\frac{2}{3} S_y$ and may reach $0.9 S_y$. An explanation of this is also given in Reference 2 and indicates that although $0.9 S_y$ may be used as S_m, the margin on the theoretical limit load is essentially not reduced.

---

*Evaluation assures protection against catastrophic failure of the pressure boundary.
†Evaluation assures protection against fatigue failure of the pressure boundary.

10.4-2 Fatigue

The development of the fatigue design curves, which are included in Appendix I of the code, is described in Part III of Reference 1. A summary of that discussion follows.

1. For low-cycle fatigue, strain-controlled data are used. Such data are applicable because the code limits on the stress combinations other than those which include the peak stress components (the effects of local structural discontinuities) assure shakedown to elastic action of the overall structure. Therefore, the plastic deformation that occurs in local regions is constrained.

2. Since for constrained plastic deformation the strains that exist are adequately predicted by an elastic analysis, the strain range imposed for testing is converted into a (fictitious) elastic stress amplitude by multiplying the strain range by $E/2$.

3. At high cycles to failure, where the actual stress range during cycling is less than twice the yield strength, strain-controlled and stress-controlled tests give the same result.

4. The fatigue data are corrected for the worst possible effects of mean stress. This is done because the residual stress which exists in a structure after fabrication and operation may not be calculable. No correction is required if the alternating stress is higher than the yield strength, because in a strain-controlled situation, shakedown will occur such that the mean stress is reduced to zero. The mean stress correction technique used varies from material to material, in accordance with available data.

5. Following correction for the worst possible effects of mean stress, two new curves are obtained by:
   (a) Dividing the number of cycles on the corrected data curve by 20.
   (b) Dividing the stress amplitude on the corrected data curve by 2.

   The fatigue curve is taken as the lower of these two curves at each number of cycles, with a faired curve between the two sometimes used to avoid cusps.

6. As is indicated by the test data contained within the criteria document, the actual factor of safety on cycles probably ranges between 1.0 and 5.0, with a mean value of about 3. Since the data defined failure as the appearance of a visual crack, about $\frac{3}{16}$ in. long, this should be considered as a factor of safety on initiation—not on failure, Environmental and some other construction effects have indicated that this factor may be significantly lower.

Prior to 1982 the code fatigue curves did not go beyond 10^6 cycles, and there was sometimes a question as to what was intended for higher number of cycles. A review of the criteria document and Footnote 10 to NB-3222.5(b) clearly indicates that the intent is that the allowable amplitude at 10^6 cycles is equal to one-half the endurance limit. Recently Section III has just approved extension of the fatigue curves for materials to 10^{11} cycles. These new curves, in addition to addressing high-cycle fatigue also accommodate the concerns expressed with these materials at cycles $> 10^6$ when significant ranges of primary stress are present.

The fatigue test specimens used for obtaining data to develop the code fatigue curves were generally $\frac{3}{16}$ in. diameter. Therefore, the data can be interpreted as representing crack initiation and crack propagation to a small but visible crack. It is for that reason that a crack of approximately $\frac{3}{16}$ in. length was used to judge the results of the vessel tests discussed in Reference 2. It also explains why the code fatigue design procedures are considered to relate to fatigue crack initiation. In most real structures, however, initial cracks are present so that the fatigue design curves are used to deal with material containing flaws of some size and type, and the crack initiation process is complete.

The first step in performing a fatigue evaluation is to determine whether the component is cycling elastically or plastically. This is determined by evaluating the range of secondary stress and comparing it to a "shakedown limit." Reference 2 provides a detailed discussion on the development of the shakedown limit. Since the preparation of Reference 2 and the development of more sophisticated analytical techniques and more test data, simplified rules have been provided in the code that allow for consideration of fatigue cycles that exceed the shakedown limit. This technique for simplified elastic-plastic analysis was first presented in ANSI-B31.7[3] and was developed by Tagart.[4] The technique that is contained in Section III is a slight modification to the Tagart work.

10.4-3 Service Limits

Once the types of stress to be considered have been determined in accordance with Section 10.6, appropriate allowable levels of stress must be established. The allowable stresses in conjunction with the types of stresses considered provide a design margin. The previous discussion on primary bending stress and fatigue establishes the basic margins for these categories and will not be repeated here.

However, the code provides for variations on these established allowables resulting from the specific service limit that is selected for a given plant condition.

A design limit and four service limits are provided in the code. Each of these limits has allowable stresses that provide various margins of protection for either catastrophic or fatigue failure or both. The code does not define what specific service limit should be used for a given plant or system event. This is the responsibility of the owner, or his designee, in conjunction with regulatory authority requirements. In order to follow this process, plant events must first be discussed, and it is noted that the code does not do this. The code simply states that the definition of plant and system operating conditions, and the determination of their significance to the design and operability of components and supports, may be derived from specific types of nuclear power system safety criteria documents and from requirements of the regulatory authorities. These plant or system operating conditions can be categorized as follows:

Normal	Operation in design power range
	Hot standby
Upset	Deviations from normal operation
	Operator error
	Loss of load
Emergency	Low probability event
Faulted	Extremely low probability event
	Postulated combinations of emergency events

The owner, or his designee, is responsible for taking the specific plant/system operating conditions and deriving from them the appropriate design and service loads to be used to meet code limits. In doing this it must be recognized that the code deals on the component level, and even though a plant/system event may have a low probability of occurrence it may be necessary to impose a service limit that would normally be used for expected events. This decision is dependent on the function of the component for the specific plant condition under consideration. For example, a specific component may be installed in a plant to function during a plant emergency or faulted event only and to remain passive for all other events while another component is passive for the emergency or faulted events and functions only for normal plant events. The selection of service limits as a function of plant events could be quite different for these two components.

Definitions of design and service loads are given in NCA-2142.[1] In order to properly specify service limits for various types of loadings, the basis for the establishment of these limits should be recognized. Service limits A and B are provided in order to evaluate the effect of operating loads on the fatigue life of the component. Additionally, protection against membrane failure is provided for service limit B. Service limit C is provided in order to evaluate the effect of operating loads on the structural integrity of a component for situations that are not anticipated to occur for a sufficient number of times to affect fatigue life and for which large deformations in areas of structural discontinuities are not objectionable. Since the occurrence of stress associated with this limit may result in damage to the function of the component requiring removal of the component for inspection or repair, the owner should review the selection of this limit for compatibility with established system safety criteria. Service limit D is provided in order to evaluate the effect of plant operating loads on the structural integrity of a component for situations in which gross general deformations, loss of dimensional stability, and damage requiring repair (excluding loss of pressure retaining function) are

TABLE 10.4-1 STRESS INTENSITY LIMITS FOR PIPING

Stress Intensity	Limit	Yield Strength	Ultimate Tensile Strength
General primary membrane $(P_m)^a$	S_m	$\leq \frac{2}{3} Sy$	$\leq \frac{1}{3} S_u$
Local primary membrane $(P_L)^a$	$1.5 S_m$	$\leq Sy$	$\leq \frac{1}{2} S_u$
Primary membrane plus primary bending (P_b)	$1.5 S_m$	$\leq Sy$	$\leq \frac{1}{2} S_u$
Shakedown limit	$3.0 S_m$	$\leq 2 Sy$	$\leq S_u$

^{a}Calculation of P_m or P_L not required for piping—design by rule is used.

TABLE 10.4-2   SERVICE LIMITS AND ALLOWABLES FOR PIPING

Service Limit	Stress Intensity	Allowable
Design	P_b	$1.5S_m$
Level A[a]	Fatigue	$U = 1.0$
Level B[a]	P_b	$1.8S_m$ but $\leq 1.5Sy$
Level C	P_b	$2.25S_m$ but $\leq 1.8Sy$
Level D	P_b	$3.0S_m$ but $\leq 2.0Sy$

[a]Levels A and B loads must be considered in conjunction with each other in determining the cumulative usage factor (U).

not objectionable. Since the occurrence of stress associated with this limit may require removal of the component from service, the owner should review the selection of this limit for compatibility with established system safety criteria. Loads due to tests beyond those allowed by the code should be classified in appropriate service limits. This is a point that is quite often missed in developing loads for service limit stress evaluation. For example, Section XI of the ASME, BPVC[5] requires periodic testing of pumps and valves in service. This results in loadings on components that must be accounted for. Many emergency core cooling systems are designed to operate only during a low or very low probability event; however, as a result of inservice testing, they are operated a significant number of times over the life of a plant. Since they are emergency-type systems, they may include rapid opening valves that can produce dynamic loading on a system, thereby increasing the concern for evaluating these effects.

The following list defines the stresses required for evaluation of piping for the code design and service limits.

Limits	Evaluation Required
Design	Minimum thickness
	Area replacement
	Primary stress
Level A	Fatigue
Level B	Fatigue
	Primary stress
Level C	Primary stress
Level D	Primary stress

The above list indicates that primary stress evaluation is required for all limits but A; however, the allowable stress is different for each. The basic stress intensity limits are discussed in Reference 1 and are included in Tables 10.4-1 and 10.4-2 below with the variations allowed for design and service limit evaluation.

Regulatory requirements exist that take established plant/system conditions and apply code service limits using the intended function of the component as a basis. The most comprehensive document available today that addresses this issue is PSRP 3.9.3.[6] The basis of the requirements of Reference 6 is the ability to shut the plant down, maintain it in the cold shutdown condition, or mitigate the consequences of an accident. Therefore, the systems and components associated with the emergency core cooling functions are the most critical. A number of techniques are allowed which involve either the use of low allowable stresses or elaborate analytical/test procedures. In many cases it is impossible to perform the testing that is required, and a combination analytical/test program can be economically prohibitive; therefore, the industry essentially is forced into using the low allowable stress approach.

10.4-4   Summary

A comparison of the code rules as established in Reference 1 (and presented in NB-3200) and the application of those rules for piping in NB-3600 is given in the modified Hopper diagram of Fig. 10.4-1 and presented in detail in Section 10.6. This figure is given to demonstrate that each stress category (except P_L) addressed for class 1 vessel design is applicable to class 1 piping design.

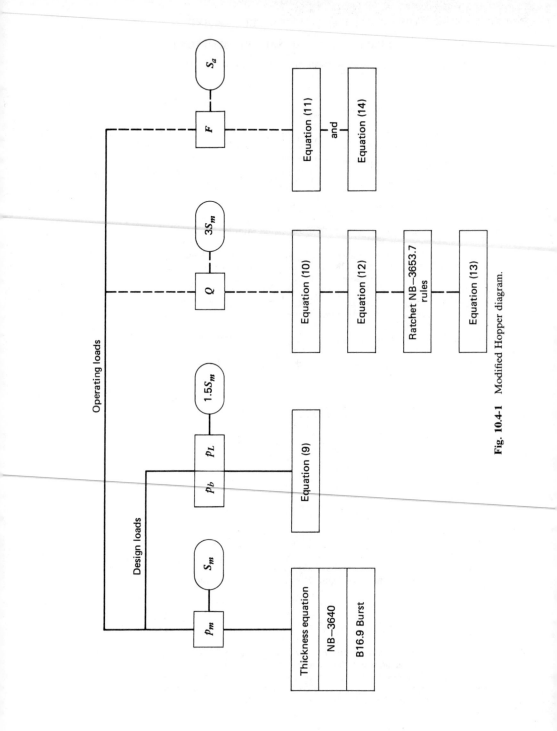

Fig. 10.4-1 Modified Hopper diagram.

REFERENCES

10.4-1 *Boiler and Pressure Vessel Code*, Section III, Nuclear Power Plant Components, Division 1, The American Society of Mechanical Engineers, New York, 1980.

10.4-2 *Criteria of the ASME Boiler and Pressure Vessel Code for Design by Analysis in Sections III and VIII*, Criteria for Design by Analysis, Division 2, The American Society of Mechanical Engineers, New York, 1969.

10.4-3 USA Standard Code for Pressure Piping, USAS B31.7, *Nuclear Power Piping* (1969).

10.4-4 S. W. Tagart, Jr., *Plastic Fatigue Analysis of Pressure Components*, The American Society of Mechanical Engineers, Paper No. 68-PVP-3.

10.4-5 *Boiler and Pressure Vessel Code*, Section XI, Division 1, Rules for Inservice Inspection of Nuclear Power Plant Components, The American Society of Mechanical Engineers, New York, 1980.

10.4-6 Preliminary Standard Review Plan, Section 3.9.3, United States Nuclear Regulatory Commission, 1981.

10.5 CLASS 2/3 CODE DESIGN MARGINS OF SAFETY—PIPING

For class 2/3 piping a different set of stress theories and limits exists. Class 2/3 piping utilizes maximum stress theory as contrasted to the maximum shear stress theory of class 1 and is essentially based on the ANSI-B31.1[1] criteria. This is important since many operating plants were designed to B31.1, and the following discussion is appropriate for those plants also. The actual application of code rules is discussed in Section 10.7.

10.5-1 Limit Load

Prior to the Winter 1981 Addenda to the code, there was a substantial difference between the approach used for B31.1, class 2/3 and class 1 piping related to providing protection against reaching the theoretical limit load. The Winter Addenda eliminates this difference and essentially provides the same rules for all classes of nuclear piping. However, in order to deal with piping designed to B31.1, some discussion of the past class 2/3 criteria in this area is necessary.

Moore and Rodabaugh[2] provide a comprehensive discussion of this topic for both class 1 and class 2/3 piping. The discussion of class 2/3 piping is also appropriate for plants designed to B31.1 since the philosophy is essentially the same. That report indicates that for limit load protection the

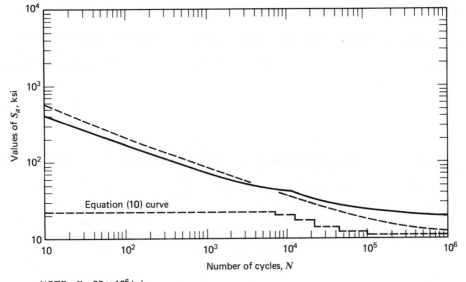

Fig. 10.5-1 Plot for stepped fatigue curve (Equation 10).

B31.1 techniques, as implemented in class 2/3, are, theoretically, not correct, unconservative as compared to class 1, and should be revised. However, it is recognized that conservative assumptions exist in the overall solution process. In addition to those associated with level of loading, damping, and so on, only a single-hinge collapse mechanism is postulated, material strain hardening is ignored, and real-time loading effects which can preclude plastic collapse are ignored. Each of these assumptions has been shown to be sufficiently conservative to indicate that the code approach to limit load protection, though theoretically not correct, provides a sufficient margin of safety.

10.5-2 Fatigue

The code requirements for class 2/3 piping do require a fatigue analysis. This analysis is quite different from that for class 1. There are essentially two fatigue equations provided in the code. [See Section 10.7, Equations (10.7-4) and (10.7-8).] The calculated stress is compared with an allowable that is essentially a point on an S/N curve represented by the factor (f) which is a function of the number of cycles. A plot of a stepped fatigue curve resulting from applying these allowables to the class 1 S/N curve for a ferritic material (SA-106, grade B at 200°F) is given in Figs. 10.5-1 and 10.5-2. Examination indicates that the Equation (10.7-4) allowable is conservative except for very high cycles. It is dramatically conservative below 7000 cycles. The Equation (10.7-8) allowable is essentially on the class 1 curve from 7000 to 24000 cycles and then becomes unconservative. Again, below 7000 cycles, the Equation (10.7-8) allowable is very conservative.

It must be recognized that the class 2/3 allowables as plotted consider that for each cycle there exists the:

1. Full range of moment from the minimum (zero or less than zero) to the maximum value.
2. Full range of pressure from zero to the design value.
3. Sustained loads are cycling.

On the other hand, it should also be recognized that the following are not considered:

1. Stresses from local temperature differences.
2. Discontinuity effects for pressure loading.
3. Any correction for plastic cycling.

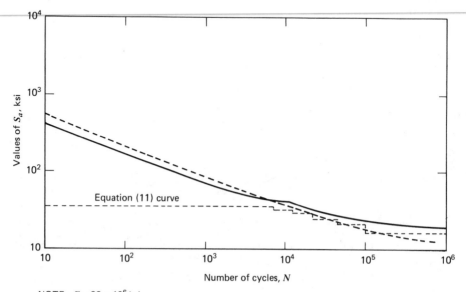

NOTE: $E = 30 \times 10^6$ ksi
--- UTS < 80.0 ksi
——— UTS 115.0-130.0 ksi
Interpolate for UTS 80.0-115.0 ksi

Fig. 10.5-2 Plot for stepped fatigue curve (Equation 11).

**TABLE 10.5-1   ALLOWABLE STRESS VALVES
FOR DESIGN AND SERVICE LIMIT
EVALUATION FOR CLASS 2/3 PIPING**

Service Limit	Stress	Allowable
Design	$S_L{}^a$	S_h
Level A	Fatigue	$S_A{}^b$
Level B	Fatigue	S_A
	S_L	$1.8S_h$ but $\leqslant 1.5Sy$
Level C	S_L	$2.25S_h$ but $\leqslant 1.8Sy$
Level D	S_L	$3.0S_h$ but $\leqslant 2.0Sy$

$^a S_L$ = maximum longitudinal stress.
$^b S_A = f(1.25S_c + 0.25S_h)$.

The code does allow the use of a higher allowable when the cycles being considered are not full temperature ranges [NC/ND 3611.2(c)(3)]. This technique has been used in the past to deal with dynamic loading that results in stresses which are a fraction of the full thermal expansion range.

10.5-3   Service Limits

The detailed discussion in Section 10.4-3 is appropriate for class 2/3 piping except that the basic stress intensity limit is defined as S_h or S_c and the variations for service limits are different. Table 10.5-1 provides the allowable stress values for design and service limit evaluation for class 2/3 piping.

REFERENCES

10.5-1   B31.1, *Power Piping*, American National Standards Institute, New York, 1980.
10.5-2   S. E. Moore and E. C. Rodabaugh, Background for the ASME Nuclear Code Simplified Method for Bounding Primary Loads in Piping Systems, *ASME J. Pressure Vessel Tech.* **104**, 351–361 (Nov. 1982).

10.6   PIPING DESIGN—NUCLEAR CLASS 1

10.6-1   General

This section presents a discussion of the design rules for class 1 piping contained in Section III.[1] Since the acceptance of ANSI-B31.7, and the subsequent adoption of nuclear piping by Section III, the piping design industry has been subjected to dramatic changes with large increases in manpower and a proliferation of computer codes. In spite of this, there still remains considerable confusion as to the proper application of these rules. This section will deal with the background of the rules, their development, and some examples of proper application.

The difference between piping and vessel criteria may appear to be major on the surface, but the following will indicate that there are little if any differences. The design procedures are certainly markedly different. In the case of vessels little guidance is given in the area of load and/or stress calculations, but considerable discussion and guidance is given concerning stress categories, allowable stress values, and criteria. In the case of piping there is negligible discussion of criteria, but equations are provided to calculate stresses. There are some important reasons for this approach in piping.

It is this understanding that allows the generic use of stress indices. Three kinds of indices are used in class 1 piping: *B*, *C*, and *K*.

In developing rules for piping design, the committee was able to take advantage of the available industry-wide standards for pipe and pipe fitting. These standards provide a control on geometry, and we can assume that a specified pipe fitting will respond to an applied load in a known manner. For example, a 20-in., schedule 80, long-radius elbow purchased to ANSI-B16.9[2] will deform in a known manner when subjected to an in-plane bending moment. It does not matter whether the elbow is in a fossil-fuel or a nuclear plant, whether it is in Massachusetts or California. We do know that the existing ANSI standards on piping fittings could provide better geometric control, particularly in the area of tees and reducers. These standards provide overall control on geometry, but do not control local geometries such as radii at intersections or changes in thickness or diameter. More will be discussed about these effects later.

10.6-2 Class 1 Piping, NB-3200 (Design By Analysis)—NB-3600 Comparison

In order to demonstrate that piping design rules of NB-3600 comply with the criteria established in design by analysis, we have to look at the modified Hopper diagram of Fig. 10.4-1. Remember that Section III provides protection against two types of failures.

Catastrophic. Protection is provided by dealing with design loads and demonstrating that membrane type stresses are less than those values for which failure may be imminent.

Fatigue. Protection is provided by dealing with all service limits A and B conditions and demonstrating that the available fatigue life of the material is not used up.

Primary Stress Considerations

NB-3200 requires calculation of general membrane, local membrane, and primary bending stress intensities and requires that they be less than some specified stress limit (S_m and $1.5S_m$, respectively).

NB-3600 provides the same kinds of protection while using different techniques. First, all piping components must meet the minimum thickness equations. For nonstandard intersections (branch connections) the reinforcement rules of NB-3600 follow those of NB-3200. For standard fittings a burst test requirement exists which ensures that the fitting will withstand the imposed design pressure without failure.

For the primary bending stress, NB-3600 requires calculation of stresses due to design pressure and design mechanical loads (weight, seismic, blowdown). Equation (10.6-2) is provided for this purpose, and the calculated stress must be within the specified limit of $1.5S_m$.

Fatigue Considerations

The first requirement is to determine whether elastic cycling or elastic-plastic cycling is occurring. NB-3200 requires calculation of the maximum range of primary plus secondary stress intensity (Q). If this value is less than $3S_m$, elastic cycling occurs and the elastic fatigue rules may be applied. If this value exceeds $3S_m$, plastic cycling occurs and the analyst must use the fatigue rules for elastic-plastic cycling in NB-3228.3.

For piping, the analyst must find the maximum range of stress intensity using Equation (10.6-3). If this calculated S_n value is less than allowable, the elastic cycling rules are followed. That is, the peak intensity stress (S_p) is calculated using Equation (10.6-6) and the alternating stress intensity (S_A) is $S_p/2$ and the rules of NB-3200 for calculation of the cumulative usage factor are used. If S_n exceeds $3S_m$, then the analyst must use the elastic-plastic rules of Equation (10.6-10) for calculating S_A. Before doing this, he must demonstrate that a hinge moment will not occur, Equation (10.6-8), the primary plus secondary membrane stress is acceptable, Equation (10.6-9), and the thermal stress ratchet rules are satisfied.

10.6-3 Pressure Design

1. Basic wall thickness. The initial task for the piping designer is to determine required wall thickness of his piping system. This procedure is the same as in the past, and for basic wall thickness requirements NB-3600 presents few new rules. The flow requirements and nominal pipe size must be given in the design specification, and from this information and the design pressure the minimum required wall thicknesses can be determined using the standard thickness equations of NB-3640.

2. Standard fittings. In the case of fittings purchased and used in accordance with the approved standards and pressure ratings of Table NB-3691.1, no minimum thickness analysis is required except the designer must provide assurance that elbows manufactured in accordance with ANSI-B16.28 shall have a minimum thickness in the crotch region 20% greater than required.

3. Pipe bends. Pipe bend wall thickness, after bending, must meet the minimum wall thickness calculation. A table of suggested minimum thickness prior to bending is given in Table NB-3642.1-1. These values are based on experience and good shop practice, but do not assure satisfaction of the minimum wall thickness requirements. The designer is cautioned to discuss this with the fabricator to ensure that the requirements are met. Failure to meet the minimum wall thickness requirements could result in the designer having to perform a detail analysis of the pipe in accordance with the rules of NB-3200 or develop stress indices for the bend to be used in the applicable equations of NB-3600.

4. Intersection. Intersections purchased in accordance with the applicable standards of Reference 1 are acceptable, all others must be analyzed in accordance with NB-3640. The rules of this subdivision are based on the area replacement technique. That is, if a portion of pipe material which carries membrane stress is removed, it must be replaced in close proximity to the area of removal. The amount of metal to be replaced, and the location or limits within which it must be provided, are described in detail for both welded branch connections and extruded outlets.

5. Special rules. Special rules are also provided for closures, blanks, reducers, and flanges. One item requiring careful attention by the designer is the flange. The rules for flange design differ from past practice, particularly as applied to bolting. The loadings that must be considered when analyzing flange bolting are preload, pressure, differential thermal expansion of the mating flanges and the bolts, and expansion moments and forces.

With respect to mitered joints, special rules are provided which pertain to minimum thickness and geometry; however, no stress indices are available, and the use of miters would require development of those indices by the designer.

10.6-4 Protection Against Membrane Failure

Stresses

Basic protection for internal pressure has been provided as discussed in Section 10.6-3. However, since membrane protection requires the consideration of all non-self-limiting loads, the solution of Equation (10.6-2) requires the analyst to have determined the moments resulting from all non-self-limiting loads. These loads must be defined in the design specification as "design loads" and include weight, inertial earthquake, water hammer, and relief valve thrusts. In examining Equation (10.6-2) one finds two terms: pressure and bending moment. It is only necessary to calculate the stress in an equivalent straight pipe:

$$\frac{PD_0}{2t} + \frac{M_i D_0}{2I} \tag{10.6-1}$$

and multiply these results by the applicable B_1 and B_2 index factors to find the stress in the fitting under consideration.

Stress Indices

The indices B_1 and B_2 are mechanical load (L) indices; this load L could be pressure or moment or some form of concentrated load. A stress index is a ratio, σ/s, of a significant stress in a piping component to some nominal stress in that component due to the same load L. The nominal stress should be proportional to the load L and should be simple to calculate.

Stress indices have been used in ANSI-B31.1 since 1955 where they were called stress intensification factors. These stress intensification factors are still used in nuclear class 2 and 3 piping. The vessel rules also use stress indices for determining pressure stresses in nozzles. The B indices have a different scope than the stress intensification factors of B31.1 and class 2 or 3 piping in that they are based on a limit load type of analysis. The use of B indices has been structured to provide a conservative limit load restriction. In addition, strain-hardening effects are ignored and for most ferritic materials allowable stresses are limited by one-third ultimate strength rather than two-thirds yield strength.

$$B_1\frac{PD_0}{2t} + B_2\frac{M_i D_0}{2I} \le 1.5S_m \tag{10.6-2}$$

where B_1, B_2 = primary stress indices for the specific product under investigation (NB-3680)
P = design pressure, psi
D_0 = outside diameter of pipe, in. (NB-3683)
t = nominal wall thickness of product, in. (NB-3683)
I = moment of inertia, in.[4] (NB-3683)
M_i = resultant moment due to a combination of design mechanical loads, in.·lb. All design mechanical loads, and combinations thereof shall be provided in the design specification. In the combination of loads all directional moment components in the same direction shall be combined before determining the resultant moment (i.e., resultant moments from different load sets shall not be used in calculating the moment M_i). If the method of analysis for earthquake or other dynamic loads is such that only magnitudes without relative algebraic signs are obtained, the most conservative combination shall be assumed.
S_m = allowable design stress intensity value, psi.

Loadings

The calculation of M_1, M_2, and M_3 should be done algebraically when the sign of the earthquake moments is known. If the method of earthquake analysis is such that only magnitudes (without relative algebraic signs) are obtained, the most disadvantageous combination must be assumed. An example of this follows for the case where relative algebraic signs are known for the earthquake load:

Moment	Weight	Earthquake Load	Total	
M_1	+13,000	±8,500	+21,500	+4,500
M_2	−14,000	±2,000	−12,000	−16,000
M_3	−1,000	±350	−650	−1,350

For the case where relative algebraic signs are not known for the earthquake load:

Moment	Weight	Earthquake Load	Total
M_1	+13,000	8,500	+21,500
M_2	−14,000	2,000	−16,000
M_3	−1,000	350	−1,350

The case of known algebraic signs for earthquake load is obvious and the value of $M_i = 24,630$ is the value to be used in Equation (10.6-2). The case where algebraic signs of earthquake load is not known is not as obvious. The requirement is that the most disadvantageous combination must be assumed. In the total column the moments are added and the signs of the weight moments are used, the value of $M_i = 26,830$. Since Equation (10.6-2) only requires the use of single amplitude of earthquake load, then taking the range of earthquake moment is not required as it will be for Equations (10.6-3) and (10.6-6).

It is important to recognize that Equation (10.6-2) is provided to prevent failure from a single application of loading. For this reason the analyst must know what loadings can cause such a failure and which of these occur together. Under the stresses column above a list of some of the loads that must be considered are given; however, the design specification must provide these and must also define the load combinations to be considered. The analysis of a given piping system could require more than one solution of Equation (10.6-2) in order to consider all the load combinations specified without being overly conservative. For example, consider a main steam system subjected to the following loadings:

Weight
Earthquake
Turbine Trip
Relief Valve Operation
    Initial
    Steady state

The analyst could consider all of these acting together and provide the required support system and piping geometry to accommodate the stress limit of Equation (10.6-2). This would be overconservative. The design specification should provide proper combinations based on the probability of an event occurring, the probability of an event triggering another event, the time over which an event occurs, and so on.

Weight is always present and must be considered at all times. Let us assume an earthquake occurs (recognize that the magnitude of the specified earthquake is probable in the life of the plant). It is probable that the earthquake will trip the turbine stop valves generating a pressure wave in the steam line, pressure will build, and the relief valves will blow. If we look at the time over which these events occur a number of combinations could be developed.

1. Weight + earthquake + turbine trip
2. Weight + turbine trip + relief valve initial
3. Weight + relief valve steady state

This specification would require the solution of Equation (10.6-2) for three different load combinations.

10.6-5   Protection Against Fatigue Failure

General

Some discussions of the philosophy behind the equations of NB-3650 and the applicable stress indices is appropriate.

In the formulation of design rules two questions had to be answered: (i) How can the design philosophy of Section III, ASME boiler and pressure vessel code, be utilized without completely confusing the average piping designer who was never required to analyze discontinuity and thermal stress by past piping codes? (ii) How can assurance be provided that the stresses in a specific fitting or pipe are adequate without requiring an expensive analysis for a relatively inexpensive piece of equipment? Both of these questions address themselves to the need for a simplified solution that will cover all variables.

The simplified solution provided covers all loading and geometry variables in a conservative manner as follows: (1) The stresses due to all loading conditions are considered to be additive in all cases; (2) the stress indices (B, C, and K factors) are assumed to have the maximum possible value for a given fitting subjected to a given load. Actually, the maximum stress resulting from two separate loading conditions could be 90° apart in a given fitting. For example, in an elbow the maximum stress due to in-plane bending occurs at the side of the elbow and the maximum stress due to pressure occurs at the crotch. The equations of NB-3650 assume that they occur at the same location in the fitting and are additive.

Simplification of analysis is provided by the use of stress indices. Stress indices are the ratio of a particular stress to a nominal stress. In the case of NB-3600, the nominal stress is that for a straight pipe having the same diameter and wall thickness of the specific component under consideration. The stress index corrects this stress to the maximum value of stress known to exist for a given component under consideration subjected to an applied load.

For example, the theoretical solution of stresses in an elbow subjected to one loading condition is an arduous task. Multiply this by the number of applied loads and the number of elbows in a system and the effort is staggering. However, since the solutions were available, it was a much simpler procedure to provide a factor which, when multiplied by the nominal stress, gives a conservative estimate of the stress in the elbow. This meant the long, theoretical solution would only have to be performed once; that is, to develop the stress index.

The C indices for moment loading are closely tied in with the ANSI-B31.1 and class 2 and 3 stress intensification factors. The C indices are close to double the intensification factors, which is nothing new since the intensification factor predicts about one-half of the elastic stress. The C indices for pressure and temperature are discussed in detail below. They are essentially predictors of the primary plus secondary elastic stress in a piping component resulting from pressure or thermal loading.

NB-3650 provides protection against two types of fatigue failure: fatigue failure in which the gross structure is subjected to elastic cycling and fatigue failure in which the gross structure is subjected to plastic cycling.

The designer must determine whether the structure cycles elastically or inelastically. This is accomplished by meeting the requirements of Equation (10.6-3). The shakedown criterion as embodied in Equation (10.6-3) states that the maximum primary plus secondary stress intensity range, exclusive of stress concentration effects, be less than $3S_m$. Briefly, the shakedown criterion requires that after a few cycles of load application, the maximum stress will cycle within the range of tensile yield strength and compressive yield strength and therefore be subjected to elastic cycling. Satisfaction of this criterion allows the calculation of stresses assuming completely elastic behavior.

$$S_n = C_1 \frac{P_0 D_0}{2t} + C_2 \frac{D_0}{2I} M_i + C_3 E_{ab} |\alpha_a T_a - \alpha_b T_b| \leq 3S_m \qquad (10.6\text{-}3)$$

where C_1, C_2, C_3 = secondary stress indices for the specific component under investigation (NB-3680)
D_0, t, I, S_m = as defined for Equation (10.6-2)
M_i = resultant range of moment which occurs when the system goes from one service load set to another, in.·lb. Service loads and combinations thereof shall be provided in the design specification. In the combination of moments from load sets, all directional moment components in the same direction shall be combined before determining the resultant moment (i.e., resultant moments from different load sets shall not be used in calculating the moment range M_i). Weight effects need not be considered in determining the loading range since they are noncyclic in character. If the method of analysis is such that only magnitudes without relative algebraic signs are obtained, the most conservative combination shall be assumed. If a combination includes earthquake effects, M_i shall be either: (1) the resultant range of moment

due to the combination of all loads considering one-half the range of the earthquake or (2) the resultant range of moment due to the full range of the earthquake alone, whichever is greater.

$T_a(T_b)$ = range of average temperature on side $a(b)$ of gross structural discontinuity or material discontinuity, °F. For generally cylindrical shapes, the averaging of T (NB-3653.2) shall be over a distance of $\sqrt{d_a t_a}$ for T_a and over a distance of $\sqrt{d_b t_b}$ for T_b.

$d_a(d_b)$ = inside diameter on side $a(b)$ of a gross structural discontinuity or material discontinuity, in.

$t_a(t_b)$ = average wall thickness through the length $\sqrt{d_a t_a}$ ($\sqrt{d_b t_b}$), in. A trial and error solution for t_a and t_b may be necessary.

$\alpha_a(\alpha_b)$ = coefficient of thermal expansion on side $a(b)$ of a gross structural discontinuity or material discontinuity, at room temperature, $1/$°F.

E_{ab} = average modulus of elasticity of the two sides of a gross structural discontinuity or material discontinuity at room temperature, psi

P_0 = range of service pressure, psi

The first two terms are familiar from Equation (10.6-2), the only differences being the B indices are replaced by C indices and the value of M_i changes as defined below. In the case of Equation (10.6-3), all terms predict the maximum range of primary and secondary stress resulting from the specific load associated with that term (P—pressure, M_i—bending moment, $T_a - T_b$—temperature). Since primary and secondary stresses are calculated here, all self-limiting loads that cycle must be considered.

It is important to recognize that Equation (10.6-3) deals with ranges of loads; that is, each term represents a load range and the summation of these load ranges represents a stress range. In design by analysis the calculation of stresses with respect to time is required; here in NB-3600 the calculation of loads with respect to time is also required. This comes about because of our stress index approach.

Looking at Fig. 10.6-1 we see the maximum stress range for a specified set of operating conditions. This is obtained by taking the difference between the algebraic maximum and the algebraic minimum stress intensity at any given time. For NB-3600 we take the values of pressure $\Delta T_1, \Delta T_2, T_a - T_b$ versus time for a specified set of operating conditions and find the difference between the algebraic maximum and the algebraic minimum values of these loads. These values are then used in Equations (10.6-3) and (10.6-6) and a stress range for each load range is calculated, so that the stress ranges added together provide a value of S_p equivalent to that obtained using NB-3200. This calculation of load ranges must be done for each and every possible combination of operating conditions. For three operating conditions this is relatively simple, that is, combine 1 and 2, 1 and 3, 2 and 3; for 20 operating conditions the task is much more difficult.

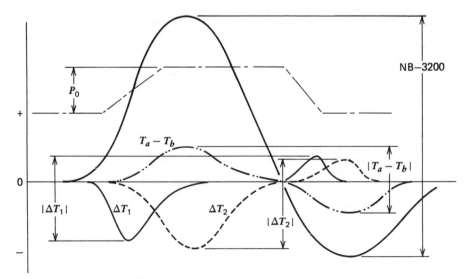

Fig. 10.6-1 Maximum stress range for a specified set of operating conditions.

Pressure

The loads that result from pressure, that are self-limiting, and are considered by the C index of Equation (10.6-3) are loads that occur at a structural discontinuity. For example, consider a piece of straight pipe joined to another pipe with dissimilar cross section. The assembly will be subjected to discontinuity stresses at the junction because the free radial deflection of one type of cylinder under pressure is different from that of the other, resulting in a restraint on free deformation. This restraint produces forces and moments at the discontinuity that are necessary to provide compatibility of the structure. The discontinuity loads are self-limiting in nature since they are a function of the relative restraint of adjacent members rather than the imposed loading which produces the deflection. This condition is shown in Fig. 10.6-2.

The loads H_1, H_2, M_1, and M_2 are the discontinuity reactions required to maintain compatibility of point A at the junction.

The calculation of the redundant loads is performed as follows:

$$u = \frac{H}{2D\lambda^3} + \frac{M}{2D\lambda^2} + \frac{PR^2}{Et}\left(1 - \frac{\nu}{2}\right) \qquad (10.6-4)$$

$$\beta = \frac{H}{2D\lambda^2} + \frac{M}{D\lambda} \qquad (10.6-5)$$

where u = radial deflection
β = rotation
H = shear load, lb/in.
M = moment, in.·lb/in.
R = inside radius, in.
t = pipewall thickness, in.
ν = Poisson's ratio = 0.3
P = internal pressure, psi
λ = pipe attenuation length

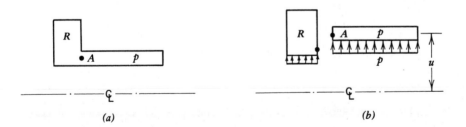

(a)   (b)

u = radial deflection for a semi−infinite cylinder

$[u = PR^2/ET\,(1 - \frac{\nu}{2})]$

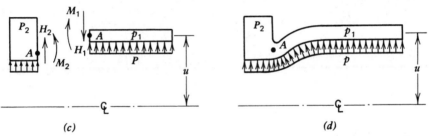

(c)   (d)

Fig. 10.6-2 Sample determinations of C_1 factors. (a) Unloaded state; (b) free deflection of each piece as though not connected; (c) loadings generated at A since P_1 and P_2 are connected; (d) actual deflected shape.

The same equations are used for both types of pipes since the deflections and rotations must be equal:

$$u \text{ pipe } 1 = u \text{ pipe } 2$$

$$\beta \text{ pipe } 1 = \beta \text{ pipe } 2$$

Solving the above two simultaneous equations will yield the two unknowns H and M. Since stress indices are provided, this type of solution is not required for standard components as the C_1 indices account for this loading.

It is interesting to note that the influence of discontinuity loads (and, in fact, all self-limiting loads) is local within the region of the discontinuity and dies away along the pipe. At a distance from the discontinuity where the pipe radial deflection is the same as that for a semi-infinite (free) cylinder (see Fig. 10.6-2), the discontinuity reactions have negligible influence.

Bending Moments

The value of M_i includes all cyclic non-self-limiting and self-limiting external moments. The discussion here will deal with the differences in the type of moments generated by non-self-limiting and self-limiting loads. Consider first the moments generated by the weight of the piping system and its contents. These loads are always present and can only be accommodated by the amount of material provided in the pipe. If there is insufficient material, the loads will continue to produce deformation of the structure until failure occurs. Changing the amount of material or the design does not change the applied load but rather changes the load's effect on the structure. However, since weight loading is a constant (noncyclic) load, it is not necessary to consider its effect on shakedown and fatigue. The effect of weight stress on the fatigue life of a structure is that of changing the mean stress about which the alternating component of stress cycles from the zero value. Changing the mean stress to some value other than zero has a deleterious effect on the fatigue life; however, the fatigue curves used in Section III are adjusted for the maximum effect of mean stress. That is, the fatigue curves used assume the maximum mean stress possible is present, whether or not the component under consideration is subjected to any mean stress. There is no way the structure can reduce or change the value of the applied load if it is non-self-limiting.

Self-limiting loads are quite different and are actually a secondary effect as in the case of loads at a discontinuity resulting from pressure loading. Consider a piping system that is heated to some temperature above ambient. In an unrestrained state the system will undergo free expansion. However, since the system is anchored, or restrained, loads are generated in the system to keep it in the desired position. The loads then are a function of the amount of restraint on the free deflection, and the system must accommodate restraint of deflection rather than applied moments.

Temperature

The temperature term of Equation (10.6-3) produces stresses resulting from self-limiting loads produced by thermal gradients.

The temperature term $[C_3 E_{ab}(\alpha_a T_a - \alpha_b T_b)]$ describes the stress resulting from temperature differences occurring in adjacent parts, or from a difference in the coefficient of thermal expansion (α) of adjacent parts, or both. An example of this phenomenon would be the cylinder–cylinder junction discussed under pressure loading. In the process of heating the system, the mean temperature of the thinner cylinder becomes 480°F, and the mean temperature of the thickest cylinder becomes 300°F and discontinuity thermal stresses will occur. The thinner cylinder wants to expand in the radial direction some quantity $R\alpha\Delta T$; the other wants to expand also, but not as much since it is 180°F cooler. The thicker cylinder imposes a restraint on the thinner, generating internal forces and moments to provide compatibility of the structure just as in the pressure case. For two different materials, at a bimetallic joint for example, the same explanation holds, even when both are at the same temperature, since the different alphas (α's) result in different values of radial deflection. These solutions are identical to those for the pressure case of Fig. 10.6-2; the only change is that the free thermal radial deflection ($R\alpha\Delta T$) replaces the $PR^2/Et(1 - \nu/2)$ term.

The C_3 factor for axial geometric discontinuities given in NB-3600 is 1.8. This value is derived from the case of a completely built-in cylinder. In reality there are no standard components that introduce this type of restraint on a pipe. The variation of the factor that actually occurs for longitudinal stress (though not recognized by the code) as a function of the thickness ratio of adjacent parts is shown in Fig. 10.6-3.

Fatigue

Fatigue failure occurs when the maximum stress from all loads and displacements is concentrated at a point and continued cycling of the stress produces a crack which propagates through the material and results in a leak.

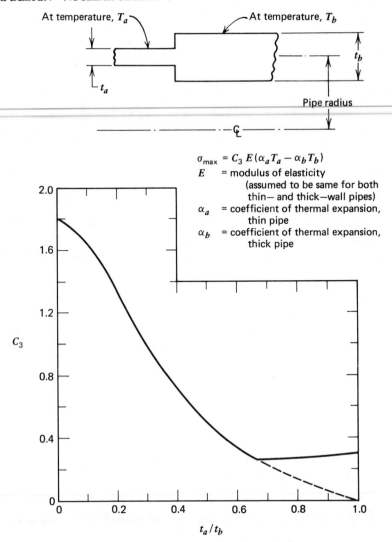

$$\sigma_{max} = C_3 E(\alpha_a T_a - \alpha_b T_b)$$

E = modulus of elasticity
(assumed to be same for both
thin— and thick—wall pipes)

α_a = coefficient of thermal expansion,
thin pipe

α_b = coefficient of thermal expansion,
thick pipe

Fig. 10.6-3 Effect of relative thickness on C_3 factor.

Elastic Fatigue Analysis

The following equation from Section III provides the designer with a tool for calculating the peak stress intensity.

$$S_p = K_1 C_1 \frac{P_0 D_0}{2t} + K_2 C_2 \frac{D_0}{2I} M_i + \frac{1}{2(1-\nu)} K_3 E\alpha|\Delta T_1|$$

$$+ K_3 C_3 E_{ab}|\alpha_a T_a - \alpha_b T_b| + \frac{1}{1-\nu} E\alpha|\Delta T_2| \tag{10.6-6}$$

where K_1, K_2, K_3 = local stress indices for the specific component under investigation (NB-3680)

$E\alpha$ = modulus of elasticity (E) times the mean coefficient of thermal expansion (α) both at room temperature, psi/°F

ν = Poisson's ratio = 0.3

$|\Delta T_2|$ = absolute value of the range for that portion of the nonlinear thermal gradient through the wall thickness not included in ΔT_1 as shown below, °F

$|\Delta T_1|$ = absolute value of the range of the temperature difference between the temperature of the outside surface T_0 and the temperature of the inside surface T_i of the piping product assuming moment generating equivalent linear temperature distribution, °F

The peak stress intensity at a point is merely the range of stress intensity calculated at that point including any local structural discontinuity (notch) effects plus any local thermal stresses that are present. Examination of Equation (10.6-6) above indicates that K values have been added to the terms that appear in Equation (10.6-3) to represent notch effects and two new temperature terms have been added to the equation. The K values are stress indices that predict the notch effects and are generally elastic stress concentration factors.

The two thermal stress terms are a function of the stress generated due to the radial gradient through the wall. ΔT_1 is the linear portion and ΔT_2 is the nonlinear portion. Some discussion concerning ΔT_1 and ΔT_2 is appropriate at this point since the NB-3600 rules revised the treatment of ΔT_1 in the Summer 1979 Addenda. The inclusion of ΔT_1 in the secondary stress category was done initially during the formulation of requirements for ANSI-B31.7.

The rules for vessels at that time (NB-3200) did not include the radial gradient in the secondary category. The vessel rules limited the magnitude of the radial gradient as it affects ratcheting and then considered it only in fatigue. All other cyclic load stresses were limited to $3S_m$ unless a detailed elastic-plastic analysis was performed. When the rules for piping were adopted in NB-3600 the vessel people agreed, although "grudgingly," to also include the linear portion of the radial gradient in the secondary stress category. This resulted in a number of situations where $3S_m$ was exceeded and a simplified approach to elastic-plastic analysis was adopted. Equations (10.6-7) and (10.6-9) of NB-3600 and NB-3228.3 provide this simplified approach.

One of the major impacts of a radial gradient on shell-type structures is that ratcheting of the structure can occur and failure from this effect is possible. Based on the magnitude of the gradient, a ratchet-type failure can occur in a fewer number of cycles than that predicted by the fatigue rules. In design by analysis, NB-3200, this is addressed in NB-3222.5 by limiting the linear and parabolic variation of temperature through the wall. In NB-3600 this rule did not exist.

NB-3213.9 provides the following definition: "Secondary stress is a normal stress or a shear stress developed by the constraint of adjacent material or by self-constraint of the structure. The basic characteristic of a secondary stress is that it is self-limiting. Local yielding and minor distortions can satisfy the conditions which cause the stress to occur and failure from one application of the stress is not to be expected. Examples of secondary stresses are:

General thermal stress (NB-3213.12(a));
Bending stress at a gross structural discontinuity."

NB-3213.9(a) requires us to review NB-3213.13(a) for thermal generated secondary stresses.

NB-3213.13(a) provides the following definition: "General thermal stress is associated with distortion of the structure in which it occurs. If a stress of this type, neglecting stress concentrations, exceeds twice the yield strength of the material, the elastic analysis may be invalid and successive thermal cycles may produce incremental distortion. Therefore this type is classified as secondary stress in Table NB-3217.1. Examples of general thermal stresses are:

1. Stress produced by an axial temperature distribution in a cylindrical shell;
2. Stress produced by the temperature difference between a nozzle and the shell to which it is attached;
3. The equivalent linear stress produced by the radial temperature distribution in a cylindrical shell."

We must recognize that item 3 above, was added to comply with the rules that placed ΔT_1 in the secondary category. That is, the definition came after the fact, not before.

The calculation of alternating stress (S_{alt}) for cases where $S_n > 3S_m$ (i.e., plastic cycling is occurring) requires the use of a K_e factor. The present K_e factors are conservative, particularly as the $S_n/3S_m$ ratio becomes larger. This resulted in extremely conservative estimates of the cumulative usage factor (U). The inclusion of ΔT_1 in the secondary stress category increases the calculated S_n, resulting in a significant number of conditions for which K_e must be applied to S_{alt}. This created some problems in the industry in the past but none worth changing requirements about, that is, until the Nuclear Regulatory Commission published pipe rupture criteria based on the calculated value of S_n and the cumulative usage factor U.

A technique that would eliminate some of the conservatism associated with the calculation of U, thereby reducing the number of postulated break locations, was deemed necessary since there is substantial feeling that pipe whip restraints can impair normal operation of piping systems, increase radiation exposure of personnel, and hinder effective in-service inspection.

The most effective way to do this, at this stage of evolution of code requirements, was to remove ΔT_1 from the secondary stress category (S_n) for piping, which is a return to the Section III requirements that existed for vessels prior to the inclusion of piping design rules in that code.

Elastic-Plastic Fatigue Analysis

When the calculated value of S_n from Equation (10.6-3) exceeds $3S_m$, the procedure of fatigue analysis must include the effects of plastic cycling. This procedure has been provided in a simplified manner in NB-3653.6.

Since stresses are allowed to exceed shakedown limits, the control of thermal expansion stresses was required. The requirements of Equation (10.6-7) provide this control and essentially limit thermal expansion stresses to a level comparative to class 2, 3, and B31.1 rules. Equation (10.6-7) requires that the resultant range of moment (M_i) from thermal expansion and anchor movements does not produce a primary plus secondary stress which exceeds $3S_m$.

$$S_e = C_2 \frac{D_0}{2I} M_i \leq 3S_m \tag{10.6-7}$$

where S_e = nominal value of expansion stress, psi
  M_i = same as M_i in Equation (10.6-3), except that it includes only moments due to thermal expansion and thermal anchor movements, in.·lb

Having satisfied the control of thermal expansion stresses, the code requires compliance with the thermal stress ratchet rules of NB-3653.7. These rules provide protection against failure due to ratcheting that could occur when large radial gradients are imposed on a structure subjected to a primary membrane stress. This ratcheting failure could occur quicker than that predicted by fatigue. The value of the range of ΔT_1 cannot exceed that calculated as follows:

$$\Delta T_1 \text{ range} \leq \frac{y' S_y}{0.7 E \alpha} C_4 \tag{10.6-8}$$

where $y' = 3.33, 2.00, 1.20,$ and 0.80 for $x = 0.3, 0.5, 0.7,$ and 0.8, respectively
  $x = (PD_0 / 2t)(1/S_y)$
  P = maximum pressure for the set of conditions under consideration
  $C_4 = 1.1$ for ferritic material
    $= 1.3$ for austenitic material
  $E\alpha$ = as defined for Equation (10.6-6)
  S_y = yield strength value, psi, taken at average fluid temperature of the transient under consideration

Next the requirements of Equation (10.6-9) must be met. These requirements limit the primary plus secondary membrane plus primary bending stress intensity, excluding any thermal bending and thermal expansion stress to $3S_m$.

$$C_1 \frac{P_0 D_0}{2t} + C_2 \frac{D_0 M_i}{2I} + C_3' E_{ab} |\alpha_a T_a - \alpha_b T_b| \leq 3S_m \tag{10.6-9}$$

where M_i = as defined in NB-3652, and all other variables are as defined in NB-3653
  C_3' = values in Table NB-3681 (a)-1

Now the alternating stress intensities are calculated using

$$S_{\text{alt}} = \tfrac{1}{2} K_e S_p \tag{10.6-10}$$

where S_{alt} = alternating stress intensity, psi
  S_p = peak stress intensity value calculated by Equation (10.6-6), NB-3653.2, psi

where

$$K_e = 1.0 + \frac{1-n}{n(m-1)} \left(\frac{S_n}{3S_m} - 1 \right) \quad \text{for} \quad 1.0 < \frac{S_n}{3S_m} < m \tag{10.6-11}$$

and

$$K_e = \frac{1}{n} \quad \text{for} \quad \frac{S_n}{3S_m} \geq m \tag{10.6-12}$$

TABLE 10.6-1   VALUES OF m, n, AND T_{max} FOR VARIOUS CLASSES OF PERMITTED MATERIALS

Materials	m	n	T_{max}, °F
Carbon steel	3.0	0.2	700
Low-alloy steel	2.0	0.2	700
Martensitic stainless steel	2.0	0.2	700
Austenitic stainless steel	1.7	0.3	800
Nickel–chromium–iron	1.7	0.3	800
Nickel–copper	1.7	0.3	800

where S_n = primary plus secondary stress intensity value calculated in Equation (10.6-3), NB-3653.1, psi

m, n = material parameters given in Table NB-3228.3(b)-1

Value of the material parameters m and n are given for the various classes of code materials in Table 10.6-1. This equation provides for the decay on fatigue life when cycling in the plastic range.

Having found the S_{alt} values for each transient condition, for those transients that produce plastic cycling, and taking $\frac{1}{2}S_p$ as calculated in Equation (10.6-6) for those transients which produce elastic cycling, the designer follows the rules for calculating the cumulative usage factor described in NB-3653.4 and NB-3653.5.

10.6-6   Application of Rules

Having discussed the rules of NB-3600, we should become familiar with their application. It is only then that we can see the detail required and the engineering disciplines needed to generate a class 1 stress report and to provide the information necessary to fabricate the piping system.

Some major points of consideration in arranging the piping system are applicable here since they affect stress levels but are not really accounted for in the code rules.

System Unbalance

Strains calculated on an elastic basis are sufficiently accurate for systems in which there are no severe plastic strain concentrations. However, elastic calculations fail to reflect the actual strain distribution in unbalanced systems where only a small length of the piping undergoes plastic strain, while the major portion of the length remains essentially elastic. In these cases, the weaker or higher stressed portions will be subjected to plastic strain concentrations because of the elastic follow-up of the stiffer or lower stressed portions of the piping. That is, the imposed deflection on the piping system will be absorbed almost entirely in the weaker portion of the system, and the remainder of the system will remain unchanged.

Such unbalance can be produced by:

1. Use of small pipe runs in series with larger or stiffer pipe with the small pipe relatively lightly stressed.
2. Local reduction in size of a cross section or local use of a weaker material.
3. In a system of uniform size, by use of a configuration for which most of the piping lies near a straight line drawn between the anchors or terminals with only a very small portion that projects away from this line and absorbs most of the expansion strain.

Conditions of this type should be avoided particularly where materials of relatively low ductility are used.

Longitudinal Loads

Direct stress resulting from longitudinal loads is not calculated in NB-3600 since, in a normal configuration, the moment loading due to the longitudinal load produces bending stresses so large that in comparison, the direct stresses become insignificant. A case could develop, however (although unlikely), in which the direct stress resulting from the longitudinal load becomes significant. For example, a straight line between two anchor points, which when heated, produces zero bending moments but very high longitudinal compressive loads. The designer is cautioned to check his end loadings to determine the magnitude and effect of the calculated loads.

Support and Vibration Control—General

There are no specific methods provided for support and/or vibration control of a piping system; however, NB-3622.3 specifies that the designer shall be responsible for ensuring that vibration of piping systems is within acceptable levels. Experience indicates that a major source of piping failure has been due to system vibration, particularly in small piping. The designer must be more than casually concerned with this problem in developing the support system and during preoperational testing should conduct on-site checking for excessive vibration. Operating personnel should report excessive piping system vibration during plant operation as soon as this condition is discovered. In addition, the following guidance with respect to support and hanger location can be helpful in resolving problems.

Support System

A procedure for performing the design of a support system is presented here. The initial flexibility analysis provides the designer with deflections of the piping system at any point. This information will be used to determine whether a spring or solid-type hanger will be used when the location of the support is determined.

Hanger Spacing

Table 10.6-2, taken from ANSI-B31.1, provides the designer with an adequate guideline for his first hanger spacing selection. Table 10.6-2 is based on the equation $S = 1.2 \, (wl^2/Z)$, where S is the maximum bending stress (psi), Z is the section modulus (in.^{3}), l is the pipe span between supports (ft), and w is the total unit weight (lb/ft).

The table values assume the value of $S = 1500$ psi, but the designer can vary this value. However, he is cautioned not to choose a high S value since it must be included with pressure and other effects in meeting the allowable stresses of the code as discussed in Section 5.2.

Hanger Location Considerations

Using Table 10.6-2, the designer has arrived at the first hanger spacing. However, the location of the actual supports includes some additional considerations such as the piping itself, adjacent structure, calculated deflections, and accessibility. Some major items of consideration are:

1. Locate supports as close as possible to heavy concentrated loads on the piping system imposed by valves, flanges, minor vessels, and so on. However, the designer must avoid attaching the support directly to a component (valve, strainer, etc.) that structurally is not his responsibility without checking with the component supplier to determine acceptability of his support.

2. Locate supports on straight runs rather than bends, elbows, or tees. The latter components are usually the most highly stressed portions of the system, and any additional restraint at these locations should be avoided. In relation to elbows and bends, the localized restraint of welded attachments will

TABLE 10.6-2   HANGER SPACING FOR PIPE SUPPORT

Normal Pipe Size (in.)	Suggested Maximum Span Between Supports (ft)	
	Water	Steam, Gas, or Air
1	7	9
2	10	13
3	12	15
4	14	17
6	17	21
8	19	24
12	23	30
16	30	35
20	32	39
24	32	42

reduce flexibility of these components and require experimental determination of the stress index and flexibility factor.

3. Ensure that the structure used as a foundation of the hanger can carry the load imposed by the piping. The ideal situation is to attach supports to large structural members such as columns or trusses. It may be necessary to provide additional intermediate steel reinforcement in order to have an adequate foundation. Additional steel reinforcement should be provided judiciously to eliminate interferences with other piping, equipment, electrical cables, and so on, and the steel should be checked by the designer to ensure its adequacy for carrying the imposed loading and to determine the deflection of the steel at the point of support attachment. In the final analysis the designer responsible for the building structure should approve all support attachments and/or additional foundations. The use of steel baseplates and expansion concrete anchor bolts requires caution by the designer. The manufacturers load capacities and spacing must be followed or the support capacity will be reduced. The use of thin, flexible plates should be avoided since the anchor bolt load will be increased by plate bending.

4. Ensure that hangers, which require examination after installation, are acceptable for the required examination. Avoid locating supports on sections of piping that require periodic removal. This will eliminate the need for temporary support of the adjacent piping during removal.

5. Whenever possible, locate unidirectional supports at points of zero or minimum deflection in the direction of the support. This will eliminate the need for springs and allow for the use of rigid supports. There obviously must be some decisions made as to this requirement, in relation to that specified in item 1 above, the most critical being the concentrated load situation of item 1. When rigid supports are used, their restraint on the deflection of the piping system in the other two directions (normal to the plane of the support) must be considered. Short strut-type hangers can provide considerable restraint in these other directions.

Once the support locations and types for the system have been determined, the weight, dynamic and thermal flexibility analyses may be performed. With respect to dynamic and flexibility analysis, it is important to recognize that the results of such analysis not only demonstrate the adequacy of the system but are of utmost importance to the pipe support designer. For this reason it is imperative that sufficient operating cases be investigated to assure that the full *range* of piping deflections have been considered. The support system must be designed for those conditions that are going to exist during plant operation, not those conditions some system designer feels like imposing for conservatism.

Criteria

There must be some interim criteria (guidelines) by which the analyst can determine the adequacy of this support system while performing each individual portion of the analysis. The analyst does not solve the system for all loads at one time, but rather deals with each condition separately, allotting a certain stress allowable for each condition so that when the last condition is dealt with the required combinations will not exceed the allowables. An example would be the use of the following guidelines, which generally ensure compliance with NC-3650 and Equation (10.6-2) of NB-3650 for a seismic category I system.

Weight stress   5,000 psi
Thermal expansion, range   20,000 psi
Seismic inertial effects   10,000 psi
Seismic end movements   7,000 psi
Dynamic loadings, $1.20 S_H$ (pressure stress and seismic stress)

Summary

Now that the system has been analyzed for all loading conditions specified, the analyst satisfies the equations of NC-3650 and Equation (10.6-2) of NB-3650 using the loading combinations specified in the design specification for all specified service limits. Having done this, the analyst can release the piping system for fabrication and transmit the required information to the support designer to initiate support fabrication. The support designer should receive as a minimum, the following:

Range of imposed forces and moments
Range of deflections and rotations
Stiffness of support and steel used in analysis
Type of support
Restrictions on pipe attachment

10.6-7 Temperature Distribution Analysis

Two general techniques are used to calculate the required values of temperature. The decision as to which technique is to be used is made by the analyst at the time of determination of ΔT's. The two methods will be described here in detail, and then some discussion as to when to use one or the other will follow.

Method 1: Simplified Analysis

This method is defined as simplified because it does not involve the use of computers and is relatively easy to apply.

A number of curves have been prepared that give the value of ΔT_1, ΔT_2, and mean temperature for both a ramp and a jump (step) change in temperature as a function of the Fourier (N_{Fo}) number and the Biot (N_{Bi}) number. The appropriate L and N values obtained from the curves are multiplied by the fluid temperature change to obtain the values of ΔT_1, ΔT_2 and mean temperature

$$N_{Fo} = \frac{\alpha t}{a^2} \tag{10.6-13}$$

and

$$N_{Bi} = \frac{ha}{k} \tag{10.6-14}$$

where $\alpha =$ thermal diffusivity of the pipe material (ft²/hr)
      $t =$ time from start of transient (hr)
      $a =$ pipe wall thickness (ft)
      $h =$ surface heat transfer coefficient (Btu/hr·ft²·°F)
      $k =$ thermal conductivity of pipe material (Btu/hr·ft·°F)

In order to make this approach more universal and simplified, it is possible to assume that the film coefficient for all situations is 5000 Btu/hr · ft² · °F which enables the plotting of Biot number versus pipe thickness for both austenitic and ferritic steels. This eliminates the need to calculate a film coefficient and subsequently a Biot number. The Fourier number may be plotted versus time for a range of pipe sizes. This is made possible by assuming that all pipe is schedule 160. The use of Btu/hr · ft² · °F for a film coefficient and schedule 160 pipe is conservative for most nuclear systems. Thermal response curves used for this "simplified analysis" are given in Figs. 10.6-4–10.6-7.

Method 2: Detail Analysis

Since the simplified procedure outlined as method 1 is conservative, it is frequently necessary to determine ΔT's more accurately. Some examples of when a detail analysis would be used are:

1. When axial heat flow is significant such as in a branch connection.
2. When the specified transient is so complex that use of the simplified procedure becomes too cumbersome. The procedure outlined in Section 10.6-7 is usually efficient when sufficient hold time is specified at either end of the transient. However, when hold time is short, followed by a second change in temperature, the simplified procedure can be burdensome.
3. When the simplified procedure results in ΔT values that produce stresses in excess of the allowable criteria.

Grid generators have been developed that prepare the model and geometric input automatically for both a one-dimensional and two-dimensional heat transfer problem. In the process of development of these generators it was determined that eight elements through the thickness were required in the region of interest to obtain accurate ΔT_1 and ΔT_2 data.

One-Dimensional Model. An example of a one-dimensional model is shown in Fig. 10.6-8. The actual field changes for a given transient are applied to boundary node 11, and the response of the pipe is a function of the pipe material properties and the heat transfer mechanism existing between the fluid (node 11) and the inside wall of the pipe (node 10). The ambient conditions are applied to boundary node 12. For "perfectly" insulated pipe the film coefficient applied to the surface node 9 is zero and no heat transfer occurs between the pipe wall and the external environment. For those cases for which there is no insulation or for which the insulation is not "perfect," a film coefficient is calculated that reflects the heat loss to the environment. This film coefficient is then applied at node 9.

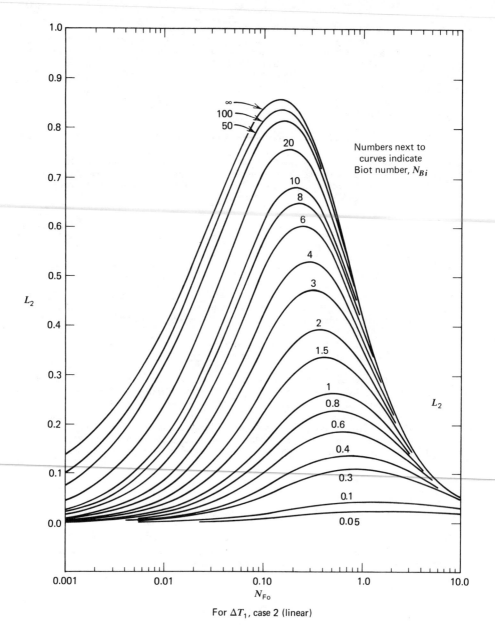

For ΔT_1, case 2 (linear)

Fig. 10.6-4 Thermal response curve for ΔT_1, linear.

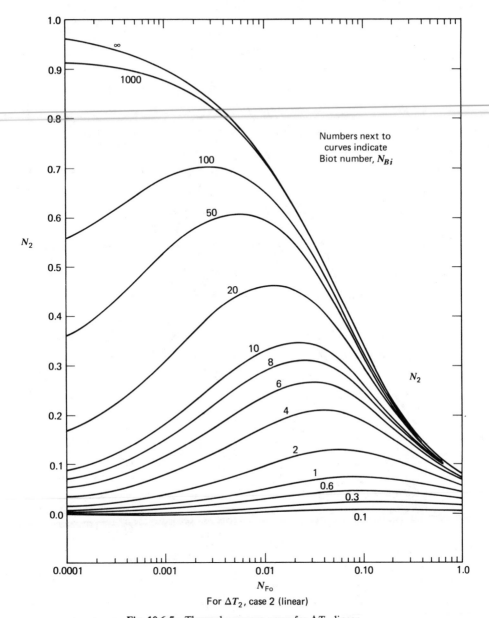

For ΔT_2, case 2 (linear)

Fig. 10.6-5 Thermal response curve for ΔT_2, linear.

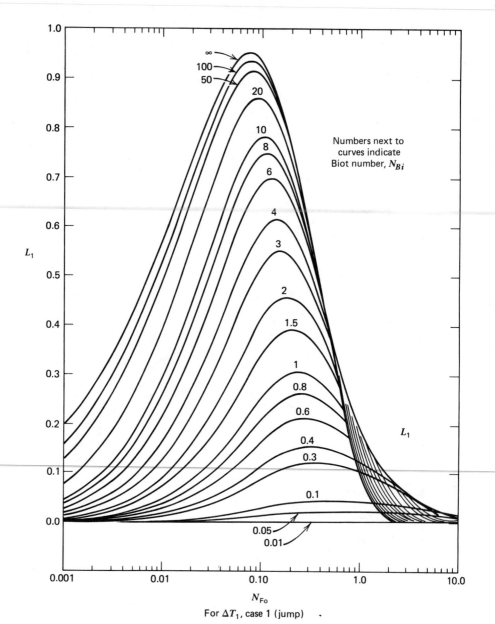

Numbers next to curves indicate Biot number, N_{Bi}

For ΔT_1, case 1 (jump)

Fig. 10.6-6 Thermal response curve for ΔT_1, jump.

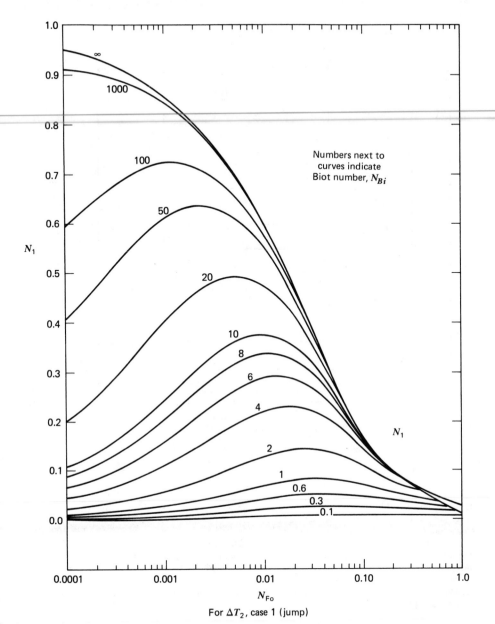

For ΔT_2, case 1 (jump)

Fig. 10.6-7 Thermal response curve for ΔT_2, jump.

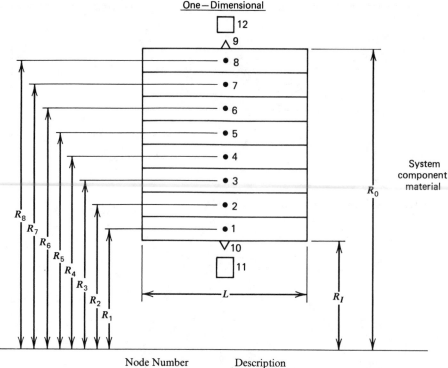

Node Number	Description
1 thru 8	Internal (pipe wall) nodes
9 and 10	Surface nodes
11 and 12	Boundary nodes

Fig. 10.6-8 Example of one-dimensional model.

Two-Dimensional Model. For the branch connection problem, two distinct models have been generated. The first model covers tees (full size and reducing), sweepolets, and fabricated branch connections (see Fig. 10.6-9). The second model covers half couplings and weldolets (see Fig. 10.6-10).

The same techniques concerning fluid changes, film coefficients, and so on are used here as in the one-dimensional model. The major differences are that the branch and run fluid can be varied independently and the effects of axial heat flow are included.

The values of ΔT_1 and ΔT_2 are taken on a diagonal through the crotch of the two-dimensional models. The values of T_a and T_b are taken in the run and branch pipe along a length equal to $\sqrt{Dt}$ as shown on both figures.

10.6-8   Results of Thermal Analysis

The procedures described for temperature distribution analysis are used to obtain appropriate ΔT_1, ΔT_2, and $T_a - T_b$ values for each of the specified thermal transients. Frequently, results for similar transients will be derived from a ratio of detailed results obtained for one of the transients. Sometimes less severe transients conservatively use ΔT values obtained for more severe transients.

REFERENCES

10.6-1 *Boiler and Pressure Vessel Code*, Section III, Nuclear Power Plant Components, Division 1, The American Society of Mechanical Engineers, New York, 1980.

10.6-2 B16.9, *Wrought Steel Butt Weld Fittings*, American National Standards Institute, New York, 1978.

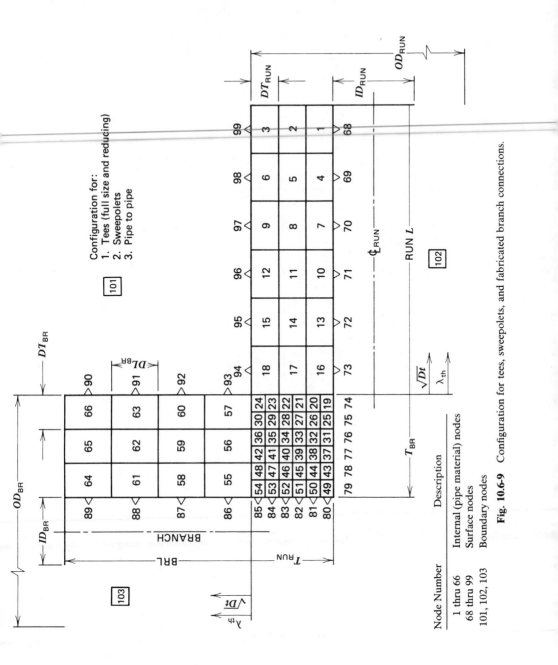

Fig. 10.6-9 Configuration for tees, sweepolets, and fabricated branch connections.

Node Number	Description
1 thru 66	Internal (pipe material) nodes
68 thru 99	Surface nodes
101, 102, 103	Boundary nodes

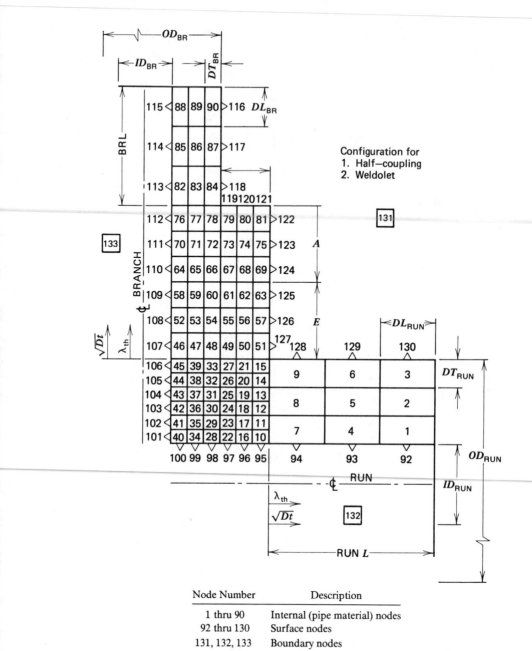

Fig. 10.6-10 Configuration for half-couplings and weldolets.

Node Number	Description
1 thru 90	Internal (pipe material) nodes
92 thru 130	Surface nodes
131, 132, 133	Boundary nodes

10.7 PIPING DESIGN—NUCLEAR CLASS 2 AND CLASS 3

10.7-1 General

This section presents a discussion of the design rules for class 2 and 3 piping contained in Section III.[1] The basis for these rules had always been the established criteria of ANSI-B31.1[2] and still is today with very few exceptions. We will, therefore, review the background of these rules, their development, and discuss some examples of proper application.

Piping analysis criteria have been in existence since the issue of B31.1 in 1955. This document established rules for the design and analysis of piping systems and associated components. The basic tools used in piping analysis are piping component standards and stress intensification factors.

The first item, component standards, is critical to piping design and analysis. These standards serve two important functions. First, they provide general geometric control so that we can determine the response of a given fitting to a given load at one time, and use this knowledge each time we have to analyze that size fitting. This is basically the stress intensification factor (i) approach. A stress intensification factor is a ratio of σ/s, of a significant (σ) in a piping component due to moment M, to some nominal stress (s). In B31.1 the nominal stress is the moment divided by the section modulus of the pipe, M/Z. The analyst should now recognize that an i factor of 2 means the significant stress in the component is twice the nominal stress in an equivalent straight pipe.

The second function served by these standards is the requirement that the fitting has passed, or it be demonstrated that it is capable of passing, a burst test. This allows the acceptance of standard fittings such as tees for pressure design which would not always pass the reinforcement rules. The basic philosophy is since the fitting is acceptable at the burst pressure of the attached pipe then it is as good as the pipe.

10.7-2 Stress Categories

The existing B31 piping codes and associated standards provide two kinds of failure protection. The first being protection against catastrophic or burst-type failure. This is provided by requiring, as a minimum, the following:

1. The use of standard fittings for which prototypes have been subject to (or can be demonstrated to meet) a pressure burst test.
2. The calculation of a minimum pipe wall thickness.
3. That certain fabricated branch connections meet established reinforcement rules.
4. That stresses due to sustained loads and occasional loads meet specific allowable stresses.

Although these requirements may not appear to be related, they most certainly are. In each case the code is requiring that sufficient material is available in the pipe or fitting wall to provide protection against a catastrophic failure of the system. It is important to recognize that protection cannot be assured unless all sustained loadings to which the system will be subjected are considered.

In terms of stresses the code is limiting the membrane stress in the pipe wall, resulting from sustained loads, to a percentage of the material yield stress. Since gross yielding of the pipe wall will not occur, it follows that protection against catastrophic failures is provided.

The second kind of protection provided is protection against fatigue or leak type failure. This is accomplished by requiring that stresses in the piping system resulting from the following be calculated: constraint of free end displacement and imposed deflections and rotations.

The code requires that moments in the system resulting from the above be calculated and the resulting expansion stress be compared with an allowable.

In terms of stress, the fatigue or cyclic stress in the pipe wall is limited to a value that is below the allowable cyclic stress level for the material. That is, a given material can be subjected to a specific stress level for a certain number of cycles before significant material cracking occurs. If the stress level is changed, the number of cycles changes. An increase in stress results in a decrease in the number of times that stress can be applied, and conversely, a decrease in stress results in an increase in allowable cycles.

10.7-3 Stress Intensification Factors

It is of utmost importance that we understand stress intensification factors, what they represent, and the meaning of the answer we get when using them.

Perhaps the greatest problem area is the lack of understanding as to what we are calculating. This, of course, is a generalization, but one can look in all areas of the piping industry, not only nuclear, and find that problem. Fortunately, the people who wrote the B31.1 rules understand, the pipe understands, so the system works. The problems arise when people start writing requirements and positions without having a basic understanding of B31.1. We must recognize that nuclear class 2 and 3 rules are identical to B31.1 in the stress analysis area and that nuclear class 1 is a totally different approach. To assume they are the same is the first error, and to assume that we can calculate the strain in a piping component using these criteria is the second problem.

To understand what one calculates for a stress it is necessary to go back to the development of the stress intensity (i) values. These values were developed as a result of cyclic bending fatigue testing of components and predict, fairly accurately, the effective fatigue response of various piping components. This work was followed by a number of theoretical approaches to the development of intensification factors and together these form the basis for the i values appearing in the codes. The i values *do not* predict the maximum elastic stress and are approximately one-half the theoretical stress index and therefore one-half the elastic stress. This is pointed out quite clearly in the code where the analyst is allowed to use $C_2 K_2/2$ values from NB-3600 for components that have no i value given in NC/ND-3600.

The basic lack of understanding is failure to recognize that i values are based on the prediction of the fatigue stresses for use with a matching fatigue design curve. To use stresses calculated with i values to evaluate collapse or to evaluate catastrophic type failure is an error unless the allowable stresses have been modified accordingly.

10.7-4 Primary Stress Protection

The Code provides for protection against catastrophic failure (i.e., failure due to a single application of load or loads) in the following manner.

Pressure

Straight Pipe. The minimum thickness equation of NC/ND-3641.1 must be satisfied for straight pipe:

$$t_m = \frac{PD_0}{2(S + Py)} + A \qquad (10.7\text{-}1)$$

where t_m = minimum required wall thickness, in. If pipe is ordered by its nominal wall thickness, the manufacturing tolerance on wall thickness must be taken into account. After the minimum pipe wall thickness t_m is determined, this minimum thickness shall be increased by an amount sufficient to provide the manufacturing tolerance allowed in the applicable pipe specification or required by the process. The next heavier commercial wall thickness shall then be selected from standard thickness schedules such as contained in ANSI-B36.10 or from manufacturers' schedules for other than standard thickness.

P = internal design pressure, psi

y = a coefficient having a value of 0.4, except that for pipe with a D_0/t_m ratio less than 6, the value of y shall be taken as

$$y = \frac{d}{d + D_0}$$

D_0 = outside diameter of pipe, in. For design calculations the outside diameter of pipe as given in tables of standards and specifications shall be used in obtaining the value of t_m. When calculating the allowable pressure of pipe on hand or in stock, the actual measured outside diameter and actual measured minimum wall thickness at the thinner end of the pipe may be used to calculate this pressure.

d = inside diameter of pipe.

S = maximum allowable stress for the material at the design temperature, psi.

A = an additional thickness to provide for material removed in threading, corrosion or erosion allowance, and material required for structural strength of the pipe during erection, as appropriate, in.

Standard Fittings. In the case of fittings purchased and used in accordance with the approved standards and pressure ratings of Table NC-3132.1, no minimum thickness analysis is required.

Pipe Bends. Pipe bend wall thickness after bending must meet the minimum wall thickness calculated by Equation (10.7-1). A table of suggested minimum thickness prior to bending is given in Table NC-3642.1(c)-1 and Fig. 10.7-1. These values are based on experience and good shop practice, but do not assure satisfaction of the minimum wall thickness requirements. The designer is cautioned to discuss this with the fabricator to ensure that the requirements are met.

Radius of Bends	Minimum Thickness Recommended Prior To Bending
6 pipe diameter or greater	$1.06t_m$
5 pipe diameter	$1.08t_m$
4 pipe diameter	$1.16t_m$
3 pipe diameter	$1.25t_m$

Fig. 10.7-1   Minimum thickness for bending.

Intersections.   Intersections that are not purchased in accordance with applicable standards must be analyzed in accordance with NC-3643. The rules of this subdivision are based on the area replacement technique. That is, if a portion of pipe material which carries membrane stress is removed, it must be replaced in close proximity to the area of removal. The amount of metal to be replaced, and the location or limits within which it must be provided, are described in detail for both welded branch connections and extruded outlets.

Expansion Joints.   Rather specific rules are provided in NC/ND-3649.1 and NC/ND-3649.2 and are appropriate for providing adequate joints. The real problem here results from the fact that these type joints are not applied properly. When we see failures of expansion joints occur, they are usually due to improper use by system designer and/or improper installation. For example:

Guides removed by installer who assumes they were for shipping only

Improper anchoring of the system

Severe vibration environment (i.e., pump discharge)

Damage to convolutions during installation (denting, weld spatter, etc.)

Improper alignment

Control of Primary Stresses

Sustained Loads.   The effects of pressure, weight, and other sustained mechanical loads must meet the following requirements:

$$S_{\text{SL}} = B_1 \frac{PD_0}{2t_n} + B_2 \frac{M_A}{Z} \leq 1.5 S_h \tag{10.7-2}$$

where B_1, B_2 = primary stress indices for the specific product under investigation (NB-3680)
P = internal design pressure, psi
D_0 = outside diameter of pipe, in.
t_n = nominal wall thickness, in.
M_A = resultant moment loading on cross section due to weight and other sustained loads, in.·lb (NC-3653.3)
Z = section modulus of pipe, in.³
S_h = basic material allowable stress at design temperature, psi

Occasional Loads.   The effects of pressure, weight, other sustained mechanical loads, and occasional loads, including earthquake, must meet the following requirements:

$$S_{\text{OL}} = B_1 \frac{P_{\text{max}} D_0}{2t_n} + B_2 \left[\frac{(M_A + M_B)}{Z} \right] \leq 1.8 S_h \tag{10.7-3}$$

but not greater than $1.5 S_y$. Terms are the same as in Equation (10.7-2), except:

P_{max} = peak pressure, psi
M_B = resultant moment loading on cross section due to occasional loads, such as thrusts from relief and safety valve loads from pressure and flow transients, and earthquake, if the design specifications require calculation of moments due to earthquake, in.·lb. For earthquake use only one-half the range. Effects of anchor displacement due to earthquake may be excluded from Equation (10.7-3) if they are included in Equations (10.7-4) and (10.7-8) (NC-3653.2)
S_y = basic material yield strength at design temperature, psi

The code is not specific here on how one treats a multiple-load event. For example, M_b represents the resultant moment loading from relief valve thrusts, flow transients, and earthquake. If a system is subjected to an earthquake event plus relief valve blow, how does one combine these loads to arrive at M_b. There are two obvious choices, absolute sum (ABS) and square root sum of the squares (SRSS).

Summary

Having satisfied the requirements for pressure and for control of primary stress, the designer has provided protection against a catastrophic failure occurring. Further discussion concerning the appropriateness of this approach will come after we look at the class 1 rules.

10.7-5   Fatigue Protection

Fatigue protection is provided by satisfying Equation (10.7-4) or (10.7-8). Thermal expansion loading must satisfy the following:

$$S_e = \frac{iM_c}{Z} \le S_a \tag{10.7-4}$$

where M_c is the range of resultant moments due to thermal expansion. Also include moments resulting for anchor displacements due to earthquake if these effects were omitted from Equation (10.7-3).

Earlier discussion indicated that the stress intensification factor i was a fatigue factor to be used with a matching fatigue curve. The term S_a defines this fatigue curve:

$$S_a = f(1.25S_c + 0.25S_h) \tag{10.7-5}$$

where S_c = basic material allowable stress at minimum (cold) temperature, psi
   S_h = basic material allowable stress at maximum (hot) temperature, psi
   f = stress range reduction factor for cyclic conditions for total number of full temperature cycles

The factor f decreases with increasing number of cycles, and one can generate an S_a versus f or S_a versus cycles curve from this data.

Piping codes always permitted the designer to increase the allowable S_a value by the amount of allowable S_h not used up for primary stress protection, Equation (10.7-2). In equation form this translated to

$$S_a = f[1.25(S_c + S_h) - S_{SL}] \tag{10.7-6}$$

where S_{SL} is as defined in Equation (10.7-2).

This indicates that, in the formulation of these fatigue rules, the committee assumed that the sustained longitudinal stress also cycled. For example, if we could design a piping system that had no pressure in it and zero weight stress, then the allowable S_a would be:

$$S_a = f(1.25S_c + 1.25S_h) \tag{10.7-7}$$

For the case where pressure and weight are equal to S_h we merely subtract $1.0S_h$ from the S_a of Equation (10.7-7) and are left with Equation (10.7-5), which is the common standard.

This discussion leads us to Equation (10.7-8) of NC/ND-3600, which is the same philosophy with a little different application:

$$S_{te} = \frac{PD_0}{4t_n} + 0.75i\frac{M_a}{Z} + i\frac{M_c}{Z} \le (S_h + S_a) \tag{10.7-8}$$

The first two terms with the allowable S_h represent Equation (10.7-2), S_{SL}, and the last term with the allowable S_a represents Equation (10.7-4), S_e.

REFERENCES

10.7-1 *Boiler and Pressure Vessel Code*, Section III, Nuclear Power Plant Components, Division 1, The American Society of Mechanical Engineers, New York, 1980.

10.7-2 B31.1, *Power Piping*, American National Standards Institute, New York, 1980.

INDEX